W. Greiner · J. Reinhardt
FIELD QUANTIZATION

Springer
*Berlin
Heidelberg
New York
Barcelona
Budapest
Hong Kong
London
Milan
Paris
Santa Clara
Singapore
Tokyo*

Greiner
Quantum Mechanics
An Introduction   3rd Edition

Greiner
Quantum Mechanics
Special Chapters

Greiner · Müller
Quantum Mechanics
Symmetries   2nd Edition

Greiner
Relativistic Quantum Mechanics
Wave Equations

Greiner · Reinhardt
Field Quantization

Greiner · Reinhardt
Quantum Electrodynamics
2nd Edition

Greiner · Schäfer
Quantum Chromodynamics

Greiner · Maruhn
Nuclear Models

Greiner · Müller
Gauge Theory of Weak Interactions
2nd Edition

Greiner
Mechanics I
(in preparation)

Greiner
Mechanics II
(in preparation)

Greiner
Electrodynamics
(in preparation)

Greiner · Neise · Stöcker
**Thermodynamics
and Statistical Mechanics**

Walter Greiner · Joachim Reinhardt

FIELD QUANTIZATION

With a Foreword by
D. A. Bromley

With 46 Figures,
and 52 Worked Examples and Problems

Springer

Professor Dr. Walter Greiner
Dr. Joachim Reinhardt

Institut für Theoretische Physik der
Johann Wolfgang Goethe-Universität Frankfurt
Postfach 11 19 32
D-60054 Frankfurt am Main
Germany

Street address:

Robert-Mayer-Strasse 8–10
D-60325 Frankfurt am Main
Germany

email:
greiner@th.physik.uni-frankfurt.de (W. Greiner)
jr@th.physik.uni-frankfurt.de (J. Reinhardt)

Title of the original German edition: *Theoretische Physik,* Ein Lehr- und Übungsbuch,
Band 7a: Feldquantisierung © Verlag Harri Deutsch, Thun 1986, 1993

Library of Congress Cataloging-in-Publication Data.
Greiner, Walter, 1935 –
[Feldquantisierung. English] Field quantization / Walter Greiner, Joachim Reinhardt: with a foreword by D. A. Bromley. p.cm. Includes bibliographical references and index.
ISBN 3-540-59179-6 (pbk.: alk. paper).
1. Quantum field theory. 2. Path integrals. I. Reinhardt, J. (Joachim), 1952– . II. Title.
QC174.45.G72 1996 530.1'43–dc20 95-45449

ISBN 3-540-59179-6 Springer-Verlag Berlin Heidelberg New York

This work is subject to copyright. All rights are reserved, whether the whole or part of the material is concerned, specifically the rights of translation, reprinting, reuse of illustrations, recitation, broadcasting, reproduction on microfilm or in any other way, and storage in data banks. Duplication of this publication or parts thereof is permitted only under the provisions of the German Copyright Law of September 9, 1965, in its current version, and permission for use must always be obtained from Springer-Verlag. Violations are liable for prosecution under the German Copyright Law.

© Springer-Verlag Berlin Heidelberg 1996
Printed and bound by Hamilton Printing Company, Rensselaer, NY.
Printed in the United States of America.

9 8 7 6 5 4 3 2

The use of general descriptive names, registered names, trademarks, etc. in this publication does not imply, even in the absence of a specific statement, that such names are exempt from the relevant protective laws and regulations and therefore free for general use.

Typesetting: Camera ready copy from the authors
Cover design; Design Concept, Emil Smejkal, Heidelberg
Copy Editor: V. Wicks
Production Editor: P. Treiber
SPIN 10469515   56/3144 - 5 4 3 2 1   - Printed on acid-free paper

Foreword to Earlier Series Editions

More than a generation of German-speaking students around the world have worked their way to an understanding and appreciation of the power and beauty of modern theoretical physics – with mathematics, the most fundamental of sciences – using Walter Greiner's textbooks as their guide.

The idea of developing a coherent, complete presentation of an entire field of science in a series of closely related textbooks is not a new one. Many older physicists remember with real pleasure their sense of adventure and discovery as they worked their ways through the classic series by Sommerfeld, by Planck and by Landau and Lifshitz. From the students' viewpoint, there are a great many obvious advantages to be gained through use of consistent notation, logical ordering of topics and coherence of presentation; beyond this, the complete coverage of the science provides a unique opportunity for the author to convey his personal enthusiasm and love for his subject.

The present five-volume set, *Theoretical Physics*, is in fact only that part of the complete set of textbooks developed by Greiner and his students that presents the quantum theory. I have long urged him to make the remaining volumes on classical mechanics and dynamics, on electromagnetism, on nuclear and particle physics, and on special topics available to an English-speaking audience as well, and we can hope for these companion volumes covering all of theoretical physics some time in the future.

What makes Greiner's volumes of particular value to the student and professor alike is their completeness. Greiner avoids the all too common "it follows that ..." which conceals several pages of mathematical manipulation and confounds the student. He does not hesitate to include experimental data to illuminate or illustrate a theoretical point and these data, like the theoretical content, have been kept up to date and topical through frequent revision and expansion of the lecture notes upon which these volumes are based.

Moreover, Greiner greatly increases the value of his presentation by including something like one hundred completely worked examples in each volume. Nothing is of greater importance to the student than seeing, in detail, how the theoretical concepts and tools under study are applied to actual problems of interest to a working physicist. And, finally, Greiner adds brief biographical sketches to each chapter covering the people responsible for the development of the theoretical ideas and/or the experimental data presented. It was Auguste Comte (1798–1857) in his *Positive Philosophy* who noted, "To understand a science it is necessary to know its history". This is all too often forgotten in modern physics teaching and the bridges that Greiner builds to the pioneering figures of our science upon whose work we build are welcome ones.

Greiner's lectures, which underlie these volumes, are internationally noted for their clarity, their completeness and for the effort that he has devoted to making physics an integral whole; his enthusiasm for his science is contagious and shines through almost every page.

These volumes represent only a part of a unique and Herculean effort to make all of theoretical physics accessible to the interested student. Beyond that, they are of enormous value to the professional physicist and to all others working with quantum phenomena. Again and again the reader will find that, after dipping into a particular volume to review a specific topic, he will end up browsing, caught up by often fascinating new insights and developments with which he had not previously been familiar.

Having used a number of Greiner's volumes in their original German in my teaching and research at Yale, I welcome these new and revised English translations and would recommend them enthusiastically to anyone searching for a coherent overview of physics.

Yale University *D. Allan Bromley*
New Haven, CT, USA Henry Ford II Professor of Physics
1989

Preface

Theoretical physics has become a many-faceted science. For the young student it is difficult enough to cope with the overwhelming amount of new scientific material that has to be learned, let alone obtain an overview of the entire field, which ranges from mechanics through electrodynamics, quantum mechanics, field theory, nuclear and heavy-ion science, statistical mechanics, thermodynamics, and solid-state theory to elementary-particle physics. And this knowledge should be acquired in just 8–10 semesters, during which, in addition, a Diploma or Master's thesis has to be worked on or examinations prepared for. All this can be achieved only if the university teachers help to introduce the student to the new disciplines as early on as possible, in order to create interest and excitement that in turn set free essential new energy.

At the Johann Wolfgang Goethe University in Frankfurt we therefore confront the student with theoretical physics immediately, in the first semester. Theoretical Mechanics I and II, Electrodynamics, and Quantum Mechanics I – An Introduction are the basic courses during the first two years. These lectures are supplemented with many mathematical explanations and much support material. After the fourth semester of studies, graduate work begins, and Quantum Mechanics II – Symmetries, Statistical Mechanics and Thermodynamics, Relativistic Quantum Mechanics, Quantum Electrodynamics, the Gauge Theory of Weak Interactions, and Quantum Chromodynamics are obligatory. Apart from these a number of supplementary courses on special topics are offered, such as Hydrodynamics, Classical Field Theory, Special and General Relativity, Many-Body Theories, Nuclear Models, Models of Elementary Particles, and Solid-State Theory.

The present volume of lectures covers the subject of field quantization which lies at the heart of many developments in modern theoretical physics. The observation by Planck and Einstein that a classical field theory – electrodynamics – had to be augmented by corpuscular and nondeterministic aspects stood at the cradle of quantum theory. At around 1930 it was recognized that not only the radiation field with its photons but also matter fields, e.g. the electrons, can be described by the same procedure of "second quantization". Within this formalism, matter is represented by operator-valued fields which are subject to certain (anti-) commutation relations. In this way one arrives at a theory describing systems of several particles (field quanta) which in particular provides a very natural way to formulate the creation and annihilation of particles. Quantum field theory has become *the* language of modern theoretical physics. It is used in particle and high-energy physics, but also the description of many-body systems encountered in solid-state, nuclear and atomic physics make use of the methods of quantum field theory.

The aim of this volume is to present an introduction to the techniques of field quantization on their own, putting less emphasis on their application to specific physical models. The book was conceived as a companion to the volume *Quantum Electrodynamics* in this series, which concentrates on the evaluation of results and applications without working out the field-theoretical foundations. The two books, however, do not rely on each other and can be studied independently. As a prerequisite for the present volume the reader only needs some familiarity with quantum mechanics, both in the nonrelativistic and in the relativistic form.

In the first two chapters we review the classical and quantum mechanical description of oscillating systems with a finite number of degrees of freedom, followed by an exposition of the classical theory of fields. Here the emphasis is laid on the important topic of symmetries and conservation laws. The main part of the book deals with the method of "canonical quantization" which is applied, step by step, to various types of fields: the nonrelativistic Schrödinger field, and the relativistic fields with spin 0, 1/2, and 1. The subsequent chapters treat the topic of interacting quantum fields and show how perturbation theory can be employed to systematically derive observable quantities, in particular the scattering matrix. This is complemented by a chapter on discrete symmetry transformations, which play an important role for the development of models of elementary particles.

The last part of the book contains the description of a formalism which in a way stands in competition with the "seasoned" method of canonical quantization. The method of quantization using path integrals, which essentially is equivalent to the canonical formalism, has gained increasing popularity over the years. Apart from their elegance and formal appeal path-integral quantization and the related functional techniques are particulary well suited to implement conditions of constraint, which is necessary for the treatment of gauge fields. Nowadays any student of theoretical physics must be familiar with both the canonical and the path-integral formalism.

We repeat that this book is not intended to provide an exhaustive introduction to all aspects of quantum field theory. In particular, the important topic of renormalization is covered only in passing. Our main goal has been to present a detailed and comprehensible introduction to the methods of field quantization themselves, leaving aside most applications. We hope to attain this goal by presenting the subject in considerable detail, explaining the mathematical tools in a rather informal way, and by including a large number of examples and worked exercises.

We would like to express our gratitude to Dr. Ch. Hofmann for his help in proofreading the German edition of the text. For the typesetting of the English edition we enjoyed the help of Mr. M. Bleicher. Once again we are pleased to acknowledge the agreeable collaboration with Dr. H.J. Kölsch and his team at Springer-Verlag, Heidelberg. The English manuscript was copy edited by Dr. Victoria Wicks.

Frankfurt am Main, *Walter Greiner*
December 1995 *Joachim Reinhardt*

Contents

Part I. Many-Body Systems and Classical Field Theory

1. **Classical and Quantum Mechanics of Particle Systems** 3
   - 1.1 Introduction 3
   - 1.2 Classical Mechanics of Mass Points 4
   - 1.3 Quantum Mechanics: The Harmonic Oscillator 6
     - 1.3.1 The Harmonic Oscillator 8
   - 1.4 The Linear Chain (Classical Treatment) 10
   - 1.5 The Linear Chain (Quantum Treatment) 18

2. **Classical Field Theory** 31
   - 2.1 Introduction 31
   - 2.2 The Hamilton Formalism 34
   - 2.3 Functional Derivatives 36
   - 2.4 Conservation Laws in Classical Field Theories 39
   - 2.5 The Generators of the Poincaré Group 49

Part II. Canonical Quantization

3. **Nonrelativistic Quantum Field Theory** 57
   - 3.1 Introduction 57
   - 3.2 Quantization Rules for Bose Particles 58
   - 3.3 Quantization Rules for Fermi Particles 65

4. **Spin-0 Fields: The Klein–Gordon Equation** 75
   - 4.1 The Neutral Klein–Gordon Field 75
   - 4.2 The Charged Klein–Gordon Field 91
   - 4.3 Symmetry Transformations 95
   - 4.4 The Invariant Commutation Relations 100
   - 4.5 The Scalar Feynman Propagator 106
   - 4.6 Supplement: The Δ Functions 109

5. **Spin-$\frac{1}{2}$ Fields: The Dirac Equation** 117
   - 5.1 Introduction 117
   - 5.2 Canonical Quantization of the Dirac Field 123

    5.3 Plane-Wave Expansion of the Field Operator 124
    5.4 The Feynman Propagator for Dirac Fields 132

6. **Spin-1 Fields: The Maxwell and Proca Equations** 141
    6.1 Introduction 141
    6.2 The Maxwell Equations 141
         6.2.1 The Lorentz Gauge 144
         6.2.2 The Coulomb Gauge 144
         6.2.3 Lagrange Density and Conserved Quantities 145
         6.2.4 The Angular-Momentum Tensor 148
    6.3 The Proca Equation 152
    6.4 Plane-Wave Expansion of the Vector Field 154
         6.4.1 The Massive Vector Field 154
         6.4.2 The Massless Vector Field 156
    6.5 Canonical Quantization of the Massive Vector Field 158

7. **Quantization of the Photon Field** 171
    7.1 Introduction 171
    7.2 The Electromagnetic Field in Lorentz Gauge 172
    7.3 Canonical Quantization in the Lorentz Gauge 176
         7.3.1 Fourier Decomposition of the Field Operator 177
    7.4 The Gupta–Bleuler Method 180
    7.5 The Feynman Propagator for Photons 185
    7.6 Supplement: Simple Rule for Deriving Feynman Propagators .. 188
    7.7 Canonical Quantization in the Coulomb Gauge 196
         7.7.1 The Coulomb Interaction 200

8. **Interacting Quantum Fields** 211
    8.1 Introduction 211
    8.2 The Interaction Picture 211
    8.3 The Time-Evolution Operator 215
    8.4 The Scattering Matrix 219
    8.5 Wick's Theorem 225
    8.6 The Feynman Rules of Quantum Electrodynamics 233
    8.7 Appendix: The Scattering Cross Section 267

9. **The Reduction Formalism** 269
    9.1 Introduction 269
    9.2 In and Out Fields 270
    9.3 The Lehmann–Källén Spectral Representation 278
    9.4 The LSZ Reduction Formula 282
    9.5 Perturbation Theory for the n-Point Function 290

10. **Discrete Symmetry Transformations** 301
    10.1 Introduction 301
    10.2 Scalar Fields 301
         10.2.1 Space Inversion 301
         10.2.2 Charge Conjugation 305
         10.2.3 Time Reversal 306

   10.3 Dirac Fields 311
       10.3.1 Space Inversion 312
       10.3.2 Charge Conjugation 313
       10.3.3 Time Reversal 315
   10.4 The Electromagnetic Field 318
   10.5 Invariance of the S Matrix 324
   10.6 The CPT Theorem 326

Part III. Quantization with Path Integrals

11. The Path-Integral Method 337
   11.1 Introduction 337
   11.2 Path Integrals in Nonrelativistic Quantum Mechanics 337
   11.3 Feynman's Path Integral 343
   11.4 The Multi-Dimensional Path Integral 350
   11.5 The Time-Ordered Product and n-Point Functions 356
   11.6 The Vacuum Persistence Amplitude $W[J]$ 360

12. Path Integrals in Field Theory 365
   12.1 The Path Integral for Scalar Quantum Fields 365
   12.2 Euclidian Field Theory 371
   12.3 The Feynman Propagator 375
   12.4 Generating Functional and Green's Function 380
   12.5 Generating Functional for Interacting Fields 384
   12.6 Green's Functions in Momentum Space 391
   12.7 One-Particle Irreducible Graphs and the Effective Action 400
   12.8 Path Integrals for Fermion Fields 408
   12.9 Generating Functional and Green's Function
       for Fermion Fields 419
   12.10 Generating Functional and Feynman Propagator
       for the Free Dirac Field 421

Index 433

Contents of Examples and Exercises

1.1 Normal Coordinates 15
1.2 The Linear Chain Subject to External Forces..................... 20
1.3 The Baker–Campbell–Hausdorff Relation 27
2.1 The Symmetrized Energy-Momentum Tensor 47
2.2 The Poincaré Algebra for Classical Fields 51
3.1 The Normalization of Fock States 68
3.2 Interacting Particle Systems: The Hartree–Fock Approximation.... 69
4.1 Commutation Relations for Creation and Annihilation Operators .. 83
4.2 Commutation Relations of the Angular-Momentum Operator...... 85
4.3 The Field Operator in the Spherical Representation 86
4.4 The Charge of a State... 94
4.5 Commutation Relations Between Field Operators and Generators.. 99
4.6 The Function $\Delta_1(\mathbf{x} - \mathbf{y})$ for Equal Time Arguments 105
5.1 The Symmetrized Dirac Lagrange Density 121
5.2 The Symmetrized Current Operator............................ 134
5.3 The Momentum Operator 135
5.4 Helicity States 136
5.5 General Commutation Relations and Microcausality............. 138
6.1 The Lagrangian of the Maxwell Field........................... 149
6.2 Coupled Maxwell and Dirac Fields 150
6.3 Fourier Decomposition of the Proca Field Operator 160
6.4 Invariant Commutation Relations
    and the Feynman Propagator of the Proca Field................. 167
7.1 The Energy Density of the Photon Field in the Lorentz Gauge 175
7.2 Gauge Transformations and Pseudo-photon States 183
7.3 The Feynman Propagator for Arbitrary Values
    of the Gauge Parameter ζ 189
7.4 The Transverse Delta Function 202
7.5 General Commutation Rules for the Electromagnetic Field 205
8.1 The Gell-Mann–Low Theorem.................................. 220
8.2 Proof of Wick's Theorem 231
8.3 Disconnected Vacuum Graphs 243
8.4 Møller Scattering and Compton Scattering 245
8.5 The Feynman Graphs of Photon–Photon Scattering 249
8.6 Scalar Electrodynamics... 251
8.7 ϕ^4 Theory 261
9.1 Derivation of the Yang–Feldman Equation 276
9.2 The Reduction Formula for Spin-$\frac{1}{2}$ Particles..................... 288

9.3   The Equation of Motion for the Operator $\hat{U}(t)$ 294
9.4   Green's Functions and the S Matrix of ϕ^4 Theory 295
10.1 The Operators $\hat{\mathcal{P}}$ and $\hat{\mathcal{C}}$ for Scalar Fields 309
10.2 The Classification of Positronium States 320
10.3 Transformation Rules for the Bilinear Covariants 330
10.4 The Relation Between Particles and Antiparticles 331
11.1 The Path Integral for the Propagation of a Free Particle 345
11.2 Weyl Ordering of Operators 347
11.3 Gaussian Integrals in D Dimensions 353
12.1 Construction of the Field-Theoretical Path Integral 368
12.2 Series Expansion of the Generating Functional 386
12.3 A Differential Equation for $W[J]$ 387
12.4 The Perturbation Series for the ϕ^4 Theory 392
12.5 Connected Green's Functions 398
12.6 Grassmann Integration 415
12.7 Yukawa Coupling 424

Part I

Many-Body Systems and Classical Field Theory

1. Classical and Quantum Mechanics of Particle Systems

1.1 Introduction

In this book we will develop the quantum-theoretical description of systems which can be described in terms of a *field theory*. Starting from the framework of classical physics, the concept of a field at first might evoke ideas about macroscopic systems, for example velocity fields or temperature fields in fluids and gases, etc. Fields of this kind will not concern us, however; they can be viewed as derived quantities which arise from an averaging of microscopic particle densities. Our subjects are the fundamental fields that describe matter on a microscopic level: it is the *quantum-mechanical wave function* $\psi(\boldsymbol{x},t)$ of a system which can be viewed as a field from which the observable quantities can be deduced. In quantum mechanics the wave function is introduced as an ordinary complex-valued function of space and time. In Dirac's terminology it has the character of a "c number". Quantum field theory goes one step further and treats the wave function itself as an object which has to undergo quantization. In this way the wave function $\psi(\boldsymbol{x},t)$ is transmuted into a *field operator* $\hat{\psi}(\boldsymbol{x},t)$, which is an operator-valued quantity (a "q number") satisfying certain commutation relations. This process, often called "second quantization", is quite analogous to the route that in ordinary quantum mechanics leads from a set of classical coordinates q_i to a set of quantum operators $\hat{q}_i$. There is one important technical difference, though, since $\psi(\boldsymbol{x},t)$ is a field, i.e., an object which depends on the coordinate $\boldsymbol{x}$. The latter plays the role of a "continuous-valued index", in contrast to the discrete index i, which labels the set q_i. Field theory therefore is concerned with systems having an infinite number of degrees of freedom.

The concept of field quantization has far-reaching consequences and is one of the cornerstones of modern physics. Field quantization provides an elegant language to describe particle systems. Moreover the theory naturally leads to the insight that there are *field quanta* which can be *created and annihilated*. These field quanta come in many guises and are found in virtually all areas of physics. In addition to the ordinary elementary particles encountered in high-energy physics, quasiparticles, i.e., collective excitations in macroscopic or microscopic systems, can also be descibed as field quanta. The list of possible examples is very rich; to name just a few there are phonons, magnons, excitons, and plasmons in solid-state physics, quantized vibrations of hadronic matter in nuclear physics, etc. Perhaps the most conspicuous example is given by the quanta of the electromagnetic field. The existence of photons and the discovery of processes involving their emission and absorption historically presented the first occasion that called for the concept of field quantization.

According to the principles of quantum theory the fields undergo *quantum fluctuations* which will be present even in the absence of particles, i.e., in the vacuum. The description of the effects of vacuum field fluctuations in interacting systems is one of the key problems in quantum field theory. In the following chapters, however, we will touch on this subject and the related problem of renormalization only in passing. Our main concern will be directed to laying the groundwork and to describing in some detail, how the transition from classical to quantized fields is made.

As a preparation for greater things to come, in this first chapter we will briefly remind the reader of the formalism of classical mechanics describing point particles. Then the method of quantization of such systems will be mentioned, with particular emphasis on the harmonic oscillator. Finally a system of coupled mass points with harmonic interaction (the oscillating chain) will be discussed on the classical and quantum levels. The continuum limit of this system leads to a simple field theory and thus sets the stage for our main subject.

1.2 Classical Mechanics of Mass Points

In highly condensed form we will collect here some basic information on the classical mechanics of point particles. The details can be found in any textbook on classical mechanics. Let us consider a structureless particle of mass m which moves in one dimension under the influence of a time-independent potential $V(q)$. The time evolution of the trajectory $q(t)$ is determined by Newton's equation of motion

$$m\ddot{q} = -\frac{\partial V}{\partial q} \,, \tag{1.1}$$

which has to be solved with certain initial conditions $q(t_0)$ and $\dot{q}(t_0)$. An alternative and more flexible description of the same system is obtained by using the Lagrange function or Lagrangian

$$L(q, \dot{q}) = T - V = \frac{1}{2}m\dot{q}^2 - V(q) \,, \tag{1.2}$$

from which the action

$$W = \int_{t_0}^{t_1} d\tau \, L(q(\tau), \dot{q}(\tau)) \,, \tag{1.3}$$

can be constructed. The equation of motion now follows from a variational problem, i.e., from Hamilton's principle of least action. According to this principle the particle will follow a trajectory for which the action is stationary (usually minimal) under variation:

$$\delta W = 0 \,. \tag{1.4}$$

The functions to be varied are $q(t)$ and $\dot{q}(t)$, subject to the boundary conditions $\delta q(t_0) = \delta q(t_1) = 0$. Using variational calculus this allows us to determine the trajectory $q(t)$ as a solution of the differential equation of Euler and Lagrange

$$\frac{d}{dt}\frac{\partial L}{\partial \dot{q}} - \frac{\partial L}{\partial q} = 0 \,. \tag{1.5}$$

Using the Lagrangian (1.2) we can see that this is equivalent to Newton's equation.

One further equivalent formulation is obtained when the canonically conjugate momentum

$$p = \frac{\partial L}{\partial \dot{q}} \tag{1.6}$$

is introduced as an independent variable instead of the velocity $\dot{q}$. Using a Legendre transformation, one switches from the pair of variables $(q, \dot{q})$ to (q, p). To do this the Hamilton function or Hamiltonian, for short,

$$H(q, p) = p\,\dot{q}(p) - L(q, \dot{q}(p)) \tag{1.7}$$

is introduced. The motion of the mass point is then described in terms of a system of two coupled differential equation of first order:

$$\dot{p} = -\frac{\partial H}{\partial q} \,, \tag{1.8a}$$

$$\dot{q} = \frac{\partial H}{\partial p} \,. \tag{1.8b}$$

This set of Hamilton equations again is equivalent to Newton's equation but can be more easily generalized to other problems.

Finally we remind the reader of the formalism of Poisson brackets. The Poisson bracket of two dynamical variables (i.e., quantities which depend on p and q) $A(p, q)$ and $B(p, q)$ is defined through

$$\{A, B\}_{\mathrm{PB}} = \frac{\partial A}{\partial q}\frac{\partial B}{\partial p} - \frac{\partial A}{\partial p}\frac{\partial B}{\partial q} \,. \tag{1.9}$$

Using Hamilton's equations (1.8) it is simple to show that the time derivative of a dynamical variable A can be expressed in terms of its Poisson bracket with the Hamiltonian H:

$$\begin{aligned}\frac{dA}{dt} &= \frac{\partial A}{\partial t} + \frac{\partial A}{\partial p}\dot{p} + \frac{\partial A}{\partial q}\dot{q} = \frac{\partial A}{\partial t} - \frac{\partial A}{\partial p}\frac{\partial H}{\partial q} + \frac{\partial A}{\partial q}\frac{\partial H}{\partial p} \\ &= \frac{\partial A}{\partial t} + \{A, H\}_{\mathrm{PB}} \,.\end{aligned} \tag{1.10}$$

As a special case of this rule the Hamiltonian itself is found to be a constant of the motion (provided that the potential and the mass do not depend on time):

$$\frac{dH}{dt} = \frac{\partial H}{\partial t} + \{H, H\}_{\mathrm{PB}} = \frac{\partial H}{\partial t} \,. \tag{1.11}$$

The mutual Poisson brackets of the coordinates read

$$\{p,p\}_{\text{PB}} = \{q,q\}_{\text{PB}} = 0,\qquad(1.12a)$$
$$\{q,p\}_{\text{PB}} = 1.\qquad(1.12b)$$

All the equations given in this section can be easily generalized to systems with an arbitrary number N of degrees of freedom. In this process the values of q and p become multicomponent vectors q_i and p_i, $i = 1,\ldots,N$ and the equations of motion are endowed with the appropriate indices. As usual the appearance of repeated indices will imply summation.

The methods of Lagrange and Hamilton provide for an elegant and flexible description of dynamical systems. They can be applied to any chosen set of generalized coordinates q_i, the only prerequisite being the knowledge of the Lagrange or Hamilton function. The change between different sets of coordinates can be elegantly described through canonical transformations.

1.3 Quantum Mechanics: The Harmonic Oscillator

The traditional way to go from a classical to the quantum description of a system involves the replacement of the classical dynamical variables, which are pure numbers, by linear operators in a Hilbert space. In the one-dimensional case we have discussed so far, one replaces $q \to \hat{q}$ and $p \to \hat{p}$. Quantum operators are objects that in general do not commute. The following *quantization postulate* relates the commutators between two operators $[\hat{A}, \hat{B}] = \hat{A}\hat{B} - \hat{B}\hat{A}$ to the classical Poisson bracket (1.9)

$$\{A, B\}_{\text{PB}} \to \frac{1}{i\hbar}[\hat{A}, \hat{B}]. \qquad (1.13)$$

A special case of this is Heisenberg's fundamental quantization rule for the position and momentum variables:

$$[\hat{q}, \hat{p}] = i\hbar, \qquad (1.14)$$

as can be seen from (1.12). The state of the quantum system is described by a vector in Hilbert space. We will use Dirac's notation and denote the elements of this space by ket vectors $|\Psi\rangle$. The space of ket vectors will be complemented by the dual vector space of adjoint vectors (bra vectors $\langle\Phi|$). The overlap between two vectors $\langle\Phi|\Psi\rangle$ is a complex number and holds the usual properties of a scalar product:

$$\begin{aligned}\langle\Phi|\alpha\Psi_1 + \beta\Psi_2\rangle &= \alpha\langle\Phi|\Psi_1\rangle + \beta\langle\Phi|\Psi_2\rangle \\ \langle\Phi|\Psi\rangle^* &= \langle\Psi|\Phi\rangle \\ \langle\Phi|\Phi\rangle = 0 &\leftrightarrow |\Phi\rangle = 0.\end{aligned} \qquad (1.15)$$

According to the probabilistic nature of quantum mechanics the squared modulus $|\langle\Phi|\Psi\rangle|^2$ is interpreted as the probability that a system that was prepared in state $|\Psi\rangle$ can be found in state $|\Phi\rangle$. Measurable quantities are associated with observable operators $\hat{O}$ which are hermitean (self-adjoint[1]), i.e., the adjoint operator $\hat{O}^\dagger$, defined through

---

[1] From a mathematical point of view self-adjointness is more restrictive than hermiticity since it requires that the domains of $\hat{O}$ and $\hat{O}^\dagger$ are equal. We will ignore this distinction, however.

1.3 Quantum Mechanics: The Harmonic Oscillator

$$\langle \Phi | \hat{O}^\dagger \Psi \rangle = \langle \hat{O} \Phi | \Psi \rangle , \qquad (1.16)$$

has to agree with $\hat{O}$. This condition ensures that measurable quantities are described by real numbers.

The statistical average of a measurable quantity is described by the expectation value of the corresponding operator, evaluated with respect to the quantum state

$$O = \langle \Psi | \hat{O} | \Psi \rangle . \qquad (1.17)$$

When repeated measurements of the quantity (in an ensemble of systems with identical preparation) are performed the results will exhibit statistical fluctuations about the mean value with a variance of

$$(\Delta O)^2 = \langle \Psi | (\hat{O} - O)^2 | \Psi \rangle = \langle \Psi | \hat{O}^2 | \Psi \rangle - \langle \Psi | \hat{O} | \Psi \rangle^2 . \qquad (1.18)$$

This variance will vanish, leading to a sharply defined measuring value, if $|\Psi\rangle$ is an eigenstate of the operator $\hat{O}$:

$$\hat{O} |\Psi\rangle = O |\Psi\rangle \qquad (1.19)$$

having eigenvalue O. $|\Psi\rangle$ cannot be the simultaneous eigenstate of two non-commuting operators $\hat{A}$ and $\hat{B}$. In this case the uncertainty relation

$$\Delta A \, \Delta B \geq \frac{1}{2} \left| \langle \Psi | [\hat{A}, \hat{B}] | \Psi \rangle \right| \qquad (1.20)$$

applies. The particular uncertainty relation for the position and momentum coordinates reads

$$\Delta p \Delta q \geq \frac{1}{2} \hbar . \qquad (1.21)$$

The eigenstates of a self-adjoint operator form a basis of the Hilbert space, i.e., each vector can be expanded in this set of states. In general the expansion consists of a summation over the discrete part and an integration over the continuous part of the spectrum.

The eigenstates of the position operator describe a particle localized at a particular point in space:

$$\hat{q} |q\rangle = q |q\rangle . \qquad (1.22)$$

Two states localized at distinct points in space will have zero overlap. The position eigenstates are conveniently normalized "on a delta function" according to

$$\langle q' | q \rangle = \delta(q - q') . \qquad (1.23)$$

They satisfy the completeness relation

$$1 = \int dq \, |q\rangle \langle q| . \qquad (1.24)$$

Projecting an arbitrary state vector $|\Psi\rangle$ onto the coordinate eigenstates leads to the Schrödinger wave function in coordinate representation:

$$\psi(q) = \langle q | \Psi \rangle . \qquad (1.25)$$

In an analogous fashion one can also construct momentum eigenstates

$$\hat{p}|p\rangle = p|p\rangle \; , \tag{1.26a}$$

$$\langle p'|p\rangle = 2\pi\hbar\,\delta(p-p') \; , \quad 1 = \int \frac{\mathrm{d}p}{2\pi\hbar}\,|p\rangle\langle p| \; . \tag{1.26b}$$

The coordinate-space representation of a momentum eigenstate represents a plane wave which, in agreement with the uncertainty relation, is completely delocalized in space:

$$\langle q|p\rangle = \mathrm{e}^{\mathrm{i}pq/\hbar} \; . \tag{1.27}$$

The factor $2\pi\hbar$ in (1.26) appears by convention. One can as well normalize the momentum eigenstates with a factor of 1 instead of $2\pi\hbar$. The plane wave in (1.27) then carries a normalization factor of $(2\pi\hbar)^{-1/2}$.

The dynamical evolution of the quantum system is described by Heisenberg's equation of motion for operators

$$\mathrm{i}\hbar\frac{\mathrm{d}\hat{O}}{\mathrm{d}t} = [\hat{O},\hat{H}] \; . \tag{1.28}$$

This equation emerges from the classical equation of motion (1.10) through the substitution (1.13). We have assumed that $\hat{O}$ is not endowed with an explicit time dependence. The state vectors are constant in time in the Heisenberg picture:

$$\frac{\mathrm{d}}{\mathrm{d}t}|\Psi\rangle = 0 \; . \tag{1.29}$$

It is well known that there exists a completely equivalent representation of quantum mechanics based on the Schrödinger picture, being characterized by time-dependent state vectors and constant operators. The transition between these two pictures is accomplished by a unitary transformation mediated by the operator of time development

$$\hat{U}(t,t_0) = \mathrm{e}^{-\mathrm{i}\hat{H}(t-t_0)/\hbar} \; . \tag{1.30}$$

We will skip the details here but later in Chap. 8 the issue will come up again.

1.3.1 The Harmonic Oscillator

Let us briefly recapitulate the quantum theory of the harmonic oscillator as it will turn out to be closely related to the formalism of field quantization. The Hamiltonian of a one-dimensional oscillator of mass m and oscillation frequency ω reads

$$\hat{H} = \frac{\hat{p}^2}{2m} + \frac{m}{2}\omega^2\hat{q}^2 \; . \tag{1.31}$$

To find the energy eigenstates of the system one may write down the Schrödinger equation in coordinate representation and solve the resulting second-order differential equation for the wave function. An elegant alternative method of solution, which is very convenient for our purpose, is based on purely algebraic means. We define the following mixed operator:

$$\hat{a} = \sqrt{\frac{m\omega}{2\hbar}}\left(\hat{q} + \mathrm{i}\frac{\hat{p}}{m\omega}\right) \; , \tag{1.32}$$

having the adjoint

$$\hat{a}^\dagger = \sqrt{\frac{m\omega}{2\hbar}} \left(\hat{q} - i\frac{\hat{p}}{m\omega} \right) . \tag{1.33}$$

Note that the factor i makes this a nonhermitean operator. These relations can be inverted, leading to

$$\hat{q} = \sqrt{\frac{\hbar}{2m\omega}} \left(\hat{a} + \hat{a}^\dagger \right) , \tag{1.34a}$$

$$\hat{p} = \frac{1}{i}\sqrt{\frac{\hbar m\omega}{2}} \left(\hat{a} - \hat{a}^\dagger \right) . \tag{1.34b}$$

The canonical commutation relation between $\hat{p}$ and $\hat{q}$ leads to

$$[\hat{a}, \hat{a}^\dagger] = 1 . \tag{1.35}$$

Expressed in terms of the new operators $\hat{a}$ and $\hat{a}^\dagger$, the Hamiltonian $\hat{H}$ acquires a very simple form:

$$\begin{aligned}\hat{H} &= -\frac{1}{2m}\frac{\hbar m\omega}{2}\left(\hat{a} - \hat{a}^\dagger\right)^2 + \frac{m}{2}\omega^2\frac{\hbar}{2m\omega}\left(\hat{a} + \hat{a}^\dagger\right)^2 \\ &= \tfrac{1}{2}\hbar\omega\left(\hat{a}^\dagger\hat{a} + \hat{a}\hat{a}^\dagger\right) = \hbar\omega\left(\hat{a}^\dagger\hat{a} + \tfrac{1}{2}\right) . \end{aligned} \tag{1.36}$$

The space of eigenstates of $\hat{H}$ can be constructed easily based on the commutation relation (1.35). The states can be labelled by an integer index n:

$$\hat{H}|n\rangle = E_n|n\rangle , \quad n = 0, 1, 2, \dots . \tag{1.37}$$

Starting from a state $|n\rangle$, one finds that $\hat{a}^\dagger|n\rangle$ and $\hat{a}|n\rangle$ also are eigenstates of the Hamiltonian, but with the shifted energy eigenvalues:

$$\begin{aligned}\hat{H}\hat{a}^\dagger|n\rangle &= \hbar\omega(\hat{a}^\dagger\hat{a} + \tfrac{1}{2})\hat{a}^\dagger|n\rangle = \hbar\omega\hat{a}^\dagger(\hat{a}^\dagger\hat{a} + \tfrac{3}{2})|n\rangle \\ &= (E_n + \hbar\omega)\hat{a}^\dagger|n\rangle \end{aligned} \tag{1.38}$$

and

$$\hat{H}\hat{a}|n\rangle = (E_n - \hbar\omega)|n\rangle . \tag{1.39}$$

The *ground state* $|0\rangle$ is defined to be the state with the lowest possible energy, which implies

$$\hat{a}|0\rangle = 0 . \tag{1.40}$$

This state has the zero-point energy $E_0 = \tfrac{1}{2}\hbar\omega$. The complete spectrum of eigenenergies is given by $E_n = \hbar\omega(n + \tfrac{1}{2})$. The objects $\hat{a}^\dagger$ and $\hat{a}$ are called the *ladder operators* or raising and lowering operators of the harmonic oscillator. They serve to raise or lower the number of excitation quanta (also called phonons in some applications). In the framework of field theory their counterparts will play the role of *creation* and *annihilation operators* for particles.

Any eigenstate of the oscillator can be constructed through repeated application of the operator $\hat{a}^\dagger$ on the ground state:

$$|n\rangle = \frac{1}{\sqrt{n!}} \left(\hat{a}^\dagger\right)^n |0\rangle . \tag{1.41}$$

The square-root factor guarantees the normalization condition $\langle n|n\rangle = 1$; a possible complex phase factor was chosen to be $+1$. The matrix elements of the ladder operators are

$$\langle m|\hat{a}^\dagger|n\rangle = \sqrt{n+1}\,\delta_{m,n+1}\,, \tag{1.42a}$$

$$\langle m|\hat{a}|n\rangle = \sqrt{n}\,\delta_{m,n-1}\,. \tag{1.42b}$$

Up to now we have ignored time. In the *Heisenberg picture* the time dependence is carried by the operators. They satisfy the equation of motion

$$i\hbar\frac{d}{dt}\hat{a}(t) = \left[\hat{a}(t),\hat{H}\right]\,, \tag{1.43}$$

or, using the commutator relation (1.35), which is valid for arbitrary (but equal) time arguments,

$$i\hbar\frac{d}{dt}\hat{a}(t) = \hbar\omega\,\hat{a}(t)\,. \tag{1.44}$$

This simple differential equation can be solved analytically at once:

$$\hat{a}(t) = e^{-i\omega t}\,\hat{a}(0)\,, \tag{1.45}$$

and accordingly

$$\hat{a}^\dagger(t) = e^{+i\omega t}\,\hat{a}^\dagger(0)\,. \tag{1.46}$$

From (1.34) the time dependence of the position and momentum operators $\hat{q}(t)$ and $\hat{p}(t)$ follows at once.

The expectation values of the coordinate and momentum operator vanish if the oscillator is in an energy eigenstate,

$$\langle n|\hat{q}|n\rangle = \langle n|\hat{p}|n\rangle = 0\,, \tag{1.47}$$

while higher moments of even order have finite expectation values. If the particle is required to move on a definite (mean) trajectory $\bar{q}(t)$ then the quantum state must be a superposition of several energy eigenstates $|n\rangle$. Particularly useful in this connection are the *coherent states* (see Example 1.2), since they allow us to make the transition to the classical limit in which the particle moves on a well-defined trajectory.

1.4 The Linear Chain (Classical Treatment)

Before tackling field theories, as a preparation we study a finite system of N discrete point masses lined up on a one-dimensional linear chain, being separated by an equilibrium distance a. Neighboring masses are connected via elastic springs.[2] We want to study *longitudinal oscillations* of the chain. As coordinates we take the excursions $q_1,\ldots,q_N$ from the equilibrium position of the point masses (see Fig. 1.1).

---

[2] As a realistic example for such a system one could think of an extended macromolecule. To understand the dynamics of such a system, quantization becomes important, which we will treat in Sect. 1.5.

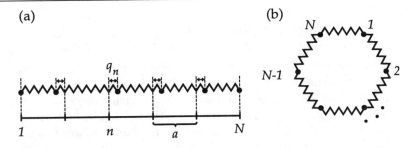

Fig. 1.1. (a) A linear chain consisting of point masses that can move in the longitudinal direction. (b) The linear chain with periodic boundary conditions corresponds to a closed ring

The Lagrangian of the system reads

$$L = T - V = \sum_{n=1}^{N} \frac{m}{2} \dot{q}_n^2 - \sum_{n=1}^{N} \frac{\kappa}{2}(q_{n+1} - q_n)^2 \,, \tag{1.48}$$

which according to (1.5) leads to the equation of motion

$$m\ddot{q}_n + \frac{\kappa}{2}\Big[2(q_n - q_{n-1}) - 2(q_{n+1} - q_n)\Big] = 0 \tag{1.49}$$

or

$$m\ddot{q}_n = \kappa(q_{n+1} + q_{n-1} - 2q_n) \,. \tag{1.50}$$

An interesting aspect of this equation of motion is its *continuum limit*. For vanishing neighbor separation, $a \to 0$, the combination on the right-hand side turns into the second-order derivative operator:

$$\frac{q_{n+1} - 2q_n + q_{n-1}}{a^2} \to \frac{\partial^2}{\partial x^2} q(x,t) \,. \tag{1.51}$$

Thus the continuum limit of (1.50) is the *wave equation*

$$\left(\frac{\partial^2}{\partial t^2} - v^2 \frac{\partial^2}{\partial x^2}\right) q(x,t) = 0 \,, \tag{1.52}$$

which describes the excitations of a continuous elastic string. The propagation velocity is $v = \sqrt{a^2\kappa/m}$.

The system (1.50) of coupled linear differential equations can be written in matrix form as

$$m\ddot{q} = \kappa A q \tag{1.53}$$

where q is an N-dimensional vector. Since only direct neighbors in the chain interact with each other the $N \times N$ coupling matrix A is tridiagonal:

$$A = \begin{pmatrix} -2 & 1 & & & \\ 1 & -2 & 1 & & \\ & \ddots & \ddots & \ddots & \\ & & 1 & -2 & 1 \\ & & & 1 & -2 \end{pmatrix} . \tag{1.54}$$

The canonical momenta are

$$p_n = \frac{\partial L}{\partial \dot{q}_n} = m\dot{q}_n \tag{1.55}$$

and the Hamiltonian reads

$$H = \sum_{n=1}^{N} \frac{1}{2m} p_n^2 + \sum_{n=1}^{N} \frac{\kappa}{2} (q_{n+1} - q_n)^2 \, . \tag{1.56}$$

Before trying to solve (1.50) we have to specify the boundary conditions. As an example, the ends of the chain could be clamped to fixed positions, $q_1 = q_N = 0$, which would mean that only solutions having nodes at the boundaries would be admitted. Wave-like excitations then would be fully reflected at the boundaries. If the finite chain is meant to model an infinitely extended system then the use of *periodic boundary conditions* is advisable. Here the two ends are identified with each other:

$$q_{N+1}(t) = q_1(t) \, , \tag{1.57}$$

which implies $q_{N+n}(t) = q_n(t)$ for any integer n. In this way the linear chain is endowed with the topology of a closed ring; see Fig. 1.1(b) (but there are, of course, no curvature effects).

The dynamics of a set of coupled oscillators is best studied by introducing *normal coordinates*. The position coordinates $q_n(t)$ are expanded with respect to a set of linearly independent *basis functions* u_n^k,

$$q_n(t) = \sum_k a_k(t) \, u_n^k \, , \tag{1.58}$$

where the index k counts the members of the basis set. In principle any complete set of functions can be chosen for the basis. A natural and convenient choice is the use of harmonic functions

$$u_n^k = \frac{1}{\sqrt{N}} \, \mathrm{e}^{ikan} \tag{1.59}$$

which makes (1.58) a discrete Fourier decomposition. The index k has the dimension of an inverse length and correponds to the *wave number* of the "plane wave" (1.59). The range of values k can assume is restricted by the imposed periodic boundary conditions (1.57) since $u_{N+n}^k = u_n^k$ implies $\mathrm{e}^{ikaN} = 1$. Thus k must satisfy

$$k = \frac{2\pi}{Na} l \, , \tag{1.60}$$

where l is an integer. To represent an arbitrary function q_n according to (1.58), values of l in the range

$$-\frac{N}{2} < l \leq \frac{N}{2} \tag{1.61}$$

are sufficient. Larger wave numbers would exceed the "resolving power" of the lattice as depicted in Fig. 1.2 and the associated functions u_n^k would no longer be linearly independent. The dimension of the basis $\{u_n^k\}$ just coincides with the number of degrees of freedom N of the system of oscillators since according to (1.61) k can take on N different values.

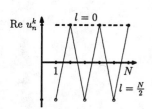

Fig. 1.2. Real part of the basis function u_n^k for the special wave numbers $k = 0$ (*dashed*) and $k = \pi/a$ (*full*)

The basis wave function (1.59) satisfies the *orthonormality condition*

$$\sum_{n=1}^{N} u_n^{k'*} u_n^k = \delta_{kk'} \, . \tag{1.62}$$

The summation over the position index n can be viewed as a scalar product between the basis functions. In this sense different basis functions are orthogonal to each other and have unit length. Furthermore the basis functions satisfy the *completeness relation*

$$\sum_k u_{n'}^{k*} u_n^k = \delta_{nn'} . \qquad (1.63)$$

Proof of (1.62) and (1.63). We have to prove

$$\sum_{n=1}^{N} e^{2\pi i(l-l')n/N} = N \delta_{ll'} . \qquad (1.64)$$

This obviously is true for $l = l'$. If l and l' are unequal, (1.64) can be summed as a geometric series leding to

$$\sum_{n=1}^{N} e^{2\pi i(l-l')n/N} = \frac{e^{2\pi i(l-l')} - 1}{e^{2\pi i(l-l')/N} - 1} = 0 . \qquad (1.65)$$

The denominator will not vanish since $l - l' \neq 0$ and $|l - l'| < N$. The completeness relation follows in a similar way by using (1.60) and (1.64):

$$\sum_k u_{n'}^{k*} u_n^k = \frac{1}{N} \sum_k e^{ika(n-n')} = \frac{1}{N} \sum_{l=-N/2+1}^{N/2} e^{2\pi i(n-n')l/N} \qquad (1.66)$$

$$= \frac{1}{N} e^{-i\pi(n-n')} \sum_{l'=1}^{N} e^{2\pi i(n-n')l'/N} = e^{-i\pi(n-n')} \delta_{nn'} = \delta_{nn'} .$$

A further important property of the chosen basis functions is their behavior under complex conjugation:

$$u_n^{k*} = u_n^{-k} . \qquad (1.67)$$

Since the coordinates have to be real quantities, $q_n^* = q_n$, this implies that the expansion coefficients in (1.58) satisfy

$$a_k^*(t) = a_{-k}(t) . \qquad (1.68)$$

The equation of motion (1.50) for the coordinates $q_n(t)$ can be transformed into a differential equation for the time evolution of the coefficients $a_k(t)$. Insertion of the expansion (1.58) and projection on u_n^k by using (1.62) leads to

$$\ddot{a}_k(t) = \frac{\kappa}{m} \sum_{k'} a_{k'}(t) \sum_n u_n^{k*} \left(u_{n+1}^{k'} + u_{n-1}^{k'} - 2u_n^{k'} \right) . \qquad (1.69)$$

At this point it turns out to be very useful that the Fourier basis functions (1.59) show a simple behavior under translations. The values of u_n^k taken at different coordinates n differ only by a phase factor

$$u_{n\pm 1}^k = e^{\pm ika} u_n^k . \qquad (1.70)$$

After using this property on the right-hand side of (1.69), we can use orthogonality relation:

Fig. 1.3. The dispersion relation of an oscillating chain. The dots mark the discrete values of k

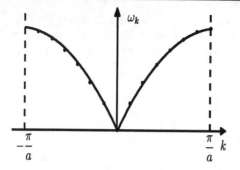

$$\ddot{a}_k(t) = \frac{\kappa}{m}\left(e^{ika} + e^{-ika} - 2\right) a_k(t) \equiv -\omega_k^2\, a_k(t)\ . \tag{1.71}$$

This is the differential equation of a harmonic oscillator with a frequency

$$\omega_k = \sqrt{2\frac{\kappa}{m}(1-\cos ka)} = 2\sqrt{\frac{\kappa}{m}}\left|\sin\frac{ka}{2}\right|\ . \tag{1.72}$$

The *dispersion relation* of the chain undergoing longitudinal oscillations is shown in Fig. 1.3. For small wave numbers, i.e., wavelengths much larger than the lattice constant a, the dispersion relation is approximately linear, $\omega_k \simeq \sqrt{\frac{\kappa}{m}}|k|a$.

In contrast to the original equation of motion (1.50), (1.71) describes a system of *uncoupled* oscillators. The achievement of such a decoupling was the reason for introducing normal coordinates. The general solution of (1.71) can be written down at once:

$$a_k(t) = b_k\, e^{-i\omega_k t} + b^*_{-k}\, e^{+i\omega_k t}\ , \tag{1.73}$$

which takes into account the constraint (1.68) ensuring real-valued coordinates. The general expansion now becomes

$$q_n(t) = \sum_k \left(b_k\, e^{-i\omega_k t} + b^*_{-k}\, e^{+i\omega_k t}\right) u_n^k\ . \tag{1.74}$$

This can be written in a more symmetric fashion if $k \to -k$ is substituted in the second term (this does not influence the range of summation) and (1.67) is employed:

$$q_n(t) = \sum_k \left(b_k\, e^{-i\omega_k t} u_n^k + b^*_k\, e^{+i\omega_k t} u_n^{k*}\right)\ , \tag{1.75}$$

or written explicitly

$$q_n(t) = \frac{1}{\sqrt{N}} \sum_k \left(b_k\, e^{-i(\omega_k t - kan)} + b^*_k\, e^{+i(\omega_k t - kan)}\right)\ . \tag{1.76}$$

The second term is the complex conjugate to the first one so that the coordinate is indeed real. We will often encounter expansions of the type (1.75) in this book.

In order to apply Hamilton's formalism to the problem of the oscillating chain we have to express the canonical momenta p_n in terms of the normal coordinates. Using (1.55) and (1.75) we find

$$p_n(t) = m \sum_k (-i\omega_k) \left(b_k \, e^{-i\omega_k t} u_n^k - b_k^* \, e^{+i\omega_k t} u_n^{k*} \right) . \tag{1.77}$$

The expansions (1.75) and (1.77) can be used to evaluate the Hamiltonian (1.56). A somewhat lengthy calculation (see Exercise 1.1), which makes use of the orthonormality of the basis functions and of the dispersion relation, leads to

$$H = \sum_k 2m\omega_k^2 \, b_k^* b_k . \tag{1.78}$$

As expected the energy of the system is constant in time. When dealing with quantum fields we will frequently encounter quadratic forms of this kind, the only difference being a replacement of the classical normal coordinate b_k by a quantum operator $\hat{b}_k$.

The Poisson brackets between the normal coordinates b_k and b_k^* take the following simple form (see Exercise 1.1c)

$$\{b_k, b_k^*\}_{PB} = \frac{-i}{2m\omega_k} \delta_{kk'} , \tag{1.79a}$$

$$\{b_k, b_k\}_{PB} = \{b_k^*, b_k^*\}_{PB} = 0 . \tag{1.79b}$$

The knowledge of the normal coordinates can be used to construct solutions to the initial-value problem for the coordinates $q_n(t)$. The starting conditions $q_n(0)$ and $\dot{q}_n(0)$ are projected on the basis u_n^k, which yields the coefficients b_k. Then according to (1.75) the motion is known for arbitrary times. As shown in Exercise 1.1b this leads to the explicit solution

$$q_n(t) = \frac{1}{N} \sum_k \sum_{n'} \left[q_{n'}(0) \cos\big(ka(n-n') - \omega_k t\big) \right.$$
$$\left. - \frac{1}{\omega_k} \dot{q}_n(0) \sin\big(ka(n-n') - \omega_k t\big) \right] . \tag{1.80}$$

EXERCISE

1.1 Normal Coordinates

Problem. (a) Derive the Hamiltonian of the linear chain, expressed in terms of the normal coordinates b_k and b_k^*.
(b) Solve the initial-value problem for the linear chain using normal coordinates, i.e., give an expression for $q_n(t)$ in terms of the initial conditions $q_n(0)$ and $\dot{q}_n(0)$.
(c) Compute the Poisson brackets between the normal coordinates b_k and b_k^*.

Solution. (a) The expansions (1.75) and (1.77) for the coordinate and momentum have to be inserted into the expression (1.56) for the Hamiltonian $H = T + V$. The part for the kinetic energy thus becomes

$$T = \frac{m}{2} \sum_n \sum_{k,k'} (-i\omega_{k'})(-i\omega_k)$$

Exercise 1.1

$$\times \left[b_{k'} b_k \, \mathrm{e}^{-\mathrm{i}(\omega_{k'}+\omega_k)t} u_n^{k'} u_n^{k} - b_{k'} b_k^* \, \mathrm{e}^{-\mathrm{i}(\omega_{k'}-\omega_k)t} u_n^{k'} u_n^{k*} \right.$$

$$\left. - b_{k'}^* b_k \, \mathrm{e}^{+\mathrm{i}(\omega_{k'}-\omega_k)t} u_n^{k'*} u_n^{k} + b_{k'}^* b_k^* \, \mathrm{e}^{+\mathrm{i}(\omega_{k'}+\omega_k)t} u_n^{k'*} u_n^{k*} \right] . \quad (1)$$

The second and third term in the sum can be simplified by using the orthonormality relation (1.62), which allows us to perform the summation over n. The same procedure can also be applied to the other two terms since according to (1.67) the relations

$$\sum_n u_n^{k'} u_n^{k} = \sum_n u_n^{-k'*} u_n^{k} = \delta_{k',-k} ,$$

$$\sum_n u_n^{k'*} u_n^{k*} = \delta_{k',-k} \quad (2)$$

hold. In this way the k' sum in (1) breaks down and (using $\omega_{-k} = \omega_k$)

$$T = \frac{m}{2} \sum_k (-\omega_k^2) \left[b_{-k} b_k \, \mathrm{e}^{-2\mathrm{i}\omega_k t} - b_k b_k^* - b_k^* b_k + b_{-k}^* b_k^* \, \mathrm{e}^{+2\mathrm{i}\omega_k t} \right] . \quad (3)$$

The potential energy

$$V = \frac{\kappa}{2} \sum_n (q_{n+1} - q_n)^2 \quad (4)$$

can be treated similarly. After insertion of the expansion for q_{n+1} and q_n, (1.70) can be used to express u_{n+1}^k in terms of u_n^k, which leads to

$$V = \frac{\kappa}{2} \sum_n \sum_{k,k'} \left[b_{k'} b_k \, \mathrm{e}^{-\mathrm{i}(\omega_{k'}+\omega_k)t} (\mathrm{e}^{\mathrm{i}k'a} - 1)(\mathrm{e}^{\mathrm{i}ka} - 1) u_n^{k'} u_n^{k} \right.$$

$$+ b_{k'} b_k^* \, \mathrm{e}^{-\mathrm{i}(\omega_{k'}-\omega_k)t} (\mathrm{e}^{\mathrm{i}k'a} - 1)(\mathrm{e}^{-\mathrm{i}ka} - 1) u_n^{k'} u_n^{k*}$$

$$+ b_{k'}^* b_k \, \mathrm{e}^{+\mathrm{i}(\omega_{k'}-\omega_k)t} (\mathrm{e}^{-\mathrm{i}k'a} - 1)(\mathrm{e}^{\mathrm{i}ka} - 1) u_n^{k'*} u_n^{k}$$

$$\left. + b_{k'}^* b_k^* \, \mathrm{e}^{+\mathrm{i}(\omega_{k'}+\omega_k)t} (\mathrm{e}^{-\mathrm{i}k'a} - 1)(\mathrm{e}^{-\mathrm{i}ka} - 1) u_n^{k'*} u_n^{k*} \right] . \quad (5)$$

Again, two of the sums can be eliminated and, using $(\mathrm{e}^{-\mathrm{i}ka} - 1)(\mathrm{e}^{+\mathrm{i}ka} - 1) = 4\sin^2 \frac{ka}{2}$, we get

$$V = \frac{m}{2} \sum_k 4 \frac{\kappa}{m} \sin^2 \frac{ka}{2} \left[b_{-k} b_k \, \mathrm{e}^{-2\mathrm{i}\omega_k t} + b_k b_k^* + b_k^* b_k + b_{-k}^* b_k^* \, \mathrm{e}^{+2\mathrm{i}\omega_k t} \right] . \quad (6)$$

The factor in front of the square bracket, according to the dispersion relation (1.72), becomes ω_k^2, which agrees with its counterpart in (3). Thus the kinetic and potential energies can be combined to yield the simple result

$$H = T + V = \sum_k m\omega_k^2 (b_k b_k^* + b_k^* b_k) = \sum_k 2m\omega_k^2 b_k^* b_k . \quad (7)$$

(b) The connection between the initial conditions and the normal coordinates is given by

$$q_n(0) = \sum_k (b_k u_n^k + b_k^* u_n^{k*}) , \quad (8a)$$

$$\dot{q}_n(0) = \sum_k (-\mathrm{i}\omega_k)(b_k u_n^k - b_k^* u_n^{k*}) . \quad (8b)$$

1.4 The Linear Chain (Classical Treatment)

Exercise 1.1

The coefficients b_k are obtained by projecting these expansions onto the set of basis functions. The initial positions lead to

$$\sum_n u_n^{k*} q_n(0) = \sum_{k'}\left(b_{k'}\sum_n u_n^{k*}u_n^{k'} + b_{k'}^*\sum_n u_n^{k*}u_n^{k'*}\right)$$
$$= \sum_{k'}\left(b_{k'}\delta_{k',k} + b_{k'}^*\delta_{k',-k}\right)$$
$$= b_k + b_{-k}^*, \tag{9}$$

where use was made of the orthogonality conditions (1.62) and (2). For the initial velocities we get

$$\sum_n u_n^{k*} \dot{q}_n(0) = -i\omega_k\left(b_k - b_{-k}^*\right). \tag{10}$$

These two equations can be solved for b_k, yielding the expansion coefficients as a discrete Fourier transform of the initial conditions:

$$b_k = \frac{1}{2}\sum_n u_n^{k*}\left(q_n(0) + \frac{i}{\omega_k}\dot{q}_n(0)\right). \tag{11}$$

The information contained in the initial conditions ($2N$ real numbers) is now encoded in the set of N complex numbers b_k. The solution of the initial-value problem is known for all times t:

$$q_n(t) = \sum_k \left(b_k e^{-i\omega_k t} u_n^k + b_k^* e^{+i\omega_k t} u_n^{k*}\right)$$
$$= \sum_k \sum_{n'} \frac{1}{2}\Big[q_{n'}(0)\left(e^{-i\omega_k t} u_n^k u_{n'}^{k*} + e^{+i\omega_k t} u_n^{k*} u_{n'}^k\right)$$
$$+ \frac{i}{\omega_k}\dot{q}_{n'}(0)\left(e^{-i\omega_k t} u_n^k u_{n'}^{k*} - e^{+i\omega_k t} u_n^{k*} u_{n'}^k\right)\Big]. \tag{12}$$

Using the abbreviation

$$G_{nn'}(t) = \sum_k e^{-i\omega_k t} u_n^k u_{n'}^{k*} = \frac{1}{N}\sum_k e^{i(ka(n-n')-\omega_k t)} \tag{13}$$

this result can be written as

$$q_n(t) = \sum_{n'}\left[q_{n'}(0)\operatorname{Re} G_{nn'}(t) - \frac{1}{\omega_k}\dot{q}_{n'}(0)\operatorname{Im} G_{nn'}(t)\right] \tag{14}$$

leading to (1.80). Because of the completeness relation (1.63), the condition $G_{nn'}(0) = \delta_{nn'}$ holds, so that (14) obviously satisfies the initial condition at $t=0$.

(c) By generalizing (11), the connection between the normal coordinates b_k and q_n or p_n at arbitrary time t reads

$$b_k = \frac{1}{2}\sum_n u_n^{k*} e^{i\omega_k t}\left(q_n(t) + \frac{i}{\omega_k m}p_n(t)\right). \tag{15}$$

It is easily checked that the right-hand side does not depend on time. The Poisson bracket

Exercise 1.1

$$\{b_k, b_{k'}^*\}_{\text{PB}} = \sum_n \left(\frac{\partial b_k}{\partial q_n} \frac{\partial b_{k'}^*}{\partial p_n} - \frac{\partial b_k}{\partial p_n} \frac{\partial b_{k'}^*}{\partial q_n} \right) \qquad (16)$$

is obtained by differentiating (15) and using the orthonormality relation

$$\{b_k, b_{k'}^*\}_{\text{PB}} = \frac{1}{4} \frac{-\mathrm{i}}{m} \mathrm{e}^{\mathrm{i}(\omega_k - \omega_{k'})t} \left(\frac{1}{\omega_{k'}} + \frac{1}{\omega_k} \right) \sum_n u_n^{k*} u_n^{k'} = \frac{-\mathrm{i}}{2m\omega_k} \delta_{kk'} \; . \quad (17)$$

A similar calculation shows that the other Poisson brackets vanish:

$$\{b_k, b_{k'}\}_{\text{PB}} = \{b_k^*, b_{k'}^*\}_{\text{PB}} = 0 \; . \qquad (18)$$

1.5 The Linear Chain (Quantum Treatment)

To make the transition from classical to quantum mechanics one replaces the position and momentum coordinates of the point masses by linear operators $\hat{q}_n$ and $\hat{p}_n$. These operators satisfy the commutation rules (for equal time arguments, which will be suppressed)

$$[\hat{q}_n, \hat{p}_{n'}] = \mathrm{i}\hbar \delta_{nn'} \; , \qquad (1.81\mathrm{a})$$
$$[\hat{q}_n, \hat{q}_{n'}] = [\hat{p}_n, \hat{p}_{n'}] = 0 \; . \qquad (1.81\mathrm{b})$$

The Hamiltonian of the system according to (1.56) is given by

$$\hat{H} = \sum_{n=1}^N \frac{1}{2m} \hat{p}_n^2 + \sum_{n=1}^N \frac{\kappa}{2} (\hat{q}_{n+1} - \hat{q}_n)^2 \; . \qquad (1.82)$$

As we did in the classical treatment we will again introduce normal coordinates, expanding the operators $\hat{q}_n$ and $\hat{p}_n$ in terms of a basis u_n^k. The expansion coefficients $\hat{b}_k$ now of course will be operators in Hilbert space. In analogy with (1.75) we expand

$$\hat{q}_n(t) = \sum_k \left(\hat{b}_k(t) u_n^k + \hat{b}_k^\dagger(t) u_n^{k*} \right) \qquad (1.83)$$

and for the momentum operator

$$\hat{p}_n(t) = \sum_k (-\mathrm{i}m\omega_k) \left(\hat{b}_k(t) u_n^k - \hat{b}_k^\dagger(t) u_n^{k*} \right) \; . \qquad (1.84)$$

These operators are hermitean by construction, $\hat{q}_n^\dagger = \hat{q}_n$ and $\hat{p}_n^\dagger = \hat{p}_n$, which guarantees that their expectation values are real. The basis expansions can be inverted as in the classical case (see (15) in Exercise 1.1):

$$\hat{b}_k(t) = \frac{1}{2} \sum_n u_n^{k*} \left(\hat{q}_n(t) + \frac{\mathrm{i}}{\omega_k m} \hat{p}_n(t) \right) \; . \qquad (1.85)$$

The time dependence of $\hat{b}_k(t)$ is very simple:

$$\hat{b}_k(t) = \mathrm{e}^{-\mathrm{i}\omega_k t} \hat{b}_k(0) \; . \qquad (1.86)$$

1.5 The Linear Chain (Quantum Treatment)

This is easily checked by inserting (1.85) and (1.82) into the Heisenberg equation and using the canonical commutation relations, which after some calculations leads to

$$i\hbar \frac{d}{dt}\hat{b}_k = [\hat{b}_k, \hat{H}] = \hbar\omega_k \hat{b}_k . \tag{1.87}$$

Equation (1.85) allows us to calculate the commutators between the operators $\hat{b}_k(t)$ and $\hat{b}_k^\dagger(t)$. Using (1.81) and the orthogonality relation (1.62) we find (equal time arguments of the operators are assumed)

$$\begin{aligned}{}[\hat{b}_k, \hat{b}_{k'}^\dagger] &= \frac{1}{4}\sum_{nn'} u_n^{k*} u_{n'}^{k'*} \left[\hat{q}_n + \frac{i}{\omega_k m}\hat{p}_n, \hat{q}_{n'} - \frac{i}{\omega_{k'} m}\hat{p}_{n'}\right] \\ &= \frac{\hbar}{2m\omega_k}\delta_{kk'} \end{aligned} \tag{1.88a}$$

and

$$[\hat{b}_k, \hat{b}_{k'}] = [\hat{b}_k^\dagger, \hat{b}_{k'}^\dagger] = 0 . \tag{1.88b}$$

Note that the prescription (1.13) relates these results to the Poisson brackets (1.79). The operators $\hat{b}_k$ carry the dimension of a length so that the commutator (1.88) contains a factor that has the dimension of (length)^2. It will be advantageous to split this factor off and introduce new *dimensionless* operators $\hat{c}_k$ by defining

$$\hat{c}_k = \sqrt{\frac{2m\omega_k}{\hbar}}\,\hat{b}_k . \tag{1.89}$$

The expansions (1.83) and (1.84) then become

$$\hat{q}_n(t) = \sum_k \sqrt{\frac{\hbar}{2m\omega_k}} \left(\hat{c}_k(t) u_n^k + \hat{c}_k^\dagger(t) u_n^{k*}\right) , \tag{1.90}$$

$$\hat{p}_n(t) = -i\sum_k \sqrt{\frac{\hbar m\omega_k}{2}} \left(\hat{c}_k(t) u_n^k - \hat{c}_k^\dagger(t) u_n^{k*}\right) , \tag{1.91}$$

and the commutation relations read

$$[\hat{c}_k, \hat{c}_{k'}^\dagger] = \delta_{kk'} , \tag{1.92a}$$
$$[\hat{c}_k, \hat{c}_{k'}] = [\hat{c}_k^\dagger, \hat{c}_{k'}^\dagger] = 0 . \tag{1.92b}$$

When transforming the Hamiltonian (1.82), the calculation of Exercise 1.1a can be used essentially unchanged. Only in the very last step does the operator property of $\hat{b}_k$ play a role. According to (7)

$$\begin{aligned}\hat{H} &= \sum_k m\omega_k^2 \left(\hat{b}_k \hat{b}_k^\dagger + \hat{b}_k^\dagger \hat{b}_k\right) = \sum_k \frac{\hbar\omega_k}{2}\left(\hat{c}_k\hat{c}_k^\dagger + \hat{c}_k^\dagger \hat{c}_k\right) \\ &= \sum_k \hbar\omega_k \left(\hat{c}_k^\dagger \hat{c}_k + \tfrac{1}{2}\right) .\end{aligned} \tag{1.93}$$

This is the Hamiltonian of a system of uncoupled harmonic oscillators with energies $\hbar\omega_k$. Equation (1.93) contains also the zero-point energy $\tfrac{1}{2}\hbar\omega_k$ per oscillator mode. The quantized vibrations of the system are called *phonons*.

The eigenstates of the Hamiltonian can be constructed in the same way as in Sect. 1.3, the only difference being the appearance of an index k which describes the N different oscillator modes. Since these modes are uncoupled the energy just is given by the sum of the various contributions. Accordingly, the state vectors are given by direct products over subspaces corresponding to the different eigenmodes. The operators $\hat{c}_k$ and $\hat{c}_k^\dagger$ are *annihilation and creation operators for phonons*. The ground state $|0\rangle$ of the chain (which of course contains zero-point motion) is characterized by the property

$$\hat{c}_k |0\rangle = 0 \qquad \text{for all } k \ . \tag{1.94}$$

Multi-phonon states containing n_k phonons in each of the excitation modes k are constructed as direct products in the following way:

$$|n\rangle \equiv |n_1, n_2, \ldots\rangle = \prod_k |n_k\rangle \ , \tag{1.95}$$

where (1.41) applies for each vibration mode:

$$|n_k\rangle = \frac{1}{\sqrt{n_k!}} \left(\hat{c}_k^\dagger\right)^{n_k} |0_k\rangle \ . \tag{1.96}$$

These states satisfy the stationary Schrödinger equation

$$(\hat{H} - E_n)|n\rangle = 0 \ , \tag{1.97}$$

with the energy eigenvalues

$$E_n = \sum_k \hbar\omega_k (n_k + \tfrac{1}{2}) \ . \tag{1.98}$$

EXAMPLE

1.2 The Linear Chain Subject to External Forces

Up to now we have studied a freely moving system of harmonically coupled point masses. Now the chain will be subjected to the influence of an externally applied force. This force $\mathcal{F}_n$ may vary in space but it is assumed to be constant in time. The force is described by an additional contribution to the potential energy which is proportional to the spatial coordinate q_n. The Hamiltonian of the total system then reads

$$\hat{H} = \hat{H}_0 + \hat{V}_1 \ , \tag{1}$$

where $\hat{H}_0$ is the unperturbed Hamiltonian (1.93) and the additional potential energy is given by

$$\hat{V}_1 = -\sum_n \mathcal{F}_n \hat{q}_n \ . \tag{2}$$

If we use the expansion (1.90) of the position operator this becomes

1.5 The Linear Chain (Quantum Treatment)

Example 1.2

$$\begin{aligned}\hat{V}_1 &= -\sum_n \mathcal{F}_n \sum_k \sqrt{\frac{\hbar}{2m\omega_k}}\left(\hat{c}_k u_n^k + \hat{c}_k^\dagger u_n^{k*}\right) \\ &= -\sum_k \left(F_k \hat{c}_k + F_k^* \hat{c}_k^\dagger\right),\end{aligned} \qquad (3)$$

where the Fourier coefficients of the force are given by

$$F_k = \sum_n \mathcal{F}_n \sqrt{\frac{\hbar}{2m\omega_k}} u_n^k = \frac{1}{\sqrt{N}}\sqrt{\frac{\hbar}{2m\omega_k}} \sum_n \mathcal{F}_n \, e^{ikan} \,. \qquad (4)$$

The complete Hamilonian thus reads

$$\hat{H} = \sum_k \left(\hbar\omega_k \left(\hat{c}_k^\dagger \hat{c}_k + \tfrac{1}{2}\right) - F_k \hat{c}_k - F_k^* \hat{c}_k^\dagger\right) . \qquad (5)$$

The interaction operator is *linear* in the creation and annihilation operators $\hat{c}_k^\dagger$ and $\hat{c}_k$. Owing to this simple structure of the interaction a transformation can be found which diagonalizes the Hamiltonian and thus again provides a decoupling of the oscillator modes. It is sufficient to introduce a *shift of the operators* by suitable c numbers α_k. We introduce a set of new creation and annihilation operators through

$$\hat{d}_k = \hat{c}_k - \alpha_k \quad , \quad \hat{d}_k^\dagger = \hat{c}_k^\dagger - \alpha_k^* \,. \qquad (6)$$

This transformation leaves the commutation relations unchanged:

$$\left[\hat{d}_k, \hat{d}_{k'}^\dagger\right] = \delta_{kk'} \,. \qquad (7)$$

Expressed in terms of the new operators, the Hamiltonian becomes

$$\begin{aligned}\hat{H} &= \sum_k \Big(\hbar\omega_k \left(\hat{d}_k^\dagger \hat{d}_k + \tfrac{1}{2}\right) - \hat{d}_k (F_k - \alpha_k^* \hbar\omega_k) - \hat{d}_k^\dagger (F_k^* - \alpha_k \hbar\omega_k) \\ &\quad + \hbar\omega_k |\alpha_k|^2 - \alpha_k F_k - \alpha_k^* F_k^*\Big) .\end{aligned} \qquad (8)$$

Obviously one has to choose

$$\alpha_k = \frac{1}{\hbar\omega_k} F_k^* \qquad (9)$$

in order to make $\hat{H}$ diagonal. The result is

$$\hat{H} = \sum_k \hbar\omega_k \left(\hat{d}_k^\dagger \hat{d}_k + \tfrac{1}{2}\right) + \Delta E \qquad (10)$$

with

$$\Delta E = -\sum_k \frac{1}{\hbar\omega_k} |F_k|^2 = -\sum_k \hbar\omega_k |\alpha_k|^2 \,. \qquad (11)$$

From this we conclude that the excitation spectrum of the linear chain under the influence of (time-independent) external forces remains essentially the same. The only influence is a simultaneous lowering of all the eigenenergies by the value ΔE, i.e.,

$$E_n = \sum_k \hbar\omega_k \left(n_k + \tfrac{1}{2}\right) + \Delta E \,. \qquad (12)$$

Example 1.2

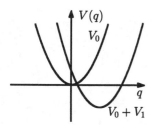

Fig. 1.4. The influence of a constant external force lowers the potential-energy curve of a harmonic oscillator but does not change its shape

This drop in energy corresponds to the work expended by the external force in order to bring the system to the new equilibrium configuration. Note that the "potential well" always gets lowered, irrespective of the strength and direction of the applied force, as demonstrated in Fig. 1.4 (for a system with one degree of freedom). Since the perturbation is linear in q the new potential $V = V_0 + V_1$ remains a parabola of unchanged shape.

The state vectors can be obtained by applying the creation operators $\hat{d}_k^\dagger$ to the new ground state $|0, \alpha\rangle$ which is defined by

$$\hat{d}_k |0, \alpha\rangle = 0 \quad \text{for all } k . \tag{13}$$

This leads to

$$|n, \alpha\rangle = \prod_k \frac{1}{\sqrt{n_k!}} (\hat{d}_k^\dagger)^{n_k} |0, \alpha\rangle . \tag{14}$$

The notation applied here explicitly takes into account that the eigenstates of the interacting system will depend on the force parameters α_k. This implies that the interacting and noninteracting ground states also do not agree $|0, \alpha\rangle \neq |0\rangle$. The explicit connection between these states will be derived later (see (37)).

The transition from the free to the interacting system accomplished through (6) can be formulated in an elegant manner by introducing a *unitary transformation* between the operators. We will construct a *shift operator* $\hat{S}(\alpha)$ which has the following property

$$\hat{S}(\alpha) \hat{c}_k \hat{S}^\dagger(\alpha) = \hat{c}_k + \alpha_k ,$$
$$\hat{S}(\alpha) \hat{c}_k^\dagger \hat{S}^\dagger(\alpha) = \hat{c}_k^\dagger + \alpha_k^* . \tag{15}$$

The unitarity condition for this operator reads

$$\hat{S}^\dagger(\alpha) = \hat{S}^{-1}(\alpha) . \tag{16}$$

This transformation operator (its explicit form will be derived soon) is chosen in such a way that it diagonalizes the Hamiltonian $\hat{H}$. Inserting $\hat{S}^{-1}\hat{S} = \hat{S}^\dagger\hat{S} = 1$ and applying $\hat{S}$ from the left the eigenvalue equation, we change

$$\hat{H}|\Psi\rangle = E|\Psi\rangle \tag{17}$$

to

$$\left(\hat{S}\hat{H}\hat{S}^\dagger\right)\hat{S}|\Psi\rangle = E\hat{S}|\Psi\rangle \quad \text{or} \quad \hat{H}'|\Psi'\rangle = E|\Psi'\rangle , \tag{18}$$

with

$$\hat{H}' = \hat{S}\hat{H}\hat{S}^\dagger , \tag{19a}$$
$$|\Psi'\rangle = \hat{S}|\Psi\rangle . \tag{19b}$$

The transformed Hamiltonian $\hat{H}'$ follows as in (8):

$$\hat{H}' = \hat{S}\hat{H}\hat{S}^\dagger = \sum_k \hbar\omega_k \left(\hat{c}_k^\dagger \hat{c}_k + \tfrac{1}{2}\right) + \Delta E , \tag{20}$$

where ΔE is the energy shift defined in (11). Thus $\hat{H}'$ has the same structure as the original Hamiltonian $\hat{H}$ in (10) but is still expressed in terms of the operators $\hat{c}_k$.

Example 1.2

The eigenstates of (20) are given by

$$|n', \alpha\rangle = \prod_k \frac{1}{\sqrt{n_k!}} (\hat{c}_k^\dagger)^{n_k} |0', \alpha\rangle \qquad (21)$$

with the ground state

$$|0', \alpha\rangle = \hat{S}(\alpha) |0, \alpha\rangle , \qquad (22)$$

which satisfies

$$\hat{c}_k |0', \alpha\rangle = 0 \quad \text{for all } k . \qquad (23)$$

The energy spectrum of course is again given by (12), remaining unchanged under the unitary transformation. The state vectors $|n', \alpha\rangle$ are related to the $|n, \alpha\rangle$ through the transformation operator $\hat{S}$. Let us check the consistency of both calculations. For the ground state the transformed equation (23) reads

$$0 = \hat{S}^\dagger \hat{c}_k \hat{S} |0, \alpha\rangle = (\hat{c}_k - \alpha_k) |0, \alpha\rangle = \hat{d}_k |0, \alpha\rangle , \qquad (24)$$

and thus agrees with (13). If we introduce factors $1 = \hat{S}\hat{S}^\dagger$, the general eigenstate (21) can be written as

$$\begin{aligned}
\hat{S}^\dagger |n', \alpha\rangle &= \prod_k \frac{1}{\sqrt{n_k!}} (\hat{S}^\dagger \hat{c}_k^\dagger \hat{S})^{n_k} \hat{S}^\dagger |0', \alpha\rangle \\
&= \prod_k \frac{1}{\sqrt{n_k!}} (\hat{d}_k^\dagger)^{n_k} |0, \alpha\rangle = |n, \alpha\rangle ,
\end{aligned} \qquad (25)$$

in agreement with (19b). Here and in the previous equation we used the inverted version of (15)

$$\begin{aligned}
\hat{S}^\dagger \hat{c}_k \hat{S} &= \hat{c}_k - \alpha_k = \hat{d}_k , \\
\hat{S}^\dagger \hat{c}_k^\dagger \hat{S} &= \hat{c}_k^\dagger - \alpha_k^* = \hat{d}_k^\dagger .
\end{aligned} \qquad (26)$$

What still is missing is an explicit contruction of a transformation operator $\hat{S}(\alpha)$ that fulfills the conditions (15) and (16). It will turn out to be useful to express this operator as the exponential function of another operator $\hat{\Lambda}(\alpha)$:

$$\hat{S}(\alpha) = e^{\hat{\Lambda}(\alpha)} . \qquad (27)$$

The exponential function of an operator can be defined via the Taylor expansion. Since $\hat{S}$ is unitary and since the general relation $(e^{\hat{A}})^\dagger = e^{\hat{A}^\dagger}$, $(e^{\hat{A}})^{-1} = e^{-\hat{A}}$ holds, the operator $\hat{\Lambda}$ must be antihermitean:

$$\hat{\Lambda}^\dagger(\alpha) = -\hat{\Lambda}(\alpha) . \qquad (28)$$

Now we demand that (15) holds, i.e., for all values of k

$$e^{\hat{\Lambda}} \hat{c}_k e^{-\hat{\Lambda}} = \hat{c}_k + \alpha_k . \qquad (29)$$

Let us investigate the Taylor expansion of the expression on the left-hand side:

Example 1.2

$$e^{\hat{\Lambda}} \hat{c}_k e^{-\hat{\Lambda}} = \left(1 + \hat{\Lambda} + \tfrac{1}{2}\hat{\Lambda}^2 \pm \ldots\right) \hat{c}_k \left(1 - \hat{\Lambda} + \tfrac{1}{2}\hat{\Lambda}^2 \mp \ldots\right)$$

$$= \hat{c}_k + \left(\hat{\Lambda}\hat{c}_k - \hat{c}_k\hat{\Lambda}\right) + \tfrac{1}{2}\left(\hat{\Lambda}^2 \hat{c}_k - 2\hat{\Lambda}\hat{c}_k\hat{\Lambda} + \hat{c}_k \hat{\Lambda}^2\right) + \ldots$$

$$= \hat{c}_k + [\hat{\Lambda}, \hat{c}_k] + \tfrac{1}{2}\left[\hat{\Lambda}, [\hat{\Lambda}, \hat{c}_k]\right] + \ldots . \tag{30}$$

This amounts to an infinite series containing multiple commutators of progressively higher order (see Exercise 1.3). The postulated equation (29) obviously is satisfied if

$$[\hat{\Lambda}, \hat{c}_k] = \alpha_k \tag{31}$$

holds. Since this commutator is a c number, all the higher multiple commutators simply vanish and the series (30) terminates. The condition (31), because of (1.92), is satisfied for all k if one chooses

$$\hat{\Lambda}(\alpha) = -\sum_k \alpha_k \hat{c}_k^\dagger . \tag{32}$$

This operator is not yet antihermitean. We can, however, enforce this property since it is possible to add any linear combination of annihilation operators $\hat{c}_k$ to (32) without violating (31). Thus we extend the operator $\hat{\Lambda}$ to the antihermitean form

$$\hat{\Lambda}(\alpha) = -\sum_k \left(\alpha_k \hat{c}_k^\dagger - \alpha_k^* \hat{c}_k\right) . \tag{33}$$

This is the final result for the transformation operator since it also satisfies the second condition (15):

$$e^{\hat{\Lambda}} \hat{c}_k^\dagger e^{-\hat{\Lambda}} = \hat{c}_k^\dagger + \alpha_k^* . \tag{34}$$

The unitary shift operator $\hat{S}(\alpha)$ thus reads explicitly

$$\hat{S}(\alpha) = \exp\left(-\sum_k (\alpha_k \hat{c}_k^\dagger - \alpha_k^* \hat{c}_k)\right) . \tag{35}$$

Finally we will study the connection between the ground state $|0, \alpha\rangle$ of the perturbed sytem and the old ground state $|0\rangle$, which describes the system in the absence of external forces. The transformed ground state $|0', \alpha\rangle$ can serve as a bridge between them. Comparing the defining relations for $|0\rangle$ and $|0', \alpha\rangle$, (1.94) and (23), we find that they coincide. Thus we can make the identification

$$|0', \alpha\rangle = |0\rangle . \tag{36}$$

The state $|0', \alpha\rangle$ thus actually does not depend on the force coefficients α_k after all. This dependence was "transformed away" by the operator $\hat{S}(\alpha)$. From (36) and (22) the perturbed ground state is given by

$$|0, \alpha\rangle = \hat{S}^\dagger(\alpha) |0\rangle = \exp\left(\sum_k (\alpha_k \hat{c}_k^\dagger - \alpha_k^* \hat{c}_k)\right) |0\rangle . \tag{37}$$

The operator acting on $|0\rangle$ contains a complicated mixture of creation and annihilation operators when expanded into a Taylor series. It would be convenient to isolate that part which consists of the annihilators only since their action on the ground state is trivial, i.e., $\hat{c}_k|0\rangle = 0$. Indeed there exists a very valuable general operator identity which can be put to use here. It allows us to factorize an exponential function of a sum of operators into the product of exponential functions. As shown in Exercise 1.3, the *Baker–Campbell–Hausdorff relation* reads

$$e^{\hat{A} + \hat{B}} = e^{-\frac{1}{2}[\hat{A}, \hat{B}]} e^{\hat{A}} e^{\hat{B}} \tag{38}$$

where $\hat{A}$ and $\hat{B}$ are operators which satisfy the condition

$$[\hat{A}, [\hat{A}, \hat{B}]] = [\hat{B}, [\hat{A}, \hat{B}]] = 0 \,. \tag{39}$$

If we define

$$\hat{A} = \sum_k \alpha_k \hat{c}_k^\dagger \quad, \quad \hat{B} = -\sum_k \alpha_k^* \hat{c}_k \quad, \quad [\hat{A}, \hat{B}] = \sum_k |\alpha_k|^2 \,, \tag{40}$$

then the condition (39) is obviously satisfied and (38) can be employed to simplify (37):

$$\begin{aligned}
|0, \alpha\rangle &= \exp\left(-\frac{1}{2} \sum_k |\alpha_k|^2\right) \exp\left(\sum_k \alpha_k \hat{c}_k^\dagger\right) \exp\left(-\sum_k \alpha_k^* \hat{c}_k\right) |0\rangle \\
&= \exp\left(-\frac{1}{2} \sum_k |\alpha_k|^2\right) \exp\left(\sum_k \alpha_k \hat{c}_k^\dagger\right) |0\rangle \,.
\end{aligned} \tag{41}$$

The BCH relation (38) also tells us that the transformation operator $\hat{S}(\alpha)$ can be factorized with respect to the various eigenmodes, since the operators $\hat{c}_k$ and $\hat{c}_k^\dagger$ belonging to different wave numbers do commute:

$$\hat{S}(\alpha) = \prod_k \hat{S}(\alpha_k) \,. \tag{42}$$

As a consequence the ground state can also be factorized according to

$$|0, \alpha\rangle = \prod_k |0, \alpha_k\rangle \,. \tag{43}$$

The perturbed ground state according to (41) is given by a *coherent superposition* of infinitely many multi-phonon states of the free system:

$$\begin{aligned}
|0, \alpha_k\rangle &= e^{-\frac{1}{2}|\alpha_k|^2} e^{\alpha_k \hat{c}_k^\dagger} |0\rangle \\
&= e^{-\frac{1}{2}|\alpha_k|^2} \sum_{n_k=0}^\infty \frac{1}{n_k!} \alpha_k^{n_k} (\hat{c}_k^\dagger)^{n_k} |0\rangle \\
&= e^{-\frac{1}{2}|\alpha_k|^2} \sum_{n_k} \frac{1}{\sqrt{n_k!}} \alpha_k^{n_k} |n_k\rangle \,.
\end{aligned} \tag{44}$$

Here $|n_k\rangle$ refers to the state consisting of n_k phonons of frequency ω_k which was introduced in (1.95). The ground state of the perturbed system is said to contain a superpositon of *virtual phonons*. The probability of finding n_k phonons of the type k is given by the squared projected amplitude

$$\left|\langle n_k|0, \alpha_k\rangle\right|^2 = e^{-|\alpha_k|^2} \frac{1}{n_k!} |\alpha_k|^{2n_k} \,. \tag{45}$$

Example 1.2

This is just the well-known *Poisson distribution*. The mean number of virtual phonons of type k is given by

$$\begin{aligned}\langle 0,\alpha_k|\hat{c}_k^\dagger \hat{c}_k|0,\alpha_k\rangle &= \sum_{n_k}\langle 0,\alpha_k|n_k\rangle\langle n_k|\hat{c}_k^\dagger \hat{c}_k|0,\alpha_k\rangle \\ &= \sum_{n_k=0}^{\infty} n_k \left|\langle n_k|0,\alpha_k\rangle\right|^2 \\ &= |\alpha_k|^2 \, e^{-|\alpha_k|^2} \sum_{n_k=1}^{\infty} \frac{1}{(n_k-1)!}|\alpha_k|^{2(n_k-1)} \\ &= |\alpha_k|^2 \,.\end{aligned} \qquad (46)$$

This result is quite plausible: the number of virtual phonons of wave number k present in the shifted ground state grows with the intensity of the corresponding Fourier component of the perturbing force.

A quantum state which is constructed as a superposition of states with varying particle number according to (44) is called a *coherent state*. In terms of a formal definition a coherent state is introduced as an *eigenstate of the annihilation operator*.[3] This is confirmed by the following calculation, which makes use of (15):

$$\begin{aligned}\hat{c}_k|0,\alpha_k\rangle &= \hat{c}_k \hat{S}^\dagger(\alpha_k)|0\rangle = \hat{S}^\dagger(\alpha_k)(\hat{c}_k+\alpha_k)|0\rangle = \alpha_k \hat{S}^\dagger(\alpha_k)|0\rangle \\ &= \alpha_k|0,\alpha_k\rangle \,.\end{aligned} \qquad (47)$$

The same result also follows immediately from (13) and (6).

For any set of complex parameters α_k a coherent state can be defined through Eqs. (43) and (44). The usefulness of coherent states is not restricted to the example of the shifted ground state of a chain of oscillators. Among other applications coherent states play an important role when the transition to the classical limit of a quantum system is studied. Whereas we know from (1.42) that for pure phonon states the expectation values of creation (annihilation) operators vanish, a coherent state leads to a finite result. With coherent states, the classical limit can then be performed by simply replacing

$$\hat{c}_k \to \alpha_k \quad \text{and} \quad \hat{c}_k^\dagger \to \alpha_k^* \,. \qquad (48)$$

This is valid provided that the operator under consideration has been *normal ordered*, which implies that product terms are constructed in such a way that creators are always positioned to the left of annihilators.

The expectation of the position operator (1.90) in the coherent state $|0,\alpha\rangle$ is given by

$$\begin{aligned}\bar{q}_n(t) &= \langle 0,\alpha|\hat{q}_n(t)|0,\alpha\rangle \\ &= \sum_k \sqrt{\frac{\hbar}{2m\omega_k}}\left(\hat{c}_k \, e^{-i\omega_k t} u_n^k + \hat{c}_k^\dagger \, e^{i\omega_k t} u_n^{k*}\right)\end{aligned}$$

---

[3] We note in passing that eigenstates of creation operators do not exist. A state $\hat{c}_k^\dagger|\Psi\rangle$ necessarily contains more phonons than the state $|\Psi\rangle$. The analogous argument applied to the annihilation operator does not work since $\hat{c}_k|0\rangle = 0$ vanishes.

$$= \sum_k \sqrt{\frac{\hbar}{2m\omega_k}} \left(\alpha_k \, e^{-i\omega_k t} u_n^k + \alpha_k^* \, e^{i\omega_k t} u_n^{k*} \right)$$

$$= \frac{1}{\sqrt{N}} \sum_k \sqrt{\frac{\hbar}{2m\omega_k}} \, 2 \, \mathrm{Re}\left[\alpha_k \, e^{i(kan - \omega_k t)} \right] . \quad (49)$$

In contrast to (1.47) this is a finite value that represents the equilibrium displacement reached under the influence of the external forces.

The energy expectation value in a coherent state is found from (5) to be

$$\langle 0, \alpha | \hat{H} | 0, \alpha \rangle = \sum_k \left[\hbar\omega_k \left(|\alpha_k|^2 + \tfrac{1}{2} \right) - F_k \alpha_k - F_k^* \alpha_k^* \right] . \quad (50)$$

This quadratic function of the admixture coefficients α_k is minimized if $\alpha_k = (1/\hbar\omega_k) F_k^*$ is chosen. This is just the result from (9) for the ground state.

Remark. A somewhat pathological special case results if the external force has a $k = 0$ component. This implies that a combined net force $\mathcal{F} = \sum_n \mathcal{F}_n \neq 0$ acts on the chain. Because of the factor $\omega_k = 0$ in the denominator of the expressions for α_k, energy, displacement, etc., divergencies appear. This is the result of the lack of a restoring force that could counter $\mathcal{F}$. Consequently the system as a whole undergoes an unlimited acceleration.

EXERCISE

1.3 The Baker–Campbell–Hausdorff Relation

Problem. Prove the following identities for the operators $\hat{A}$ and $\hat{B}$:

(a) $\quad e^{\hat{A}} \hat{B} e^{-\hat{A}} = \hat{B} + [\hat{A}, \hat{B}] + \frac{1}{2!} [\hat{A}, [\hat{A}, \hat{B}]] + \ldots ,\quad (1)$

(b) $\quad e^{\hat{A}+\hat{B}} = e^{\hat{A}} e^{\hat{B}} e^{-\frac{1}{2}[\hat{A},\hat{B}]} \quad (2)$

$\quad\quad$ povided that $\quad [[\hat{A}, \hat{B}], \hat{A}] = [[\hat{A}, \hat{B}], \hat{B}] = 0 . \quad (3)$

Hint: instead of a term-by-term inspection of the Taylor series, a more elegant treament proceeds as follows. To prove (1) introduce a continuous auxiliary parameter x and study the generalized expression

$$\hat{U}(x) = e^{x\hat{A}} \hat{B} e^{-x\hat{A}} . \quad (4)$$

$\hat{U}(x)$ satisfies an integral equation whose iterative solution, taken at the point $x = 1$, coincides with (1). Similarly (2) can be generalized to

$$e^{x(\hat{A}+\hat{B})} = e^{x\hat{A}} \hat{Q}(x) e^{x\hat{B}} . \quad (5)$$

Differentiation with respect to x leads to a differential equation for the operator-valued function $\hat{Q}(x)$ which can be solved easily, provided that (3) holds.

Solution. (a) To find an integral equation for the operator $\hat{U}(x)$ we differentiate (4) with respect to x:

$$\frac{d\hat{U}}{dx} = \hat{A} \, e^{x\hat{A}} \hat{B} e^{-x\hat{A}} - e^{x\hat{A}} \hat{B} e^{-x\hat{A}} \hat{A} = [\hat{A}, \hat{U}(x)] . \quad (6)$$

Exercise 1.3

The solution is fixed uniquely through the initial condition $\hat{U}(0) = \hat{B}$. As one immediately verifies by differentiation the following integral equation of Fredholm type is equivalent to (6):

$$\hat{U}(x) = \hat{B} + \int_0^x dy \left[\hat{A}, \hat{U}(y)\right] . \tag{7}$$

This integral equation can be solved iteratively, which amounts to constructing the Neumann series. In the first step one identifies

$$\hat{U}^{(0)}(x) = \hat{B} , \tag{8}$$

and subsequently

$$\hat{U}^{(1)}(x) = \hat{B} + \int_0^x dy \left[\hat{A}, \hat{U}^{(0)}(y)\right] = \hat{B} + [\hat{A}, \hat{B}]x , \tag{9}$$

followed by

$$\begin{aligned}\hat{U}^{(2)}(x) &= \hat{B} + \int_0^x dy \left[\hat{A}, \hat{U}^{(1)}(y)\right] = \hat{B} + \int_0^x dy \left[\hat{A}, \hat{B} + [\hat{A}, \hat{B}]y\right] \\ &= \hat{B} + [\hat{A}, \hat{B}]x + [\hat{A},[\hat{A}, \hat{B}]] \tfrac{1}{2}x^2 .\end{aligned}$$

This procedure can be repeated n times

$$\hat{U}^{(n)}(x) = \hat{B} + [\hat{A}, \hat{B}]x + \ldots + [\hat{A},[\ldots,[\hat{A}, \hat{B}]]] \tfrac{1}{n!}x^n . \tag{10}$$

In the limit $n \to \infty$ this leads to (1) when we set $x = 1$.

(b) The differential equation for the operator $\hat{Q}(x)$ is obtained by deriving (5) with respect to the parameter x:

$$(\hat{A} + \hat{B}) e^{x(\hat{A}+\hat{B})} = \hat{A} e^{x\hat{A}} \hat{Q}(x) e^{x\hat{B}} + e^{x\hat{A}} \frac{d\hat{Q}}{dx} e^{x\hat{B}} + e^{x\hat{A}} \hat{Q}(x) \hat{B} e^{x\hat{B}} \tag{11}$$

or

$$\hat{B} e^{x\hat{A}} \hat{Q} e^{x\hat{B}} = e^{x\hat{A}} \frac{d\hat{Q}}{dx} e^{x\hat{B}} + e^{x\hat{A}} \hat{Q} \hat{B} e^{x\hat{B}} . \tag{12}$$

After multiplication by $e^{-x\hat{A}}$ from the left and by $e^{-x\hat{B}}$ from the right, one arrives at the differential equation

$$\frac{d\hat{Q}}{dx} = e^{-x\hat{A}} \hat{B} e^{x\hat{A}} \hat{Q} - \hat{Q} \hat{B} . \tag{13}$$

Now we can make use of the operator identity (1). Since the multiple commutators involving $\hat{A}$ vanish according to (3), the series is seen to terminate and (11) becomes

$$\frac{d\hat{Q}}{dx} = (\hat{B} - x[\hat{A}, \hat{B}])\hat{Q} - \hat{Q}\hat{B} = -x[\hat{A}, \hat{B}]\hat{Q} + [\hat{B}, \hat{Q}] . \tag{14}$$

Assuming that the operator $\hat{Q}$ commutes with $\hat{B}$ this reduces to the simple differential equation

$$\frac{d\hat{Q}}{dx} = -x[\hat{A}, \hat{B}]\hat{Q} \quad \text{where} \quad \hat{Q}(0) = 1 . \tag{15}$$

Exercise 1.3

The solution reads
$$\hat{Q}(x) = e^{-\frac{1}{2}x^2[\hat{A},\hat{B}]} .\qquad(16)$$

This solution also justifies the assumption made when going from (14) to (15) since the operator $\hat{Q}$ commutes with $\hat{B}$ provided that $[[\hat{A},\hat{B}],\hat{B}] = 0$, which is true by assumption (3). If we set $x = 1$, the result (2) is obtained.

Note: If the condition (3) is not satisfied then (2) will contain an infinite series of multiple commutators. The general result has the form[4]

$$\begin{aligned}e^{\hat{A}}e^{\hat{B}} &= \exp\Big\{ \hat{A} + \hat{B} + \tfrac{1}{2!}[\hat{A},\hat{B}] + \tfrac{1}{3!}\left(\tfrac{1}{2}[[\hat{A},\hat{B}],\hat{B}] + \tfrac{1}{2}[\hat{A},[\hat{A},\hat{B}]]\right) \\ &\quad + \tfrac{1}{4!}[[\hat{A},[\hat{A},\hat{B}]],\hat{B}] + \ldots \Big\} .\end{aligned}\qquad(17)$$

---

[4] This result dates back to the mathematicians Campbell, Baker and Hausdorff. The original publications are J.E. Campbell: Proc. Lond. Math. Soc. **29**, 14 (1898); H.F. Baker: Proc. Lond Math. Soc. Ser. **3**, 24 (1904); F. Hausdorff: Ber. Verh. Sächs. Akad. Wiss. Leipzig, Math.-Naturwiss. Klasse **58**, 19 (1906).

2. Classical Field Theory

2.1 Introduction

Up to now we have been concerned with systems of point masses characterized by a discrete set of coordinates $q_i(t)$. Now we will turn to the study of *fields*. Each point of a (finite or infinite) region in space will be associated with some continuous field variable which we will denote by the generic name $\phi(\boldsymbol{x}, t)$. This obviously constitutes a system with an infinite number of degrees of freedom. The dynamical variables of the theory are now the values of the field $\phi(\boldsymbol{x})$ *at each point of space* instead of the discrete set of coordinates q_i encountered previously. The space coordinate $\boldsymbol{x}$ plays the role of a "continuous-valued index"

The Lagrange function now is a *functional* of the field, i.e., a mapping from a space of functions to the real numbers. It is customary to denote a functional dependence by square brackets:

$$L(t) = L[\phi(\boldsymbol{x}, t), \dot{\phi}(\boldsymbol{x}, t)] \ . \tag{2.1}$$

Being a functional, $L(t)$ simultaneously depends on the values of ϕ and $\dot{\phi}$ at all points in space. Note that the functional does not depend on the coordinate $\boldsymbol{x}$ itself.

In order to apply Hamilton's principle we define the *variation of a functional* $F[\phi(\boldsymbol{x})]$ as

$$\begin{aligned} \delta F[\phi] &= F[\phi + \delta\phi] - F[\phi] \\ &:= \int d^3x \, \frac{\delta F[\phi]}{\delta \phi(\boldsymbol{x})} \delta\phi(\boldsymbol{x}) \ . \end{aligned} \tag{2.2}$$

In the second line the *functional derivative* $\delta F/\delta\phi(\boldsymbol{x})$ of the functional $F[\phi]$ with respect to the function ϕ at the space point $\boldsymbol{x}$ was introduced. This tells us how the value of the functional is changed when the value of the function ϕ is varied at the point $\boldsymbol{x}$. As we will discuss in Sect. 2.3 in some more detail, the functional derivative obeys many of the rules of ordinary differential calculus.

Let us apply (2.2) to the Lagrangian (2.1), which depends on the functions ϕ and $\dot{\phi}$,

$$\delta L[\phi, \dot{\phi}] = \int d^3x \left(\frac{\delta L}{\delta \phi(\boldsymbol{x})} \delta\phi(\boldsymbol{x}) + \frac{\delta L}{\delta \dot{\phi}(\boldsymbol{x})} \delta\dot{\phi}(\boldsymbol{x}) \right) \ . \tag{2.3}$$

Both sides of this equation have an additional time dependence which is not marked. Integration of the Lagrangian leads to the action $W[\phi, \dot{\phi}]$ which is also a functional of the functions ϕ and $\dot{\phi}$. By integration over a time interval

$t_1 \ldots t_2$ (often $t_1 = -\infty$ and $t_2 = +\infty$ are chosen), the *variation of the action* follows as

$$\begin{aligned}
\delta W &= \delta \int_{t_1}^{t_2} dt\, L[\phi, \dot\phi] \\
&= \int_{t_1}^{t_2} dt\, d^3x \left(\frac{\delta L}{\delta \phi(\boldsymbol{x},t)} \delta\phi(\boldsymbol{x},t) + \frac{\delta L}{\delta \dot\phi(\boldsymbol{x},t)} \delta\dot\phi(\boldsymbol{x},t) \right) \\
&= \int_{t_1}^{t_2} dt\, d^3x \left(\frac{\delta L}{\delta \phi(\boldsymbol{x},t)} - \frac{\partial}{\partial t} \frac{\delta L}{\delta \dot\phi(\boldsymbol{x},t)} \right) \delta\phi(\boldsymbol{x},t)\,.
\end{aligned} \qquad (2.4)$$

Here an integration by parts was performed and the boundary condition $\delta\phi(\boldsymbol{x},t_1) = \delta\phi(\boldsymbol{x},t_2) = 0$ was used, together with the relation $\delta\dot\phi = \partial/\partial t\, \delta\phi$. *Hamilton's principle* of stationary action

$$\delta W[\phi, \dot\phi] = \delta \int_{t_1}^{t_2} dt\, L[\phi, \dot\phi] = 0 \qquad (2.5)$$

thus leads to

$$\frac{\delta L}{\delta \phi} - \frac{\partial}{\partial t} \frac{\delta L}{\delta \dot\phi} = 0\,, \qquad (2.6)$$

which is just the Euler–Lagrange equation (1.5) generalized to field theory.

To get a better understanding of the somwhat "abstract" notion of functional derivatives let us for a while return to a discretized description. We assume that space is divided into small cells of size ΔV_i. With each cell we associate the respective average value of the function $\phi(\boldsymbol{x},t)$, i.e.,

$$\phi_i(t) := \frac{1}{\Delta V_i} \int_{\Delta V_i} d^3x\, \phi(\boldsymbol{x},t)\,. \qquad (2.7)$$

Now L depends on the discrete set of "coordinates" $\phi_i, \dot\phi_i$. The variation of the Lagrangian then can be written as

$$\begin{aligned}
\delta L(\phi_i, \dot\phi_i) &= \sum_i \left(\frac{\delta L}{\delta \phi_i} \delta\phi_i + \frac{\delta L}{\delta \dot\phi_i} \delta\dot\phi_i \right) \\
&= \sum_i \left(\frac{1}{\Delta V_i} \frac{\delta L}{\delta \phi_i} \delta\phi_i + \frac{1}{\Delta V_i} \frac{\delta L}{\delta \dot\phi_i} \delta\dot\phi_i \right) \Delta V_i\,.
\end{aligned} \qquad (2.8)$$

Since variations at different spatial points are assumed to be independent, a comparison of (2.8) and (2.3) stipulates the identification

$$\begin{aligned}
\frac{\delta L(t)}{\delta \phi(\boldsymbol{x},t)} &= \lim_{\Delta V_i \to 0} \frac{1}{\Delta V_i} \frac{\delta L(t)}{\delta \phi_i(t)}\,, \\
\frac{\delta L(t)}{\delta \dot\phi(\boldsymbol{x},t)} &= \lim_{\Delta V_i \to 0} \frac{1}{\Delta V_i} \frac{\delta L(t)}{\delta \dot\phi_i(t)}\,,
\end{aligned} \qquad (2.9)$$

where $\boldsymbol{x}$ is located in the cell ΔV_i. The functional derivative thus essentially means differentiation with respect to the value the field takes at the point $\boldsymbol{x}$. To be more specific, in the following we will consider *local field theories* in which the Lagrangian L can be written as a volume integral over a density function $\mathcal{L}$:

$$L(t) = \int \mathrm{d}^3 x\, \mathcal{L}\big(\phi(\boldsymbol{x},t), \boldsymbol{\nabla}\phi(\boldsymbol{x},t), \dot{\phi}(\boldsymbol{x},t)\big) \,. \tag{2.10}$$

The Lagrange density may depend on the field function ϕ, on its time derivative $\dot{\phi}$, and also on the gradient $\boldsymbol{\nabla}\phi$. In principle $\mathcal{L}$ could depend also on higher derivatives of ϕ. This, however, would have undesirable consequences, since the resulting equations of motion would be of higher than second order. Furthermore we will not study *nonlocal* theories in which the Lagrange density at $\boldsymbol{x}$ has an additional dependence on the value of the field at other space points $\boldsymbol{y} \neq \boldsymbol{x}$. The restriction to local Lagrange densities containing first-order derivatives has been found to be sufficiently general to form the basis for all present-day field theories.

When constructing a Lagrange density, one starts with considerable freedom. The choices get narrowed down by imposing the symmetry and invariance properties the theory under consideration is expected to satisfy. As the most basic requirement for any relativistic field theory the Lagrange density must be a *Lorentz scalar*.

Using the Lagrange density $\mathcal{L}$, the variation of $L(t)$ can be written as

$$\begin{aligned}
\delta L(t) &= \int \mathrm{d}^3 x \left(\frac{\partial \mathcal{L}}{\partial \phi(\boldsymbol{x},t)} \delta\phi(\boldsymbol{x},t) + \frac{\partial \mathcal{L}}{\partial (\boldsymbol{\nabla}\phi(\boldsymbol{x},t))} \delta\boldsymbol{\nabla}\phi(\boldsymbol{x},t) \right. \\
&\qquad \left. + \frac{\partial \mathcal{L}}{\partial \dot{\phi}(\boldsymbol{x},t)} \delta\dot{\phi}(\boldsymbol{x},t) \right) \\
&= \int \mathrm{d}^3 x \left(\left(\frac{\partial \mathcal{L}}{\partial \phi(\boldsymbol{x},t)} - \boldsymbol{\nabla} \frac{\partial \mathcal{L}}{\partial (\boldsymbol{\nabla}\phi(\boldsymbol{x},t))} \right) \delta\phi(\boldsymbol{x},t) \right. \\
&\qquad \left. + \frac{\partial \mathcal{L}}{\partial \dot{\phi}(\boldsymbol{x},t)} \delta\dot{\phi}(\boldsymbol{x},t) \right) \,.
\end{aligned} \tag{2.11}$$

Here the relation $\delta\boldsymbol{\nabla}\phi = \boldsymbol{\nabla}\delta\phi$ was used, followed by an integration by parts. To justify this procedure the fields and their derivatives on the surface have to approach zero fast enough. Comparing this with (2.3) we arrive at explicit expressions for the functional derivatives:

$$\frac{\delta L(t)}{\delta \phi(\boldsymbol{x},t)} = \frac{\partial \mathcal{L}}{\partial \phi(\boldsymbol{x},t)} - \boldsymbol{\nabla} \frac{\partial \mathcal{L}}{\partial (\boldsymbol{\nabla}\phi(\boldsymbol{x},t))} \,, \tag{2.12a}$$

$$\frac{\delta L(t)}{\delta \dot{\phi}(\boldsymbol{x},t)} = \frac{\partial \mathcal{L}}{\partial \dot{\phi}(\boldsymbol{x},t)} \,. \tag{2.12b}$$

Expressed in terms of the Lagrange density the *Euler–Lagrange equation* (2.6) reads

$$\frac{\partial \mathcal{L}}{\partial \phi(\boldsymbol{x},t)} - \boldsymbol{\nabla} \frac{\partial \mathcal{L}}{\partial (\boldsymbol{\nabla}\phi(\boldsymbol{x},t))} - \frac{\partial}{\partial t} \frac{\partial \mathcal{L}}{\partial \dot{\phi}(\boldsymbol{x},t)} = 0 \tag{2.13}$$

or, using relativistic (covariant) notation[1] with $x^\mu = (x^0, \boldsymbol{x}) = (t, \boldsymbol{x})$,

$$\frac{\partial \mathcal{L}}{\partial \phi(x)} - \frac{\partial}{\partial x^\mu} \frac{\partial \mathcal{L}}{\partial (\partial_\mu \phi)} = 0 \,. \tag{2.14}$$

Here the four-dimensional gradient was written as usual as $\partial \phi / \partial x^\mu \equiv \partial_\mu \phi = (\partial_t, \boldsymbol{\nabla})\phi$.

---

[1] Where not stated otherwise we will always use relativistic units, i.e., we set $c = 1$.

As soon as the Lagrange density of a physical theory is known the field equations can be written down, taking the form of partial differential equations in the variables x and t. This will be worked out later in great detail for the Schrödinger, Klein–Gordon, Dirac, and Maxwell fields. Since we have assumed that $\mathcal{L}$ does not depend on derivatives of higher than first order, the fields satisfy differential equations of at most second order.

If the Lagrange function depends on several independent fields $\phi_r, r = 1, \ldots, N$, then (2.14) can be generalized as

$$\frac{\partial \mathcal{L}}{\partial \phi_r} - \frac{\partial}{\partial x^\mu} \frac{\partial \mathcal{L}}{\partial (\partial_\mu \phi_r)} = 0 \ . \tag{2.15}$$

Problems can arise if the components ϕ_r are mutually dependent, as will be the case for particles with nonzero spin.

We finally remark that the field eqution can also be deduced directly, i.e., without the formal introduction of the functional derivative, from the variational principle

$$\delta \int_{t_1}^{t_2} \mathrm{d}t \int \mathrm{d}^3 x \ \mathcal{L}\big(\phi(x), \partial_\mu \phi(x)\big) = 0 \tag{2.16}$$

by using integration by parts in four dimensions.

2.2 The Hamilton Formalism

To apply Hamilton's formalism to a field theory, one first has to define a "momentum" that is canonically conjugate to the field variable. In analogy to (1.6) we define the *canonically conjugate field* by the functional derivative

$$\pi(\boldsymbol{x}, t) = \frac{\delta L(t)}{\delta \dot{\phi}(\boldsymbol{x}, t)} \ . \tag{2.17}$$

According to the results of the last section this reduces to an ordinary derivative of the Lagrange density

$$\pi(\boldsymbol{x}, t) = \lim_{\Delta V_i \to 0} \frac{1}{\Delta V_i} \frac{\delta L(t)}{\delta \dot{\phi}_i(t)} = \frac{\partial \mathcal{L}}{\partial \dot{\phi}(\boldsymbol{x}, t)} \ . \tag{2.18}$$

The time derivative of π follows from the Euler–Lagrange equation (2.6)

$$\dot{\pi}(\boldsymbol{x}, t) = \frac{\delta L}{\delta \phi(\boldsymbol{x}, t)} \ . \tag{2.19}$$

Now the *Hamiltonian* is introduced through the Legendre transformation

$$H(t) = \int \mathrm{d}^3 x \ \pi(\boldsymbol{x}, t) \dot{\phi}(\boldsymbol{x}, t) - L(t) \ . \tag{2.20}$$

This can also be written as an integral over the Hamilton density $\mathcal{H}(\boldsymbol{x}, t)$:

$$H(t) = \int \mathrm{d}^3 x \, \mathcal{H}(x) \quad \text{where} \quad \mathcal{H}(x) = \pi(x) \dot{\phi}(x) - \mathcal{L}(x) \ . \tag{2.21}$$

Hamilton's equations of motion take a form reminiscent of the earlier result from ordinary mechanics; see (1.8):

$$\dot{\phi} = \frac{\delta H}{\delta \pi} \quad , \quad \dot{\pi} = -\frac{\delta H}{\delta \phi} . \tag{2.22}$$

These equations are derived by taking the variation of H and using the Legendre transformation (2.20):

$$\delta H = \int d^3x \left(\dot{\phi}\,\delta\pi + \pi\,\delta\dot{\phi} \right) - \delta L = \int d^3x \left(\dot{\phi}\,\delta\pi - \dot{\pi}\,\delta\phi \right) , \tag{2.23}$$

since

$$\delta L = \int d^3x \left(\frac{\delta L}{\delta \phi}\,\delta\phi + \frac{\delta L}{\delta \dot{\phi}}\,\delta\dot{\phi} \right) = \int d^3x \left(\dot{\pi}\,\delta\phi + \pi\,\delta\dot{\phi} \right) . \tag{2.24}$$

This directly yields (2.22).

Since the functional $H[\phi, \pi]$ can depend on ϕ, π and their gradients $\nabla\phi, \nabla\pi$ the functional derivatives in (2.22) can be expressed explicitly in terms of the Hamilton density $\mathcal{H}$ as

$$\frac{\delta H}{\delta \phi} = \frac{\partial \mathcal{H}}{\partial \phi} - \nabla \frac{\partial \mathcal{H}}{\partial (\nabla \phi)} , \tag{2.25a}$$

$$\frac{\delta H}{\delta \pi} = \frac{\partial \mathcal{H}}{\partial \pi} - \nabla \frac{\partial \mathcal{H}}{\partial (\nabla \pi)} . \tag{2.25b}$$

Finally let us study the role of the Poisson brackets in field theory. Given two functionals $F[\phi, \pi]$ and $G[\phi, \pi]$ we define

$$\{F, G\}_{\mathrm{PB}} = \int d^3x \left(\frac{\delta F}{\delta \phi(x)} \frac{\delta G}{\delta \pi(x)} - \frac{\delta F}{\delta \pi(x)} \frac{\delta G}{\delta \phi(x)} \right) . \tag{2.26}$$

Because of Hamilton's equations of motion, (2.22), the time evolution of a functional satisfies

$$\dot{F}(t) = \int d^3x \left(\frac{\delta F(t)}{\delta \phi(x)} \dot{\phi}(x) + \frac{\delta F(t)}{\delta \pi(x)} \dot{\pi}(x) \right) = \{F, H\}_{\mathrm{PB}} \tag{2.27}$$

provided that there is no explicit time dependence.

There is a seemingly trivial but very important special case which allows the immediate evaluation of a functional derivative. This is based on the observation that a function can be written as a functional depending on itself:

$$\phi(\boldsymbol{x}, t) = \int d^3x'\, \phi(\boldsymbol{x}', t)\delta^3(\boldsymbol{x} - \boldsymbol{x}') , \tag{2.28}$$

where $\boldsymbol{x}$, like t, can be viewed as a parameter of the functional. The functional derivative with respect to $\phi(\boldsymbol{x}', t)$ then reads

$$\frac{\delta \phi(\boldsymbol{x}, t)}{\delta \phi(\boldsymbol{x}', t)} = \delta^3(\boldsymbol{x} - \boldsymbol{x}') . \tag{2.29}$$

This plausible result follows from the defining relation for the functional derivative (2.2) through

$$\delta\phi(\boldsymbol{x}, t) = \int d^3x'\, \frac{\delta \phi(\boldsymbol{x}, t)}{\delta \phi(\boldsymbol{x}', t)} \delta\phi(\boldsymbol{x}', t). \tag{2.30}$$

The result also can be understood in the framework of the discretized formulation involving volume elements. Since $\phi(\boldsymbol{x}, t)$ is then localized in a spatial cell taking the value $\phi_j(t)$ the derivative becomes

$$\frac{\delta\phi(\boldsymbol{x},t)}{\delta\phi(\boldsymbol{x}',t)} = \lim_{\Delta V_i \to 0} \frac{1}{\Delta V_i}\frac{\partial\phi_j}{\partial\phi_i} = \lim_{\Delta V_i \to 0} \frac{\delta_{ij}}{\Delta V_i} = \delta^3(\boldsymbol{x}-\boldsymbol{x}') \;. \tag{2.31}$$

Obviously the relations

$$\frac{\delta\pi(\boldsymbol{x},t)}{\delta\pi(\boldsymbol{x}',t)} = \delta^3(\boldsymbol{x}-\boldsymbol{x}') \tag{2.32}$$

and

$$\frac{\delta\pi(\boldsymbol{x},t)}{\delta\phi(\boldsymbol{x}',t)} = \frac{\delta\phi(\boldsymbol{x},t)}{\delta\pi(\boldsymbol{x}',t)} = 0 \tag{2.33}$$

also hold since ϕ and π are independent functionals.

With the use of (2.31)-(2.33) the Poisson brackets of ϕ and π with the Hamilton function can be evaluated:

$$\begin{aligned}\dot\phi(\boldsymbol{x},t) &= \{\phi(\boldsymbol{x},t), H(t)\}_{\text{PB}} \\ &= \int\!\mathrm{d}^3 x' \,\frac{\delta\phi(\boldsymbol{x},t)}{\delta\phi(\boldsymbol{x}',t)}\frac{\delta H(t)}{\delta\pi(\boldsymbol{x}',t)} = \frac{\delta H}{\delta\pi(\boldsymbol{x},t)} \;,\end{aligned} \tag{2.34}$$

and similarly

$$\begin{aligned}\dot\pi(\boldsymbol{x},t) &= \{\pi(\boldsymbol{x},t), H(t)\}_{\text{PB}} \\ &= -\int\!\mathrm{d}^3 x' \,\frac{\delta\pi(\boldsymbol{x},t)}{\delta\pi(\boldsymbol{x}',t)}\frac{\delta H(t)}{\delta\phi(\boldsymbol{x}',t)} = -\frac{\delta H(t)}{\delta\phi(\boldsymbol{x},t)} \;.\end{aligned} \tag{2.35}$$

These are Hamilton's equations of motion (2.22).

The mutual Poisson brackets of the fields will turn out to be of special interest. For these we find

$$\begin{aligned}\{\phi(\boldsymbol{x},t), \pi(\boldsymbol{x}',t)\}_{\text{PB}} &= \int\!\mathrm{d}^3 x'' \,\frac{\delta\phi(\boldsymbol{x},t)}{\delta\phi(\boldsymbol{x}'',t)}\frac{\delta\pi(\boldsymbol{x}',t)}{\delta\pi(\boldsymbol{x}'',t)} \\ &= \int\!\mathrm{d}^3 x'' \,\delta^3(\boldsymbol{x}-\boldsymbol{x}'')\delta^3(\boldsymbol{x}'-\boldsymbol{x}'') \\ &= \delta^3(\boldsymbol{x}-\boldsymbol{x}')\end{aligned} \tag{2.36}$$

and because of (2.33)

$$\{\phi(\boldsymbol{x},t), \phi(\boldsymbol{x}',t)\}_{\text{PB}} = \{\pi(\boldsymbol{x},t), \pi(\boldsymbol{x}',t)\}_{\text{PB}} = 0 \;. \tag{2.37}$$

These simple relations of course only apply if both fields are taken at equal time values $t' = t$.

A field theory can be quantized in the same way as a discrete mechanical system by replacing the Poisson brackets by commutation relations between operators in Hilbert space. This will be treated in detail later.

MATHEMATICAL SUPPLEMENT

2.3 Functional Derivatives

The discussion of field theories makes ample use of the functional derivative, i.e., the differentiation of a functional with respect to its argument. Here we

2.3 Functional Derivatives

Supplement 2.3

want to present a brief and nonrigorous discussion of the definition and some properties of this mathematical operation.

Let $F[\phi]$ be a functional, i.e., a mapping from a normed linear space of functions (a Banach space) $M = \{\phi(x) : x \in \mathbb{R}\}$ to the field of real or complex numbers, $F : M \to \mathbb{R}$ or $\mathbb{C}$. The object $\delta F[\phi]/\delta\phi(x)$ tells how the value of the functional changes if the function $\phi(x)$ is changed at the point x. Thus the functional derivative (also known as the Fréchet derivative) itself is an ordinary function depending on x. As its defining relation we will use

$$\delta F[\phi] = \int \mathrm{d}x \, \frac{\delta F[\phi]}{\delta\phi(x)} \, \delta\phi(x) \,, \tag{1}$$

which implies that the total change in F upon variation of the function $\phi(x)$ is a linear superposition of the local changes summed over the whole range of x values. Figure 2.1 illustrates this construction for the simplest possible case of a scalar function in one dimension.

As in ordinary differentiation, the functional derivative can also be represented as the limit of divided differences. To see this we construct a variation of the "independent variable", i.e., the function $\phi(x)$ which is localized at the point y having strength ϵ:

$$\delta\phi(x) = \epsilon \, \delta(x - y) \,. \tag{2}$$

Inserted into (1) this leads to

$$\delta F[\phi] = F[\phi + \epsilon \, \delta(x - y)] - F[\phi] = \int \mathrm{d}x \, \frac{\delta F[\phi]}{\delta\phi(x)} \, \epsilon \, \delta(x - y) = \epsilon \, \frac{\delta F}{\delta\phi(y)} \tag{3}$$

or in the limit of vanishing ϵ

$$\frac{\delta F[\phi]}{\delta\phi(y)} = \lim_{\epsilon \to 0} \frac{F[\phi + \epsilon \delta(x - y)] - F[\phi]}{\epsilon} \,. \tag{4}$$

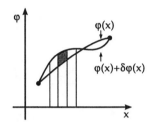

Fig. 2.1. Illustration of the functional derivative of a functional $F[\phi]$

This relation is equivalent to the definition (1) and is well suited for the practical evaluation of functional derivatives. To arrive at unique results we must specify the order of the mathematical operations. We introduce the rule that the limit $\epsilon \to 0$ has to be taken first, before other possible limiting operations. Note that the x dependence on the right-hand side of (4) is only a formal one; it might have been preferable to write $\epsilon(\,\cdot\, - y)$ for the variational function, in keeping with the notation $\phi(\cdot)$ instead of $\phi(x)$, which is used sometimes in the mathematical literature if x is a "silent" argument.

Most of the rules of ordinary differential calculus also apply to functional derivatives. Obviously we deal with a *linear* operation. For the product of two functionals $G[\phi]$ and $H[\phi]$ the *product rule* applies. The derivative of the combined functional $F[\phi] = G[\phi] H[\phi]$ is given by

$$\frac{\delta F[\phi]}{\delta\phi(x)} = \frac{\delta G[\phi]}{\delta\phi(x)} H[\phi] + G[\phi] \frac{\delta H[\phi]}{\delta\phi(x)} \,. \tag{5}$$

Similarly the *chain rule* can be applied to the functional of a functional:

$$\frac{\delta}{\delta\phi(y)} F[G[\phi]] = \int \mathrm{d}x \, \frac{\delta F[G]}{\delta G(x)} \frac{\delta G[\phi]}{\delta\phi(y)} \,. \tag{6}$$

The proof of (5) and (6) mutatis mutandis can be taken over from ordinary differential calculus. In a special case of the chain rule the functional $G[\phi]$

Supplement 2.3

may be an ordinary function $g(\phi)$ being localized at the point x. Then the integral in (6) disappears and we have

$$\frac{\delta}{\delta\phi(y)}F[g(\phi)] = \frac{\delta F}{\delta g(\phi(y))}\frac{\mathrm{d}g(\phi)}{\mathrm{d}\phi(y)} \,. \tag{7}$$

Now let us illustrate the functional derivative through a few specific examples. Note that all integrals in the following will extend over the whole real axis.

1. $\displaystyle F[\phi] = \int \mathrm{d}x \, \big(\phi(x)\big)^n \,. \tag{8}$

Equation (4) leads to the functional derivative

$$\begin{aligned}
\frac{\delta F[\phi]}{\delta\phi(y)} &= \lim_{\epsilon\to 0}\frac{1}{\epsilon}\left(\int \mathrm{d}x\,\big(\phi(x)+\epsilon\delta(x-y)\big)^n - \int \mathrm{d}x\,\big(\phi(x)\big)^n\right) \\
&= \int \mathrm{d}x\, n\big(\phi(x)\big)^{n-1}\delta(x-y) = n\big(\phi(y)\big)^{n-1} \,.
\end{aligned} \tag{9}$$

In this case functional differentiation cancels the integration over x. It should be noted that here the limit $\epsilon \to 0$ had to be taken before the integration since otherwise indefinite expressions of the type $(\delta(x-y))^2$ would have appeared. The result (9) can immediately be generalized to any function $g(\phi(x))$ in the integrand, which has a Taylor series expansion. Each term of the power series can then be treated as in (9), which leads to the result

$$\frac{\delta}{\delta\phi(y)}\int \mathrm{d}x\, g(\phi(x)) = g'(\phi(y)) \,. \tag{10}$$

The prime denotes the derivative of g with respect to its argument. The same result would also have followed from the chain rule (7).

2. $\displaystyle F[\phi] = \int \mathrm{d}x \left(\frac{\mathrm{d}\phi(x)}{\mathrm{d}x}\right)^n \,. \tag{11}$

The prescription (4) yields

$$\begin{aligned}
\frac{\delta F[\phi]}{\delta\phi(y)} &= \lim_{\epsilon\to 0}\frac{1}{\epsilon}\left(\int \mathrm{d}x\left(\frac{\mathrm{d}}{\mathrm{d}x}[\phi(x)+\epsilon\delta(x-y)]\right)^n - \int \mathrm{d}x\left(\frac{\mathrm{d}}{\mathrm{d}x}\phi(x)\right)^n\right) \\
&= \lim_{\epsilon\to 0}\frac{1}{\epsilon}\int \mathrm{d}x\left(\left(\frac{\mathrm{d}\phi}{\mathrm{d}x}\right)^n + n\epsilon\left(\frac{\mathrm{d}\phi}{\mathrm{d}x}\right)^{n-1}\frac{\mathrm{d}}{\mathrm{d}x}\delta(x-y) + O(\epsilon^2) - \left(\frac{\mathrm{d}\phi}{\mathrm{d}x}\right)^n\right) \\
&= n\int \mathrm{d}x\left(\frac{\mathrm{d}\phi}{\mathrm{d}x}\right)^{n-1}\frac{\mathrm{d}}{\mathrm{d}x}\delta(x-y) = -n\frac{\mathrm{d}}{\mathrm{d}x}\left(\frac{\mathrm{d}\phi}{\mathrm{d}x}\right)^{n-1}\bigg|_y \,.
\end{aligned} \tag{12}$$

Here in the last step an integration by parts was performed. This result can also be extended to general functions $h(\mathrm{d}\phi/\mathrm{d}x)$:

$$\frac{\delta}{\delta\phi(y)}\int \mathrm{d}x\, h\!\left(\frac{\mathrm{d}\phi}{\mathrm{d}x}\right) = -\frac{\mathrm{d}}{\mathrm{d}x}\frac{\mathrm{d}h}{\mathrm{d}\!\left(\frac{\mathrm{d}\phi}{\mathrm{d}x}\right)}\bigg|_y \,. \tag{13}$$

3. $F_y[\phi] = \int dx' \, K(y, x')\phi(x')$. (14)

As indicated by the index, this functional carries an additional dependence on the coordinate y. The derivative with respect to $\phi(x)$ simply gives the integral kernel:

$$\frac{\delta F_y[\phi]}{\delta \phi(x)} = K(y, x) \,.$$ (15)

4. $F_x[\phi] = \phi(x)$. (16)

This "trivial" functional associates the function ϕ with its value at the point x, i.e., it has a parametric coordinate dependence as had the previous example. The functional derivative gets very simple:

$$\frac{\delta F_x[\phi]}{\delta \phi(y)} = \lim_{\epsilon \to 0} \frac{1}{\epsilon} \Big(\phi(x) + \epsilon \delta(x-y) - \phi(x) \Big) = \delta(x-y) \,.$$ (17)

Because of (1) this result is intuitively clear since the value of ϕ at x can only change if the variation $\delta\phi$ acts on exactly the same location.

5. $F_x[\phi] = \nabla \phi(x)$. (18)

We obtain

$$\frac{\delta F_x[\phi]}{\delta \phi(y)} = \lim_{\epsilon \to 0} \frac{1}{\epsilon} \Big(\nabla_x (\phi(x) + \epsilon \delta(x-y)) - \nabla_x \phi(x) \Big) = \nabla_x \delta(x-y) \,.$$ (19)

In this example the difference between functional differentiation and ordinary partial differentiation is clearly evident. In the latter case $\nabla \phi$ and ϕ are viewed as independent functions leading to the result

$$\frac{\partial F_x}{\partial \phi} = 0 \,.$$ (20)

2.4 Conservation Laws in Classical Field Theories

Conservation laws, i.e., the existence of quantities which do not change in time, independent of the dynamical evolution of a system, play an all-important role in theoretical physics. The conservation of energy, momentum, and angular momentum are fundamental laws that any theory has to guarantee if it is to give a valid description of nature. In addition to these basic properties, physical systems often possess further additional conserved quantities, such as charge, isospin, or generalizations thereof. From a fundamental point of view the conservation laws are natural consequences of the *symmetry properties* of a system. For each continuous transformation of the coordinates and/or the fields under which "physics does not change" the existence of a conserved quantity can be deduced. For instance, the conservation of energy, momentum, and angular momentum are based on the invariance of a theory under temporal and spatial translations and under rotations in space. Similarly the conservation of charge follows from an invariance under phase transformations.

The mathematical foundation of this connection was elucidated systematically at a the beginning of the century. The link between symmetry properties and conservation laws is know as *Noether's theorem*.[2] In the following we will derive this theorem for a general classical field theory and general symmetry transformations. Subsequently the various conserved quantities mentioned earlier will be derived as special applications of the theorem.

Noether's Theorem. Let us first consider the case that the action integral does not change if the *coordinates* are subject to a continuous transformation. It will be sufficient to study *infinitesimal transformations* of the kind

$$x'_\mu = x_\mu + \delta x_\mu \,. \tag{2.38}$$

Let the corresponding change in the field $\phi_r(x)$ be

$$\phi'_r(x') = \phi_r(x) + \delta\phi_r(x) \,, \tag{2.39}$$

with the resulting change in the Lagrange density of

$$\mathcal{L}'(x') = \mathcal{L}(x) + \delta\mathcal{L}(x) \,. \tag{2.40}$$

Here for brevity's sake the functional dependance of the Lagrange density $\mathcal{L}(x) \equiv \mathcal{L}\big(\phi(x), \partial\phi/\partial x_\mu(x)\big)$ was omitted. $\mathcal{L}'(x')$ is obtained by inserting the primed quantities into the original Lagrange densitiy $\mathcal{L}'(x') = \mathcal{L}\big(\phi'(x'), \partial\phi'/\partial x'_\mu(x')\big)$. It is important to understand that the variation defined above consists of two ingredients, namely the transformation of the coordinates from x to x' and furthermore the change of the "shape" of the field function from ϕ to ϕ'. As an obvious example, think of a vector field that changes its direction if the coordinates system is rotated. Therefore it is useful to define a *modified variation*

$$\tilde\delta\phi_r(x) = \phi'_r(x) - \phi_r(x) \,, \tag{2.41}$$

which keeps the value of the coordinate x fixed and only takes into account the change of shape of the field.[3] The two types of variations are related through

$$\begin{aligned}
\tilde\delta\phi_r(x) &= \phi'_r(x) - \phi'_r(x') + \phi'_r(x') - \phi_r(x) \\
&= \delta\phi_r(x) - \big(\phi'_r(x') - \phi'_r(x)\big) = \delta\phi_r(x) - \frac{\partial\phi'_r(x)}{\partial x_\mu}\delta x_\mu \\
&= \delta\phi_r(x) - \frac{\partial\phi_r}{\partial x_\mu}\delta x_\mu \,.
\end{aligned} \tag{2.42}$$

In the second but last step the first term of the Taylor expansion was inserted and finally, in lowest order, $\phi'_r(x)$, was replaced by $\phi_r(x)$. Thus this equation

---

[2] E. Noether: *Invariante Variationsprobleme*, Nachr. d. Kgl. Ges. d. Wiss. (Math. phys. Klasse), Göttingen (1918), 235.

[3] The quantity $\tilde\delta\phi_r(x)$ is sometimes called a *total variation*, whereas $\delta\phi_r(x)$ defined in (2.39) is the *local variation* since here x' and x *refer to the same point*, which is represented differently only in terms of the two sets of coordinates.

and many of the following results are only valid up to first order in the variations. This is sufficient since the whole treatment in any case only applies to infinitesimal variations.

The modified variation $\tilde{\delta}$ quite conveniently has the property to commute with differentiation $\partial/\partial x_\mu$:

$$\frac{\partial}{\partial x_\mu}\tilde{\delta}\phi_r(x) = \tilde{\delta}\Big(\frac{\partial \phi_r(x)}{\partial x_\mu}\Big) \,. \tag{2.43}$$

This is immediately obvious from the definition (2.41). Variations of the δ type do not share this property. When calculating the gradient one finds an additional term

$$\begin{aligned}
\frac{\partial}{\partial x_\mu}\big(\delta\phi_r(x)\big) &= \frac{\partial}{\partial x_\mu}\phi'_r(x') - \frac{\partial}{\partial x_\mu}\phi_r(x) \\
&= \Big(\frac{\partial \phi'_r(x')}{\partial x'_\mu} - \frac{\partial \phi_r(x)}{\partial x_\mu}\Big) + \frac{\partial \phi'_r(x')}{\partial x_\mu} - \frac{\partial \phi'_r(x')}{\partial x'_\mu} \\
&= \tilde{\delta}\Big(\frac{\partial \phi_r(x)}{\partial x_\mu}\Big) + \frac{\partial x'^\nu}{\partial x_\mu}\frac{\partial \phi'_r(x')}{\partial x'^\nu} - \frac{\partial \phi'_r(x')}{\partial x'_\mu} \\
&= \tilde{\delta}\Big(\frac{\partial \phi_r(x)}{\partial x_\mu}\Big) + \frac{\partial \phi'_r(x')}{\partial x'^\nu}\frac{\partial \delta x^\nu}{\partial x_\mu} \\
&= \tilde{\delta}\Big(\frac{\partial \phi_r(x)}{\partial x_\mu}\Big) + \frac{\partial \phi_r(x)}{\partial x^\nu}\frac{\partial \delta x^\nu}{\partial x_\mu} \,,
\end{aligned} \tag{2.44}$$

where, according to (2.38), the identity $\partial x'^\nu/\partial x_\mu = g^{\nu\mu} + \partial \delta x^\nu/\partial x_\mu$ was used and the last step is valid only to first order.

Now we study the consequences that follow if the transformations (2.38) and (2.39) *leave the action integral invariant*, i.e., we demand

$$\delta W \equiv \int_{\Omega'} d^4x' \, \mathcal{L}'(x') - \int_\Omega d^4x \, \mathcal{L}(x) \stackrel{!}{=} 0 \,, \tag{2.45}$$

where Ω' denotes the same volume of integration as Ω, being expressed in terms of the new coordinates x'. We introduce the variation of the Lagrange density into: (2.45)

$$\delta W = \int_{\Omega'} d^4x' \, \delta\mathcal{L}(x) + \int_{\Omega'} d^4x' \, \mathcal{L}(x) - \int_\Omega d^4x \, \mathcal{L}(x) \,. \tag{2.46}$$

Transformation of the volume of integration in (2.46) introduces a Jacobi determinant that in first order reduces to

$$\begin{aligned}
d^4x' &= \Big|\frac{\partial(x'^\mu)}{\partial(x^\nu)}\Big| d^4x = \begin{vmatrix} 1+\frac{\partial \delta x_0}{\partial x_0} & \frac{\partial \delta x_0}{\partial x_1} & \cdots & \cdots \\ \frac{\partial \delta x_1}{\partial x_0} & 1+\frac{\partial \delta x_1}{\partial x_1} & \cdots & \cdots \\ \vdots & \vdots & \ddots & \cdots \\ \vdots & \vdots & \vdots & 1+\frac{\partial \delta x_3}{\partial x_3} \end{vmatrix} d^4x \\
&= \Big(1+\frac{\partial \delta x^\mu}{\partial x^\mu}\Big) d^4x \,.
\end{aligned} \tag{2.47}$$

All terms involving mixed derivatives of the variation $\delta x'^\mu$ are of higher order. If we use the modified variation of type (2.42), (2.46) is simplified in first order:

$$\begin{aligned}\delta W &= \int_\Omega d^4x\, \delta\mathcal{L}(x) + \int_\Omega d^4x\, \mathcal{L}(x)\frac{\partial \delta x^\mu}{\partial x^\mu} \\ &= \int_\Omega d^4x\left(\tilde\delta\mathcal{L}(x) + \frac{\partial \mathcal{L}(x)}{\partial x^\mu}\delta x^\mu\right) + \int_\Omega d^4x\, \mathcal{L}(x)\frac{\partial \delta x^\mu}{\partial x^\mu} \\ &= \int_\Omega d^4x\left(\tilde\delta\mathcal{L}(x) + \frac{\partial}{\partial x^\mu}(\mathcal{L}(x)\delta x^\mu)\right).\end{aligned} \qquad (2.48)$$

Now we express the total variation $\tilde\delta\mathcal{L}(x)$ in terms of the variations of the fields and their derivatives:

$$\begin{aligned}\tilde\delta\mathcal{L}(x) &= \frac{\partial \mathcal{L}(x)}{\partial \phi_r}\tilde\delta\phi_r(x) + \frac{\partial \mathcal{L}(x)}{\partial(\partial^\mu\phi_r)}\tilde\delta\left(\frac{\partial \phi_r(x)}{\partial x_\mu}\right) \\ &= \left[\frac{\partial \mathcal{L}(x)}{\partial \phi_r}\tilde\delta\phi_r(x) - \frac{\partial}{\partial x_\mu}\left(\frac{\partial \mathcal{L}(x)}{\partial(\partial^\mu\phi_r)}\right)\tilde\delta\phi_r(x)\right] \\ &\quad + \frac{\partial}{\partial x_\mu}\left(\frac{\partial \mathcal{L}(x)}{\partial(\partial^\mu\phi_r)}\right)\tilde\delta\phi_r(x) + \frac{\partial \mathcal{L}(x)}{\partial(\partial^\mu\phi_r)}\frac{\partial}{\partial x_\mu}(\tilde\delta\phi_r(x)) \\ &= \left[\frac{\partial \mathcal{L}(x)}{\partial \phi_r} - \frac{\partial}{\partial x_\mu}\left(\frac{\partial \mathcal{L}(x)}{\partial(\partial^\mu\phi_r)}\right)\right]\tilde\delta\phi_r(x) \\ &\quad + \frac{\partial}{\partial x_\mu}\left[\frac{\partial \mathcal{L}(x)}{\partial(\partial^\mu\phi_r)}\tilde\delta\phi_r(x)\right].\end{aligned} \qquad (2.49)$$

Here (2.43) was used, i.e., the fact that variation and differentiation can be interchanged. We use the summation convention not only for the Minkowski indices but also for the component index r. Thus for fields ϕ_r with several degrees of freedom, $r = 1,\ldots,N$, a summation over r is implied whenever the index occurs twice in an expression.

Now we nearly have reached our goal. Since the range of integration Ω can be chosen arbitrarily, the integrand of (2.48) itself has to vanish if the action integral is to be invariant, as postulated in (2.45). Using (2.49) the integrand reads

$$\left[\frac{\partial \mathcal{L}(x)}{\partial \phi_r} - \frac{\partial}{\partial x_\mu}\frac{\partial \mathcal{L}(x)}{\partial(\partial^\mu\phi_r)}\right]\tilde\delta\phi_r(x) + \frac{\partial}{\partial x_\mu}\left[\frac{\partial \mathcal{L}(x)}{\partial(\partial^\mu\phi_r)}\tilde\delta\phi_r(x) + \mathcal{L}(x)\delta x^\mu\right] = 0. \qquad (2.50)$$

The first term is recognized as the Euler–Lagrange equation (2.14). This term therefore vanishes, provided that the field ϕ_r satisfies the equation of motion. We are left with an expression with vanishing four-divergence that, using (2.42), can be written as

$$\frac{\partial}{\partial x_\mu}\left[\frac{\partial \mathcal{L}(x)}{\partial(\partial^\mu\phi_r)}\left(\delta\phi_r(x) - \frac{\partial \phi_r}{\partial x_\nu}\delta x_\nu\right) + \mathcal{L}(x)\delta x_\mu\right] = 0. \qquad (2.51)$$

This is an *equation of continuity* for the vector field defined by the terms in the square bracket, i.e.,

$$\frac{\partial}{\partial x_\mu}f_\mu(x) = 0 \qquad (2.52)$$

with the "current" density

$$f_\mu(x) = \frac{\partial \mathcal{L}(x)}{\partial(\partial^\mu \phi_r)} \delta\phi_r(x) - \left(\frac{\partial \mathcal{L}(x)}{\partial(\partial^\mu \phi_r)} \frac{\partial \phi_r}{\partial x^\nu} - g_{\mu\nu} \mathcal{L}(x) \right) \delta x^\nu . \qquad (2.53)$$

As is well known, an equation of continuity is just the expression of a *conservation law* in terms of a differential equation. This becomes obvious if (2.52) is integrated over three-dimensional space and the theorem of Gauss is used:

$$\begin{aligned} 0 &= \int_V d^3x \, \frac{\partial}{\partial x_\mu} f_\mu(x) = \int_V d^3x \, \frac{\partial}{\partial x_0} f_0(x) + \int_V d^3x \, \boldsymbol{\nabla} \cdot \boldsymbol{f}(x) \\ &= \frac{d}{dx_0} \int_V d^3x \, f_0(x) + \oint_{\partial V} d\boldsymbol{o} \cdot \boldsymbol{f}(x) . \end{aligned} \qquad (2.54)$$

The value of the integral over the surface ∂V vanishes since the fields and their derivatives are assumed to fall off sufficiently fast at infinity. Therefore

$$G := \int_V d^3x \, f_0(x) \qquad (2.55)$$

is a *conserved quantity* having a value constant in time. This is the essential result of *Noether's theorem*:

- Each continuous symmetry transformation leads to a conservation law. The conserved quantity G can be obtained from the Lagrange density through the use of (2.53) and (2.55).

Let us now study several important applications of Noether's theorem.

(1) Invariance Under Translation. The invariance under translations

$$x'^\mu = x^\mu + \epsilon^\mu , \qquad (2.56)$$

follows from the homogeneity of space-time. Since the shape of the fields is supposed not to change under translations,

$$\phi'_r(x') = \phi_r(x) , \qquad (2.57)$$

the local variation vanishes, $\delta\phi_r = 0$, and the conserved "Noether current" takes a simple form. The differential conservation law (2.52) reads (after splitting off a constant factor ϵ^ν)

$$\frac{\partial}{\partial x_\mu} \Theta_{\mu\nu} = 0 \qquad (2.58)$$

with the *canonical energy-momentum tensor*

$$\Theta_{\mu\nu} = \frac{\partial \mathcal{L}}{\partial(\partial^\mu \phi_r)} \frac{\partial \phi_r}{\partial x^\nu} - g_{\mu\nu} \mathcal{L} . \qquad (2.59)$$

Because $\nu = 0, \ldots, 3$ this implies four conserved quantities, which are the energy E and the momentum vector $\boldsymbol{P}$ of the field. In four-dimensional notation

$$P^\nu = (E/c, \boldsymbol{P}) = \frac{1}{c} \int_V d^3x \, \Theta^{0\nu}(x) = \text{const} . \qquad (2.60)$$

Let us remark that the tensor $\Theta_{\mu\nu}$ defined in (2.59) in some cases turns out not to be symmetric. It is possible, however, to go over to a *symmetrical energy-momentum tensor* $T_{\mu\nu} = T_{\nu\mu}$ by adding a suitable term with vanishing four-divergence. The new tensor also satisfies the conservation law (2.58) and yields the same value of P^μ. We will come back to this question in Example 2.1.

(2) Lorentz Invariance. Four-dimensional space-time is assumed to be not only homogeneous under translations but also *isotropic* with respect to rotations. Therefore the action is required to be *invariant under Lorentz transformations*. These (proper) Lorentz transformations are known to include ordinary three-dimensional rotations in space as well as velocity transformations (Lorentz boosts) which can be viewed as rotations in mixed spatio-temporal hyperplanes of Minkowski space. A general infinitesimal rotation is given by

$$x'^\mu = x^\mu + \delta\omega^{\mu\nu} x_\nu \ . \tag{2.61}$$

The matrix $\delta\omega^{\mu\nu}$ depends on the rotation angles (in four dimensions) and is antisymmetric

$$\delta\omega^{\mu\nu} = -\delta\omega^{\nu\mu} \ . \tag{2.62}$$

This ensures that the length of the vector x^μ, measured with respect to the Minkowski metric, remains invariant under the transformation. This is true since in the lowest order

$$\begin{aligned}
x'^\mu x'_\mu &= \left(x^\mu + \delta\omega^{\mu\sigma} x_\sigma\right)\left(x_\mu + \delta\omega_\mu{}^\tau x_\tau\right) \\
&= x^\mu x_\mu + \delta\omega^{\mu\sigma} x_\sigma x_\mu + \delta\omega_\mu{}^\tau x^\mu x_\tau \\
&= x^\mu x_\mu + 2 x_\mu x_\nu \delta\omega^{\mu\nu} = x^\mu x_\mu + x_\mu x_\nu \left(\delta\omega^{\mu\nu} + \delta\omega^{\nu\mu}\right) \ ,
\end{aligned} \tag{2.63}$$

which leads to the requirement (2.62).

The transformed field function $\phi'_r(x')$ will show a linear dependece on the rotation angles and on the values $\phi_r(x)$. We describe this dependence through the general ansatz

$$\phi'_r(x') = \phi_r(x) + \frac{1}{2}\delta\omega_{\mu\nu} \left(I^{\mu\nu}\right)_{rs} \phi_s(x) \ . \tag{2.64}$$

The quantities $I^{\mu\nu}$ are the *infinitesimal generators of the Lorentz transformation*.[4] The physical fields ϕ_r transform according to an irreducible representation of the Lorentz group. Therefore the $\left(I^{\mu\nu}\right)_{rs}$ are the elements of the *matrix representation* of the corresponding infinitesimal generator. They describe a mixing of the various components of a multicomponent field (e.g., spinor, vector, ...). The infinitesimal generators can be chosen to be antisymmetric with respect to the Lorentz indices μ and ν, $I^{\mu\nu} = -I^{\nu\mu}$, since a symmetric part does not contribute to (2.64) because of (2.62). Thus there are *six independent generators*. Three of them, $(\mu, \nu) = (1,2), (1,3), (2,3)$, correspond to spatial rotations whereas the remaining three, $(\mu, \nu) = (0,1), (0,2), (0,3)$, describe Lorentz boosts, i.e., velocity transformations along the various coordinate axes.

For completeness we remark that the infinitesimal generators satisfy the following commutator relations:

$$\left[I_{\mu\nu}, I_{\sigma\tau}\right] = +g_{\nu\sigma} I_{\mu\tau} + g_{\mu\tau} I_{\nu\sigma} - g_{\nu\tau} I_{\mu\sigma} - g_{\mu\sigma} I_{\nu\tau} \ . \tag{2.65}$$

---

[4] For more details see the volume W. Greiner: *Relativistic Quantum Mechanics* (Springer, Berlin, Heidelberg, 1990).

This *Lie algebra* defines the structure of the Lorentz group and is valid in any representation. The matrix indices $r, s, \ldots$ have been omitted for brevity in (2.65).

The transformation relations for the coordinates (2.61) and for the field (2.64) can now be inserted into Noether's theorem. According to (2.53) this leads to the conserved field

$$f_\mu(x) = \frac{\partial \mathcal{L}}{\partial(\partial^\mu \phi_r)} \frac{1}{2} \delta\omega_{\nu\lambda} (I^{\nu\lambda})_{rs} \phi_s(x) - \Theta_{\mu\nu} \delta\omega^{\nu\lambda} x_\lambda \,, \tag{2.66}$$

where $\Theta_{\mu\nu}$ is the energy-momentum tensor introduced in (2.59). If we use the antisymmetry of $\delta\omega^{\nu\lambda}$, the last term can be written as

$$\Theta_{\mu\nu} \delta\omega^{\nu\lambda} x_\lambda = \frac{1}{2} \delta\omega^{\nu\lambda} (\Theta_{\mu\nu} x_\lambda - \Theta_{\mu\lambda} x_\nu) \,, \tag{2.67}$$

so that (2.66) becomes

$$f_\mu(x) = \frac{1}{2} \delta\omega^{\nu\lambda} M_{\mu\nu\lambda}(x) \tag{2.68}$$

with the abbreviation

$$M_{\mu\nu\lambda}(x) = \Theta_{\mu\lambda} x_\nu - \Theta_{\mu\nu} x_\lambda + \frac{\partial \mathcal{L}}{\partial(\partial^\mu \phi_r)} (I_{\nu\lambda})_{rs} \phi_s(x) \,. \tag{2.69}$$

The corresponding integral conservation law tells us that the antisymmetric tensor

$$M_{\nu\lambda} = \int \mathrm{d}^3 x \left[\Theta_{0\lambda} x_\nu - \Theta_{0\nu} x_\lambda + \frac{\partial \mathcal{L}}{\partial(\partial^0 \phi_r)} (I_{\nu\lambda})_{rs} \phi_s(x) \right] \tag{2.70}$$

is a constant of motion. Here $M_{\nu\lambda}$ plays the role of the *tensor of angular momentum*. This becomes obvious for the spatial rotations, i.e., when taking the values $1, 2, 3$ for ν and λ. Then (2.70) is split into the following two parts:

$$M_{nl} = L_{nl} + S_{nl} \tag{2.71}$$

where

$$\begin{aligned} L_{nl} &= \int \mathrm{d}^3 x \left(x_n \Theta_{0l} - x_l \Theta_{0n} \right) \\ &= \int \mathrm{d}^3 x \frac{\partial \mathcal{L}}{\partial(\partial^0 \phi_r)} \left(x_n \frac{\partial}{\partial x^l} - x_l \frac{\partial}{\partial x^n} \right) \phi_r(x) \,, \end{aligned} \tag{2.72}$$

and

$$S_{nl} = \int \mathrm{d}^3 x \frac{\partial \mathcal{L}}{\partial(\partial^0 \phi_r)} (I^{nl})_{rs} \phi_s(x) \,. \tag{2.73}$$

The first part, L_{nl}, contains one component of the cross product between the position vector and the momentum (or rather the momentum density). Therefore L_{nl} has the meaning of the *orbital angular momentum* orthogonal to the plane spanned by the x_n and x_l. The second contribution, S_{nl}, contains the generators $(I^{nl})_{rs}$ and thus depends on the intrinsic transformation properties of the field ϕ_r. It describes the internal or *spin angular momentum* and will look very differently for, scalar, spinor, or vector fields, for example. In addition to describing the angular momentum the tensor $M_{\nu\lambda}$ gives three further conserved properties with mixed spatial-temporal indices. These quantities are related to the relativistic generalization of the *center of mass*.

The space component of the angular-momentum tensor, M_{ln}, has three independent components and can be mapped to the three-dimensional angular-momentum vector $\boldsymbol{J}$. This is achieved by contraction with the antisymmetric unit tensor

$$M_{nl} = \epsilon_{nlk} J^k \; . \tag{2.74}$$

The cartesian components of $\boldsymbol{J}$ are then given by

$$J^1 = M_{23} \quad , \quad J^2 = M_{31} \quad , \quad J^3 = M_{12} \tag{2.75}$$

or, written in closed form,

$$J^m = \tfrac{1}{2}\epsilon_{mnl} M_{nl} \; . \tag{2.76}$$

This follows from the contraction rule for the three-dimensional Levi–Civita tensor

$$\epsilon_{nlk} \epsilon_{nlm} = 2\,\delta_{km} \; . \tag{2.77}$$

(Note that here the summation convention was applied to three-vectors for which no distinction between covariant and contravariant components was needed.)

(3) Internal Symmetries. The symmetries investigated so far have all been related coordinate transformations (four-dimensional translation and rotation). This has led to conservation laws which are of fundamental importance and have to be valid for any physical system. In addition, the Lagrange density of a given theory can exhibit further invariance properties not related to space and time. This can happen for fields with an *internal structure*, which is indicated by the presence of several components ϕ_r. An infinitesimal *internal symmetry transformation* amounts to mixing these components. It can be written as

$$\phi'_r(x) = \phi_r(x) + \mathrm{i}\epsilon \sum_s \lambda_{rs} \phi_s(x) \; . \tag{2.78}$$

If this transformation leaves the action variable invariant, Noether's theorem at once leads to a conserved vector field. If we use

$$\delta x_\mu = 0 \quad , \quad \delta\phi_r(x) = \mathrm{i}\epsilon \sum_s \lambda_{rs} \phi_s(x) \; , \tag{2.79}$$

the "Noether current" reads

$$f_\mu = \sum_r \frac{\partial \mathcal{L}}{\partial(\partial^\mu \phi_r)} \delta\phi_r = \mathrm{i}\epsilon \sum_{rs} \frac{\partial \mathcal{L}}{\partial(\partial^\mu \phi_r)} \lambda_{rs} \phi_s(x) \; . \tag{2.80}$$

The conserved quantity obtained by integration is called a Noether charge

$$Q = \int \mathrm{d}^3 x \sum_{rs} \pi_r(x) \lambda_{rs} \phi_s(x) \; . \tag{2.81}$$

As an important application of this concept, let us study a *complex field* described by a Lagrange density which is *invariant under phase transformation*:

$$\phi' = \phi + i\epsilon\phi \quad , \quad \phi^{*\prime} = \phi^* - i\epsilon\phi^* . \tag{2.82}$$

This relation also can be written in the notation of (2.78) by taking the real and imaginary part of ϕ as the two components of a real field (ϕ_1, ϕ_2) so that $\lambda_{12} = 1, \lambda_{21} = -1$. It is more convenient, however, to adopt a complex field and to take $\delta\phi$ and $\delta\phi^*$ as independent variations. This leads to the conserved current

$$j_\mu(x) = (-i)\left(\frac{\partial \mathcal{L}}{\partial(\partial^\mu\phi)}\phi(x) - \frac{\partial \mathcal{L}}{\partial(\partial^\mu\phi^*)}\phi^*(x)\right) \tag{2.83}$$

where a convenient normalization factor was introduced. The transformation coefficients with respect to the "components" ϕ and ϕ^* are given by $\lambda_{11} = +1$, $\lambda_{22} = -1$. The Noether current j_μ turns out to be identical with the usual *electromagnetic current density*, as we will demonstrate later.

EXAMPLE

2.1 The Symmetrized Energy-Momentum Tensor

Noether's theorem leads to equations of continuity, i.e., to conservation laws in differential form. The densities and currents obtained in this way are not fixed uniquely since it is possible to add certain four-dimensional divergence terms without influencing the equation of continuity. Let us illustate this for the *canonical energy-momentum tensor* $\Theta_{\mu\nu}$. We define a *modified tensor* through

$$T_{\mu\nu} = \Theta_{\mu\nu} + \partial^\sigma \chi_{\sigma\mu\nu} , \tag{1}$$

where the tensor $\chi_{\sigma\mu\nu}$ is arbitrary except for the requirement of being antisymmetric with respect to the first two indices:

$$\chi_{\sigma\mu\nu} = -\chi_{\mu\sigma\nu} . \tag{2}$$

This condition guarantees that the conservation law remains unchanged:

$$\begin{aligned}\partial^\mu T_{\mu\nu} &= \partial^\mu \Theta_{\mu\nu} + \partial^\mu \partial^\sigma \chi_{\sigma\mu\nu} \\ &= \partial^\mu \Theta_{\mu\nu} + \frac{1}{2}\partial^\mu \partial^\sigma (\chi_{\sigma\mu\nu} + \chi_{\mu\sigma\nu}) = \partial^\mu \Theta_{\mu\nu} = 0 .\end{aligned} \tag{3}$$

Also the total energy and momentum are not affected by the transformation (1):

$$\tilde{P}_\nu = \int d^3x\, T_{0\nu} = \int d^3x \left(\Theta_{0\nu} + \partial^0 \chi_{00\nu} + \partial^k \chi_{0k\nu}\right) = \int d^3x\, \Theta_{0\nu} = P_\nu . \tag{4}$$

Here $\chi_{00\nu}$ vanishes because of (2) and we have assumed that $\chi_{0k\nu}$ falls off fast enough at large distances to assure that the surface integral in Gauss' theorem can be neglected.

The additional freedom of choice expressed by (1) allows the construction of a modified energy-momentum tensor $T_{\mu\nu}$ which is *symmetric* under permutation of the indices:[5]

$$T_{\mu\nu} = T_{\nu\mu} . \tag{5}$$

---

[5] J. Belinfante: Physica **6**, 887 (1939).

Example 2.1

As a side effect, the energy-momentum tensor constructed in this way can be used to find a simpler formulation of the conservation law for angular momentum. We define the *modified angular-momentum tensor*

$$\widetilde{M}_{\mu\nu\lambda} = T_{\mu\lambda}x_\nu - T_{\mu\nu}x_\lambda \,, \tag{6}$$

which does not contain the additional term contained in (2.69). Then the following differential conservation law for $\widetilde{M}_{\mu\nu\lambda}$ is derived immediately:

$$\partial^\mu \widetilde{M}_{\mu\nu\lambda} = (\partial^\mu T_{\mu\lambda})x_\nu + g^\mu{}_\nu T_{\mu\lambda} - (\partial^\mu T_{\mu\nu})x_\lambda - g^\mu{}_\lambda T_{\mu\nu} = T_{\nu\lambda} - T_{\lambda\nu} = 0 \,. \tag{7}$$

The tensor $\widetilde{M}_{\mu\nu\lambda}$ again has to agree with the canonical angular-momentum tensor $M_{\mu\nu\lambda}$ up to a four-divergence:

$$\widetilde{M}_{\mu\nu\lambda} = M_{\mu\nu\lambda} + \partial^\sigma \eta_{\sigma\mu\nu\lambda} \,, \tag{8}$$

where η has to be antisymmetric in the first two indices

$$\eta_{\sigma\mu\nu\lambda} = -\eta_{\mu\sigma\nu\lambda} \,. \tag{9}$$

The same argument as in (4) shows that the conserved quantity remains unchanged:

$$\widetilde{M}_{\nu\lambda} = \int d^3x \, \widetilde{M}_{0\nu\lambda} = \int d^3x \, M_{0\nu\lambda} = M_{\nu\lambda} \,. \tag{10}$$

Now we will construct the transformation function $\chi_{\mu\nu\lambda}$. Inserting the explicit expressions for $\widetilde{M}_{\mu\nu\lambda}$ and $M_{\mu\nu\lambda}$ into (8) gives

$$(\Theta_{\mu\nu} + \partial^\sigma \chi_{\sigma\mu\lambda})x_\nu - (\Theta_{\mu\nu} + \partial^\sigma \chi_{\sigma\mu\nu})x_\lambda$$
$$= \Theta_{\mu\lambda}x_\nu - \Theta_{\mu\nu}x_\lambda + \frac{\partial \mathcal{L}}{\partial(\partial^\mu \phi_r)}(I_{\nu\lambda})_{rs}\phi_s + \partial^\sigma \eta_{\sigma\mu\nu\lambda} \,. \tag{11}$$

Now make a special choice for the function $\eta_{\sigma\mu\nu\lambda}$ in such a way that the expression (11) get simplified:

$$\eta_{\sigma\mu\nu\lambda} = x_\nu \chi_{\sigma\mu\lambda} - x_\lambda \chi_{\sigma\mu\nu} \,, \tag{12}$$

which, because of (2), satisfies the condition of antisymmetry (9). Equation (11) becomes

$$(\partial^\sigma \chi_{\sigma\mu\lambda})x_\nu - (\partial^\sigma \chi_{\sigma\mu\nu})x_\lambda - \partial^\sigma (x_\nu \chi_{\sigma\mu\lambda} - x_\lambda \chi_{\sigma\mu\nu}) = \frac{\partial \mathcal{L}}{\partial(\partial^\mu \phi_r)}(I_{\nu\lambda})_{rs}\phi_s \tag{13}$$

or

$$\chi_{\nu\mu\lambda} - \chi_{\lambda\mu\nu} = -\frac{\partial \mathcal{L}}{\partial(\partial^\mu \phi_r)}(I_{\nu\lambda})_{rs}\phi_s \,. \tag{14}$$

The quantities $I_{\nu\lambda}$ are antisymmetric in ν, λ: $I_{\nu\lambda} = -I_{\lambda\nu}$. The relation (14) serves to determine only that part of $\chi_{\nu\mu\lambda}$ which is antisymmetric with respect to $\nu \leftrightarrow \lambda$. Thus the general solution of (14) reads

$$\chi_{\nu\mu\lambda} = -\frac{1}{2}\frac{\partial \mathcal{L}}{\partial(\partial^\mu \phi_r)}(I_{\nu\lambda})_{rs}\phi_s + a_{\nu\mu\lambda} \,, \tag{15}$$

which contains an arbitrary part symmetric in ν, λ:

$$a_{\nu\mu\lambda} = a_{\lambda\mu\nu} \, . \tag{16}$$

This freedom can now be put to use to satisfy the original requirement (2). We choose

$$\chi_{\nu\mu\lambda} = \frac{1}{2}\left[-\frac{\partial \mathcal{L}}{\partial(\partial^\mu \phi_r)}(I_{\nu\lambda})_{rs} + \frac{\partial \mathcal{L}}{\partial(\partial^\nu \phi_r)}(I_{\mu\lambda})_{rs} + \frac{\partial \mathcal{L}}{\partial(\partial^\lambda \phi_r)}(I_{\mu\nu})_{rs}\right]\phi_s \, , \tag{17}$$

where the added term obviously satisfies the condition (16). The required antisymmetry with respect to the interchange $\nu \leftrightarrow \mu$ is easily checked:

$$\begin{aligned}\chi_{\mu\nu\lambda} &= \frac{1}{2}\left[-\frac{\partial \mathcal{L}}{\partial(\partial^\nu \phi_r)}(I_{\mu\lambda})_{rs} + \frac{\partial \mathcal{L}}{\partial(\partial^\mu \phi_r)}(I_{\nu\lambda})_{rs} + \frac{\partial \mathcal{L}}{\partial(\partial^\lambda \phi_r)}(I_{\nu\mu})_{rs}\right]\phi_s \\ &= -\chi_{\nu\mu\lambda} \, ,\end{aligned} \tag{18}$$

by using $(I_{\mu\nu})_{rs} = -(I_{\nu\mu})_{rs}$ and interchanging the first terms in (17).

Thus we have found that although the canonical energy-momentum tensor in general (the exception is the scalar field) is not symmetric it can be *symmetrized* by adding the divergence of (17). This has the added benefit of leading to a *modified angular-momentum tensor* having the simple form (6). The symmetrized form of the energy-momentum tensor, $T_{\mu\nu}$, is often viewed as more "fundamental" than the canonical tensor $\Theta_{\mu\nu}$. We remark that it is also possible to deduce $T_{\mu\nu}$ directly from a variational principle.[6]

2.5 The Generators of the Poincaré Group

As shown in (2.55), for each continuous symmetry there exists a conserved quantity

$$G = \int d^3x \left(\pi_r \delta\phi_r - (\pi_r \partial_\nu \phi_r - g_{0\nu}\mathcal{L})\delta x^\nu \right) \, . \tag{2.84}$$

Not surprisingly these conserved quantities play a central role in the formulation of a field theory. In particular the knowledge of G allows us to determine the corresponding (total) variation $\tilde{\delta}\phi_r(x) = \phi'_r(x) - \phi_r(x)$. For this, one has to calculate the Poisson bracket:

$$\tilde{\delta}\phi_r(x) = \{\phi_r(x), G\}_{\text{PB}} \, . \tag{2.85}$$

Because of this connection, G is called the *generator of the symmetry transformation* (which should not be confused with the infinitesimal generators $I^{\mu\nu}$ of the Lorentz group).

Proof of (2.85). We first show that the Poisson bracket can be reduced to the functional derivative with respect to the conjugate field $\pi_r(x)$:

$$\begin{aligned}\{\phi_r(x), G\}_{\text{PB}} &= \int d^3x' \left(\frac{\delta\phi_r(x)}{\delta\phi_s(x')}\frac{\delta G}{\delta\pi_s(x')} - \frac{\delta\phi_r(x)}{\delta\pi_s(x')}\frac{\delta G}{\delta\phi_s(x')}\right) \\ &= \frac{\delta G}{\delta\pi_r(x)} \, .\end{aligned} \tag{2.86}$$

---

[6] See A.O. Barut: *Electrodynamics and Classical Theory of Fields and Particles* (Dover Publications, New York 1980).

This derivative is easily evaluated if there is no time translation, i.e., for $\nu = k \neq 0$ in (2.84):

$$\begin{aligned}\frac{\delta G}{\delta \pi_r(x)} &= \frac{\delta}{\delta \pi_r(x)} \int d^3x' \, \pi_s(x') \big(\delta\phi_s(x') - \partial_k \phi_s(x') \, \delta x^k\big) \\ &= \delta\phi_r(x) - \delta x^k \partial_k \phi_r(x) \, .\end{aligned} \quad (2.87)$$

According to (2.42) this just coincides with the total variation $\tilde{\delta}\phi_r(x)$. In the case of a time translation, $\nu = 0$, the evaluation of the functional derivative is not quite as easy. We have to take into account that the "velocity" field $\partial_0 \phi(x) \equiv \dot\phi(x)$ depends on $\pi(x)$. We assume that this function (without the indices),

$$\pi = \frac{\partial \mathcal{L}(\phi, \nabla\phi, \dot\phi)}{\partial \dot\phi} =: \Pi(\phi, \nabla\phi, \dot\phi) \, , \quad (2.88)$$

can be inverted as

$$\dot\phi = \Pi^{-1}(\phi, \nabla\phi, \pi) \, . \quad (2.89)$$

Then when forming the derivative of (2.84) we can use the functional chain rule:

$$\begin{aligned}\frac{\delta G}{\delta \pi_r(x)} &= \frac{\delta}{\delta \pi_r(x)} \int d^3x' \, \Big(\pi_s \delta\phi_s - \big(\pi_s \dot\phi_s - \mathcal{L}\big)\delta x^0\Big) \\ &= \delta\phi_r(x) - \dot\phi_r(x)\delta x^0 \\ &\quad - \int d^3x'' \, \frac{\delta \dot\phi_t(x'')}{\delta \pi_r(x)} \frac{\delta}{\delta \dot\phi_t(x'')} \int d^3x' \, \Big(\pi_s \dot\phi_s - \mathcal{L}(\phi, \nabla\phi, \dot\phi)\Big) \, .\end{aligned} \quad (2.90)$$

In the last term the derivative is meant to act only on $\dot\phi_s$ and not on π_s. This term is found to cancel since

$$\frac{\delta}{\delta \dot\phi_t(x'')} \int d^3x' \, \big(\pi_s \dot\phi_s - \mathcal{L}\big) = \pi_t(x'') - \frac{\partial \mathcal{L}}{\partial \dot\phi_t(x'')} = 0 \quad (2.91)$$

according to the definition of the conjugate field. By the way, this can be derived at once from the functional Hamilton equation (2.22) for the field π. Thus we again find

$$\frac{\delta G}{\delta \pi_r(x)} = \delta\phi_r(x) - \delta x^0 \partial_0 \phi_r(x) = \tilde{\delta}\phi_r(x) \quad (2.92)$$

and the relation (2.85) for the generator has been proven in both cases.

Now we want to study in more detail the properties of the generators G in the case of the fundamental relativistic symmetries. The homogeneity and isotropy of space-time lead to the conservation law for the momentum vector (2.60) and the generalized angular-momentum tensor (2.70). The set of translations, spatial rotations, and Lorentz boosts together form the inhomogeneous Lorentz group, also known as the *Poincaré group*. This is a Lie group with $6+4 = 10$ parameters (3 rotation angles, 3 boost velocities, and 3 spatial and 1 temporal shift). Using (2.60) and (2.70) we can write the generators of this group as

$$G = -\epsilon_\mu P^\mu + \frac{1}{2}\delta\omega_{\mu\nu} M^{\mu\nu} \, . \quad (2.93)$$

According to (2.85) this allows us to determine the variation of the field under the coordinate transformation $x'^\mu = \delta\omega^\mu{}_\nu x^\nu + \epsilon^\mu$. For translations, (2.56) and (2.57) lead to

$$\{\phi_r(x), P^\mu\}_{PB} = \partial^\mu \phi_r(x) , \tag{2.94}$$

and for Lorentz transformations, (2.61), (2.64), and (2.42) yield

$$\{\phi_r(x), M^{\mu\nu}\}_{PB} = (I^{\mu\nu})_{rs}\phi_s(x) + x^\mu \partial^\nu \phi_r - x^\nu \partial^\mu \phi_r . \tag{2.95}$$

The generators P^μ and $M^{\mu\nu}$ are connected through a set of relations which are known as the *Poincaré algebra*. In the framework of classical field theory this algebra is realized in terms of Poisson brackets. The Poincaré algebra reads

$$\{P_\mu, P_\nu\}_{PB} = 0 , \tag{2.96a}$$
$$\{M_{\mu\nu}, P_\lambda\}_{PB} = g_{\nu\lambda}P_\mu - g_{\mu\lambda}P_\nu , \tag{2.96b}$$
$$\{M_{\mu\nu}, M_{\sigma\tau}\}_{PB} = g_{\nu\sigma}M_{\mu\tau} + g_{\mu\tau}M_{\nu\sigma} - g_{\nu\tau}M_{\mu\sigma} - g_{\mu\sigma}M_{\nu\tau} . \tag{2.96c}$$

The Poincaré algebra is an expression of fundamental properties of space and time. Therefore relations similar to those in (2.96) will also hold for quantum fields, except that the Poisson brackets will be replaced by commutators (see Sect. 4.3).

EXERCISE

2.2 The Poincaré Algebra for Classical Fields

Problem. Show the validity of the Poincaré algebra (2.96a-c).
Hint: Use Hamilton's equations and the equation of continuity for the momentum density.

Solution. (a) To evaluate the Poisson bracket

$$\{P_\mu, P_\nu\}_{PB} = \int d^3x \left(\frac{\delta P_\mu}{\delta \phi_s(x)} \frac{\delta P_\nu}{\delta \pi_s(x)} - \frac{\delta P_\mu}{\delta \pi_s(x)} \frac{\delta P_\nu}{\delta \phi_s(x)} \right) , \tag{1}$$

the functional derivative of the four-momentum

$$P_\mu = \int d^3x' \left(\pi_r(x') \partial'_\mu \phi_r(x') - g_{0\mu} \mathcal{L}(\phi(x'), \dot\phi(x')) \right) \tag{2}$$

has to be formed. For $\mu = i \neq 0$ we find, using integration by parts,

$$\frac{\delta P_i}{\delta \phi_s(x)} = \frac{\delta}{\delta \phi_s(x)} \int d^3x' \, \pi_r(x') \partial'_i \phi_r(x') = -\partial_i \pi_s(x) . \tag{3}$$

In the case $i = 0$, P_μ agrees with the Hamiltonian H and Hamilton's equation leads to

$$\frac{\delta P_0}{\delta \phi_s(x)} = \frac{\delta H}{\delta \phi_s(x)} = -\dot\pi_s(x) . \tag{4}$$

Equations (3) and (4) can be combined. Similarly the derivative of the conjugate field can be found, leading to

$$\frac{\delta P_\mu}{\delta \phi_s(x)} = -\partial_\mu \pi_s(x) , \quad \frac{\delta P_\mu}{\delta \pi_s(x)} = \partial_\mu \phi_s(x) . \tag{5}$$

The Poisson bracket (1) then reads

Exercise 2.2

$$\{P_\mu, P_\nu\}_{\text{PB}} = -\int d^3x \left(\partial_\mu \pi_s \, \partial_\nu \phi_s - \partial_\mu \phi_s \, \partial_\nu \pi_s\right). \tag{6}$$

When evaluating this expression, one has to distinguish between space and time indices. If both indices are spatial, $\mu = i$, $\nu = j$, then (6) can be transformed into a surface term, which as usual is assumed to vanish.

$$\begin{aligned}\{P_i, P_j\}_{\text{PB}} &= -\int d^3x \left(\partial_i \pi_s \, \partial_j \phi_s - \partial_i \phi_s \, \partial_j \pi_s\right) \\ &= -\int d^3x \left(\partial_i(\pi_s \partial_j \phi_s) - \pi_s \partial_i \partial_j \phi_s - \partial_j(\pi_s \partial_i \phi_s) + \pi_s \partial_j \partial_i \phi_s\right) \\ &= 0. \end{aligned} \tag{7}$$

In the case of mixed indices, $\mu = i$, $\nu = 0$, the Poisson bracket can be transformed to the time derivative of the conserved quantity P_i:

$$\begin{aligned}\{P_i, P_0\}_{\text{PB}} &= -\int d^3x \left(\dot\phi_s \, \partial_i \pi_s - \dot\pi_s \, \partial_i \phi_s\right) = \int d^3x \left(\pi_s \, \partial_i \dot\phi_s + \dot\pi_s \, \partial_i \phi_s\right) \\ &= \partial_0 \int d^3x \, \pi_s \, \partial_i \phi_s = \partial_0 P_i = 0. \end{aligned} \tag{8}$$

Since $P_0 = H$ the result (8) just agrees with the equation of motion (2.27) for P_i. Thus we could have skipped the detailed derivation.

(b)

$$\{M_{\mu\nu}, P_\lambda\}_{\text{PB}} = \int d^3x \left(\frac{\delta M_{\mu\nu}}{\delta \phi_s(x)} \frac{\delta P_\lambda}{\delta \pi_s(x)} - \frac{\delta M_{\mu\nu}}{\delta \pi_s(x)} \frac{\delta P_\lambda}{\delta \phi_s(x)}\right). \tag{9}$$

For the case of spatial indices the functional derivatives of the angular momentum tensor (2.70) reduce to

$$\frac{\delta M_{nl}}{\delta \phi_s} = -x_n \partial_l \pi_s + x_l \partial_n \pi_s + \pi_r (I_{nl})_{rs}, \tag{10a}$$

$$\frac{\delta M_{nl}}{\delta \pi_s} = x_n \partial_l \phi_s - x_l \partial_n \phi_s + (I_{nl})_{sr} \phi_r, \tag{10b}$$

The mixed case of space and time indices produces to a supplementary term when the ϕ_s derivative is performed:

$$\begin{aligned}\frac{\delta}{\delta \phi} \int d^3x \, \mathcal{L}(\phi, \nabla\phi, \dot\phi) x_n &= \frac{\partial \mathcal{L}}{\partial \phi} x_n - \partial_k\left(\frac{\partial \mathcal{L}}{\partial(\partial_k \phi)} x_n\right) \\ &= \partial_0 \frac{\partial \mathcal{L}}{\partial(\partial_0 \phi)} x_n - \frac{\partial \mathcal{L}}{\partial(\partial^n \phi)}, \end{aligned} \tag{11}$$

where the Euler–Lagrange equation was used. This leads to

$$\frac{\delta M_{n0}}{\delta \phi_s} = -x_n \partial_0 \pi_s + x_0 \partial_n \pi_s + \pi_r (I_{n0})_{rs} + \frac{\partial \mathcal{L}}{\partial(\partial^n \phi_s)}, \tag{12a}$$

$$\frac{\delta M_{n0}}{\delta \pi_s} = x_n \partial_0 \phi_s - x_0 \partial_n \phi_s + (I_{n0})_{sr} \phi_r. \tag{12b}$$

The Poisson bracket (9) calls for the study of four different cases. Their evaluation makes repeated use of integration by parts. For brevity we will no longer write down the component indices r, s.

Case 1.

$$\begin{aligned}\{M_{nl}, P_i\}_{\text{PB}} &= \int d^3x \left[\left(-x_n\partial_l\pi + x_l\partial_n\pi + \pi I_{nl}\right)\partial_i\phi \right.\\ &\quad \left. + \partial_i\pi\left(x_n\partial_l\phi - x_l\partial_n\phi + I_{nl}\phi\right)\right] \\ &= \int d^3x \left(-\pi\partial_l\phi\, g_{in} + \pi\partial_n\phi\, g_{il}\right) = g_{il}P_n - g_{in}P_l\,. \quad (13)\end{aligned}$$

Case 2.

$$\begin{aligned}\{M_{nl}, P_0\}_{\text{PB}} &= \int d^3x \left[\pi(x_n\partial_l - x_l\partial_n + I_{nl})\dot\phi + \dot\pi(x_n\partial_l - x_l\partial_n + I_{nl})\phi\right] \\ &= \partial_0 M_{nl} = 0\,. \quad (14)\end{aligned}$$

Case 3. One might guess that the Poisson bracket of M_{n0} and $P_0 = H$ is zero, as in (14), since the time derivative vanishes. This is not quite true, however, since M_{n0} has an *explicit* time dependence which has to be taken into account when we form the derivative:

$$\{M_{n0}, P_0\}_{\text{PB}} = \frac{d}{dx_0}M_{n0} - \frac{\partial}{\partial x_0}M_{n0} = -\frac{\partial}{\partial x_0}M_{n0}\,. \quad (15)$$

Here $\partial/\partial x_0$ acts only on the time coordinate explicitly contained in M_{n0}, leading to

$$\{M_{n0}, P_0\}_{\text{PB}} = -\frac{\partial}{\partial x_0}\int d^3x\,(-\pi\partial_n\phi)x_0 = \int d^3x\, \pi\partial_n\phi = P_n \quad (16)$$

The same result also is obtained from an explicit calculation using (12).

Case 4. The remaining Poisson bracket is evaluated as follows

$$\begin{aligned}\{M_{n0}, P_i\}_{\text{PB}} &= \int d^3x \left[\left(-x_n\dot\pi + x_0\partial_n\pi + \pi I_{n0} + \frac{\partial \mathcal{L}}{\partial(\partial^n\phi)}\right)(\partial_i\phi) \right.\\ &\quad \left. + (\partial_i\pi)\left(x_n\dot\phi - x_0\partial_n\phi + I_{n0}\phi\right)\right] \\ &= \int d^3x \left[-\partial_0(\pi\partial_i\phi)x_n - g_{in}\pi\dot\phi + \frac{\partial\mathcal{L}}{\partial(\partial^n\phi)}\partial_i\phi\right]\,. \quad (17)\end{aligned}$$

The last term under the integral can be expressed through the momentum density

$$\frac{\partial\mathcal{L}}{\partial(\partial^n\phi)}\partial_i\phi = \Theta_{ni} + g_{ni}\mathcal{L}\,. \quad (18)$$

If we use this result and $\Theta_{0i} = \pi\partial_i\phi$, (17) becomes

$$\{M_{n0}, P_i\}_{\text{PB}} = \int d^3x \left[-g_{in}(\pi\dot\phi - \mathcal{L}) - \partial^0\Theta_{0i}x_n + \Theta_{ni}\right]\,. \quad (19)$$

Because of the equation of continuity for the momentum current $\partial^0\Theta_{0i} + \partial^k\Theta_{ki} = 0$, the last two terms in (19) can be combined into a divergence term which does not contribute to the integral:

$$\begin{aligned}\{M_{n0}, P_i\}_{\text{PB}} &= -g_{in}P_0 + \int d^3x\left[(\partial^k\Theta_{ki})x_n + \Theta_{ni}\right] \\ &= -g_{in}P_0 + \int d^3x\,\partial^k(\Theta_{ki}x_n) = -g_{in}P_0\,. \quad (20)\end{aligned}$$

Exercise 2.2

This is the last special case needed to complete the proof of (2.96b).

(c) The proof of (2.96c) proceeds along the same lines but is rather lengthy. For example the case of the spatial angular-momentum components is treated by using (10). After several integrations by parts, the Poisson bracket becomes

$$\{M_{ij}, M_{kl}\}_{\text{PB}} = \int d^3x \left(\frac{\delta M_{ij}}{\delta \phi} \frac{\delta M_{kl}}{\delta \pi} - \frac{\delta M_{ij}}{\delta \pi} \frac{\delta M_{kl}}{\delta \phi} \right)$$

$$= \int d^3x \, \pi \Big[g_{ik}(-x_j \partial_l + x_l \partial_j) + g_{il}(x_j \partial_k - x_k \partial_j)$$

$$+ g_{jk}(x_i \partial_l - x_l \partial_i) + g_{jl}(x_i \partial_k - x_k \partial_i) + I_{ij} I_{kl} - I_{ij} I_{kl} \Big] \phi \,, \quad (21)$$

which leads to (2.96c) provided that the Lie algebra (2.65) of the infinitesimal generators of the Poincaré group holds. The mixed combinations of space and time indices can be treated similarly.

Part II

Canonical Quantization

3. Nonrelativistic Quantum Field Theory

3.1 Introduction

To introduce the formalism of field quantization we will first concentrate on the familiar example of nonrelativistic quantum mechanics. In this framework the "motion" of a particle is described in terms of the complex Schrödinger wavefunction $\psi(\boldsymbol{x},t)$. The square $\psi^*\psi$ gives the probability of finding the particle at position $\boldsymbol{x}$. Allowing for the presence of an external potential $V(\boldsymbol{x},t)$, the wave function $\psi(\boldsymbol{x},t)$ satisfies the Schrödinger equation

$$i\hbar \frac{\partial \psi}{\partial t} = -\frac{\hbar^2}{2m}\nabla^2 \psi + V(\boldsymbol{x},t)\psi \; . \tag{3.1}$$

The step of setting up this equation often is called *first quantization* which is followed by the step of *second quantization*. This terminology may be misleading to some degree. However, in the context of quantum field theory one can view the wave function $\psi(\boldsymbol{x},t)$ as a classical field and (3.1) as a classical field equation for describing nonrelativistic particles.

To make use of the techniques of classical field theory developed in the last chapter we have to know the Lagrange density of the Schrödinger field. This can be constructed quite easily by "inverting" the Euler–Lagrange equation (2.13). We will verify that the following Lagrange density belongs to (3.1)

$$\mathcal{L}(\psi, \boldsymbol{\nabla}\psi, \dot{\psi}) = i\hbar \psi^* \frac{\partial \psi}{\partial t} - \frac{\hbar^2}{2m}\boldsymbol{\nabla}\psi^* \cdot \boldsymbol{\nabla}\psi - V(\boldsymbol{x},t)\psi^*\psi \; . \tag{3.2}$$

When the variation of the action integral is carried out, the functions $\psi(\boldsymbol{x},t)$ and $\psi^*(\boldsymbol{x},t)$ can be treated as two independent fields. This leads to the two Euler–Lagrange equations

$$\frac{\partial \mathcal{L}}{\partial \psi} - \boldsymbol{\nabla}\frac{\partial \mathcal{L}}{\partial(\boldsymbol{\nabla}\psi)} - \frac{\partial}{\partial t}\frac{\partial \mathcal{L}}{\partial \dot{\psi}} = 0 \tag{3.3}$$

and

$$\frac{\partial \mathcal{L}}{\partial \psi^*} - \boldsymbol{\nabla}\frac{\partial \mathcal{L}}{\partial(\boldsymbol{\nabla}\psi^*)} - \frac{\partial}{\partial t}\frac{\partial \mathcal{L}}{\partial \dot{\psi}^*} = 0 \; . \tag{3.4}$$

One immediately verifies that the ansatz (3.2) for the Lagrange density $\mathcal{L}$ produces the Schrödinger equation (3.1) and its complex-conjugate counterpart. The *canonically conjugate field* associated with $\psi(\boldsymbol{x},t)$ is

$$\pi(\boldsymbol{x},t) = \frac{\partial \mathcal{L}}{\partial \dot{\psi}} = i\hbar \psi^*(\boldsymbol{x},t) \; , \tag{3.5}$$

whereas the field conjugate to $\psi^*(\boldsymbol{x},t)$ is found to vanish. Therefore there are only two independent fields, $\psi(\boldsymbol{x},t)$ and $\pi(\boldsymbol{x},t)$. The Hamilton density is

$$\mathcal{H} = \pi \frac{\partial \psi}{\partial t} - \mathcal{L} = \frac{\hbar^2}{2m} \boldsymbol{\nabla}\psi^* \cdot \boldsymbol{\nabla}\psi + V(\boldsymbol{x},t)\,\psi^*\psi \,, \tag{3.6}$$

which, after an integration by parts, leads to the Hamiltonian

$$H = \int \mathrm{d}^3 x\, \mathcal{H}(x) = \int \mathrm{d}^3 x\, \psi^*(x)\Big(-\frac{\hbar^2}{2m}\boldsymbol{\nabla}^2 + V(\boldsymbol{x},t)\Big)\psi(x) \,. \tag{3.7}$$

The Poisson brackets of the fields ψ and π have already been derived in (2.36) and (2.37):

$$\big\{\psi(\boldsymbol{x},t),\,\pi(\boldsymbol{x}',t)\big\}_{\mathrm{PB}} = \delta^3(\boldsymbol{x}-\boldsymbol{x}') \,, \tag{3.8a}$$

$$\big\{\psi(\boldsymbol{x},t),\,\psi(\boldsymbol{x}',t)\big\}_{\mathrm{PB}} = \big\{\pi(\boldsymbol{x},t),\,\pi(\boldsymbol{x}',t)\big\}_{\mathrm{PB}} = 0 \,. \tag{3.8b}$$

3.2 Quantization Rules for Bose Particles

In the process of *canonical field quantization* the classical fields $\psi(\boldsymbol{x},t)$ and $\pi(\boldsymbol{x},t)$ are replaced by operators $\hat{\psi}(\boldsymbol{x},t)$ and $\hat{\pi}(\boldsymbol{x},t) = \mathrm{i}\hbar\hat{\psi}^\dagger(\boldsymbol{x},t)$. The compex conjugate wave function $\psi^*(\boldsymbol{x},t)$ is replaced by the hermitean adjoint field operator $\hat{\psi}^\dagger(\boldsymbol{x},t)$. As in (1.13) these operators are postulated to satisfy the *equal-time commutation relations* (ETCR):

$$[\hat{\psi}(\boldsymbol{x},t),\hat{\psi}^\dagger(\boldsymbol{x}',t)] = \delta^3(\boldsymbol{x}-\boldsymbol{x}') \,, \tag{3.9a}$$

$$[\hat{\psi}(\boldsymbol{x},t),\hat{\psi}(\boldsymbol{x}',t)] = [\hat{\psi}^\dagger(\boldsymbol{x},t),\hat{\psi}^\dagger(\boldsymbol{x}',t)] = 0 \,. \tag{3.9b}$$

Because $\hat{\psi}^\dagger$ was used instead of $\hat{\pi}$ the factor $\mathrm{i}\hbar$ has dropped out. In Sect. 3.3 we will find that (3.9) is not the only possible rule of quantization. As we will soon discuss, quantization according to (3.9) implies that the particles satisfy Bose statistics.

The dynamics of the field operators is determined by Heisenberg's equation of motion

$$\dot{\hat{\psi}} = \frac{1}{\mathrm{i}\hbar}\big[\hat{\psi},\hat{H}\big] \quad,\quad \dot{\hat{\pi}} = \frac{1}{\mathrm{i}\hbar}\big[\hat{\pi},\hat{H}\big] \,. \tag{3.10}$$

The second equation is seen to be the hermitean conjugate of the first one since

$$\begin{aligned}\dot{\hat{\pi}} &= \mathrm{i}\hbar\dot{\hat{\psi}}^\dagger = -[\hat{\psi},\hat{H}]^\dagger = +(\hat{\psi}^\dagger\hat{H}^\dagger - \hat{H}^\dagger\hat{\psi}^\dagger) = +[\hat{\psi}^\dagger,\hat{H}^\dagger]\\ &= \frac{1}{\mathrm{i}\hbar}[\hat{\pi},\hat{H}] \,,\end{aligned} \tag{3.11}$$

where we use the fact that the Hamilton operator is hermitean, $\hat{H}^\dagger = \hat{H}$. With the help of the commutation relations (3.9) one finds that the equation of motion for $\hat{\psi}(\boldsymbol{x},t)$ leads back to the original Schrödinger equation. To verify this we use $\mathcal{D}_x$ as an abbreviation for the Schrödinger differential operator taken from (3.7)

$$\mathcal{D}_x = -\frac{\hbar^2}{2m}\nabla^2 + V(\boldsymbol{x}, t), \tag{3.12}$$

and we find

$$\begin{aligned}
[\hat{\psi}(x), \hat{H}] &= \int \mathrm{d}^3 x' \, [\hat{\psi}(x), \hat{\psi}^\dagger(x')\mathcal{D}_{x'}\hat{\psi}(x')] \\
&= \int \mathrm{d}^3 x' \left([\hat{\psi}(x), \hat{\psi}^\dagger(x')]\mathcal{D}_{x'}\hat{\psi}(x') + \hat{\psi}^\dagger(x')\mathcal{D}_{x'}[\hat{\psi}(x), \hat{\psi}(x')] \right) \\
&= \mathcal{D}_x \hat{\psi}(x). \tag{3.13}
\end{aligned}$$

Thus the field operator also satisfies the time-dependent Schrödinger equation

$$i\hbar \frac{\partial \hat{\psi}}{\partial t} = -\frac{\hbar^2}{2m}\nabla^2 \hat{\psi} + V(\boldsymbol{x},t)\hat{\psi}. \tag{3.14}$$

Note, however, that this is not a fundamental but rather a derived relation. The fundamental law of motion is given by the Heisenberg equation (3.10). It is not guaranteed that the field operator $\hat{\psi}$ always satisfies the same differential equations as does the classical field function ψ.

In the process of field quantization we have introduced the objects $\hat{\psi}$, $\hat{\psi}^\dagger$ which are operators in an abstract space of state vectors $|\Phi\rangle$. If one wants to understand and put to use this formalism it is expedient to construct an explicit representation of these operators. For a start we choose a complete set of "classical" wave function $u_i(\boldsymbol{x})$. The field operator can be represented in terms of a generalized Fourier decomposition by expanding it with respect to this set of functions:

$$\begin{aligned}
\hat{\psi}(\boldsymbol{x}, t) &= \sum_i \hat{a}_i(t)\, u_i(\boldsymbol{x}), & (3.15\mathrm{a}) \\
\hat{\psi}^\dagger(\boldsymbol{x}, t) &= \sum_i \hat{a}_i^\dagger(t)\, u_i^*(\boldsymbol{x}). & (3.15\mathrm{b})
\end{aligned}$$

The operator property of $\hat{\psi}(\boldsymbol{x}, t)$ is now carried by the time-dependent expansion coefficients $\hat{a}_i(t)$, while the $u_i(\boldsymbol{x})$ are ordinary complex-valued functions. These functions are assumed to form a complete and orthogonal system, i.e.,

$$\int \mathrm{d}^3 x\, u_i^*(\boldsymbol{x})\, u_j(\boldsymbol{x}) = \delta_{ij}, \tag{3.16}$$

$$\sum_i u_i(\boldsymbol{x})\, u_i^*(\boldsymbol{x}') = \delta^3(\boldsymbol{x} - \boldsymbol{x}'). \tag{3.17}$$

The expansion (3.15) in general implies summation over a set of discrete states as well as integration over a continuum part that may be characterized by a real parameter λ. In the latter case, δ_{ij} in (3.16) is to be interpreted as a delta function with respect to the continuous parameter, $\delta(\lambda - \lambda')$. Insertion of the expansion (3.15) into the equal-time commutation relations (3.9) leads to

$$\sum_{ij} u_i(\boldsymbol{x})\, u_j^*(\boldsymbol{x}')\, [\hat{a}_i(t), \hat{a}_j(t)^\dagger] = \delta^3(\boldsymbol{x} - \boldsymbol{x}'). \tag{3.18}$$

From (3.17) it is obvious that the operators $\hat{a}_i(t)$ have to satisfy the *commutation relations*

$$[\hat{a}_i(t), \hat{a}_j^\dagger(t)] = \delta_{ij} \tag{3.19a}$$

and

$$[\hat{a}_i(t), \hat{a}_j(t)] = [\hat{a}_i^\dagger(t), \hat{a}_j^\dagger(t)] = 0 \,. \tag{3.19b}$$

We can check this by projecting out the "Fourier coefficients" from (3.15a) and (3.15b),

$$\hat{a}_i(t) = \int d^3x \, u_i^*(\boldsymbol{x}) \, \hat{\psi}(\boldsymbol{x}, t) \,, \tag{3.20a}$$

$$\hat{a}_i^\dagger(t) = \int d^3x \, u_i(\boldsymbol{x}) \, \hat{\psi}^\dagger(\boldsymbol{x}, t) \,, \tag{3.20b}$$

and inserting them into (3.19).

The interpretation of the coefficients $\hat{a}_i$ becomes particularly simple if the system has no explicit time dependence, i.e., if the potential reads $V(\boldsymbol{x}, t) = V(\boldsymbol{x})$. Then it is quite natural to use the *eigensolutions of the stationary Schrödinger equation* as the basis for expanding the field operator:

$$\left(-\frac{\hbar^2}{2m}\nabla^2 + V(\boldsymbol{x})\right) u_i(\boldsymbol{x}) = \epsilon_i \, u_i(\boldsymbol{x}) \,. \tag{3.21}$$

This immediately leads to the *Hamilton operator* (for which we will continue to use the name Hamiltonian)

$$\hat{H} = \int d^3x \, \hat{\psi}^\dagger(\boldsymbol{x}, t) \left(-\frac{\hbar^2}{2m}\nabla^2 + V(\boldsymbol{x})\right) \hat{\psi}(\boldsymbol{x}, t) = \sum_i \hat{a}_i^\dagger \, \hat{a}_i \, \epsilon_i \,. \tag{3.22}$$

The time dependence of the operators $\hat{a}_i(t)$ is determined by

$$i\hbar \dot{\hat{a}}_i(t) = [\hat{a}_i(t), \hat{H}] = \sum_j \epsilon_j [\hat{a}_i(t), \hat{a}_j^\dagger(t)\hat{a}_j(t)] = \epsilon_i \hat{a}_i(t) \,, \tag{3.23}$$

which is solved by

$$\hat{a}_i(t) = e^{-i\epsilon_i t/\hbar} \hat{a}_i(0) \equiv e^{-i\epsilon_i t/\hbar} \hat{a}_i \,. \tag{3.24}$$

The use of the eigenfunction basis thus has the effect that the time dependence of the operators $a_i(t)$ becomes trivial, being characterized by a simple phase factor. In the following we will construct the state vectors, making use of the time-independent operators $\hat{a}_i$. This is possible within the *Heisenberg picture* which implies that the operators are time dependent while the state vectors are constant.

The Hamiltonian (3.22) obviously describes the total energy of a collection of particles distributed over various excited states u_i having energies ϵ_i. The *particle-number operator* is defined as

$$\hat{N} = \sum_i \hat{n}_i = \sum_i \hat{a}_i^\dagger \, \hat{a}_i \,. \tag{3.25}$$

The particle-number operators $\hat{n}_i$ for different states i commute among each other since the commutation relations (3.19) imply

$$\begin{aligned}
[\hat{n}_i, \hat{n}_j] &= [\hat{a}_i^\dagger \hat{a}_i, \hat{a}_j^\dagger \hat{a}_j] \\
&= \hat{a}_i^\dagger \hat{a}_j^\dagger [\hat{a}_i, \hat{a}_j] + \hat{a}_i^\dagger [\hat{a}_i, \hat{a}_j^\dagger] \hat{a}_j + \hat{a}_j^\dagger [\hat{a}_i^\dagger, \hat{a}_j] \hat{a}_i + [\hat{a}_i^\dagger, \hat{a}_j^\dagger] \hat{a}_i \hat{a}_j \\
&= 0 \,.
\end{aligned} \tag{3.26}$$

Therefore the particle-number operator $\hat{N}$ commutes with the Hamiltonian $\hat{H}$ and thus is a constant:

$$\dot{\hat{N}} = \frac{1}{i\hbar}[\hat{N}, \hat{H}] = \frac{1}{i\hbar}\sum_{ij}[\hat{a}_i^\dagger \hat{a}_i, \hat{a}_j^\dagger \hat{a}_j]\,\epsilon_j = 0\,. \tag{3.27}$$

According to the rules of quantum mechanics it is possible to find a common set of eigenstates for all these commuting operators. These states are fully characterized by specifying the particle numbers n_i (i.e., the eigenvalues of the operators $\hat{n}_i$) for all the states i:

$$\hat{n}_i\,|n_1, n_2, \ldots, n_i, \ldots\rangle = n_i\,|n_1, n_2, \ldots, n_i, \ldots\rangle\,. \tag{3.28}$$

The state vectors $|n_1, n_2, \ldots\rangle$ form a *basis of the Hilbert space of second quantization* in the "particle-number representation". This Hilbert space is also called the *Fock space*. The scalar product in Fock space is defined as

$$\langle n_1', n_2', \ldots | n_1, n_2, \ldots\rangle = \delta_{n_1', n_1}\,\delta_{n_2', n_2}\cdots\,. \tag{3.29}$$

One immediately finds for the total number of particles n:

$$\hat{N}\,|n_1, n_2, \ldots\rangle = \left(\sum_i n_i\right)|n_1, n_2, \ldots\rangle = n\,|n_1, n_2, \ldots\rangle\,. \tag{3.30}$$

The application of the operator $\hat{a}_i^\dagger$ to $|n_1, n_2, \ldots\rangle$ produces a new state vector. Let us calculate the particle number in the new state $\hat{a}_i^\dagger\,|n_1, n_2, \ldots\rangle$ using the commutation relations (3.19):

$$\begin{aligned}
\hat{N}\hat{a}_i^\dagger\,|n_1,\ldots\rangle &= \sum_j \hat{a}_j^\dagger \hat{a}_j \hat{a}_i^\dagger\,|n_1,\ldots\rangle = \sum_j \hat{a}_j^\dagger(\delta_{ij} + \hat{a}_i^\dagger \hat{a}_j)\,|n_1,\ldots\rangle \\
&= \sum_j(\delta_{ij}\hat{a}_j^\dagger + \hat{a}_i^\dagger \hat{a}_j^\dagger \hat{a}_j)\,|n_1,\ldots\rangle = \hat{a}_i^\dagger(1 + \hat{N})\,|n_1,\ldots\rangle \\
&= (n+1)\,\hat{a}_i^\dagger\,|n_1,\ldots\rangle\,.
\end{aligned} \tag{3.31}$$

Similarly one finds

$$\hat{N}\hat{a}_i\,|n_1,\ldots\rangle = (n-1)\hat{a}_i\,|n_1,\ldots\rangle\,. \tag{3.32}$$

The operators $\hat{a}_i^\dagger$ and $\hat{a}_i$ have the effect of increasing or decreasing the particle number n_i in the state u_i and therefore are called *creation* and *annihilation operators*. The whole set of state vectors can be built up based on a single state which does not contain any particles. In the context of quantum field theory this state is called the *vacuum* $|0\rangle$. The vacuum is defined as a state which is destroyed by the application of any annihilation operator, i.e.,

$$\hat{a}_i\,|0\rangle = 0 \quad \text{for all } i\,. \tag{3.33}$$

Many-body states in the *occupation-number representation* are constructed by applying the appropriate numbers $n_1, n_2, \ldots$ of creation operators to the vacuum:

$$|n_1, n_2, \ldots\rangle = C_{n_1, n_2, \ldots}\,(\hat{a}_1^\dagger)^{n_1}(\hat{a}_2^\dagger)^{n_2}\cdots|0\rangle\,. \tag{3.34}$$

The normalization condition (3.29) for the state vector is satisfied if the constant $C_{n_1,\ldots}$ in (3.34) is chosen as

$$C_{n_1,n_2,\ldots} = \frac{1}{\sqrt{n_1! n_2! \cdots}} \,. \tag{3.35}$$

This can be checked if one starts from the scalar product (3.29) and lets the operators $\hat{a}_i^\dagger$ contained in the "bra" vector act to the right. As shown in Exercise 3.1, repeated use of the commutation relations (3.19) leads to (3.35).

Finally we find the *matrix representation* of the creation and annihilation operators:

$$\langle n_1', \ldots, n_i', \ldots | \hat{a}_i | n_1, \ldots, n_i, \ldots \rangle = \sqrt{n_i}\, \delta_{n_1', n_1} \cdots \delta_{n_i', n_i - 1} \cdots, \tag{3.36a}$$

$$\langle n_1', \ldots, n_i' \ldots | \hat{a}_i^\dagger | n_1, \ldots, n_i, \ldots \rangle = \sqrt{n_i + 1}\, \delta_{n_1', n_1} \cdots \delta_{n_i', n_i + 1} \cdots \tag{3.36b}$$

To investigate the connection between the Fock-space theory and ordinary quantum mechanics we now construct the *wave function in position space*. We start with the definition of a new type of n-particle state vector

$$|\boldsymbol{x}_1, \boldsymbol{x}_2, \ldots, \boldsymbol{x}_n; t\rangle = \frac{1}{\sqrt{n!}} \hat{\psi}^\dagger(\boldsymbol{x}_1, t) \cdots \hat{\psi}^\dagger(\boldsymbol{x}_n, t) |0\rangle \,, \tag{3.37}$$

which depends on space and time. This state describes a system of n particles that, at a given time t, are *localized in coordinate space* at the points $\boldsymbol{x}_1, \ldots, \boldsymbol{x}_n$. This interpretation can be confirmed if one introduces the *local particle-number operator*

$$\hat{N}_V(t) = \int_V \mathrm{d}^3 x\, \hat{\psi}^\dagger(\boldsymbol{x}, t) \hat{\psi}(\boldsymbol{x}, t) \,. \tag{3.38}$$

Note that in contrast to $\hat{N}$ this operator in general will depend on time, in view of the possible spreading of the wave function. The index V indicates that the integration extends over a limited test volume (which in the limit can be taken to be infinitesimally small). Because of (3.9) the operator $\hat{N}_V$ satisfies the commutator relation

$$[\hat{N}_V, \hat{\psi}^\dagger(\boldsymbol{x}, t)] = \int_V \mathrm{d}^3 x'\, \hat{\psi}^\dagger(\boldsymbol{x}', t) \delta^3(\boldsymbol{x} - \boldsymbol{x}') = \begin{cases} \hat{\psi}^\dagger(\boldsymbol{x}, t) & \boldsymbol{x} \in V \\ 0 & \boldsymbol{x} \notin V \end{cases}. \tag{3.39}$$

Using (3.37) and $\hat{N}_V |0\rangle = 0$ it follows that $|\boldsymbol{x}_1, \boldsymbol{x}_2, \ldots, \boldsymbol{x}_n; t\rangle$ is an eigenstate of the operator $\hat{N}_V$. Its eigenvalue is 0 if none of the vectors $\boldsymbol{x}_i$ lies in the test volume V, it is 1 if exactly one $\boldsymbol{x}_i$ is contained in V, and so on.

The localized state vectors (3.37) also form a basis of the Hilbert space. They are constructed in such a way that the following orthonormality relation is valid:

$$\langle \boldsymbol{x}_1', \ldots, \boldsymbol{x}_n'; t | \boldsymbol{x}_1, \ldots, \boldsymbol{x}_n; t \rangle$$
$$= \frac{1}{n!} \sum_{\text{Permut}} P\left[\delta^3(\boldsymbol{x}_1 - \boldsymbol{x}_1') \cdots \delta^3(\boldsymbol{x}_n - \boldsymbol{x}_n')\right] \,. \tag{3.40}$$

Here P denotes the permutation operator which interchanges the ordering of the indices. The sum over the permutations of the spatial coordinats arises since the ordering of the arguments $\boldsymbol{x}_1, \ldots, \boldsymbol{x}_n$ in the state vector (3.37) has no meaning. Our formalism describes indistinguishable particles.

As we know from the eigenstates of the position operator in ordinary quantum mechanics the state $|\boldsymbol{x},t\rangle$ is not normalizable in the strict sense since $\langle \boldsymbol{x},t|\boldsymbol{x},t\rangle = \delta^3(\boldsymbol{0}) = \infty$. If this is annoying, one can construct wavepacket states according to

$$|\Phi_g\rangle = \int d^3x\, g(\boldsymbol{x})\,|\boldsymbol{x}\rangle \tag{3.41}$$

with a square-intregrable function $g(\boldsymbol{x})$. These states can be normalized as

$$\langle\Phi_g|\Phi_g\rangle = \int d^3x'd^3x\, g^*(\boldsymbol{x}')g(\boldsymbol{x})\,\langle\boldsymbol{x}'|\boldsymbol{x}\rangle = 1\,. \tag{3.42}$$

The same can be done for the many-particle states $|\boldsymbol{x}_1,\boldsymbol{x}_2,\ldots,\boldsymbol{x}_n\rangle$.

The completeness relation for the localized state vectors reads

$$\begin{aligned}\mathbb{1} &= |0\rangle\langle 0| + \int d^3x_1\,|\boldsymbol{x}_1,t\rangle\langle\boldsymbol{x}_1,t| \\ &\quad + \int d^3x_1 d^3x_2\,|\boldsymbol{x}_1,\boldsymbol{x}_2,t\rangle\langle\boldsymbol{x}_1,\boldsymbol{x}_2,t| + \ldots\,.\end{aligned} \tag{3.43}$$

Let us inspect the effect of the field operators in some more detail. Obviously $\hat{\psi}^\dagger(\boldsymbol{x})$ has the meaning of a *creation operator for a particle at position* $\boldsymbol{x}$. Its action changes the n-particle state into an $(n+1)$-particle state. From (3.37) we can deduce immediately that

$$\hat{\psi}^\dagger(\boldsymbol{x})|\boldsymbol{x}_1,\ldots,\boldsymbol{x}_n\rangle = \sqrt{n+1}\,|\boldsymbol{x},\boldsymbol{x}_1,\ldots,\boldsymbol{x}_n\rangle\,. \tag{3.44}$$

Correspondingly the field operator $\hat{\psi}(\boldsymbol{x})$ has the effect of removing a particle at position $\boldsymbol{x}$. The zero-particle vacuum state gets destroyed:

$$\hat{\psi}(\boldsymbol{x})|0\rangle = 0\,. \tag{3.45}$$

If there is a finite number of particles n present, one of them can be destroyed by $\hat{\psi}(\boldsymbol{x})$. Since this fate can strike any of the particles, this leads to the sum

$$\hat{\psi}(\boldsymbol{x})|\boldsymbol{x}_1,\ldots,\boldsymbol{x}_n\rangle = \frac{1}{\sqrt{n}}\sum_{i=1}^n \delta^3(\boldsymbol{x}_i-\boldsymbol{x})\,|\boldsymbol{x}_1,\ldots,\boldsymbol{x}_{i-1},\boldsymbol{x}_{i+1}\ldots\boldsymbol{x}_n\rangle\,, \tag{3.46}$$

as one easily confirms by n-fold application of the commutation relation (3.9).

Now we employ the localized state vector to analyze the Fock-space states in coordinate space. To achieve this we define the probability amplitude

$$\Phi^{(n)}_{[n_1 n_2\ldots]}(\boldsymbol{x}_1,\boldsymbol{x}_2,\ldots,\boldsymbol{x}_n;t) = \langle\boldsymbol{x}_1,\boldsymbol{x}_2,\ldots,\boldsymbol{x}_n;t\,|\,n_1,n_2,\ldots\rangle \tag{3.47}$$

with $n = n_1 + n_2 + \ldots$. Mathematically speaking this is the transformation coefficient between the occupation-number basis of the Hilbert space and the basis of localized n-particle states. The square $|\Phi^{(n)}_{[n_1 n_2\ldots]}|^2$ of this amplitude clearly tells us the probability of finding one particle each at the positions $\boldsymbol{x}_1,\ldots,\boldsymbol{x}_n$ provided that one knows that the level u_1 is occupied by n_1 particles, etc. The correct normalization is ensured by the completeness relation (3.43) and by (3.29):

$$\int d^3x_1 \ldots d^3x_n \, \Phi^{(n)*}_{[n'_1 n'_2 \ldots]}(\boldsymbol{x}_1, \ldots, \boldsymbol{x}_n; t) \, \Phi^{(n)}_{[n_1 n_2 \ldots]}(\boldsymbol{x}_1, \ldots, \boldsymbol{x}_n; t)$$

$$= \int d^3x_1 \ldots d^3x_n \, \langle n'_1, n'_2, \ldots \, | \, \boldsymbol{x}_1, \ldots, \boldsymbol{x}_n \rangle \langle \boldsymbol{x}_1, \ldots, \boldsymbol{x}_n \, | \, n'_1, n'_2, \ldots \rangle$$

$$= \delta_{n_1 n'_1} \delta_{n_2 n'_2} \ldots \,. \tag{3.48}$$

The quantity $\Phi^{(n)}_{[n_1 n_2 \ldots]}(\boldsymbol{x}_1, \ldots, \boldsymbol{x}_n; t)$ therefore can be identified with the *wave function of an n-particle system in coordinate space* known from ordinary quantum mechanics. The equation of motion (3.14) for the field operator

$$i\hbar \dot{\hat{\psi}}(\boldsymbol{x}, t) = \left(-\frac{\hbar^2}{2m} \boldsymbol{\nabla}^2 + V(\boldsymbol{x}) \right) \hat{\psi}(\boldsymbol{x}, t) \tag{3.49}$$

implies that our many-body wave function $\Phi^{(n)}$ satisfies the *many-body Schrödinger equation*. This follows by applying the product rule of differentiation

$$i\hbar \frac{\partial}{\partial t} \Phi^{(n)}_{[n_1 n_2 \ldots]}(\boldsymbol{x}_1, \ldots, \boldsymbol{x}_n; t) = i\hbar \frac{\partial}{\partial t} \langle \boldsymbol{x}_1, \ldots, \boldsymbol{x}_n, t | n_1, n_2 \ldots \rangle$$

$$= i\hbar \frac{\partial}{\partial t} \frac{1}{\sqrt{n!}} \langle 0 | \hat{\psi}(\boldsymbol{x}_n, t) \cdots \hat{\psi}(\boldsymbol{x}_1, t) | n_1, n_2, \ldots \rangle$$

$$= \sum_i \frac{1}{\sqrt{n!}} \langle 0 | \hat{\psi}(\boldsymbol{x}_n, t) \cdots$$

$$\times \left(-\frac{\hbar^2}{2m} \boldsymbol{\nabla}_i^2 + V(\boldsymbol{x}_i) \right) \hat{\psi}(\boldsymbol{x}_i, t) \cdots \hat{\psi}(\boldsymbol{x}_1, t) | n_1, n_2, \ldots \rangle$$

$$= \sum_i \left(-\frac{\hbar^2}{2m} \boldsymbol{\nabla}_i^2 + V(\boldsymbol{x}_i) \right) \Phi^{(n)}_{[n_1 n_2 \ldots]}(\boldsymbol{x}_1, \ldots, \boldsymbol{x}_n; t) \,. \tag{3.50}$$

The symmetry property of the many-body wave function is of particular interest: since all creation operators $\hat{\psi}^\dagger(\boldsymbol{x}_i, t)$ commute among each other the wave function $\Phi^{(n)}_{[n_1 n_2 \ldots]}(\boldsymbol{x}_1, \ldots, \boldsymbol{x}_n; t)$ according to (3.9) and (3.37) is *symmetric under permutation of the coordinates*. This means that the Schrödinger field subject to the quantization condition (3.9) describes a *collection of indistinguishable particles that obey Bose–Einstein statistics*.

How does the wave function (3.47) look explicitly when it is expressed in terms of the basis functions $u_i(\boldsymbol{x})$? Using the expansion (3.15) the one-particle state with level k occupied is found as[1]

$$\begin{aligned} \Phi^{(1)}_k(\boldsymbol{x}_1) &= \langle \boldsymbol{x}_1 | 0, \ldots, 0, 1, 0, \ldots \rangle = \langle 0 | \hat{\psi}(\boldsymbol{x}_1) \hat{a}^\dagger_k | 0 \rangle \\ &= \sum_i u_i(\boldsymbol{x}_1) \langle 0 | \hat{a}_i \hat{a}^\dagger_k | 0 \rangle = u_k(\boldsymbol{x}_1) \,. \end{aligned} \tag{3.51}$$

The two-particle state with particles in the levels k_1 and k_2 (assuming $k_1 \neq k_2$) has the wave function

$$\Phi^{(2)}_{k_1 k_2}(\boldsymbol{x}_1, \boldsymbol{x}_2) = \langle \boldsymbol{x}_1 \boldsymbol{x}_2 | \hat{a}^\dagger_{k_1} \hat{a}^\dagger_{k_2} | 0 \rangle$$

---

[1] Here we change the notation. In contrast to (3.47) the state $\Phi^{(n)}$ is now labelled with reference to the n occupied states k_i instead of the infinite (but mostly zero) set of occupation numbers n_i.

$$= \frac{1}{\sqrt{2!}} \sum_{i,j} u_i(\boldsymbol{x}_1) u_j(\boldsymbol{x}_2) \langle 0 | \hat{a}_j \hat{a}_i \hat{a}_{k_1}^\dagger \hat{a}_{k_2}^\dagger | 0 \rangle$$

$$= \frac{1}{\sqrt{2!}} \left(u_{k_1}(\boldsymbol{x}_1) u_{k_2}(\boldsymbol{x}_2) + u_{k_2}(\boldsymbol{x}_1) u_{k_1}(\boldsymbol{x}_2) \right) . \quad (3.52)$$

Otherwise, if both particles occupy the same level $k_1 = k_2$, the normalization factor (3.35) becomes important, with the result

$$\Phi^{(2)}_{k_1 k_1}(\boldsymbol{x}_1, \boldsymbol{x}_2) = \langle \boldsymbol{x}_1 \boldsymbol{x}_2 | \frac{1}{\sqrt{2!}} \hat{a}_{k_1}^\dagger \hat{a}_{k_1}^\dagger | 0 \rangle = \frac{1}{\sqrt{2!}} \frac{1}{\sqrt{2!}} 2\, u_{k_1}(\boldsymbol{x}_1) u_{k_1}(\boldsymbol{x}_2)$$

$$= u_{k_1}(\boldsymbol{x}_1) u_{k_1}(\boldsymbol{x}_2) . \quad (3.53)$$

This can be generalized to the case of an arbitrary n-particle state with particles in the levels $k_1, k_2, \ldots, k_n$ (some of which may have multiple occupation):

$$\Phi^{(n)}_{k_1 \ldots k_n}(\boldsymbol{x}_1, \ldots, \boldsymbol{x}_n) = \frac{1}{\sqrt{n_1! n_2! \cdots}} \frac{1}{\sqrt{n!}} \sum_{\text{Permut}} P\left[u_{k_1}(\boldsymbol{x}_1) \cdots u_{k_n}(\boldsymbol{x}_n) \right] . \quad (3.54)$$

Clearly this is a *completely symmetrized product wave function*. The structure of (3.54) is strongly reminiscent of the determinant of the matrix constructed from the values $u_{k_i}(x_j)$, $i, j = 1, \ldots, n$. In contrast to the determinant here all terms enter with a positive sign. This mathematical construction is sometimes called the *permanent* of the matrix.

3.3 Quantization Rules for Fermi Particles

It is a well-established empirical fact that many of the particles found in nature are fermions, which obey the *Pauli exclusion principle*. Their wave functions are *antisymmetric*, i.e., they change their sign when the particles are interchanged. This clearly stands in contrast to the results we obtained in the last section and thus our quantization rule is inadequate to describe particles of this type. Fortunately it is possible to modify the formalism of field quantization in such a way that it is applicable to fermions too. This is achieved by replacing the commutation relations (3.9) among the field operators by *anticommutation relations*. The fermionic quantization conditions, which were first formulated by P. Jordan and E. Wigner,[2] are

$$\{ \hat{\psi}(\boldsymbol{x}, t), \hat{\psi}^\dagger(\boldsymbol{x}, t) \} = \delta^3(\boldsymbol{x} - \boldsymbol{x}') , \quad (3.55a)$$

$$\{ \hat{\psi}(\boldsymbol{x}, t), \hat{\psi}(\boldsymbol{x}', t) \} = \{ \hat{\psi}^\dagger(\boldsymbol{x}, t), \hat{\psi}^\dagger(\boldsymbol{x}', t) \} = 0 . \quad (3.55b)$$

The curly brackets denote the anticommutator, i.e., $\{\hat{A}, \hat{B}\} := \hat{A}\hat{B} + \hat{B}\hat{A}$. Apart from (3.55) the general formalism of field quantization, in particular the equation of motion (3.10), is supposed to be equally valid for both kinds of particles.

Let us first consider which consequence the condition (3.55b) has on the symmetry of the localized states (3.37):

---

[2] P. Jordan, E.P. Wigner: Z. Physik **47**, 631 (1928).

$$|\boldsymbol{x}_1,\boldsymbol{x}_2,\ldots,\boldsymbol{x}_n;t\rangle = \frac{1}{\sqrt{n!}}\hat{\psi}^\dagger(\boldsymbol{x}_1,t)\cdots\hat{\psi}^\dagger(\boldsymbol{x}_n,t)|0\rangle\,. \tag{3.56}$$

Obviously the product of operators $\hat{\psi}^\dagger(\boldsymbol{x}_1,t)\ldots\hat{\psi}^\dagger(\boldsymbol{x}_n,t)$ and as a consequence also the wave function (3.47) are *completely antisymmetric*, i.e.

$$\Phi^{(n)}_{[n_1 n_2\ldots]}(\boldsymbol{x}_1,\boldsymbol{x}_2,\ldots,\boldsymbol{x}_n;t) = \text{sgn}\,P\,\Phi^{(n)}_{[n_1 n_2\ldots]}(\boldsymbol{x}_{i_1},\boldsymbol{x}_{i_2},\ldots,\boldsymbol{x}_{i_n};t)\,, \tag{3.57}$$

where the sign, $\text{sgn}\,P = \pm 1$, distinguishes between even and odd permutations of the indices $(i_1, i_2, \ldots, i_n) = P(1, 2, \ldots, n)$. The validity of (3.57) follows from the anticommutation relation (3.55b) alone. If we want to interpret (3.56) as a meaningful n-particle state vector, (3.55a) must also be valid. To substantiate this claim we look at the commutator of the local particle-number operator $\hat{N}_V$ defined in (3.38) with the field operator. We employ the following simple but very useful identity, which allows the transformation between commutators and anticommutators:

$$[\hat{A},\hat{B}\hat{C}] = \{\hat{A},\hat{B}\}\hat{C} - \hat{B}\{\hat{A},\hat{C}\} \tag{3.58a}$$

or

$$[\hat{A}\hat{B},\hat{C}] = \hat{A}\{\hat{B},\hat{C}\} - \{\hat{A},\hat{C}\}\hat{B}\,. \tag{3.58b}$$

This leads to

$$\begin{aligned}
[\hat{N}_V,\hat{\psi}^\dagger(\boldsymbol{x},t)] &= \int_V d^3 x'\,[\hat{\psi}^\dagger(\boldsymbol{x}',t)\hat{\psi}(\boldsymbol{x}',t),\hat{\psi}^\dagger(\boldsymbol{x},t)] \\
&= \int_V d^3 x'\,\Big(\hat{\psi}^\dagger(\boldsymbol{x}',t)\{\hat{\psi}(\boldsymbol{x}',t),\hat{\psi}^\dagger(\boldsymbol{x},t)\} \\
&\quad - \{\hat{\psi}^\dagger(\boldsymbol{x}',t),\hat{\psi}^\dagger(\boldsymbol{x},t)\}\hat{\psi}(\boldsymbol{x}',t)\Big) \\
&= \int_V d^3 x'\,\hat{\psi}^\dagger(\boldsymbol{x}',t)\{\hat{\psi}(\boldsymbol{x}',t),\hat{\psi}^\dagger(\boldsymbol{x},t)\}\,. \end{aligned} \tag{3.59}$$

Equation (3.39) and the ensuing reasoning are thus only valid if the anticommutator between $\hat{\psi}(\boldsymbol{x},t)$ and $\hat{\psi}^\dagger(\boldsymbol{x}',t)$ reduces to a delta function, as postulated in (3.55a).

The basis expansion (3.15) as well as the Hamiltonian (3.22) and the time dependence (3.24) of the operators $\hat{a}_i(t)$ are the same for both quantization conditions. The many-body Schrödinger equation (3.50) also remains unchanged. This is so because using (3.58) we find that the equation of motion for the field operator $\hat{\psi}(\boldsymbol{x},t)$ again is

$$\begin{aligned}
i\hbar\dot{\hat{\psi}}(\boldsymbol{x},t) &= [\hat{\psi},\hat{H}] = \int d^3 x'\,\Big[\hat{\psi}(\boldsymbol{x},t),\hat{\psi}^\dagger(\boldsymbol{x}',t)\mathcal{D}_{\boldsymbol{x}'}\hat{\psi}(\boldsymbol{x}',t)\Big] \\
&= \int d^3 x'\,\Big(\{\hat{\psi}(\boldsymbol{x},t),\hat{\psi}^\dagger(\boldsymbol{x}',t)\}\mathcal{D}_{\boldsymbol{x}'}\hat{\psi}(\boldsymbol{x}',t) \\
&\quad - \hat{\psi}^\dagger(\boldsymbol{x}',t)\mathcal{D}_{\boldsymbol{x}'}\{\hat{\psi}(\boldsymbol{x},t),\hat{\psi}(\boldsymbol{x}',t)\}\Big) \\
&= \Big(-\frac{1}{2m}\boldsymbol{\nabla}^2 + V(\boldsymbol{x})\Big)\hat{\psi}(\boldsymbol{x},t)\,, \end{aligned} \tag{3.60}$$

where $\mathcal{D}_x$ denotes the Schrödinger differential operator defined in (3.12).

Now let us investigate the consequences of the Jordan–Wigner quantization rule (3.55) for the construction of the Fock space. The basis expansion (3.15) of the field operator immediately leads to anticommutation relations for the creation and annihilation operators:

$$\{\hat{a}_i, \hat{a}_j^\dagger\} = \delta_{ij}, \quad (3.61a)$$

$$\{\hat{a}_i, \hat{a}_j\} = \{\hat{a}_i^\dagger, \hat{a}_j^\dagger\} = 0. \quad (3.61b)$$

As a special case the square of a creation or annihilation operator is seen to vanish: $\hat{a}_i^2 = (\hat{a}_i^\dagger)^2 = 0$. From this it is obvious that the operator $\hat{n}_i$ for the particle number in a level i can have only eigenvalues n_i that are 0 or 1 since

$$(\hat{n}_i)^2 = \hat{a}_i^\dagger (\hat{a}_i \hat{a}_i^\dagger) \hat{a}_i = \hat{a}_i^\dagger (1 - \hat{a}_i^\dagger \hat{a}_i) \hat{a}_i = \hat{a}_i^\dagger \hat{a}_i = \hat{n}_i, \quad (3.62)$$

and thus $n_i^2 = n_i$. The Fock space for fermions thus has a simple structure: according to (3.34) it consists of the state vectors

$$|n_1, n_2, \ldots\rangle = (\hat{a}_1^\dagger)^{n_1} (\hat{a}_2^\dagger)^{n_2} \cdots |0\rangle \quad (3.63)$$

where the numbers $n_1, n_2, \ldots$ are either 0 or 1. The creation operator $\hat{a}_i^\dagger$ increases the occupation number of the level i by 1. If the state was empty this results in

$$\hat{a}_i^\dagger |n_1, n_2, \ldots, 0_i, \ldots\rangle = (-1)^{\sum_{k=1}^{i-1} n_k} (\hat{a}_1^\dagger)^{n_1} \ldots (\hat{a}_i^\dagger) \ldots |0\rangle$$

$$= (-1)^{\sum_{k=1}^{i-1} n_k} |n_1, n_2, \ldots, 1_i, \ldots\rangle. \quad (3.64)$$

If, on the other hand, the level i was already occupied then the application of $\hat{a}_i^\dagger$ destroys the state vector:

$$\hat{a}_i^\dagger |n_1, n_2, \ldots, 1_i, \ldots\rangle = (-1)^{\sum_{k=1}^{i-1} n_k} (\hat{a}_1^\dagger)^{n_1} \ldots (\hat{a}_i^\dagger)^2 \ldots |0\rangle = 0. \quad (3.65)$$

In this way no two particles are allowed to occupy the same level and the Pauli principle is satisfied. Note that (3.64) contains a phase factor $(-1)^{\sum_{k=1}^{i-1} n_k}$ which has no physical significance as it depends on the labelling of the states i, which is arbitrary.

The many-particle wave function in coordinate space, which we constructed in the last section for Bose particles, can be easily extended to the case of fermions. The two-particle wave function (3.52) becomes

$$\Phi^{(2)}_{k_1 k_2}(x_1, x_2) = \frac{1}{\sqrt{2!}} \left(u_{k_1}(x_1) u_{k_2}(x_2) - u_{k_2}(x_1) u_{k_1}(x_2) \right). \quad (3.66)$$

which can be extended to the n-particle wave function

$$\Phi^{(n)}_{k_1 \ldots k_n}(x_1, \ldots, x_n) = \frac{1}{\sqrt{n!}} \sum_{\text{Permut}} \text{sgn} P \, P\left[u_{k_1}(x_1) \cdots u_{k_n}(x_n) \right]. \quad (3.67)$$

Owing to the alternating sign, (3.67) can be written as a determinant:

$$\Phi^{(n)}_{k_1 \ldots k_n}(x_1, \ldots, x_n) = \frac{1}{\sqrt{n!}} \begin{vmatrix} u_{k_1}(x_1) & u_{k_1}(x_2) & \cdots & u_{k_1}(x_n) \\ u_{k_2}(x_1) & u_{k_2}(x_2) & \cdots & u_{k_2}(x_n) \\ \vdots & \vdots & \ddots & \vdots \\ u_{k_n}(x_1) & u_{k_n}(x_2) & \cdots & u_{k_n}(x_n) \end{vmatrix}. \quad (3.68)$$

This is the famous *Slater determinant* for describing the wave function of many-fermion systems.

In the present and the last sections we have seen that there are two alternative procedures used to quantize a field, which lead to quite different physical consequences. At the present stage we do not know from theoretical arguments which of the two alternatives (Bose–Einstein or Fermi–Dirac statistics) has to be chosen; both of them lead to mathematically consistent theories. Later on, however, we will find that this freedom of choice exists only within the framework of nonrelativistic Schrödinger theory. The quantization of relativistic fields can be done consistently with only one of the two alternatives: Bose statistics applies to particles with integer spin and Fermi statistics to particles with half-integer spin. The opposite choice would lead to the violation of basic principles; see Sects. 4.4 and 5.3.

EXERCISE

3.1 The Normalization of Fock States

Problem. Show that the normalization factor for the states in Fock space has to be chosen as

$$C_{n_1,n_2,\ldots} = \frac{1}{\sqrt{n_1!n_2!\cdots}} \ . \tag{1}$$

Solution. The conjecture will be proven by mathematical induction. Equation (1) is trivial for the zero-particle state (the vacuum). For one-particle states we find at once

$$\begin{aligned}\langle 0,\ldots,1_i,\ldots|0,\ldots,1_i,\ldots\rangle &= \langle 0|\hat{a}_i\hat{a}_i^\dagger|0\rangle = \langle 0|\hat{a}_i^\dagger\hat{a}_i + 1|0\rangle \\ &= \langle 0|0\rangle = 1 \ .\end{aligned} \tag{2}$$

We assume that (1) is the correct normalization factor for a general many-particle state $|n_1, n_2, \ldots\rangle$ and evaluate the norm of the "next higher" state

$$\begin{aligned}&\langle n_1,\ldots,n_i+1,\ldots|n_1,\ldots,n_i+1,\ldots\rangle \\ &= |C_{n_1\ldots n_i+1\ldots}|^2 \langle 0|\ldots(\hat{a}_i)^{n_i+1}\ldots(\hat{a}_1)^{n_1}(\hat{a}_1^\dagger)^{n_1}\ldots(\hat{a}_i^\dagger)^{n_i+1}\ldots|0\rangle \ .\end{aligned} \tag{3}$$

The extra annihilation operator $\hat{a}_i$ is now repeatedly commuted to the right until it meets with its counterpart $\hat{a}_i^\dagger$. Additional commutation steps lead to

$$\begin{aligned}\hat{a}_i(\hat{a}_i^\dagger)^{n_i+1} &= (\hat{a}_i^\dagger\hat{a}_i + 1)(\hat{a}_i^\dagger)^{n_i} = \hat{a}_i^\dagger(\hat{a}_i^\dagger\hat{a}_i+1+1)(\hat{a}_i^\dagger)^{n_i-1} = \ldots \\ &= (\hat{a}_i^\dagger)^{n_i}(n_i+1+\hat{a}_i^\dagger\hat{a}_i) \ .\end{aligned} \tag{4}$$

The last term in the sum does not contribute since the operator $\hat{a}_i$ can be moved further to the right and finally will annihilate the vacuum $\hat{a}_i|0\rangle = 0$. Thus we arrive at the condition

$$\begin{aligned}1 &= (n_i+1)|C_{n_1\ldots n_i+1\ldots}|^2 \langle 0|\ldots(\hat{a}_i)^{n_i}\ldots(\hat{a}_1)^{n_1}(\hat{a}_1^\dagger)^{n_1}\ldots(\hat{a}_i^\dagger)^{n_i}\ldots|0\rangle \\ &= (n_i+1)|C_{n_1\ldots n_i+1\ldots}|^2 \cdot |C_{n_1\ldots n_i\ldots}|^{-2} \ .\end{aligned} \tag{5}$$

Choosing a phase factor $+1$, we find that the square root of this result confirms the normalization constant (1).

EXAMPLE ▬▬▬▬▬▬▬▬

3.2 Interacting Particle Systems: The Hartree–Fock Approximation

So far in this chapter we have been mainly concerned with systems of non-interacting particles. Here we want to describe a system of fermions that are coupled to each other via a *two-body interaction*. The Hamiltonian of such a system can be expressed in terms of the field operator through the following ansatz:

$$\hat{H} = \hat{H}_0 + \hat{H}_1 \tag{1}$$

with the familiar free Hamiltonian

$$\hat{H}_0 = \int d^3x\, \hat{\psi}^\dagger(\boldsymbol{x},t)\mathcal{D}_x\hat{\psi}(\boldsymbol{x},t) , \tag{2}$$

where $\mathcal{D}_x$ is the "free" Schrödinger operator (3.12), which may include an external potential $V(\boldsymbol{x})$ (e.g., the Coulomb field of an atomic nucleus). The two-body interaction Hamiltonian is given by

$$\hat{H}_1 = \frac{1}{2}\int d^3x'd^3x\, \hat{\psi}^\dagger(\boldsymbol{x}',t)\hat{\psi}^\dagger(\boldsymbol{x},t)U(\boldsymbol{x},\boldsymbol{x}')\hat{\psi}(\boldsymbol{x},t)\hat{\psi}(\boldsymbol{x}',t) . \tag{3}$$

The function $U(\boldsymbol{x},\boldsymbol{x}')$ describes the interaction potential, which is assumed to be real and symmetric, i.e.,

$$U(\boldsymbol{x},\boldsymbol{x}') = U(\boldsymbol{x}',\boldsymbol{x}) . \tag{4}$$

As an example, this could stand for the mutual Coulomb interaction of a system of electrons:

$$U(\boldsymbol{x},\boldsymbol{x}') = \frac{e^2}{|\boldsymbol{x}-\boldsymbol{x}'|} . \tag{5}$$

The factor $\frac{1}{2}$ in (3) was included to remove double counting of the interaction energy.

Let us set up the equation of motion for the field operator $\hat{\psi}(\boldsymbol{x},t)$. Its general form is given by Heisenberg's equation

$$i\hbar\partial_0\hat{\psi}(\boldsymbol{x},t) = \left[\hat{\psi}(\boldsymbol{x},t), \hat{H}\right] . \tag{6}$$

The commutator contains two parts. The free contribution gives[3]

$$\left[\hat{\psi}(x), \hat{H}_0\right] = \int d^3x\, \left[\hat{\psi}(x), \hat{\psi}^\dagger \mathcal{D}_{x'}\hat{\psi}(x')\right] = \mathcal{D}_x\hat{\psi}(x) \tag{7}$$

as in (3.60). Making use of the identity (3.58a) we can write the interaction term in such a way that the quantization rules (3.55) can be applied:

---

[3] For brevity we write $\hat{\psi}(\boldsymbol{x},t) \equiv \hat{\psi}(x)$ implying the same time argument for all operators.

Example 3.2

$$[\hat{\psi}(x), \hat{H}_1] = \frac{1}{2}\int d^3x' \, d^3x'' \, U(\boldsymbol{x}',\boldsymbol{x}'') \, [\hat{\psi}(x), \hat{\psi}^\dagger(x'')\hat{\psi}^\dagger(x')\hat{\psi}(x')\hat{\psi}(x'')]$$

$$= \frac{1}{2}\int d^3x' \, d^3x'' \, U(\boldsymbol{x}',\boldsymbol{x}'') \Big(\hat{\psi}^\dagger(x'')\hat{\psi}^\dagger(x') \, [\hat{\psi}(x), \hat{\psi}(x')\hat{\psi}(x'')]$$
$$+ [\hat{\psi}(x), \hat{\psi}^\dagger(x'')\hat{\psi}^\dagger(x')] \, \hat{\psi}(x')\hat{\psi}(x'') \Big)$$

$$= \frac{1}{2}\int d^3x' \, d^3x'' \, U(\boldsymbol{x}',\boldsymbol{x}'')$$
$$\times \Big[\hat{\psi}^\dagger(x'')\hat{\psi}^\dagger(x') \Big(\{\hat{\psi}(x), \hat{\psi}(x')\}\hat{\psi}(x'') - \hat{\psi}(x')\{\hat{\psi}(x), \hat{\psi}(x'')\} \Big)$$
$$+ \Big(\{\hat{\psi}(x), \hat{\psi}^\dagger(x'')\}\hat{\psi}^\dagger(x') - \hat{\psi}^\dagger(x'')\{\hat{\psi}(x), \hat{\psi}^\dagger(x')\} \Big) \hat{\psi}(x')\hat{\psi}(x'') \Big]$$

$$= \frac{1}{2}\int d^3x' \, d^3x'' \, U(\boldsymbol{x}',\boldsymbol{x}'')$$
$$\times \Big(\delta^3(\boldsymbol{x}-\boldsymbol{x}'')\hat{\psi}^\dagger(x') - \hat{\psi}^\dagger(x'')\delta^3(\boldsymbol{x}-\boldsymbol{x}') \Big) \hat{\psi}(x')\hat{\psi}(x'')$$

$$= \frac{1}{2}\int d^3x' \, \Big(U(\boldsymbol{x}',\boldsymbol{x}) + U(\boldsymbol{x},\boldsymbol{x}') \Big) \hat{\psi}^\dagger(x')\hat{\psi}(x')\hat{\psi}(x) \, . \tag{8}$$

Because of the symmetry condition (4) the commutator becomes

$$[\hat{\psi}(x), \hat{H}_1] = \frac{1}{2}\int d^3x' \, \hat{\psi}^\dagger(x')\hat{\psi}(x') \, U(\boldsymbol{x}',\boldsymbol{x}) \, \hat{\psi}(x) \, . \tag{9}$$

The *equation of motion for the field operator* thus reads

$$i\hbar\partial_0\hat{\psi}(x) - \mathcal{D}_x\hat{\psi}(x) - \int d^3x' \, \hat{\psi}^\dagger(x')\hat{\psi}(x') \, U(\boldsymbol{x}',\boldsymbol{x}) \, \hat{\psi}(x) = 0 \, . \tag{10}$$

This is a nonlinear partial integro-differential equation. The complicated structure of this equation defeats any hope of finding an exact solution, so that one has to take recourse to approximation methods. A major problem lies in the fact that the objects $\hat{\psi}(\boldsymbol{x},t)$ are not ordinary c-number functions – this would allow the application of standard numerical integration techniques – but are complicated noncommuting operators in Hilbert space.

To cast (10) into a more easily solvable form we will try to replace the field operators by classical functions. Without loosing generality we can expand the field operator into a complete orthonormal basis of single-particle functions $\varphi_i(\boldsymbol{x})$ as in (3.15):

$$\hat{\psi}(\boldsymbol{x},t) = \sum_i \hat{a}_i(t)\,\varphi_i(\boldsymbol{x}) \, , \tag{11}$$

leading to the standard anticommutation relations (3.61). When expressed in terms of creation and annihilation operators the Hamiltonian now has a more complicated structure. The "free" part is

$$\hat{H}_0 = \sum_{i,j} d_{ij}\, \hat{a}_i^\dagger(t)\hat{a}_j(t) \, , \tag{12}$$

where

$$d_{ij} = \int d^3x \, \varphi_i^*(\boldsymbol{x})\mathcal{D}_x\varphi_j(\boldsymbol{x}) \, . \tag{13}$$

3.3 Quantization Rules for Fermi Particles

Example 3.2

The interaction term is given by

$$\hat{H}_1 = \sum_{i,j,k,l} u_{ijkl}\, \hat{a}_i^\dagger(t)\hat{a}_j^\dagger(t)\hat{a}_k(t)\hat{a}_l(t)\,, \tag{14}$$

where

$$u_{ijkl} = \int d^3 x'\, d^3 x\, \varphi_i^*(\boldsymbol{x}')\varphi_j^*(\boldsymbol{x})\, U(\boldsymbol{x},\boldsymbol{x}')\, \varphi_k(\boldsymbol{x})\varphi_l(\boldsymbol{x}')\,. \tag{15}$$

Because of the interaction Hamiltonian (14) the time development of the Heisenberg operator $\hat{a}_i(t)$ is no longer given by a simple phase factor as in (3.24). The formal solution of the Heisenberg equation

$$\hat{a}_i(t) = e^{+i\hat{H}t/\hbar}\, \hat{a}_i(0)\, e^{-i\hat{H}t/\hbar} \tag{16}$$

indicates that the operator character of $\hat{a}_i$ gets mixed up as a consequence of the interaction. Only for the free Hamiltonian $\hat{H}_0$ with a diagonal coupling matrix $d_{ij} = \epsilon_i\delta_{ij}$ will the evaluation of (16) lead to the simple time dependence (3.24). This can be easily checked by using the operator identity (1) derived in Exercise 1.3. The commutator $[\hat{H},\hat{a}_i]$ involving the full Hamiltonian, however, contains product terms of the form $\hat{a}_i\hat{a}_j^\dagger\hat{a}_l$. As a consequence the operator $\hat{a}_i(t)$, which at time $t = 0$ started out as a pure one-particle annihilator, at later times *will develop into a complicated superposition of creation and annihilation operators*!

Therefore there is no reason to expect that the interacting system is exactly describable by an n-particle state vector like

$$|\Psi\rangle = |n_1, n_2, \ldots\rangle = (\hat{a}_1^\dagger)^{n_1}(\hat{a}_2^\dagger)^{n_2}\cdots|0\rangle\,. \tag{17}$$

Expressed in coordinate space this implies that the true wave function of an interacting system *cannot be fully described by a Slater determinant* like (3.68). As illustrated in Fig. 3.1 the true ground state of a many-fermion system contains admixtures of *particle-hole excitations* described by operators of the type $\hat{a}_i^\dagger\hat{a}_j$ and higher powers.

To arrive at an *approximate* description of the system we will now neglect these admixtures and assume that the time development (16) can be replaced by the simple relation

$$\hat{a}_i(t) = e^{-i\epsilon_i t/\hbar}\hat{a}_i\,. \tag{18}$$

Further we assume that the state vector is given by a pure product as in (17) (a Slater determinant). The aim of our calculation will be to choose the single-particle basis $\varphi_i(\boldsymbol{x})$ in (11) in such a way that the error introduced by these approximations is as small as possible.

For this purpose we form the matrix elements of the equation of motion (10) for the field operators, taken between the n-particle state vector $|\Psi\rangle$ and the $(n-1)$-particle vector $\hat{a}_l|\Psi\rangle$:

$$\langle\Psi|\hat{a}_l^\dagger\hat{O}|\Psi\rangle = 0\,. \tag{19}$$

Here we have used the symbolic notation $\hat{O} = 0$ for the operator equation (10). Of course equation (10) in principle should hold in general and not only for this special type of matrix element (19). It is impossible to achieve this, however, with the state constructed in (17).

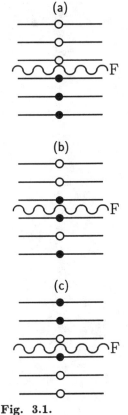

Fig. 3.1.
(a) The ground state of an interacting fermion system can be approximately described by single-particle levels φ_i, which are occupied up to the "Fermi level" F. The true state vector, however, contains an infinite admixture of contributions having one-particle-one-hole (b) two-particle-two-hole (c) etc. character. These contributions are neglected in the Hartree-Fock approximation

Example 3.2

The condition (19) leads to an equation which can be used to determine the basis functions $\varphi_i(\boldsymbol{x})$. To derive this equation we have to evaluate the matrix elements of the three terms in (10). For the time-derivative term the ansatz (18) leads to

$$\langle\Psi|\, \hat{a}_l^\dagger i\hbar\partial_0 \hat{\psi}\,|\Psi\rangle = \sum_i \epsilon_i \varphi_i(\boldsymbol{x}) \langle\Psi|\, \hat{a}_l^\dagger \hat{a}_i\,|\Psi\rangle \,. \tag{20}$$

Since $|\Psi\rangle$ is assumed to have the form of (17) the matrix element (20) will vanish unless the indices l and i are "paired". For $i = l$ we find

$$\langle\Psi|\, \hat{a}_l^\dagger i\hbar\partial_0 \hat{\psi}\,|\Psi\rangle = n_l \epsilon_l \varphi_l(\boldsymbol{x}) \,. \tag{21}$$

For the noninteracting part in (10) we find

$$\langle\Psi|\, \hat{a}_l^\dagger \mathcal{D}_x \hat{\psi}(x)\,|\Psi\rangle = n_l \mathcal{D}_x \varphi_l(\boldsymbol{x}) \,. \tag{22}$$

The interaction term leads to a somewhat more complicated result:

$$\langle\Psi|\hat{a}_l^\dagger \int\!\mathrm{d}^3x'\, \hat{\psi}^\dagger(x')\hat{\psi}(x')\, U(\boldsymbol{x}',\boldsymbol{x})\, \hat{\psi}(x)|\Psi\rangle$$
$$= \int\!\mathrm{d}^3x'\, \sum_{i,j,k} U(\boldsymbol{x}',\boldsymbol{x})\, \varphi_i^*(\boldsymbol{x}')\varphi_j(\boldsymbol{x}')\varphi_k(\boldsymbol{x})\, \langle\Psi|\, \hat{a}_l^\dagger \hat{a}_i^\dagger \hat{a}_j \hat{a}_k\,|\Psi\rangle \,. \tag{23}$$

As before, the indices have to be paired off, but this can be achieved in two different ways:

$$\langle\Psi|\, \hat{a}_l^\dagger \hat{a}_i^\dagger \hat{a}_j \hat{a}_k\,|\Psi\rangle = \delta_{lj}\delta_{ik}\langle\Psi|\,\hat{a}_l^\dagger \hat{a}_i^\dagger \hat{a}_l \hat{a}_i\,|\Psi\rangle + \delta_{lk}\delta_{ij}\langle\Psi|\,\hat{a}_l^\dagger \hat{a}_i^\dagger \hat{a}_i \hat{a}_l\,|\Psi\rangle$$
$$= \delta_{lj}\delta_{ik}\langle\Psi|-\hat{n}_l\hat{n}_i + \delta_{li}\hat{n}_l|\Psi\rangle + \delta_{lk}\delta_{ij}\langle\Psi|+\hat{n}_l\hat{n}_i - \delta_{li}\hat{n}_l|\Psi\rangle$$
$$= n_l n_i \bigl(\delta_{lk}\delta_{ij} - \delta_{lj}\delta_{ik}\bigr) \,. \tag{24}$$

After insertion of the three parts and removal of the common factor n_l, (19) takes the form

$$\mathcal{D}_x \varphi_l(\boldsymbol{x}) + \int\!\mathrm{d}^3x' \sum_i n_i \Bigl[\varphi_i^*(\boldsymbol{x}')\varphi_i(\boldsymbol{x}')\, U(\boldsymbol{x}',\boldsymbol{x})\, \varphi_l(\boldsymbol{x})$$
$$-\varphi_i^*(\boldsymbol{x}')\varphi_l(\boldsymbol{x}')\, U(\boldsymbol{x}',\boldsymbol{x})\, \varphi_i(\boldsymbol{x})\Bigr] = \epsilon_l \varphi_l(\boldsymbol{x}) \,. \tag{25}$$

This is the well-known *Hartree–Fock equation*. It can be written a bit more compactly in terms of the density matrix

$$\rho(\boldsymbol{x}',\boldsymbol{x}) = \sum_i n_i\, \varphi_i^*(\boldsymbol{x}')\varphi_i(\boldsymbol{x}) = \rho^*(\boldsymbol{x},\boldsymbol{x}') \tag{26}$$

and the density function, which is the diagonal part of the density matrix,

$$\rho(\boldsymbol{x}) = \rho(\boldsymbol{x},\boldsymbol{x}) = \sum_i n_i\, \varphi_i^*(\boldsymbol{x})\varphi_i(\boldsymbol{x}) \,. \tag{27}$$

The Hartree–Fock equation in this notation reads

$$\mathcal{D}_x \varphi_l(\boldsymbol{x}) + \int\!\mathrm{d}^3x'\, \rho(\boldsymbol{x}')\, U(\boldsymbol{x}',\boldsymbol{x})\, \varphi_l(\boldsymbol{x})$$
$$- \int\!\mathrm{d}^3x'\, \rho(\boldsymbol{x}',\boldsymbol{x})\, U(\boldsymbol{x}',\boldsymbol{x})\, \varphi_l(\boldsymbol{x}') = \epsilon_l \varphi_l(\boldsymbol{x}) \,. \tag{28}$$

3.3 Quantization Rules for Fermi Particles

Example 3.2

The Hartree–Fock equation contains two separate interaction terms. The *direct term* looks just as one would have expected: the particle in orbital l "feels" the potential which originates from the density distribution $\rho(x')$ of all particles in the system (this includes the self-interaction term $i = l$). According to Fermi–Dirac statistics (antisymmetry under particle exchange) an additional *exchange term* with negative sign arises. This term has a more complicated structure than the direct term; it is *nonlocal*, depending on the values $\varphi_l(x')$ for all values of x' and thus does not have the interpretation of an interaction mediated by a potential. As an agreeable side effect we notice that the exchange term contains a part which cancels the self interaction.

In contrast to the operator equation (10) the relation (28) is a system of coupled nonlinear integro-differential equations for ordinary c-number-valued functions. Such a problem can be attacked sucessfully by using interative numerical methods. One starts from a set of starting solutions $\varphi_l^{(0)}(x)$, which are guessed more or less accurately, and constructs the densities. Subsequently (25) is solved by numerical integration which leads to an improved solution $\varphi_l^{(1)}(x)$. This procedure is repeated until, hopefully, convergence is achieved.

Projection of the Hartree–Fock equation (25) onto $\int d^3x \, \varphi_k^*(x)\ldots$, using (13) and (15), yields

$$d_{kl} + \sum_i n_i (u_{kiil} - u_{kili}) = \epsilon_l \int d^3x \, \varphi_k^*(x)\varphi_l(x) \ . \tag{29}$$

Furthermore projection of the complex conjugate equation (25) onto $\int d^3x \ldots \varphi_l(x)$ leads to

$$d_{lk}^* + \sum_i n_i (u_{kiil} - u_{kili}) = \epsilon_k \int d^3x \, \varphi_k^*(x)\varphi_l(x) \ . \tag{30}$$

Since $d_{lk}^* = d_{kl}$, subtraction of these two equations leads to the conclusion that the single-particle wave functions for different eigenvalues remain orthogonal also in the interacting system, allowing us to impose the condition

$$\int d^3x \, \varphi_k^*(x)\varphi_l(x) = \delta_{kl} \ . \tag{31}$$

The single-particle energies ϵ_i do not directly correspond to observable quantities. The *total energy of the n-particle system* follows from (1)–(3), (17), and (24):

$$E = \langle \Psi | \hat{H} | \Psi \rangle = \sum_i n_i d_{ii} + \frac{1}{2} \sum_{ij} n_i n_j (u_{ijji} - u_{ijij}) \ . \tag{32}$$

The total energy thus results from a summation of the kinetic and potential energies of all particles, taking into account the exchange interaction. It is also possible to find an expression for E in terms of the single-particle energies ϵ_i. Using (29) we can eliminate the diagonal elements d_{ii}, leading to

$$E = \sum_i n_i \epsilon_i - \frac{1}{2} \sum_{ij} n_i n_j (u_{ijji} - u_{ijij}) \ . \tag{33}$$

This result is quite plausible: the total energy results from adding up all the single-particle energies of the occupied orbitals. The interation energy of the particles has to be subtracted from this value. This is necessary since otherwise the interaction energy would be double counted because it is contained in the energies of both partners of the interaction.

The Hartree–Fock method is one of the cornerstones of many-body theory which allows the calculation of various properties of systems in areas such as atomic, solid-state, and nuclear physics. Often, however, its accuracy is not high enough and one has to go beyond the ansatz of a product state (Slater determinant) (17) and take into account *particle correlations*.

Remark: The derivation starting from the equation of motion of Heisenberg field operators is not the most common way to derive the Hartree–Fock equation. Usually the Schrödinger picture is used and a variational principle is employed that minimizes the energy, i.e. the expectation value of the Hamiltonian with respect to a trial state vector $|\Psi\rangle$ of type (17):

$$\delta\Big(\langle\Psi|\hat{H}|\Psi\rangle - \lambda_{ij}\int\mathrm{d}^3x\,\varphi_i^*(\boldsymbol{x})\varphi_j(\boldsymbol{x})\Big) = 0\,. \tag{34}$$

The variation is performed with respect to the single-particle basis orbitals $\varphi_l^*(\boldsymbol{x})$, the Lagrange parameters λ_{ij} are introduced to enforce the orthonormality of the set φ_i. This variational principle leads to the same Hartree–Fock equation (22).

4. Spin-0 Fields: The Klein–Gordon Equation

4.1 The Neutral Klein–Gordon Field

The simplest example of a relativistic field theory deals with spin-0 particles described by the Klein–Gordon equation. In this section we will describe those aspects that are relevant for the quantization of this field. For more information on the Klein–Gordon equation the reader can consult the volume *Relativistic Quantum Mechanics* in this series. We will consecutively treat real and complex-valued fields.

The Lagrange density of a *real spin-0 field* $\phi(x) = \phi(\mathbf{x}, t)$ with mass m reads

$$\mathcal{L}(x) = \frac{\hbar^2}{2} \frac{\partial \phi}{\partial x_\mu} \frac{\partial \phi}{\partial x^\mu} - \frac{1}{2} m^2 c^2 \phi^2 \ . \tag{4.1}$$

In the following we will apply *natural units* of measurement, as is customary in field theory, i.e., we will set $\hbar = c = 1$. This serves to simplify the equations and does no harm since, if desired, the appropriate powers of c and $\hbar$ can be re-inserted, if we know the physical dimension of the quantity of interest. It is important to be aware of the dimension of the scalar field ϕ: Since the Lagrange density has the dimension $\dim [\mathcal{L}] = \text{energy} \times \text{length}^{-3}$ we deduce $\dim [\phi] = \text{energy}^{1/2} \times \text{length}^{-3/2} \times \text{momentum}^{-1} = \text{mass}^{-1/2} \times \text{length}^{-3/2}$. In the natural system of units there is only one type of dimension: $\dim[\text{energy}] = \dim[\text{mass}] = \dim[\text{length}^{-1}]$. Therefore the scalar field ϕ has the "natural dimension" $d = +1$.

The Euler–Lagrange equation $\frac{\partial \mathcal{L}}{\partial \phi} = \frac{\partial}{\partial x_\mu} \frac{\partial \mathcal{L}}{\partial (\partial^\mu \phi)}$ immediately leads to the Klein–Gordon equation

$$(\Box + m^2)\phi(x) = 0 \ , \tag{4.2}$$

where $\Box = \partial^\mu \partial_\mu$ is the four-dimensional Laplace operator. The *canonically conjugate field* is

$$\pi(x) = \frac{\partial \mathcal{L}}{\partial \dot\phi(x)} = \dot\phi(x) \ , \tag{4.3}$$

which leads to the Hamilton density

$$\mathcal{H}(x) = \pi(x)\dot\phi(x) - \mathcal{L}(x) = \frac{1}{2}\left(\pi(x)^2 + \left(\boldsymbol{\nabla}\phi(x)\right)^2 + m^2 \phi(x)^2\right) \ . \tag{4.4}$$

Now we follow the prescription of field quantization: the fields $\phi(\boldsymbol{x},t)$ and $\pi(\boldsymbol{x},t)$ are replaced by operators $\hat{\phi}(\boldsymbol{x},t)$ and $\hat{\pi}(\boldsymbol{x},t)$ for which the *equal-time commutation relations* (ETCR) are[1]

$$[\hat{\phi}(\boldsymbol{x},t), \hat{\pi}(\boldsymbol{x}',t)] = \mathrm{i}\delta^3(\boldsymbol{x}-\boldsymbol{x}'), \tag{4.5a}$$

$$[\hat{\phi}(\boldsymbol{x},t), \hat{\phi}(\boldsymbol{x}',t)] = [\hat{\pi}(\boldsymbol{x},t), \hat{\pi}(\boldsymbol{x}',t)] = 0. \tag{4.5b}$$

Here we have chosen the ordinary commutator (with a minus sign) and consequently the field quanta will satisfy Bose–Einstein statistics. In Sect. 4.4 we will show that the choice of the anticommutator leads to inconsistencies.

Using the quantized *Hamilton operator*

$$\hat{H} = \int \mathrm{d}^3x\, \frac{1}{2}\Big(\hat{\pi}(\boldsymbol{x},t)^2 + (\boldsymbol{\nabla}\hat{\phi}(\boldsymbol{x},t))^2 + m^2\hat{\phi}(\boldsymbol{x},t)^2\Big) \tag{4.6}$$

and the commutation relations (4.5) we can show through a brief calculation that Hamilton's equations of motion lead to

$$\dot{\hat{\phi}}(\boldsymbol{x},t) = -\mathrm{i}\big[\hat{\phi}(\boldsymbol{x},t), \hat{H}\big] = \hat{\pi}(\boldsymbol{x},t) \tag{4.7a}$$

and

$$\dot{\hat{\pi}}(\boldsymbol{x},t) = -\mathrm{i}\big[\hat{\pi}(\boldsymbol{x},t), \hat{H}\big] = (\boldsymbol{\nabla}^2 - m^2)\, \hat{\phi}(\boldsymbol{x},t). \tag{4.7b}$$

To obtain the operator $\boldsymbol{\nabla}^2$ in (4.7b) the equation (4.5a) was differentiated, $[\hat{\pi}(\boldsymbol{x},t), \boldsymbol{\nabla}'\hat{\phi}(\boldsymbol{x}',t)] = \boldsymbol{\nabla}'[\hat{\pi}(\boldsymbol{x},t), \hat{\phi}(\boldsymbol{x}',t)] = -\mathrm{i}\boldsymbol{\nabla}'\delta^3(\boldsymbol{x}-\boldsymbol{x}')$, followed by an integration by parts. Thus we have found that the field operator of the quantized theory still satisfies the original Klein–Gordon equation [2]

$$\ddot{\hat{\phi}}(\boldsymbol{x},t) = (\boldsymbol{\nabla}^2 - m^2)\, \hat{\phi}(\boldsymbol{x},t). \tag{4.8}$$

Now the field operator $\hat{\phi}(\boldsymbol{x},t)$ will be expanded with respect to a basis. For this we use the set of plane waves

$$u_{\boldsymbol{p}}(\boldsymbol{x}) = N_{\boldsymbol{p}}\, \mathrm{e}^{\mathrm{i}\boldsymbol{p}\cdot\boldsymbol{x}}, \tag{4.9}$$

which means

$$\hat{\phi}(\boldsymbol{x},t) = \int \mathrm{d}^3p\, N_{\boldsymbol{p}}\, \mathrm{e}^{\mathrm{i}\boldsymbol{p}\cdot\boldsymbol{x}}\, \hat{a}_{\boldsymbol{p}}(t). \tag{4.10}$$

The normalization constant $N_{\boldsymbol{p}}$ will be fixed later. Insertion of (4.8) immediately yields the equation of motion for the operators $\hat{a}_{\boldsymbol{p}}(t)$:

$$\ddot{\hat{a}}_{\boldsymbol{p}}(t) = -(\boldsymbol{p}^2 + m^2)\, \hat{a}_{\boldsymbol{p}}(t). \tag{4.11}$$

The general solution of (4.11) is obviously given by

$$\hat{a}_{\boldsymbol{p}}(t) = \hat{a}_{\boldsymbol{p}}^{(1)}\mathrm{e}^{-\mathrm{i}\omega_{\boldsymbol{p}}t} + \hat{a}_{\boldsymbol{p}}^{(2)}\mathrm{e}^{+\mathrm{i}\omega_{\boldsymbol{p}}t}, \tag{4.12}$$

---

[1] Although we deal with a relativistic theory the quantization rule treats the time coordinate in a special way since the vectors x' and x are supposed to have the same time component, which is not a Lorentz-invariant notion. A covariant generalization of the commutation relations (4.5) to the case of unequal times $t' \neq t$ will be discussed in Sect. 4.4.

[2] This is not a trivial conclusion. In general the equations of motion for the classical fields and for the quantized field operators do not necessarily agree.

where the operators $\hat{a}_{\boldsymbol{p}}^{(1)}$ and $\hat{a}_{\boldsymbol{p}}^{(2)}$ are constant in time and the frequency is defined to be (relativistic dispersion relation)

$$\omega_p = +\sqrt{\boldsymbol{p}^2 + m^2} \, . \tag{4.13}$$

Since we started from a real-valued classical field, $\phi^* = \phi$, the corresponding field operator should be *hermitean*, $\hat{\phi}^\dagger = \hat{\phi}$. This constraint allows us to express one of the two operators in (4.12) in terms of its counterpart

$$\left(\hat{a}_{\boldsymbol{p}}^{(1)}\right)^\dagger = \hat{a}_{-\boldsymbol{p}}^{(2)} \, , \tag{4.14}$$

and the basis expansion (4.10) becomes

$$\hat{\phi}(\boldsymbol{x},t) = \int \mathrm{d}^3 p \, N_p \left(\hat{a}_{\boldsymbol{p}} \, \mathrm{e}^{\mathrm{i}(\boldsymbol{p}\cdot\boldsymbol{x}-\omega_p t)} + \hat{a}_{\boldsymbol{p}}^\dagger \, \mathrm{e}^{-\mathrm{i}(\boldsymbol{p}\cdot\boldsymbol{x}-\omega_p t)}\right) , \tag{4.15}$$

where $\hat{a}_{\boldsymbol{p}} \equiv \hat{a}_{\boldsymbol{p}}^{(1)}$. Because $\hat{\pi} = \dot{\hat{\phi}}$, the basis expansion of the conjugate field reads

$$\hat{\pi}(\boldsymbol{x},t) = \int \mathrm{d}^3 p \, N_p (-\mathrm{i}\omega_p) \left(\hat{a}_{\boldsymbol{p}} \, \mathrm{e}^{\mathrm{i}(\boldsymbol{p}\cdot\boldsymbol{x}-\omega_p t)} - \hat{a}_{\boldsymbol{p}}^\dagger \, \mathrm{e}^{-\mathrm{i}(\boldsymbol{p}\cdot\boldsymbol{x}-\omega_p t)}\right) . \tag{4.16}$$

The inevitable emergence of *two contributions with positive and negative frequency* in (4.15) and (4.16) is a decisive difference between relativistic field theories and the nonrelativistic Schrödinger field in (3.15) and (3.24).

Now the commutation relations for $\hat{a}_{\boldsymbol{p}}$ and $\hat{a}_{\boldsymbol{p}}^\dagger$ have to be found. We can expect that these operators fulfill the algebra (3.19) typical for creation and annihilation operators, i.e.,

$$[\hat{a}_{\boldsymbol{p}}, \hat{a}_{\boldsymbol{p}'}^\dagger] = \delta^3(\boldsymbol{p}-\boldsymbol{p}') , \tag{4.17a}$$

and

$$[\hat{a}_{\boldsymbol{p}}, \hat{a}_{\boldsymbol{p}'}] = [\hat{a}_{\boldsymbol{p}}^\dagger, \hat{a}_{\boldsymbol{p}'}^\dagger] = 0 \, . \tag{4.17b}$$

To check whether this is true we evaluate the commutator

$$\begin{aligned}
[\hat{\phi}(\boldsymbol{x},t), \hat{\pi}(\boldsymbol{x}',t)] = \int \mathrm{d}^3 p \int \mathrm{d}^3 p' \, N_p N_{p'} (-\mathrm{i}\omega_{p'}) \\
\times \Big([\hat{a}_{\boldsymbol{p}}, \hat{a}_{\boldsymbol{p}'}] \, \mathrm{e}^{-\mathrm{i}(p\cdot x - p'\cdot x')} - [\hat{a}_{\boldsymbol{p}}, \hat{a}_{\boldsymbol{p}'}^\dagger] \, \mathrm{e}^{-\mathrm{i}(p\cdot x + p'\cdot x')} \\
+ [\hat{a}_{\boldsymbol{p}}^\dagger, \hat{a}_{\boldsymbol{p}'}] \, \mathrm{e}^{+\mathrm{i}(p\cdot x - p'\cdot x')} - [\hat{a}_{\boldsymbol{p}}^\dagger, \hat{a}_{\boldsymbol{p}'}^\dagger] \, \mathrm{e}^{+\mathrm{i}(p\cdot x + p'\cdot x')}\Big) ,
\end{aligned} \tag{4.18}$$

where the relativistic notation $p\cdot x = \omega_p t - \boldsymbol{p}\cdot\boldsymbol{x}$, $t' \equiv t$ is used. The application of (4.17) leads to

$$\begin{aligned}
[\hat{\phi}(\boldsymbol{x},t), \hat{\pi}(\boldsymbol{x}',t)] &= \int \mathrm{d}^3 p \int \mathrm{d}^3 p' \, N_p N_{p'} (+\mathrm{i}\omega_{p'}) \delta^3(\boldsymbol{p}-\boldsymbol{p}') \\
&\quad \times \left(\mathrm{e}^{-\mathrm{i}(p\cdot x - p'\cdot x')} + \mathrm{e}^{\mathrm{i}(p\cdot x - p'\cdot x')}\right) \\
&= \mathrm{i} \int \mathrm{d}^3 p \, N_p^2 \, \omega_p \left(\mathrm{e}^{\mathrm{i}\boldsymbol{p}\cdot(\boldsymbol{x}-\boldsymbol{x}')} + \mathrm{e}^{-\mathrm{i}\boldsymbol{p}\cdot(\boldsymbol{x}-\boldsymbol{x}')}\right) .
\end{aligned} \tag{4.19}$$

Clearly a wise choice for the normalization factor N_p is[3]

---

[3] Some authors use the normalization $N_p = (2\pi)^{-3}(2\omega_p)^{-1}$. With this choice the commutation relation (4.17a) reads $[\hat{a}_{\boldsymbol{p}}, \hat{a}_{\boldsymbol{p}'}^\dagger] = (2\pi)^3 2\omega_p \delta^3(\boldsymbol{p}-\boldsymbol{p}')$.

$$N_p = \frac{1}{\sqrt{2\omega_p(2\pi)^3}} . \tag{4.20}$$

This leads to the desired result

$$[\hat{\phi}(\boldsymbol{x},t), \hat{\pi}(\boldsymbol{x}',t)] = i\int \frac{d^3p}{(2\pi)^3} e^{i\boldsymbol{p}\cdot(\boldsymbol{x}-\boldsymbol{x}')} = i\delta^3(\boldsymbol{x}-\boldsymbol{x}') . \tag{4.21}$$

Along the same lines, the equations (4.5b) also can be derived from the commutation relations (4.17) for the creation and annihilation operators.

The commutation (4.17) could also have been derived directly. If we project onto the basis wave functions (the plane waves), the Fourier expansion (4.15) of the field operator can be inverted. Let us define the normalized time-dependent plane waves

$$u_p(\boldsymbol{x},t) = N_p \, e^{-i p \cdot x} = \frac{1}{\sqrt{2\omega_p(2\pi)^3}} e^{-i(\omega_p t - \boldsymbol{p}\cdot\boldsymbol{x})} . \tag{4.22}$$

These functions form a complete set of solutions of the Klein–Gordon equation

$$(\partial_0^2 - \nabla^2 + m^2) u_p(\boldsymbol{x},t) = 0 . \tag{4.23}$$

The *scalar product of two Klein–Gordon wave functions* ϕ and χ (which is not strictly a scalar product since it is not positive definite) is defined as

$$\begin{aligned}(\phi, \chi) &= i\int d^3x \, \phi^*(\boldsymbol{x},t) \overset{\leftrightarrow}{\partial_0} \chi(\boldsymbol{x},t) \\ &\equiv i\int d^3x \left[\phi^*(\boldsymbol{x},t) \frac{\partial \chi(\boldsymbol{x},t)}{\partial t} - \frac{\partial \phi^*(\boldsymbol{x},t)}{\partial t} \chi(\boldsymbol{x},t)\right] .\end{aligned} \tag{4.24}$$

Inserting (4.22) we easily verify that the plane waves form an orthonormal set with respect to the scalar product (4.24):

$$(u_{p'}, u_p) = i\int d^3x \, u_{p'}^*(\boldsymbol{x},t) \overset{\leftrightarrow}{\partial_0} u_p(\boldsymbol{x},t) = \delta^3(\boldsymbol{p}-\boldsymbol{p}') , \tag{4.25a}$$

$$(u_{p'}^*, u_p^*) = -\delta^3(\boldsymbol{p}-\boldsymbol{p}') . \tag{4.25b}$$

Similarly the plane waves with opposite signs for the frequency all are orthogonal:

$$(u_{p'}, u_p^*) = (u_{p'}^*, u_p) = 0 . \tag{4.26}$$

Projection of the field operator

$$\hat{\phi}(\boldsymbol{x},t) = \int d^3p \left(\hat{a}_p \, u_p(\boldsymbol{x},t) + \hat{a}_p^\dagger \, u_p^*(\boldsymbol{x},t)\right) \tag{4.27}$$

on u_p or u_p^* gives the "Fourier coefficients"

$$\hat{a}_p = (u_p, \hat{\phi}) = i\int d^3x \, u_p^*(\boldsymbol{x},t) \overset{\leftrightarrow}{\partial_0} \hat{\phi}(\boldsymbol{x},t) , \tag{4.28a}$$

$$\hat{a}_p^\dagger = -(u_p^*, \hat{\phi}) = -i\int d^3x \, u_p(\boldsymbol{x},t) \overset{\leftrightarrow}{\partial_0} \hat{\phi}(\boldsymbol{x},t) . \tag{4.28b}$$

Using the equations of motion (4.8) for $\hat{\phi}$ and (4.23) for u_p one immediately verifies, through an integration by parts, that the operators $\hat{a}_p$ and $\hat{a}_p^\dagger$ are constant in time. Insertion of (4.28) allows the direct evaluation of the commutators (4.17) (see Exercise 4.1).

Remark: We have employed the continuum normalization when constructing the plane waves (4.22). Sometimes it is advantageous to use the *box normalization* instead. Here periodic boundary conditions are imposed at the surface of a cube having a (sufficiently large) volume $V = L^3$. As a consequence the momentum can take on only the *discrete values*

$$p_l = \frac{2\pi}{L} l , \tag{4.29}$$

where l is a tripel of three integer numbers

$$l = (l_x, l_y, l_z) \quad \text{with} \quad l_i = 0, \pm 1, \pm 2, \ldots . \tag{4.30}$$

The corresponding plane waves within the finite box can be normalized to unity,

$$(u_{p_{l'}}, u_{p_l}) = \delta_{p_{l'}, p_l} \tag{4.31}$$

if the normalization factor

$$N_p = \frac{1}{\sqrt{2\omega_p L^3}} \tag{4.32}$$

is chosen. The continuous integrals over momentum are replaced by discrete sums in the case of box normalization:

$$\int \frac{d^3p}{\sqrt{(2\pi)^3}} \leftrightarrow \frac{1}{L^3} \sum_l . \tag{4.33}$$

The commutation relation (4.17a) gets replaced by

$$[\hat{a}_{p_l}, \hat{a}^\dagger_{p_{l'}}] = \delta_{ll'} . \tag{4.34}$$

The commutation relations (4.17) or equivalently (4.34) are identical to those we found for the quantized nonrelativistic Schrödinger field by using plane waves as an expansion basis. Therefore the construction of the Fock space presented in Sect. 3.1 can immediately be taken over to the relativistic Klein–Gordon field. The operators $\hat{a}^\dagger_p$ and $\hat{a}_p$ again have the function to *create* and *annihilate* particles, described by the wave functions $u_p(x,t)$. It is not essential that momentum eigenstates (plane waves) have been chosen as the expansion basis. It is possible as well (and sometimes advantageous) to choose, for example, angular-momentum eigenstates (spherical waves); see Exercise 4.3.

Now we will express the *Hamiltonian* in terms of creation and annihilation operators. Insertion of (4.27) and the conjugate field operator

$$\hat{\pi}(x,t) = -i \int d^3p\, \omega_p \left(\hat{a}_p u_p(x,t) - \hat{a}^\dagger_p u^*_p(x,t) \right) \tag{4.35}$$

leads to

$$\hat{H} = \frac{1}{2} \int d^3x \left(\hat{\pi}^2 + (\nabla \hat{\phi})^2 + m^2 \hat{\phi}^2 \right)$$

$$= \frac{1}{2} \int d^3x \left[-\int d^3p'\, \omega_{p'} \left(\hat{a}_{p'} u_{p'} - \hat{a}^\dagger_{p'} u^*_{p'} \right) \int d^3p\, \omega_p \left(\hat{a}_p u_p - \hat{a}^\dagger_p u^*_p \right) \right.$$

$$-\int d^3p'\, p'\left(\hat{a}_{p'}\,u_{p'} - \hat{a}^\dagger_{p'}\,u^*_{p'}\right)\cdot\int d^3p\, p\left(\hat{a}_p\,u_p - \hat{a}^\dagger_p\,u^*_p\right)$$
$$+\int d^3p'\, m\left(\hat{a}_{p'}\,u_{p'} + \hat{a}^\dagger_{p'}\,u^*_{p'}\right)\int d^3p\, m\left(\hat{a}_p\,u_p + \hat{a}^\dagger_p\,u^*_p\right)\Bigg]. \tag{4.36}$$

The integration over x can be carried out using (4.22) and gives delta functions:

$$\int d^3x\, u^*_{p'}(\boldsymbol{x},t)u_p(\boldsymbol{x},t) = \frac{1}{2\omega_p}\delta^3(\boldsymbol{p}-\boldsymbol{p}'), \tag{4.37a}$$

$$\int d^3x\, u_{p'}(\boldsymbol{x},t)u_p(\boldsymbol{x},t) = \frac{1}{2\omega_p}e^{-2i\omega_p t}\delta^3(\boldsymbol{p}+\boldsymbol{p}'). \tag{4.37b}$$

We are left with a single momentum integration

$$\hat{H} = \frac{1}{2}\Bigg[-\int d^3p\,\frac{\omega_p^2}{2\omega_p}\left(\hat{a}_{-p}\hat{a}_p e^{-2i\omega_p t} - \hat{a}^\dagger_p\hat{a}_p - \hat{a}_p\hat{a}^\dagger_p + \hat{a}^\dagger_{-p}\hat{a}^\dagger_p e^{2i\omega_p t}\right)$$
$$-\int d^3p\,\frac{\boldsymbol{p}^2}{2\omega_p}\left(-\hat{a}_{-p}\hat{a}_p e^{-2i\omega_p t} - \hat{a}^\dagger_p\hat{a}_p - \hat{a}_p\hat{a}^\dagger_p - \hat{a}^\dagger_{-p}\hat{a}^\dagger_p e^{2i\omega_p t}\right)$$
$$+\int d^3p\,\frac{m^2}{2\omega_p}\left(\hat{a}_{-p}\hat{a}_p e^{-2i\omega_p t} + \hat{a}^\dagger_p\hat{a}_p + \hat{a}_p\hat{a}^\dagger_p + \hat{a}^\dagger_{-p}\hat{a}^\dagger_p e^{2i\omega_p t}\right)\Bigg]. \tag{4.38}$$

The terms involving $\hat{a}\hat{a}$ and $\hat{a}^\dagger\hat{a}^\dagger$ are multiplied by a factor $(-\omega_p^2 + \boldsymbol{p}^2 + m^2)$ and therefore are found to vanish. The remaining expression for the Hamiltonian simply is

$$\hat{H} = \frac{1}{2}\int d^3p\,\omega_p\left(\hat{a}^\dagger_p\hat{a}_p + \hat{a}_p\hat{a}^\dagger_p\right). \tag{4.39}$$

Obviously this operator is beset with a problem: because of $\hat{a}_p\hat{a}^\dagger_p = \hat{a}^\dagger_p\hat{a}_p + \delta^3(\boldsymbol{0})$ its expectation value with respect to any state, including the vacuum, is infinite! To identify the root of the problem we switch back, for a while, from the continuum formulation to the discretized version of the theory by confining the system in a large but finite box. This leads to discrete values p_l for the momentum, and the Hamiltonian becomes

$$\hat{H} = \frac{1}{2}\sum_l \omega_{p_l}\left(\hat{a}^\dagger_{p_l}\hat{a}_{p_l} + \hat{a}_{p_l}\hat{a}^\dagger_{p_l}\right)$$
$$= \sum_l \omega_{p_l}\left(\hat{a}^\dagger_{p_l}\hat{a}_{p_l} + \frac{1}{2}\right) \tag{4.40}$$

because of the commutation relation $[\hat{a}_{p_l},\hat{a}^\dagger_{p_{l'}}] = \delta_{ll'}$. This Hamiltonian is easy to interpret. Each momentum state p_l is occupied by particles whose number is determined by the expectation values of the operator $\hat{n}_{p_l} = \hat{a}^\dagger_{p_l}\hat{a}_{p_l}$. Each of them contributes an energy quantum of ω_{p_l} to the total energy. In addition there is a *zero-point energy*, which is independent of the occupation number. This has the same origin as the zero-point energy of a quantum mechanical harmonic oscillator. The quantum field corresponds to a system of (infinitely many!) harmonic oscillators, where each term in the expansion

(4.40) describes one normal mode. Since there is an infinite number of modes the *vacuum energy*

$$E_0 = \sum_l \frac{1}{2} \omega_{p_l} \qquad (4.41)$$

is strongly divergent. Fortunately, from a practical point of view this infinity (unlike other infinities that crop up in quantum field theory) is rather harmless. Since physical observables involve energy differences and not the absolute value of the energy, the constant zero-point value E_0 always drops out. This can be formally accounted for by *defining* the Hamiltonian in such a way that the vacuum energy is removed:

$$\hat{\tilde{H}} = \hat{H} - E_0 \, . \qquad (4.42)$$

A mathematical trick can be used to achive such a "trivial renormalization". For this the field operator (4.27) is split into two parts containing positive and negative frequencies:

$$\hat{\phi}(\boldsymbol{x},t) = \hat{\phi}^{(+)}(\boldsymbol{x},t) + \hat{\phi}^{(-)}(\boldsymbol{x},t) \, . \qquad (4.43)$$

The first part comprises the annihilation operators and the second part the creation operators. Now we introduce the important notion of *normal ordering* of two operators $\hat{\phi}, \hat{\chi}$. The normal product is defined as a product in which the parts with negative frequency generally stand to the left of the parts with positive frequency (remember that "positive frequency" refers to a time evolution with the phase $\exp(-i\omega t), \omega > 0$):

$$:\hat{\phi}\hat{\chi}: \; = \; \hat{\phi}^{(-)}\hat{\chi}^{(-)} + \hat{\phi}^{(-)}\hat{\chi}^{(+)} + \hat{\chi}^{(-)}\hat{\phi}^{(+)} + \hat{\phi}^{(+)}\hat{\chi}^{(+)} \, . \qquad (4.44)$$

Thus if the Hamiltonian is defined as the normal-ordered product of the field operators,

$$\begin{aligned}\hat{\tilde{H}} &= \frac{1}{2}\int d^3x \; :\!\left(\hat{\pi}^2 + (\boldsymbol{\nabla}\hat{\phi})^2 + m^2\hat{\phi}^2\right)\!: \\ &= \int d^3p \, \omega_p \, \hat{a}_{\boldsymbol{p}}^\dagger \hat{a}_{\boldsymbol{p}} \, , \end{aligned} \qquad (4.45)$$

the creation operators are automatically moved to the left of the annihilation operators and the commutator (4.40) which gave rise to the vacuum energy does not arise from the beginning.

However, this is a rather formal argument which should not induce us to dismiss the vacuum energy as a pure artefact of the theory. It is possible to find conditions that allow the measurement of *energy differences between different vacuum configurations*:

$$\Delta E = E_0[1] - E_0[2] \, . \qquad (4.46)$$

Here the arguments 1 and 2 refer to *different boundary conditions* imposed on the field ϕ. Depending on the circumstances, the value of ΔE can be finite and nonzero, leading to experimentally observable effects. The most famous example is the case of two conducting plates which "lock up" the zero-point fluctuations of the electromagnetic field. The value of ΔE depends on the distance between the plates, which are thus found to exert a (weak but measurable) force on each other although they do not carry an electric charge. This is known as the Casimir effect[4].

Now let us turn to the *momentum* of our quantum field. The classical expression for the momentum follows from Noether's theorem as

$$P_\mu = \int d^3x \, \Theta_{0\mu} = \int d^3x \left(\pi \frac{\partial \phi}{\partial x^\mu} - g_{0\mu} \mathcal{L} \right), \qquad (4.47)$$

and in particular

$$\boldsymbol{P} = -\int d^3x \, \pi \boldsymbol{\nabla} \phi. \qquad (4.48)$$

The momentum operator of the quantum field therefore could be defined as

$$\hat{\boldsymbol{P}} = -\int d^3x \, \hat{\pi}(\boldsymbol{x}, t) \boldsymbol{\nabla} \hat{\phi}(\boldsymbol{x}, t). \qquad (4.49)$$

There is no reason, however, why the noncommuting operators $\hat{\pi}$ and $\hat{\phi}$ should be multiplied just in this order. This certainly cannot be deduced from the classical theory since the odering problem does not arise there. A more natural choice would be to *symmetrize* the momentum operator according to the definition

$$\hat{\boldsymbol{P}} = -\frac{1}{2} \int d^3x \left(\hat{\pi}(\boldsymbol{x},t) \boldsymbol{\nabla} \hat{\phi}(\boldsymbol{x},t) + \boldsymbol{\nabla} \hat{\phi}(\boldsymbol{x},t) \hat{\pi}(\boldsymbol{x},t) \right). \qquad (4.50)$$

This choice also guarantees that $\hat{\boldsymbol{P}}$ is a hermitean operator. Now the quantum fields again will be expanded into plane waves and (4.27), (4.35), and (4.37) are used:

$$\begin{aligned}
\hat{\boldsymbol{P}} &= -\frac{1}{2} \Big[-\mathrm{i} \int d^3 p' \omega_{p'} (\hat{a}_{p'} u_p - \hat{a}_{p'}^\dagger u_{p'}^*) \int d^3 p \, \mathrm{i} \, \boldsymbol{p} \, (\hat{a}_p u_p - \hat{a}_p^\dagger u_p^*) \\
&\quad + \text{Exchange} \Big] \\
&= -\frac{1}{2} \int d^3 p \, \frac{1}{2\omega_p} \omega_p \\
&\quad \times \Big[\left(\boldsymbol{p} \, \hat{a}_{-p} \hat{a}_p \mathrm{e}^{-2\mathrm{i}\omega_p t} - \boldsymbol{p} \, \hat{a}_p^\dagger \hat{a}_p - \boldsymbol{p} \, \hat{a}_p \hat{a}_p^\dagger + \boldsymbol{p} \, \hat{a}_{-p}^\dagger \hat{a}_p^\dagger \mathrm{e}^{2\mathrm{i}\omega_p t} \right) \\
&\quad + \left(-\boldsymbol{p} \, \hat{a}_{-p} \hat{a}_p \mathrm{e}^{-2\mathrm{i}\omega_p t} - \boldsymbol{p} \, \hat{a}_p^\dagger \hat{a}_p - \boldsymbol{p} \, \hat{a}_p \hat{a}_p^\dagger - \boldsymbol{p} \, \hat{a}_{-p}^\dagger \hat{a}_p^\dagger \mathrm{e}^{2\mathrm{i}\omega_p t} \right) \Big] \\
&= \frac{1}{2} \int d^3 p \, \boldsymbol{p} \left(\hat{a}_p^\dagger \hat{a}_p + \hat{a}_p \hat{a}_p^\dagger \right). \qquad (4.51)
\end{aligned}$$

Let us remark that the contributions involving $\hat{a}_{-p} \hat{a}_p$ and $\hat{a}_{-p}^\dagger \hat{a}_p^\dagger$ would have dropped out even without imposing the symmetrization prescription (4.50) since the integrand is an odd function in $\boldsymbol{p}$. The momentum operator (4.51) also appears to suffer from a divergence. Using the discretized formulation, however, it is seen that

$$\hat{\boldsymbol{P}} = \sum_l \boldsymbol{p}_l \, \hat{a}_{p_l}^\dagger \hat{a}_{p_l} + \frac{1}{2} \sum_l \boldsymbol{p}_l = \sum_l \boldsymbol{p}_l \, \hat{a}_{p_l}^\dagger \hat{a}_{p_l}. \qquad (4.52)$$

---

[4] The original publication is H.B.G. Casimir: Proc. Netherl. Acad. Wetensch. **51**, 793 (1948); see also G. Plunien, B. Müller, W. Greiner: Physics Reports **134**, 87 (1986).

Since all spatial directions are equivalent the contributions to the sum over p_l cancel each other and the vacuum momentum necessarily vanishes due to symmetry reason. The additional imposition of the normal ordering prescription is not required to define the momentum operator but, of course, it can do no harm.

Our expansion of the field operator of the Klein–Gordon theory as in (4.15) has lead to a formalism closely similar to the quantized Schrödinger field of Chap. 3. Again it is possible to construct the *Fock space* by applying the creation operators $\hat{a}_p^\dagger$ to the vacuum state $|0\rangle$. The field quanta carry momentum p and energy $\omega_p = (p^2 + m^2)^{1/2}$ and they are counted by the number operator $\hat{n}_p = \hat{a}_p^\dagger \hat{a}_p$.

The states constructed in this way do not have a well-defined *angular momentum*. It is possible to show that the operator of orbital angular momentum (there is no spin, of course) derived from Noether's theorem is nondiagonal with respect to the Fock space generated by the operators $\hat{a}_p$. When we introduce symmetrization and normal ordering the angular-momentum operator reads

$$\hat{L} = -\frac{1}{2}\int d^3x : \left(\hat{\pi}\, \boldsymbol{x} \times \boldsymbol{\nabla}\hat{\phi} + (\boldsymbol{x} \times \boldsymbol{\nabla}\hat{\phi})\, \hat{\pi}\right):$$
$$= i\int d^3p\, \hat{a}_p^\dagger \left(\boldsymbol{p} \times \boldsymbol{\nabla}_p\right) \hat{a}_p \ . \tag{4.53}$$

This will be derived in Exercise 4.3d.

As an alternative to the procedure presented above, the Fock space can be constructed from field quanta that carry other sets of quantum numbers. As an example the field operator can be expanded with respect to *spherical waves* as basis functions. In this case the field quanta have good angular momentum and energy; the momentum operator, however, obtains a complicated nondiagonal structure. The sperical representation will be investigated in Exercise 4.3.

EXERCISE

4.1 Commutation Relations for Creation and Annihilation Operators

Problem. Derive the commutation relations for the Fourier coefficients $\hat{a}_p$ and $\hat{a}_p^\dagger$ in the expansion (4.27) of the field operator

$$\hat{\phi}(\boldsymbol{x},t) = \int d^3p \left(\hat{a}_p\, u_p(\boldsymbol{x},t) + \hat{a}_p^\dagger\, u_p^*(\boldsymbol{x},t)\right) \ . \tag{1}$$

Solution. (a) Calculation of $[\hat{a}_p, \hat{a}_{p'}]$.
We start by inserting (4.28a) into the commutator:

$$[\hat{a}_p, \hat{a}_{p'}] = i^2 \int d^3x\, d^3x' \left[u_p^*(\boldsymbol{x},t)\, \overset{\leftrightarrow}{\partial_0}\, \hat{\phi}(\boldsymbol{x},t),\ u_{p'}^*(\boldsymbol{x}',t)\, \overset{\leftrightarrow}{\partial_0}\, \hat{\phi}(\boldsymbol{x}',t)\right] \ . \tag{2}$$

This is transformed using the definition (4.24). The functions $u_p(\boldsymbol{x},t)$ are c numbers and commute with the field operators

Exercise 4.1

$$\left[u_p^*(x,t) \stackrel{\leftrightarrow}{\partial}_0 \hat{\phi}(x,t), u_{p'}^*(x',t) \stackrel{\leftrightarrow}{\partial}_0 \hat{\phi}(x',t)\right] \quad (3)$$

$$= u_p^*(x,t)u_{p'}^*(x',t)\left[\dot{\hat{\phi}}(x,t),\dot{\hat{\phi}}(x',t)\right] - u_p^*(x,t)\dot{u}_{p'}^*(x',t)\left[\dot{\hat{\phi}}(x,t),\hat{\phi}(x',t)\right]$$
$$- \dot{u}_p^*(x,t)u_{p'}^*(x',t)\left[\hat{\phi}(x,t),\dot{\hat{\phi}}(x',t)\right] + \dot{u}_p^*(x,t)\dot{u}_{p'}^*(x',t)\left[\hat{\phi}(x,t),\hat{\phi}(x',t)\right].$$

Inserting the commutation relations (4.5), we find that the first and fourth term on the right-hand side vanish and the remaining commutators reduce to delta functions:

$$\begin{aligned}[\hat{a}_p, \hat{a}_{p'}] &= -\mathrm{i}\int \mathrm{d}^3 x \left(u_p^*(x,t)\dot{u}_{p'}^*(x,t) - \dot{u}_p^*(x,t)u_{p'}^*(x,t)\right)\\
&= -\mathrm{i}\int \mathrm{d}^3 x \, u_p^*(x,t) \stackrel{\leftrightarrow}{\partial}_0 u_{p'}^*(x,t) = -\left(u_p, u_{p'}^*\right).\end{aligned} \quad (4)$$

This expression vanishes because of the orthogonality relation (4.26) and thus

$$[\hat{a}_p, \hat{a}_{p'}] = 0. \quad (5)$$

An analogous calculation applies to the commutator of creation operators. Because of (4.28) one simply has to take the complex conjugate of $u_p(x,t)$. The result is

$$[\hat{a}_p^\dagger, \hat{a}_{p'}^\dagger] = 0. \quad (6)$$

(b) Calculation of $[\hat{a}_p, \hat{a}_{p'}^\dagger]$.
As above, the use of (4.28) leads to

$$[\hat{a}_p, \hat{a}_{p'}^\dagger] = -\mathrm{i}^2 \int \mathrm{d}^3 x \mathrm{d}^3 x' \left[u_p^*(x,t) \stackrel{\leftrightarrow}{\partial}_0 \hat{\phi}(x,t), u_{p'}(x',t) \stackrel{\leftrightarrow}{\partial}_0 \hat{\phi}(x',t)\right]. \quad (7)$$

Insertion of (4.24) gives

$$\left[u_p^*(x,t) \stackrel{\leftrightarrow}{\partial}_0 \hat{\phi}(x,t), u_{p'}(x',t) \stackrel{\leftrightarrow}{\partial}_0 \hat{\phi}(x',t)\right] \quad (8)$$

$$= u_p^*(x,t)u_{p'}(x',t)\left[\dot{\hat{\phi}}(x,t),\dot{\hat{\phi}}(x',t)\right] - u_p^*(x,t)\dot{u}_{p'}(x',t)\left[\dot{\hat{\phi}}(x,t),\hat{\phi}(x',t)\right]$$
$$- \dot{u}_p^*(x,t)u_{p'}(x',t)\left[\hat{\phi}(x,t),\dot{\hat{\phi}}(x',t)\right] + \dot{u}_p^*(x,t)\dot{u}_{p'}(x',t)\left[\hat{\phi}(x,t),\hat{\phi}(x',t)\right].$$

Again the first and fourth terms vanish and the remaining commutators reduce to delta functions:

$$[\hat{a}_p, \hat{a}_{p'}^\dagger] = \mathrm{i}\int \mathrm{d}^3 x \, u_p^*(x,t) \stackrel{\leftrightarrow}{\partial}_0 u_{p'}(x,t) = \left(u_p, u_{p'}\right). \quad (9)$$

From (4.25) one finds

$$[\hat{a}_p, \hat{a}_{p'}^\dagger] = \delta^3(p - p'). \quad (10)$$

Equations (5), (6) and (10) are the commutation relations for creation and annihilations operators known from (4.17).

EXERCISE

4.2 Commutation Relations of the Angular-Momentum Operator

Problem. (a) Show that the quantized angular-momentum operator $\hat{\boldsymbol{J}}$ of the spin-0 field satisfies the standard commutation relations

$$[\hat{J}_i, \hat{J}_j] = i\epsilon_{ijk}\hat{J}_k \ . \tag{1}$$

Here ϵ_{ijk} is the completely antisymmetric unit tensor (the Levi–Civita tensor) in three dimensions.

(b) Confirm the commutation relations between the momentum and angular-momentum operators

$$[\hat{P}_i, \hat{J}_j] = i\epsilon_{ijk}\hat{P}_k \ . \tag{2}$$

Solution. (a) The general expression for the angular-momentum tensor has been derived from Noether's theorem in (2.70):

$$M_{ij} = \int d^3x \, \frac{\partial \mathcal{L}}{\partial(\partial_0\phi_r)} \left((x_i\partial_j - x_j\partial_i)\phi_r + (I^{ij})_{rs}\phi_s \right) . \tag{3}$$

In the case of a scalar field the generators of Lorentz transformations $(I^{ij})_{rs}$ drop out and we are left with the orbital angular momentum

$$\boldsymbol{J} = \boldsymbol{L} = -\int d^3x \, \pi(x)(\boldsymbol{x} \times \boldsymbol{\nabla})\phi(x) \ . \tag{4}$$

The sign (which does not follow from Noether's theorem) was chosen to be consistent with the standard definition of angular momentum. If the metric is taken into account the signs of (4) and (2.72) agree.

Quantization is achieved by making the changes $\phi \to \hat{\phi}$ and $\pi \to \hat{\pi}$ in (4). Keep in mind, however, that this introduces an arbitrary choice of the ordering when products of field operators $\hat{\phi}$ and $\hat{\pi}$ are involved. This can be avoided by symmetrizing the products of operators. We will adopt this prescription here since it would not change the course of the derivation and would not affect the final result.

Let us study the commutator

$$[\hat{L}_i, \hat{L}_j] = \int d^3x \, d^3x' \left[\hat{\pi}(x)(\boldsymbol{x} \times \boldsymbol{\nabla})_i \hat{\phi}(x), \, \hat{\pi}(x')(\boldsymbol{x}' \times \boldsymbol{\nabla}')_j \hat{\phi}(x') \right] , \tag{5}$$

which is to be evaluated for equal time arguments $x_0 = x'_0$. The equal-time commutation relations (4.5) can be applied, in particular $[\hat{\phi}(\boldsymbol{x},t), \hat{\pi}(\boldsymbol{x}',t)] = i\delta^3(\boldsymbol{x} - \boldsymbol{x}')$. The commutator in (5) can be expanded into four terms, two of which are found to vanish because they contain $[\hat{\pi}(x), \hat{\pi}(x')] = [\hat{\phi}(x), \hat{\phi}(x')] = 0$. We are left with

$$[\hat{L}_i, \hat{L}_j] = \int d^3x \, d^3x' \Big(\hat{\pi}(x)(\boldsymbol{x} \times \boldsymbol{\nabla})_i \, [\hat{\phi}(x), \hat{\pi}(x')] \, (\boldsymbol{x}' \times \boldsymbol{\nabla}')_j \hat{\phi}(x')$$
$$+ \hat{\pi}(x')(\boldsymbol{x}' \times \boldsymbol{\nabla}')_j \, [\hat{\pi}(x), \hat{\phi}(x')] \, (\boldsymbol{x} \times \boldsymbol{\nabla})_i \hat{\phi}(x) \Big)$$

Exercise 4.2

$$
= \int d^3x\, d^3x' \Big(\hat{\pi}(x)(x \times \nabla)_i\, i\delta^3(x-x')\, (x' \times \nabla')_j \hat{\phi}(x')
$$
$$
+ \hat{\pi}(x')(x' \times \nabla')_j(-i)\delta^3(x-x')\, (x \times \nabla)_i \hat{\phi}(x) \Big) . \tag{6}
$$

The delta functions can be isolated through an integration by parts with respect to x and x', and the spatial integration collapses:

$$
\begin{aligned}
[\hat{L}_i, \hat{L}_j] &= -i \int d^3x\, d^3x'\, \delta^3(x-x') \Big((x \times \nabla)_i \hat{\pi}(x)(x' \times \nabla')_j \hat{\phi}(x') \\
&\quad - (x' \times \nabla')_j \hat{\pi}(x')(x \times \nabla)_i \hat{\phi}(x) \Big) \\
&= i \int d^3x\, \Big(\hat{\pi}(x)(x \times \nabla)_i (x \times \nabla)_j \hat{\phi}(x) \\
&\quad - \hat{\pi}(x)(x \times \nabla)_j (x \times \nabla)_i \hat{\phi}(x) \Big) .
\end{aligned} \tag{7}
$$

The last step involves a further integration by parts. The result (7) can be attributed to the action of the ∇ operator on the coordinate vector x. This is the same algebra that is responsible for the angular-momentum commutator in ordinary quantum mechanics:

$$
(x \times \nabla)_i\, (x \times \nabla)_j - (x \times \nabla)_j\, (x \times \nabla)_i = x_j \partial_i - x_i \partial_j = -\epsilon_{ijk}(x \times \nabla)_k . \tag{8}
$$

(b) If we use the momentum operator (again in its unsymmetrized form)

$$
\hat{P} = -\int d^3x\, \hat{\pi}(x)\nabla\hat{\phi}(x) , \tag{9}
$$

the calculation proceeds exactly as in **(a)**. The commutator reads

$$
\begin{aligned}
[\hat{P}_i, \hat{L}_j] &= \int d^3x\, d^3x' \left[\hat{\pi}(x)\nabla_i\hat{\phi}(x),\, \hat{\pi}(x')(x' \times \nabla')_j \hat{\phi}(x') \right] \\
&= \int d^3x\, d^3x' \Big(\hat{\pi}(x)[\nabla_i\hat{\phi}(x), \hat{\pi}(x')](x' \times \nabla')_j \hat{\phi}(x') \\
&\quad + \hat{\pi}(x')[\hat{\pi}(x), (x' \times \nabla')_j \hat{\phi}(x')]\nabla_i\hat{\phi}(x) \Big) \\
&= -i \int d^3x\, \Big((\nabla_i\hat{\pi}(x))(x \times \nabla)_j \hat{\phi}(x) - ((x \times \nabla)_j \hat{\pi}(x))\nabla_i\hat{\phi}(x) \Big) \\
&= i \int d^3x\, \hat{\pi}(x) \Big(\nabla_i(x \times \nabla)_j - (x \times \nabla)_j \nabla_i \Big) \hat{\phi}(x) \\
&= i\epsilon_{ijk}\hat{P}_k .
\end{aligned} \tag{10}
$$

We have integrated by parts twice and at the end used the identity

$$
\nabla_i(x \times \nabla)_j - (x \times \nabla)_j \nabla_i = -\epsilon_{ijk}\nabla_k . \tag{11}
$$

EXERCISE ▰

4.3 The Field Operator in the Spherical Representation

Problem. The field operator $\hat{\phi}(x, t)$ can be expanded into spherical waves instead of plane waves. Then the field modes are classified according to their angular momentum instead of their longitudinal momentum. This is to be worked out for the case of the neutral Klein–Gordon field.

Exercise 4.3

(a) Set up a basis of stationary spherical waves which solve the radial Klein–Gordon equation and find the corresponding creation and annihilation operators.
(b) Find the connection with the expansion into the plane-wave basis.
(c) Derive expressions for the Hamiltonian and for the angular-momentum operator in the spherical-wave representation.
(d) Compute the angular-momentum operator $\hat{\boldsymbol{L}}$ in terms of the plane-wave operators $\hat{a}_p$ and $\hat{a}_p^\dagger$.

Solution. (a) We want to expand the field operators in the basis of stationary spherical waves. These are solutions of the Klein–Gordon equation characterized by the absolute value of the radial momentum p and by the angular-momentum quantum numbers l and m:

$$\phi_{plm}(\boldsymbol{x}) = N_p \, u_{pl}(r) \, Y_{lm}(\Omega) \,. \tag{1}$$

The basis functions $u_{pl}(r)$ satisfy the radial equation (we split the Laplacian according to $\Delta = \Delta_r + \Delta_\Omega$)

$$\left(\Delta_r - \frac{l(l+1)}{r^2} + p^2\right) u_{pr}(r) = \left(\frac{d^2}{dr^2} + \frac{2}{r}\frac{d}{dr} - \frac{l(l+1)}{r^2} + p^2\right) u_{pl}(r) = 0 \,. \tag{2}$$

The real-valued solution to this differential equation which is regular at the origin is given by

$$u_{pl}(r) = \sqrt{\frac{2}{\pi}} \, p \, j_l(pr) \,. \tag{3}$$

This is a spherical Bessel function $j_l(pr)$, which was "normalized to a delta function". It is known from the mathematical theory of differential equations (Eq. (2) belongs to the class of Sturm-Liouville problems) that solutions having different eigenvalues p are mutually orthogonal. The normalization factor can be confirmed by inserting the asymptotic expansion of the spherical Bessel function $j_l(x) \to \sin(x - l\pi/2)/x$, which leads to

$$\int_0^\infty dr \, r^2 \, j_l(pr) \, j_l(p'r) = \frac{\pi}{2} \frac{1}{p^2} \delta(p - p') \,. \tag{4}$$

The spherical harmonics $Y_{lm}(\Omega)$ satisfy the differential equation

$$\left(\Delta_\Omega + \frac{l(l+1)}{r^2}\right) Y_{lm}(\Omega) = 0 \,, \tag{5}$$

where Δ_Ω is the angular part of the Laplacian. The basis functions fulfill the orthogonality relation

$$\int d^3x \, \phi_{plm}^* \, \phi_{p'l'm'} = N_p^2 \, \delta(p - p') \, \delta_{ll'} \, \delta_{mm'} \tag{6}$$

and the completeness relation

$$\int_0^\infty dp \sum_{l=0}^\infty \sum_{m=-l}^{+l} u_{pl}(r) Y_{lm}^*(\Omega_r) \, u_{pl}(r') Y_{lm}(\Omega_{r'}) = \delta^3(\boldsymbol{r} - \boldsymbol{r}') \,. \tag{7}$$

Exercise 4.3

The factor N_p was introduced with the aim of obtaining a unit delta function in the canonical commutation relation to be discussed below.

The field operator $\hat{\phi}(\boldsymbol{x},t)$ can be expanded as

$$\hat{\phi}(\boldsymbol{x},t) = \int_0^\infty dp \sum_{l=0}^\infty \sum_{m=-l}^{+l} N_p\, u_{pl}(r) Y_{lm}(\Omega)\, \hat{a}_{plm}(t) \,. \tag{8}$$

In the following, the integration and summation together will be denoted by $\sum_{plm}$. The time dependence can be easily split off and solved. According to (4.8) we know that

$$\begin{aligned}\ddot{\hat{\phi}}(\boldsymbol{x},t) &= (\Delta - m^2)\, \hat{\phi}(\boldsymbol{x},t) \\ &= \sum_{plm} N_p \left(\Delta_r u_{pl}(r) + u_{pl}(r)\Delta_\Omega - m^2 u_{pl}(r)\right) Y_{lm}(\Omega)\, \hat{a}_{plm}(t) \,.\end{aligned} \tag{9}$$

Inserting (5) and (2) leads to

$$\ddot{\hat{\phi}}(\boldsymbol{x},t) = \sum_{plm} N_p \left(-p^2 - m^2\right) u_{pl}(r) Y_{lm}(\Omega)\, \hat{a}_{plm}(t) \tag{10}$$

and thus

$$\ddot{\hat{a}}_{plm}(t) = -(p^2 + m^2)\, \hat{a}_{plm}(t) \,, \tag{11}$$

which is solved by

$$\hat{a}_{plm}(t) = \hat{a}^{(+)}\, e^{-i\omega_p t} + \hat{a}^{(-)}\, e^{+i\omega_p t} \,, \tag{12}$$

where $\omega_p = \sqrt{p^2 + m^2}$. Since the field operator is required to be hermitean (remember that we discuss a neutral Klein–Gordon field) its expansion into spherical waves reads

$$\hat{\phi}(\boldsymbol{x},t) = \sum_{plm} N_p\, u_{pl}(r) \left(Y_{lm}(\Omega)\, \hat{a}_{plm}\, e^{-i\omega_p t} + Y_{lm}^*(\Omega)\, \hat{a}_{plm}^\dagger\, e^{+i\omega_p t}\right) \,. \tag{13}$$

A similar result applies for the canonically conjugate field operator

$$\hat{\pi}(\boldsymbol{x},t) = \sum_{plm} N_p\, u_{pl}(r)\, (-i\omega_p) \left(Y_{lm}(\Omega)\, \hat{a}_{plm}\, e^{-i\omega_p t} - Y_{lm}^*(\Omega)\, \hat{a}_{plm}^\dagger\, e^{+i\omega_p t}\right) \,. \tag{14}$$

Now we can surmise that $\hat{a}_{plm}$ and $\hat{a}_{plm}^\dagger$ play the role of annihilation and creation operators, satisfying the familiar commutation relations

$$\left[\hat{a}_{plm},\, \hat{a}_{p'l'm'}^\dagger\right] = \delta(p-p')\, \delta_{ll'}\, \delta_{mm'} \,, \tag{15a}$$

$$\left[\hat{a}_{plm},\, \hat{a}_{p'l'm'}\right] = \left[\hat{a}_{plm}^\dagger,\, \hat{a}_{p'l'm'}^\dagger\right] = 0 \,. \tag{15b}$$

To check this proposition we use these relations to evaluate the equal-time commutator $[\hat{\phi}, \hat{\pi}]$ and compare it with the expected result. Using (15) and the completeness relation (7) we find

$$\left[\hat{\phi}(\boldsymbol{x},t),\dot{\hat{\phi}}(\boldsymbol{x}',t)\right] = \sum_{plm}\sum_{p'l'm'} N_p N_{p'}\, u_{pl}(r) u_{p'l'}(r)\,(\mathrm{i}\omega_{p'})$$

$$\times \Big(\delta(p-p')\delta_{ll'}\delta_{mm'} Y_{lm}(\Omega) Y^*_{l'm'}(\Omega')\,\mathrm{e}^{-\mathrm{i}(\omega_p-\omega_{p'})t}$$

$$+\,\delta(p-p')\delta_{ll'}\delta_{mm'} Y^*_{lm}(\Omega) Y_{l'm'}(\Omega')\,\mathrm{e}^{+\mathrm{i}(\omega_p-\omega_{p'})t}\Big)$$

$$= \sum_{plm} N_p^2\,(\mathrm{i}\omega_p) u_{pl}(r) u_{pl}(r')\Big(Y_{lm}(\Omega)Y^*_{lm}(\Omega') + Y^*_{lm}(\Omega)Y_{lm}(\Omega')\Big)$$

$$= 2\omega_p N_p^2\,\mathrm{i}\,\delta^3(\boldsymbol{x}-\boldsymbol{x}') = \mathrm{i}\,\delta^3(\boldsymbol{x}-\boldsymbol{x}')\;, \qquad (16)$$

if the normalization factor is chosen as $N_p = (2\omega_p)^{-1/2}$. Similarly the other ETCR are recovered, thus confirming the validity of (15). It is also possible to evaluate the commutators (15) directly, by solving (13) and (14) for the operators $\hat{a}_{plm}$ and $\hat{a}^\dagger_{plm}$ and proceeding as in Exercise 4.1.

(b) The connection between the representation in terms of plane waves and in terms of spherical waves can be found by equating the expansions of the field operator (13) and (4.15), i.e.,

$$\int_0^\infty \mathrm{d}p \sum_{lm} \frac{1}{\sqrt{2\omega_p}}\, u_{pl}(r) Y_{lm}(\Omega_r)\,\hat{a}_{plm}\,\mathrm{e}^{-\mathrm{i}\omega_p t}$$

$$= \int_0^\infty \mathrm{d}p\, p^2 \int \mathrm{d}\Omega_p\, \frac{1}{\sqrt{(2\pi)^3 2\omega_p}}\, \mathrm{e}^{\mathrm{i}\boldsymbol{p}\cdot\boldsymbol{x}}\,\hat{a}_{\boldsymbol{p}}\,\mathrm{e}^{-\mathrm{i}\omega_p t}\;. \qquad (17)$$

The plane wave has the following "Rayleigh expansion" into spherical waves[5]

$$\mathrm{e}^{\mathrm{i}\boldsymbol{p}\cdot\boldsymbol{x}} = 4\pi \sum_{lm} \mathrm{i}^l\, j_l(pr)\, Y^*_{lm}(\Omega_p) Y_{lm}(\Omega_r)\;. \qquad (18)$$

Inserting this into (17), comparing the coefficients, leads to the following tranformation between the two types of annihilation operators

$$\hat{a}_{plm} = \mathrm{i}^l\, p \int \mathrm{d}\Omega_p\, Y^*_{lm}(\Omega_p)\,\hat{a}_{\boldsymbol{p}}\;. \qquad (19)$$

(c) Integrating by parts and using (4.8), we can write the normal-ordered *Hamiltonian* (4.45) as

$$\hat{H} = \frac{1}{2}\int \mathrm{d}^3 x\, :\!\Big(\hat{\pi}^2 - \hat{\phi}\ddot{\hat{\phi}}\Big)\!:\;. \qquad (20)$$

Insertion of the expansions (13) and (14) gives

$$\hat{H} = \frac{1}{2}\int \mathrm{d}^3 x \sum_{p'l'm'}\sum_{plm} :\!\Big[(\omega_p^2 - \omega_p\omega_{p'})\phi_{p'l'm'}\phi_{plm}\,\hat{a}_{p'l'm'}\hat{a}_{plm}\,\mathrm{e}^{-\mathrm{i}(\omega_{p'}+\omega_p)t}$$

$$+ (\omega_p^2 - \omega_p\omega_{p'})\phi^*_{p'l'm'}\phi^*_{plm}\,\hat{a}^\dagger_{p'l'm'}\hat{a}^\dagger_{plm}\,\mathrm{e}^{+\mathrm{i}(\omega_{p'}+\omega_p)t}$$

$$+ (\omega_p^2 + \omega_p\omega_{p'})\phi_{p'l'm'}\phi^*_{plm}\,\hat{a}_{p'l'm'}\hat{a}^\dagger_{plm}\,\mathrm{e}^{-\mathrm{i}(\omega_{p'}-\omega_p)t}$$

$$+ (\omega_p^2 + \omega_p\omega_{p'})\phi^*_{p'l'm'}\phi_{plm}\,\hat{a}^\dagger_{p'l'm'}\hat{a}_{plm}\,\mathrm{e}^{+\mathrm{i}(\omega_{p'}-\omega_p)t}\Big]\!:$$

---

[5] see, e.g., C. Joachain: *Quantum Collision Theory*, (North Holland, Amsterdam 1975).

Exercise 4.3

$$= \frac{1}{2}\sum_{plm} 2\omega_p^2 \frac{1}{2\omega_p} : \left(\hat{a}_{plm}^\dagger \hat{a}_{plm} + \hat{a}_{plm}\hat{a}_{plm}^\dagger\right) : . \tag{21}$$

Here the orthonormality relation (6) was used, as well as the fact that integrals of the type $\phi_{p'l'm'}\phi_{plm}$ (no complex conjugation) vanish for $p \neq p'$. To be more specific, because of $Y_{lm}^* = (-1)^l Y_{l,-m}$ and (6) the following relation holds:

$$\int d^3x\, \phi_{plm}\, \phi_{p'l'm'} = N_p^2\, \delta(p-p')\,(-1)^l \delta_{ll'}\, \delta_{m,-m'} . \tag{22}$$

As a consequence the contributions containing the factor $\omega_p^2 - \omega_p\omega_{p'}$ drop out. The resulting expression for the normal-ordered Hamiltonian

$$\hat{H} = \int_0^\infty dp \sum_{lm} \omega_p\, \hat{a}_{plm}^\dagger \hat{a}_{plm} \tag{23}$$

has the expected structure. It describes a system of field quanta which carry the energy ω_p and are described by the quantum numbers plm.

The symmetrized and normal-ordered operator for the *angular momentum* of the Klein–Gordon field is (cf. (2.72))

$$\hat{\boldsymbol{L}} = -\frac{1}{2}\int d^3x : \left(\hat{\pi}\,\boldsymbol{x}\times\boldsymbol{\nabla}\hat{\phi} + (\boldsymbol{x}\times\boldsymbol{\nabla}\phi)\,\hat{\pi}\right): . \tag{24}$$

Conveniently the basis functions ϕ_{plm} are eigenfunctions of the angular-momentum differential operator $L_3 = -i\,(\boldsymbol{x}\times\boldsymbol{\nabla})_3$ since $(-i\,\boldsymbol{x}\times\boldsymbol{\nabla})_3 Y_{lm} = m\, Y_{lm}$. Thus, inserting the expansion of the field operators, we obtain

$$\begin{aligned}\hat{L}_3 &= \frac{1}{2}\int d^3x \sum_{p'l'm'}\sum_{plm}(\omega_{p'}m + \omega_p m') \\ &\times : \Big[-\phi_{p'l'm'}\phi_{plm}\, \hat{a}_{p'l'm'}\, \hat{a}_{plm}\, e^{-i(\omega_{p'}+\omega_p)t} \\ &\quad + \phi_{p'l'm'}^*\phi_{plm}^*\, \hat{a}_{p'l'm'}^\dagger \hat{a}_{plm}^\dagger\, e^{+i(\omega_{p'}+\omega_p)t} \\ &\quad + \phi_{p'l'm'}\phi_{plm}^*\, \hat{a}_{p'l'm'}\, \hat{a}_{plm}^\dagger\, e^{-i(\omega_{p'}-\omega_p)t} \\ &\quad - \phi_{p'l'm'}^*\phi_{plm}\, \hat{a}_{p'l'm'}^\dagger \hat{a}_{plm}\, e^{+i(\omega_{p'}-\omega_p)t}\Big]: .\end{aligned} \tag{25}$$

Because of (22) the factor $\omega_{p'}m + \omega_p m'$ in the first and last terms drop out and, after performing normal ordering, the result

$$\hat{L}_3 = \int_0^\infty dp \sum_{lm} m\, \hat{a}_{plm}^\dagger \hat{a}_{plm} \tag{26}$$

is obtained. Obviously the field quanta carry the well-defined angular-momentum projection m.

(d) In order to express the angular-momentum operator (24) in terms of creation and annihilation operators of field quanta having well-defined momentum, the expansion of the field operator

$$\hat{\phi}(\boldsymbol{x},t) = \int d^3p\, \frac{1}{\sqrt{2\omega_p(2\pi)^3}}\left(\hat{a}_p\, e^{-i(\omega_p t - \boldsymbol{p}\cdot\boldsymbol{x})} + \hat{a}_p^\dagger\, e^{+i(\omega_p t - \boldsymbol{p}\cdot\boldsymbol{x})}\right) \tag{27}$$

and $\hat{\pi}(\boldsymbol{x},t) = \dot{\hat{\phi}}(\boldsymbol{x},t)$ are substituted into (24). The d^3x integration can be simplified if we substitute $\boldsymbol{x}\times\boldsymbol{\nabla}_x \to -i\boldsymbol{\nabla}_p\times i\boldsymbol{p} = \boldsymbol{\nabla}_p\times\boldsymbol{p}$. This is justified since the gradient in position space as well as in momentum space acts on plane waves $\exp(\pm i\boldsymbol{p}\cdot\boldsymbol{x})$ which are symmetric with respect to the interchange of $\boldsymbol{x}$ and $\boldsymbol{p}$. The symmetrized angular-momentum operator then becomes

Exercise 4.3

$$\begin{aligned}\hat{\boldsymbol{L}} = {}& \frac{\mathrm{i}}{2}\int \mathrm{d}^3 p' \mathrm{d}^3 p\, \frac{1}{(2\pi)^3\sqrt{2\omega_{p'}2\omega_p}}\int \mathrm{d}^3 x \\ & \times : \Big[\omega_{p'}\Big(\hat{a}_{\boldsymbol{p}'}\,\mathrm{e}^{-\mathrm{i}(\omega_{p'}t-\boldsymbol{p}'\cdot\boldsymbol{x})} - \hat{a}^{\dagger}_{\boldsymbol{p}'}\,\mathrm{e}^{+\mathrm{i}(\omega_{p'}t-\boldsymbol{p}'\cdot\boldsymbol{x})}\Big) \\ & \times (\boldsymbol{\nabla}_p\times\boldsymbol{p})\Big(\hat{a}_{\boldsymbol{p}}\,\mathrm{e}^{-\mathrm{i}(\omega_p t-\boldsymbol{p}\cdot\boldsymbol{x})} + \hat{a}^{\dagger}_{\boldsymbol{p}}\,\mathrm{e}^{+\mathrm{i}(\omega_p t-\boldsymbol{p}\cdot\boldsymbol{x})}\Big) \\ & + (\boldsymbol{\nabla}_{p'}\times\boldsymbol{p}')\Big(\hat{a}_{\boldsymbol{p}'}\,\mathrm{e}^{-\mathrm{i}(\omega_{p'}t-\boldsymbol{p}'\cdot\boldsymbol{x})} + \hat{a}^{\dagger}_{\boldsymbol{p}'}\,\mathrm{e}^{+\mathrm{i}(\omega_{p'}t-\boldsymbol{p}'\cdot\boldsymbol{x})}\Big) \\ & \times \omega_p\Big(\hat{a}_{\boldsymbol{p}}\,\mathrm{e}^{-\mathrm{i}(\omega_p t-\boldsymbol{p}\cdot\boldsymbol{x})} - \hat{a}^{\dagger}_{\boldsymbol{p}}\,\mathrm{e}^{+\mathrm{i}(\omega_p t-\boldsymbol{p}\cdot\boldsymbol{x})}\Big)\Big] : .\end{aligned} \qquad (28)$$

The $\mathrm{d}^3 x$ integration yields delta functions $\delta^3(\boldsymbol{p}\pm\boldsymbol{p}')$. Only the mixed products of $\hat{a}_{\boldsymbol{p}}$ and $\hat{a}^{\dagger}_{\boldsymbol{p}}$ survive. For example, the contribution containing two annihilation operators vanishes,

$$\hat{a}_{\boldsymbol{p}'}\hat{a}_{\boldsymbol{p}}\,\mathrm{e}^{-\mathrm{i}(\omega_{p'}+\omega_p)t}\Big(\omega_{p'}\boldsymbol{\nabla}_p\times\boldsymbol{p} + \omega_p\boldsymbol{\nabla}_{p'}\times\boldsymbol{p}'\Big)\delta^3(\boldsymbol{p}+\boldsymbol{p}') = 0 , \qquad (29)$$

because of $\boldsymbol{p}' = -\boldsymbol{p}$ and $\boldsymbol{\nabla}_{p'}\delta^3(\boldsymbol{p}+\boldsymbol{p}') = \boldsymbol{\nabla}_p\delta^3(\boldsymbol{p}+\boldsymbol{p}')$. The mixed term $\hat{a}^{\dagger}_{\boldsymbol{p}'}\hat{a}_{\boldsymbol{p}}$ leads to the integrand

$$\begin{aligned}&-\hat{a}^{\dagger}_{\boldsymbol{p}'}\hat{a}_{\boldsymbol{p}}\,\mathrm{e}^{\mathrm{i}(\omega_{p'}-\omega_p)t}\Big(\omega_{p'}\boldsymbol{\nabla}_p\times\boldsymbol{p} - \omega_p\boldsymbol{\nabla}_{p'}\times\boldsymbol{p}'\Big)\delta^3(\boldsymbol{p}-\boldsymbol{p}') \\ &= -2\,\hat{a}^{\dagger}_{\boldsymbol{p}'}\hat{a}_{\boldsymbol{p}}\,\omega_p\boldsymbol{\nabla}_p\times\boldsymbol{p}\,\delta^3(\boldsymbol{p}-\boldsymbol{p}') ,\end{aligned} \qquad (30)$$

and similarly for $\hat{a}\,\hat{a}^{\dagger}$. Now one of the momentum integrations in (28) can be done, leading to

$$\hat{\boldsymbol{L}} = \frac{\mathrm{i}}{2}\int \mathrm{d}^3 p :\Big(\hat{a}^{\dagger}_{\boldsymbol{p}}\,\boldsymbol{p}\times\boldsymbol{\nabla}_p\hat{a}_{\boldsymbol{p}} + (\boldsymbol{p}\times\boldsymbol{\nabla}_p\hat{a}_{\boldsymbol{p}})\,\hat{a}^{\dagger}_{\boldsymbol{p}}\Big): \qquad (31)$$

or, taking into account normal ordering,

$$\hat{\boldsymbol{L}} = \mathrm{i}\int \mathrm{d}^3 p\,\hat{a}^{\dagger}_{\boldsymbol{p}}\Big(\boldsymbol{p}\times\boldsymbol{\nabla}_p\Big)\hat{a}_{\boldsymbol{p}} . \qquad (32)$$

The presence of the operator $\boldsymbol{\nabla}_p$ demonstrates that the angular-momentum operator in plane-wave representation is *nondiagonal*. This implies that the field quanta created by $\hat{a}^{\dagger}_{\boldsymbol{p}}$ do not possess good angular-momentum.

4.2 The Charged Klein–Gordon Field

The real-valued Klein–Gordon field was found to describe a collection of spin-0 particles of identical type. It is easy to generalize this to particles having an *internal degree of freedom*. The simplest generalization of this type introduces a doublet of particles and antiparticles that can be described by going over to *complex* fields $\phi \neq \phi^*$ (and consequently nonhermitean field operators $\hat{\phi} \neq \hat{\phi}^{\dagger}$).

The Lagrange density, which of course remains a real-valued function, can be described by the ansatz

$$\mathcal{L} = \frac{\partial \phi^*}{\partial x_\mu} \frac{\partial \phi}{\partial x^\mu} - m^2 \phi^* \phi , \tag{4.54}$$

where ϕ and ϕ^* can be treated as independent fields.[6] The two canonically conjugate fields are

$$\pi = \frac{\partial \mathcal{L}}{\partial \dot\phi} = \dot\phi^* , \tag{4.55a}$$

$$\pi^* = \frac{\partial \mathcal{L}}{\partial \dot\phi^*} = \dot\phi , \tag{4.55b}$$

and the Hamiltonian becomes

$$\begin{aligned} H &= \int d^3 x \left(\pi \partial_0 \phi + \pi^* \partial_0 \phi^* - \mathcal{L} \right) \\ &= \int d^3 x \left(\pi^* \pi + \boldsymbol{\nabla}\phi^* \cdot \boldsymbol{\nabla}\phi + m^2 \phi^* \phi \right) . \end{aligned} \tag{4.56}$$

Note the absence of the factor $\frac{1}{2}$ in the expressions for $\mathcal{L}$ and H. Quantization of the theory is again achieved by going over to field operators $\hat\phi$, $\hat\phi^\dagger$, $\hat\pi$, $\hat\pi^\dagger$. The field operators are required to satisfy the *commutation relations* at equal time,

$$\left[\hat\phi(\boldsymbol{x},t), \hat\pi(\boldsymbol{x}',t)\right] = \left[\hat\phi^\dagger(\boldsymbol{x},t), \hat\pi^\dagger(\boldsymbol{x}',t)\right] = i\delta^3(\boldsymbol{x}-\boldsymbol{x}') , \tag{4.57}$$

while the commutators involving other combinations of fields are required to vanish. The Fourier decomposition (4.27) of the field operator now is replaced by

$$\hat\phi(\boldsymbol{x},t) = \int d^3 p \left(\hat{a}_{\boldsymbol{p}} u_{\boldsymbol{p}}(\boldsymbol{x},t) + \hat{b}^\dagger_{\boldsymbol{p}} u^*_{\boldsymbol{p}}(\boldsymbol{x},t) \right) , \tag{4.58a}$$

$$\hat\phi^\dagger(\boldsymbol{x},t) = \int d^3 p \left(\hat{a}^\dagger_{\boldsymbol{p}} u^*_{\boldsymbol{p}}(\boldsymbol{x},t) + \hat{b}_{\boldsymbol{p}} u_{\boldsymbol{p}}(\boldsymbol{x},t) \right) . \tag{4.58b}$$

Since the field operator is no longer hermitean, $\hat\phi^\dagger \neq \hat\phi$, the coefficients $\hat{b}^\dagger_{\boldsymbol{p}}$ now cannot be expressed in terms of $\hat{a}_{\boldsymbol{p}}$ as in (4.14). Rather, there are two independent sets of creation and annihilation operators. These operators again obey simple commmutations relations that can be derived in the same way as in the case of the real Klein–Gordon field in the last section. The result is

$$\begin{aligned} \left[\hat{a}_{\boldsymbol{p}}, \hat{a}^\dagger_{\boldsymbol{p}'}\right] &= \left[\hat{b}_{\boldsymbol{p}}, \hat{b}^\dagger_{\boldsymbol{p}'}\right] = \delta^3(\boldsymbol{p}-\boldsymbol{p}') , \\ \left[\hat{a}_{\boldsymbol{p}}, \hat{a}_{\boldsymbol{p}'}\right] &= \left[\hat{b}_{\boldsymbol{p}}, \hat{b}_{\boldsymbol{p}'}\right] = \left[\hat{a}^\dagger_{\boldsymbol{p}}, \hat{a}^\dagger_{\boldsymbol{p}'}\right] = \left[\hat{b}^\dagger_{\boldsymbol{p}}, \hat{b}^\dagger_{\boldsymbol{p}'}\right] = 0 , \\ \left[\hat{a}_{\boldsymbol{p}}, \hat{b}_{\boldsymbol{p}'}\right] &= \left[\hat{a}_{\boldsymbol{p}}, \hat{b}^\dagger_{\boldsymbol{p}'}\right] = \left[\hat{a}^\dagger_{\boldsymbol{p}}, \hat{b}_{\boldsymbol{p}'}\right] = \left[\hat{a}^\dagger_{\boldsymbol{p}}, \hat{b}^\dagger_{\boldsymbol{p}'}\right] = 0 . \end{aligned} \tag{4.59}$$

The Hamiltonian and the momentum operator are easily expressed in terms of these operators. The *normal-ordered Hamiltonian* is found to be

---

[6] In case of doubt one can employ a set of two independent real fields $\phi_1 = \frac{1}{\sqrt{2}}(\phi+\phi^*)$ and $\phi_2 = \frac{-i}{\sqrt{2}}(\phi-\phi^*)$ instead of ϕ and ϕ^* and transform back to the complex fields at the end of the calculation.

$$\begin{aligned}\hat{H} &= \boldsymbol{:}\!\int\!\mathrm{d}^3p\,\omega_p(\hat{a}_{\boldsymbol{p}}\hat{a}_{\boldsymbol{p}}^\dagger + \hat{b}_{\boldsymbol{p}}^\dagger\hat{b}_{\boldsymbol{p}})\boldsymbol{:}\\ &= \int\!\mathrm{d}^3p\,\omega_p(\hat{a}_{\boldsymbol{p}}^\dagger\hat{a}_{\boldsymbol{p}} + \hat{b}_{\boldsymbol{p}}^\dagger\hat{b}_{\boldsymbol{p}}) \equiv \int\!\mathrm{d}^3p\,\omega_p(\hat{n}_{\boldsymbol{p}}^{(a)} + \hat{n}_{\boldsymbol{p}}^{(b)})\,. \end{aligned} \qquad (4.60)$$

The derivation of this expression can be copied from the last section, replacing half of the $\hat{a}$ operators by $\hat{b}$ operators in (4.36) and (4.38). Similarly the momentum operator becomes

$$\hat{\boldsymbol{P}} = \int\!\mathrm{d}^3p\,\boldsymbol{p}(\hat{a}_{\boldsymbol{p}}^\dagger\hat{a}_{\boldsymbol{p}} + \hat{b}_{\boldsymbol{p}}^\dagger\hat{b}_{\boldsymbol{p}}) \equiv \int\!\mathrm{d}^3p\,\boldsymbol{p}\,(\hat{n}_{\boldsymbol{p}}^{(a)} + \hat{n}_{\boldsymbol{p}}^{(b)})\,. \qquad (4.61)$$

and the angular-momentum operator reads

$$\hat{\boldsymbol{L}} = \mathrm{i}\int\!\mathrm{d}^3p\,\left[\hat{a}_{\boldsymbol{p}}^\dagger(\boldsymbol{p}\times\boldsymbol{\nabla}_{\boldsymbol{p}})\hat{a}_{\boldsymbol{p}} + \hat{b}_{\boldsymbol{p}}^\dagger(\boldsymbol{p}\times\boldsymbol{\nabla}_{\boldsymbol{p}})\hat{b}_{\boldsymbol{p}}\right]\,. \qquad (4.62)$$

Obviously our new theory describes *two independent types of particles* a and b having the same mass m. Again a Fock space can be constructed, starting from the vacuum state which is defined to contain neither a nor b particles:

$$\hat{a}_{\boldsymbol{p}}|0\rangle = \hat{b}_{\boldsymbol{p}}|0\rangle = 0 \quad \text{for all} \quad \boldsymbol{p}\,. \qquad (4.63)$$

General many-particle states are now characterized by two sets of occupation numbers $\{n_i^{(a)}\}$ and $\{n_i^{(b)}\}$. Closer inspection of the Lagrange density (4.54) uncovers a connection between the two types of particles. It exhibits a *symmetry under phase transformations*

$$\phi' = \phi\,\mathrm{e}^{\mathrm{i}\alpha}\,,\quad \phi^{*\prime} = \phi^*\,\mathrm{e}^{-\mathrm{i}\alpha} \qquad (4.64)$$

with real phases α. Noether's theorem (see Chap. 2) tells us that this continuous symmetry transformation leads to a conserved quantity we will call the *charge*

$$\begin{aligned}Q &= \int\!\mathrm{d}^3x\,j^0(x) = -\mathrm{i}\int\!\mathrm{d}^3x\,\left(\frac{\partial\mathcal{L}}{\partial\pi^*}\phi - \frac{\partial\mathcal{L}}{\partial\pi}\phi^*\right)\\ &= -\mathrm{i}\int\!\mathrm{d}^3x\,\left(\pi\phi - \pi^*\phi^*\right)\,. \end{aligned}\qquad (4.65)$$

Because of the connection $\pi = \dot{\phi}^*$, $\pi^* = \dot{\phi}$ this can also be written in the compact form

$$Q = \mathrm{i}\int\!\mathrm{d}^3x\,\phi^*\,\overleftrightarrow{\partial}_0\,\phi = \left(\phi,\phi\right)\,, \qquad (4.66)$$

using the notation of (4.24).

Within the quantized theory the charge becomes an operator:

$$\hat{Q} = -\mathrm{i}\int\!\mathrm{d}^3x\,\boldsymbol{:}\!(\hat{\pi}\hat{\phi} - \hat{\pi}^\dagger\hat{\phi}^\dagger)\boldsymbol{:}\,. \qquad (4.67)$$

Again the ordering of the operators is not determined from classical theory. By adopting the prescription of normal ordering we have eliminated an infinite but unobservable vacuum charge. Insertion of (4.58) and the corresponding expansions of the conjugate field operators $\hat{\pi}$, $\hat{\pi}^\dagger$ leads to

$$\hat{Q} = :\frac{1}{2}\int d^3p\left(\hat{a}_p^\dagger \hat{a}_p + \hat{a}_p \hat{a}_p^\dagger - \hat{b}_p^\dagger \hat{b}_p - \hat{b}_p \hat{b}_p^\dagger\right):$$

$$= \int d^3p\left(\hat{a}_p^\dagger \hat{a}_p - \hat{b}_p^\dagger \hat{b}_p\right) \equiv \int d^3p\left(\hat{n}_p^{(a)} - \hat{n}_p^{(b)}\right). \quad (4.68)$$

It is immediately seen that the charge remains a conserved quantity also in the quantized theory: the charge operator satisfies the equation of motion

$$\dot{\hat{Q}} = -i\left[\hat{Q}, \hat{H}\right] = 0. \quad (4.69)$$

Collecting our results, we find that the operator $\hat{a}_p^\dagger$ creates a particle having energy ω_p, momentum p and charge $+1$, whereas $\hat{b}_p^\dagger$ creates a particle with the same energy, the same momentum but *opposite charge* -1. (Of course the absolute sign of the charge is just a matter of convention.) The b particles therefore are called the *antiparticles* of the a particles.

Particles described by a real-valued Klein–Gordon field are known as *strictly neutral*. They do not possess antiparticles (one may as well say that they "are their own antiparticles") and of course are electrically neutral. A well-known example is the neutral pion π^0. The distinction is somewhat subtle since there are also particles that carry no electric charge but are not strictly neutral and have antiparticles. This is possible since the notion of "charge" is not restricted to the electrical charge. Other types of charge can arise from internal symmetries. An example for a particle which is not strictly neutral is the K meson[7] which comes in two types, the K^0 and the $\bar{K}^0$.

EXERCISE

4.4 The Charge of a State

Problem. What happens to the charge of a state $|\alpha\rangle$ if the field operator $\hat{\phi}^\dagger$ is applied?

Solution. We assume that $|\alpha\rangle$ is an eigenstate of the charge operator $\hat{Q}$ with eigenvalue q. We want to determine the action of $\hat{Q}$ on the state $\hat{\phi}^\dagger|\alpha\rangle$. Using the expansions

$$\hat{Q} = \int d^3p\left(\hat{a}_p^\dagger \hat{a}_p - \hat{b}_p^\dagger \hat{b}_p\right) \quad (1)$$

and

$$\hat{\phi}^\dagger = \int d^3p'\left(\hat{a}_{p'}^\dagger u_{p'}^* + \hat{b}_{p'} u_{p'}\right), \quad (2)$$

we find that the product of the operators $\hat{Q}$ and $\hat{\phi}^\dagger$ becomes

$$\hat{Q}\hat{\phi}^\dagger = \int d^3p \int d^3p' \left[\left(\hat{a}_p^\dagger \hat{a}_p \hat{a}_{p'}^\dagger - \hat{b}_p^\dagger \hat{b}_p \hat{a}_{p'}^\dagger\right)u_{p'}^* + \left(\hat{a}_p^\dagger \hat{a}_p \hat{b}_{p'} - \hat{b}_p^\dagger \hat{b}_p \hat{b}_{p'}\right)u_{p'}\right]. \quad (3)$$

---

[7] See the volume W. Greiner, B. Müller: Quantum Mechanics – Symmetries (Springer, Berlin, Heidelberg 1994)

This expression can be related to the product $\hat{\phi}^\dagger\hat{Q}$. For this the creation and annihilation operators carrying the index p' have to be commuted to the left by using (4.59):

Exercise 4.4

$$\hat{Q}\hat{\phi}^\dagger = \int d^3p \int d^3p' \left[\left(\hat{a}^\dagger_{p'}\hat{a}^\dagger_p\hat{a}_p + \hat{a}^\dagger_p\delta^3(p-p') - \hat{a}^\dagger_{p'}\hat{b}^\dagger_p\hat{b}_p\right)u^*_{p'}\right.$$
$$\left. + \left(\hat{b}_{p'}\hat{a}^\dagger_p\hat{a}_p - \hat{b}_{p'}\hat{b}^\dagger_p\hat{b}_p + \delta^3(p-p')\hat{b}_{p'}\right)u_{p'}\right]. \quad (4)$$

The integrals containing delta functions drop out and the remaining terms are seen to agree with the operator $\hat{\phi}^\dagger\hat{Q}$:

$$\hat{Q}\hat{\phi}^\dagger = \hat{\phi}^\dagger\hat{Q} + \int d^3p\left(\hat{a}^\dagger_p u^*_p + \hat{b}_p u_p\right) = \hat{\phi}^\dagger\hat{Q} + \hat{\phi}^\dagger = \hat{\phi}^\dagger(\hat{Q}+1). \quad (5)$$

Thus if $|\alpha\rangle$ is an eigenstate of the charge operator then this is true also for the state $\hat{\phi}^\dagger|\alpha\rangle$:

$$\hat{Q}\hat{\phi}^\dagger|\alpha\rangle = \hat{\phi}^\dagger(\hat{Q}+1)|\alpha\rangle = (q+1)\hat{\phi}^\dagger|\alpha\rangle. \quad (6)$$

Thus the field operator $\hat{\phi}^\dagger$ increases the charge of a state by one unit. Similarly one can show that $\hat{\phi}$ reduces the charge by one unit. The result can be written as

$$[\hat{Q},\hat{\phi}(x)] = -\hat{\phi}(x)\quad,\quad [\hat{Q},\hat{\phi}^\dagger(x)] = \hat{\phi}^\dagger(x). \quad (7)$$

4.3 Symmetry Transformations

In Sect. 2.4 we have studied in some detail the transformation properties of classical fields and the connection between continuous symmetries and conservation laws. These considerations will now be extended to the case of quantum fields. We will be guided by the general correspondence principle and postulate that the expectation values of quantum operators have the same transformation properties as the related classical quantities.

Consider a general transformation that depends on a set of parameters ω and changes the coordinates according to

$$x \to x' = L(\omega, x). \quad (4.70)$$

The classical field is transformed as

$$\phi(x) \to \phi'(x') = \Lambda(\omega)\phi(x). \quad (4.71)$$

In the case of a multicomponent field, $\Lambda(\omega)$ of course is a matrix. A state vector in Hilbert space $|\alpha\rangle$ in the old system gets transformed into the vector $|\alpha'\rangle$ in the new system through the *linear transformation*[8]

$$|\alpha'\rangle = \hat{\mathcal{U}}(\omega)|\alpha\rangle. \quad (4.72)$$

---

[8] From a mathematical point of view also *antilinear* transformations are admissible. These are found to arise in the case of time-reflection transformations, $(x_0, \boldsymbol{x}) \to (-x_0, \boldsymbol{x})$, see Chap. 10.

To preserve the norm of the state, $\hat{\mathcal{U}}(\omega)$ must be a *unitary operator*

$$\hat{\mathcal{U}}^\dagger(\omega) = \hat{\mathcal{U}}^{-1}(\omega) \ . \tag{4.73}$$

Now we postulate the following connection between the primed and unprimed matrix elements of the field operator $\hat{\phi}(x)$:

$$\langle \beta'|\hat{\phi}(x')|\alpha'\rangle = \Lambda(\omega)\langle\beta|\hat{\phi}(x)|\alpha\rangle \ . \tag{4.74}$$

Note: the field operator is *not* primed. This would amount to double counting since the new system is already reached by using the transformed state vectors. Indeed $\hat{\phi}(x)$ and $\hat{\phi}(x')$ are *different operators* if $x \neq x'$. The degrees of freedom of the quantum field are described completely by specifying the set of field operators $\hat{\phi}(x)$ at all points x of Minkowski space. Since (4.74) is meant to apply to all states, using (4.72) and (4.73) we deduce the following *transformation law of the field operator*

$$\hat{\mathcal{U}}^{-1}(\omega)\hat{\phi}(x')\hat{\mathcal{U}}(\omega) = \Lambda(\omega)\hat{\phi}(x) \tag{4.75}$$

or

$$\hat{\mathcal{U}}^{-1}(\omega)\hat{\phi}(x)\hat{\mathcal{U}}(\omega) = \Lambda(\omega)\hat{\phi}\big(L^{-1}(\omega,x)\big) \ . \tag{4.76}$$

The set of all transformations forms a group, where group multiplication is defined as the subsequent application of the elements. The corresponding operators $\hat{\mathcal{U}}$ then form a *unitary representation* of the transformation group in Hilbert space.

In the case of a *continuous* transformation[9] it is reasonable to start with *infinitesimal* transformations $x' = x + \delta\omega$. The operator $\hat{\mathcal{U}}$ will be close to the identity and we write

$$\hat{\mathcal{U}}(\delta\omega) = 1 - \mathrm{i}\,\hat{G}(\delta\omega) \tag{4.77}$$

where $\hat{G}(\delta\omega)$ is a *hermitean operator* that is linear (to lowest order) in the small "rotation angles" $\delta\omega$. Then (4.76) to lowest order becomes

$$\mathrm{i}\big[\hat{\phi}(x),\hat{G}\big] = -\partial_\mu\hat{\phi}(x)\,\delta x^\mu + \big(\Lambda(\delta\omega) - 1\big)\hat{\phi}(x) \ . \tag{4.78}$$

Now let us study the example of an *infinitesimal Poincaré transformation* (translation plus Lorentz transformation)

$$\delta x^\mu = \epsilon^\mu + \delta\omega^\mu{}_\nu x^\nu \ . \tag{4.79}$$

For the transformation operator (4.77) we make the general linear ansatz

$$\hat{G}(\epsilon^\mu, \delta\omega^{\mu\nu}) = -\epsilon_\mu \hat{P}^\mu + \frac{1}{2}\delta\omega_{\mu\nu}\hat{M}^{\mu\nu} \ . \tag{4.80}$$

This is precisely the form of the generator of the Poincaré transformation (2.93) introduced in classical field theory. There the quantities P^μ and $M^{\mu\nu}$ had been identified as the momentum vector and the angular-momentum tensor, respectively. Now $\hat{P}^\mu$ and $\hat{M}^{\mu\nu}$ have become operators in Hilbert space.

In the special case of pure *translations* we have $\Lambda = 1$ and (4.78) takes on the form

---

[9] The important case of *discrete transformations* (reflections) will be treated in Chap. 10.

$$i[\hat{\phi}(x),\hat{P}^\mu] = -\partial^\mu\hat{\phi}(x). \tag{4.81}$$

This makes it clear that the momentum operator $\hat{P}^\mu$ is related to space-time translations of the field operator. This connection becomes even more obvious when *finite translations* $x'_\mu = x_\mu + a_\mu$ are studied. The unitary operator (4.77) is generalized as

$$\hat{\mathcal{U}}(a) = e^{ia\cdot\hat{P}} \tag{4.82}$$

and from (4.78) the following important rule for the change of the field operator under translations is obtained:

$$\hat{\phi}(x+a) = e^{ia\cdot\hat{P}}\,\hat{\phi}(x)\,e^{-ia\cdot\hat{P}}. \tag{4.83}$$

This can be shown by using the general operator identity (cf. Exercise 1.3)

$$e^{\hat{A}}\hat{B}e^{-\hat{A}} = \hat{B} + [\hat{A},\hat{B}] + \tfrac{1}{2!}[\hat{A},[\hat{A},\hat{B}]] + \ldots. \tag{4.84}$$

From the commutator relation (4.81) it can be seen that the right-hand side of (4.83) is just the Taylor expansion of the function $\hat{\phi}(x)$:

$$\begin{aligned}
e^{ia\cdot\hat{P}}\hat{\phi}(x)e^{-ia\cdot\hat{P}} &= \hat{\phi}(x) + [ia\cdot\hat{P},\hat{\phi}(x)] + [ia\cdot\hat{P},[ia\cdot\hat{P},\hat{\phi}(x)]] + \ldots \\
&= \hat{\phi}(x) + a_\mu\partial^\mu\hat{\phi}(x) + \tfrac{1}{2!}a_\mu a_\nu \partial^\mu\partial^\nu\hat{\phi}(x) + \ldots \\
&= \hat{\phi}(x+a).
\end{aligned} \tag{4.85}$$

Similarly, (4.81) and thus (4.85) also hold for the canonically conjugate field $\hat{\pi}(x)$. Therefore the identity

$$\hat{F}(x+a) = e^{ia\cdot\hat{P}}\,\hat{F}(x)\,e^{-ia\cdot\hat{P}} \tag{4.86}$$

holds for any operator that can be expanded in powers of $\hat{\phi}$ and $\hat{\pi}$. As a special case, for a purely temporal translation with $a=(t,\mathbf{0})$ and $x=0$ we have

$$\hat{F}(t) = e^{i\hat{H}t}\,\hat{F}(0)\,e^{-i\hat{H}t}, \tag{4.87}$$

from which the equation motion

$$\partial_0\hat{F}(t) = i\hat{H}e^{i\hat{H}t}\hat{F}(0)e^{-i\hat{H}t} - ie^{i\hat{H}t}\hat{F}(0)e^{-i\hat{H}t}\hat{H} = i[\hat{H},\hat{F}(t)] \tag{4.88}$$

can be deduced. This is Heisenberg's equation of motion for the operator $\hat{F}$. Replacing the temporal by a general translation we arrive at the *four-dimensional generalization of Heisenberg's equation*

$$\partial_\mu\hat{F}(x) = i[\hat{P}_\mu,\hat{F}(x)]. \tag{4.89}$$

In the case of infinitesimal *Lorentz transformations* (boosts or spatial rotations), (4.78) leads to a commutator relation between the field operator and the angular-momentum tensor. As in (2.64) we write

$$\Lambda(\delta\omega) = 1 + \frac{1}{2}\delta\omega_{\mu\nu}I^{\mu\nu}, \tag{4.90}$$

where $\delta\omega_{\mu\nu}$ denotes the boost parameters or rotation angles and $I^{\mu\nu}$ are the infinitesimal generators of the Lorentz transformation. For each combination of μ and ν the generator $I^{\mu\nu}$ is a matrix in the space of the components of the field operator. In contrast to (2.64) we will not denote this explicitly here. Taking into account that $\hat{M}^{\mu\nu}$ has to be antisymmetric, (4.78) becomes

$$\mathrm{i}\big[\hat{\phi}(x)\,,\,\hat{M}^{\mu\nu}\big] = \big(x^\nu\partial^\mu - x^\mu\partial^\nu\big)\hat{\phi}(x) - I^{\mu\nu}\hat{\phi}(x)\,. \tag{4.91}$$

The contributions of orbital and spin angular momenta are clearly separated. This result is the quantized version of the Poisson-bracket relation (2.95).

The relations (4.81) and (4.91) are of fundamental importance for quantum field theory. A precise definition of the operators $\hat{P}^\mu$ and $\hat{M}^{\mu\nu}$, however, has not yet been given. We can expect that these operators emerge from the corresponding classical expressions deduced from Noether's theorem by the substitution $\phi(x) \to \hat{\phi}(x)$. The operators $\hat{P}^\mu$ und $\hat{M}^{\mu\nu}$ formed within this recipe satisfy equations (4.81) and (4.91). This can be shown with the use of the canonical commutation relations for the field operator. In Exercise 4.5 this will be checked for the case of a charged Klein–Gordon field.

The generators $\hat{P}_\mu$ and $\hat{M}_{\mu\nu}$ satisfy the *Poincaré algebra* which we have already discussed in the framework of classical field theory. We have to replace the Poisson brackets in (2.96) by commutators, $\{\,,\,\}_{\mathrm{PB}} \to -\mathrm{i}[\,,\,]$, which gives

$$[\hat{P}_\mu\,,\,\hat{P}_\nu] = 0\,, \tag{4.92a}$$

$$[\hat{M}_{\mu\nu}\,,\,\hat{P}_\lambda] = \mathrm{i}\big(g_{\mu\lambda}\hat{P}_\nu - g_{\nu\lambda}\hat{P}_\mu\big)\,, \tag{4.92b}$$

$$[\hat{M}_{\mu\nu}\,,\,\hat{M}_{\sigma\tau}] = \mathrm{i}\big(g_{\nu\tau}\hat{M}_{\mu\sigma} - g_{\nu\sigma}\hat{M}_{\mu\tau} - g_{\mu\tau}\hat{M}_{\nu\sigma} + g_{\mu\sigma}\hat{M}_{\nu\tau}\big)\,. \tag{4.92c}$$

The special cases of spatial momentum and angular momentum operators have already been checked explicitly in Exercise 4.2.

In addition to the transformations involving space and time we can also study the *internal degrees of freedom* of the field. For infinitesimal transformations (2.78)

$$\Lambda_{rs}(\epsilon) = \delta_{rs} + \mathrm{i}\epsilon\lambda_{rs}\,, \tag{4.93}$$

with the notation $\hat{G} = -\epsilon\hat{Q}$ for the charge operator as a generator, the following commutation relation is deduced:

$$\big[\hat{\phi}_r(x)\,,\,\hat{Q}\big] = \lambda_{rs}\hat{\phi}_s(x)\,. \tag{4.94}$$

In the case of a *symmetry transformation* the related generator $\hat{G}$ will be a *conserved* quantity. According to (4.88) this implies that the generator has to commute with the Hamiltonian

$$[\hat{H}\,,\,\hat{G}] = 0\,. \tag{4.95}$$

For the momentum and angular-momentum operators this condition is a special case of the Poincaré algebra (4.92a) and (4.92b). The canonical commutation relations can be used to check whether (4.95) is satisfied in a given theory.

EXERCISE

4.5 Commutation Relations Between Field Operators and Generators

Problem. Prove the validity of the commutation relations (4.81), (4.91), and (4.94) for the case of a charged scalar quantum field.

Solution. The Lagrangian of a free charged Klein–Gordon field is given by (4.54). The time and space components of the momentum operator expressed in terms of the field operators $\hat{\phi}$, $\hat{\phi}^\dagger$, $\hat{\pi} = \partial_0 \hat{\phi}^\dagger$, and $\hat{\pi}^\dagger = \partial_0 \hat{\phi}$ are

$$\hat{P}_k = \int d^3x \left(\hat{\pi} \partial_k \hat{\phi} + \hat{\pi}^\dagger \partial_k \hat{\phi}^\dagger \right), \tag{1a}$$

$$\hat{P}_0 = \int d^3x \left(\hat{\pi}^\dagger \hat{\pi} + \boldsymbol{\nabla} \hat{\phi}^\dagger \cdot \boldsymbol{\nabla} \hat{\phi} + m^2 \hat{\phi}^\dagger \hat{\phi} \right). \tag{1b}$$

The angular-momentum tensor is similarly given by

$$\hat{M}_{nl} = \int d^3x \left(x_l (\hat{\pi} \partial_n \hat{\phi} + \hat{\pi}^\dagger \partial_n \hat{\phi}^\dagger) \right.$$
$$\left. - x_n (\hat{\pi} \partial_l \hat{\phi} + \hat{\pi}^\dagger \partial_l \hat{\phi}^\dagger) \right), \tag{2a}$$

$$\hat{M}_{n0} = \int d^3x \left(x_0 (\hat{\pi} \partial_n \hat{\phi} + \hat{\pi}^\dagger \partial_n \hat{\phi}^\dagger) - x_n (\hat{\pi}^\dagger \hat{\pi} + \boldsymbol{\nabla} \hat{\phi}^\dagger \cdot \boldsymbol{\nabla} \hat{\phi} + m^2 \hat{\phi}^\dagger \hat{\phi}) \right). \tag{2b}$$

According to the canonical quantization rules (4.58), in the calculation the commutators of $\phi(x)$ with these operators only terms of the type $[\hat{\phi}, \hat{\pi}]$ and $[\hat{\phi}^\dagger, \hat{\pi}^\dagger]$ will contribute. Since the resulting delta function cancels the integration we immediately obtain the desired results

$$[\hat{\phi}(x), \hat{P}_k] = i \partial_k \hat{\phi}, \tag{3a}$$

$$[\hat{\phi}(x), \hat{P}_0] = i \hat{\pi}^\dagger = i \partial_0 \hat{\phi}, \tag{3b}$$

$$[\hat{\phi}(x), \hat{M}_{nl}] = i \left(x_l \partial_n \hat{\phi} - x_n \partial_l \hat{\phi} \right), \tag{3c}$$

$$[\hat{\phi}(x), \hat{M}_{n0}] = i \left(x_0 \partial_n \hat{\phi} - x_n \hat{\pi}^\dagger \right) = i \left(x_0 \partial_n \hat{\phi} - x_n \partial_0 \hat{\phi} \right), \tag{3d}$$

The charge operator is given by (2.83):

$$\hat{Q} = -i \int d^3x \left(\hat{\pi}(x) \hat{\phi}(x) - \hat{\pi}^\dagger(x) \hat{\phi}^\dagger(x) \right). \tag{4}$$

This leads to the commutation relations (see Exercise 4.4)

$$[\hat{\phi}(x), \hat{Q}] = -i \int d^3x' \, [\hat{\phi}(x), \hat{\pi}(x')] \hat{\phi}(x') = \hat{\phi}(x), \tag{5a}$$

$$[\hat{\phi}^\dagger(x), \hat{Q}] = +i \int d^3x' \, [\hat{\phi}^\dagger(x), \hat{\pi}^\dagger(x')] \hat{\phi}^\dagger(x') = -\hat{\phi}^\dagger(x). \tag{5b}$$

The choice of operator ordering (e.g., normal ordering) does not affect the commutations relations (3) and (5).

4.4 The Invariant Commutation Relations

Up to now our discussion of canonical field quantization has been based on the commutation relations between field operators at two different points having arbitrary spatial separation but equal time arguments. Although this proves sufficient for us to carry out quantization it would be gratifying to have a formulation that does not single out the time coordinate but is relativistically invariant. In the case of *free fields* this can be achieved very easily. Since we know the explicit time dependence of the field operators, expressed in terms of creation and annihilation operators, the commutators can be evaluated for arbitrary values of the coordinates.

In the following we will study a complex Klein–Gordon field, starting with the commutator between the field operators $\hat{\phi}(\boldsymbol{x}, x_0)$ and $\hat{\phi}^\dagger(\boldsymbol{y}, y_0)$ for *arbitrary, possibly unequal* times x_0, y_0:

$$i\Delta(x-y) := \left[\hat{\phi}(x), \hat{\phi}^\dagger(y)\right] . \tag{4.96}$$

This function was first studied by W. Pauli and P. Jordan. Sometimes it is also associated with the name of J. Schwinger. The notation $\Delta(x-y)$ already takes into account the homogeneous character of space-time which implies that the commutator can depend only on the difference $x-y$ of the coordinates, not on their absolute values. To find an explicit expression for the function $\Delta(x-y)$ we insert the expansion (4.58) of the field operator with respect to the plane-wave basis (4.58)

$$\hat{\phi}(x) = \int d^3p \left(\hat{a}_{\boldsymbol{p}} u_{\boldsymbol{p}}(x) + \hat{b}_{\boldsymbol{p}}^\dagger u_{\boldsymbol{p}}^*(x)\right) = \hat{\phi}^{(+)}(x) + \hat{\phi}^{(-)}(x) \tag{4.97}$$

and the hermitean adjoint of this equation. The expression gets simplified through the use of the commutation relations (4.59) between the creation and annihilation operators:

$$\begin{aligned}
i\Delta(x-y) &= \int d^3p' \, d^3p \left(u_{\boldsymbol{p}'}(x) u_{\boldsymbol{p}}^*(y) \left[\hat{a}_{\boldsymbol{p}'}, \hat{a}_{\boldsymbol{p}}^\dagger\right] + u_{\boldsymbol{p}'}^*(x) u_{\boldsymbol{p}}(y) \left[\hat{b}_{\boldsymbol{p}'}^\dagger, \hat{b}_{\boldsymbol{p}}\right]\right) \\
&= \int d^3p \left(u_{\boldsymbol{p}}(x) u_{\boldsymbol{p}}^*(y) - u_{\boldsymbol{p}}^*(x) u_{\boldsymbol{p}}(y)\right) \\
&= \int \frac{d^3p}{(2\pi)^3} \frac{1}{2\omega_{\boldsymbol{p}}} \left(e^{-ip\cdot(x-y)} - e^{+ip\cdot(x-y)}\right) \\
&\equiv i\Delta^{(+)}(x-y) + i\Delta^{(-)}(x-y) ,
\end{aligned} \tag{4.98}$$

where the frequency factor in the exponent for both terms is given by $p_0 = \omega_{\boldsymbol{p}} \equiv +\sqrt{\boldsymbol{p}^2 + m^2}$. Equation (4.98) can also be written as

$$\Delta(x-y) = -\int \frac{d^3p}{(2\pi)^3} \frac{\sin p \cdot (x-y)}{\omega_{\boldsymbol{p}}} . \tag{4.99}$$

Now let us study the properties of the commutation function $\Delta(x-y)$.

1. $\Delta(x-y)$ is a *Lorentz-invariant* function. This could have been expected as the definition did not single out a specific frame of reference. To underline its Lorentz invariance the Pauli–Jordan function can be written in a manifestly four-dimensional form. The three-dimensional momentum integration in

(4.98) will be extended to four dimensions. For this we make the substitution $p \to -p$ and write (abbreviating $z \equiv x - y$)

$$\int \frac{d^3p}{(2\pi)^3} \frac{1}{2\omega_p} \left(e^{-i(\omega_p z_0 - p \cdot z)} - e^{i(\omega_p z_0 - p \cdot z)} \right)$$

$$= \int \frac{d^4p}{(2\pi)^3} \frac{1}{2\omega_p} \left(\delta(p_0 - \omega_p) - \delta(p_0 + \omega_p) \right) e^{-i(p_0 z_0 - p \cdot z)}$$

$$= \int \frac{d^4p}{(2\pi)^3} \frac{\epsilon(p_0)}{2\omega_p} \left(\delta(p_0 - \omega_p) + \delta(p_0 + \omega_p) \right) e^{-ip \cdot z} . \quad (4.100)$$

Here

$$\epsilon(p_0) \equiv \text{sgn}(p_0) = \begin{cases} +1 & \text{für} \quad p_0 > 0 \\ -1 & \text{für} \quad p_0 < 0 \end{cases}$$

is the sign function. Both delta functions can be cleverly combined:

$$\frac{1}{2\omega_p} \left(\delta(p_0 - \omega_p) + \delta(p_0 + \omega_p) \right) = \delta\big((p_0 - \omega_p)(p_0 + \omega_p)\big) = \delta(p_0^2 - \omega_p^2)$$

$$= \delta(p_0^2 - \mathbf{p}^2 - m^2)$$

$$= \delta(p^2 - m^2) , \quad (4.101)$$

and therefore

$$i\Delta(x - y) = \int \frac{d^4p}{(2\pi)^3} \epsilon(p_0) \delta(p^2 - m^2) e^{-ip \cdot (x-y)} . \quad (4.102)$$

This expression obviously is Lorentz invariant since it is composed entirely of scalar quantities. This is true also for the factor $\epsilon(p_0)$, since the sign of p_0 does not change under arbitrary proper (orthochronous) Lorentz transformations: time-like ($p^2 = m^2 > 0$) momentum vectors with $p_0 > 0$ always lie in the forward light cone and those with $p_0 < 0$ in the backward light cone

2. Δ is an *odd* function:

$$\Delta(x - y) = -\Delta(y - x) . \quad (4.103)$$

This is an immediate consequence of the definition.

3. The following *boundary conditions at vanishing time difference* hold:

$$\Delta(0, \mathbf{x}) = 0 \quad (4.104)$$

and

$$\left. \frac{\partial}{\partial x_0} \Delta(x_0, \mathbf{x}) \right|_{x_0 = 0} = -\delta^3(\mathbf{x}) . \quad (4.105)$$

The first condition (4.104) can be read from (4.99) since the integrand is an odd function in $\mathbf{p}$. Equation (4.105) can be verified by differentiating (4.98) with respect to time:

$$\left. \frac{\partial}{\partial x_0} \Delta(x_0, \mathbf{x}) \right|_{x_0 = 0} = \left. \frac{1}{i} \int \frac{d^3p}{(2\pi)^3} \frac{1}{2\omega_p} \left(-i\omega_p e^{-ip \cdot x} - (i\omega_p) e^{+ip \cdot x} \right) \right|_{x_0 = 0}$$

$$= -\int \frac{d^3p}{(2\pi)^3} \frac{1}{2} \left(e^{+ip \cdot x} + e^{-ip \cdot x} \right) = -\delta^3(\mathbf{x}) . \quad (4.106)$$

We further note that the spatial derivative of $\Delta(x)$ vanishes at $x_0 = 0$ owing to the antisymmetry of the integrand:

$$\nabla \Delta(x_0, \boldsymbol{x})|_{x_0=0} = \int \frac{d^3p}{(2\pi)^3} \frac{\boldsymbol{p}\cos(\boldsymbol{p}\cdot\boldsymbol{x})}{\omega_p} = 0 \ . \tag{4.107}$$

This symmetry argument can be easily generalized to the case of higher derivatives. At $x_0 = 0$ the following derivatives of $\Delta(x)$ can be shown to vanish: (i) all spatial derivatives; (ii) all time derivatives of even order.

As a corrollary to (4.106) the equal-time commutation relations can be regained since according to $\hat{\pi}(x) = \dot{\hat{\phi}}^\dagger(x)$

$$\left[\hat{\phi}(x), \hat{\pi}(y)\right] = \left[\hat{\phi}(x), \dot{\hat{\phi}}^\dagger(y)\right] = \frac{\partial}{\partial y_0}\left[\hat{\phi}(x), \hat{\phi}^\dagger(y)\right] = i\frac{\partial}{\partial y_0}\Delta(x-y) \ , \tag{4.108}$$

and for equal time arguments

$$\left[\hat{\phi}(\boldsymbol{x},t), \hat{\pi}(\boldsymbol{y},t)\right] = i\frac{\partial}{\partial y_0}\Delta(x-y)\bigg|_{x_0 \to y_0} = i\delta^3(\boldsymbol{x}-\boldsymbol{y}) \ . \tag{4.109}$$

The commutation relations

$$\left[\hat{\phi}(x), \hat{\phi}(y)\right] = \left[\hat{\pi}(x), \hat{\pi}(y)\right] = 0 \tag{4.110}$$

turn out to be valid for arbitrary times x_0, y_0. After the insertion of the expansion (4.97) of the free field operator, only those combinations of creation and annihilation operators turn up that have a vanishing commutator according to (4.59).

4. The function $\Delta(x)$ satisfies the homogeneous Klein–Gordon equation

$$(\Box + m^2)\,\Delta(x) = 0 \ . \tag{4.111}$$

This can be seen from (4.96) written in the form

$$\Delta(x) = \frac{1}{i}\left[\hat{\phi}(x), \hat{\phi}^\dagger(0)\right] \ , \tag{4.112}$$

since the field operator $\hat{\phi}(x)$ according to (4.8) satisfies the Klein–Gordon equation: $(\Box + m^2)\hat{\phi}(x) = 0$.

We note that the properties **3** and **4** are sufficient to provide a unique definition of the function $\Delta(x)$. Mathematical theory shows that the solution of a hyperbolic differential operator like $(\Box + m^2)$ is completely specified by giving the initial values of $\Delta(x)$ and $\partial/\partial x_0 \Delta(x)$ on a space-like hypersurface $x_0 = 0$.

5. A very fundamental property of quantum fields can be deduced from **3**:

$$\Delta(x-y) = 0 \quad \text{for} \quad (x-y)^2 < 0 \ , \tag{4.113}$$

i.e. the invariant function $\Delta(x-y)$ vanishes if the argument is a *space-like four-vector* as illustrated in Fig. 4.1. According to (4.104) this property is valid in the special case $x_0 = y_0$. It must be valid also for any four-vector $x - y$ which can be reached by a proper Lorentz transformation.

In quantum theory the vanishing of the commutator between two operators has the important consequence that the corresponding observable quantities

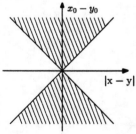

Fig. 4.1. The invariant function $\Delta(x-y)$ vanishes outside of the light cone, i.e., for space-like distances $(x-y)^2 < 0$. On the surface of the light cone $\Delta(x-y)$ is a singular function

can be measured independently. In our case this implies that *measurements at two points that have a space-like separation*, i.e., those which cannot get into contact through the transmission of light signals, *do not influence each other*. To put it in another way: disturbances cannot propagate with superluminal velocity. This is one of the most fundamental demands to be imposed on a physical theory. It is also known as the condition of *microcausality*. On a macroscopic scale the absence of faster-than-light transmission of signals has been well tested. Nothing indicates that the causality principle might break down at the scale of atoms or elementary particles. Therefore it is generally accepted that microcausality describes an essential property of the fabric of space and time.

We have to admit that the discussion so far has been a bit careless since the fields ϕ themselves do not correspond to physical observables.[10] More correctly one has to form *bilinear forms* of field operators (e.g., currents). As the following argumentation shows, however, the conclusions are not modified if this refinement is taken into account.

The general form of an operator that corresponds to a local observable can be written as

$$\hat{O}(x) = \hat{\phi}^\dagger(x)\, O(x)\, \hat{\phi}(x)\,, \tag{4.114}$$

where $O(x)$ is a c-number-valued function or a differential operator. The property of microcausality is determined by the commutator of two observables:[11]

$$[\hat{O}(x),\, \hat{O}(y)] = O(x)O(y)\left[\hat{\phi}^\dagger(x)\hat{\phi}(x),\, \hat{\phi}^\dagger(y)\hat{\phi}(y)\right]\,. \tag{4.115}$$

The commutator involving bilinear form can be rewritten as follows

$$\begin{aligned}
&[\hat{\phi}^\dagger(x)\hat{\phi}(x),\, \hat{\phi}^\dagger(y)\hat{\phi}(y)] \\
&= \hat{\phi}^\dagger(x)\hat{\phi}^\dagger(y)\left[\hat{\phi}(x),\hat{\phi}(y)\right] + \hat{\phi}^\dagger(x)\left[\hat{\phi}(x),\hat{\phi}^\dagger(y)\right]\hat{\phi}(y) \\
&\quad + \hat{\phi}^\dagger(y)\left[\hat{\phi}^\dagger(x),\hat{\phi}(y)\right]\hat{\phi}(x) + \left[\hat{\phi}^\dagger(x),\hat{\phi}^\dagger(y)\right]\hat{\phi}(y)\hat{\phi}(x) \\
&= \hat{\phi}^\dagger(x)\,\mathrm{i}\Delta(x-y)\hat{\phi}(y) + \hat{\phi}^\dagger(y)(-\mathrm{i})\Delta(y-x)\hat{\phi}(x)\,.
\end{aligned} \tag{4.116}$$

We have used the commutation relation (4.96) as well as (cf. (4.110))

$$[\hat{\phi}(x),\, \hat{\phi}(y)] = [\hat{\phi}^\dagger(x),\, \hat{\phi}^\dagger(y)] = 0\,. \tag{4.117}$$

Thus the commutator of observables taken at different points in space-time reads

$$[\hat{O}(x),\, \hat{O}(y)] = O(x)O(y)\left(\hat{\phi}^\dagger(x)\hat{\phi}(y) + \hat{\phi}^\dagger(y)\hat{\phi}(x)\right)\mathrm{i}\Delta(x-y)\,. \tag{4.118}$$

As we see, the property of microcausality obeyed by the Pauli–Jordan function, which vanishes for space-like separations, translates into the commutator of bilinear observables:

$$[\hat{O}(x),\, \hat{O}(y)] = 0 \quad \text{for} \quad (x-y)^2 < 0\,. \tag{4.119}$$

---

[10] A careful discussion of the measuring process and commutation relations in field theory was carried out in N. Bohr and L. Rosenfeld: Phys. Rev. **78**, 794 (1950).

[11] If $O(x)$ is a differential operator one has to be careful on which argument it acts.

The condition of microcausality can be used to explain why spin-0 particles cannot be made to satisfy the Fermi–Dirac statistics, i.e., why *the Klein–Gordon field cannot be quantized by using anticommutators*. Let us attempt to adopt these "wrong" commutation rules:

$$\{\hat{a}_{\boldsymbol{p}}, \hat{a}_{\boldsymbol{p}'}^\dagger\} = \{\hat{b}_{\boldsymbol{p}}, \hat{b}_{\boldsymbol{p}'}^\dagger\} = \delta^3(\boldsymbol{p}-\boldsymbol{p}') \qquad \text{etc.} \ . \tag{4.120}$$

The anticommutator between field operators would then be

$$\begin{aligned}
\mathrm{i}\Delta_1(x-y) &:= \{\hat{\phi}(x), \hat{\phi}^\dagger(y)\} \\
&= \int \mathrm{d}^3 p' \, \mathrm{d}^3 p \left(u_{\boldsymbol{p}'}(x) u_{\boldsymbol{p}}^*(y) \{\hat{a}_{\boldsymbol{p}'}, \hat{a}_{\boldsymbol{p}}^\dagger\} + u_{\boldsymbol{p}'}^*(x) u_{\boldsymbol{p}}(y) \{\hat{b}_{\boldsymbol{p}'}^\dagger, \hat{b}_{\boldsymbol{p}}\} \right) \\
&= \int \mathrm{d}^3 p \left(u_{\boldsymbol{p}}(x) u_{\boldsymbol{p}}^*(y) + u_{\boldsymbol{p}}^*(x) u_{\boldsymbol{p}}(y) \right) \\
&= \mathrm{i}\Delta^{(+)}(x-y) - \mathrm{i}\Delta^{(-)}(x-y) \ ,
\end{aligned} \tag{4.121}$$

which differs from (4.98) in the relative sign of the two terms. As in (4.99), combining the exponential functions leads to

$$\mathrm{i}\Delta_1(x-y) = \int \frac{\mathrm{d}^3 p}{(2\pi)^3} \frac{\cos[\boldsymbol{p}\cdot(\boldsymbol{x}-\boldsymbol{y})]}{\omega_{\boldsymbol{p}}} \ . \tag{4.122}$$

The function $\Delta_1(x-y)$ is also relativistically invariant and satisfies the homogeneous Klein–Gordon equation (cf. (4.6)). There is one decisive difference, however: in contrast to $\Delta(x-y)$, the function $\Delta_1(x-y)$ *does not vanish for space-like separations*. Indeed for the special case of equal times $x_0 = y_0$ the solution of the integral in spherical coordinates (see Exercise 4.6) yields

$$\int \frac{\mathrm{d}^3 p}{(2\pi)^3} \frac{\cos[\boldsymbol{p}\cdot(\boldsymbol{x}-\boldsymbol{y})]}{\sqrt{\boldsymbol{p}^2+m^2}} = \frac{m}{2\pi^2 |\boldsymbol{x}-\boldsymbol{y}|} \mathrm{K}_1(m|\boldsymbol{x}-\boldsymbol{y}|) \neq 0 \ . \tag{4.123}$$

The modified Bessel function (MacDonald function) $\mathrm{K}_1(x)$ drops off exponentially for large arguments. However, it certainly does not vanish in a region of the size of the Compton wavelength of the particle. Using this property we now proceed to show that the quantization prescription (4.120) would *violate the principle of microcausality*, i.e., we will show that the commutator between two observables located at space-like distance does not vanish. For the observable we again adopt an expression bilinear in the field operator, as in (4.114). The required commutator is transformed using the identities $[\hat{A}, \hat{B}\hat{C}] = [\hat{A}, \hat{B}]\hat{C} + \hat{B}[\hat{A}, \hat{C}]$ and $[\hat{A}, \hat{B}\hat{C}] = \{\hat{A}, \hat{B}\}\hat{C} - \hat{B}\{\hat{A}, \hat{C}\}$ as follows

$$\begin{aligned}
&[\hat{\phi}^\dagger(x)\hat{\phi}(x), \hat{\phi}^\dagger(y)\hat{\phi}(y)] \\
&= -\hat{\phi}^\dagger(x)\hat{\phi}^\dagger(y)\{\hat{\phi}(x),\hat{\phi}(y)\} + \hat{\phi}^\dagger(x)\{\hat{\phi}(x),\hat{\phi}^\dagger(y)\}\hat{\phi}(y) \\
&\quad -\hat{\phi}^\dagger(y)\{\hat{\phi}^\dagger(x),\hat{\phi}(y)\}\hat{\phi}(x) + \{\hat{\phi}^\dagger(x),\hat{\phi}^\dagger(y)\}\hat{\phi}(y)\hat{\phi}(x) \\
&= \left(\hat{\phi}^\dagger(x)\hat{\phi}(y) - \hat{\phi}^\dagger(y)\hat{\phi}(x)\right)\mathrm{i}\Delta_1(x-y) \ .
\end{aligned} \tag{4.124}$$

This implies that on the scale of the particle's Compton wavelength microcausality would be violated if we were to adopt fermionic quantization

$$[\hat{O}(x), \hat{O}(y)] \neq 0 \qquad \text{for} \quad (x-y)^2 < 0 \ . \tag{4.125}$$

This shows that the Klein–Gordon field may not be quantized by using the rule for fermions. This is a special case of the general *spin-statistics theorem*, which goes back to W. Pauli[12].

EXERCISE

4.6 The Function $\Delta_1(x-y)$ for Equal Time Arguments

Problem. Calculate the propagation function $\Delta_1(x-y)$ for the special case of equal time arguments $x_0 = y_0$, i.e., for space-like separations.

Solution. For $x_0 = y_0$, (4.122) simplifies according to

$$i\Delta_1(0, \boldsymbol{x} - \boldsymbol{y}) = \int \frac{d^3p}{(2\pi)^3} \frac{\cos(\boldsymbol{p} \cdot (\boldsymbol{x} - \boldsymbol{y}))}{\sqrt{\boldsymbol{p}^2 + m^2}} . \qquad (1)$$

We express the vector $\boldsymbol{p}$ in spherical coordinates, taking the polar axis to point into the direction $\boldsymbol{r} = \boldsymbol{x} - \boldsymbol{y}$:

$$i\Delta_1(0, \boldsymbol{x} - \boldsymbol{y}) = \frac{1}{(2\pi)^3} \int_0^{2\pi} d\phi \int_0^\pi d\theta \sin\theta \int_0^\infty dp \frac{p^2 \cos(pr\cos\theta)}{\sqrt{p^2 + m^2}} . \qquad (2)$$

The ϕ integration can be done immediately, leading to a factor 2π. To solve the θ integration we substitute $z = pr\cos\theta$:

$$\begin{aligned}
i\Delta_1(0, \boldsymbol{x} - \boldsymbol{y}) &= \frac{1}{(2\pi)^2} \int_0^\infty \frac{p^2 dp}{\sqrt{p^2 + m^2}} \int_0^\pi d\theta \sin\theta \cos(pr\cos\theta) \\
&= \frac{1}{(2\pi)^2} \int_0^\infty \frac{p^2 dp}{\sqrt{p^2 + m^2}} \int_{-pr}^{pr} dz \frac{1}{pr} \cos z \\
&= \frac{1}{2\pi^2 r} \int_0^\infty dp \frac{p \sin pr}{\sqrt{p^2 + m^2}} . \qquad (3)
\end{aligned}$$

The denominator in this integral can be eliminated through the transformation $p = m\sinh t$ and thus $dp = \sqrt{p^2+m^2}\, dt$, leading to

$$i\Delta_1(0, \boldsymbol{x} - \boldsymbol{y}) = \frac{m}{2\pi^2 r} \int_0^\infty dt\, \sinh t\, \sin(mr \sinh t) . \qquad (4)$$

This turns out to be an integral representation of the modified Bessel function (MacDonald function)[13] $K_1(mr)$. The resulting special case for the invariant function thus becomes

$$i\Delta_1(0, \boldsymbol{x} - \boldsymbol{y}) = \frac{m}{2\pi^2} \frac{1}{|\boldsymbol{x} - \boldsymbol{y}|} K_1(m|\boldsymbol{x} - \boldsymbol{y}|) . \qquad (5)$$

---

[12] W. Pauli: Phys. Rev. **58**, 716 (1940); M. Fierz: Helv. Phys. Acta **12**, 3 (1939).
[13] I.S. Gradshteyn, I.M. Ryzhik: *Table of Integrals, Series, and Products*, (Academic Press, New York 1965) no. 3.996.1

4.5 The Scalar Feynman Propagator

The functions $\Delta(x-y)$ and $\Delta_1(x-y)$ are just two members of a whole family of invariant functions that solve the Klein–Gordon equation with special boundary conditions. Perhaps the most important member of this family is the *Feynman propagator* $\Delta_F(x-y)$ which will turn out to be an essential ingredient for the treatment of interacting field theories by using perturbation theory (see Chap. 8). The Feynman propagator is defined as a particular expectation value of a product of field operators:

$$i\Delta_F(x-y) = \langle 0 | T(\hat{\phi}(x)\hat{\phi}^\dagger(y)) | 0 \rangle . \tag{4.126}$$

The symbol T denotes the *time-ordered product* of the operators $\hat{\phi}(x)$ and $\hat{\phi}^\dagger(y)$. Given two time-dependent vectors $\hat{A}$ and $\hat{B}$ we define the time-ordered product as

$$T(\hat{A}(x)\hat{B}(y)) = \hat{A}(x)\hat{B}(y)\Theta(x_0-y_0) \pm \hat{B}(y)\hat{A}(x)\Theta(y_0-x_0) . \tag{4.127}$$

The factors are put into chronological order so that the operator having the later time argument is put first. When reordering is carried out, a plus (minus) sign is used, depending on the bosonic (fermionic) character of the operators. The definition (4.127) of the T product is easily generalized to the case of more than two factors. The operators are then ordered in such a way the time arguments decrease monotonically from left to right. If two time arguments happen to be equal, problems might arise since the ordering then is not well defined. In the present case this poses no problem for finite spatial distances $\boldsymbol{x} \neq \boldsymbol{y}$ since the operators $\hat{\phi}(x)$ and $\hat{\phi}^\dagger(y)$ are then known to commute

To evaluate the Feynman propagator (4.126) we first consider the case $x_0 > y_0$:

$$\begin{aligned} \langle 0 | T(\hat{\phi}(x)\hat{\phi}^\dagger(y)) | 0 \rangle &= \langle 0 | \hat{\phi}(x)\hat{\phi}^\dagger(y) | 0 \rangle \\ &= \langle 0 | \hat{\phi}(x)^{(+)}\hat{\phi}^{\dagger(-)}(y) | 0 \rangle \quad \text{for } x_0 > y_0 , \end{aligned} \tag{4.128}$$

where the superscript $(+)$ or $(-)$ denotes those parts of the field operator (4.58) that have positive or negative frequency, respectively:

$$\hat{\phi}^{(+)}(x) = \int d^3p \, \hat{a}_{\boldsymbol{p}} \, u_{\boldsymbol{p}}(x) \quad , \quad \hat{\phi}^{\dagger(+)}(x) = \int d^3p \, \hat{b}_{\boldsymbol{p}} \, u_{\boldsymbol{p}}(x) , \tag{4.129a}$$

$$\hat{\phi}^{(-)}(x) = \int d^3p \, \hat{b}_{\boldsymbol{p}}^\dagger \, u_{\boldsymbol{p}}^*(x) \quad , \quad \hat{\phi}^{\dagger(-)}(x) = \int d^3p \, \hat{a}_{\boldsymbol{p}}^\dagger \, u_{\boldsymbol{p}}^*(x) . \tag{4.129b}$$

Three terms could be dropped in (4.128) because of

$$\hat{\phi}^{\dagger(+)}(y) | 0 \rangle = \langle 0 | \hat{\phi}^{(-)}(x) = 0 . \tag{4.130}$$

Insertion of the expansion (4.129) yields

$$\begin{aligned} i\Delta_F(x-y) &= \int d^3p \, u_{\boldsymbol{p}}(x) u_{\boldsymbol{p}}^*(y) = \int \frac{d^3p}{(2\pi)^3} \frac{1}{2\omega_{\boldsymbol{p}}} e^{-ip\cdot(x-y)} \\ &\equiv i\Delta^{(+)}(x-y) \quad \text{for } x_0 > y_0 . \end{aligned} \tag{4.131}$$

An analogous derivation for the opposite time ordering yields

$$i\Delta_F(x-y) = \langle 0|\hat{\phi}^\dagger(y)\hat{\phi}(x)|0\rangle = \langle 0|\hat{\phi}^{\dagger(+)}(y)\hat{\phi}^{(-)}(x)|0\rangle$$

$$= \int d^3 p\, u_p(y) u_p^*(x) = \int \frac{d^3 p}{(2\pi)^3} \frac{1}{2\omega_p} e^{+ip\cdot(x-y)}$$

$$\equiv -i\Delta^{(-)}(x-y) \quad \text{for} \quad x_0 < y_0 \,. \tag{4.132}$$

Combining these results we find

$$i\Delta_F(x-y) = \Theta(x_0-y_0)\, i\Delta^{(+)}(x-y) - \Theta(y_0-x_0)\, i\Delta^{(-)}(x-y)$$

$$= \int \frac{d^3 p}{(2\pi)^3} \frac{1}{2\omega_p} \Big[\Theta(x_0-y_0)\, e^{-ip\cdot(x-y)} + \Theta(y_0-x_0)\, e^{+ip\cdot(x-y)}\Big].$$

$$\tag{4.133}$$

The function $\Delta_F(x-y)$ can be represented in a form that is very useful for practical calculations by introducing a further integration of the variable p_0. Going to the complex p_0 plane, we can interpret both terms in (4.133) as the residues of a common integrand. If we make the substitution $\boldsymbol{p} \to -\boldsymbol{p}$ in the second term we can combine them into

$$\frac{1}{2\omega_p}\Big[\Theta(x_0-y_0)\, e^{-i\omega_p(x_0-y_0)} + \Theta(y_0-x_0)\, e^{+i\omega_p(x_0-y_0)}\Big]$$

$$= -\int_{C_F} \frac{dp_0}{2\pi i} \frac{e^{-ip_0(x_0-y_0)}}{(p_0-\omega_p)(p_0+\omega_p)} \,. \tag{4.134}$$

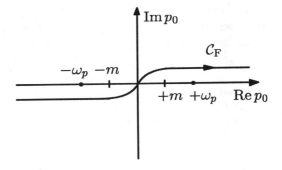

Fig. 4.2. The integration contour C_F used to define the Feynman propagator Δ_F

Here the *integration contour* C_F has to be used, which circumnavigates the poles of the integrand in the complex plan as depicted in Fig. 4.2. Equation (4.134) immediately follows from Cauchy's theorem of residues if one closes the path of integration by a large semicircle in the upper half plane (if $x_0 < y_0$) or in the lower half plane (if $x_0 > y_0$). Thus the Feynman propagator can be written in the compact form

$$\Delta_F(x-y) = \int_{C_F} \frac{d^4 p}{(2\pi)^4} \frac{e^{-ip\cdot(x-y)}}{p^2-m^2} \,. \tag{4.135}$$

The same effect can be achieved if one integrates along the real p_0 axis instead of the contour C_F and avoids hitting the singularities by shifting them by an infinitesimal amount in the complex plane. The correct signs of this shift are obtained if a term $+i\epsilon$ is added in the denominator. Thus the Feynman propagator can also be expressed as

$$\Delta_F(x-y) = \int \frac{d^4p}{(2\pi)^4} e^{-ip\cdot(x-y)} \Delta_F(p)$$

$$= \int \frac{d^4p}{(2\pi)^4} e^{-ip\cdot(x-y)} \frac{1}{p^2 - m^2 + i\epsilon} \qquad (4.136)$$

which shows that in momentum space the Feynman propagator is a simple rational function. The integral representation can be used immediately to show that $\Delta_F(x-y)$ is a *Green's function of the Klein–Gordon equation*. Application of the Klein–Gordon wave operator to (4.136) gives

$$(\Box_x + m^2)\Delta_F(x-y) = \int \frac{d^4p}{(2\pi)^4}(-p^2+m^2)\frac{e^{-ip\cdot(x-y)}}{p^2-m^2+i\epsilon}$$

$$= -\int \frac{d^4p}{(2\pi)^4} e^{-ip\cdot(x-y)} = -\delta^4(x-y), \qquad (4.137)$$

i.e., the function $\Delta_F(x-y)$ is a solution of the inhomogeneous Klein–Gordon equation containing a delta function as a source term.

The Feynman propagator $\Delta_F(x_2 - x_1)$ defined through (4.126) will later be seen to emerge naturally in the perturbation series for interacting field theories. We can illustrate its meaning already at the present stage. According to Exercise 4.4 the operator $\hat{\phi}^\dagger(\boldsymbol{x},t)$ increases the charge Q of a state by one unit, whereas $\hat{\phi}(\boldsymbol{x},t)$ decreases it by the same amount. Application of the adjoint field operator on the vacuum $|0\rangle$ gives a state $\hat{\phi}^\dagger(\boldsymbol{x}_1,t_1)|0\rangle$ which containes a single particle, described by the creation operator $\hat{a}^\dagger_{\boldsymbol{p}}$. Projecting this state onto $(\hat{\phi}^\dagger(\boldsymbol{x}_2,t_2)|0\rangle)^\dagger = \langle 0|\hat{\phi}(\boldsymbol{x}_2,t_2)$ leads to a probability amplitude that describes the process by which a particle starts at the point $\boldsymbol{x}_1, t_1$ in space-time and propagates to the point $\boldsymbol{x}_2, t_2$. This is given by

$$i\Delta_F(x_2-x_1)\Theta(t_2-t_1) = \langle 0|\hat{\phi}(\boldsymbol{x}_2,t_2)\hat{\phi}^\dagger(\boldsymbol{x}_1,t_1)|0\rangle \Theta(t_2-t_1). \qquad (4.138)$$

One can say as well that a particle is created at point 1 (i.e., $\boldsymbol{x}_1, t_1$) and gets destroyed at point 2 (i.e., $\boldsymbol{x}_2, t_2$). There is a closely related second process in which a unit of charge moves from point 1 to point 2: instead of the annihilation of a particle at 2 an antiparticle (described by $\hat{b}_{\boldsymbol{p}}$) can be created at this point. This can move to 1 where it is annihilated, thereby increasing the charge Q at this point by one unit, just as it was increased by the creation of the particle in the first process. The amplitude for the propagation of an antiparticle is given by

$$i\Delta_F(x_2-x_1)\Theta(t_1-t_2) = \langle 0|\hat{\phi}^\dagger(\boldsymbol{x}_1,t_1)\hat{\phi}(\boldsymbol{x}_2,t_2)|0\rangle \Theta(t_1-t_2). \qquad (4.139)$$

Obviously the Feynman propagator $\Delta_F(x_2-x_1)$ includes both processes of particle and antiparticle propagation, depending on the time ordering of t_1 and t_2. Figure 4.3 illustrates this for the example of the propagation of charged pions (π^- = particle, π^+ = antiparticle). Both processes can be combined in an illustrative manner by using Feynman's interpretation, which considers *an antiparticle to be a particle that moves backward in time*. If this language is used one has to invert the direction of the arrow in Fig. 4.3(b) and at the same time change the label from π^+ to π^-.

In the volume *Quantum Electrodynamics* in this series of books, the discussion of the Feynman propagator is the starting point for developing the

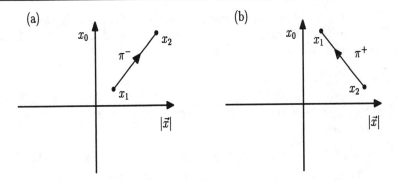

Fig. 4.3. Space-time diagram showing the propagation of a particle (**a**) and of an antiparticle (**b**), both being described by the Feynman propagator $\Delta_F(x_2 - x_1)$

theory of quantum electrodynamics, without making use of the methods of field quantization. Central to this presentation is the intuitive physical picture of particle and antiparticle propagation developed by Stückelberg and Feynman: particles propagate forward in time; *antiparticles are viewed as particles with negative energy that move backward in time.*

4.6 Supplement: The Δ Functions

In the last sections, several important functions have been constructed, starting from a scalar field. Now we want to present a complete collection of these functions and to derived their mutual relationships. The whole family of functions will be denoted by the letter Δ, each member being distinguished by an additional subscript. Table 4.1 contains a listing of the Δ functions commonly used.[14]

Table 4.1. A collection of the invariant commutation and propagation functions of a scalar field

Symbol	Name	Contour	Relation
$\Delta(x)$	Pauli–Jordan function	$\mathcal{C}$	–
$\Delta^{(+)}(x)$	positive frequency	$\mathcal{C}^{(+)}$	$\frac{1}{2}(\Delta + \Delta_1)$
$\Delta^{(-)}(x)$	negative frequency	$\mathcal{C}^{(-)}$	$\frac{1}{2}(\Delta - \Delta_1)$
$\Delta_1(x)$	anticommutator	$\mathcal{C}_1$	–
$\Delta_F(x)$	Feynman propagator	$\mathcal{C}_F$	$\frac{1}{2}(\epsilon(x_0)\Delta + \Delta_1)$
$\Delta_R(x)$	retarded propagator	$\mathcal{C}_R$	$\Theta(x_0)\Delta$
$\Delta_A(x)$	advanced propagator	$\mathcal{C}_A$	$-\Theta(-x_0)\Delta$
$\Delta_D(x)$	Dyson propagator	$\mathcal{C}_D$	$\frac{1}{2}(\epsilon(x_0)\Delta - \Delta_1)$
$\bar{\Delta}(x)$	principal-part propagator	$\bar{\mathcal{C}}$	$\frac{1}{2}\epsilon(x_0)\Delta$

---

[14] The literature contains a multitude of different conventions that differ in the names and by multiplicative factors.

These functions can be classified according to the result obtained when applying the Klein–Gordon operator. The functions $\Delta, \Delta^{(+)}, \Delta^{(-)}, \Delta_1$ satisfy the *homogeous Klein-Gordon equation*

$$(\Box + m^2)\, \Delta(x) = 0 \quad \text{etc.} . \tag{4.140}$$

In contrast, the functions $\Delta_F, \Delta_R, \Delta_A, \Delta_D, \bar{\Delta}$ are Green's functions, i.e., they solve the *inhomogeneous Klein-Gordon equation* with a delta function as the source term:

$$(\Box + m^2)\, \Delta_F(x) = -\delta^4(x) \quad \text{etc.} . \tag{4.141}$$

In the following, we will call these two classes the *commutation functions* and the *propagation functions*.

The Commutation Functions

As the most important representative of this type, in Sect. 4.4 we have become acquainted with the *Pauli–Jordan function*, which was defined as the commutator of two scalar field operators:

$$\mathrm{i}\Delta(x) = \left[\hat{\phi}(x),\, \hat{\phi}(y)\right] . \tag{4.142}$$

The Pauli–Jordan function has the three-dimensional Fourier representation (using $p_0 = \omega_p \equiv +\sqrt{\mathbf{p}^2 + m^2}$)

$$\Delta(x) = -\mathrm{i}\int \frac{\mathrm{d}^3 p}{(2\pi)^3}\, \frac{1}{2\omega_p} \left(\mathrm{e}^{-\mathrm{i}p\cdot x} - \mathrm{e}^{+\mathrm{i}p\cdot x}\right) \tag{4.143}$$

or

$$\Delta(x) = -\int \frac{\mathrm{d}^3 p}{(2\pi)^3}\, \mathrm{e}^{\mathrm{i}\mathbf{p}\cdot\mathbf{x}}\, \frac{\sin \omega_p x_0}{\omega_p} . \tag{4.144}$$

Equivalently we have deduced the four-dimensional Fourier representation

$$\Delta(x) = -\mathrm{i}\int \frac{\mathrm{d}^4 p}{(2\pi)^3}\, \mathrm{e}^{-\mathrm{i}p\cdot x}\, \epsilon(p_0)\, \delta(p^2 - m^2) , \tag{4.145}$$

where $\epsilon(p_0) = \mathrm{sgn}(p_0)$ denotes the sign function.

There is another way to represent $\Delta(x)$ in terms of a four-dimensional integral in momentum space. Both terms in (4.143) can be interpreted as the residues of an integral over $\mathrm{d}p_0$ having poles at $p_0 = \pm\omega_p$. To generate these residues one has to integrate over a closed contour C that encompasses both poles (cf. Fig. 4.4a).

This is confirmed using the theorem of residues for the p_0 integration according to

$$\begin{aligned}
\int_C \frac{\mathrm{d}p_0}{2\pi}\, \frac{\mathrm{e}^{-\mathrm{i}p_0 x_0}}{p^2 - m^2} &= \int_C \frac{\mathrm{d}p_0}{2\pi}\, \frac{\mathrm{e}^{-\mathrm{i}p_0 x_0}}{p_0^2 - \omega_p^2} \\
&= -2\pi\mathrm{i}\left(\mathrm{Res}\Big|_{p_0=+\omega_p} + \mathrm{Res}\Big|_{p_0=-\omega_p}\right) \\
&= -\mathrm{i}\frac{1}{2\omega_p}\left(\mathrm{e}^{-\mathrm{i}\omega_p x_0} - \mathrm{e}^{+\mathrm{i}\omega_p x_0}\right) .
\end{aligned} \tag{4.146}$$

A comparison with (4.143) reveals that

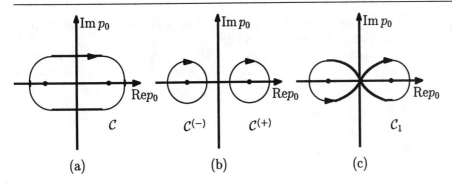

Fig. 4.4. The integration contours that define the commutation functions

$$\Delta(x) = \int_C \frac{d^4p}{(2\pi)^4} \frac{e^{-ip\cdot x}}{p^2 - m^2}, \qquad (4.147)$$

where in the second term of (4.143) p was replaced by $-p$.

The Pauli–Jordan function contains two parts, one with positive and onw with negative frequency:

$$\Delta(x) = \Delta^{(+)}(x) + \Delta^{(-)}(x). \qquad (4.148)$$

Obviously each of these parts is generated from one of the two residues in the integral (4.146). One can obtain these parts by using the contours $C^{(+)}$ or $C^{(-)}$, which circle around one of the poles as defined in Fig. 4.4b. The functions $\Delta^{(\pm)}(x)$ can be written in the invariant form (4.147) by using the identities (see (4.101))

$$\frac{1}{2\omega_p} \delta(p_0 \mp \omega_p) = \Theta(\pm p_0)\, \delta(p^2 - m^2), \qquad (4.149)$$

which lead to

$$\Delta^{(\pm)}(x) = \mp i \int \frac{d^4p}{(2\pi)^3} e^{-ip\cdot x}\, \Theta(\pm p_0)\, \delta(p^2 - m^2). \qquad (4.150)$$

One further type of commutation function can be constructed by surrounding the two poles in opposite directions, using a contour C_1 shaped like the figure eight (see Fig. 4.4c):

$$\Delta_1(x) = \Delta^{(+)}(x) - \Delta^{(-)}(x). \qquad (4.151)$$

The function $\Delta_1(x)$ defined in this way[15] has the three-dimensional Fourier representation

$$\Delta_1(x) = \int_{C_1} \frac{d^4p}{(2\pi)^4} \frac{e^{-ip\cdot x}}{p^2 - m^2} = -i \int \frac{d^3p}{(2\pi)^3} \frac{1}{2\omega_p} \left(e^{-ip\cdot x} + e^{+ip\cdot x}\right)$$

$$= -i \int \frac{d^3p}{(2\pi)^3} e^{i\mathbf{p}\cdot\mathbf{x}} \frac{\cos\omega_p x_0}{\omega_p}. \qquad (4.152)$$

An alternative representation follows from (4.150)

$$\Delta_1(x) = -i \int \frac{d^4p}{(2\pi)^3} e^{-ip\cdot x} \delta(p^2 - m^2). \qquad (4.153)$$

---

[15] Often the definition of $\Delta_1(x)$ is given with an additional factor i which makes it a real-valued function.

We have encountered the function $i\Delta_1(x-y)$ in (4.121) where it was identified with the anticommutator of the field operators $\hat{\phi}(x)$ and $\hat{\phi}(y)$.

The commutation functions have the following properties under reflection and complex conjugation, which are easily verified from their integral representations:

$$\begin{aligned}
\Delta^{(\pm)}(-x) &= -\Delta^{(\mp)}(x) , \\
\Delta(-x) &= -\Delta(x) , \\
\Delta_1(-x) &= \Delta_1(x) , \\
\Delta^{(\pm)*}(x) &= \Delta^{(\mp)}(x) , \\
\Delta^*(x) &= \Delta(x) \quad \text{real} , \\
\Delta_1^*(x) &= -\Delta_1(x) \quad \text{imaginary} .
\end{aligned} \qquad (4.154)$$

The Propagation Functions

All the functions introduced in the last section satisfy the homogeneous Klein–Gordon equation. This becomes immediately obvious from the four-dimensional Fourier representations (4.145), (4.150), and (4.153)) since $(p^2 - m^2)\delta(p^2 - m^2) = 0$. The Green's functions, on the other hand, can be constructed by solving the Fourier integral (4.147) with *open integration contours* that extend to infinity. Upon application of the operator $\Box + m^2$ the poles are canceled and we are left with an integral from $-\infty$ to $+\infty$. Since the integrand is the exponential function $\exp(-ip_0 x_0)$ this leads to a delta function.

The most important representative of this class of functions is the *Feynman propagator*, which was introduced in Sect. 4.5:

$$i\Delta_F(x-y) = \langle 0 | T(\hat{\phi}(x)\,\hat{\phi}^\dagger(y)) | 0 \rangle . \qquad (4.155)$$

This functions results from the integration

$$\Delta_F(x) = \int_{C_F} \frac{d^4p}{(2\pi)^4} \frac{e^{-ip\cdot x}}{p^2 - m^2} \qquad (4.156)$$

extended over the Feynman contour C_F defined in Fig. 4.5a (see also Sect. 4.5).

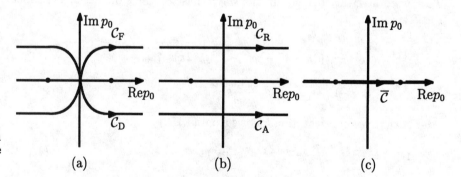

Fig. 4.5. The integration contours that define the propagation functions

The contour C_F is open and extends to infinity. Since the integrand is a holomorphic function, the p_0 integration can be closed by adding a "half-circle at infinity". Here the sign of the time coordinate x_0 makes an inportant difference: the contour can be closed in the upper (lower) half plane provided that x_0 is negative (positive). The integration contour extended in this way encircles only one of the two poles. It can be deformed continuously without changing the value of the integral and is thus seen to be topologically equivalent either to the contour $C^{(+)}$ or to the contour $-C^{(-)}$ from Fig. 4.4b. This obviously implies the relation

$$\Delta_F(x) = \Theta(x_0)\Delta^{(+)}(x) - \Theta(-x_0)\Delta^{(-)}(x) \, . \qquad (4.157)$$

Therefore the Feynman propagator corresponds to a sum of the positive-frequency Pauli–Jordan function for positive time arguments and of the negative-frequency Pauli–Jordan function for negative time arguments, both being combined with a negative relative sign. This construction of the Feynman propagator ensures the correct implementation of the condition of causality in the study of the propagation of perturbations. Here in the spirit of Stückelberg and Feynman antiparticles are treated as particles with negative frequency (energy) that move backward in time (see also the discussion in the volume *Quantum Electrodynamics*).

Alternative propagation functions can be defined through the open integration contours C_A and C_R in Fig. 4.5b, which pass both poles on the same side. Closing these contours can lead to the Pauli–Jordan function $\Delta(x)$ from Fig. 4.4a. If the time coordinate has the opposite sign, the integral vanishes since no poles are encircled. Thus we have

$$\Delta_R(x) = \Theta(x_0)\,\Delta(x) \, , \qquad (4.158\text{a})$$
$$\Delta_A(x) = -\Theta(-x_0)\,\Delta(x) \, . \qquad (4.158\text{b})$$

These are the *retarded* and the *advanced* Green's function of the Klein–Gordon equation. Δ_R (Δ_A) is nonvanishing only for a positive (negative) time argument. Looking at (4.158) we find that the Pauli–Jordan function $\Delta(x)$ can be written as the difference between the retarded and the advanced propagators:

$$\Delta(x) = \Delta_R(x) - \Delta_A(x) \, . \qquad (4.159)$$

This is quite natural since a comparison of Figs. 4.4a and 4.5b reveals that by pasting together the contours C_R and $-C_A$ at infinity, we generate the integration contour C.

For completeness' sake we mention the "anticausal" propagator (also known as the Dyson propagator) $\Delta_D(x)$, which encircles the poles in the opposite way compared to the Feynman propagator (see Fig. 4.5a):

$$\Delta_D(x) = \Theta(x_0)\,\Delta^{(-)}(x) - \Theta(-x_0)\,\Delta^{(+)} \, . \qquad (4.160)$$

Finally the principal-part propagator $\bar{\Delta}(x)$ can be introduced, which is generated if the p_0 integration on the real axis runs through the poles at $p_0 = \pm\omega_p$ as sketched in Fig. 4.5c. The integration over the singularities is interpreted as a principal-part integral. This prescription can be interpreted as the arithmetic mean of two integrals along contours that pass the pole to the left or to the right. This implies

$$\bar{\Delta}(x) = \frac{1}{2}\big(\Delta_{\mathrm{R}}(x) + \Delta_{\mathrm{A}}(x)\big) \ . \tag{4.161}$$

According to (4.158) the principal-part propagator has a simple connection with the Pauli–Jordan function:

$$\bar{\Delta}(x) = \frac{1}{2}\,\epsilon(x_0)\,\Delta(x) \ . \tag{4.162}$$

A connection to the Feynman propagator is given through

$$\Delta_{\mathrm{F}}(x) = \bar{\Delta}(x) + \frac{1}{2}\,\Delta_1(x) \ . \tag{4.163}$$

A further useful connection between the commutation and propagation functions is

$$\Delta_1(x) = \Delta_{\mathrm{F}}(x) - \Delta_{\mathrm{D}}(x) \ . \tag{4.164}$$

All the Δ functions can be expressed in terms of two independent "basic" functions (since the integrand in the Fourier integral has two poles). A reasonable choice for these basic functions is $\Delta(x)$ and $\Delta_1(x)$. The last column of Table 4.1 on page 109 shows how the other Δ functions can be obtained.

Obviously all the propagation functions contain a product of the function $\Delta(x)$ with a unit step function in time $[\Theta(x_0)$ or $\tfrac{1}{2}\epsilon(x_0)]$. It is this step function that gives rise to the delta function when the Klein–Gordon operator is applied. For example the Klein–Gordon operator acts on the Feynman propagator as follows:

$$\begin{aligned}(\Box + m^2)\Delta_{\mathrm{F}}(x) &= (\Box + m^2)\frac{1}{2}\big(\epsilon(x_0)\,\Delta(x) + \Delta_1(x)\big) \\ &= \big(\partial_0^2 - \nabla^2 + m^2\big)\frac{1}{2}\epsilon(x_0)\,\Delta(x) \\ &= \big(\partial_0\delta(x_0)\big)\Delta(x) + 2\,\delta(x_0)\big(\partial_0\Delta(x)\big) + \frac{1}{2}\epsilon(x_0)(\Box + m^2)\Delta(x) \ .\end{aligned} \tag{4.165}$$

The last contribution vanishes since $\Delta_1(x)$ and $\Delta(x)$ solve the homoneneous Klein–Gordon equation. The first term containing the derivative of the delta function is equivalent to $-\delta(x_0)\big(\partial_0\Delta(x)\big)$. This is true since this expression has to be viewed as a distribution. Thus it makes sense only when it gets multiplied by a (sufficiently smooth) test function $f(x_0)$ and integrated over x_0. Then an integration by parts leads to

$$\begin{aligned}\int &\mathrm{d}x_0\,\big(\partial_0\delta(x_0)\big)\,\Delta(x)f(x_0) \\ &= -\int \mathrm{d}x_0\,\delta(x_0)\big(\partial_0\Delta(x)\big)f - \int \mathrm{d}x_0\,\delta(x_0)\Delta(x)(\partial_0 f) \\ &= -\partial_0\Delta(x)\big|_{x_0=0}f - \Delta(0,\boldsymbol{x})\,\partial_0 f\big|_{x_0=0} \\ &= -\partial_0\Delta(x)\big|_{x_0=0}f \ ,\end{aligned} \tag{4.166}$$

since $\Delta(0,\boldsymbol{x}) = 0$ according to (4.104). The second boundary condition (see (4.105)), $\partial_0\Delta(x)|_{x_0=0} = -\delta^3(\boldsymbol{x})$, can be applied and (4.165) becomes

$$(\Box + m^2)\, \Delta_\mathrm{F}(x) = -\delta^4(x)\,, \qquad (4.167)$$

as expected.

The functions $\Delta(x)$ and $\Delta_1(x)$, and as a consequence the whole family of Δ functions, are expressed in terms of Fourier integrals that can be solved in closed form. Unfortunately the resulting expressions in coordinate space are quite unwieldy and are not very suitable for practical calculations. Usually it is much more convenient to work in momentum space. Nevertheless let us quote the resulting functions in coordinate space:[16]

$$\Delta(x) = -\frac{1}{2\pi}\epsilon(x_0)\,\delta(x^2) + \frac{m}{4\pi\sqrt{x^2}}\epsilon(x_0)\,\Theta(x^2)\,\mathrm{J}_1\!\left(m\sqrt{x^2}\right), \qquad (4.168\mathrm{a})$$

$$\mathrm{i}\Delta_1(x) = \frac{m}{4\pi\sqrt{x^2}}\Theta(x^2)\,\mathrm{N}_1\!\left(m\sqrt{x^2}\right) + \frac{m}{2\pi^2\sqrt{-x^2}}\Theta(-x^2)\,\mathrm{K}_1\!\left(m\sqrt{-x^2}\right). \qquad (4.168\mathrm{b})$$

These expressions contain the Bessel function J_1, the Neumann function N_1, and the MacDonald function K_1 (which is a Hankel function of an imaginary argument), all of first order. For large values of the argument z the functions $\mathrm{J}_1(z)$ and $\mathrm{N}_1(z)$ are oscillatory, whereas $\mathrm{K}_1(z)$ falls off exponentially. From (4.168) we learn that the Pauli–Jordan function $\Delta(x)$ vanishes identically for space-like separations ($x^2 < 0$). In Sect. 4.4 this property was found to guarantee the condition of microcausality. In contrast to this, the function $\Delta_1(x)$ and thus also the Feynman propagator $\Delta_\mathrm{F}(x)$ extend into the space-like region, dropping off on the scale of the Compton wavelength $1/m$.

All the Δ functions show *singular behavior on the light cone*. Taking into account the singularities of the Bessel functions at argument zero, we find the functional dependence in the vicinity of the light cone:

$$\Delta(x) \simeq -\frac{1}{2\pi}\epsilon(x_0)\,\delta(x^2) + \frac{m^2}{8\pi}\epsilon(x_0)\,\Theta(x^2)\,, \qquad (4.169\mathrm{a})$$

$$\mathrm{i}\Delta_1(x) \simeq -\frac{1}{2\pi^2 x^2} + \frac{m^2}{4\pi^2}\ln\frac{m\sqrt{|x^2|}}{2}\,. \qquad (4.169\mathrm{b})$$

Four different types of singularities are seen to arise at the light cone: $\delta(x^2)$, $\Theta(x^2)$, $1/x^2$, and $\ln|x^2|$.

---

[16] See the volume *Quantum Electrodynamics* (Springer, Berlin, Heidelberg, 1994) and also N.N. Bogoliubov, D.V. Shirkov, *Introduction to the Theory of Quantized Fields*, (Wiley, New York 1980), Chap. 16.

5. Spin-$\frac{1}{2}$ Fields: The Dirac Equation

5.1 Introduction

Having treated the quantization of the Klein–Gordon field, we now turn to the study of particles with spin $\frac{1}{2}$, which are described by the Dirac equation. We will assume that readers have some familiarity with this equation and refer them to the volume *Relativistic Quantum Mechanics* in this series of books for details. As in the previous chapters we will again start by considering the Dirac wave function $\psi(\boldsymbol{x}, t)$ as a "classical" field and derive the corresponding wave equation from a Lagrange function. Subsequently this field will be subjected to canonical quantization.

With the covariant relativistic notation, the Dirac equation for massive spin-$\frac{1}{2}$ particles reads ($\hbar = c = 1$)

$$(i\gamma^\mu \partial_\mu - m)\psi = 0 \qquad (5.1)$$

where the four Dirac matrices $\gamma^\mu, \mu = 0, \ldots, 3$, satisfy the algebra

$$\gamma^\mu \gamma^\nu + \gamma^\nu \gamma^\mu = 2g^{\mu\nu} . \qquad (5.2)$$

The wave function ψ has four components and satisfies the the transformation laws of a relativistic spinor. The *dimension* of the Dirac field is $\dim[\psi] = \text{length}^{-3/2}$, i.e., it has the natural dimension $d = +3/2$.

What will be the Lagrangian that leads to (5.1) as an equation of motion? The Lagrange density will be a bilinear function composed of the fields $\psi, \dot{\psi}, \boldsymbol{\nabla}\psi$ and the hermitean conjugate fields $\psi^\dagger, \dot{\psi}^\dagger, \boldsymbol{\nabla}\psi^\dagger$. It has to transform as a Lorentz scalar density and can contain only derivatives of first order since the Dirac equation itself is of first order. This strongly narrows down the choice of possible candidates for the Lagrange density. As a first attempt we choose the ansatz

$$\mathcal{L} = \bar{\psi}(i\gamma^\mu \partial_\mu - m)\psi = i\psi^\dagger \dot{\psi} + i\psi^\dagger \boldsymbol{\alpha} \cdot \boldsymbol{\nabla}\psi - m\psi^\dagger \beta \psi . \qquad (5.3)$$

Here in the second step the matrices $\beta = \gamma^0$, $\beta^2 = 1$, and $\boldsymbol{\alpha} = \gamma^0 \boldsymbol{\gamma}$ were used and the adjoint spinor $\bar{\psi} = \psi^\dagger \gamma^0$ was introduced. We will treat the spinors ψ and $\psi^\dagger$ as independent fields, each having four components. Differentiation of $\mathcal{L}$ with respect to these fields and to their time and space derivatives gives

$$\frac{\partial \mathcal{L}}{\partial \psi} = \mathrm{i}\psi^\dagger \dot{}, \qquad \frac{\partial \mathcal{L}}{\partial \dot\psi^\dagger} = 0,$$

$$\frac{\partial \mathcal{L}}{\partial(\boldsymbol{\nabla}\psi)} = \mathrm{i}\psi^\dagger \boldsymbol{\alpha}, \qquad \frac{\partial \mathcal{L}}{\partial(\boldsymbol{\nabla}\psi^\dagger)} = 0, \qquad (5.4)$$

$$\frac{\partial \mathcal{L}}{\partial \psi} = -m\psi^\dagger \beta, \qquad \frac{\partial \mathcal{L}}{\partial \psi^\dagger} = \mathrm{i}\dot\psi + \mathrm{i}\boldsymbol{\alpha}\cdot\boldsymbol{\nabla}\psi - m\beta\psi.$$

Of course these equations are to be interpreted as matrix equations. For example the last but one equation reads explicitely

$$\frac{\partial \mathcal{L}}{\partial \psi_\rho} = -m \sum_\tau \psi_\tau^\dagger \beta_{\tau\rho}. \qquad (5.5)$$

Fortunately in most situations the spinor indices can be dropped without causing confusion. In general we will use this convention since it makes for a much more compact notation. Variation of the action with respect to $\psi^\dagger$ leads to the Euler–Lagrange equation

$$\frac{\partial}{\partial t}\frac{\partial \mathcal{L}}{\partial \dot\psi^\dagger} = \frac{\partial \mathcal{L}}{\partial \psi^\dagger} - \boldsymbol{\nabla}\frac{\partial \mathcal{L}}{\partial(\boldsymbol{\nabla}\psi^\dagger)}, \qquad (5.6)$$

which by use of (5.4) simply becomes

$$\mathrm{i}\dot\psi + \mathrm{i}\boldsymbol{\alpha}\cdot\boldsymbol{\nabla}\psi - m\beta\psi = 0. \qquad (5.7)$$

Multiplying by $\gamma^0 = \beta$, we obtain the standard form of the Dirac equation

$$(\mathrm{i}\gamma^\mu \partial_\mu - m)\psi = 0, \qquad (5.8)$$

as desired. Variation with respect to ψ using (5.4) leads to the differential equation

$$\mathrm{i}\dot\psi^\dagger = -m\,\psi^\dagger \beta - \mathrm{i}\boldsymbol{\nabla}\psi^\dagger \boldsymbol{\alpha}. \qquad (5.9)$$

This can be identified as the hermitean conjugate of (5.7)

$$\bar\psi(\mathrm{i}\gamma^\mu \overleftarrow{\partial}_\mu + m) = 0. \qquad (5.10)$$

The arrow indicates that the partial derivative acts on the function to the left.

As a side remark we note that the Lagrange density (5.3) and thus also the action W of the Dirac field happens to vanish, $\mathcal{L}=0$, if we insert solutions of the Dirac equation for the wave function ψ. This property has no particular consequence, however, since what matters is not the numerical value of the action but rather its response to variations.

To proceed further we need the canonically conjugate fields π_ψ and $\pi_{\psi^\dagger}$. As in the case of the Schrödinger field in Chap. 3 these turn out to dependent quantities. We find

$$\pi_\psi = \frac{\partial \mathcal{L}}{\partial \dot\psi} = \mathrm{i}\psi^\dagger \quad, \quad \pi_{\psi^\dagger} = \frac{\partial \mathcal{L}}{\partial \dot\psi^\dagger} = 0. \qquad (5.11)$$

Thus there are two independent degrees of freedom of the Dirac field, which are given by ψ and $\psi^\dagger$.

As usual the *Hamilton density* is obtained through the Legendre transformation

$$\begin{aligned}\mathcal{H} &= \pi_\psi \dot\psi + \pi_{\psi^\dagger}\dot\psi^\dagger - \mathcal{L} = i\psi^\dagger \dot\psi - i\psi^\dagger\dot\psi - i\psi^\dagger\boldsymbol{\alpha}\cdot\boldsymbol{\nabla}\psi + m\,\psi^\dagger\beta\psi \\ &= \psi^\dagger(-i\boldsymbol{\alpha}\cdot\boldsymbol{\nabla} + \beta m)\psi\;.\end{aligned} \quad (5.12)$$

The Hamiltonian is thus given by the expectation value of Dirac's differential operator $H_D = \boldsymbol{\alpha}\cdot\boldsymbol{p} + \beta m$:

$$H = \int d^3x\,\psi^\dagger(\boldsymbol{x},t)\bigl(-i\boldsymbol{\alpha}\cdot\boldsymbol{\nabla} + \beta m\bigr)\psi(\boldsymbol{x},t)\;. \quad (5.13)$$

Finally let us derive the conserved quantities that follow from Noether's theorem. The canonical *energy-momentum tensor* $\Theta_{\mu\nu}$ reads

$$\begin{aligned}\Theta_{\mu\nu} &= \frac{\partial\mathcal{L}}{\partial(\partial^\mu\psi)}\partial_\nu\psi + \frac{\partial\mathcal{L}}{\partial(\partial^\mu\psi^\dagger)}\partial_\nu\psi^\dagger - g_{\mu\nu}\mathcal{L} \\ &= \bar\psi i\gamma_\mu\partial_\nu\psi - g_{\mu\nu}\bar\psi(i\gamma^\sigma\partial_\sigma - m)\psi\;,\end{aligned} \quad (5.14)$$

which leads to the conserved *four-momentum vector*

$$P_\nu = \int d^3x\,\Theta_{0\nu} = \int d^3x\,\Bigl(\bar\psi i\gamma_0\partial_\nu\psi - g_{0\nu}\bar\psi(i\gamma^\sigma\partial_\sigma - m)\psi\Bigr)\;. \quad (5.15)$$

The time component of this vector is the energy

$$\begin{aligned}P_0 &= \int d^3x\,\bar\psi\Bigl(i\gamma_0\partial_0 - i\gamma^0\partial_0 - i\boldsymbol{\gamma}\cdot\boldsymbol{\nabla} + m\Bigr)\psi \\ &= \int d^3x\,\psi^\dagger(-i\boldsymbol{\alpha}\cdot\boldsymbol{\nabla} + \beta m)\psi\;.\end{aligned} \quad (5.16)$$

As expected this is just the Hamiltonian H of (5.13). The spatial momentum vector reads

$$\boldsymbol{P} = -i\int d^3x\,\psi^\dagger\boldsymbol{\nabla}\psi\;. \quad (5.17)$$

The transformation law of Dirac spinors under infinitesimal Lorentz transformations is given by (see *Relativistic Quantum Mechanics*, Chap. 3)

$$\psi'(x') = \psi(x) - \tfrac{i}{4}\delta\omega_{\mu\nu}\sigma^{\mu\nu}\psi(x)\;, \quad (5.18)$$

where

$$\sigma^{\mu\nu} = \tfrac{i}{2}\bigl[\gamma^\mu,\gamma^\nu\bigr]\;. \quad (5.19)$$

Comparing this with (2.64) we read off the infinitesimal generators

$$\bigl(I^{\mu\nu}\bigr)_{\alpha\beta} = -\tfrac{i}{2}\bigl(\sigma^{\mu\nu}\bigr)_{\alpha\beta}\;, \quad (5.20)$$

where $\mu,\nu = 0,\ldots,3$ are Lorentz indices and $\alpha,\beta = 1,\ldots,4$ are Dirac indices. In particular we find

$$\sigma^{0i} = \tfrac{i}{2}\bigl(\gamma^0\gamma^i - \gamma^i\gamma^0\bigr) = i\gamma^0\gamma^i = i\alpha^i\;, \quad (5.21a)$$

$$\sigma^{ij} = \epsilon^{ijk}\Sigma_k\;. \quad (5.21b)$$

The generalized *angular-momentum tensor* (2.70) of the Dirac field has the form

$$M_{\nu\lambda} = L_{\nu\lambda} + S_{\nu\lambda}, \tag{5.22a}$$

$$L_{\nu\lambda} = \int d^3x \left(\Theta_{0\lambda}x_\nu - \Theta_{0\nu}x_\lambda\right), \tag{5.22b}$$

$$S_{\nu\lambda} = \int d^3x \frac{\partial \mathcal{L}}{\partial(\partial^0\psi)} I_{\nu\lambda}\psi = \tfrac{1}{2}\int d^3x\, \psi^\dagger \sigma_{\nu\lambda}\psi. \tag{5.22c}$$

The three-dimensional vectors of orbital and spin angular momentum are

$$\boldsymbol{L} = -\mathrm{i}\int d^3x\, \psi^\dagger\, \boldsymbol{x} \times \boldsymbol{\nabla}\psi, \tag{5.23a}$$

$$\boldsymbol{S} = \tfrac{1}{2}\int d^3x\, \psi^\dagger\, \boldsymbol{\Sigma}\psi. \tag{5.23b}$$

In the standard representation of the Dirac matrices, $\boldsymbol{\Sigma}$ is the "double" Pauli matrix

$$\boldsymbol{\Sigma} = \begin{pmatrix} \boldsymbol{\sigma} & 0 \\ 0 & \boldsymbol{\sigma} \end{pmatrix}. \tag{5.24}$$

The Lagrange density (5.3) is invariant under global phase transformations $\psi \to \psi\,\mathrm{e}^{\mathrm{i}\chi}$ and $\psi^\dagger \to \psi^\dagger\,\mathrm{e}^{-\mathrm{i}\chi}$. This implies, according to (2.83), the existence of a conserved *current-density vector* $j_\mu(\boldsymbol{x},t)$:

$$j_\mu = -\mathrm{i}e\left(\frac{\partial \mathcal{L}}{\partial(\partial^\mu\psi)}\psi - \frac{\partial \mathcal{L}}{\partial(\partial^\mu\psi^\dagger)}\psi^\dagger\right) = e\bar{\psi}\gamma_\mu\psi. \tag{5.25}$$

We have included the electrical elementary charge e as a factor in the definition of j_μ since expression (5.25) is just the electrical current density of the Dirac field. The resulting conserved quantity is the *total charge*

$$Q = \int d^3x\, j_0(\boldsymbol{x},t) = e\int d^3x\, \psi^\dagger\psi. \tag{5.26}$$

The constancy of Q is seen to coincide with the conservation of the norm of the Dirac field.

Our ansatz for the Lagrangian of the Dirac fields looks very natural and so far has lead to reasonable results. The expression (5.3) has one annoying feature, however. It is easy to show that $\mathcal{L}$ is not a real number since

$$\begin{aligned}
\mathcal{L}^* \equiv \mathcal{L}^\dagger &= \left[\bar{\psi}(\mathrm{i}\gamma^\mu\overrightarrow{\partial}_\mu - m)\psi\right]^\dagger = \psi^\dagger\left(-\mathrm{i}\gamma^{\mu\dagger}\overleftarrow{\partial}_\mu - m\right)\gamma^{0\dagger}\psi \\
&= \psi^\dagger\left(-\mathrm{i}\gamma^0\gamma^\mu\gamma^0\overleftarrow{\partial}_\mu - m\right)\gamma^0\psi = \bar{\psi}\left(-\mathrm{i}\gamma^\mu\overleftarrow{\partial}_\mu - m\right)\psi,
\end{aligned} \tag{5.27}$$

where the properties of the γ matrices $\gamma^{0\dagger} = \gamma^0$ and $\gamma^0\gamma^{\mu\dagger}\gamma^0 = \gamma^\mu$ have been used. The expression (5.27) in general will not agree with (5.3), $\mathcal{L}^* \neq \mathcal{L}$. This deficiency, however, can be cured quite easily by defining the *real Lagrange density* (which will become a hermitean operator after quantization) as

$$\begin{aligned}
\mathcal{L}' := \tfrac{1}{2}(\mathcal{L} + \mathcal{L}^\dagger) &= \tfrac{1}{2}\bar{\psi}\left(\mathrm{i}\gamma^\mu\overrightarrow{\partial}_\mu - m\right)\psi + \tfrac{1}{2}\bar{\psi}\left(-\mathrm{i}\gamma^\mu\overleftarrow{\partial}_\mu - m\right)\psi \\
&= \tfrac{1}{2}\bar{\psi}\mathrm{i}\gamma^\mu\overleftrightarrow{\partial}_\mu\psi - m\bar{\psi}\psi.
\end{aligned} \tag{5.28}$$

This *symmetrized Lagrange density* essentially leads to the same equations of motion and to the same conserved quantities as the simpler ansatz (5.3). This is studied in Exercise 5.1.

EXERCISE

5.1 The Symmetrized Dirac Lagrange Density

Problem. Show that the symmetrized Lagrange density (5.28) leads to the same equations of motion and the same conservation laws as the unsymmetrized expression (5.3).

Solution. It is easy to verify that (5.28) produces the familiar Dirac equation. This can be seen even without performing the explicit calculation since $\mathcal{L}'$ and $\mathcal{L}$ are identical up to a four-divergence:

$$\mathcal{L}' = \mathcal{L} + \delta\mathcal{L} = \mathcal{L} + \frac{1}{2}(-\mathrm{i})\partial^\mu(\bar\psi\gamma_\mu\psi) \,. \tag{1}$$

The additional term $\delta\mathcal{L}$ only leads to a surface term in the action integral and thus does not show up in the Euler–Lagrange equation.

To compute the conserved quantities the canonically conjugate fields are needed:

$$\pi_\psi = \frac{\partial\mathcal{L}}{\partial\dot\psi} = \frac{\mathrm{i}}{2}\psi^\dagger \equiv \pi \quad \text{and} \quad \pi_{\psi^\dagger} = \frac{\partial\mathcal{L}}{\partial\dot\psi^\dagger} = -\frac{\mathrm{i}}{2}\psi \equiv \pi^\dagger \,. \tag{2}$$

The energy-momentum tensor then becomes

$$\begin{aligned}
\Theta'_{0\nu} &= \pi\partial_\nu\psi + \pi^\dagger\partial_\nu\psi^\dagger - g_{0\nu}\mathcal{L}' \\
&= \frac{\mathrm{i}}{2}\psi^\dagger\overset{\leftrightarrow}{\partial}_\nu \psi - g_{0\nu}\left(\frac{\mathrm{i}}{2}\bar\psi\gamma^\mu\overset{\leftrightarrow}{\partial}_\mu\psi - m\bar\psi\psi\right) \,.
\end{aligned} \tag{3}$$

The difference between $\Theta'_{0\nu}$ and the original result $\Theta_{0\nu}$ given in (5.14) is

$$\delta\Theta_{0\nu} = \Theta'_{0\nu} - \Theta_{0\nu} = -\frac{\mathrm{i}}{2}\partial_\nu(\psi^\dagger\psi) + \frac{\mathrm{i}}{2}g_{0\nu}\partial_\mu(\bar\psi\gamma^\mu\psi) \,. \tag{4}$$

This implies that there is no difference in the field energy

$$\begin{aligned}
\delta P_0 &= \int\mathrm{d}^3x\,\Theta_{00} = \int\mathrm{d}^3x\left[-\frac{\mathrm{i}}{2}\partial_0(\psi^\dagger\psi) + \frac{\mathrm{i}}{2}\partial_0(\bar\psi\gamma^0\psi) + \frac{\mathrm{i}}{2}\boldsymbol{\nabla}\cdot(\bar\psi\boldsymbol{\gamma}\psi)\right] \\
&= \frac{\mathrm{i}}{2}\int\mathrm{d}^3x\,\boldsymbol{\nabla}\cdot(\psi^\dagger\boldsymbol{\alpha}\psi) = 0 \,.
\end{aligned} \tag{5}$$

In the same way the space components ($n = 1, 2, 3$) of the momentum differ only by a vanishing surface term

$$\delta P_n = -\frac{\mathrm{i}}{2}\int\mathrm{d}^3x\,\partial_n(\psi^\dagger\psi) = 0 \,. \tag{6}$$

The generalized angular-momentum tensor becomes

$$M'_{\nu\lambda} = L'_{\nu\lambda} + S'_{\nu\lambda} \tag{7}$$

where

$$\begin{aligned}
L'_{\nu\lambda} &= \int\mathrm{d}^3x\,\left(\Theta'_{0\lambda}x_\nu - \Theta'_{0\nu}x_\lambda\right) \\
&= L_{\nu\lambda} + \int\mathrm{d}^3x\,\left(\delta_{0\lambda}x_\nu - \delta\Theta_{0\nu}x_\lambda\right) = L_{\nu\lambda} + \delta L_{\nu\lambda} \,.
\end{aligned} \tag{8}$$

Exercise 5.1

The difference in the spatial orbital angular-momentum tensor ($\nu = n, \lambda = l$) again vanishes:

$$\begin{aligned}\delta L_{nl} &= \int d^3x \left[-\frac{i}{2}\partial_l(\psi^\dagger\psi)x_n + \frac{i}{2}\partial_n(\psi^\dagger\psi)x_l\right] \\ &= \int d^3x \left[-\frac{i}{2}\partial_l(\psi^\dagger\psi x_n) + -\frac{i}{2}\psi^\dagger\psi\, g_{ln} + \frac{i}{2}\partial_n(\psi^\dagger\psi x_l) + \frac{i}{2}\psi^\dagger\psi\, g_{nl}\right] \\ &= 0 \,. \end{aligned} \quad (9)$$

The mixed space-time components, however, do not vanish:

$$\begin{aligned}\delta L_{0l} &= \int d^3x \left(\delta\Theta_{0l}x_0 - \delta\Theta_{00}x_l\right) = -\frac{i}{2}\int d^3x \left[\partial_l(\psi^\dagger\psi)x_0 + \partial_k(\bar\psi\gamma^k\psi)x_l\right] \\ &= -\frac{i}{2}\int d^3x \left[\partial_k(\bar\psi\gamma^k\psi x_l) - \bar\psi\gamma^k\psi\, g_{kl}\right] = \frac{i}{2}\int d^3x\, \bar\psi\gamma_l\psi \neq 0 \,. \end{aligned} \quad (10)$$

To compute the spin contribution $S'_{\nu\lambda}$ in (7), the hermitean conjugate of (5.18) is needed:

$$\psi'^\dagger(x') = \psi^\dagger(x) + \frac{1}{2}\Delta\omega^{\mu\nu}\psi^\dagger(x)I^\dagger_{\mu\nu} \,, \quad (11)$$

which according to (5.20) leads to

$$S'_{\nu\lambda} = \int d^3x \left(\pi I_{\nu\lambda}\psi + \psi^\dagger I^\dagger_{\nu\lambda}\pi^\dagger\right) = \frac{1}{4}\int d^3x \left(\psi^\dagger\sigma_{\nu\lambda}\psi + \psi^\dagger\sigma^\dagger_{\nu\lambda}\pi\right) \,. \quad (12)$$

The space components satisfy

$$\sigma^\dagger_{nl} = \left[\frac{i}{2}(\gamma_n\gamma_l - \gamma_l\gamma_n)\right]^\dagger = -\frac{i}{2}(\gamma^\dagger_l\gamma^\dagger_n - \gamma^\dagger_n\gamma^\dagger_l) = \sigma_{nl} \,, \quad (13)$$

which implies

$$S'_{nl} = \frac{1}{2}\int d^3x\, \psi^\dagger\sigma_{nl}\psi = S_{nl} \,. \quad (14)$$

The mixed components, on the other hand, vanish identically since

$$\sigma^\dagger_{0l} = \left[\frac{i}{2}(\gamma_0\gamma_l - \gamma_l\gamma_0)\right]^\dagger = -\sigma_{0l} \quad (15)$$

and thus

$$S'_{0l} = S_{0l} + \delta S_{0l} = 0 \,. \quad (16)$$

Therefore the differences in the orbital and spin angular momentum are found to cancel each other:

$$\delta M_{0l} = \delta L_{0l} + \delta S_{0l} = \frac{i}{2}\int d^3x\, \bar\psi\gamma_l\psi - \frac{1}{2}\int d^3x\, \psi^\dagger i\gamma_0\gamma_l\psi = 0 \,. \quad (17)$$

5.2 Canonical Quantization of the Dirac Field

Quantization of the Dirac field is achieved by replacing the spinors $\psi(\boldsymbol{x},t)$ and $\psi^\dagger(\boldsymbol{x},t)$ by field operators $\hat{\psi}(\boldsymbol{x},t)$ and $\hat{\psi}^\dagger(\boldsymbol{x},t)$. At this stage a decision has to be made whether to use commutation or anticommutation rules. Of course we already know the answer, since electrons, which are described by the Dirac equation, are found empirically to satisfy the Pauli exlusion principle. Therefore we will follow the "correct" path and use the quantization rules of Jordan and Wigner. The deeper reason for this choice will become clear later.

Using (5.11) we postulate that the Dirac field operators satisfy the *equal-time anticommutation rules*

$$\{\hat{\psi}_\alpha(\boldsymbol{x},t), \hat{\psi}_\beta^\dagger(\boldsymbol{x}',t)\} = \delta_{\alpha\beta}\,\delta^3(\boldsymbol{x}-\boldsymbol{x}')\,, \tag{5.29a}$$

$$\{\hat{\psi}_\alpha(\boldsymbol{x},t), \hat{\psi}_\beta(\boldsymbol{x}',t)\} = \{\hat{\psi}_\alpha^\dagger(\boldsymbol{x},t), \hat{\psi}_\beta^\dagger(\boldsymbol{x}',t)\} = 0\,. \tag{5.29b}$$

This choice guarantees that the particles (field quanta) are governed by Fermi–Dirac statistics. The alternate choice of Bose–Einstein statistics (minus-sign commutators) contradicts the empirical observation that spin-$\frac{1}{2}$ particles satisfy the Pauli principle. At the end of Sect. 5.3 we will see that this wrong choice of quantization would also lead to internal inconsistencies of the theory.

Heisenberg's equation of motion for the field operator $\hat{\psi}(\boldsymbol{x},t)$ reads

$$\begin{aligned}\dot{\hat{\psi}}(\boldsymbol{x},t) &= -\mathrm{i}\,[\hat{\psi}(\boldsymbol{x},t), \hat{H}] \\ &= -\int \mathrm{d}^3 x'\, \Big[\hat{\psi}(\boldsymbol{x},t)\,,\, \hat{\psi}^\dagger(\boldsymbol{x}',t)\,\boldsymbol{\alpha}\cdot\boldsymbol{\nabla}'\hat{\psi}(\boldsymbol{x}',t) \\ &\quad + \mathrm{i}m\,\hat{\psi}^\dagger(\boldsymbol{x}',t)\beta\hat{\psi}(\boldsymbol{x}',t)\Big]\,. \end{aligned} \tag{5.30}$$

The commutators required in this expression can be rewritten by using the identity

$$[\hat{A}, \hat{B}\hat{C}] = \{\hat{A},\hat{B}\}\hat{C} - \hat{B}\{\hat{A},\hat{C}\} \tag{5.31}$$

in such a way that the anticommutation rules (5.29) can be applied. To avoid confusion for the time being the spinor indices will be written out explicitly:

$$\begin{aligned}\dot{\hat{\psi}}_\sigma(\boldsymbol{x},t) &= -\int \mathrm{d}^3 x'\, \Big(\{\hat{\psi}_\sigma(\boldsymbol{x},t), \hat{\psi}_\alpha^\dagger(\boldsymbol{x}',t)\}\,\boldsymbol{\alpha}_{\alpha\beta}\cdot\boldsymbol{\nabla}'\hat{\psi}_\beta(\boldsymbol{x}',t) \\ &\quad - \hat{\psi}_\alpha^\dagger(\boldsymbol{x}',t)\,\boldsymbol{\alpha}_{\alpha\beta}\cdot\boldsymbol{\nabla}'\{\hat{\psi}_\sigma(\boldsymbol{x},t), \hat{\psi}_\beta(\boldsymbol{x}',t)\} \\ &\quad + \mathrm{i}m\{\hat{\psi}_\sigma(\boldsymbol{x},t), \hat{\psi}_\alpha^\dagger(\boldsymbol{x}',t)\}\,\beta_{\alpha\beta}\,\hat{\psi}_\beta(\boldsymbol{x}',t) \\ &\quad - \mathrm{i}m\,\hat{\psi}_\alpha^\dagger(\boldsymbol{x}',t)\,\beta_{\alpha\beta}\{\hat{\psi}_\sigma(\boldsymbol{x},t), \hat{\psi}_\beta(\boldsymbol{x}',t)\}\Big) \\ &= -\int \mathrm{d}^3 x'\, \Big(\delta_{\sigma\alpha}\delta^3(\boldsymbol{x}-\boldsymbol{x}')\,\boldsymbol{\alpha}_{\alpha\beta}\cdot\boldsymbol{\nabla}'\hat{\psi}_\beta(\boldsymbol{x}',t) \\ &\quad + \mathrm{i}m\,\delta_{\sigma\alpha}\delta^3(\boldsymbol{x}-\boldsymbol{x}')\,\beta_{\alpha\beta}\,\hat{\psi}_\beta(\boldsymbol{x}',t)\Big) \\ &= \left(-\boldsymbol{\alpha}\cdot\boldsymbol{\nabla} - \mathrm{i}m\beta\right)_{\sigma\beta}\hat{\psi}_\beta(\boldsymbol{x},t)\,. \end{aligned} \tag{5.32}$$

The quantized field operator is thus found to satisfy the free Dirac equation (5.7)

$$i\dot{\hat{\psi}} = (-i\boldsymbol{\alpha}\cdot\boldsymbol{\nabla} + \beta m)\hat{\psi}\ .\tag{5.33}$$

5.3 Plane-Wave Expansion of the Field Operator

The field operator $\hat{\psi}(\boldsymbol{x},t)$ will now be expanded in a complete set of "classical" wave functions. For this purpose the plane-wave solutions of the Dirac equation are a natural choice (for details see the volume *Relativistic Quantum Mechanics*). They are given by[1]

$$\psi_{\boldsymbol{p}}^{(r)}(\boldsymbol{x},t) = (2\pi)^{-3/2}\sqrt{\frac{m}{\omega_p}}\,w_r(\boldsymbol{p})\,\mathrm{e}^{-\mathrm{i}\epsilon_r(\omega_p t - \boldsymbol{p}\cdot\boldsymbol{x})}\ .\tag{5.34}$$

The index r enumerates the four independent solutions. $r = 1, 2$ denotes the solutions with positive energy $E = +\omega_p = +\sqrt{\boldsymbol{p}^2 + m^2}$ and $r = 3, 4$ those with negative energy $E = -\omega_p = -\sqrt{\boldsymbol{p}^2 + m^2}$. This is expressed by the sign function $\epsilon_r = +1$ for $r = 1, 2$ and $\epsilon_r = -1$ for $r = 3, 4$. The plane waves (5.34) satisfy the Dirac equation

$$\left(\mathrm{i}\gamma^\mu\partial_\mu - m\right)\psi_{\boldsymbol{p}}^{(r)}(\boldsymbol{x},t) = 0\ .\tag{5.35}$$

As a consequence the Dirac unit spinors $w_r(\boldsymbol{p})$ satisfy the algebraic equation

$$\left(\gamma^\mu p_\mu - \epsilon_r m\right)w_r(\boldsymbol{p}) = 0\ .\tag{5.36}$$

The Dirac unit spinors possess the following orthogonality and completeness properties:

$$w_{r'}^\dagger(\epsilon_{r'}\boldsymbol{p})\,w_r(\epsilon_r\boldsymbol{p}) = \frac{\omega_p}{m}\delta_{rr'}\ ,\tag{5.37a}$$

$$\overline{w}_{r'}(\boldsymbol{p})\,w_r(\boldsymbol{p}) = \epsilon_r\delta_{rr'}\ ,\tag{5.37b}$$

$$\sum_{r=1}^{4}w_{r\alpha}(\epsilon_r\boldsymbol{p})\,w_{r\beta}^\dagger(\epsilon_r\boldsymbol{p}) = \frac{\omega_p}{m}\delta_{\alpha\beta}\ ,\tag{5.37c}$$

$$\sum_{r=1}^{4}\epsilon_r\,w_{r\alpha}(\boldsymbol{p})\,\overline{w}_{r\beta}(\boldsymbol{p}) = \delta_{\alpha\beta}\ .\tag{5.37d}$$

Note the negative sign of the momentum for the solutions of the negative continuum ($r = 3, 4$) in (5.37a) and (5.37c).

Equation (5.37a) guarantees that the plane waves (5.34) have the correct "normalization to delta functions":

$$\int\mathrm{d}^3x\,\psi_{\boldsymbol{p}'}^{(r')\dagger}(x)\psi_{\boldsymbol{p}}^{(r)}(x)$$

$$= \int\mathrm{d}^3x\,\frac{1}{(2\pi)^3}\sqrt{\frac{m^2}{\omega_{p'}\omega_p}}\,\mathrm{e}^{-\mathrm{i}(\epsilon_r\omega_p - \epsilon_{r'}\omega_{p'})t}\,\mathrm{e}^{+\mathrm{i}(\epsilon_r\boldsymbol{p}-\epsilon_{r'}\boldsymbol{p}')\cdot\boldsymbol{x}}\,w_{r'}^\dagger(\boldsymbol{p}')w_r(\boldsymbol{p})$$

$$= \frac{m}{\omega_p}\,\mathrm{e}^{-\mathrm{i}(\epsilon_r\omega_p - \epsilon_{r'}\omega_{p'})t}\,\delta^3(\epsilon_r\boldsymbol{p} - \epsilon_{r'}\boldsymbol{p}')\,w_{r'}^\dagger(\boldsymbol{p}')w_r(\boldsymbol{p})$$

---

[1] As in the case of the Klein–Gordon field (cf. the footnote on page 77) one often uses an alternative normalization factor $(2\pi)^{-3}(m/\omega_p)^{-1/2}$. This introduces an additional factor of $(2\pi)^3(\omega_p/m)$ at the r.h.s. of the anticommutation rules (5.45) or (5.61).

$$= \frac{m}{\omega_p} e^{-i(\epsilon_r \omega_p - \epsilon_{r'} \omega_{p'})t} \delta^3(\mathbf{p}' - \epsilon_r \epsilon_{r'} \mathbf{p}) w_{r'}^\dagger(\epsilon_r \epsilon_{r'} \mathbf{p}) w_r(\mathbf{p})$$
$$= \delta_{rr'} \delta^3(\mathbf{p} - \mathbf{p}') \tag{5.38}$$

where $\epsilon_r^2 = +1$ was used. Equation (5.38) is valid for all time arguments t.

The plane-wave expansion of the field operator now reads

$$\hat{\psi}(\mathbf{x}, t) = \sum_{r=1}^{4} \int d^3p \, \hat{a}(\mathbf{p}, r) \psi_\mathbf{p}^{(r)}(\mathbf{x}, t)$$
$$= \sum_{r=1}^{4} \int \frac{d^3p}{(2\pi)^{3/2}} \sqrt{\frac{m}{\omega_p}} \, \hat{a}(\mathbf{p}, r) \, w_r(\mathbf{p}) \, e^{-i\epsilon_r p \cdot x} \ . \tag{5.39}$$

For the hermitean conjugate field operator we have

$$\hat{\psi}^\dagger(\mathbf{x}, t) = \sum_{r=1}^{4} \int d^3p \, \hat{a}^\dagger(\mathbf{p}, r) \psi_\mathbf{p}^{(r)\dagger}(\mathbf{x}, t)$$
$$= \sum_{r=1}^{4} \int \frac{d^3p}{(2\pi)^{3/2}} \sqrt{\frac{m}{\omega_p}} \, \hat{a}^\dagger(\mathbf{p}, r) \, \overline{w}_r(\mathbf{p}) \gamma^0 \, e^{+i\epsilon_r p \cdot x} \ . \tag{5.40}$$

Now the anticommutation rules for the creation and annihilation operators $\hat{a}(\mathbf{p}, r)$ and $\hat{a}^\dagger(\mathbf{p}, r)$ have to be derived. To find them we invert the expansion (5.39) by projecting on a plane wave, using (5.38):

$$\int d^3x \, \psi_\mathbf{p}^{(r)\dagger}(\mathbf{x}, t) \hat{\psi}(\mathbf{x}, t) = \sum_{r'=1}^{4} \int d^3p' \, \hat{a}(\mathbf{p}', r') \int d^3x \, \psi_\mathbf{p}^{(r)\dagger}(\mathbf{x}, t) \psi_{\mathbf{p}'}^{(r')}(\mathbf{x}, t)$$
$$= \sum_{r'=1}^{4} \int d^3p' \, \hat{a}(\mathbf{p}', r') \delta_{rr'} \delta^3(\mathbf{p} - \mathbf{p}')$$
$$= \hat{a}(\mathbf{p}, r) \tag{5.41}$$

or

$$\hat{a}(\mathbf{p}, r) = \int \frac{d^3x}{(2\pi)^{3/2}} \sqrt{\frac{m}{\omega_p}} \, e^{+i\epsilon_r p \cdot x} w_r^\dagger(\mathbf{p}) \, \hat{\psi}(\mathbf{x}, t) \ . \tag{5.42}$$

Similarly the hermitean conjugate operator becomes

$$\hat{a}^\dagger(\mathbf{p}, r) = \int d^3x \, \hat{\psi}^\dagger(\mathbf{x}, t) \psi_\mathbf{p}^{(r)}(\mathbf{x}, t)$$
$$= \int \frac{d^3x}{(2\pi)^{3/2}} \sqrt{\frac{m}{\omega_p}} \, e^{-i\epsilon_r p \cdot x} \, \hat{\psi}^\dagger(\mathbf{x}, t) w_r(\mathbf{p}) \ . \tag{5.43}$$

From these results the anticommutation rules can be computed, using (5.29):

$$\{\hat{a}(\mathbf{p}, r), \hat{a}^\dagger(\mathbf{p}', r')\} = \int d^3x \, d^3x' \, \psi_{\mathbf{p}\alpha}^{(r)\dagger}(\mathbf{x}, t) \psi_{\mathbf{p}'\beta}^{(r')}(\mathbf{x}', t)$$
$$\times \underbrace{\{\hat{\psi}_\alpha(\mathbf{x}, t), \hat{\psi}_\beta^\dagger(\mathbf{x}', t)\}}_{=\delta_{\alpha\beta} \delta^3(\mathbf{x} - \mathbf{x}')}$$
$$= \int d^3x \, \psi_{\mathbf{p}\alpha}^{(r)\dagger}(\mathbf{x}, t) \psi_{\mathbf{p}'\alpha}^{(r')}(\mathbf{x}, t) \ . \tag{5.44}$$

Since the plane waves are orthogonal, see (5.38), the creation and annihilation operators are seen to satisfy the *anticommutation relations*

$$\{\hat{a}(\boldsymbol{p},r), \hat{a}^\dagger(\boldsymbol{p}',r')\} = \delta^3(\boldsymbol{p}-\boldsymbol{p}')\,\delta_{rr'}\;. \tag{5.45}$$

Similarly the remaining anticommutation relations can be deduced:

$$\{\hat{a}(\boldsymbol{p},r), \hat{a}(\boldsymbol{p}',r')\} = \{\hat{a}^\dagger(\boldsymbol{p},r), \hat{a}^\dagger(\boldsymbol{p}',r')\} = 0\;. \tag{5.46}$$

This result is not surprising. The same relations were also obtained for the "expansion coefficients" $\hat{a}$, $\hat{a}^\dagger$ of the Klein–Gordon field, except for the sign of the commutator, of course.

Now the *Hamilton operator* of the quantized Dirac theory will be derived. We start from the quantized version of (5.13)

$$\hat{H} = \int d^3x\, \hat{\psi}^\dagger(\boldsymbol{x},t)\left(-i\boldsymbol{\alpha}\cdot\boldsymbol{\nabla} + \beta m\right)\hat{\psi}(\boldsymbol{x},t) \tag{5.47}$$

and insert the expansion (5.39) and (5.40) of the field operator:

$$\hat{H} = \sum_{rr'}\int d^3p'\int d^3p\, \hat{a}^\dagger(\boldsymbol{p}',r')\hat{a}(\boldsymbol{p},r)\int d^3x\, \psi_{\boldsymbol{p}'}^{(r')\dagger}(-i\boldsymbol{\alpha}\cdot\boldsymbol{\nabla}+\beta m)\psi_{\boldsymbol{p}}^{(r)}. \tag{5.48}$$

Since the plane waves obey the Dirac equation (5.35),

$$(-i\boldsymbol{\alpha}\cdot\boldsymbol{\nabla}+\beta m)\psi_{\boldsymbol{p}}^{(r)}(x) = i\partial_0 \psi_{\boldsymbol{p}}^{(r)}(x) = \epsilon_r \omega_p \psi_{\boldsymbol{p}}^{(r)}(x)\;, \tag{5.49}$$

and the orthogonality relation (5.38) can be used the Hamilton operator reads

$$\begin{aligned}\hat{H} &= \sum_{rr'}\int d^3p'\int d^3p\, \hat{a}^\dagger(\boldsymbol{p}',r')\hat{a}(\boldsymbol{p},r)\,\epsilon_r\omega_p\int d^3x\,\psi_{\boldsymbol{p}'}^{(r')\dagger}(x)\psi_{\boldsymbol{p}}^{(r)}(x)\;. \\ &= \sum_{r=1}^{4}\int d^3p\,\epsilon_r\omega_p\,\hat{a}^\dagger(\boldsymbol{p},r)\hat{a}(\boldsymbol{p},r)\;.\end{aligned} \tag{5.50}$$

Separating the contributions of positive and negative frequency this becomes

$$\hat{H} = \int d^3p\left(\sum_{r=1}^{2}\omega_p\,\hat{a}^\dagger(\boldsymbol{p},r)\hat{a}(\boldsymbol{p},r) - \sum_{r=3}^{4}\omega_p\,\hat{a}^\dagger(\boldsymbol{p},r)\hat{a}(\boldsymbol{p},r)\right)\;. \tag{5.51}$$

If one introduces the *particle-number operator* related to the state $\psi_{\boldsymbol{p}}^{(r)}$ as

$$\hat{n}_{\boldsymbol{p},r} = \hat{a}^\dagger(\boldsymbol{p},r)\hat{a}(\boldsymbol{p},r)\;, \tag{5.52}$$

the Hamiltonian (5.51) appears to be useless: as the number of particles in the "lower continuum" ($r=3,4$) grows, the expectation value of $\hat{H}$, i.e., the total energy of the system, can drop to negative values beyond any bound.

To circumvent this inacceptable conclusion *Dirac's hole picture* comes to the rescue. According to this concept in the vacuum state all the levels of the lower continuum (energy $E < -m$) are occupied by particles, as depicted in Fig. 5.1. These particles, which fill up the "Dirac sea", are always present and are distributed homogeneously (in the absence of electromagnetic fields) all over space. Therefore they cannot be observed in any experiment. Their energy and charge can be removed by a simple subtraction.

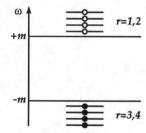

Fig. 5.1. The spectrum of the free Dirac equation. According to Dirac's hole picture all states of the lower continuum are filled

5.3 Plane-Wave Expansion of the Field Operator

It is convenient to perform this step using the box normalization for the wave functions. The integral over the momentum variable then becomes a sum over a discrete lattice of momentum values $p = p_l$ where l is a counting index. The anticommutation relations (5.45) and (5.46) assume the form

$$\{\hat{a}(\boldsymbol{p},r), \hat{a}^\dagger(\boldsymbol{p}',r')\} = \delta_{\boldsymbol{p}\boldsymbol{p}'}\,\delta_{rr'}\,, \tag{5.53a}$$

$$\{\hat{a}(\boldsymbol{p},r), \hat{a}(\boldsymbol{p}',r')\} = \{\hat{a}^\dagger(\boldsymbol{p},r), \hat{a}^\dagger(\boldsymbol{p}',r')\} = 0\,. \tag{5.53b}$$

The nonobservable *vacuum energy* is given by the sum over all energy eigenvalues in the lower continuum

$$E_0 = -\sum_{\boldsymbol{p}}\sum_{r=3}^{4} \omega_p\,, \tag{5.54}$$

which obviously is a divergent quantity. We have already had to deal with a similar problem when quantizing the Klein–Gordon field. There we encountered the zero-point energy of the field, each oscillator mode contributing an amount of $\frac{1}{2}\omega_p$ to the energy of the vacuum (cf. (4.41)). Again we resort to a subtraction procedure and define the modified Hamiltonian

$$\begin{aligned}
\hat{H}' &= \hat{H} - E_0 \\
&= \sum_{\boldsymbol{p}}\left(\sum_{r=1}^{2} \omega_p\,\hat{a}^\dagger(\boldsymbol{p},r)\hat{a}(\boldsymbol{p},r) + \sum_{r=3}^{4} \omega_p\left(1 - \hat{a}^\dagger(\boldsymbol{p},r)\hat{a}(\boldsymbol{p},r)\right)\right) \\
&= \sum_{\boldsymbol{p}}\left(\sum_{r=1}^{2} \omega_p\,\hat{a}^\dagger(\boldsymbol{p},r)\hat{a}(\boldsymbol{p},r) + \sum_{r=3}^{4} \omega_p\,\hat{a}(\boldsymbol{p},r)\hat{a}^\dagger(\boldsymbol{p},r)\right) \\
&= \sum_{\boldsymbol{p}}\left(\sum_{r=1}^{2} \omega_p\,\hat{n}_{\boldsymbol{p}r} + \sum_{r=3}^{4} \omega_p\,\hat{\bar{n}}_{\boldsymbol{p}r}\right)\,.
\end{aligned} \tag{5.55}$$

In the second term the *number operator for holes* in the state $\psi_{\boldsymbol{p}}^{(r)}$, $r = 3, 4$, was introduced. According to (5.53a) this operator satisfies

$$\hat{\bar{n}}_{\boldsymbol{p}r} = 1 - \hat{a}^\dagger(\boldsymbol{p},r)\hat{a}(\boldsymbol{p},r) = \hat{a}(\boldsymbol{p},r)\hat{a}^\dagger(\boldsymbol{p},r)\,. \tag{5.56}$$

$\hat{H}'$ is a well-defined positive-definite Hamiltonian. This mathematical construction is endowed with physical meaning if the holes are interpreted as *antiparticles*, for example as positrons. The physical *vacuum* is defined to be the state which contains neither particles nor antiparticles:

$$\hat{n}_{\boldsymbol{p}r}|0\rangle = 0 \quad \text{for} \quad r = 1, 2\,, \tag{5.57a}$$

$$\hat{\bar{n}}_{\boldsymbol{p}r}|0\rangle = 0 \quad \text{for} \quad r = 3, 4\,. \tag{5.57b}$$

Expressed in terms of the operators $\hat{a}$ and $\hat{a}^\dagger$ this implies

$$\hat{a}(\boldsymbol{p},r)|0\rangle = 0 \quad \text{for} \quad r = 1, 2\,, \tag{5.58a}$$

$$\hat{a}^\dagger(\boldsymbol{p},r)|0\rangle = 0 \quad \text{for} \quad r = 3, 4\,. \tag{5.58b}$$

Thus $\hat{a}(\boldsymbol{p},r)$ will be called the *annihilation operator for particles* ($r = 1, 2$) and $\hat{a}^\dagger(\boldsymbol{p},r)$ is the *annihilation operator for antiparticles* ($r = 3, 4$). Similarly $\hat{a}^\dagger(\boldsymbol{p},r)$, $r = 1, 2$, and $\hat{a}(\boldsymbol{p},r)$, $r = 3, 4$, are to be interpreted as *creation operators* for particles or antiparticles, respectively. The double role played by the operators $\hat{a}$ and $\hat{a}^\dagger$ is a bit confusing. Therefore it is customary to introduce separate notations for the particle and the hole operators. We take this opportunity also to change the notation for the wave functions. The names $u(p,s)$ and $v(p,s)$ will be introduced for the unit Dirac spinors of the upper and lower continuum. They are related to the spinors $w_r(\bm{p})$ as follows:[2]

$$\begin{aligned}
w_1(\bm{p}) &= u(p,+s) , \\
w_2(\bm{p}) &= u(p,-s) , \\
w_3(\bm{p}) &= v(p,-s) , \\
w_4(\bm{p}) &= v(p,+s) .
\end{aligned} \qquad (5.59)$$

Note the inverted sign of the spin in the case of the antiparticle solutions. Fitting to the spinors (5.59), the following new operators are introduced:

$$\begin{aligned}
\hat{b}(p,+s) &= \hat{a}(p,1) , \\
\hat{b}(p,-s) &= \hat{a}(p,2) , \\
\hat{d}^\dagger(p,-s) &= \hat{a}(p,3) , \\
\hat{d}^\dagger(p,+s) &= \hat{a}(p,4) .
\end{aligned} \qquad (5.60)$$

The solutions of the upper continuum are thus simply renamed $\hat{a} \to \hat{b}$. For the antiparticle solutions the direction of the spin is inverted, as suggested by the definition of the spionors (5.59) and at the same time the operator is replaced by the hermitean conjugate: $\hat{a} \to \hat{d}^\dagger$.

It is interesting to note that the transformation (5.60) does not change the form of the anticommutation relations (i.e., it is a *canonical transformation*) since $\hat{a}$ and $\hat{a}^\dagger$ are treated on an equal footing. Expressed in terms of the new operators, the anticommutation relations read

$$\{\hat{b}(p,s),\hat{b}^\dagger(p',s')\} = \delta^3(\bm{p}-\bm{p}')\delta_{ss'} , \qquad (5.61a)$$

$$\{\hat{d}(p,s),\hat{d}^\dagger(p',s')\} = \delta^3(\bm{p}-\bm{p}')\delta_{ss'} , \qquad (5.61b)$$

whereas the anticommutators involving the remaining eight combinations of operators $\hat{b},\hat{b}^\dagger,\hat{d},\hat{d}^\dagger$ vanish.

In the new language the field operator is expanded as (reverting to continuum normalization again)

$$\hat{\psi}(\bm{x},t) = \sum_s \int \frac{\mathrm{d}^3 p}{(2\pi)^{3/2}} \sqrt{\frac{m}{\omega_p}} \left(\hat{b}(p,s)u(p,s)\,\mathrm{e}^{-\mathrm{i}p\cdot x} + \hat{d}^\dagger(p,s)v(p,s)\,\mathrm{e}^{+\mathrm{i}p\cdot x} \right) . \qquad (5.62)$$

For practical calculations the following explicit form of the unit spinor can be useful:

$$u(p,s) = \frac{\not{p}+m}{\sqrt{2m(\omega_p+m)}} u(0,s) , \qquad (5.63a)$$

---

[2] s denotes the covariantly generalized spin vector. The identification (5.59) assumes that the spin is oriented in the direction of the z axis, $s = u_z = (0,0,0,1)$, which usually is taken as the quantization axis for the solutions $w_r(\bm{p})$. The use of $u(p,s)$ and $v(p,s)$ allows for an arbitrary orientation of the spin, however. See also Exercise 5.4.

$$v(p,s) = \frac{-\slashed{p}+m}{\sqrt{2m(\omega_p+m)}}\, v(0,s)\,, \tag{5.63b}$$

using the "slash notation" $\slashed{p} = p^\mu \gamma_\mu$. $u(0,s)$ and $v(0,s)$ are the unit spinors in the particle rest frame with $p=(m,\mathbf{0})$ which have only upper or only lower components. Equation (5.63) can be derived using the Lorentz boost operator (cf. Chap. 3 in *Relativistic Quantum Mechanics*). The unit spinors satisfy the following versions of the free Dirac equation:

$$(\slashed{p}-m)\,u(p,s) = 0\,, \quad (\slashed{p}+m)\,v(p,s) = 0\,, \tag{5.64a}$$

and also

$$\bar{u}(p,s)(\slashed{p}-m) = 0\,, \quad \bar{v}(\slashed{p}+m) = 0\,. \tag{5.64b}$$

The *Hamiltonian* is found to read

$$\hat{H}' = \hat{H} - E_0 = \sum_s \int d^3 p\, \omega_p \left(\hat{b}^\dagger(p,s)\hat{b}(p,s) + \hat{d}^\dagger(p,s)\hat{d}(p,s) \right). \tag{5.65}$$

Equations (5.65) and (5.61) imply the following interpretation of the new operators

$\hat{b}^\dagger$: creator of a particle,
$\hat{b}$: annihilator of a particle,
$\hat{d}^\dagger$: creator of an antiparticle,
$\hat{d}$: annihilator of an antiparticle.

Using these operators the Fock space can be constructed, starting from the *vacuum* state $|0\rangle$ defined by (cf. (5.58))

$$\hat{b}(p,s)|0\rangle = 0\,, \quad \hat{d}(p,s)|0\rangle = 0\,. \tag{5.66}$$

Repeated application of the creation operators $\hat{b}^\dagger(p,s)$ and $\hat{d}^\dagger(p,s)$ to the vacuum $|0\rangle$ allows the construction of states containing an arbitrary number of particles and antiparticles. According to Sect. 3.3 these states automatically respect the Pauli exclusion principle.

An important lesson can be learned from the above procedure: *it is not admissible to quantize the Dirac field using Bose–Einstein statistics*. We could have attempted to use minus commutators instead of (5.29). This would have worked quite well, eventually leading to the same form of the Hamiltonian

$$\hat{H} = \sum_s \int d^3 p\, \omega_p \left(\hat{b}^\dagger(p,s)\hat{b}(p,s) - \hat{d}(p,s)\hat{d}^\dagger(p,s) \right). \tag{5.67}$$

At this point, however, we are stuck. Using bosonic quantization the operators $\hat{d}, \hat{d}^\dagger$ satisfy the commutation relations

$$[\hat{d}^\dagger(p,s), \hat{d}(p',s')] = \delta^3(\mathbf{p}-\mathbf{p}')\delta_{ss'}\,. \tag{5.68}$$

Therefore reordering of the operators can *not* be invoked to invert the sign of the second term in (5.67):

$$\hat{H}' = \sum_s \int d^3 p\, \omega_p \left(\hat{b}^\dagger(p,s)\hat{b}(p,s) - \hat{d}^\dagger(p,s)\hat{d}(p,s) \right) + \text{const}\,. \tag{5.69}$$

The Hamiltonian of the Dirac field thus cannot be made a positive-definite operator if the "wrong" (i.e., bosonic) quantization prescription is employed.

By adding more and more antiparticles the system could attain an arbitrarily large negative energy which obviously is unphysical and would lead to a collapse (as soon as interactions are taken into account). Dirac's concept of a fully occupied lower continuum (which finds its mathematical expression in (5.60), (5.66)) obviously can only work for particles that satisfy the Pauli principle.

The condition of positive energies thus has been used to derive another special case of the *spin-statistics theorem* (spin = $\frac{1}{2}$ → Fermi–Dirac statistics). The corresponding argument for the Klein–Gordon field (spin = 0 → Bose–Einstein statistics) is based on the principle of microcausality. As shown in Exercise 5.5 the latter reasoning does not work for the Dirac field.

For completeness let us check the quantum numbers of the many-body states that make up the Fock space of the Dirac field. The *charge operator* according to (5.26) reads

$$\hat{Q} = e \int d^3x \, \hat{\psi}^\dagger(x)\hat{\psi}(x) \,. \tag{5.70}$$

Insertion of the expansion (5.62) and the corresponding expression for the hermitean conjugate field operator leads to

$$\begin{aligned}
\hat{Q} &= e \int d^3x \sum_s \sum_{s'} \int \frac{d^3p'}{(2\pi)^{3/2}} \int \frac{d^3p}{(2\pi)^{3/2}} \sqrt{\frac{m^2}{\omega_p \omega_{p'}}} \\
&\quad \times \left(\hat{b}^\dagger(p',s') u^\dagger(p',s') e^{ip'\cdot x} + \hat{d}(p',s') v^\dagger(p',s') e^{-ip'\cdot x} \right) \\
&\quad \times \left(\hat{b}(p,s) u(p,s) e^{-ip\cdot x} + \hat{d}^\dagger(p,s) v(p,s) e^{+ip\cdot x} \right) \\
&= e \sum_{s,s'} \int d^3p \, \frac{m}{\omega_p} \Big(\hat{b}^\dagger(p,s')\hat{b}(p,s) u^\dagger(p,s')u(p,s) \\
&\quad + \hat{d}(p,s')\hat{d}^\dagger(p,s) v^\dagger(p,s')v(p,s) \\
&\quad + \hat{b}^\dagger(-p,s')\hat{d}^\dagger(p,s) u^\dagger(-p,s')v(p,s) e^{2i\omega_p t} \\
&\quad + \hat{d}(-p,s')\hat{b}(p,s) v^\dagger(-p,s')u(p,s) e^{-2i\omega_p t} \Big) \,.
\end{aligned} \tag{5.71}$$

The spinors u and v satisfy the following orthogonality relation (cf. (5.37a))

$$u^\dagger(p,s')u(p,s) = v^\dagger(p,s')v(p,s) = \frac{\omega_p}{m} \delta_{ss'} \,, \tag{5.72a}$$

$$u^\dagger(-p,s')v(p,s) = v^\dagger(-p,s')u(p,s) = 0 \,. \tag{5.72b}$$

This can be used to simplify the *charge operator* as follows

$$\hat{Q} = e \sum_e \int d^3p \left(\hat{b}^\dagger(p,s)\hat{b}(p,s) + \hat{d}(p,s)\hat{d}^\dagger(p,s) \right) \,. \tag{5.73}$$

In order to express $\hat{Q}$ in terms of the number operators $\hat{n}(p,s)$ the operators $\hat{d}(p,s)$ and $\hat{d}^\dagger(p,s)$ have to be interchanged in the second term. The anticommutator (5.61b) obviously introduces the charge of the particles in the Dirac sea. The total charge of the Dirac vacuum is a constant (although divergent) number Q_0 which is not observable. The physical charge operator then is obtained by subtracting this vacuum charge from (5.73):

$$\hat{Q}' = \hat{Q} - Q_0 = e \sum_s \int d^3p \left(\hat{b}^\dagger(p,s)\hat{b}(p,s) - \hat{d}^\dagger(p,s)\hat{d}(p,s) \right). \quad (5.74)$$

As expected the particles and antiparticles carry opposite charges $+e$ and $-e$.

The *momentum operator* of the quantum field consists of contributions from particles and antiparticles with equal sign. According to (5.17) the operator reads

$$\hat{\boldsymbol{P}} = -i \int d^3x\, \hat{\psi}^\dagger(x) \boldsymbol{\nabla} \hat{\psi}(x). \quad (5.75)$$

As demonstrated in Exercise 5.3 the plane-wave expansion of the field operator leads to

$$\begin{aligned}\hat{\boldsymbol{P}} &= \sum_s \int d^3p\, \boldsymbol{p} \left(\hat{b}^\dagger(p,s)\hat{b}(p,s) - \hat{d}(p,s)\hat{d}^\dagger(p,s) \right) \\ &= \sum_s \int d^3p\, \boldsymbol{p} \left(\hat{b}^\dagger(p,s)\hat{b}(p,s) + \hat{d}^\dagger(p,s)\hat{d}(p,s) \right),\end{aligned} \quad (5.76)$$

where the minus sign results from differentiating the plane wave $\exp(\pm i p \cdot x)$. The change of operator ordering in the last line makes no contribution since the integral $\int d^3p\, \boldsymbol{p}$ has to vanish from symmetry arguments. In contrast to the case of energy and charge the isotropy of space guarantees that there can be no "momentum of the vacuum" $\boldsymbol{P}_0$ which would have to be subtracted.

Finally, in analogy to (5.65), (5.74), and (5.76) also the *angular-momentum operator* can be expanded with respect to particle number. As in the case of the Klein–Gordon theory (cf. (4.53)) the Dirac plane waves do not carry well-defined angular momentum, which is reflected in the nondiagonal form of the angular-momentum operator. However, a simple result is obtained for *helicity states* where the quantisation axis is directed along the momentum vector. In this case the *spin operator* becomes (see Exercise 5.4)

$$\begin{aligned}\hat{S}_p &= \frac{1}{2}\Big(\hat{b}^\dagger(p,+s)\hat{b}(p,+s) - \hat{b}^\dagger(p,-s)\hat{b}(p,-s) \\ &\quad + \hat{d}^\dagger(p,+s)\hat{d}(p,+s) - \hat{d}^\dagger(p,-s)\hat{d}(p,-s) \Big).\end{aligned} \quad (5.77)$$

This very nicely splits into contributions from particles and antiparticles, each carrying spin projections $+\frac{1}{2}$ and $-\frac{1}{2}$. This result serves to justify the redefinition of the spin direction for solutions with negative energy, which was carried out in (5.60): a missing particle in the lower continuum having spin up corresponds to an antiparticle with spin down and vice versa.

When deriving the operators for the energy $\hat{H}$ and charge $\hat{Q}$ of the Dirac field we have encountered the ubiquitous problem of nonvanishing (and, indeed, divergent) vacuum contributions. Arguing that these quantities are not observable we introduced "physical" operators by subtracting the energy and charge of the vacuum, E_0 and Q_0, which were constant numbers. From a mathematical point of view this is a rather objectionable procedure since it involves the handling of divergent expressions. The same effect can be achieved in a formally more appealing way by postulating the prescription of *normal ordering*. As in the case of the scalar field (cf. (4.15)) the field operator gets split into two parts involving positive and negative frequencies:

$$\hat{\psi}(\boldsymbol{x},t) = \hat{\psi}^{(+)}(\boldsymbol{x},t) + \hat{\psi}^{(-)}(\boldsymbol{x},t) \ . \tag{5.78}$$

The explicit form of these contributions, assuming an expansion into plane waves, is given in (5.62). An important property is the action on the vacuum state according to (5.66):

$$\hat{\psi}^{(+)}(\boldsymbol{x},t)|0\rangle = 0 \quad , \quad \hat{\bar{\psi}}^{(-)\dagger}(\boldsymbol{x},t)|0\rangle = 0 \ . \tag{5.79}$$

Normal ordering now means that all contributions involving positive frequency (i.e., annihilation operators) are moved to the right. Because of the anticommuting character of the Dirac field each reordering step goes along with a change of sign. For example the normal product of $\hat{\bar{\psi}}_\alpha$ and $\hat{\psi}_\beta$ reads

$$:\hat{\bar{\psi}}_\alpha \hat{\psi}_\beta: = \hat{\bar{\psi}}_\alpha^{(+)} \hat{\psi}_\beta^{(+)} + \hat{\bar{\psi}}_\alpha^{(-)} \hat{\psi}_\beta^{(-)} + \hat{\bar{\psi}}_\alpha^{(-)} \hat{\psi}_\beta^{(+)} - \hat{\psi}_\beta^{(-)} \hat{\bar{\psi}}_\alpha^{(+)} \ . \tag{5.80}$$

If the products of field operators are taken according to the prescription of normal ordering all vacuum contributions vanish because of the property (5.79). The "physical" operators for energy and charge then read

$$\hat{H}' = \int d^3x :\hat{\psi}^\dagger(-i\boldsymbol{\alpha}\cdot\boldsymbol{\nabla} + \beta m)\hat{\psi}: \ , \tag{5.81}$$

$$\hat{Q}' = e \int d^3x :\hat{\psi}^\dagger \hat{\psi}: \ , \tag{5.82}$$

which agrees with (5.65) and (5.74).

5.4 The Feynman Propagator for Dirac Fields

In Sect. 4.5 we have introduced the Feynman propagator Δ_F of the Klein–Gordon field. An analogous construction is possible also for spin-$\frac{1}{2}$ fields. The Feynman propagator will be defined as the vacuum expectation value of the time-ordered product of field operators taken at different points in space-time:

$$\mathrm{i} S_{\mathrm{F}\alpha\beta}(x-y) = \langle 0|T(\hat{\psi}_\alpha(x)\hat{\bar{\psi}}_\beta(y))|0\rangle \ . \tag{5.83}$$

The minus sign accounts for the fermionic (anticommuting) character of the field operator[3]

$$T(\hat{\psi}_\alpha(x)\hat{\bar{\psi}}_\beta(y)) = \begin{cases} \hat{\psi}_\alpha(x)\hat{\bar{\psi}}_\beta(y) & \text{for } x_0 > y_0 \\ -\hat{\bar{\psi}}_\beta(y)\hat{\psi}_\alpha(x) & \text{for } y_0 > x_0 \end{cases} \ . \tag{5.84}$$

Thus (5.83) becomes

---

[3] If we left out this minus sign the time ordered product would not have a Lorentz-invariant meaning. If the points x and y are separated by a space-like distance then the time ordering of x_0 and y_0 can be reversed by going to a different inertial frame. The operators $\hat{\psi}_\alpha(x)$ and $\hat{\bar{\psi}}_\beta(y)$ are known to anticommute for space-like separations (cf. Exercise 5.5). Thus, by including a minus sign in its definition (5.84), the value of the time-ordered product is not affected by such a Lorentz transformation.

$$iS_{F\alpha\beta}(x-y) = \Theta(x_0-y_0)\langle 0|\hat{\psi}_\alpha(x)\hat{\bar{\psi}}_\beta(y)|0\rangle$$
$$-\Theta(y_0-x_0)\langle 0|\hat{\bar{\psi}}_\beta(y)\hat{\psi}_\alpha(x)|0\rangle. \qquad (5.85)$$

Insertion of the Fourier decomposition (5.62) of the Dirac field operator $\hat{\psi}(\boldsymbol{x},t)$ gives

$$iS_{F\alpha\beta}(x-y) = \int \frac{\mathrm{d}^3 p}{(2\pi)^3}\frac{m}{\omega_p}\bigg(\Theta(x_0-y_0)\,\mathrm{e}^{-\mathrm{i}p\cdot(x-y)}\sum_s u_\alpha(p,s)\bar{u}_\beta(p,s)$$
$$-\Theta(y_0-x_0)\,\mathrm{e}^{+\mathrm{i}p\cdot(x-y)}\sum_s v_\alpha(p,s)\bar{v}_\beta(p,s)\bigg), \qquad (5.86)$$

where the anticommutation relations (5.61) and the definition of the vacuum state (5.66) were used. The spin summations lead to the *projection operator* (see *Relativistic Quantum Mechanics* Chap. 7.)

$$\sum_s u_\alpha(p,s)\bar{u}_\beta(p,s) = \Big(\frac{\slashed{p}+m}{2m}\Big)_{\alpha\beta}, \quad \sum_s v_\alpha(p,s)\bar{v}_\beta(p,s) = \Big(\frac{\slashed{p}-m}{2m}\Big)_{\alpha\beta}, \qquad (5.87)$$

which implies

$$iS_{F\alpha\beta}(x-y) = \int \frac{\mathrm{d}^3 p}{(2\pi)^3}\frac{1}{2\omega_p}\bigg(\Theta(x_0-y_0)\,\mathrm{e}^{-\mathrm{i}p\cdot(x-y)}(\slashed{p}+m)_{\alpha\beta}$$
$$-\Theta(y_0-x_0)\,\mathrm{e}^{+\mathrm{i}p\cdot(x-y)}(\slashed{p}-m)_{\alpha\beta}\bigg)$$
$$= \Theta(x_0-y_0)(\mathrm{i}\slashed{\nabla}+m)_{\alpha\beta}\Delta^{(+)}(x-y)$$
$$-\Theta(y_0-x_0)(\mathrm{i}\slashed{\nabla}+m)_{\alpha\beta}\Delta^{(-)}(x-y). \qquad (5.88)$$

The functions $\Delta^{(+)}$ and $\Delta^{(-)}$ have been introduced in Sect. 4.5. Obviously the spin-$\frac{1}{2}$ propagator $S_{F\alpha\beta}$ can be obtained by applying the Dirac differential operator to the scalar Feynman propagator Δ_F, (4.133):

$$S_{F\alpha\beta}(x-y) = (\mathrm{i}\slashed{\nabla}+m)_{\alpha\beta}\Big[\Theta(x_0-y_0)\Delta^{(+)}(x-y)$$
$$-\Theta(y_0-x_0)\Delta^{(-)}(x-y)\Big]$$
$$= (\mathrm{i}\slashed{\nabla}+m)_{\alpha\beta}\Delta_F(x-y). \qquad (5.89)$$

Remark: It might appear that this argument is in error since the time derivative contained in the differential operator also acts on the step functions in (5.89). The two additional terms that arise in this way, however, are found to cancel each other. They are given by

$$\mathrm{i}\gamma^0\,\delta(x_0-y_0)\,\Delta^{(+)}(x-y) - \mathrm{i}\gamma^0\,(-\delta(x_0-y_0))\,\Delta^{(-)}(x-y)$$
$$= \mathrm{i}\gamma^0\,\delta(x_0-y_0)\,\Big(\Delta^{(+)}(x-y)+\Delta^{(-)}(x-y)\Big)$$
$$= \mathrm{i}\gamma^0\,\delta(x_0-y_0)\,\Delta(x-y) = 0. \qquad (5.90)$$

This is true since (4.104) tells us that the Pauli–Jordan function $\Delta(x-y)$ vanishes at $x_0-y_0=0$.

The four-dimensional Fourier representation of (5.89) reads

$$S_{F\alpha\beta}(x-y) = \int \frac{d^4p}{(2\pi)^4}\, e^{-ip\cdot(x-y)} \frac{\slashed{p}+m}{p^2-m^2+i\epsilon}\,, \tag{5.91}$$

therefore the *Feynman propagator in momentum space* has the compact form

$$S_{F\alpha\beta}(p) = \frac{1}{\slashed{p}-m+i\epsilon}\,. \tag{5.92}$$

EXERCISE

5.2 The Symmetrized Current Operator

Problem. Show that the normal-ordered current operator of the Dirac field also can be obtained by symmetrizing the product of field operators according to

$$\hat{j}'_\mu = e : \hat{\bar{\psi}} \gamma_\mu \hat{\psi} : = \frac{e}{2}\left[\hat{\bar{\psi}}, \gamma_\mu \hat{\psi}\right]. \tag{1}$$

Solution. We insert the plane-wave expansion (5.62) of the field operator $\hat{\psi}(\boldsymbol{x},t)$ into (1):

$$\frac{e}{2}\left[\hat{\bar{\psi}}, \gamma_\mu \hat{\psi}\right] = \frac{e}{2}(\gamma_\mu)_{\alpha\beta}\left(\hat{\bar{\psi}}_\alpha \hat{\psi}_\beta - \hat{\psi}_\beta \hat{\bar{\psi}}_\alpha\right)$$

$$= \frac{e}{2}\sum_{s,s'}\int \frac{d^3p'}{(2\pi)^{3/2}}\int \frac{d^3p}{(2\pi)^{3/2}}\sqrt{\frac{m^2}{\omega_p\omega_{p'}}}$$

$$\times \Big(\bar{u}'\gamma_\mu u(\hat{b}^{\dagger\prime}\hat{b} - \hat{b}\hat{b}^{\dagger\prime})\, e^{i(p'-p)\cdot x} + \bar{v}'\gamma_\mu v(\hat{d}'\hat{d}^\dagger - \hat{d}^\dagger \hat{d}')\, e^{-i(p'-p)\cdot x}$$

$$+ \bar{u}'\gamma_\mu v(\hat{b}^{\dagger\prime}\hat{d}^\dagger - \hat{d}^\dagger \hat{b}^{\dagger\prime})\, e^{i(p'+p)\cdot x} + \bar{v}'\gamma_\mu u(\hat{d}'\hat{b} - \hat{b}\hat{d}')\, e^{-i(p'+p)\cdot x}\Big). \tag{2}$$

For brevity the spin and momentum arguments have been omitted, $\hat{b} \equiv \hat{b}(p,s)$, $\hat{b}' \equiv \hat{b}(p',s')$, etc. Using the canonical anticommutation relations (5.45) and (5.46) Eq. (2) becomes

$$\frac{e}{2}\left[\hat{\bar{\psi}}, \gamma_\mu \hat{\psi}\right] = e\sum_{s,s'}\int \frac{d^3p'}{(2\pi)^{3/2}}\int \frac{d^3p}{(2\pi)^{3/2}}\sqrt{\frac{m^2}{\omega_p\omega_{p'}}}$$

$$\times \Big(\bar{u}'\gamma_\mu u\, \hat{b}^{\dagger\prime}\hat{b}\, e^{i(p'-p)\cdot x} - \bar{v}'\gamma_\mu v\, \hat{d}^\dagger \hat{d}'\, e^{-i(p'-p)\cdot x} +$$

$$+ \bar{u}'\gamma_\mu v\, \hat{b}^{\dagger\prime}\hat{d}^\dagger\, e^{i(p'+p)\cdot x} + \bar{v}'\gamma_\mu u\, \hat{d}'\hat{b}\, e^{-i(p'+p)\cdot x} + R_\mu\Big). \tag{3}$$

Here the creation operators stand to the left of the annihilation operators, i.e., the first four terms in (3) constitute the plane-wave expansion of the normal-ordered current operator $e : \hat{\bar{\psi}}\gamma_\mu\hat{\psi} :$. The remainder

$$R_\mu = -\bar{u}\gamma_\mu u\, \delta^3(\boldsymbol{p}'-\boldsymbol{p})\delta_{ss'} + \bar{v}\gamma_\mu v\, \delta^3(\boldsymbol{p}'-\boldsymbol{p})\delta_{ss'} \tag{4}$$

can be shown to vanish if the spin summation is carried out as required in (3). Using the completeless relation (5.37d) of the unit spinors u and v we find

$$-e\sum_s\int\frac{\mathrm{d}^3p}{(2\pi)^3}\frac{m}{\omega_p}\Big(\bar{u}\gamma_\mu u-\bar{v}\gamma_\mu v\Big)$$

Exercise 5.2

$$=-e\int\frac{\mathrm{d}^3p}{(2\pi)^3}\frac{m}{\omega_p}(\gamma_\mu)_{\alpha\beta}\underbrace{\sum_s\Big(\bar{u}_\alpha(p,s)u_\beta(p,s)-\bar{v}_\alpha(p,s)v_\beta(p,s)\Big)}_{=\delta_{\alpha\beta}}$$

$$=-e\int\frac{\mathrm{d}^3p}{(2\pi)^3}\frac{m}{\omega_p}\operatorname{Sp}\gamma_\mu=0\,,\tag{5}$$

since the gamma matrices are traceless. This proves that the two forms of the current operator are equivalent as claimed in (1).

EXERCISE ▮▮▮▮▮▮▮▮▮▮▮▮▮▮▮▮▮▮▮▮▮▮▮▮

5.3 The Momentum Operator

Problem. Derive the momentum operator $\hat{\boldsymbol{P}}$ of the Dirac field, expressed in terms of plane-wave creation and annihilation operators.

Solution. Insertion of the plane-wave expansion (5.62) into the expression for the momentum operator $\hat{\boldsymbol{P}}$ yields

$$\begin{aligned}\hat{\boldsymbol{P}}&=-\mathrm{i}\int\mathrm{d}^3x\,\hat{\psi}^\dagger(x)\boldsymbol{\nabla}\hat{\psi}(x)\\&=-\mathrm{i}\int\mathrm{d}^3x\int\frac{\mathrm{d}^3p}{(2\pi)^{3/2}}\int\frac{\mathrm{d}^3p'}{(2\pi)^{3/2}}\sum_{s,s'}\sqrt{\frac{m}{\omega_p}}\sqrt{\frac{m}{\omega_{p'}}}\\&\quad\times\Big(\hat{b}^\dagger(p,s)u^\dagger(p,s)\mathrm{e}^{\mathrm{i}p\cdot x}+\hat{d}(p,s)v^\dagger(p,s)\mathrm{e}^{-\mathrm{i}p\cdot x}\Big)\\&\quad\times\mathrm{i}\boldsymbol{p}'\Big(\hat{b}(p',s')u(p',s')\mathrm{e}^{-\mathrm{i}p'\cdot x}-\hat{d}^\dagger(p',s')v(p',s')\mathrm{e}^{\mathrm{i}p'\cdot x}\Big)\,.\end{aligned}\tag{1}$$

After carrying out the multiplication this becomes

$$\begin{aligned}\hat{\boldsymbol{P}}=\int\mathrm{d}^3x\int\frac{\mathrm{d}^3p}{(2\pi)^{3/2}}\int\frac{\mathrm{d}^3p'}{(2\pi)^{3/2}}\sum_s\sum_{s'}\frac{m}{\sqrt{\omega_p\omega_{p'}}}\boldsymbol{p}'\\\times\Big(\hat{b}^\dagger(p,s)\hat{b}(p',s')u^\dagger(p,s)\,u(p',s')\,\mathrm{e}^{\mathrm{i}(p-p')\cdot x}\\-\hat{b}^\dagger(p,s)\hat{d}^\dagger(p',s')u^\dagger(p,s)\,v(p',s')\,\mathrm{e}^{\mathrm{i}(p+p')\cdot x}\\+\hat{d}(p,s)\hat{b}(p',s')v^\dagger(p,s)\,u(p',s')\,\mathrm{e}^{-\mathrm{i}(p+p')\cdot x}\\-\hat{d}(p,s)\hat{d}^\dagger(p',s')v^\dagger(p,s)\,v(p',s')\,\mathrm{e}^{-\mathrm{i}(p-p')\cdot x}\Big)\,.\end{aligned}\tag{2}$$

The integration over d^3x leads to delta functions that cancel the d^3p' integration and we are left with

$$\begin{aligned}\hat{\boldsymbol{P}}=\int\mathrm{d}^3p\sum_s\sum_{s'}\frac{m}{\omega_p}\boldsymbol{p}\Big(&\hat{b}^\dagger(p,s)\hat{b}(p,s')\,u^\dagger(p,s)u(p,s')\\&+\hat{b}^\dagger(p,s)\hat{d}^\dagger(-p,s')\,u^\dagger(p,s)v(-p,s')\,\mathrm{e}^{2\mathrm{i}\omega_pt}\\&-\hat{d}(p,s)\hat{b}(-p,s')\,v^\dagger(p,s)u(-p,s')\,\mathrm{e}^{-2\mathrm{i}\omega_pt}\end{aligned}$$

Exercise 5.3

$$-\hat{d}(p,s)\hat{d}^\dagger(p,s')\,v^\dagger(p,s)v(p,s')\Big)\,. \tag{3}$$

Use of the orthogonality relations (5.72) for the unit spinors u and v gives

$$\begin{aligned}\hat{P} &= \int d^3p \sum_s \sum_{s'} \frac{m}{\omega_p} \boldsymbol{p}\left(\hat{b}^\dagger(p,s)\hat{b}(p,s') - \hat{d}(p,s)\hat{d}^\dagger(p,s')\right)\frac{\omega_p}{m}\delta_{ss'} \\ &= \int d^3p \sum_s \boldsymbol{p}\left(\hat{b}^\dagger(p,s)\hat{b}(p,s) + \hat{d}^\dagger(p,s)\hat{d}(p,s)\right)\,.\end{aligned} \tag{4}$$

The commutator term does not contribute since it has the form $\int d^3p\,\boldsymbol{p}$ which has to vanish owing to symmetry reasons (there is no preferred direction in space).

EXERCISE

5.4 Helicity States

Problem. Derive the operator for the projection of the spin of a particle projected onto the direction of motion. Use an expansion of the field operator with respect to helicity states (see Chap. 7 in *Relativistic Quantum Mechanics* for more details on helicity).

Solution. The general construction of the unit spinors $u(p,s)$ and $v(p,s)$ used in the expansion (5.62) makes them eigenfunctions of the operator $\gamma_5 \slashed{s}$:

$$\gamma_5\slashed{s}\,u(p,\pm s) = \pm u(p,\pm s)\,, \tag{1a}$$
$$\gamma_5\slashed{s}\,v(p,\pm s) = \pm v(p,\pm s)\,, \tag{1b}$$

Here s is an arbitrary space-like unit vector, $s^2 = -1$ being orthogonal to the four-momentum vector p, i.e., $p\cdot s = 0$. This vector has an immediate physical interpretation in the *rest frame* of the particle. In this frame $s = (0, \boldsymbol{s}')$ only has space components and $\boldsymbol{s}'$ defines the direction of spin. In general ($\boldsymbol{p}\neq 0$), the spinors (1) are not eigenfunctions of the spin operator $\boldsymbol{\Sigma} = \gamma_5\gamma_0\boldsymbol{\gamma}$. *Helicity eigenstates* are an important special case. They are defined such that the spin vector in the rest frame points into the direction of the momentum, $\boldsymbol{s}' = \boldsymbol{p}/|\boldsymbol{p}|$. Application of a Lorentz boost that transforms into the system where the particle has momentum $\boldsymbol{p}$ turns $\boldsymbol{s}'$ into a four-vector. The components of this spin vector are found to be

$$s = \left(\frac{|\boldsymbol{p}|}{m},\,\frac{E}{m}\frac{\boldsymbol{p}}{|\boldsymbol{p}|}\right)\,. \tag{2}$$

This helicity vector satisfies the useful relation

$$\gamma_5\slashed{s} = \boldsymbol{\Sigma}\cdot\frac{\boldsymbol{p}}{|\boldsymbol{p}|}\frac{\slashed{p}}{m}\,, \tag{3}$$

which is easily checked by using $(\boldsymbol{\gamma}\cdot\boldsymbol{p})^2 = -\boldsymbol{p}^2$:

$$\boldsymbol{\Sigma}\cdot\frac{\boldsymbol{p}}{|\boldsymbol{p}|}\frac{\slashed{p}}{m} = \gamma_5\gamma_0\boldsymbol{\gamma}\cdot\frac{\boldsymbol{p}}{|\boldsymbol{p}|}\frac{1}{m}(E\gamma_0 - \boldsymbol{p}\cdot\boldsymbol{\gamma})$$

$$= -\gamma_5(\gamma_0)^2 \gamma \cdot \frac{p}{|p|} \frac{E}{m} - \gamma_5 \gamma_0 \frac{1}{|p|m} (\gamma \cdot p)^2$$

$$= \gamma_5 \left(\gamma_0 \frac{|p|}{m} - \gamma \cdot \frac{E}{m} \frac{p}{|p|} \right) = \gamma_5 (\gamma_0 s_0 - \gamma \cdot s)$$

$$= \gamma_5 \slashed{s} \,. \tag{4}$$

The unit spinors u and v are eigenstates of $\slashed{p}$:

$$(\slashed{p} - m) u(p,s) = 0 \quad , \quad (\slashed{p} + m) v(p,s) = 0 \,. \tag{5}$$

Using (3) and (5) we find from (1) that

$$S \cdot \frac{p}{|p|} u(p, \pm s) = \pm u(p, \pm s) \,, \tag{6a}$$

$$S \cdot \frac{p}{|p|} v(p, \pm s) = \mp v(p, \pm s) \,. \tag{6b}$$

Note that this simple relation only applies to helicity states and will not be valid for arbitrary spin vectors s. The inverted sign in (6b) can be traced back to the redefinition (5.59).

The *operator of the spin projection in the direction of motion* takes a simple form if it is expressed in terms of the helicity basis. We essentially can copy the computation that lead from (5.70) to (5.73) for the momentum operator:

$$\hat{S} \cdot \frac{p}{|p|} = \frac{1}{2} \int d^3x \, \hat{\psi}^\dagger(x,t) \, \Sigma \, \hat{\psi}(x,t)$$

$$= \frac{1}{2} \sum_{\pm s, \pm s'} \int d^3p \, \frac{m}{\omega_p} \Big(\hat{b}^\dagger(p,s') \hat{b}(p,s) \, u^\dagger(p,s') \Sigma \cdot \frac{p}{|p|} u(p,s)$$

$$+ \hat{d}(p,s') \hat{d}^\dagger(p,s) \, v^\dagger(p,s') \Sigma \cdot \frac{p}{|p|} v(p,s)$$

$$+ \ldots \Big)$$

$$= \frac{1}{2} \int d^3p \, \Big(\hat{b}^\dagger(p,s) \hat{b}(p,s) - \hat{b}^\dagger(p,-s) \hat{b}(p,-s)$$

$$- \hat{d}(p,s) \hat{d}^\dagger(p,s) + \hat{d}(p,-s) \hat{d}^\dagger(p,-s) \Big) \,. \tag{7}$$

Subtracting the fictitious "total spin of the Dirac sea", we find that the operator of the observable spin projection becomes

$$\hat{S} \cdot \frac{p}{|p|} = \frac{1}{2} \int d^3p \, \Big(\hat{b}^\dagger(p,s) \hat{b}(p,s) + \hat{d}^\dagger(p,s) \hat{d}(p,s)$$

$$+ \hat{b}^\dagger(p,-s) \hat{b}(p,-s) - \hat{d}^\dagger(p,-s) \hat{d}(p,-s) \Big) \,. \tag{8}$$

The operators $\hat{b}^\dagger(p,\pm s)\hat{b}(p,\pm s)$ and $\hat{d}^\dagger(p,\pm s)\hat{d}(p,\pm s)$ play the role of number operators for particles and antiparticles of positive (negative) helicity. The redefinition of the antiparticle spinors accounts for the fact that a missing particle of negative helicity corresponds to an antiparticle of positive helicity, and vice versa.

Exercise 5.5

EXERCISE

5.5 General Commutation Relations and Microcausality

Problem. (a) Compute the anticommutators between free Dirac field operators $\hat{\psi}(x)$ and $\hat{\bar{\psi}}(y)$ at arbitrary space-time points x and y.
(b) Show that operators describing observable quantities satisfy the condition of microcausality, i.e., they commute for space-like distances.
(c) What happens to microcausality if one attempts to use bosonic quantization rules for the Dirac field?

Solution. (a) The anticommutator of the field operators $\hat{\psi}(x)$ and $\hat{\bar{\psi}}(y)$ at arbitrary time arguments x_0 and y_0, expressed in terms of the plane-wave expansion, is

$$\{\hat{\psi}_\alpha(x), \hat{\bar{\psi}}_\beta(y)\} = \int \frac{d^3p'}{(2\pi)^{3/2}} \int \frac{d^3p}{(2\pi)^{3/2}} \sqrt{\frac{m}{\omega_{p'}}} \sqrt{\frac{m}{\omega_p}} \sum_{s,s'}$$
$$\times \Big\{ \hat{b}(p',s')u_\alpha(p',s')\,e^{-ip'\cdot x} + \hat{d}^\dagger(p',s')v_\alpha(p',s')\,e^{+ip'\cdot x},$$
$$\hat{b}^\dagger(p,s)\bar{u}_\beta(p,s)\,e^{+ip\cdot y} + \hat{d}(p,s)\bar{v}_\beta(p,s)\,e^{-ip\cdot y} \Big\}. \quad (1)$$

Only the anticommutators involving $\{\hat{b}, \hat{b}^\dagger\}$ and $\{\hat{d}^\dagger, \hat{d}\}$ contribute:

$$\{\hat{\psi}_\alpha(x), \hat{\bar{\psi}}_\beta(y)\} = \int \frac{d^3p}{(2\pi)^3} \frac{m}{\omega_p} \sum_s \Big(u_\alpha(p,s)\bar{u}_\beta(p,s)\,e^{-ip\cdot(x-y)}$$
$$+ v_\alpha(p,s)\bar{v}_\beta(p,s)\,e^{+ip\cdot(x-y)} \Big). \quad (2)$$

Summation over the unit spinors produces the projection operators (5.87), which implies

$$\{\hat{\psi}_\alpha(x), \hat{\bar{\psi}}_\beta(y)\} = \int \frac{d^3p}{(2\pi)^3} \frac{1}{2\omega_p}$$
$$\times \Big((\slashed{p}+m)_{\alpha\beta}\,e^{-ip\cdot(x-y)} - (-\slashed{p}+m)_{\alpha\beta}\,e^{+ip\cdot(x-y)} \Big)$$
$$= (i\slashed{\nabla}+m)_{\alpha\beta} \int \frac{d^3p}{(2\pi)^3} \frac{1}{2\omega_p} \Big(e^{-ip\cdot(x-y)} - e^{+ip\cdot(x-y)} \Big)$$
$$= (i\slashed{\nabla}+m)_{\alpha\beta}\,i\Delta(x-y) \equiv i S_{\alpha\beta}(x-y). \quad (3)$$

$\Delta(x-y')$ denotes the Lorentz invariant Pauli–Jordan function introduced in Sect. 4.4. It vanishes outside the light cone, i.e., for space-like separations. The relation (3) contains the equal-time quantization condition as a special case, which can be seen by using (4.104), (4.106), and (4.107):

$$\{\hat{\psi}_\alpha(x), \hat{\bar{\psi}}_\beta(y)\}\Big|_{x_0=y_0} = (-\gamma^0\partial_0 - \gamma^i\partial_i + im)_{\alpha\beta}\,\Delta(x-y)\Big|_{x_0=y_0}$$
$$= \gamma^0{}_{\alpha\beta}\,\delta^3(\mathbf{x}-\mathbf{y}). \quad (4)$$

This coincides with (5.29a) after multiplication by γ^0. The remaining anticommutators are seen to vanish,

$$\{\hat{\psi}_\alpha(x), \hat{\psi}_\beta(y)\} = \{\hat{\bar{\psi}}_\alpha(x), \hat{\bar{\psi}}_\beta(y)\} = 0 , \tag{5}$$

since the "wrong" combinations of creation and annihilation operators are involved.

Exercise 5.5

(b) Observables are constructed as bilinear combinations of the field operator

$$\hat{O}(x) = \hat{\bar{\psi}}_\alpha(x) \, O_{\alpha\beta}(x) \hat{\psi}_\beta(x) , \tag{6}$$

were $O_{\alpha\beta}(x)$ is a Dirac matrix consisting of c numbers (and, possibly, differentiation operators). As an example, take the Dirac current operator for which $O^\mu_{\alpha\beta} = e\gamma^\mu_{\alpha\beta}$. The resulting commutator between two operators describing an observable quantity at two different points in space-time reads

$$\left[\hat{O}(x), \hat{O}(y)\right] = O_{\alpha\beta}(x) O_{\gamma\delta}(y) \left[\hat{\bar{\psi}}_\alpha(x)\hat{\psi}_\beta(x), \hat{\bar{\psi}}_\gamma(y)\hat{\psi}_\delta(y)\right] . \tag{7}$$

Using the identities $[\hat{A}, \hat{B}\hat{C}] = [\hat{A}, \hat{B}]\hat{C} + \hat{B}[\hat{A}, \hat{C}]$ and $[\hat{A}, \hat{B}\hat{C}] = \{\hat{A}, \hat{B}\}\hat{C} - \hat{B}\{\hat{A}, \hat{C}\}$, we can write the commutator in (7) as

$$\begin{aligned}
&\left[\hat{\bar{\psi}}_\alpha(x)\hat{\psi}_\beta(x), \hat{\bar{\psi}}_\gamma(y)\hat{\psi}_\delta(y)\right] \\
&= \hat{\bar{\psi}}_\alpha(x)\Big(\{\hat{\psi}_\beta(x), \hat{\bar{\psi}}_\gamma(y)\}\hat{\psi}_\delta(y) - \hat{\bar{\psi}}_\gamma(y)\{\hat{\psi}_\beta(x), \hat{\psi}_\delta(y)\}\Big) \\
&\quad + \Big(\{\hat{\bar{\psi}}_\alpha(x), \hat{\bar{\psi}}_\gamma(y)\}\hat{\psi}_\delta(y) - \hat{\bar{\psi}}_\gamma(y)\{\hat{\bar{\psi}}_\alpha(x), \hat{\psi}_\delta(y)\}\Big)\hat{\psi}_\beta(x) .
\end{aligned} \tag{8}$$

Using (3) and (5) this becomes

$$\begin{aligned}
\left[\hat{\bar{\psi}}_\alpha(x)\hat{\psi}_\beta(x), \hat{\bar{\psi}}_\gamma(x)\hat{\psi}_\delta(x)\right] &= \mathrm{i}S_{\beta\gamma}(x-y)\hat{\bar{\psi}}_\alpha(x)\hat{\psi}_\delta(y) \\
&\quad - \mathrm{i}S_{\delta\alpha}(y-x)\hat{\bar{\psi}}_\gamma(y)\hat{\psi}_\beta(x) .
\end{aligned} \tag{9}$$

Since $\Delta(x-y)$ and consequently $S_{\alpha\beta}(x-y)$ are known to vanish for space-like separations, the microcausality condition is guaranteed:

$$\left[\hat{O}(x), \hat{O}(y)\right] = 0 \quad \text{for} \quad (x-y)^2 < 0 . \tag{10}$$

(c) Contrary to the case of the Klein–Gordon field (see Sect. 4.4), microcausality cannot be used to rule out the "wrong" quantization prescription. If the calculation of part (a) is repeated using minus commutators one obtains

$$\begin{aligned}
\left[\hat{\psi}_\alpha(x), \hat{\bar{\psi}}_\beta(y)\right] &= \int \frac{\mathrm{d}^3 p'}{(2\pi)^{3/2}} \int \frac{\mathrm{d}^3 p}{(2\pi)^{3/2}} \sqrt{\frac{m}{\omega_{p'}}}\sqrt{\frac{m}{\omega_p}} \sum_{s,s'} \\
&\quad \times \Big([\hat{b}(p',s'), \hat{b}^\dagger(p,s)]\, u_\alpha(p',s')\bar{u}_\beta(p,s)\, \mathrm{e}^{-\mathrm{i}p'\cdot x + \mathrm{i}p\cdot y} \\
&\qquad + [\hat{d}^\dagger(p',s'), \hat{d}(p,s)]\, v_\alpha(p',s')\bar{v}_\beta(p,s)\, \mathrm{e}^{+\mathrm{i}p'\cdot x - \mathrm{i}p\cdot y} \\
&\qquad + \text{mixed terms}\Big) .
\end{aligned} \tag{11}$$

The bosonic quantization condition would read

$$[\hat{a}(p',s'), \hat{a}^\dagger(p,s)] = \delta^3(\boldsymbol{p}-\boldsymbol{p}')\,\delta_{ss'} . \tag{12}$$

If we use the redefinition of operators according to (5.60) this becomes (compare (5.68))

Exercise 5.5

$$\left[\hat{b}(p',s'),\,\hat{b}^\dagger(p,s)\right] = \left[\hat{d}^\dagger(p',s'),\,\hat{d}(p,s)\right] = \delta^3(\boldsymbol{p}-\boldsymbol{p}')\,\delta_{ss'}\;. \tag{13}$$

This implies that the relative sign of the terms in (11) remains unchanged and the commutator again reduces to (2). Thus we have found that when we use bosonic quantization, the commutator between fields,

$$\left[\hat{\psi}_\alpha(x),\,\hat{\bar{\psi}}_\beta(y)\right] = \mathrm{i}\,S_{\alpha\beta}(x-y)\;, \tag{14}$$

vanishes in the same way for space-like separations as did the anticommutator (3) when fermionic quantization was used. The derivation of the spin-statistics theorem for Dirac fields therefore can be only based on the positivity condition of the energy, as explained in Sect. 5.3.

6. Spin-1 Fields: The Maxwell and Proca Equations

6.1 Introduction

In field theory, particles of spin 1 are described as the quanta of vector fields. Such *vector bosons* play a central role as the mediators of interactions in particle physics. The important examples are the gauge fields of the electromagnetic (massless photons), the weak (massive $W^\pm$ and Z^0 bosons), and the strong (massless gluons) interactions. On a somewhat less fundamental level vector fields can also be used to describe spin-1 mesons, for example the ρ and the ω meson.

When treating vector fields one encounters a characteristic problem: the theory seems to be "overdetermined", that is to say the vector field nominally has *four degrees of freedom* which is more than is required to describe a spin-1 particle. Here a distinction has to be made between massive and massless particles. In the former case there are *three* degrees of freedom, whereas in the massless case the number is further reduced to *two* degrees of freedom. We all know from electrodynamics that the photon comes in only two polarization states, both of of which are transverse. Massive vector fields allow for an additional longitudinal polarization state. The reduction of the degrees of freedom has to be taken into account through as a *constraint* when quantizing the theory.

In the following sections we will first discuss the wave equations for the massless and the massive classical spin-1 fields, i.e., the Maxwell and the Proca equations. This will mainly serve to define the notation. Subsequently several approaches to the canonical quantization of these fields will be studied. Owing to its central importance, the quantization of the electromagnetic field will be treated separately in the next chapter.

6.2 The Maxwell Equations

The electromagnetic phenomena in vacuum can be described by two three-dimensional vector fields: the electric and magnetic field strength, $\boldsymbol{E}(\boldsymbol{x},t)$ and $\boldsymbol{B}(\boldsymbol{x},t)$. They satisfy the Maxwell equations

$$\boldsymbol{\nabla} \cdot \boldsymbol{E} = \rho \;, \tag{6.1a}$$

$$\boldsymbol{\nabla} \cdot \boldsymbol{B} = 0 \;, \tag{6.1b}$$

$$\nabla \times \boldsymbol{B} - \frac{\partial \boldsymbol{E}}{\partial t} = \boldsymbol{j} \,, \tag{6.1c}$$

$$\nabla \times \boldsymbol{E} + \frac{\partial \boldsymbol{B}}{\partial t} = 0 \,. \tag{6.1d}$$

Specifically, these are the laws of Gauss (6.1a) and Ampere (6.1c), Faraday's law of induction (6.1d) and the condition of vanishing magnetic charge (6.1b). Note: we use the "rationalized" Lorentz–Heaviside system of units; in the Gaussian system additional factors of 4π would appear. Furthermore the speed of light was omitted, i.e., we set $c = 1$.

It is well known that the set of equations (6.1) can be cast into a manifestly covariant form by introducing an antisymmetric tensor of rank 2, the *field-strength tensor* $F^{\mu\nu}$ defined as

$$F^{\mu\nu} = \begin{pmatrix} 0 & -E^1 & -E^2 & -E^3 \\ E^1 & 0 & -B^3 & B^2 \\ E^2 & B^3 & 0 & -B^1 \\ E^3 & -B^2 & B^1 & 0 \end{pmatrix} \,. \tag{6.2}$$

The *dual field-strength tensor* $^*F^{\mu\nu}$ is obtained by contracting $F^{\mu\nu}$ with the completely antisymmetric unit tensor (the Levi–Civita tensor) $\epsilon^{\mu\nu\rho\sigma}$:

$$^*F^{\mu\nu} = \frac{1}{2}\epsilon^{\mu\nu\rho\sigma} F_{\rho\sigma} = F^{\mu\nu}(\boldsymbol{E} \to \boldsymbol{B}, \boldsymbol{B} \to -\boldsymbol{E}) \,. \tag{6.3}$$

The inhomogeneous and the homogeneous Maxwell equations, (6.1a) and (6.1c) as well as (6.1b) and (6.1d), can be combined into

$$\partial_\mu F^{\mu\nu} = j^\nu \,, \tag{6.4}$$
$$\partial^\lambda F^{\mu\nu} + \partial^\nu F^{\lambda\mu} + \partial^\mu F^{\nu\lambda} = 0 \tag{6.5}$$

where $j^\mu = (\rho, \boldsymbol{j})$ is the four-current density.

Contracting (6.5) with the Levi–Civita tensor and renaming the summation indices we arrive at

$$\partial_\mu {}^*F^{\mu\nu} = 0 \,, \tag{6.6}$$

since

$$\begin{aligned} 0 &= \epsilon_{\lambda\nu\rho\sigma}\left(\partial^\lambda F^{\rho\sigma} + \partial^\sigma F^{\lambda\rho} + \partial^\rho F^{\sigma\lambda}\right) \\ &= \left(\epsilon_{\lambda\nu\rho\sigma} + \epsilon_{\rho\nu\sigma\lambda} + \epsilon_{\sigma\nu\lambda\rho}\right)\partial^\lambda F^{\rho\sigma} = 6\,\partial^\lambda\, {}^*F_{\lambda\nu} \,. \end{aligned}$$

Taking the four-divergence of (6.4) immediately leads to the *continuity equation* for the electric current vector

$$\partial_\nu j^\nu = 0 \,. \tag{6.7}$$

This is a compatibility condition since $F^{\mu\nu}$ is constructed as an antisymmetric tensor.

The field-strength tensor can be used to construct scalar and pseudoscalar bilinear functions

$$F_{\mu\nu} F^{\mu\nu} = -2(\boldsymbol{E}^2 - \boldsymbol{B}^2) \,, \tag{6.8a}$$
$$F_{\mu\nu} {}^*F^{\mu\nu} = -4\boldsymbol{E} \cdot \boldsymbol{B} \,. \tag{6.8b}$$

In the framework of quantum theory the field strength, $\boldsymbol{E}$ and $\boldsymbol{B}$ are usually treated as derived quantities. For the fundamental dynamical variable the *vector potential* $A^\mu = (A^0, \boldsymbol{A})$ is employed.[1] They are connected to the field strength through

$$\boldsymbol{B} = \nabla \times \boldsymbol{A} ,$$
$$\boldsymbol{E} = -\frac{\partial \boldsymbol{A}}{\partial t} - \nabla A_0 , \qquad (6.9)$$

or

$$F^{\mu\nu} = \partial^\mu A^\nu - \partial^\nu A^\mu . \qquad (6.10)$$

This construction automatically guarantees that the field-strength tensor satisfies the homogeneous Maxwell equation (6.5): the six terms involving derivatives of second order cancel each other. The inhomogeneous Maxwell equation leads to a second-order wave equation

$$\Box A^\nu - \partial^\nu (\partial_\mu A^\mu) = j^\nu , \qquad (6.11)$$

or, with the space and time components written separately,

$$\Box \boldsymbol{A} + \nabla \left(\frac{\partial}{\partial t} A_0 + \nabla \cdot \boldsymbol{A} \right) = \boldsymbol{j} , \qquad (6.12a)$$

$$-\Delta A_0 - \frac{\partial}{\partial t} (\nabla \cdot \boldsymbol{A}) = \rho . \qquad (6.12b)$$

Taking the four-divergence of (6.11) immediately shows that *the electromagnetic current is conserved*, i.e., it satisfies the continuity equation (6.7).

It is well known that the potential $A^\mu(\boldsymbol{x}, t)$ is not a directly observable quantity and that it is not uniquely determined. A *local gauge transformation*, which adds the four-gradient of an arbitrary scalar function $\Lambda(\boldsymbol{x}, t)$ to the potential,

$$A'^\mu(x) = A^\mu(x) + \partial^\mu \Lambda(x) , \qquad (6.13)$$

will not modify the values of the field strengths (6.10) and will preserve the wave equation (6.11). The property of gauge invariance leads to technical complications in the study of the electromagnetic field. To calculate a quantity of interest one often subjects the potential to a particular gauge condition, i.e., one "fixes the gauge". The details of the ensuing calculation can depend strongly on the chosen gauge condition, although of course the observable quantities derived at the end have to be gauge invariant. The problems are aggravated when *nonabelian* gauge theories are studied, which are of central importance for modern particle physics. In this book, however, we will restrict our attention to abelian fields.

Several gauge conditions are used frequently and have special names. Among these are

---

[1] We mention in passing that it is nevertheless also possible to apply the canonical quantization procedure directly to the field strengths. This approach dates back to W. Pauli.

$$\begin{aligned}
\text{Lorentz gauge:} \quad & \partial_\mu A^\mu = 0 \;, \\
\text{Coulomb gauge:} \quad & \boldsymbol{\nabla} \cdot \boldsymbol{A} = 0 \;, \\
\text{temporal gauge:} \quad & A_0 = 0 \;, \\
\text{axial gauge:} \quad & A_3 = 0 \;.
\end{aligned} \qquad (6.14)$$

For the electromagnetic field the first two gauges are used most frequently.

6.2.1 The Lorentz Gauge

The gauge condition (6.14a) is the only one that enjoys the privilege of being manifestly Lorentz covariant. In Sect. 7.3 the quantization scheme for photons will be based on the Lorentz gauge.

Any given potential can be made to satisfy the Lorentz condition by applying the transformation (6.13) with a function $\Lambda(x)$, which satisfies the equation

$$\Box\, \Lambda = -\partial_\mu A^\mu \;. \qquad (6.15)$$

After this step, however, the field A^μ is still not defined uniquely. There is an infinitely large class of gauge-equivalent fields describing the same physics and all satisfying the Lorentz condition. They are related to each other by transformation functions $\tilde{\Lambda}(x)$ which solve the homogeneous wave equation

$$\Box\, \tilde{\Lambda} = 0 \;. \qquad (6.16)$$

From the observation that the potential A^μ is still not uniquely fixed even after the impositon of the constraint $\partial \cdot A = 0$, we can deduce that it has only *two independent degrees of freedom*, instead of four, which naively would be expected for a four-dimensional vector field. In the next chapter we will explicitly uncover these degrees of freedom.

6.2.2 The Coulomb Gauge

This gauge requires that the three-dimensional divergence $\boldsymbol{\nabla} \cdot \boldsymbol{A} = 0$ vanishes, i.e., that the vector potential $\boldsymbol{A}(\boldsymbol{x}, t)$ is a spatially transverse field, with polarization vectors orthogonal to the direction of propagation. Therefore one also uses the name *transverse gauge*. An arbitrary potential can be made to satisfy the Coulomb gauge condition if the transfomation (6.13) is chosen in such a way that $\Lambda(x)$ is a solution of the Poisson equation

$$\Delta \Lambda = \boldsymbol{\nabla} \cdot \boldsymbol{A} \;. \qquad (6.17)$$

Explicitly the re-gauged potential then reads

$$\boldsymbol{A}' = \boldsymbol{A} - \boldsymbol{\nabla} \int d^3 x'\, G(\boldsymbol{x} - \boldsymbol{x}')\, \boldsymbol{\nabla}' \cdot \boldsymbol{A}(\boldsymbol{x}', t) \;. \qquad (6.18)$$

Here $G(\boldsymbol{x} - \boldsymbol{x}')$ is the Green's function of the Laplacian, defined by

$$\Delta\, G(\boldsymbol{x} - \boldsymbol{x}') = \delta^3(\boldsymbol{x} - \boldsymbol{x}') \;, \qquad (6.19)$$

which has the explicit form

$$G(\boldsymbol{x} - \boldsymbol{x}') = -\frac{1}{4\pi}\frac{1}{|\boldsymbol{x} - \boldsymbol{x}'|} \;. \qquad (6.20)$$

One arrives at the same result by splitting the potential $\boldsymbol{A}$ into its *transverse* and *longitudinal* parts:[2]

$$\boldsymbol{A} = \boldsymbol{A}_\perp + \boldsymbol{A}_\parallel = (P_\perp + P_\parallel)\boldsymbol{A}, \qquad (6.21)$$

so that

$$\nabla \cdot \boldsymbol{A}_\perp = 0 \quad \text{and} \quad \nabla \times \boldsymbol{A}_\parallel = 0. \qquad (6.22)$$

The *projection operators* in (6.21) can be written as

$$(P_\perp)_{ij} = \delta_{ij} - \partial_i \frac{1}{\Delta} \partial_j, \qquad (6.23a)$$

$$(P_\parallel)_{ij} = \partial_i \frac{1}{\Delta} \partial_j. \qquad (6.23b)$$

The symbol $1/\Delta$ denotes the inverse Laplacian. The action of this operator on a vector field $\boldsymbol{V}(\boldsymbol{x})$ is given by

$$\frac{1}{\Delta} \boldsymbol{V}(\boldsymbol{x}) = \int \mathrm{d}^3 x'\, G(\boldsymbol{x}-\boldsymbol{x}')\,\boldsymbol{V}(\boldsymbol{x}'), \qquad (6.24)$$

where the integral kernel consists of the Green's function (6.20). With (6.23a) the transverse part of the vector potential $\boldsymbol{A}_\perp$ becomes (6.18). Clearly the projection operators $P_\perp$ and $P_\parallel$ are *nonlocal* operators in coordinate space.

The field equations (6.12) in the Coulomb gauge get simplified:

$$\Box \boldsymbol{A} + \frac{\partial}{\partial t} \nabla A_0 = \boldsymbol{j}, \qquad (6.25a)$$

$$\Delta A_0 = -\rho. \qquad (6.25b)$$

The potential A_0 simply satisfies the Poisson equation, which no longer contains a time derivative. As a consequence $A_0(\boldsymbol{x},t)$ can be determined directly from the charge distribution $\rho(\boldsymbol{x},t)$ taken at the same time:

$$A_0(\boldsymbol{x},t) = \int \mathrm{d}^3 x'\, \frac{1}{4\pi}\, \frac{\rho(\boldsymbol{x}',t)}{|\boldsymbol{x}-\boldsymbol{x}'|}. \qquad (6.26)$$

This *instantaneous Coulomb potential* obviously no longer plays the role of a dynamical degree of freedom of the theory. Since the longitudinal degree of freedom of $\boldsymbol{A}(\boldsymbol{x},t)$ has also been eliminated, $\boldsymbol{A}_\parallel = 0$, the Coulomb gauge should be very suitable for describing the photon field with its two transverse degrees of freedom. The drawback of this gauge lies in the violation of (manifest) relativistic covariance. The condition (6.14b) will be satisfied only in a particular preferred inertial frame and the whole calculation will be restricted to this frame. We will present the quantization of the electromagnetic field in the Coulomb gauge in Sect. 7.7.

6.2.3 Lagrange Density and Conserved Quantities

The Lagrange density of the electromagnetic field interacting with a charged source $j_\mu(x)$ reads (cf. Exercise 6.1)

---

[2] Note that this splitting is not relativistically invariant. Both components will get mixed under Lorentz transformations.

$$\mathcal{L} = -\frac{1}{4} F_{\mu\nu} F^{\mu\nu} - j_\mu A^\mu \,. \tag{6.27}$$

This leads to the action integral

$$W = \int \mathrm{d}^4 x \, \mathcal{L}(x) = \int \mathrm{d}^4 x \left(\frac{1}{2}(\boldsymbol{E}^2 - \boldsymbol{B}^2) - \rho A_0 + \boldsymbol{j} \cdot \boldsymbol{A} \right) \,. \tag{6.28}$$

We remind the reader of the connection between gauge invariance and current conservation. The field-strength tensor $F_{\mu\nu}$ is gauge invariant (in abelian field theories) but this does not hold automatically for the interaction term. In order to guarantee the invariance of the action the integrand at most may obtain an additional divergence term under local gauge transformations (6.13). This is satisfied provided that the current j_μ satisfies the *continuity equation* (6.7)

$$\partial^\mu j_\mu = 0 \,, \tag{6.29}$$

so that

$$\begin{aligned} j_\mu A'^\mu &= j_\mu A^\mu + j_\mu \partial^\mu \Lambda = j_\mu A^\mu + \partial^\mu (j_\mu \Lambda) - \Lambda (\partial^\mu j_\mu) \\ &\to j_\mu A^\mu + \text{surface term} \,. \end{aligned}$$

Since the gauge function $\Lambda(x)$ is arbitrary, the current conservation law $\partial^\mu j_\mu = 0$ must be satisfied everywhere in order to leave the action invariant.

Knowing the Lagrange density, we can evaluate the *canonical energy-momentum tensor* $\Theta^{\mu\nu}$. According to Sect. 2.4 it is

$$\Theta^{\mu\nu} = \frac{\partial \mathcal{L}}{\partial(\partial_\mu A_\sigma)} \partial^\nu A_\sigma - g^{\mu\nu} \mathcal{L} \,. \tag{6.30}$$

Using the identity

$$\frac{\partial(F_{\alpha\beta} F^{\alpha\beta})}{\partial(\partial_\mu A_\sigma)} = 4 F^{\mu\sigma} \,, \tag{6.31}$$

which is easily verified, we obtain

$$\Theta^{\mu\nu} = \frac{1}{4} g^{\mu\nu} F_{\alpha\beta} F^{\alpha\beta} - F^{\mu\sigma} \partial^\nu A_\sigma + g^{\mu\nu} j_\sigma A^\sigma \,. \tag{6.32}$$

In the absence of a current, $j_\mu = 0$, the divergence of this tensor vanishes, which implies a conservation law for energy and momentum. The Lagrangian (6.27) also contains the option that a current j_μ is coupled to the field. This current is treated as an externally given source and thus the translation invariance of the system is broken. Therefore the continuity equation for energy and momentum density will obtain an extra term.

Using the equation of motion (6.4) the four-divergence of the canonical energy-momentum tensor is

$$\begin{aligned} \partial_\mu \Theta^{\mu\nu} &= \frac{1}{2}(\partial^\nu F_{\alpha\beta}) F^{\alpha\beta} - (\partial_\mu F^{\mu\sigma}) \partial^\nu A_\sigma - F^{\mu\sigma} \partial_\mu \partial^\nu A_\sigma \\ &\quad + (\partial^\nu j_\sigma) A^\sigma + j_\sigma \partial^\nu A^\sigma \\ &= -\frac{1}{2} \partial^\nu (\partial_\alpha A_\beta + \partial_\beta A_\alpha) F^{\alpha\beta} + (\partial^\nu j_\sigma) A^\sigma \,. \end{aligned} \tag{6.33}$$

Since $F^{\alpha\beta}$ is antisymmetric this reduces to

$$\partial_\mu \Theta^{\mu\nu} = (\partial^\nu j_\sigma) A^\sigma \ . \tag{6.34}$$

Thus energy conservation ($\nu=0$) is violated if the source current $j_\mu(\boldsymbol{x},t)$ depends on time. Similarly the electromagnetic field can aquire or lose momentum ($\nu = 1,2,3$) provided that $j_\mu(\boldsymbol{x},t)$ is inhomogeneous and has a spatial gradient. Of course this violation of energy and momentum conservation is just a consequence of the ansatz which treats $j_\mu(\boldsymbol{x},t)$ as a given function. In a complete theory the current will be caused by other fields, which also have to be treated as dynamical variables. These fields will contribute to the total energy-momentum tensor on the left-hand side of (6.34), whereas the right-hand side will vanish (see Exercise 6.2).

The canonical energy-momentum tensor introduced in (6.32) has some disagreeable properties: it is *not symmetric* and also *not gauge invariant*. The latter property follows if the gauge transformation (6.13) is applied:

$$\Theta'^{\mu\nu} = \Theta^{\mu\nu} - F^{\mu\sigma} \partial^\nu \partial_\sigma \Lambda + g^{\mu\nu} j^\sigma \partial_\sigma \Lambda \ . \tag{6.35}$$

Now we swap the derivative ∂_σ and use the equation of motion (6.4) and the continuity equation (6.7):

$$\begin{aligned}\Theta'^{\mu\nu} &= \Theta^{\mu\nu} - \partial_\sigma(F^{\mu\sigma} \partial^\nu \Lambda - g^{\mu\nu} j^\sigma \Lambda) + (\partial_\sigma F^{\mu\sigma}) \partial^\nu \Lambda - g^{\mu\nu}(\partial_\sigma j^\sigma) \Lambda \\ &= \Theta^{\mu\nu} - \partial_\sigma(F^{\mu\sigma} \partial^\nu \Lambda - g^{\mu\nu} j^\sigma \Lambda) - j^\mu \partial^\nu \Lambda \ .\end{aligned} \tag{6.36}$$

Thus $\Theta^{\mu\nu}$ is not gauge invariant, even in the absence of external currents, $j_\mu = 0$. However, in the latter case the change of gauge only leads to a divergence, which upon integration produces a surface term, making no contribution to the total energy and total momentum of the electromagnetic field. As discussed in Example 2.1 it is allowed to add to the energy-momentum tensor the divergence of an arbitrary tensor of rank 3, being antisymmetric in the first two indices, $\chi^{\sigma\mu\nu} = -\chi^{\mu\sigma\nu}$,

$$\tilde{\Theta}^{\mu\nu} = \Theta^{\mu\nu} + \partial_\sigma \chi^{\sigma\mu\nu} \ , \tag{6.37}$$

without changing the conserved quantities. This is true since Gauss' theorem implies

$$\begin{aligned}\tilde{P}^\nu &= \int\!\mathrm{d}^3x\, \tilde{\Theta}^{0\nu} = \int\!\mathrm{d}^3x\, \left(\Theta^{0\nu} + \partial_\sigma \chi^{\sigma 0\nu}\right) \\ &= P^\nu + \int\!\mathrm{d}^3x\, \partial_0 \chi^{00\nu} + \text{surface term} = P^\nu\end{aligned} \tag{6.38}$$

provided that surface terms at spatial infinity can be neglected.

We can make use of this fact in order to construct a "physical" *modified energy-momentum tensor* $T^{\mu\nu}$. Following the pattern of (6.37) we add the four-divergence of a tensor $\chi^{\sigma\mu\nu}$ which is chosen such that it cancels the second term in (6.36) under gauge transformations

$$T^{\mu\nu} = \Theta^{\mu\nu} + \partial_\sigma (F^{\mu\sigma} A^\nu) \ . \tag{6.39}$$

Employing the Maxwell equation we obtain

$$T^{\mu\nu} = \frac{1}{4} g^{\mu\nu} F_{\alpha\beta} F^{\alpha\beta} + F^{\mu\sigma} F_\sigma{}^\nu + g^{\mu\nu} j_\sigma A^\sigma - j^\mu A^\nu \ . \tag{6.40}$$

The gauge invariance in the absence of sources, $j_\mu = 0$, is obvious from the fact that (6.40), in contrast to (6.32), depends only on the (gauge invariant) field-strength tensor $F^{\mu\nu}$. Furthermore the redefinition (6.39) has transformed $T^{\mu\nu}$ into a *symmetric* tensor (in the absence of sources). We remark that the transition from $\Theta_{\mu\nu}$ to $T_{\mu\nu}$ precisely coincides with the general procedure of Belinfante for symmetrizing the energy-momentum tensor discussed in Example 2.1. The extra term in (6.39) is obtained from the general equation (17) in this example if the transformation coefficients $(I_{\alpha\beta})_{\mu\nu}$, to be derived below, (see (6.47)) are inserted.

The apparent violation of gauge invariance in the presence of a current $j_\mu(x)$ is no reason for concern since we are dealing with an open system with a externally prescribed source. The argument applying to the apparent violation of energy and momentum conservation can be repeated; gauge invariance is restored if the source is included in the system and bestowed with appropriate transformation properties (see Exercise 6.2).

Explicitly we find from (6.40) the *energy density*

$$\begin{aligned} w &= T^{00} = \frac{1}{4} F_{\alpha\beta} F^{\alpha\beta} + F^{0\sigma} F_\sigma{}^0 + j_\sigma A^\sigma - j^0 A_0 \\ &= \frac{1}{2}(\boldsymbol{B}^2 + \boldsymbol{E}^2) - \boldsymbol{j} \cdot \boldsymbol{A} \,, \end{aligned} \qquad (6.41)$$

and for the *momentum density* (i.e., the Poynting vector)

$$\begin{aligned} p^k &= T^{0k} = F^{0\sigma} F_\sigma{}^k - j^0 A^k \,, \\ \boldsymbol{p} &= \boldsymbol{E} \times \boldsymbol{B} - j^0 \boldsymbol{A} \,. \end{aligned} \qquad (6.42)$$

6.2.4 The Angular-Momentum Tensor

Noether's theorem enables us to calculate the angular momentum of a given field configuration $A_\mu(x)$. For this we have to know the behavior of the vector potential under infinitesimal Lorentz transformations

$$x'^\mu = x^\mu + \delta\omega^{\mu\nu} x_\nu \,. \qquad (6.43)$$

The general ansatz for the transformation of multicomponent fields was given in (2.64):

$$A'^\mu(x') = A^\mu(x) + \frac{1}{2}\delta\omega_{\alpha\beta} (I^{\alpha\beta})^{\mu\nu} A_\nu(x) \,. \qquad (6.44)$$

The infinitesimal generators $(I^{\alpha\beta})^{\mu\nu}$ of the Lorentz group can be easily determined from the transformation properties. Clearly the field $A^\mu(x)$ must be a four-vector, following the same infinitesimal transformation rule as the coordinate vector (6.43):

$$A'^\mu(x') = A^\mu(x) + \delta\omega^{\mu\nu} A_\nu(x) \,. \qquad (6.45)$$

This implies

$$\delta\omega_{\alpha\beta}\left[\frac{1}{2}(I^{\alpha\beta})^{\mu\nu} - g^{\alpha\mu}g^{\beta\nu}\right] = 0 \,. \qquad (6.46)$$

Furthermore the generators can be chosen to be antisymmetric, $(I^{\alpha\beta})^{\mu\nu} = -(I^{\beta\alpha})^{\mu\nu}$ since a symmetric part will drop out after contraction with the antisymmetric matrix $\delta\omega_{\alpha\beta}$. Therefore (6.46) is solved by

$$(I^{\alpha\beta})^{\mu\nu} = g^{\alpha\mu}g^{\beta\nu} - g^{\alpha\nu}g^{\beta\mu} \,. \tag{6.47}$$

This matrix determines the properties of a spin-1 field under Lorentz transformations. The objects (6.47) satisfy the Lie algebra of the Lorentz group (2.65). The tensor of *angular-momentum density* $M^{\mu\nu\lambda}$ is then obtained from the general formula (2.70) as

$$\begin{aligned}
M^{\mu\nu\lambda} &= \Theta^{\mu\lambda}x^\nu - \Theta^{\mu\nu}x^\lambda + \frac{\partial \mathcal{L}}{\partial(\partial_\mu A^\sigma)}(I^{\nu\lambda})^{\sigma\tau}A_\tau \\
&= \Theta^{\mu\lambda}x^\nu - \Theta^{\mu\nu}x^\lambda - F^\mu{}_\sigma(g^{\nu\sigma}g^{\lambda\tau} - g^{\nu\tau}g^{\lambda\sigma})A_\tau \\
&= \Theta^{\mu\lambda}x^\nu - \Theta^{\mu\nu}x^\lambda + (F^{\mu\lambda}A^\nu - F^{\mu\nu}A^\lambda) \,.
\end{aligned} \tag{6.48}$$

The last term describes the intrinsic angular momentum (the spin) of the vector field. To be more precise the latter is obtained from the $\mu = 0$ component, taking spatial components ν and λ:

$$S^{nl} = \int d^3x \, (F^{0l}A^n - F^{0n}A^l) \,. \tag{6.49}$$

Written in terms of three-vectors, the *spin of the electromagnetic field* reads

$$\boldsymbol{S} = \int d^3x \, \boldsymbol{E} \times \boldsymbol{A} \,. \tag{6.50}$$

EXERCISE

6.1 The Lagrangian of the Maxwell Field

Problem. Derive the most general form of the Lagrange density of the massless spin-1 field, which produces the field equation (6.11). Show that the result agrees with

$$\mathcal{L}_0 = -\frac{1}{4}F_{\mu\nu}F^{\mu\nu} - j_\mu A^\mu \tag{1}$$

up to a four-divergence term.

Solution. The Lagrange density $\mathcal{L}$ is a Lorentz scalar that has to be constructed from the vector field A^μ and its derivatives. Since the field equation for A^μ is supposed to be linear, $\mathcal{L}$ will consist of second-order terms. We make the following general ansatz:

$$\mathcal{L} = \alpha(\partial_\mu A^\mu)^2 + \beta(\partial_\mu A^\nu)(\partial^\mu A_\nu) + \gamma(\partial_\mu A^\nu)(\partial_\nu A^\mu) + \delta A_\mu A^\mu + \epsilon\, j_\mu A^\mu \,. \tag{2}$$

A possible further term, $(\partial_\mu\partial_\nu A^\nu)A^\mu$, was left out since it agrees with the first term up to a divergence. The interaction with the current j_μ can be determined from the field equation (6.11). Variation of the action with respect to the field A_μ leads to the Euler–Lagrange equation

$$\frac{\partial \mathcal{L}}{\partial A_\mu} - \partial^\nu \frac{\partial \mathcal{L}}{\partial(\partial^\nu A_\mu)} = 0 \,. \tag{3}$$

We need the derivatives

Exercise 6.1

$$\frac{\partial \mathcal{L}}{\partial A_\mu} = 2\delta A^\mu + \epsilon j^\mu , \tag{4}$$

$$\frac{\partial \mathcal{L}}{\partial(\partial^\nu A_\mu)} = \alpha\, 2 g^\mu{}_\nu (\partial_\sigma A^\sigma) + \beta\, 2\partial_\nu A^\mu + \gamma\, 2\partial^\mu A_\nu . \tag{5}$$

Inserting this into (3) gives

$$\begin{aligned} \delta A^\mu + \tfrac{1}{2}\epsilon j^\mu &= \partial^\nu \left[\alpha g^\mu{}_\nu (\partial_\sigma A^\sigma) + \beta \partial_\nu A^\mu + \gamma \partial^\mu A_\nu \right] \\ &= (\alpha+\gamma)\partial^\mu(\partial_\sigma A^\sigma) + \beta \,\Box\, A^\mu . \end{aligned} \tag{6}$$

This is required to agree with the field equation

$$\Box A^\mu - \partial^\mu(\partial_\nu A^\nu) = j^\mu . \tag{7}$$

Obviously the choice $\delta = 0$, $\beta = \tfrac{1}{2}\epsilon$, $\alpha + \gamma = -\tfrac{1}{2}\epsilon$ will do the trick. Of course $\mathcal{L}$ in this way can be determined only up to the constant factor ϵ. The value of ϵ can be fixed using the interaction term. The potential energy of a charge q experiencing a potential A_0 should be $V = qA_0$, entering the Lagrangian with a negative sign $(L = T - V)$. From this we deduce $\epsilon = -1$. The general expression for the Lagrange density then reads

$$\mathcal{L} = \alpha(\partial_\mu A^\mu)^2 - \tfrac{1}{2}(\partial_\mu A^\nu)(\partial^\mu A_\nu) + \left(\tfrac{1}{2} - \alpha\right)(\partial_\mu A^\nu)(\partial_\nu A^\mu) - j_\mu A^\mu . \tag{8}$$

A comparison with the expression

$$\mathcal{L}_0 = -\tfrac{1}{4}F_{\mu\nu}F^{\mu\nu} - j_\mu A^\mu = -\tfrac{1}{2}(\partial_\mu A_\nu)(\partial^\mu A^\nu) + \tfrac{1}{2}(\partial_\mu A_\nu)(\partial^\nu A^\mu) - j_\mu A^\mu . \tag{9}$$

exhibits the difference

$$\mathcal{L} - \mathcal{L}_0 = \alpha \left[(\partial_\mu A^\mu)(\partial_\nu A^\nu) - (\partial_\mu A^\nu)(\partial_\nu A^\mu) \right] , \tag{10}$$

where α is an arbitrary parameter. The term in square brackets is a four-divergence which will not contribute to the action integral. Using the product rule and renumbering a few indices the content of the square bracket can be written as

$$\begin{aligned} [\;\;] &= \left(\partial_\mu(A^\mu \partial_\nu A^\nu) - A^\mu \partial_\mu \partial_\nu A^\nu \right) - \left(\partial_\mu(A^\nu \partial_\nu A^\mu) - A^\nu \partial_\mu \partial_\nu A^\mu \right) \\ &= \partial_\mu \left(A^\mu \partial_\nu A^\nu - A^\nu \partial_\nu A^\mu \right) . \end{aligned} \tag{11}$$

Therefore without restricting generality, α can be chosen to vanish.

EXERCISE

6.2 Coupled Maxwell and Dirac Fields

Problem. Quantum electrodynamics describes a coupled system of a Maxwell field and an electrically charged Dirac field. Their coupling is driven by the Dirac current $j^\mu = e\bar\psi \gamma^\mu \psi$. Derive the resulting classical equations of motion. Show that the energy-momentum vector of the total system is gauge invariant and that energy and momentum are conserved.

Solution. The Lagrangian of the system is

$$\begin{aligned}\mathcal{L} &= \mathcal{L}_{\text{Dirac}} + \mathcal{L}_{\text{e.m.}} + \mathcal{L}_{\text{int}} \\ &= \bar{\psi}(i\gamma^\mu\partial_\mu - m)\psi - \frac{1}{4}F_{\mu\nu}F^{\mu\nu} - e\bar{\psi}\gamma^\mu\psi A_\mu \,.\end{aligned} \quad (1)$$

This leads to the following set of coupled equations of motions for the fields ψ, $\bar{\psi}$, and A^μ:

$$\left[\gamma^\mu i(\partial_\mu + ieA_\mu) - m\right]\psi = 0 \,, \quad (2a)$$

$$\bar{\psi}\left[\gamma^\mu i(\overleftarrow{\partial}_\mu - ieA_\mu) + m\right] = 0 \,, \quad (2b)$$

$$\Box A^\mu - \partial^\mu(\partial\cdot A) = e\bar{\psi}\gamma^\mu\psi \,. \quad (2c)$$

The Lagrangian (1) and the field equations (2) are invariant under local gauge transformations

$$A'_\mu(x) = A_\mu(x) + \partial_\mu\Lambda(x) \,, \quad (3a)$$

$$\psi'(x) = \exp\left[-ie\Lambda(x)\right]\psi(x) \,, \quad \bar{\psi}'(x) = \exp\left[+ie\Lambda(x)\right]\bar{\psi}(x) \,. \quad (3b)$$

This was achieved by the prescription of *minimal coupling*, replacing the partial derivative by the "gauge-covariant" derivative, $\partial_\mu \to D_\mu = \partial_\mu + ieA_\mu$. The canonical energy-momentum tensor resulting from (1) is

$$\Theta^{\mu\nu} = \Theta^{\mu\nu}_{\text{Dirac}} + \Theta^{\mu\nu}_{\text{e.m.}} + \Theta^{\mu\nu}_{\text{int}} \,, \quad (4)$$

where

$$\Theta^{\mu\nu}_{\text{Dirac}} = \bar{\psi}i\gamma^\mu\partial^\nu\psi - g^{\mu\nu}\bar{\psi}(i\gamma^\sigma\partial_\sigma - m)\psi \,, \quad (5a)$$

$$\Theta^{\mu\nu}_{\text{e.m.}} = -F^{\mu\sigma}\partial^\nu A_\sigma + g^{\mu\nu}\frac{1}{4}F_{\sigma\tau}F^{\sigma\tau} \,, \quad (5b)$$

$$\Theta^{\mu\nu}_{\text{int}} = g^{\mu\nu}e\bar{\psi}\gamma^\sigma\psi A_\sigma \,. \quad (5c)$$

Application of the gauge transformation (3) leads to the following extra terms

$$\Delta\Theta^{\mu\nu}_{\text{Dirac}} = e\bar{\psi}\gamma^\mu\psi\,\partial^\nu\Lambda - g^{\mu\nu}e\bar{\psi}\gamma^\sigma\psi\,\partial_\sigma\Lambda \,, \quad (6a)$$

$$\Delta\Theta^{\mu\nu}_{\text{e.m.}} = -F^{\mu\sigma}\partial^\nu\partial_\sigma\Lambda \,, \quad (6b)$$

$$\Delta\Theta^{\mu\nu}_{\text{int}} = g^{\mu\nu}e\bar{\psi}\gamma^\sigma\psi\,\partial_\sigma\Lambda \,. \quad (6c)$$

The sum of these contributions reduces to

$$\begin{aligned}\Delta\Theta^{\mu\nu} &= e\bar{\psi}\gamma^\mu\psi\,\partial^\nu\Lambda - F^{\mu\sigma}\partial^\nu\partial_\sigma\Lambda \\ &= \left(e\bar{\psi}\gamma^\mu\psi + \partial_\sigma F^{\mu\sigma}\right)\partial^\nu\Lambda - \partial_\sigma\left(F^{\mu\sigma}\partial^\nu\Lambda\right) \\ &= -\partial_\sigma\left(F^{\mu\sigma}\partial^\nu\Lambda\right) \,, \end{aligned} \quad (7)$$

where in the last step the field equation (2c) has been used. The energy-momentum four-vector proves to be gauge invariant since its change reduces to a surface integral:

$$\begin{aligned}\Delta P^\nu &= \int d^3x\,\Delta\Theta^{0\nu} = -\int d^3x\,\partial_\sigma\left(F^{0\sigma}\partial^\nu\Lambda\right) = \int d^3x\,\partial_i\left(E^i\partial^\nu\Lambda\right) \\ &= 0 \,.\end{aligned} \quad (8)$$

Exercise 6.2

Using the field equations (2), we can write the four-divergences of the contributions to the electromagnetic field-strength tensor as

$$\partial_\mu \Theta^{\mu\nu}_{\text{Dirac}} = -\partial^\nu (\bar{\psi}\gamma^\sigma \psi) A_\sigma \,, \tag{9a}$$

$$\partial_\mu \Theta^{\mu\nu}_{\text{e.m.}} = -(e\bar{\psi}\gamma^\sigma \psi)(\partial^\nu A_\sigma) \,, \tag{9b}$$

$$\partial_\mu \Theta^{\mu\nu}_{\text{int}} = \partial^\nu (e\bar{\psi}\gamma^\sigma \psi) A_\sigma + (e\bar{\psi}\gamma^\sigma \psi)(\partial^\nu A_\sigma) \,. \tag{9c}$$

Energy-momentum conservation in the combined systems follows from summing up the three contributions

$$\partial_\mu \Theta^{\mu\nu}_{\text{Dirac}} + \partial_\mu \Theta^{\mu\nu}_{\text{e.m.}} + \partial_\mu \Theta^{\mu\nu}_{\text{int}} = \partial_\mu \Theta^{\mu\nu} = 0 \,. \tag{10}$$

This also holds for the symmetrized energy-momentum tensor $T^{\mu\nu}$.

6.3 The Proca Equation

The transition from the massless electromagnetic field to a field of *massive vector bosons* is easily done: we simply add a quadratic mass term to the Lagrangian. Its form can be deduced by comparing with the massive Klein–Gordon theory:

$$\mathcal{L} = -\frac{1}{4} F_{\mu\nu} F^{\mu\nu} + \frac{1}{2} m^2 A_\mu A^\mu - j_\mu A^\mu \,. \tag{6.51}$$

We have assumed that A^μ is a real-valued *neutral spin-1 field*. For a *charged field*, A_μ is complex and the mass term in (6.51) has to be replaced by $m^2 A^*_\mu A^\mu$.

The sign of the mass term has to be treated with care. At first sight it might appear to disagree with the sign of the Klein–Gordon mass term. One has to keep in mind, however, that the scalar product $A_\sigma A^\sigma = (A^0)^2 - (\boldsymbol{A})^2$ consists of four different contributions. The sign was chosen such that it agrees with the Klein–Gordon case for the *spatial components* of $A^\mu(x)$. It will turn out that the time component $A^0(x)$ play a special role and cannot be identified with a degree of freedom of a real physical particle (see Sect. 6.5).

The Euler–Lagrange equation emerging from (6.51) reads

$$\partial_\mu F^{\mu\nu} + m^2 A^\nu = j^\nu \,, \tag{6.52}$$

which can be written in terms of the potential A^μ as

$$\Box A^\nu - \partial^\nu (\partial_\mu A^\mu) + m^2 A^\nu = j^\nu \,. \tag{6.53}$$

This is the *Proca equation*.

An essential difference compared to the massless case is encountered when taking the four-divergence of this equation. Since the first two terms cancel each other we find, assuming $m \neq 0$

$$\partial_\nu A^\nu = \frac{1}{m^2} \partial_\nu j^\nu \,. \tag{6.54}$$

In the following we will assume that either there are no sources, $j^\nu = 0$, or that the source current is conserved,[3] $\partial_\nu j^\nu = 0$. This simplifies the condition (6.54):

$$\partial_\nu A^\nu = 0 \,. \tag{6.55}$$

In contrast to the electromagnetic field, the massive spin-1 field will *automatically satisfy the Lorentz condition* (6.55) (or more generally (6.54)). The Lorentz condition has become a condition of consistency for the Proca field. As a consequence the field equation (6.53) gets simplified:

$$(\Box + m^2) A^\nu = j^\nu \,. \tag{6.56}$$

Thus the four components of the spin-1 field separately satisfy the Klein–Gordon equation. This is not surprising since it is an expression of the relativistic energy momentum relation. Simultaneously, the constraint condition (6.54) has also to be satisfied. It can be shown that in the absence of this constraint the field A^μ would describe a mixture of independent spin-0 and spin-1 particles, which is not what we intend.

Obviously the Proca theory *cannot be gauge invariant*: the mass term in (6.51) upon re-gauging of the field transforms according to

$$A_\sigma A^\sigma \to (A_\sigma + \partial_\sigma \Lambda)(A^\sigma + \partial^\sigma \Lambda) = A_\sigma A^\sigma + 2 A_\sigma \partial^\sigma \Lambda + \partial_\sigma \Lambda \partial^\sigma \Lambda \,, \tag{6.57}$$

and there is no way that the gauge-dependent terms could be expressed as a four-divergence. Also the field equation (6.56) immediately is seen to lack gauge invariance.

The *energy-momentum tensor* of the Proca theory can be copied from electrodynamics. The only difference is an additional mass term, i.e., (6.32) is generalized to

$$\Theta^{\mu\nu} = \frac{1}{4} g^{\mu\nu} F_{\alpha\beta} F^{\alpha\beta} - F^{\mu\sigma} \partial^\nu A_\sigma - \frac{1}{2} m^2 g^{\mu\nu} A_\sigma A^\sigma + g^{\mu\nu} j_\sigma A^\sigma \,. \tag{6.58}$$

Again this energy-momentum tensor can be symmetrized by adding a divergence as in (6.39), leading to a result that is similar to (6.40):

$$\begin{aligned} T^{\mu\nu} &= \frac{1}{4} g^{\mu\nu} F_{\alpha\beta} F^{\alpha\beta} + F^{\mu\sigma} F_\sigma{}^\nu - \frac{1}{2} m^2 g^{\mu\nu} A_\sigma A^\sigma + m^2 A^\mu A^\nu \\ &\quad + g^{\mu\nu} j_\sigma A^\sigma - j^\mu A^\nu \,. \end{aligned} \tag{6.59}$$

This implies for the energy and momentum density:

$$w = \frac{1}{2}(\boldsymbol{B}^2 + \boldsymbol{E}^2) + \frac{1}{2} m^2 (A_0^2 + \boldsymbol{A}^2) - \boldsymbol{j} \cdot \boldsymbol{A} \,, \tag{6.60a}$$

$$\boldsymbol{p} = \boldsymbol{E} \times \boldsymbol{B} - m^2 A^0 \boldsymbol{A} - j^0 \boldsymbol{A} \,. \tag{6.60b}$$

---

[3] This depends on the particular nature of the source. In contrast to the electromagnetic case the current is no longer conserved automatically.

6.4 Plane-Wave Expansion of the Vector Field

Drawing on the experience from the Klein–Gordon and the Dirac field, we can expect that a decomposition of the vector field $A^\mu(x)$ in terms of plane waves will be very useful. The general form of a field mode will look like

$$A_\mu(\boldsymbol{k},\lambda;x) = N_k\, e^{-i(\omega_k t - \boldsymbol{k}\cdot\boldsymbol{x})}\, \epsilon_\mu(\boldsymbol{k},\lambda) \qquad (6.61)$$

where $\omega_k = +\sqrt{\boldsymbol{k}^2 + m^2}$. Here $\epsilon_\mu(\boldsymbol{k},\lambda)$ denotes a set of four-dimensional *polarization vectors*, which play a similar role as the unit spinors u and v in the plane-wave decomposition of the Dirac field. Since we deal with Lorentz vectors there should be four linearly independent polarization vector ϵ^μ, three of them being space-like and one time-like. Things get simplified if the polarization vectors are defined with respect to the direction of the momentum vector, thus leading to particle states with well-defined helicity (this choice is not mandatory, however; see the footnote on page 155). Without restricting generality we demand that the polarization vectors form a *four-dimensional orthonormal system* satisfying

$$\epsilon_\mu(\boldsymbol{k},\lambda)\epsilon^\mu(\boldsymbol{k},\lambda') = g_{\lambda\lambda'}\,. \qquad (6.62)$$

The following discussion will depend on whether the mass m of the vector field is finite (Proca field) or vanishing (Maxwell field). Therefore these two cases will be treated separately.

6.4.1 The Massive Vector Field

To construct the set of polarization vectors $\epsilon_\mu(\boldsymbol{k},\lambda)$ we choose a frame of reference (which is arbitrary but will be kept fixed) in which the plane wave has momentum $\boldsymbol{k}$. Now we choose two space-like *transverse polarization vectors*

$$\epsilon(\boldsymbol{k},1) = \big(0, \boldsymbol{\epsilon}(\boldsymbol{k},1)\big)\,, \qquad (6.63a)$$
$$\epsilon(\boldsymbol{k},2) = \big(0, \boldsymbol{\epsilon}(\boldsymbol{k},2)\big)\,, \qquad (6.63b)$$

imposing the conditions

$$\boldsymbol{\epsilon}(\boldsymbol{k},1)\cdot\boldsymbol{k} = \boldsymbol{\epsilon}(\boldsymbol{k},2)\cdot\boldsymbol{k} = 0 \qquad (6.64)$$

and

$$\boldsymbol{\epsilon}(\boldsymbol{k},i)\cdot\boldsymbol{\epsilon}(\boldsymbol{k},j) = \delta_{ij}\,. \qquad (6.65)$$

The third polarization vector ($\lambda = 3$) is constructed such that its spatial component points in the direction of the momentum $\boldsymbol{k}$ while being normalized according to (6.62). Its zero component is not yet specified. We will adopt the further condition that the four-vector $\epsilon(\boldsymbol{k},3)$ is orthogonal to the momentum four-vector,

$$k^\mu \epsilon_\mu(\boldsymbol{k},3) = 0\,. \qquad (6.66)$$

Taking this equation and the normalization condition (6.62) for $\lambda = \lambda' = 3$, we find the components of the thus constructed *longitudinal polarization vector*:

$$\epsilon(\boldsymbol{k},3) = \left(\frac{|\boldsymbol{k}|}{m}, \frac{\boldsymbol{k}}{|\boldsymbol{k}|}\frac{k_0}{m}\right)\,. \qquad (6.67)$$

The normalization condition $(\epsilon(\boldsymbol{k},3))^2 = -1$ is satisfied since $k^2/m^2 - k_0^2/m^2 = -m^2/m^2 = -1$. It is worth noting that in the massless case it would not have been possible to construct a vector that is transverse in four dimensions, (6.66), and at the same time has a nonzero norm: (6.67) is not well defined in the limit $m \to 0$. This problem will be addressed in the next section.

The spatial components of the three polarization vectors $\lambda = 1, 2, 3$ in the chosen special Lorentz frame form an orthogonal dreibein as sketched in Fig. 6.1. This, of course, is not a Lorentz invariant property since it refers to three-vectors. The four-dimensional orthogonality condition (6.62), however, by construction is valid in every inertial frame.

To complete the vector basis in Minkowski space a fourth *time-like polarization vector* with index $\lambda = 0$ is needed. For this we can simply use the momentum vector k, namely

$$\epsilon(\boldsymbol{k},0) = \frac{1}{m} k . \tag{6.68}$$

The factor $1/m$ ensures the normalization according to (6.62). Also, it is obvious that (6.68) is orthogonal to the three space-like polarization vectors $\epsilon(\boldsymbol{k},\lambda)$. Let us write down the four-dimensional scalar product of our set of polarization vectors[4] with the momentum vector:

$$k \cdot \epsilon(\boldsymbol{k},1) = k \cdot \epsilon(\boldsymbol{k},2) = k \cdot \epsilon(\boldsymbol{k},3) = 0 \tag{6.70a}$$

$$k \cdot \epsilon(\boldsymbol{k},0) = m . \tag{6.70b}$$

It can be expected that the four polarization vectors satisfy a *completeness relation*. It is not difficult to guess its form:

$$\sum_{\lambda=0}^{3} g_{\lambda\lambda}\, \epsilon_\mu(\boldsymbol{k},\lambda)\, \epsilon_\nu(\boldsymbol{k},\lambda) = g_{\mu\nu} . \tag{6.71}$$

Here the factor $g_{\lambda\lambda}$ on the left-hand side is not meant to designate a tensor, it simply stands for the sign factor $+1$ or -1. Thus repetition of the index λ here, as an exception to the general rule, does not imply a summation. The conjecture (6.71) is most easily checked in the rest frame of the particle where only the $\lambda = 0$ vector has a time-like component. We find

$$\sum_{\lambda=0}^{3} g_{\lambda\lambda}\, \epsilon_\mu(\boldsymbol{k},\lambda)\, \epsilon_\nu(\boldsymbol{k},\lambda)$$

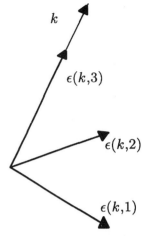

Fig. 6.1. The directions of the longitudinal and the two transverse polarization vectors

---

[4] It is also possible to construct a set of *general polarization vectors* which do not give special meaning to the momentum vector $\boldsymbol{k}$. To achieve this, one starts from an arbitrarily chosen dreibein of unit vectors $\epsilon^{(1)}$, $\epsilon^{(2)}$, $\epsilon^{(3)}$. These vectors are extended into four-vectors and are subject to a Lorentz boost transformation going to the frame of reference in which the particle moves with momentum $\boldsymbol{k}$. The result is given by

$$\epsilon(\boldsymbol{k},\lambda) = \left(\frac{\boldsymbol{k}\cdot\boldsymbol{\epsilon}^{(\lambda)}}{m},\, \boldsymbol{\epsilon}^{(\lambda)} + \frac{\boldsymbol{k}\cdot\boldsymbol{\epsilon}^{(\lambda)}}{m(k_0+m)} \boldsymbol{k} \right), \quad \lambda = 1,2,3 . \tag{6.69}$$

In combination with (6.68) for $\lambda = 0$ this prescription leads to a set of general polarization vectors which satisfy the conditions (6.62) and (6.70). The previous vectors (6.63) and (6.67) can be viewed as a special case of (6.69).

$$= \begin{pmatrix} 1 \\ 0 \end{pmatrix}_\mu \begin{pmatrix} 1 \\ 0 \end{pmatrix}_\nu - \sum_{\lambda=1}^{3} \begin{pmatrix} 0 \\ \epsilon(\boldsymbol{k},\lambda) \end{pmatrix}_\mu \begin{pmatrix} 0 \\ \epsilon(\boldsymbol{k},\lambda) \end{pmatrix}_\nu . \tag{6.72}$$

If $\mu = 0$ and $\nu = 0$ the result is $+1$ and (6.71) is satisfied. Furthermore the mixed spatial and temporal components of (6.72) vanish. If both indices i, j are spatial we have to prove that

$$\sum_{\lambda=1}^{3} \epsilon_i(\boldsymbol{k},\lambda)\,\epsilon_j(\boldsymbol{k},\lambda) = \delta_{ij} . \tag{6.73}$$

This is just the ordinary completeness relation for an orthogonal dreibein. It is important to include the sign factor $g_{\lambda\lambda}$ in (6.71). In the rest frame also the relation

$$\sum_{\lambda=0}^{3} \epsilon_\mu(\boldsymbol{k},\lambda)\,\epsilon_\nu(\boldsymbol{k},\lambda) = \delta_{\mu\nu} \tag{6.74}$$

would be valid. This, however, is not a covariant equation since the left-hand side consists of a contravariant tensor of rank 2, while the right-hand side is not a tensor. The correct form of the completeness relation, valid in all Lorentz frames, therefore is given by (6.71).

In order to understand the physical meaning of the plane-wave solutions we have to insert the ansatz (6.61) into the Proca equation. The field equation (6.56) for the free ($j^\mu = 0$) vector field simply leads to the energy-momentum relation $k^2 = m^2$. The constraint (6.55) turns into the condition of four-dimensional transversality

$$k \cdot \epsilon = 0 . \tag{6.75}$$

This constraint implies that the massive vector field has only *three polarization states* ($\lambda = 1, 2, 3$). The fourth state ($\lambda = 0$) contradicts the consistency condition and thus cannot be used to describe a (free) particle.

Nevertheless we have to note that the three "physical" polarization states on their own do not form a complete set in the mathematical sense. Rather, the completeness relation will contain an extra term since (6.71) can be written in the form

$$\sum_{\lambda=1}^{3} \epsilon_\mu(\boldsymbol{k},\lambda)\,\epsilon_\nu(\boldsymbol{k},\lambda) = -\left(g_{\mu\nu} - \frac{1}{m^2}k_\mu k_\nu\right) . \tag{6.76}$$

6.4.2 The Massless Vector Field

When constructing the polarization states of the massless vector field we can begin as in the massive case and introduce two *transverse polarization vectors* $\lambda = 1, 2$ (cf. (6.63)), which refer to a fixed Lorentz system. However, now the momentum vector k can no longer be used as a basis vector: k cannot be normalized to 1 since the dispersion relation now reads $k^2 = 0$. In addition we recall that the longitudinal polarization vector (6.67) is not defined for $m = 0$. In the massless case it is impossible to construct a third polarization vector which is normalizable and at the same time transverse (in four dimensions).

To avoid this problem we arbitrarily define a time-like unit vector which in the chosen special Lorentz frame simply is given by

$$n = (1, 0, 0, 0) \quad \text{where} \quad n^2 = +1 . \tag{6.77}$$

The *longitudinal polarization vector* can then be written in covariant form as[5]

$$\epsilon(\boldsymbol{k}, 3) = \frac{k - n(k \cdot n)}{\left[(k \cdot n)^2 - k^2\right]^{1/2}} \tag{6.78}$$

This vector indeed has the correct normalization

$$\epsilon(\boldsymbol{k}, 3) \cdot \epsilon(\boldsymbol{k}, 3) = \frac{k^2 - 2(k \cdot n)^2 + n^2 (k \cdot n)^2}{(k \cdot n)^2 - k^2} = -1 . \tag{6.79}$$

Furthermore we have $\epsilon(\boldsymbol{k}, 3) \cdot n = 0$. In the special Lorentz frame, (6.78) is reduced to

$$\epsilon(\boldsymbol{k}, 3) = \left(0, \frac{\boldsymbol{k}}{|\boldsymbol{k}|}\right) . \tag{6.80}$$

(Note the difference compared to the polarization vector $\epsilon(\boldsymbol{k}, 3)$ of the massive vector field which contained a time-like component.) The vector (6.80) thus is orthogonal to the transverse vectors (6.63). To add a fourth member to our vector basis we simply can use the unit vector n as a *time-like polarization vector*

$$\epsilon(\boldsymbol{k}, 0) = n , \tag{6.81}$$

which by definition is normalized to $+1$ and stands orthogonal to the other polarization vectors ($\lambda = 1, 2, 3$).

The polarization basis in an arbitrary inertial frame can be obtained via a Lorentz transformation. Of course this will give a more complicated appearance to the vectors $\epsilon^\mu(\boldsymbol{k}, \lambda)$ which in general will consist of a mixture of space and time components.

The four-dimensional scalar product of the basis vectors and the momentum vector reads

$$k \cdot \epsilon(\boldsymbol{k}, 1) = k \cdot \epsilon(\boldsymbol{k}, 2) = 0 , \tag{6.82a}$$
$$k \cdot \epsilon(\boldsymbol{k}, 0) = -k \cdot \epsilon(\boldsymbol{k}, 3) = k \cdot n , \tag{6.82b}$$

which of course is valid in any frame of reference. This has to be compared with (6.70) for the massive vector field. The *completeness relation* (6.71) remains valid also for the massless case:

$$\sum_{\lambda=0}^{3} g_{\lambda\lambda}\, \epsilon_\mu(\boldsymbol{k}, \lambda)\, \epsilon_\nu(\boldsymbol{k}, \lambda) = g_{\mu\nu} . \tag{6.83}$$

The proof contained in (6.72) can be taken over, although now the rest frame does not exist for a massless particle. Our special Lorentz frame will do as well, however, since the only property used in the proof was the absence of

---

[5] Free photons satisfy the dispersion relation $k^2 = 0$ and the normalizing denominator in (6.78) is simplified to $k \cdot n$.

time-like components of the $\lambda = 1, 2, 3$ vectors. For free photons ($k^2 = 0$) the completeness relation can be written as

$$\sum_{\lambda=1}^{2} \epsilon_\mu(\boldsymbol{k}, \lambda)\, \epsilon_\nu(\boldsymbol{k}, \lambda) = -g_{\mu\nu} - \frac{k_\mu k_\nu}{(k \cdot n)^2} + \frac{k_\mu n_\nu + k_\nu n_\mu}{k \cdot n} \,. \qquad (6.84)$$

Here we have isolated the transverse modes on the left-hand side. The general expression applicable also for virtual photons ($k^2 \neq 0$) looks a bit more complex:

$$\sum_{\lambda=1}^{2} \epsilon_\mu(\boldsymbol{k}, \lambda)\, \epsilon_\nu(\boldsymbol{k}, \lambda) = -g_{\mu\nu} - \frac{k_\mu k_\nu - (k_\mu n_\nu + k_\nu n_\mu) k \cdot n + n_\mu n_\nu k^2}{(k \cdot n)^2 - k^2} \,. \qquad (6.85)$$

6.5 Canonical Quantization of the Massive Vector Field

The canonical quantization procedure amounts to the imposition of equal-time commutation relations (ETCR) for the field variables and their canonically conjugate "momenta". For the spin-1 field this is not so trivial since one has to isolate the "relevant" degrees of freedom first. We will start with the massive vector field (Proca field), which can be quantized quite easily. The more involved case of the photon field will be treated separately in Chap. 7.

For a start, we have to introduce the Hamilton formulation of the Proca theory. The field variable is a four-dimensional vector $A^\mu(x)$. For each component $\mu = 0, \ldots, 3$ a canonically conjugate field can be obtained by differentiation of the Lagrange density $\mathcal{L}$ given in (6.51). According to the identity (6.31) one finds[6]

$$\pi^\mu = \frac{\partial \mathcal{L}}{\partial(\partial_0 A_\mu)} = -F^{0\mu} \,, \qquad (6.86)$$

or

$$\pi^0 = 0 \quad \text{and} \quad \pi^i = E^i \,. \qquad (6.87)$$

We are confronted with the new phenomenon that the 0 component of the vector potential A^0 *does not possess a canonically conjugate field* π^0 since the Lagrangian $\mathcal{L}$ does not contain the time derivative $\partial_0 A_0$. It might appear that the nonexistence of π_0 spells deep trouble for any attempt to quantize the theory as it is impossible to postulate a canonical commutation relation for A^0. Fortunately, however, such a relation is not needed since the Proca field has only three independent degrees of freedom, as we know from the consistency condition (6.55). Thus *the field $A_0(x)$ is a dependent variable*. The independent dynamical variables of the theory are the three-vector fields $\boldsymbol{A}$ and $\boldsymbol{\pi} = \boldsymbol{E}$. If they are known, the value of A_0 follows automatically. Taking the Proca equation (6.52), $\partial_\mu F^{\mu 0} + m^2 A^0 = 0$, we have the connection

$$A^0 = -\frac{1}{m^2} \boldsymbol{\nabla} \cdot \boldsymbol{E} \,, \qquad (6.88)$$

---

[6] Note that A^μ and π_μ are the conjugate fields. Thus the space components obtain a minus sign if contravariant vectors $\boldsymbol{A}$ and $\boldsymbol{\pi}$ are used for both fields.

6.5 Canonical Quantization of the Massive Vector Field

which proves that A^0 is a dependent quantity and not a dynamical variable of its own.

Let us now derive the Hamiltonian of the Proca field. For the Hamilton density we find

$$\begin{aligned}\mathcal{H} &= \pi_\mu \dot{A}^\mu - \mathcal{L} = -F_{0\mu}\dot{A}^\mu + \frac{1}{4}F_{\mu\nu}F^{\mu\nu} - \frac{1}{2}m^2 A_\mu A^\mu \\ &= -\boldsymbol{E}\cdot\dot{\boldsymbol{A}} - \frac{1}{2}(\boldsymbol{E}^2 - \boldsymbol{B}^2) - \frac{1}{2}m^2(A_0^{\ 2} - \boldsymbol{A}^2). \end{aligned} \quad (6.89)$$

The right-hand side has to be expressed in terms of the canonical variables $\boldsymbol{A}$ and $\boldsymbol{E}$. To achieve this we can make use of $\boldsymbol{E} = -\nabla A_0 - \partial_0 \boldsymbol{A}$ and

$$\boldsymbol{E}\cdot\nabla A_0 = \nabla\cdot(\boldsymbol{E} A_0) - A_0(\nabla\cdot\boldsymbol{E}) = \nabla\cdot(\boldsymbol{E} A_0) + m^2 A_0^2, \quad (6.90)$$

as well as (6.88). This leads to

$$\begin{aligned}\mathcal{H} &= -\boldsymbol{E}\cdot(-\boldsymbol{E} - \nabla A_0) + \frac{1}{2}(-\boldsymbol{E}^2 + \boldsymbol{B}^2 + m^2\boldsymbol{A}^2) - \frac{1}{2}m^2 A_0^2 \\ &= \frac{1}{2}(\boldsymbol{E}^2 + \boldsymbol{B}^2 + m^2\boldsymbol{A}^2 + m^2 A_0^2) + \nabla\cdot(\boldsymbol{E} A_0). \end{aligned} \quad (6.91)$$

The divergence term yields only a surface contribution upon integration over d^3x and can be discarded. The *Hamiltonian* thus reads

$$H = \int d^3x \frac{1}{2}\left(\boldsymbol{E}^2 + (\nabla\times\boldsymbol{A})^2 + m^2\boldsymbol{A}^2 + \frac{1}{m^2}(\nabla\cdot\boldsymbol{E})^2\right). \quad (6.92)$$

The same result is obtained from the integrated energy-momentum tensor $\Theta^{00}(x)$ or $T^{00}(x)$ (see (6.60a)).

Canonical quantization of the Proca field is achieved as usual by imposing the ETCR for the canonically conjugate field variables A^i and π_j (keep in mind the sign change when going to contravariant vectors):

$$[\hat{A}^i(\boldsymbol{x},t), \hat{E}^j(\boldsymbol{x}',t)] = -\mathrm{i}\delta_{ij}\delta^3(\boldsymbol{x}-\boldsymbol{x}'), \quad (6.93\mathrm{a})$$

$$[\hat{A}^i(\boldsymbol{x},t), \hat{A}^j(\boldsymbol{x}',t)] = [\hat{E}^i(\boldsymbol{x},t), \hat{E}^j(\boldsymbol{x}',t)] = 0. \quad (6.93\mathrm{b})$$

The three spatial components of $\hat{\boldsymbol{A}}$ are treated as independent fields, which finds expression in the factor δ_{ij} on the right-hand side of (6.93a). We have used ordinary commutators with a minus sign for quantization since the spin-1 particles are bosons. Attempts to use fermionic quantization rules as in the spin-0 case would lead to a violation of microcausality. The arguments of Sect. 4.4 can be applied to prove this.

The Hamilton operator of the (neutral) massive vector field is the field-quantized version of (6.92):

$$\hat{H} = \int d^3x \frac{1}{2}\left(\hat{\boldsymbol{E}}^2 + (\nabla\times\hat{\boldsymbol{A}})^2 + m^2\hat{\boldsymbol{A}}^2 + \frac{1}{m^2}(\nabla\cdot\hat{\boldsymbol{E}})^2\right). \quad (6.94)$$

Using (6.93) one easily confirms that the Heisenberg equations of motion for the operators $\hat{\boldsymbol{A}}(\boldsymbol{x},t)$ and $\hat{\boldsymbol{E}}(\boldsymbol{x},t)$ become

$$\dot{\hat{\boldsymbol{A}}} = -\hat{\boldsymbol{E}} + \frac{1}{m^2}\nabla(\nabla\cdot\hat{\boldsymbol{E}}), \quad (6.95\mathrm{a})$$

$$\dot{\hat{\boldsymbol{E}}} = -\nabla^2\hat{\boldsymbol{A}} + \nabla(\nabla\cdot\hat{\boldsymbol{A}}) + m^2\hat{\boldsymbol{A}}. \quad (6.95\mathrm{b})$$

This again agrees with the Proca equation of the classical theory, (6.53) with (6.55), expressed in terms of $\boldsymbol{A}$ and $\boldsymbol{E}$.

The quantization rule (6.93) appears to have the unpleasant property of not being manifestly covariant, since it is expressed in terms of the three-vectors $\hat{\boldsymbol{A}}$ and $\hat{\boldsymbol{E}}$. It would have been much nicer to use four-vectors and to quantize the vector potential A^μ. This can be achieved if one *defines* the field operator $\hat{A}^0(x)$ to be

$$\hat{A}^0 = -\frac{1}{m^2}\boldsymbol{\nabla}\cdot\hat{\boldsymbol{E}}\,, \tag{6.96}$$

corresponding to the classical relation (6.88). Using (6.93), we find that this operator satisfies the commutation relations

$$\left[\hat{\boldsymbol{A}}(\boldsymbol{x},t),\,\hat{A}^0(\boldsymbol{x}',t)\right] = \mathrm{i}\frac{1}{m^2}\boldsymbol{\nabla}\delta^3(\boldsymbol{x}-\boldsymbol{x}')\,, \tag{6.97a}$$

$$\left[\hat{A}^0(\boldsymbol{x},t),\,\hat{A}^0(\boldsymbol{x}',t)\right] = 0\,. \tag{6.97b}$$

Because the space and time components of $\hat{A}^\mu$ are treated differently our quantization procedure appears not to be covariant. In Exercise 6.4 it will be shown, however, that (6.93) and (6.97) are special cases of the following *covariant commutation relation*, which is valid for arbitrary space-time coordinates x and y:

$$\left[\hat{A}^\mu(x),\,\hat{A}^\nu(y)\right] = -\mathrm{i}\left(g^{\mu\nu} + \frac{1}{m^2}\partial^\mu\partial^\nu\right)\Delta(x-y)\,. \tag{6.98}$$

Here $\Delta(x-y)$ is the invariant Pauli–Jordan–Schwinger function introduced in Sect. 4.4. The form of (6.98) thus demonstrates that the canonical quantization procedure is covariant also for the massive spin-1 field.

EXAMPLE

6.3 Fourier Decomposition of the Proca Field Operator

The plane-wave decomposition of the vector field operator makes use of the basis states $A^\mu(\boldsymbol{k},\lambda)$ having well-defined polarization ($\lambda=1,2,3$), which were introduced in Sect. 6.4. The field operator is written as

$$\hat{A}^\mu(x) = \int\mathrm{d}^3k\sum_{\lambda=1}^{3}\left(\hat{a}_{\boldsymbol{k}\lambda}\,A^\mu(\boldsymbol{k},\lambda;x) + \hat{a}_{\boldsymbol{k}\lambda}^\dagger\,A^{\mu*}(\boldsymbol{k},\lambda;x)\right) \tag{1a}$$

$$= \int\frac{\mathrm{d}^3k}{\sqrt{2\omega_k(2\pi)^3}}\sum_{\lambda=1}^{3}\left(\hat{a}_{\boldsymbol{k}\lambda}\,\epsilon^\mu(\boldsymbol{k},\lambda)\,\mathrm{e}^{-\mathrm{i}k\cdot x} + \hat{a}_{\boldsymbol{k}\lambda}^\dagger\,\epsilon^{\mu*}(\boldsymbol{k},\lambda)\,\mathrm{e}^{\mathrm{i}k\cdot x}\right). \tag{1b}$$

The normalization factor $N_k = (2\omega_k(2\pi)^3)^{-1/2}$, which agrees with that used in the spin-0 case, ensures that the expansion coefficients $\hat{a}_{\boldsymbol{k}\lambda}$ satisfy simple commutation relations. The field operator in (1b) by construction is hermitean, which in the classical theory corresponds to a real-valued field. As a consequence our theory describes a *neutral spin-1 field*. If one wants to describe a *charged spin-1 field*, (1) has to be replaced by the expansion

Example 6.3

$$\hat{A}^\mu(x) = \int d^3k \sum_{\lambda=1}^{3} \left(\hat{a}_{k\lambda}\, A^\mu(k,\lambda;x) + \hat{b}^\dagger_{k\lambda}\, A^{\mu*}(k,\lambda;x) \right). \tag{2}$$

The operators $\hat{a}_{k\lambda}$ and $\hat{b}_{k\lambda}$ separately describe particles and antiparticles (cf. the treatment of the charged Klein–Gordon field in Chap. 4). In the following we will concentrate on the neutral field described by (1), but the discussion can be easily carried over to the charged case.

The expansion (1) sums over the two transverse ($\lambda = 1, 2$) and the longitudinal ($\lambda = 3$) polarization state. Explicit expression for the polarization vectors $\epsilon^\mu(k, \lambda)$ have been given in (6.63) and (6.67).[7]

The operator of the three-dimensional vector potential has the following expansion:

$$\hat{\mathbf{A}}(x) = \int \frac{d^3k}{\sqrt{2\omega_k(2\pi)^3}} \sum_{\lambda=1}^{3} \boldsymbol{\epsilon}(k,\lambda) \left(\hat{a}_{k\lambda}\, e^{-ik\cdot x} + \hat{a}^\dagger_{k\lambda}\, e^{ik\cdot x} \right). \tag{3}$$

The corresponding field-strength vector (i.e., the canonically conjugate field) is given by

$$\begin{aligned}
\hat{\mathbf{E}}(x) &= -\partial_0\, \hat{\mathbf{A}} - \boldsymbol{\nabla}\hat{A}_0 \\
&= \int \frac{d^3k}{\sqrt{2\omega_k(2\pi)^3}} \sum_{\lambda=1}^{3} \left(i\omega_k\boldsymbol{\epsilon}(k,\lambda) - i\mathbf{k}\,\epsilon^0(k,\lambda) \right)\left(\hat{a}_{k\lambda}\, e^{-ik\cdot x} - \hat{a}^\dagger_{k\lambda}\, e^{ik\cdot x} \right) \\
&= i\int \frac{d^3k}{\sqrt{2\omega_k(2\pi)^3}} \sum_{\lambda=1}^{3} \omega_k\tilde{\boldsymbol{\epsilon}}(k,\lambda)\left(\hat{a}_{k\lambda}\, e^{-ik\cdot x} - \hat{a}^\dagger_{k\lambda}\, e^{ik\cdot x} \right). \tag{4}
\end{aligned}$$

We have introduced the following modified polarization vectors:

$$\tilde{\boldsymbol{\epsilon}}(k,\lambda) := \boldsymbol{\epsilon}(k,\lambda) - \frac{1}{\omega_k}\mathbf{k}\,\epsilon^0(k,\lambda) \tag{5a}$$

$$= \boldsymbol{\epsilon}(k,\lambda) - \frac{k^2}{\omega_k^2}\mathbf{k}\cdot\boldsymbol{\epsilon}(k,\lambda), \tag{5b}$$

where in the second line the transversality condition $\omega_k\epsilon^0 - \mathbf{k}\cdot\boldsymbol{\epsilon} = 0$ has been used. These vectors have the property

$$\mathbf{k}\cdot\tilde{\boldsymbol{\epsilon}}(k,\lambda) = \left(1 - \frac{k^2}{\omega_k^2}\right)\mathbf{k}\cdot\boldsymbol{\epsilon}(k,\lambda) = \frac{m^2}{\omega_k^2}\mathbf{k}\cdot\boldsymbol{\epsilon}(k,\lambda). \tag{6}$$

For the special choice of polarization vectors used in Sect. 6.4.1 we get explicitly

$$\tilde{\boldsymbol{\epsilon}}(k,\lambda) = f_\lambda \boldsymbol{\epsilon}(k,\lambda) \quad \text{with} \quad f_\lambda = \begin{cases} 1 & \text{for } \lambda = 1,2 \\ m^2/\omega_k^2 & \text{for } \lambda = 3 \end{cases}. \tag{7}$$

---

[7] To describe planar linear polarization these vectors can be chosen to be real-valued, $\epsilon^{\mu*}(k,\lambda) = \epsilon^\mu(k,\lambda)$. For circular polarization see (38).

Example 6.3

The Commutation Relations

We can expect that $\hat{a}^\dagger_{k\lambda}$ and $\hat{a}_{k\lambda}$ as usual will play the role of creation and annihilation operators for spin-1 bosons. To check their commutation relations we express them by the "inverse Fourier representation" in terms of the field operators $\hat{A}(x)$ and $\hat{E}(x)$. We follow the analogous treatment of the Klein–Gordon theory given in Sect. 4. The *scalar product* of two Proca vectors will be defined in analogy to (4.24) as

$$\left(A(x), A'(x)\right) = \mathrm{i} \int \mathrm{d}^3 x \, A^{\mu*}(x) \stackrel{\leftrightarrow}{\partial_0} A'_\mu(x) \,, \tag{8}$$

where $A \stackrel{\leftrightarrow}{\partial_0} B = A(\partial_0 B) - (\partial_0 A)B$.

The scalar product of two plane waves then becomes

$$\begin{aligned}
\left(A(\boldsymbol{k}', \lambda'), A(\boldsymbol{k}, \lambda)\right) &= \mathrm{i} \int \mathrm{d}^3 x \, \frac{1}{\sqrt{2\omega_{k'}(2\pi)^3}} \frac{1}{\sqrt{2\omega_k(2\pi)^3}} \\
&\quad \times \epsilon^\mu(\boldsymbol{k}', \lambda') \, \mathrm{e}^{\mathrm{i}k'\cdot x} \stackrel{\leftrightarrow}{\partial_0} \epsilon_\mu(\boldsymbol{k}, \lambda) \, \mathrm{e}^{-\mathrm{i}k\cdot x} \\
&= \delta^3(\boldsymbol{k}' - \boldsymbol{k}) \, \epsilon^\mu(\boldsymbol{k}, \lambda')\epsilon_\mu(\boldsymbol{k}, \lambda) \,.
\end{aligned} \tag{9}$$

The orthogonality of the four-dimensional polarization vectors then implies

$$\left(A(\boldsymbol{k}', \lambda'), A(\boldsymbol{k}, \lambda)\right) = \delta^3(\boldsymbol{k}' - \boldsymbol{k}) \, g_{\lambda\lambda'} \,. \tag{10}$$

Similarly we find

$$\left(A^*(\boldsymbol{k}', \lambda'), A^*(\boldsymbol{k}, \lambda)\right) = -\delta^3(\boldsymbol{k}' - \boldsymbol{k}) \, g_{\lambda\lambda'} \tag{11}$$

and

$$\left(A(\boldsymbol{k}', \lambda'), A^*(\boldsymbol{k}, \lambda)\right) = \left(A^*(\boldsymbol{k}', \lambda'), A(\boldsymbol{k}, \lambda)\right) = 0 \,. \tag{12}$$

If we apply these relations to (1) the creation and annihilation operators can be projected out. Keeping in mind $g_{\lambda\lambda} = -1$, we obtain

$$\hat{a}_{k\lambda} = \left(A(\boldsymbol{k}, \lambda), \hat{A}(x)\right) = -\mathrm{i} \int \mathrm{d}^3 x \, A^{\mu*}(\boldsymbol{k}, \lambda) \stackrel{\leftrightarrow}{\partial_0} \hat{A}_\mu(x) \tag{13}$$

and a similar equation for $\hat{a}^\dagger_{k\lambda}$. When we insert the plane wave, (13) becomes

$$\begin{aligned}
\hat{a}_{k\lambda} &= -\mathrm{i} \int \mathrm{d}^3 x \left(A^{\mu*}(\boldsymbol{k}, \lambda) \, \partial_0 \hat{A}_\mu(x) - \partial_0 A^{\mu*}(\boldsymbol{k}, \lambda) \, \hat{A}_\mu(x) \right) \\
&= -\mathrm{i} \int \mathrm{d}^3 x \, \frac{\mathrm{e}^{\mathrm{i}k\cdot x}}{\sqrt{2\omega_k(2\pi)^3}} \, \epsilon^\mu(\boldsymbol{k}, \lambda) \left(\partial_0 \hat{A}_\mu(x) - \mathrm{i}\omega_k \hat{A}_\mu(x) \right) \,.
\end{aligned} \tag{14}$$

This is to be expressed in terms of the three-dimensional fields $\hat{\boldsymbol{E}}$ and $\hat{\boldsymbol{A}}$:

$$\begin{aligned}
\hat{a}_{k\lambda} &= -\mathrm{i} \int \mathrm{d}^3 x \, \frac{\mathrm{e}^{\mathrm{i}k\cdot x}}{\sqrt{2\omega_k(2\pi)^3}} \left(\epsilon^0 \partial_0 \hat{A}_0 - \boldsymbol{\epsilon} \cdot \partial_0 \hat{\boldsymbol{A}} - \mathrm{i}\omega_k \epsilon^0 \hat{A}_0 + \mathrm{i}\omega_k \boldsymbol{\epsilon} \cdot \hat{\boldsymbol{A}} \right) \\
&= -\mathrm{i} \int \mathrm{d}^3 x \, \frac{\mathrm{e}^{\mathrm{i}k\cdot x}}{\sqrt{2\omega_k(2\pi)^3}} \\
&\quad \times \left(-\epsilon^0 \boldsymbol{\nabla} \cdot \hat{\boldsymbol{A}} + \boldsymbol{\epsilon} \cdot \hat{\boldsymbol{E}} + \boldsymbol{\epsilon} \cdot \boldsymbol{\nabla} \hat{A}_0 - \mathrm{i}\omega_k \epsilon^0 \hat{A}_0 + \mathrm{i}\omega_k \boldsymbol{\epsilon} \cdot \hat{\boldsymbol{A}} \right) \,,
\end{aligned} \tag{15}$$

where we have used $\partial \hat{A}_0 = -\boldsymbol{\nabla}\cdot\hat{\boldsymbol{A}}$ and $-\partial_0 \hat{\boldsymbol{A}} = \hat{\boldsymbol{E}} + \boldsymbol{\nabla}\hat{A}_0$. Equation (15) can be simplified by integrating by parts:

$$-\epsilon^0 \boldsymbol{\nabla}\cdot\hat{\boldsymbol{A}} \to -\mathrm{i}\epsilon^0 \, \boldsymbol{k}\cdot\hat{\boldsymbol{A}} \,. \tag{16}$$

Using $k\cdot\epsilon = 0$ we also find

$$\boldsymbol{\epsilon}\cdot\boldsymbol{\nabla}\hat{A}_0 \omega_k\, \epsilon^0 \hat{A}_0 \to \left(\mathrm{i}\boldsymbol{\epsilon}\cdot\boldsymbol{k} - \mathrm{i}\omega_k \epsilon^0\right) \hat{A}_0 = 0 \,. \tag{17}$$

Thus the expansion coefficients of the field operator are given by

$$\begin{aligned}
\hat{a}_{\boldsymbol{k}\lambda} &= \int\!\mathrm{d}^3 x \,\frac{\mathrm{e}^{\mathrm{i}\boldsymbol{k}\cdot\boldsymbol{x}}}{\sqrt{2\omega_k(2\pi)^3}} \Big((\omega_k \boldsymbol{\epsilon} - \epsilon^0 \boldsymbol{k})\cdot\hat{\boldsymbol{A}} - \mathrm{i}\boldsymbol{\epsilon}\cdot\hat{\boldsymbol{E}}\Big) \\
&= \int\!\mathrm{d}^3 x \,\frac{\mathrm{e}^{\mathrm{i}\boldsymbol{k}\cdot\boldsymbol{x}}}{\sqrt{2\omega_k(2\pi)^3}} \Big(\omega_k \widetilde{\boldsymbol{\epsilon}}(\boldsymbol{k},\lambda)\cdot\hat{\boldsymbol{A}}(x) - \mathrm{i}\boldsymbol{\epsilon}(\boldsymbol{k},\lambda)\cdot\hat{\boldsymbol{E}}(x)\Big) \,,
\end{aligned} \tag{18}$$

where the modified polarization vector $\widetilde{\boldsymbol{\epsilon}}(\boldsymbol{k},\lambda)$ from (5) has been used. With the aid of this inverse Fourier representation, the commutation relations for the creation and annihilation operators can be derived immediately. We find

$$\begin{aligned}
\left[\hat{a}_{\boldsymbol{k}'\lambda'}, \hat{a}^\dagger_{\boldsymbol{k}\lambda}\right] &= \int\!\mathrm{d}^3 x'\,\mathrm{d}^3 x \,\frac{\mathrm{e}^{\mathrm{i}\boldsymbol{k}'\cdot\boldsymbol{x}'}}{\sqrt{2\omega_{k'}(2\pi)^3}}\,\frac{\mathrm{e}^{-\mathrm{i}\boldsymbol{k}\cdot\boldsymbol{x}}}{\sqrt{2\omega_k(2\pi)^3}} \\
&\quad \times \left[\omega_{k'}\widetilde{\boldsymbol{\epsilon}}'\cdot\hat{\boldsymbol{A}}' - \mathrm{i}\boldsymbol{\epsilon}'\cdot\hat{\boldsymbol{E}}',\, \omega_k \widetilde{\boldsymbol{\epsilon}}\cdot\hat{\boldsymbol{A}} + \mathrm{i}\boldsymbol{\epsilon}\cdot\hat{\boldsymbol{E}}\right] \\
&= \int\!\mathrm{d}^3 x \,\frac{\mathrm{e}^{\mathrm{i}(\omega_{k'} - \omega_k)t - \mathrm{i}(\boldsymbol{k}' - \boldsymbol{k})\cdot\boldsymbol{x}}}{2(2\pi)^3 \sqrt{\omega_{k'}\omega_k}} \left(\omega_{k'}\widetilde{\boldsymbol{\epsilon}}'\cdot\boldsymbol{\epsilon} + \omega_k \boldsymbol{\epsilon}'\cdot\widetilde{\boldsymbol{\epsilon}}\right) \\
&= \delta^3(\boldsymbol{k}' - \boldsymbol{k})\tfrac{1}{2}\big(\widetilde{\boldsymbol{\epsilon}}(\boldsymbol{k},\lambda')\cdot\boldsymbol{\epsilon}(\boldsymbol{k},\lambda) + \boldsymbol{\epsilon}(\boldsymbol{k},\lambda')\cdot\widetilde{\boldsymbol{\epsilon}}(\boldsymbol{k},\lambda)\big),
\end{aligned} \tag{19}$$

where the equal-time commutation relations (6.93) have been used. The vectors $\widetilde{\boldsymbol{\epsilon}}$ and $\boldsymbol{\epsilon}$ satisfy the following orthogonality relation (using (5a) and (6.62)):

$$\begin{aligned}
\widetilde{\boldsymbol{\epsilon}}(\boldsymbol{k},\lambda')\cdot\boldsymbol{\epsilon}(\boldsymbol{k},\lambda) &= \boldsymbol{\epsilon}(\boldsymbol{k},\lambda')\cdot\boldsymbol{\epsilon}(\boldsymbol{k},\lambda) - \frac{1}{\omega_k}\epsilon^0(\boldsymbol{k},\lambda')\boldsymbol{k}\cdot\boldsymbol{\epsilon}(\boldsymbol{k},\lambda) \\
&= \boldsymbol{\epsilon}(\boldsymbol{k},\lambda')\cdot\boldsymbol{\epsilon}(\boldsymbol{k},\lambda) - \epsilon^0(\boldsymbol{k},\lambda')\epsilon^0(\boldsymbol{k},\lambda) \\
&= -\epsilon(\boldsymbol{k},\lambda')\cdot\epsilon(\boldsymbol{k},\lambda) = -g_{\lambda'\lambda} = \delta_{\lambda'\lambda} \,.
\end{aligned} \tag{20}$$

This leads to

$$\left[\hat{a}_{\boldsymbol{k}'\lambda'}, \hat{a}^\dagger_{\boldsymbol{k}\lambda}\right] = \delta^3(\boldsymbol{k}' - \boldsymbol{k})\,\delta_{\lambda\lambda'} \,, \tag{21}$$

which is just the expected result. The other commutation relations of interest,

$$\left[\hat{a}_{\boldsymbol{k}'\lambda'}, \hat{a}_{\boldsymbol{k}\lambda}\right] = \left[\hat{a}^\dagger_{\boldsymbol{k}'\lambda'}, \hat{a}^\dagger_{\boldsymbol{k}\lambda}\right] = 0 \,, \tag{22}$$

can be derived similarly. Thus we have confirmed that the $\hat{a}^\dagger_{\boldsymbol{k}\lambda}$ and $\hat{a}_{\boldsymbol{k}\lambda}$ have the familiar algebraic properties of creation and annihilation operators.

The Hamilton Operator

In the following, we show that the quantized Hamiltonian $\hat{H}$ can be written as the familiar sum over particle-number operators $\hat{n}_{\boldsymbol{k},\lambda} = \hat{a}^\dagger_{\boldsymbol{k}\lambda}\hat{a}_{\boldsymbol{k}\lambda}$ of the individual field modes multiplied by $\hbar\omega_k$. This simple result is reached by an elementary but somewhat lengthy calculation, which we will sketch here. One

Example 6.3

Example 6.3

has to insert the expansions (3) and (4) of the field operators $\hat{\boldsymbol{A}}$ and $\hat{\boldsymbol{E}}$ into the expression for the Hamiltonian (6.94). To eliminate possible (divergent) vacuum contributions, the Hamiltonian will be taken in its normal-ordered form:

$$\hat{H} = \int d^3x \, d\frac{1}{2} : \left(\hat{\boldsymbol{E}}^2 + m^2 \hat{\boldsymbol{A}}^2 + (\boldsymbol{\nabla} \times \hat{\boldsymbol{A}})^2 + \frac{1}{m^2}(\boldsymbol{\nabla} \cdot \hat{\boldsymbol{E}})^2 \right) : , \qquad (23)$$

After inserting the expansions for $\hat{\boldsymbol{A}}$ and $\hat{\boldsymbol{E}}$ and squaring them, one arrives at a sum over sixteen terms under the integral (23) containing various combinations of $\hat{a}$ and $\hat{a}^\dagger$. As an example let us inspect the term containing $\hat{a}^\dagger \hat{a}$, i.e.,

$$\frac{1}{2}\int d^3x \sum_{\lambda'\lambda} \int \frac{d^3k'}{\sqrt{2\omega_{k'}(2\pi)^3}} \frac{d^3k}{\sqrt{2\omega_k(2\pi)^3}} \left(\omega_{k'}\omega_k \widetilde{\boldsymbol{\epsilon}}' \cdot \widetilde{\boldsymbol{\epsilon}} + (\boldsymbol{k}' \times \boldsymbol{\epsilon}') \cdot (\boldsymbol{k} \times \boldsymbol{\epsilon}) \right.$$
$$\left. + m^2 \boldsymbol{\epsilon}' \cdot \boldsymbol{\epsilon} + \frac{1}{m^2} \omega_{k'}\omega_k (\boldsymbol{k} \cdot \widetilde{\boldsymbol{\epsilon}}')(\boldsymbol{k} \cdot \widetilde{\boldsymbol{\epsilon}}) \right) e^{i(k'-k)\cdot x} \hat{a}^\dagger_{k'\lambda'} \hat{a}_{k\lambda}$$
$$= \frac{1}{2} \sum_{\lambda'\lambda} \int \frac{d^3k}{2\omega_k} \left(\omega_k^2 \widetilde{\boldsymbol{\epsilon}}' \cdot \widetilde{\boldsymbol{\epsilon}} + (\boldsymbol{k} \times \boldsymbol{\epsilon}') \cdot (\boldsymbol{k} \times \boldsymbol{\epsilon}) \right.$$
$$\left. + m^2 \boldsymbol{\epsilon}' \cdot \boldsymbol{\epsilon} + \frac{\omega_k^2}{m^2} (\boldsymbol{k} \cdot \widetilde{\boldsymbol{\epsilon}}')(\boldsymbol{k} \cdot \widetilde{\boldsymbol{\epsilon}}) \right) \hat{a}^\dagger_{k\lambda'} \hat{a}_{k\lambda} . \qquad (24)$$

Using the identity $(\boldsymbol{k} \times \boldsymbol{\epsilon}') \cdot (\boldsymbol{k} \times \boldsymbol{\epsilon}) = k^2 \boldsymbol{\epsilon}' \cdot \boldsymbol{\epsilon} - (\boldsymbol{k} \cdot \boldsymbol{\epsilon}')(\boldsymbol{k} \cdot \boldsymbol{\epsilon})$ and (5a), (6) and $\boldsymbol{k} \cdot \boldsymbol{\epsilon} = \omega_k \epsilon^0$, the polarization-dependent terms in (24) become

$$\begin{aligned}(\ldots) &= \omega_k^2 \boldsymbol{\epsilon}' \cdot \boldsymbol{\epsilon} + \omega_k^2 \widetilde{\boldsymbol{\epsilon}}' \cdot \widetilde{\boldsymbol{\epsilon}} - (\boldsymbol{k} \cdot \boldsymbol{\epsilon}')(\boldsymbol{k} \cdot \boldsymbol{\epsilon}) + \frac{\omega_k^2}{m^2}(\boldsymbol{k} \cdot \widetilde{\boldsymbol{\epsilon}}')(\boldsymbol{k} \cdot \widetilde{\boldsymbol{\epsilon}}) \\ &= 2\omega_k^2 (\boldsymbol{\epsilon}' \cdot \boldsymbol{\epsilon} - \epsilon^{0'}\epsilon^0) = -2\omega_k^2 g_{\lambda'\lambda} = 2\omega_k^2 \delta_{\lambda'\lambda} .\end{aligned} \qquad (25)$$

The same coefficients also stand in front of the term containing $\hat{a}\hat{a}^\dagger$. Because in (23) normal ordering was prescribed, both terms can be combined into one. A similar calculation will show that the cofficients of the terms involving $\hat{a}\,\hat{a}$ and $\hat{a}^\dagger\hat{a}^\dagger$ vanish. Thus the *Hamiltonian of the massive spin-1 field* in second quantization indeed have the expected form

$$\hat{H} = \sum_{\lambda=1}^{3} \int d^3k \, \omega_k \, \hat{a}^\dagger_{k\lambda} \hat{a}_{k\lambda} . \qquad (26)$$

In a similar fashion the *momentum operator* can be derived, starting from the energy-momentum tensor T^{0i}, leading to

$$\hat{\boldsymbol{P}} = \sum_{\lambda=1}^{3} \int d^3k \, \boldsymbol{k} \, \hat{a}^\dagger_{k\lambda} \hat{a}_{k\lambda} . \qquad (27)$$

Thus also for the vector field the quanta are found to carry energy ω_k and momentum $\boldsymbol{k}$.

The Spin Operator

Since the vector field carries an intrinsic angular momentum of 1 there must exist a spin operator that is added to the ordinary orbital angular momentum operator. The spin term follows from Noether's theorem (Sect. 2.4) and we already have derived the relevant classical expression (6.49). The spin operator is given by

$$\hat{S} = \int d^3x : \hat{E} \times \hat{A} : , \qquad (28)$$

where again we have adopted the normal-ordering prescription. The Fourier decomposition of the field operators (3) and (4) leads to

$$\begin{aligned}
\hat{S} &= \int d^3x \int \frac{d^3k'}{\sqrt{2\omega_{k'}(2\pi)^3}} \int \frac{d^3k}{\sqrt{2\omega_k(2\pi)^3}} \\
&\quad \times \sum_{\lambda,\lambda'=1}^{3} i\omega_{k'}\, \tilde{\epsilon}(k',\lambda') \times \epsilon(k,\lambda) \\
&\quad :\left(\hat{a}_{k',\lambda'}\,e^{-ik'\cdot x} - \hat{a}^\dagger_{k',\lambda'}\,e^{ik'\cdot x}\right)\left(\hat{a}_{k,\lambda}\,e^{-ik\cdot x} + \hat{a}^\dagger_{k,\lambda}\,e^{ik\cdot x}\right): .
\end{aligned} \qquad (29)$$

This expression can be simplified by an elementary calculation, leading to

$$\begin{aligned}
\hat{S} &= \frac{i}{2}\int d^3k \sum_{\lambda,\lambda'=1}^{3} \Big[\tilde{\epsilon}(k,\lambda') \times \epsilon(k,\lambda)\left(\hat{a}^\dagger_{k\lambda}\hat{a}_{k\lambda'} - \hat{a}^\dagger_{k\lambda'}\hat{a}_{k\lambda}\right) \\
&\quad + \tilde{\epsilon}(-k,\lambda') \times \epsilon(k,\lambda)\left(\hat{a}_{-k\lambda'}\hat{a}_{k\lambda}\,e^{-2i\omega_k t} - \hat{a}^\dagger_{-k\lambda'}\hat{a}^\dagger_{k\lambda}\,e^{2i\omega_k t}\right)\Big] .
\end{aligned} \qquad (30)$$

As we have already observed for the spin-$\frac{1}{2}$ case, the spin S and orbital angular momentum L in general cannot be fixed separately for vector fields. However, the *helicity*, i.e., the projection $\Lambda = S \cdot k/|k|$ of the spin in the direction of motion, is a well-defined quantity. Indeed in this case the expression (30) is simplified considerably. The longitudinal polarization, $\lambda = 3$, does not contribute since $\tilde{\epsilon}(k,3)$ points in the direction of k and the cross products are orthogonal to the momentum vector. For the transverse polarizations, $\lambda = 1, 2$, we get $\tilde{\epsilon}(k,\lambda) = \epsilon(k,\lambda)$. The second contribution in (30) vanishes since it changes its sign when we relabel the variables $\lambda \leftrightarrow \lambda'$, $k \leftrightarrow -k$. We are left with

$$\hat{\Lambda} = \frac{i}{2}\int d^3k \sum_{\lambda,\lambda'=1}^{2} e_k \cdot \epsilon(k,\lambda') \times \epsilon(k,\lambda)\left(\hat{a}^\dagger_{k\lambda}\hat{a}_{k\lambda'} - \hat{a}^\dagger_{k\lambda'}\hat{a}_{k\lambda}\right) . \qquad (31)$$

Since the unit vectors $\epsilon(k,1)$, $\epsilon(k,2)$, and $e_k = k/|k|$ form a (right-handed) orthogonal dreibein, their triple product is trivial and we get

$$\hat{\Lambda} = i\int d^3k \left(\hat{a}^\dagger_{k2}\hat{a}_{k1} - \hat{a}^\dagger_{k1}\hat{a}_{k2}\right) . \qquad (32)$$

This operator is nondiagonal, i.e., it does not consist of particle-number operators $\hat{n}_{k\lambda} = \hat{a}^\dagger_{k\lambda}\hat{a}_{k\lambda}$. Thus states of the type $\hat{a}^\dagger_{k\lambda}|0\rangle$ are not eigenstates of the spin operator. This can be achieved, however, by going to the *spherical basis*. We define new set of operators by the linear transformation

Example 6.3

$$\hat{a}_{k+} = \tfrac{1}{\sqrt{2}}\left(\hat{a}_{k1} - \mathrm{i}\,\hat{a}_{k2}\right),$$

$$\hat{a}_{k-} = \tfrac{1}{\sqrt{2}}\left(\hat{a}_{k1} + \mathrm{i}\,\hat{a}_{k2}\right), \tag{33}$$

$$\hat{a}_{k0} = \hat{a}_{k3},$$

which can be inverted as

$$\hat{a}_{k1} = \tfrac{1}{\sqrt{2}}\left(\hat{a}_{k+} + \hat{a}_{k-}\right),$$

$$\hat{a}_{k2} = \tfrac{\mathrm{i}}{\sqrt{2}}\left(\hat{a}_{k+} - \hat{a}_{k-}\right), \tag{34}$$

$$\hat{a}_{k3} = \hat{a}_{k0}.$$

Equations (33) (and the corresponding equations for the hermitean conjugate operators $\hat{a}^\dagger$) describe a canonical transformation since the commutation relations are not affected,

$$\left[\hat{a}_{k'+}, \hat{a}_{k+}^\dagger\right] = \left[\hat{a}_{k'-}, \hat{a}_{k-}^\dagger\right] = \left[\hat{a}_{k'0}, \hat{a}_{k0}^\dagger\right] = \delta^3(\boldsymbol{k} - \boldsymbol{k}'), \tag{35}$$

which follows from (21)–(22). All the remaining commutators vanish. The set of operators $\hat{a}_{k+}, \hat{a}_{k-}, \hat{a}_{k0}$ and their hermitean conjugate partners thus satisfy the algebra that characterizes creation and annihilation operators. The field quanta are *circularly polarized* spin-1 particles. The operator (32) is now diagonal:

$$\hat{\Lambda} = \int \mathrm{d}^3 k \left(\hat{a}_{k+}^\dagger \hat{a}_{k+} - \hat{a}_{k-}^\dagger \hat{a}_{k-}\right). \tag{36}$$

Thus the spin projection in the direction of motion (which here is chosen to be parallel to the z axis) is carried by quanta with positive helicity which contribute $+1$ unit and quanta with negative helicity which contribute -1 unit (the units of course are $\hbar$). The triplet of spin states is completed by the longitudinal quanta ($\lambda = 3$) which carry spin projection zero. The operators for energy and momentum remain diagonal also with respect to the helicity basis. Thus (26) with the use of (34) becomes

$$\hat{H} = \sum_{\sigma=-,0,+} \int \mathrm{d}^3 k\, \omega_k\, \hat{a}_{k,\sigma}^\dagger \hat{a}_{k,\sigma}. \tag{37}$$

Of course we could have started from the outset with circular instead of linear polarizations in the expansion of the field operator (1), summing over an index $\lambda = \{-, 0, +\}$. The *circular polarization vectors* (helicity vectors) are given by

$$\epsilon^\mu(\boldsymbol{k}, \pm) = \tfrac{1}{\sqrt{2}}\left(\epsilon^\mu(\boldsymbol{k}, 1) \pm \mathrm{i}\,\epsilon^\mu(\boldsymbol{k}, 2)\right), \tag{38a}$$

$$\epsilon^\mu(\boldsymbol{k}, 0) = \epsilon^\mu(\boldsymbol{k}, 3). \tag{38b}$$

When using these basis vectors one has to take into account that they have complex components.

EXERCISE

6.4 Invariant Commutation Relations and the Feynman Propagator of the Proca Field

Problem. (a) Derive the commutators between the field operators $\hat{A}^\mu(x)$ and $\hat{A}^\nu(y)$ of the free Proca field at arbitrary space-time points x and y. Show that the equal-time commutation relations (6.93) follow as a special case. Check whether the result is consistent with the Lorentz condition $\partial_\mu \hat{A}^\mu = 0$.
(b) Construct the Feynman propagator of the Proca field, i.e., the vacuum expectation value of the time-ordered product of two field operators.

Solution. (a) We insert the plane-wave expansion of the field operator (1) given in Example 6.3 and use the commutation rules (21) and (22):

$$
\begin{aligned}
[\hat{A}^\mu(x), \hat{A}^\nu(y)] &= \int \frac{d^3k'}{\sqrt{2\omega_{k'}(2\pi)^3}} \int \frac{d^3k}{\sqrt{2\omega_k(2\pi)^3}} \sum_{\lambda,\lambda'=1}^{3} \epsilon^\mu(k',\lambda')\epsilon^\nu(k,\lambda) \\
&\quad \times \left([\hat{a}_{k'\lambda'}, \hat{a}^\dagger_{k\lambda}] \, e^{-i(k'\cdot x - k\cdot y)} + [\hat{a}^\dagger_{k'\lambda'}, \hat{a}_{k\lambda}] \, e^{i(k'\cdot x - k\cdot y)} \right) \\
&= \int \frac{d^3k}{2\omega_k(2\pi)^3} \sum_{\lambda=1}^{3} \epsilon^\mu(k,\lambda)\epsilon^\nu(k,\lambda) \left(e^{-ik\cdot(x-y)} - e^{ik\cdot(x-y)} \right).
\end{aligned}
\tag{1}
$$

The polarization vectors satisfy the modified completeness relation (6.76), which leads to

$$
\begin{aligned}
[\hat{A}^\mu(x), \hat{A}^\nu(y)] &= \int \frac{d^3k}{2\omega_k(2\pi)^3} \left(-g^{\mu\nu} + \frac{k^\mu k^\nu}{m^2} \right) \left(e^{-ik\cdot(x-y)} - e^{ik\cdot(x-y)} \right) \\
&= \left(-g^{\mu\nu} - \frac{1}{m^2} \partial^\mu \partial^\nu \right) \int \frac{d^3k}{2\omega_k(2\pi)^3} \left(e^{-ik\cdot(x-y)} - e^{ik\cdot(x-y)} \right).
\end{aligned}
\tag{2}
$$

The Fourier integral is just the invariant Pauli–Jordan function introduced in Sect. 4.4:

$$
i\Delta(x-y) = -i \int \frac{d^3k}{(2\pi)^3} \frac{\sin k \cdot (x-y)}{\omega_k}.
\tag{3}
$$

Thus the general commutator between two field operators of the massive spin-1 theory reads

$$
[\hat{A}^\mu(x), \hat{A}^\nu(y)] = -i \left(g^{\mu\nu} + \frac{1}{m^2} \partial^\mu \partial^\nu \right) \Delta(x-y).
\tag{4}
$$

As shown in Sect. 4.4 this function for vanishing time diffference $x_0 - y_0 = 0$ has the properties

$$
\Delta(0, \boldsymbol{x} - \boldsymbol{y}) = 0
\tag{5}
$$

and

Exercise 6.4

$$\left.\frac{\partial}{\partial x_0}\Delta(x_0 - y_0, \boldsymbol{x} - \boldsymbol{y})\right|_{x_0 - y_0 = 0} = -\delta^3(\boldsymbol{x} - \boldsymbol{y})\,. \tag{6}$$

The *equal-time commutation relations* between $\hat{\boldsymbol{A}}$ and the conjugate field $\hat{\boldsymbol{E}}$ can be easily obtained by using (4) and (6). Since $\hat{E}^j = \partial^j \hat{A}^0 - \partial^0 \hat{A}^j$ one finds for example:

$$\begin{aligned}
\left[\hat{A}^i(x), \hat{E}^j(y)\right] &= \partial_y^j \left[\hat{A}^i(x), \hat{A}^0(y)\right] - \partial_y^0 \left[\hat{A}^i(x), \hat{A}^j(y)\right] \\
&= (-\mathrm{i})\left[\partial_y^j \frac{1}{m^2}\partial_x^i \partial_x^0 - \partial_y^0\left(g^{ij} + \frac{1}{m^2}\partial_x^i \partial_x^j\right)\right]\Delta(x - y) \\
&= -\mathrm{i}\,g^{ij}\partial_x^0 \Delta(x - y) \\
&\xrightarrow{x_0 \to y_0} \mathrm{i}\,g^{ij}\delta^3(\boldsymbol{x} - \boldsymbol{y}) = -\mathrm{i}\delta_{ij}\delta^3(\boldsymbol{x} - \boldsymbol{y})\,. \tag{7}
\end{aligned}$$

The remaining commutation relations can be confirmed similarly:

$$\left[\hat{A}^i(x), \hat{A}^j(y)\right] = -\mathrm{i}\left(g^{ij} + \frac{1}{m^2}\partial^i\partial^j\right)\Delta(x - y) \to 0\,, \tag{8}$$

$$\left[\hat{E}^i(x), \hat{E}^j(y)\right] = -\mathrm{i}\left(-\partial^i\partial^j - g^{ij}\partial^0\partial^0\right)\Delta(x - y) \to 0\,. \tag{9}$$

This comes about since the spatial derivatives and also the time derivatives of even order of the function $\Delta(x - y)$ vanish in the limit $x_0 \to y_0$ (see the discussion following (4.107) in Sect. 4.4).

The commutation relation (4) can be shown to be *consistent with the Lorentz condition* $\partial_\mu \hat{A}^\mu = 0$:

$$\begin{aligned}
\left[\partial_\mu \hat{A}^\mu(x), \hat{A}^\nu(y)\right] &= (-\mathrm{i})\partial_\mu\left(g^{\mu\nu} + \frac{1}{m^2}\partial^\mu\partial^\nu\right)\Delta(x - y) \\
&= (-\mathrm{i})\frac{1}{m^2}\partial^\nu(m^2 + \Box)\Delta(x - y) = 0\,, \tag{10}
\end{aligned}$$

since the function $\Delta(x - y)$ is a solution of the homogeneous Klein–Gordon equation (cf. (4.111)).

(b) The vacuum expectation value of the time-ordered product of two field operators can be evaluated with the use of the Fourier expansion (1) from Example 6.3:

$$\begin{aligned}
\mathrm{i}\Delta_\mathrm{F}^{\mu\nu}(x - y) &:= \left\langle 0\left|T(\hat{A}^\mu(x)\hat{A}^\nu(y))\right|0\right\rangle \\
&= \int \frac{\mathrm{d}^3k'}{\sqrt{2\omega_{k'}}(2\pi)^3}\int\frac{\mathrm{d}^3k}{\sqrt{2\omega_k}(2\pi)^3}\sum_{\lambda,\lambda'=1}^{3}\epsilon^\mu(\boldsymbol{k}',\lambda')\epsilon^\nu(\boldsymbol{k},\lambda) \\
&\quad \times\left(\Theta(x_0 - y_0)\,\mathrm{e}^{-\mathrm{i}(k'\cdot x - k\cdot y)}\left\langle 0\left|\hat{a}_{\boldsymbol{k}'\lambda'}\hat{a}_{\boldsymbol{k}\lambda}^\dagger\right|0\right\rangle\right. \\
&\quad \left.+\Theta(y_0 - x_0)\,\mathrm{e}^{+\mathrm{i}(k'\cdot x - k\cdot y)}\left\langle 0\left|\hat{a}_{\boldsymbol{k}\lambda}\hat{a}_{\boldsymbol{k}'\lambda'}^\dagger\right|0\right\rangle\right) \\
&= \int\frac{\mathrm{d}^3k}{2\omega_k(2\pi)^3}\sum_{\lambda=1}^{3}\epsilon^\mu(\boldsymbol{k},\lambda)\,\epsilon^\nu(\boldsymbol{k},\lambda) \\
&\quad \times\left(\Theta(x_0 - y_0)\,\mathrm{e}^{-\mathrm{i}k\cdot(x - y)} + \Theta(y_0 - x_0)\,\mathrm{e}^{+\mathrm{i}k\cdot(x - y)}\right)
\end{aligned}$$

6.5 Canonical Quantization of the Massive Vector Field

$$= -\int \frac{d^3k}{2\omega_k(2\pi)^3} \left(g^{\mu\nu} - \frac{1}{m^2}k^\mu k^\nu\right)$$
$$\times \left(\Theta(x_0-y_0)\,e^{-ik\cdot(x-y)} + \Theta(y_0-x_0)\,e^{+ik\cdot(x-y)}\right). \quad (11)$$

Exercise 6.4

This expression agrees with the scalar Feynman propagator $i\Delta_F(x-y)$ except for the factor $-\left(g^{\mu\nu} - k^\mu k^\nu/m^2\right)$ under the integral which has its origin in the completeness relation of the polarization vectors.

The Feynman propagator (11) in coordinate space can be directly expressed in terms of $\Delta_F(x-y)$ by converting the k-dependent factors into gradient operators, $k^\mu \to i\partial_x^\mu$, which can be moved before the integral. This procedure, however, encounters an obstacle: the time derivative pulled before the integral will also act on the step function in (11). To compensate for this an extra term has to be subtracted. We have

$$i\partial_x^\mu\, i\partial_x^\nu \left[\Theta(x_0-y_0)\,e^{-ik\cdot(x-y)} + \Theta(y_0-x_0)\,e^{+ik\cdot(x-y)}\right]$$
$$= k^\mu k^\nu \left(\Theta(x_0-y_0)\,e^{-ik\cdot(x-y)} + \Theta(y_0-x_0)\,e^{+ik\cdot(x-y)}\right)$$
$$+ i(g^{\mu 0}k^\nu + g^{\nu 0}k^\mu)\,\delta(x_0-y_0)\left(e^{-ik\cdot(x-y)} + e^{ik\cdot(x-y)}\right)$$
$$- g^{\mu 0}g^{\nu 0}\,\delta'(x_0-y_0)\left(e^{-ik\cdot(x-y)} - e^{ik\cdot(x-y)}\right). \quad (12)$$

This can be solved for the $k^\mu k^\nu$ term needed in (11). The terms for $\mu,\nu \neq 0$ will not contribute in the second term on the right-hand side of (12) since the d^3k integration extends over an odd function in $\mathbf{k}$. This is true since the sum of exponential functions in $k\cdot(x-y)$ is even in $\mathbf{k}$, since $k\cdot(x-y) = -\mathbf{k}\cdot(\mathbf{x}-\mathbf{y})$ for $x_0 = y_0$. The integral for $\mu = \nu = 0$ leads to $2i(2\pi)^3\delta^3(\mathbf{x}-\mathbf{y})$. To evaluate the last term involving the derivative of the delta function we use

$$\delta'(x_0-y_0)\,f(x) \to -\delta(x_0-y_0)\,f'(x), \quad (13)$$

where the prime denotes derivation with respect to time x_0. The d^3k integral then becomes $-i(2\pi)^3\delta^3(\mathbf{x}-\mathbf{y})$. Collecting these results we find for the *Feynman propagator of the massive spin-1 field* in coordinate space, expressed in terms of the scalar Feynman propagator $\Delta_F(x-y)$,

$$i\Delta_F^{\mu\nu}(x-y) = \langle 0|T(\hat{A}^\mu(x)\hat{A}^\nu(y))|0\rangle \quad (14)$$
$$= -\left(g^{\mu\nu} + \frac{1}{m^2}\partial^\mu\partial^\nu\right)i\Delta_F(x-y) - \frac{i}{m^2}g^{\mu 0}g^{\nu 0}\delta^4(x-y).$$

The appearance of the extra term in (14) is rather unpleasant since it singles out the time-like components and thus seems to destroy the Lorentz covariance of the theory. In practice, however, this extra term can simply be dropped when one performs perturbative calculations using the Feynman propagator. Its contribution is exactly cancelled by so-called *normal-dependent terms*, which appear in the derivation of the perturbation series. This is further discussed in Sect. 8.6.

Exercise 6.4

The Feynman propagator (14) can also be written down in momentum space. Using (4.136) from Sect. 4.5 we find[8]

$$\Delta_F^{\mu\nu}(k) = \frac{-\left(g^{\mu\nu} - \frac{1}{m^2}k^\mu k^\nu\right)}{k^2 - m^2 + i\epsilon} - \frac{1}{m^2}g^{\mu 0}g^{\nu 0} \,. \tag{16}$$

Compared to the corresponding expression (4.136) for the spin-0 Feynman propagator the numerator contains a quadratic expression originating from the sum over polarizations. Furthermore, (16) again contains a "normal-dependent term", which will turn out to be irrelevant, however.

---

[8] This result can also be obtained directly from (11), avoiding the detour to coordinate space. In the $\mu = \nu = 0$ term one has to be careful not to put the factor $(k^0)^2$ before the integral since in (11) $k^0 \equiv \omega_k = \sqrt{\mathbf{k}^2 + m^2}$ is a fixed number. However, one can use

$$\frac{\omega_k^2}{k^2 - m^2 + i\epsilon} = \frac{\omega_k^2 - k_0^2 + k_0^2}{k_0^2 - \omega_k^2 + i\epsilon} = \frac{k_0^2}{k_0^2 - \omega_k^2 + i\epsilon} - 1 \,. \tag{15}$$

This identity can be inserted in the integral representation in order to transform the d^3k integral into a d^4k integral, having (16) in the integrand.

7. Quantization of the Photon Field

7.1 Introduction

When quantizing the massive spin-1 field within the canonical formalism, we encountered the problem that the 0 component of the potential, $A_0(x)$, does not possess a conjugate field $\pi^0(x)$ since the Lagrangian $\mathcal{L}$ does not depend on the time derivative of A_0. This did not have harmful consequences, however, since according to the Proca equation the time-like component of the field can be expressed as a dependent quantity:

$$A_0 = -\frac{1}{m^2} \boldsymbol{\nabla} \cdot \boldsymbol{E} . \qquad (7.1)$$

Therefore it was sufficient to quantize the three space-like components of the vector field, in agreement with the three physical polarization states of a massive vector particle. Contrary to superficial evidence this quantization procedure proved to be relativistically covariant (see (6.98)).

The quantization of the electromagnetic field is more delicate since the photon is massless.[1] In the limit $m \to 0$ the description of the vector field undergoes a qualitative change. Equation (7.1) obviously can no longer be applied. The vector field $A^\mu(x)$ now contains *two spurious degrees of freedom*, which in one way or another have to be eliminated. This is related to the gauge invariance of the theory according to which different fields $A^\mu(x)$ are equivalent if they are related to each other through a gauge transformation. It is not a trivial task to ensure that the quantized field operators reflect this property. In fact several implementations of the canonical quantization of the electromagnetic field have been developed. In the following, we will discuss the two most important procedures.

1. *Noncovariant gauge-fixing.* As one variation of this method, the *Coulomb gauge* $\boldsymbol{\nabla} \cdot \hat{\boldsymbol{A}} = 0$ can be used as a constraint that eliminates the unwanted longitudinal degree of freedom. Canonical quantization can then be carried through consistently and leads to plausible and correct results. This procedure, however, suffers from the lack of manifest Lorentz covariance which makes it rather unsuited for the systematic investigation of the perturbation series. Quantization in the Coulomb gauge will be discussed later in Sect. 7.7.

2. *Covariant quantization.* All components of the field $A^\mu(x)$ are treated on an equal footing. The Lagrangian of the electromagnetic field is modified in

---

[1] The present experimental upper limit for the mass of the photon is $m_\gamma < 10^{-27}$ eV $\simeq 10^{-60}$ g.

such a way that A^0 also obtains a canonically conjugate field. In order to eliminate the extraneous degrees of freedom a constraint will be imposed on the allowable state vectors in Hilbert space. This procedure will be discussed in Sect. 7.3.

7.2 The Electromagnetic Field in Lorentz Gauge

For the Maxwell Lagrangian (6.27) the demand for manifest covariance and the principle of the canonical quantization prescription are not compatible. One would need a term containing the time derivative of A^0 in order to ensure the existence of the canonically conjugate field π^0. It occurred to E. Fermi that this can be achieved by introducing an innocuous *modification of the Lagrangian*.[2] He added the term

$$\mathcal{L}' = -\frac{1}{4}F_{\mu\nu}F^{\mu\nu} - \frac{1}{2}\zeta(\partial_\sigma A^\sigma)^2 \,, \tag{7.2}$$

where ζ is a parameter that, in principle, can be chosen freely. Within the classical framework $\mathcal{L}'$ agrees with the original Lagrangian $\mathcal{L}$ provided that the potential satisfies the *Lorentz condition*

$$\partial_\mu A^\mu = 0 \,. \tag{7.3}$$

The extra part in (7.2) is called a *gauge-fixing term*. Adding the ζ term in (7.2) amounts to the introduction of a Lagrange multiplier in order to impose a constraint, i.e., the Lorentz condition, upon the field. We will see, however, that the quantized theory does not permit one to impose the Lorentz condition upon the field operator $\hat{A}^\mu(x)$; it will be necessary to find a weaker formulation of the constraint.

The *canonically conjugate field* resulting from the modified Lagrangian (7.2) reads

$$\pi_\mu = \frac{\partial \mathcal{L}'}{\partial(\partial_0 A^\mu)} = -F_{0\mu} - \zeta g_{\mu 0}(\partial_\sigma A^\sigma) \,. \tag{7.4}$$

Provided that $\zeta \neq 0$, A^0 now also has its canonically conjugate partner π_0:[3]

$$\pi_0 = -\zeta(\partial_\sigma A^\sigma) \,. \tag{7.5}$$

The Euler–Lagrange equation resulting from $\mathcal{L}'$ of course also get modified. Using

$$\frac{\partial}{\partial(\partial^\nu A_\mu)}\frac{1}{2}\zeta(\partial_\sigma A^\sigma)^2 = \zeta g_\nu{}^\mu(\partial_\sigma A^\sigma) \tag{7.6}$$

we obtain

$$\Box A^\mu - (1-\zeta)\partial^\mu(\partial_\sigma A^\sigma) = 0 \,. \tag{7.7}$$

---

[2] E. Fermi: Rev. Mod. Phys. 4, 125 (1932).
[3] Within the classical theory the r.h.s. would vanish in the Lorentz gauge. As we will see, however, the four-divergence of the quantized field operator and consequently also the operator $\hat{\pi}_0$ do not vanish.

7.2 The Electromagnetic Field in Lorentz Gauge

Within the framework of classical theory, (7.7) is equivalent to the ordinary Maxwell equations in the Lorentz gauge. Note, by the way, that it is sufficient to impose the Lorentz condition at one particular point in time. Namely, differentiation of (7.7) with respect to x_μ leads to the weaker condition

$$\zeta \Box (\partial_\sigma A^\sigma) = 0 \,. \tag{7.8}$$

This can be viewed as a Cauchy initial-value problem for the function $\partial_\sigma A^\sigma$. If the Lorentz condition is satisfied at time t_0,

$$\partial_\sigma A^\sigma(\boldsymbol{x}, t_0) = 0 \quad \text{and also} \quad \partial_0 \big(\partial_\sigma A^\sigma(\boldsymbol{x}, t_0) \big) = 0 \,, \tag{7.9}$$

then the time-evolution equation (7.8) guarantees that it will be satisfied for all times t.

If one makes the special choice $\zeta = 1$, or if the Lorentz condition is fulfilled, the potential satisfies the wave equation (d'Alembert equation)

$$\Box A^\mu = 0 \,, \tag{7.10}$$

which can be viewed as the massless version of the Klein–Gordon equation, imposed separately on all of the four components of the vector field.

In the following discussion we will choose the *special value $\zeta = 1$ for the gauge-fixing parameter* since this will simplify the calculations. It is customary to call this choice the *Feynman gauge* (sometimes also the name 'Fermi gauge' is used). For the choice $\zeta = 1$ the Lagrangian can be brought into a particularly simple form upon integration by parts in the action integral. Equation (7.2) can be transformed into

$$\begin{aligned}
\mathcal{L}' &= -\frac{1}{2}\partial_\mu A_\nu \partial^\mu A^\nu + \frac{1}{2}\partial_\mu A_\nu \partial^\nu A^\mu - \frac{1}{2}\partial_\mu A^\mu \partial_\nu A^\nu \\
&= -\frac{1}{2}\partial_\mu A_\nu \partial^\mu A^\nu + \frac{1}{2}\partial_\mu \big[A_\nu (\partial^\nu A^\mu) - (\partial_\nu A^\nu) A^\mu \big] \,.
\end{aligned} \tag{7.11}$$

The last term is a four-divergence which has no influence on the field equations. Thus the dynamics of the electromagetic field (in the Lorentz gauge) can be described by the simple Lagrangian

$$\mathcal{L}'' = -\frac{1}{2}\partial_\mu A_\nu \partial^\mu A^\nu \,. \tag{7.12}$$

The canonically conjugate field π_μ belonging to A^μ now reads

$$\pi_\mu = \frac{\partial \mathcal{L}''}{\partial(\partial_0 A^\mu)} = -\partial^0 A_\mu \,. \tag{7.13}$$

Obviously the time and space components of the A^μ field appear on an equal footing.

The *Hamilton density* of the electromagnetic field reads

$$\begin{aligned}
\mathcal{H}'' &= \pi^\mu \partial_0 A_\mu - \mathcal{L}'' = -\pi^\mu \pi_\mu + \frac{1}{2}\partial_\mu A_\nu \partial^\mu A^\nu \\
&= -\pi^\mu \pi_\mu + \frac{1}{2}\pi^\mu \pi_\mu + \frac{1}{2}\partial_k A_\nu \partial^k A^\nu \\
&= -\frac{1}{2}\pi^\mu \pi_\mu + \frac{1}{2}\partial_k A_\nu \partial^k A^\nu \,,
\end{aligned} \tag{7.14}$$

where the index k only covers the spatial components $k = 1, 2, 3$. Written out in components, this expression is

$$\mathcal{H} = \frac{1}{2} \sum_{k=1}^{3} \left[(\dot{A}^k)^2 + (\boldsymbol{\nabla} A^k)^2 \right] - \frac{1}{2} \left[(\dot{A}^0)^2 + (\boldsymbol{\nabla} A^0)^2 \right] . \tag{7.15}$$

Thus $\mathcal{H}''$ looks like the Hamilton density of four independent neutral massless Klein–Gordon fields. There is one essential difference, however, as the "scalar" part A^0 carries a negative sign! The Hamiltonian appears *not to be positive definite* and thus to have no reasonable physical interpretation. As we will see, the gauge condition will fix this problem and ensure that the electromagnetic field always carries positive energy.

The canonical *energy-momentum tensor* is given by (7.12)

$$\begin{aligned}\Theta^{\mu\nu} &= \frac{\partial \mathcal{L}''}{\partial(\partial_\mu A_\sigma)} \partial^\nu A_\sigma - g^{\mu\nu} \mathcal{L}'' \\ &= -\partial^\mu A^\sigma \partial^\nu A_\sigma + \frac{1}{2} g^{\mu\nu} \partial^\rho A^\sigma \partial_\rho A_\sigma .\end{aligned} \tag{7.16}$$

This tensor is already symmetrical so that no further symmetrization procedure, $\Theta^{\mu\nu} \to T^{\mu\nu}$ as in (6.39), is required.

It is easy to show that the expression for the *energy density* (cf. (7.14)),

$$\Theta^{00} = -\partial^0 A^\sigma \partial^0 A_\sigma + \frac{1}{2} \partial^\rho A^\sigma \partial_\rho A_\sigma , \tag{7.17}$$

can be cast into the familiar form

$$\begin{aligned}\Theta'^{00} &= \frac{1}{2} (\partial^0 \boldsymbol{A} + \boldsymbol{\nabla} A^0)^2 + \frac{1}{2} (\boldsymbol{\nabla} \times \boldsymbol{A})^2 \\ &= \frac{1}{2} (\boldsymbol{E}^2 + \boldsymbol{B}^2)\end{aligned} \tag{7.18}$$

by adding a suitable divergence term (keeping the total energy unchanged). This transformation relies on the Lorentz gauge condition (see Exercise 7.1). In contrast to (7.17) the energy density given by the expression (7.18) obviously is positive definite.

The *momentum density* is found to be

$$p^i = \Theta^{0i} = -\partial^0 A^\sigma \partial^i A_\sigma \tag{7.19}$$

or

$$\boldsymbol{p} = -\sum_{k=1}^{3} \dot{A}^k \boldsymbol{\nabla} A^k + \dot{A}^0 \boldsymbol{\nabla} A^0 . \tag{7.20}$$

EXERCISE

7.1 The Energy Density of the Photon Field in the Lorentz Gauge

Problem. Show that the expressions (7.17) and (7.18) for the energy density of the photon field in the Lorentz gauge are equivalent.

Solution. We will show that the expressions for Θ^{00} and Θ'^{00} agree up to a divergence term. The expression (7.17) reads (latin indices running from 1 to 3):

$$\begin{aligned}\Theta^{00} &= -\partial^0 A^\sigma \partial^0 A_\sigma + \frac{1}{2}\partial^\rho A^\sigma \partial_\rho A_\sigma \\ &= -\frac{1}{2}(\partial^0 A^0)^2 + \frac{1}{2}(\partial^0 \boldsymbol{A})^2 - \frac{1}{2}(\boldsymbol{\nabla} A^0)^2 + \frac{1}{2}\partial^i A^j \partial_i A_j \ . \end{aligned} \quad (1)$$

On the other hand (7.18) reads

$$\begin{aligned}\Theta'^{00} &= \frac{1}{2}(\boldsymbol{E}^2 + \boldsymbol{B}^2) \\ &= \frac{1}{2}(-\partial^0 \boldsymbol{A} - \boldsymbol{\nabla} A^0)^2 + \frac{1}{2}(\boldsymbol{\nabla} \times \boldsymbol{A})^2 \\ &= \frac{1}{2}(\partial^0 \boldsymbol{A})^2 + (\partial^0 \boldsymbol{A}) \cdot (\boldsymbol{\nabla} A^0) \\ &\quad + \frac{1}{2}(\boldsymbol{\nabla} A^0)^2 + \frac{1}{2}\partial^i A^j \partial_i A_j - \frac{1}{2}\partial^i A^j \partial_j A_i \ . \end{aligned} \quad (2)$$

The difference is given by

$$X \equiv \Theta'^{00} - \Theta^{00} = (\partial^0 \boldsymbol{A}) \cdot (\boldsymbol{\nabla} A^0) + (\boldsymbol{\nabla} A^0)^2 - \frac{1}{2}\partial^i A^j \partial_j A_i + \frac{1}{2}(\partial^0 A^0)^2 \ . \quad (3)$$

To write this as a three-dimensional divergence, the derivative operators have to be "reshuffled", making use of the Lorentz condition

$$\partial_\mu A^\mu = \partial_0 A^0 + \boldsymbol{\nabla} \cdot \boldsymbol{A} = 0 \ . \quad (4)$$

The first two terms in (3) give, using (4),

$$\begin{aligned}X_1 + X_2 &= (\partial^0 \boldsymbol{A}) \cdot (\boldsymbol{\nabla} A^0) + (\boldsymbol{\nabla} A^0) \cdot (\boldsymbol{\nabla} A^0) \\ &= \boldsymbol{\nabla} \cdot [(\partial^0 \boldsymbol{A}) A^0] - (\partial^0 \boldsymbol{\nabla} \cdot \boldsymbol{A}) A^0 + \boldsymbol{\nabla} \cdot [(\boldsymbol{\nabla} A^0) A^0] - A^0 \boldsymbol{\nabla}^2 A^0 \\ &= \boldsymbol{\nabla} \cdot [(\partial^0 \boldsymbol{A}) A^0 + (\boldsymbol{\nabla} A^0) A^0] + A^0 (\partial^0 \partial^0 - \boldsymbol{\nabla}^2) A^0 \ . \end{aligned} \quad (5)$$

The offending nondivergence term vanishes since the components of the vector potential in the Lorentz gauge satisfy the wave equation

$$(\partial^0 \partial^0 - \boldsymbol{\nabla}^2) A^\mu = 0 \ . \quad (6)$$

The third and fourth term in (3) can be combined into

$$\begin{aligned}X_3 + X_4 &= -\frac{1}{2}\partial^i A^j \partial_j A_i + \frac{1}{2}(\partial^0 A^0)^2 \\ &= \frac{1}{2}\partial^i [A_i(\partial^j A_j) - A_j(\partial^j A_i)] - \frac{1}{2}(\partial^i A_i)^2 + \frac{1}{2}(\partial^0 A^0)^2 \ . \end{aligned} \quad (7)$$

Again the nondivergence terms vanish because of the Lorentz condition. Thus the two expressions for the energy density agree up to a divergence and lead to the same total energy.

7.3 Canonical Quantization in the Lorentz Gauge

After the preparations given in the last sections the stage now is set for the canonical quantization of the electromagnetic field. As usual we postulate the following equal-time commutation relations (ETCR) for the field operators $\hat{A}^\mu$ and $\hat{\pi}^\mu$

$$\left[\hat{A}^\mu(\boldsymbol{x},t),\,\hat{\pi}^\nu(\boldsymbol{x}',t)\right] = ig^{\mu\nu}\delta^3(\boldsymbol{x}-\boldsymbol{x}')\,, \tag{7.21a}$$

$$\left[\hat{A}^\mu(\boldsymbol{x},t),\,\hat{A}^\nu(\boldsymbol{x}',t)\right] = \left[\hat{\pi}^\mu(\boldsymbol{x},t),\,\hat{\pi}^\nu(\boldsymbol{x}',t)\right] = 0\,. \tag{7.21b}$$

Using (7.13), $\hat{\pi}^\mu = -\partial^0 \hat{A}^\mu$, we can write (7.21a) as

$$\left[\hat{A}^\mu(\boldsymbol{x},t),\,\dot{\hat{A}}^\nu(\boldsymbol{x}',t)\right] = -ig^{\mu\nu}\delta^3(\boldsymbol{x}-\boldsymbol{x}')\,. \tag{7.22}$$

This relation can be compared to the corresponding quantization conditions for the neutral Klein–Gordon field (4.5):

$$\left[\hat{\phi}(\boldsymbol{x},t),\,\dot{\hat{\phi}}(\boldsymbol{x}',t)\right] = i\delta^3(\boldsymbol{x}-\boldsymbol{x}')\,. \tag{7.23}$$

The ETCR (7.21)–(7.22) apparently describe three independent scalar fields $\hat{A}^k(x)$ which are subject to the ordinary bosonic quantization scheme. The fourth field described by the operator $\hat{A}^0(x)$, however, satisfies *a commutation relation with the wrong sign*! This is unavoidable if the ETCR are written in covariant form since the right-hand side must contain the metric tensor that contains both signs. The changed sign has substantial consequences for the structure of the Hilbert space of the photon field: it will be endowed with an *indefinite metric* which implies the existence of states having negative norm. Fortunately this turns out to be only a technical complication and the formalism has a meaningful interpretation if the gauge condition is taken into account. This will be discussed in Sect. 7.4.

One immediately observes that the Lorentz gauge condition $\partial_\mu A^\mu = 0$ cannot be carried over to the field operator $\hat{A}^\mu(x)$,

$$\partial_\mu \hat{A}^\mu(x) \neq 0\,. \tag{7.24}$$

Using the ETCR (7.21)–(7.22), we find the commutator of $\partial_\mu \hat{A}^\mu(\boldsymbol{x},t)$ with the field operator $\hat{A}^\nu$ at the space point $\boldsymbol{x}'$:

$$\begin{aligned}
\left[\partial_\mu \hat{A}^\mu(\boldsymbol{x},t),\,\hat{A}^\nu(\boldsymbol{x}',t)\right] &= \left[\partial_0 \hat{A}^0 + \boldsymbol{\nabla}\cdot\hat{\boldsymbol{A}}(\boldsymbol{x},t),\,\hat{A}^\nu(\boldsymbol{x}',t)\right] \\
&= -\left[\hat{\pi}^0(\boldsymbol{x},t),\,\hat{A}^\nu(\boldsymbol{x}',t)\right] + \boldsymbol{\nabla}\cdot\left[\hat{\boldsymbol{A}}(\boldsymbol{x},t),\,\hat{A}^\nu(\boldsymbol{x}',t)\right] \\
&= ig^{\nu 0}\delta^3(\boldsymbol{x}-\boldsymbol{x}') \neq 0\,.
\end{aligned} \tag{7.25}$$

7.3.1 Fourier Decomposition of the Field Operator

To characterize the quanta of the electromagnetic field, viz. the photons, the operator $\hat{A}^\mu(x)$ can be expanded into plane waves. This discussion can be kept short since it largely coincides with the treatment of the massive vector field, which was discussed in Example 6.3.

With the polarization vectors of Sect. 6.4.2, the Fourier expansion of the field operator reads[5]

$$\hat{A}^\mu(x) = \int \frac{d^3k}{\sqrt{2\omega_k (2\pi)^3}} \sum_{\lambda=0}^{3} \left(\hat{a}_{k\lambda}\, \epsilon^\mu(k,\lambda)\, e^{-ik\cdot x} + \hat{a}_{k\lambda}^\dagger\, \epsilon^\mu(k,\lambda)\, e^{ik\cdot x} \right). \quad (7.26)$$

Note that $\omega_k = k_0 = |\mathbf{k}|$ because of the vanishing photon mass. In contrast to the case in (1) in Example 6.3, the summation now runs over all *four polarization states* $\lambda = 0, \ldots, 3$. The reduction to the two physical degrees of freedom (transverse photons) will follow later.

If we use (7.26), the canonically conjugate field $\hat{\pi}^\mu = -\dot{\hat{A}}^\mu$ is given by

$$\begin{aligned}
\hat{\pi}^\mu(x) &= i \int \frac{d^3k}{\sqrt{2\omega_k (2\pi)^3}} \omega_k \sum_{\lambda=0}^{3} \\
&\quad \times \left(\hat{a}_{k\lambda}\, \epsilon^\mu(k,\lambda)\, e^{-ik\cdot x} - \hat{a}_{k\lambda}^\dagger\, \epsilon^\mu(k,\lambda)\, e^{ik\cdot x} \right).
\end{aligned} \quad (7.27)$$

The commutation relations for the operators $\hat{a}_{k\lambda}$ follow from the ETCR (7.21) when we use the inverse Fourier expansion (see (13) in Example 6.3)

$$\hat{a}_{k\lambda} = i g_{\lambda\lambda} \int d^3x\, A^{\mu*}(k,\lambda)\, \overset{\leftrightarrow}{\partial_0}\, \hat{A}_\mu(x). \quad (7.28)$$

This leads to

$$\hat{a}_{k\lambda} = i g_{\lambda\lambda} \int d^3x\, \frac{e^{ik\cdot x}}{\sqrt{2\omega_k(2\pi)^3}}\, \epsilon^\mu(k,\lambda) \left(\dot{\hat{A}}_\mu(x) - i\omega_k \hat{A}_\mu(x) \right) \quad (7.29)$$

and a similar relation for $\hat{a}_{k\lambda}^\dagger$. A simple calculation as in Sect. 6.4, using (7.21) and (7.22), leads to

$$\left[\hat{a}_{k'\lambda'}, \hat{a}_{k\lambda}^\dagger \right] = -\delta^3(k'-k)\, g_{\lambda\lambda} g_{\lambda'\lambda'}\, \epsilon^\mu(k,\lambda')\epsilon_\mu(k,\lambda), \quad (7.30)$$

which exactly agrees with (19) in Example 6.3. The orthogonality relation of the four-dimensional polarization vectors (6.62)

$$\epsilon^\mu(k,\lambda)\epsilon_\mu(k,\lambda') = g_{\lambda\lambda'} \quad (7.31)$$

---

[4] Compare this to the relation analogous to (7.25) for the massive Proca field. In this case the commutator vanishes as shown in (10) in Exercise 6.4.

[5] Here we assume the polarization vectors to be real valued. The generalization to complex polarization vectors ϵ^μ (circular polarization) is obvious.

leads to
$$[\hat{a}_{\mathbf{k}'\lambda'}, \hat{a}^\dagger_{\mathbf{k}\lambda}] = -g_{\lambda\lambda'}\delta^3(\mathbf{k}' - \mathbf{k}) \ . \tag{7.32a}$$

Similarly one derives the remaining commutation relations
$$[\hat{a}_{\mathbf{k}'\lambda'}, \hat{a}_{\mathbf{k}\lambda}] = [\hat{a}^\dagger_{\mathbf{k}'\lambda'}, \hat{a}^\dagger_{\mathbf{k}\lambda}] = 0 \ . \tag{7.32b}$$

For the polarizations $\lambda = 1, 2, 3$ these relations agree with the corresponding result for the massive vector field: (21)–(22) in Exercise 6.3. In contrast, the operators for the new "scalar" polarization ($\lambda = 0$) satisfy commutation relations with the wrong sign.

Using (7.32), we can be easily express the Hamiltonian of the electromagnetic field in terms of particle-number operators for photons $\hat{a}^\dagger_{\mathbf{k}\lambda}\hat{a}_{\mathbf{k}\lambda}$. The integral over the quantized normal-ordered Hamilton density (7.14) reads

$$\hat{H} = \int \mathrm{d}^3x \, \hat{\mathcal{H}}(x) = -\frac{1}{2}\int \mathrm{d}^3x \, {:}\!\left(\pi^\mu\pi_\mu + \boldsymbol{\nabla} A^\mu \cdot \boldsymbol{\nabla} A_\mu\right)\!{:} \ . \tag{7.33}$$

The Fourier expansions (7.26) and (7.27) imply

$$\begin{aligned}
\hat{H} =\ & \frac{1}{2}\int\mathrm{d}^3x \int\frac{\mathrm{d}^3k'}{\sqrt{2\omega_{k'}(2\pi)^3}}\int\frac{\mathrm{d}^3k}{\sqrt{2\omega_k(2\pi)^3}} \\
& \times \sum_{\lambda\lambda'=0}^{3}\epsilon^\mu(\mathbf{k}',\lambda')\epsilon_\mu(\mathbf{k},\lambda)\left(\omega_{k'}\omega_k + \mathbf{k}'\cdot\mathbf{k}\right) \\
& \times \left(\hat{a}_{\mathbf{k}'\lambda'}\hat{a}_{\mathbf{k}\lambda}\,\mathrm{e}^{-\mathrm{i}(k'+k)\cdot x} + \hat{a}^\dagger_{\mathbf{k}'\lambda'}\hat{a}^\dagger_{\mathbf{k}\lambda}\,\mathrm{e}^{\mathrm{i}(k'+k)\cdot x}\right. \\
& \left. -\hat{a}^\dagger_{\mathbf{k}\lambda}\hat{a}_{\mathbf{k}'\lambda'}\,\mathrm{e}^{-\mathrm{i}(k'-k)\cdot x} - \hat{a}^\dagger_{\mathbf{k}'\lambda'}\hat{a}_{\mathbf{k}\lambda}\,\mathrm{e}^{\mathrm{i}(k'-k)\cdot x}\right) \ .
\end{aligned} \tag{7.34}$$

The integration over x yields a delta function $(2\pi)^3\delta^3(\mathbf{k}' \pm \mathbf{k})$. Therefore the factor $\omega_{k'}\omega_k + \mathbf{k}'\cdot\mathbf{k} \to \omega_k^2 \mp \mathbf{k}^2$ vanishes for the upper sign. Using the orthogonality relation (7.31) of the polarization vectors we finally obtain

$$\begin{aligned}
\hat{H} =\ & \int\mathrm{d}^3k\,\omega_k \sum_{\lambda=0}^{3}(-g_{\lambda\lambda})\,\hat{a}^\dagger_{\mathbf{k}\lambda}\hat{a}_{\mathbf{k}\lambda} \\
=\ & \int\mathrm{d}^3k\,\omega_k\left(\sum_{\lambda=1}^{3}\hat{a}^\dagger_{\mathbf{k}\lambda}\hat{a}_{\mathbf{k}\lambda} - \hat{a}^\dagger_{\mathbf{k}0}\hat{a}_{\mathbf{k}0}\right) \ .
\end{aligned} \tag{7.35}$$

Similarly from (7.20) the momentum operator is found to be

$$\begin{aligned}
\hat{\mathbf{P}} =\ & \int\mathrm{d}^3x\,{:}\dot{\hat{A}}^\mu\boldsymbol{\nabla}\hat{A}_\mu{:} \\
=\ & \int\mathrm{d}^3k\,\mathbf{k}\left(\sum_{\lambda=1}^{3}\hat{a}^\dagger_{\mathbf{k}\lambda}\hat{a}_{\mathbf{k}\lambda} - \hat{a}^\dagger_{\mathbf{k}0}\hat{a}_{\mathbf{k}0}\right) \ .
\end{aligned} \tag{7.36}$$

Now one might attempt to interpret $\hat{a}^\dagger_{\mathbf{k}\lambda}$ and $\hat{a}_{\mathbf{k}\lambda}$ as creation and annihilation operators of four different types of photons. This poses no problems for the transverse ($\lambda = 1, 2$) or for the longitudinal ($\lambda = 3$) polarization states, but the "scalar photons" ($\lambda = 0$) cause trouble.

7.3 Canonical Quantization in the Lorentz Gauge

The problem becomes evident if one tries to construct the *Fock space for photons*. Let us assume that a vacuum state $|0\rangle$ exists which has the property

$$\hat{a}_{k\lambda}|0\rangle = 0 \quad \text{for all} \quad k, \lambda \tag{7.37}$$

and is normalized

$$\langle 0|0\rangle = 1 \,. \tag{7.38}$$

As usual the eigenstates of the occupation-number operators can be constructed by applying the creation operators $\hat{a}^\dagger_{k\lambda}$ to the vacuum.[6] The state vector containing $n_{k,\lambda}$ photons reads

$$|n_{k,\lambda}\rangle = \frac{1}{\sqrt{n_{k,\lambda}!}} \left(\hat{a}^\dagger_{k\lambda}\right)^{n_{k,\lambda}} |0\rangle \,. \tag{7.39}$$

The norm of the one-photon state is

$$\begin{aligned}\langle 1_{k,\lambda}|1_{k,\lambda}\rangle &= \langle 0|\hat{a}_{k\lambda}\hat{a}^\dagger_{k\lambda}|0\rangle = \langle 0| - g_{\lambda\lambda}\delta^3(k-k) + \hat{a}^\dagger_{k\lambda}\hat{a}_{k\lambda}|0\rangle \\ &= -g_{\lambda\lambda}\delta^3(\mathbf{0})\langle 0|0\rangle \,. \end{aligned} \tag{7.40}$$

Thus *the norm of the state containing a scalar photon is negative*!

Remark: The presence of the delta function $\delta^3(\mathbf{0})$ signals that (7.40) is divergent. Of course this has its root in the expansion of the field operator into plane waves that extend over an infinite volume. Using the box normalization we would have to replace the delta function $\delta^3(\mathbf{0})$ by $\delta_{k,k} = 1$ (with discretized vectors k). In order to work with strictly normalizable states in the continuum limit the momentum eigenstates have to be smeared out by constructing *wave packets*. The one-photon state is then replaced by

$$|1_{k,\lambda}\rangle = \int d^3k' \, F_k(k') \hat{a}^\dagger_{k'\lambda} |0\rangle \,, \tag{7.41}$$

where $F_k(k')$ is a smooth "smearing function". The detailed form of this function is not important, provided that it is concentrated at a mean momentum k and normalized to unity:

$$\int d^3k' \, |F_k(k')|^2 = 1 \,. \tag{7.42}$$

With this language, (7.41) becomes

$$\begin{aligned}\langle 1_{k,\lambda}|1_{k,\lambda}\rangle &= \int d^3k'' \, d^3k' \, F^*_k(k'') F_k(k') \left(-g_{\lambda\lambda}\delta^3(k''-k')\right) \langle 0|0\rangle \\ &= -g_{\lambda\lambda} \int d^3k' \, |F_k(k')|^2 \langle 0|0\rangle \\ &= -g_{\lambda\lambda}\langle 0|0\rangle \,, \end{aligned} \tag{7.43}$$

a result which also could have been obtained by using the simple box normalization.

Similarly one finds that the energy of a photon, defined as the expectation value

$$\langle 1_{k,\lambda}|\hat{H}|1_{k,\lambda}\rangle = \omega_k \langle 1_{k,\lambda}|1_{k,\lambda}\rangle \,, \tag{7.44}$$

---

[6] For simplicity we restrict our attention to a single field mode k, λ; the treatment of several modes would only clog up the notation.

becomes *negative* for the scalar polarization $\lambda = 0$ owing to the negative value of the norm.

In general the action of the ladder operators $\hat{a}$ and $\hat{a}^\dagger$ on an N-particle state is given by (using box normalization)

$$\hat{a}_{k\lambda} |n_{k,\lambda}\rangle = (-g_{\lambda\lambda})\sqrt{n_{k,\lambda}} |n_{k,\lambda} - 1\rangle , \qquad (7.45a)$$

$$\hat{a}^\dagger_{k\lambda} |n_{k,\lambda}\rangle = \sqrt{n_{k,\lambda} + 1} |n_{k,\lambda} + 1\rangle . \qquad (7.45b)$$

As in Chap. 1 these results can be immediately deduced from the commutation relations. The sign factor in (7.45a) has the consequence that the *number operator* for "scalar photons" ($\lambda = 0$) obtains a minus sign:

$$\hat{n}_{k,\lambda} = \hat{a}^\dagger_{k\lambda}\hat{a}_{k\lambda} \quad \text{for } \lambda = 1, 2, 3, \quad \text{but} \quad \hat{n}_{k0} = -\hat{a}^\dagger_{k0}\hat{a}_{k0} . \qquad (7.46)$$

As a consequence the *eigenvalue* of the Hamiltonian is positive even for a scalar photon:

$$\hat{H} |1_{k0}\rangle = +\omega_k \hat{n}_{k0} |1_{k0}\rangle = +\omega_k . \qquad (7.47)$$

However, the *expectation value* of the energy is negative owing to the indefinite metric.[7]

7.4 The Gupta–Bleuler Method

The theory of the quantized photon field developed up to now is obviously missing a vital ingredient. Apparently arbitrary amounts of energy could be gained by creating "scalar photons". In reality, of course, only transverse photons are observed in experiments; the scalar and longitudinal photons are mathematical constructions that have to drop out in the calculation of observable quantities. It is the Lorentz gauge condition we have not yet taken into account that takes care of the problems. As shown in (7.24) the Lorentz condition cannot be enforced as an operator identity. Instead we will use it as a *condition for the state vectors in Hilbert space* $|\Phi\rangle$. Only those state vectors are admitted for which the *expectation value of the gauge condition* is satisfied:

$$\langle \Phi | \partial^\mu \hat{A}_\mu(x) | \Phi \rangle = 0 . \qquad (7.48)$$

Instead of (7.48) we will use a somewhat stronger condition, which is cast in the form of an eigenvalue equation. We split the field operator into two parts, the positive- and negative-frequency parts $\hat{A}^{(+)}(x)$ and $\hat{A}^{(-)}(x)$, which consist of annihilation and of creation operators, respectively:

$$\hat{A}(x) = \hat{A}^{(+)}(x) + \hat{A}^{(-)}(x) , \qquad (7.49)$$

---

[7] One could attempt to exchange the role of the $\hat{a}$ operators, interpreting $\hat{a}_{k0}$ as a creator and $\hat{a}^\dagger_{k0}$ as an annihilator of photons. This was the procedure that led to a successful interpretation of the negative-energy solutions of the Dirac equation. In the photon case, the problem of the negative norm might be circumvented in this way. However, because of the bosonic quantization the expectation value (and now also the eigenvalue) of the energy of a scalar photon would remain negative.

where $\hat{A}^{(+)\dagger} = \hat{A}^{(-)}$. This splitting, by the way, is relativistically invariant. Now the following condition for the annihilation part of the Lorentz gauge constraint is postulated:[8]

$$\partial^\mu \hat{A}^{(+)}_\mu(x)|\Phi\rangle = 0, \tag{7.50}$$

and the corresponding adjoint condition

$$\langle\Phi|\,\partial^\mu \hat{A}^{(-)}_\mu(x) = 0. \tag{7.51}$$

Obviously (7.50) and (7.51) are sufficient to guarantee (7.48) since

$$\langle\Phi|\partial^\mu \hat{A}_\mu|\Phi\rangle = \langle\Phi|[\partial^\mu \hat{A}^{(+)}_\mu|\Phi\rangle] + [\langle\Phi|\partial^\mu \hat{A}^{(-)}_\mu]|\Phi\rangle = 0. \tag{7.52}$$

The theory of the quantized electromagnetic field, described in terms of a Hilbert space with an *indefinite metric* and with use of the constraint (7.50) on the state vectors, was first suggested by S. Gupta and extended by K. Bleuler.[9] Here we will not explore this formalism in its mathematical details but will discuss only its main features.

What are the consequences of the constraint (7.50) in terms of the photon creation and annihilation operators? Using the positive-frequency part of the field operator expanded according to (7.26), we find the constraint

$$\int \frac{d^3k}{\sqrt{2\omega_k(2\pi)^3}}\, e^{-ik\cdot x} \sum_{\lambda=0}^{3} k\cdot\epsilon(\boldsymbol{k},\lambda)\,\hat{a}_{\boldsymbol{k}\lambda}|\Phi\rangle = 0. \tag{7.53}$$

This has to hold for all Fourier modes so that the sum over polarizations in (7.53) has to vanish. The consequences of this become clear if we recall the four-dimensional transversality properties (6.82) and (6.83) of the polarization vectors, which read $k\cdot\epsilon(\boldsymbol{k},1) = k\cdot\epsilon(\boldsymbol{k},2) = 0$ and $k\cdot\epsilon(\boldsymbol{k},0) = -k\cdot\epsilon(\boldsymbol{k},3)$. Thus (7.53) implies the constraint

$$\hat{L}_{\boldsymbol{k}}|\Phi\rangle \equiv (\hat{a}_{\boldsymbol{k}0} - \hat{a}_{\boldsymbol{k}3})|\Phi\rangle = 0 \quad \text{for all } \boldsymbol{k} \tag{7.54}$$

and similarly

$$\langle\Phi|\hat{L}^\dagger_{\boldsymbol{k}} \equiv \langle\Phi|(\hat{a}^\dagger_{\boldsymbol{k}0} - \hat{a}^\dagger_{\boldsymbol{k}3}) = 0. \tag{7.55}$$

This condition provides a link between the longitudinal and the scalar photons. Obviously the equations (7.54) and (7.55) imply that *the expectation values of the numbers of longitudinal and scalar photons are equal for each admissible state* $|\Phi\rangle$:

$$\langle\Phi|\,\hat{a}^\dagger_{\boldsymbol{k}0}\hat{a}_{\boldsymbol{k}0}\,|\Phi\rangle = \langle\Phi|\,\hat{a}^\dagger_{\boldsymbol{k}3}\hat{a}_{\boldsymbol{k}3}\,|\Phi\rangle. \tag{7.56}$$

In this way we get rid of the problem of negative energies in an elegant way. The expectation value of the energy accoding to (7.35) is

---

[8] The more intuitive condition $\partial^\mu \hat{A}_\mu|\Phi\rangle = 0$ does not work. Even the vacuum state leads to $\partial^\mu \hat{A}^{(-)}_\mu|0\rangle \neq 0$, since $\hat{A}^{(-)}_\mu$ consists of creation operators.
[9] S. Gupta: Proc. Roy. Soc. **63A**, 681 (1950);
K. Bleuler: Helv. Phys. Acta **23**, 567 (1950).

$$\langle\Phi|\hat{H}|\Phi\rangle = \int d^3k\,\omega_k\left(\sum_{\lambda=1}^{2}\langle\Phi|\hat{a}^\dagger_{k\lambda}\hat{a}_{k\lambda}|\Phi\rangle + \langle\Phi|\,\hat{a}^\dagger_{k3}\hat{a}_{k3} - \hat{a}^\dagger_{k0}\hat{a}_{k0}\,|\Phi\rangle\right)$$

$$= \int d^3k\,\omega_k\sum_{\lambda=1}^{2} n_{k,\lambda}\,. \tag{7.57}$$

Thus *only the transverse photons make a net contribution to the energy*. The negative energy of the scalar "pseudo photons" is exactly cancelled by the opposite contribution from the longitudinal photons! The same effect will hold true for the other observable quantities, in particular momentum and angular momentum.

When constructing the *Fock space* of photons, one has considerable freedom: *all states having the same configuration of transverse photons are physically equivalent* even if they differ in the admixture of longitudinal and scalar photons (which cancel each other). As we will discuss in more detail in Exercise 7.2 this is a consequence of the gauge invariance of the electromagnetic field.

A *general state vector in Fock space* $|\Phi_c\rangle$ can be expressed by the ansatz

$$|\Phi_c\rangle = \hat{R}_c|\Phi_T\rangle\,, \tag{7.58}$$

where $|\Phi_T\rangle$ is a state which contains only transverse photons. The operator $\hat{R}_c$ creates the admixture of longitudinal and scalar pseudo photons. Its general form is given by

$$\hat{R}_c = 1 + \int d^3k\,c(\boldsymbol{k})\hat{L}^\dagger_{\boldsymbol{k}} + \int d^3k\,d^3k'\,c(\boldsymbol{k},\boldsymbol{k}')\hat{L}^\dagger_{\boldsymbol{k}}\hat{L}^\dagger_{\boldsymbol{k}'} + \ldots \tag{7.59}$$

where the creation operator $\hat{L}^\dagger_{\boldsymbol{k}}$ has been defined in (7.55). The admixture coefficients $c(\boldsymbol{k})$ etc. can be freely chosen. The ansatz (7.59) is motivated by the observation that the commutator of $\hat{L}_{\boldsymbol{k}}$ and $\hat{L}^\dagger_{\boldsymbol{k}'}$ vanishes:

$$\begin{aligned}[\hat{L}_{\boldsymbol{k}},\hat{L}^\dagger_{\boldsymbol{k}'}] &= [\hat{a}_{k0}-\hat{a}_{k3},\hat{a}^\dagger_{k'0}-\hat{a}^\dagger_{k'3}] \\ &= [\hat{a}_{k0},\hat{a}^\dagger_{k'0}] + [\hat{a}_{k3},\hat{a}^\dagger_{k'3}] - [\hat{a}_{k3},\hat{a}^\dagger_{k'0}] - [\hat{a}_{k0},\hat{a}^\dagger_{k'3}] \\ &= -\delta^3(\boldsymbol{k}-\boldsymbol{k}') + \delta^3(\boldsymbol{k}-\boldsymbol{k}') = 0\,.\end{aligned} \tag{7.60}$$

In this way the validity of the constraint (7.54) is guaranteed for arbitrary states (7.58). Since the operator $\hat{R}_c$ consists of products of operators $\hat{L}^\dagger_{\boldsymbol{k}}$ it will commute with $\hat{L}_{\boldsymbol{k}}$. As a consequence

$$\hat{L}_{\boldsymbol{k}}|\Phi_c\rangle = \hat{L}_{\boldsymbol{k}}\hat{R}_c|\Phi_T\rangle = \hat{R}_c\hat{L}_{\boldsymbol{k}}|\Phi_T\rangle = 0\,, \tag{7.61}$$

because $|\Phi_T\rangle$ does not contain pseudo photons and thus trivially satisfies the gauge condition:

$$\hat{L}_{\boldsymbol{k}}|\Phi_T\rangle = 0\quad,\quad \langle\Phi_T|\hat{L}^\dagger_{\boldsymbol{k}} = 0\,. \tag{7.62}$$

Equation (7.60) also leads to the commutators between $\hat{R}^\dagger_c$ and $\hat{R}_c$ vanishing:

$$[\hat{R}^\dagger_c,\hat{R}_{c'}] = 0 \tag{7.63}$$

for arbitrarily chosen sets of admixture coefficients c, which also may be different. As a consequence the norm of a state, and more generally any scalar products of states, is determined solely by the transverse sector of Hilbert space:

$$\langle \Phi'_{c'} | \Phi_c \rangle = \langle \Phi'_T | \hat{R}^\dagger_{c'} \hat{R}_c | \Phi_T \rangle = \langle \Phi'_T | \hat{R}_c \hat{R}^\dagger_{c'} | \Phi_T \rangle = \langle \Phi'_T | \Phi_T \rangle . \quad (7.64)$$

Pseudo photons do not contribute to the norm.

The states $|\Phi_c\rangle$ defined in (7.58) for a given $|\Phi_T\rangle$ form an *equivalence class*. For simplicity all the coefficients c can be set equal to zero and $|\Phi_T\rangle$ can be used as the representative of this class. Thus effectively the pseudo photons can be simply ignored. As shown in Exercise 7.2 this corresponds to a special choice of the gauge of the electromagnetic field.

The situation is depicted schematically in Fig. 7.1. The large box corresponds to the full Hilbert space of photon states, most of which, however, are unphysical (shaded region) because they do not respect the Lorentz condition. The admissible space (unshaded) contains gauge-equivalent classes of states. In the figure they are symbolized by the fibers emanating from the thick line, which corresponds to a special choice of the gauge.

To close the discussion a word of caution is in order. The restriction to transverse photons is justified only with respect to photons which asymptotically are observed as free particles. When dealing with intermediate states (*virtual photons*) the longitudinal and scalar degrees of freedom *are* of importance and have to be taken into account. This will be analyzed in Sect. 7.4 in the context of the Feynman propagator for photons.

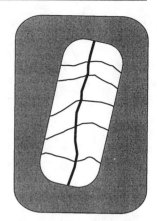

Fig. 7.1. Symbolic picture of the Hilbert space of photons. In the shaded region the Lorentz gauge condition is violated. States on the same fiber (*thin lines*) are gauge-equivalent

EXERCISE

7.2 Gauge Transformations and Pseudo-photon States

Problem. Show that the state vectors $|\Phi_c\rangle$ and $|\Phi_T\rangle$ in (7.58) differ by a gauge transformation. Show that the expectation values of the potential are related by

$$\langle \Phi_c | \hat{A}_\mu(x) | \Phi_c \rangle = \langle \Phi_T | \hat{A}_\mu(x) + \partial_\mu \Lambda(x) | \Phi_T \rangle = \langle \Phi_T | \hat{A}_\mu(x) | \Phi_T \rangle + \partial_\mu \Lambda(x) \quad (1)$$

and derive an expression for the function $\Lambda(x)$.

Solution. The state $|\Phi_c\rangle$ is defined by

$$|\Phi_c\rangle = \hat{R}_c |\Phi_T\rangle \quad (2)$$

with the creation operator of a "pseudo-photon" configuration

$$\hat{R}_c = 1 + \int d^3k\, c(\boldsymbol{k}) \hat{L}^\dagger_{\boldsymbol{k}} + \int d^3k\, d^3k'\, c(\boldsymbol{k}, \boldsymbol{k}') \hat{L}^\dagger_{\boldsymbol{k}} \hat{L}^\dagger_{\boldsymbol{k}'} + \ldots, \quad (3)$$

where $\hat{L}_{\boldsymbol{k}} = \hat{a}_{\boldsymbol{k}0} - \hat{a}_{\boldsymbol{k}3}$. We have to evaluate

$$A^c_\mu(x) \equiv \langle \Phi_T | \Big(1 + \int d^3k\, c^*(\boldsymbol{k}) \hat{L}_{\boldsymbol{k}} + \ldots \Big) \hat{A}_\mu(x) \Big(1 + \int d^3k\, c(\boldsymbol{k}) \hat{L}^\dagger_{\boldsymbol{k}} + \ldots \Big) | \Phi_T \rangle . \quad (4)$$

To simplify this expression the operator $\hat{L}_{\boldsymbol{k}}$ can be commutated to the right so that it acts on the state $|\Phi_T\rangle$ and similarly for $\hat{L}^\dagger_{\boldsymbol{k}}$ which is moved to the left. Then the constraints

Exercise 7.2

$$\hat{L}_{\boldsymbol{k}}\left|\varPhi_T\right\rangle = 0 \quad \text{and} \quad \left\langle\varPhi_T\right|\hat{L}_{\boldsymbol{k}}^\dagger = 0 \tag{5}$$

can be used. It is evident that the commutators $[\hat{L}_{\boldsymbol{k}}, \hat{A}_\mu]$ and $[\hat{A}_\mu, \hat{L}_{\boldsymbol{k}}^\dagger]$ are pure c numbers since $\hat{L}$ and $\hat{A}$ are each linear combinations of photon creation and annihilation operators. Since also

$$[\hat{L}_{\boldsymbol{k}}, \hat{L}_{\boldsymbol{k}'}^\dagger] = 0 \tag{6}$$

it is obvious without a detailed calculation that the terms of higher order in (4) do not contribute. Thus we are left with

$$\begin{aligned}\left\langle \varPhi_c\right|\hat{A}_\mu(x)\left|\varPhi_c\right\rangle &= \left\langle \varPhi_T\right|\hat{A}_\mu(x)\left|\varPhi_T\right\rangle + \left\langle \varPhi_T\right|\int \mathrm{d}^3k\, c(\boldsymbol{k})\left[\hat{A}_\mu(x), \hat{L}_{\boldsymbol{k}}^\dagger\right]\left|\varPhi_{\boldsymbol{k}}\right\rangle \\ &\quad + \left\langle \varPhi_T\right|\int \mathrm{d}^3k\, c^*(\boldsymbol{k})\left[\hat{L}_{\boldsymbol{k}}, \hat{A}_\mu(x)\right]\left|\varPhi_{\boldsymbol{k}}\right\rangle.\end{aligned} \tag{7}$$

The commutators in (7) can be readily evaluated using the Fourier decomposition of the field operator (7.26) and the canonical commutation relations (7.32). From

$$\begin{aligned}\left[\hat{a}_{\boldsymbol{k}'\lambda}, \hat{L}_{\boldsymbol{k}}^\dagger\right] &= \left[\hat{a}_{\boldsymbol{k}'\lambda}, \hat{a}_{\boldsymbol{k}0}^\dagger - \hat{a}_{\boldsymbol{k}3}^\dagger\right] = -\delta^3(\boldsymbol{k}-\boldsymbol{k}')(g_{\lambda 0} - g_{\lambda 3})\\&= -\delta^3(\boldsymbol{k}-\boldsymbol{k}')(\delta_{\lambda 0} + \delta_{\lambda 3})\end{aligned}\tag{8}$$

we obtain

$$\begin{aligned}\left[\hat{A}_\mu(x), \hat{L}_{\boldsymbol{k}}^\dagger\right] &= \int \frac{\mathrm{d}^3k'}{\sqrt{2\omega_{\boldsymbol{k}'}(2\pi)^3}}\, \mathrm{e}^{-\mathrm{i}k'\cdot x}\sum_{\lambda=0}^{3}\epsilon_\mu(\boldsymbol{k}',\lambda)\left[\hat{a}_{\boldsymbol{k}'\lambda}, \hat{L}_{\boldsymbol{k}}^\dagger\right]\\&= -\frac{1}{\sqrt{2\omega_{\boldsymbol{k}}(2\pi)^3}}\, \mathrm{e}^{-\mathrm{i}k\cdot x}\sum_{\lambda=0,3}\epsilon_\mu(\boldsymbol{k},\lambda)\end{aligned}\tag{9}$$

and similarly

$$\left[\hat{L}_{\boldsymbol{k}}, \hat{A}_\mu(x)\right] = -\frac{1}{\sqrt{2\omega_{\boldsymbol{k}}(2\pi)^3}}\, \mathrm{e}^{\mathrm{i}k\cdot x}\sum_{\lambda=0,3}\epsilon_\mu(\boldsymbol{k},\lambda). \tag{10}$$

The sum over polarizations can be rewritten using the explicit form of the polarization vectors (6.78), (6.81) (for $k^2 = 0$)

$$\sum_{\lambda=0,3}\epsilon(\boldsymbol{k},\lambda) = \epsilon(\boldsymbol{k},0) + \epsilon(\boldsymbol{k},3) = n + \frac{k - n(k\cdot n)}{k\cdot n} = \frac{1}{k\cdot n}\, k. \tag{11}$$

Thus the sum is proportional to the four-momentum vector k_μ. Then the extra term in (7) becomes

$$-\left\langle\varPhi_T|\varPhi_T\right\rangle \int \frac{\mathrm{d}^3k}{\sqrt{2\omega_{\boldsymbol{k}}(2\pi)^3}}\, \frac{1}{k\cdot n}\, k_\mu\left(c(\boldsymbol{k})\, \mathrm{e}^{-\mathrm{i}k\cdot x} + c^*(\boldsymbol{k})\, \mathrm{e}^{\mathrm{i}k\cdot x}\right). \tag{12}$$

Since the integrand contains the factor k_μ, (12) in coordinate space can be written as the gradient of a function $\Lambda(x)$. This proves (1) with a real-valued gauge function

$$\Lambda(x) = -\mathrm{i}\int \frac{\mathrm{d}^3k}{\sqrt{2\omega_{\boldsymbol{k}}(2\pi)^3}}\, \frac{1}{k\cdot n}\left(c(\boldsymbol{k})\, \mathrm{e}^{-\mathrm{i}k\cdot x} - c^*(\boldsymbol{k})\, \mathrm{e}^{\mathrm{i}k\cdot x}\right). \tag{13}$$

Because $k^2 = 0$, (13) is compatible with the Lorentz condition $\partial^\mu A_\mu = 0$ since

$$\Box \Lambda(x) = 0 \, . \tag{14}$$

The gauge function (13) does not depend on higher terms of the expansion (3) with coefficiens like $c(\boldsymbol{k},\boldsymbol{k}')$ etc. Such terms will enter when one studies the gauge transformation of more complex functions, such as products of the type $\hat{A}_\mu(x)\hat{A}_\nu(y)$.

7.5 The Feynman Propagator for Photons

The Feynman propagator for the massless photon field can be obtained in a similar way as its counterpart for the massive vector field. The second line of equation (11) in Exercise 6.4 can essentially be taken over. However, now the λ summation extends over the full set of four polarization states and it contains a sign factor $-g_{\lambda\lambda}$ originating from the commutation relations:

$$\begin{aligned}
\mathrm{i} D_\mathrm{F}^{\mu\nu}(x-y) &:= \langle 0|T\big(\hat{A}^\mu(x)\hat{A}^\nu(y)\big)|0\rangle \\
&= \int \frac{\mathrm{d}^3 k}{2\omega_k (2\pi)^3} \left(\sum_{\lambda=0}^{3}(-g_{\lambda\lambda})\epsilon^\mu(\boldsymbol{k},\lambda)\epsilon^\nu(\boldsymbol{k},\lambda)\right) \\
&\quad \times \Big(\Theta(x_0-y_0)\,\mathrm{e}^{-\mathrm{i} k\cdot(x-y)} + \Theta(y_0-x_0)\,\mathrm{e}^{+\mathrm{i} k\cdot(x-y)}\Big) \, .
\end{aligned} \tag{7.65}$$

The name $D_\mathrm{F}^{\mu\nu}$ instead of $\Delta_\mathrm{F}^{\mu\nu}$ is used for the massless propagator. The λ summation just corresponds to the completeness relation (6.83) of the polarization vectors

$$\sum_{\lambda=0}^{3} g_{\lambda\lambda}\, \epsilon^\mu(\boldsymbol{k},\lambda)\, \epsilon^\nu(\boldsymbol{k},\lambda) = g^{\mu\nu} \, . \tag{7.66}$$

Thus using (4.136) we obtain

$$D_\mathrm{F}^{\mu\nu}(x-y) = \int \frac{\mathrm{d}^4 k}{(2\pi)^4}\, \mathrm{e}^{-\mathrm{i} k\cdot(x-y)} D_\mathrm{F}^{\mu\nu}(k) \tag{7.67}$$

with the *momentum-space Feynman propagator*

$$D_\mathrm{F}^{\mu\nu}(k) = \frac{-g_{\mu\nu}}{k^2 + \mathrm{i}\epsilon} \, . \tag{7.68}$$

Remark: Equation (7.68) is the photon propagator in the "Feynman gauge", i.e., it is valid for the special choice of the gauge fixing parameter $\zeta = 1$. The general result looks a bit more complicated:

$$D_\mathrm{F}^{\mu\nu}(k) = \frac{-g_{\mu\nu}}{k^2 + \mathrm{i}\epsilon} + \frac{\zeta - 1}{\zeta}\frac{k^\mu k^\nu}{(k^2 + \mathrm{i}\epsilon)^2} \, . \tag{7.69}$$

This will be derived in Exercise 7.3. As well as the Feynman gauge, the special case $\zeta \to \infty$ also enjoys popularity. The propagator (7.69) in this limit becomes

$$D_\mathrm{F}^{\mu\nu}(k) = \frac{k^\mu k^\nu - g^{\mu\nu} k^2}{(k^2 + \mathrm{i}\epsilon)^2} \tag{7.70}$$

Exercise 7.2

and has the useful property of being transverse in four dimensions, $k_\mu D_F^{\mu\nu}(k) = 0$, which is useful for some calculations. This choice of the parameter ζ is called the *Landau gauge*.

The photon propagator could only take on the simple form (7.68) because the summation in (7.66) was not restricted to the "physical" transverse polarizations but extends over all four polarization states. To get a better understanding of this we will inspect the contributions from the degrees of freedom of the electromagnetic field separately. We write (cf. the polarization vectors (6.78) and (6.81))

$$\begin{aligned}
D_F^{\mu\nu}(k) &= \frac{1}{k^2 + i\epsilon}\left(\sum_{\lambda=1}^{2}\epsilon_\mu(\boldsymbol{k},\lambda)\epsilon_\nu(\boldsymbol{k},\lambda)\right.\\
&\qquad\left. + \epsilon_\mu(\boldsymbol{k},3)\epsilon_\nu(\boldsymbol{k},3) - \epsilon_\mu(\boldsymbol{k},0)\epsilon_\nu(\boldsymbol{k},0)\right) \\
&= \frac{1}{k^2 + i\epsilon}\left(\sum_{\lambda=1}^{2}\epsilon_\mu(\boldsymbol{k},\lambda)\epsilon_\nu(\boldsymbol{k},\lambda)\right.\\
&\qquad\left. + \frac{(k_\mu - n_\mu(k\cdot n))(k_\nu - n_\nu(k\cdot n))}{(k\cdot n)^2 - k^2} - n_\mu n_\nu\right) \\
&= \frac{1}{k^2 + i\epsilon}\left(\sum_{\lambda=1}^{2}\epsilon_\mu(\boldsymbol{k},\lambda)\epsilon_\nu(\boldsymbol{k},\lambda) + \frac{k^2 n_\mu n_\nu}{(k\cdot n)^2 - k^2}\right.\\
&\qquad\left. + \frac{k_\mu k_\nu - (k_\mu n_\nu + k_\nu n_\mu)(k\cdot n)}{(k\cdot n)^2 - k^2}\right). \quad (7.71)
\end{aligned}$$

Here the contributions of the longitudinal and scalar photons have been suitably combined to arrive at the following decomposition of the propagator:

$$D_F^{\mu\nu}(k) = D_{F(\text{trans})}^{\mu\nu}(k) + D_{F(\text{Coul})}^{\mu\nu}(k) + D_{F(\text{resid})}^{\mu\nu}(k), \quad (7.72)$$

with the transverse part

$$D_{F(\text{trans})}^{\mu\nu}(k) = \frac{1}{k^2 + i\epsilon}\sum_{\lambda=1}^{2}\epsilon_\mu(\boldsymbol{k},\lambda)\epsilon_\nu(\boldsymbol{k},\lambda), \quad (7.73)$$

the Coulomb part (the k^2 factor drops out here)

$$D_{F(\text{Coul})}^{\mu\nu}(k) = \frac{n_\mu n_\nu}{(k\cdot n)^2 - k^2}, \quad (7.74)$$

and the remainder

$$D_{F(\text{resid})}^{\mu\nu}(k) = \frac{1}{k^2 + i\epsilon}\frac{k_\mu k_\nu - (k_\mu n_\nu + k_\nu n_\mu)(k\cdot n)}{(k\cdot n)^2 - k^2}. \quad (7.75)$$

The *Coulomb propagator* (7.74) is particularly interesting. In the special Lorentz frame where the time-like reference vector n reads $n = (1,0,0,0)$ we have

$$D_{F(\text{Coul})}^{\mu\nu}(k) = \frac{\delta_{\mu 0}\delta_{\nu 0}}{|\boldsymbol{k}|^2}. \quad (7.76)$$

This is just the well-known Fourier transformed Coulomb potential! In coordinate space we find

$$\begin{aligned}
D_{\text{F(Coul)}}^{\mu\nu}(x-y) &= \int \frac{\mathrm{d}^4 k}{(2\pi)^4} \, \mathrm{e}^{-\mathrm{i}k\cdot(x-y)} \, D_{\text{F(Coul)}}^{\mu\nu}(k) \\
&= \delta(x_0 - y_0)\, \delta_{\mu 0} \delta_{\nu 0} \int \frac{\mathrm{d}^3 k}{(2\pi)^3} \frac{\mathrm{e}^{\mathrm{i}\boldsymbol{k}\cdot(\boldsymbol{x}-\boldsymbol{y})}}{|\boldsymbol{k}|^2} \\
&= \delta_{\mu 0}\delta_{\nu 0} \frac{\delta(x_0 - y_0)}{4\pi|\boldsymbol{x}-\boldsymbol{y}|} \,.
\end{aligned} \qquad (7.77)$$

This propagator describes an *instantaneous static Coulomb interaction*. It couples only to the 0 component of the electromagnetic current, i.e., to the charge density. Thus *the Coulomb interaction arises from the combined exchange of longitudinal and scalar photons*. Leaving out these degrees of freedom from the polarization sum would remove the effects of the Coulomb interaction. By simply using the *covariant photon propagator* (7.68), however, one automatically takes into account all four types of (virtual) photons, including the Coulomb term.

The role of the remainder term (7.75) has yet to be clarified. The terms in $D_{\text{F(resid)}}^{\mu\nu}(k)$ are proportional to the momentum vector k^μ or k^ν. The electromagnetic field couples to conserved currents which satisfy

$$\partial^\mu j_\mu(x) = 0 \quad \text{or} \quad k^\mu j_\mu(k) = 0 \,. \qquad (7.78)$$

Therefore the scalar product of $D_{\text{F(resid)}}^{\mu\nu}(k)$ with a conserved current produces a vanishing result. As shown in Chap. 8 the electromagnetic interaction of two (transition) currents $j^{(1)}(x)$ and $j^{(2)}(y)$ is described by

$$\begin{aligned}
& e^2 \int \mathrm{d}^4 y \int \mathrm{d}^4 x \, j_\mu^{(2)}(y) \, D_{\text{F}}^{\mu\nu}(y-x) \, j_\nu^{(1)}(x) \\
&= e^2 \int \frac{\mathrm{d}^4 k}{(2\pi)^4} \, j_\mu^{(2)}(-k) \, D_{\text{F}}^{\mu\nu}(k) \, j_\nu^{(1)}(k) \,.
\end{aligned} \qquad (7.79)$$

We conclude that the residual term does not contribute in the calculation of observable quantities and thus it can be simply left out. The same argument also applies to the gauge-dependent extra term in (7.69).

In Fig. 7.2 the interaction between two charged particles is depicted using the language of Feynman graphs. According to the decomposition (7.72), transverse as well as "Coulomb photons" are exchanged. Note, however, that this separation depends on the gauge and on the frame of reference. A well-defined unique result is obtained only by adding up the contributions.

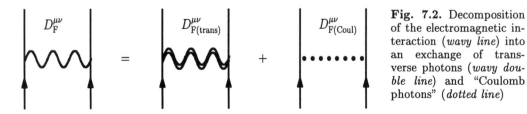

Fig. 7.2. Decomposition of the electromagnetic interaction (*wavy line*) into an exchange of transverse photons (*wavy double line*) and "Coulomb photons" (*dotted line*)

7.6 Supplement: Simple Rule for Deriving Feynman Propagators

The systematic derivation of results like (7.69) from the vacuum expectation value of time-ordered products of field operators may be somewhat laborious. There is a simple shortcut, however, by which the Feynman propagator can be directly read off from the Lagrangian of a theory.

- The Feynman propagator in momentum space is obtained by inverting the Fourier-transformed differential operator contained in the Lagrangian.

We check this explicitly for the various cases discussed so far. The Lagrangian for the *Klein–Gordon* theory

$$\mathcal{L} = \frac{1}{2}\partial_\mu \phi \partial^\mu \phi - \frac{1}{2}m^2 \phi^2 \equiv \frac{1}{2}\phi \mathcal{D}(x)\phi \tag{7.80}$$

contains the differential operator

$$\mathcal{D}(x) = \overleftarrow{\partial}_\mu \overrightarrow{\partial}^\mu - m^2 \ . \tag{7.81}$$

If we make the changes $\overrightarrow{\partial}^\mu \to -ik^\mu$ and $\overleftarrow{\partial}_\mu \to +ik_\mu$, this leads to the multiplication operator in momentum space

$$\mathcal{D}(k) = k^2 - m^2 \ . \tag{7.82}$$

The Feynman propagator in momentum space is simply obtained by inverting this

$$\Delta_F(k) = \mathcal{D}^{-1}(k) = \frac{1}{k^2 - m^2 + i\epsilon} \ , \tag{7.83}$$

where the poles are shifted as usual by adding a small negative imaginary part to the mass in order to satisfy the causality condition.

The Feynman propagator for *Dirac particles* $S_F(k)$ discussed in Sect. 5.4 is obtained in the same way.

For the *Proca theory* described by the Lagrangian

$$\mathcal{L} = -\frac{1}{4}F_{\mu\nu}F^{\mu\nu} + \frac{m^2}{2}A_\mu A^\mu \equiv \frac{1}{2}A^\mu \mathcal{D}_{\mu\nu}(x)A^\nu \tag{7.84}$$

the differential operator reads

$$\mathcal{D}_{\mu\nu}(x) = -g_{\mu\nu}\overleftarrow{\partial}_\sigma \overrightarrow{\partial}^\sigma + \overleftarrow{\partial}_\nu \overrightarrow{\partial}_\mu + m^2 g_{\mu\nu} \ . \tag{7.85}$$

In momentum space this leads to

$$\mathcal{D}_{\mu\nu}(k) = -g_{\mu\nu}(k^2 - m^2) + k_\mu k_\nu \ . \tag{7.86}$$

It is easily verified that this matrix is inverted by the Proca Feynman propagator (16) in Exercise 6.4 (leaving out the noncovariant extra term):

$$\left(-g_{\mu\nu}(k^2 - m^2) + k_\mu k_\nu\right)\left(\frac{-g^{\nu\sigma} + \frac{1}{m^2}k^\nu k^\sigma}{k^2 - m^2}\right) = g_\mu{}^\sigma \ . \tag{7.87}$$

In the case of the *Maxwell Lagrangian* (7.2) with the gauge-fixing term

$$\mathcal{L} = -\frac{1}{4}F_{\mu\nu}F^{\mu\nu} - \frac{1}{2}\zeta(\partial_\sigma A^\sigma)^2 , \qquad (7.88)$$

the differential operator is given by

$$\mathcal{D}_{\mu\nu}(x) = -g_{\mu\nu}\overleftarrow{\partial_\sigma}\overrightarrow{\partial^\sigma} + \overleftarrow{\partial_\nu}\overrightarrow{\partial_\mu} - \zeta\overleftarrow{\partial_\mu}\overrightarrow{\partial_\nu} , \qquad (7.89)$$

and in momentum space we get

$$\mathcal{D}_{\mu\nu}(k) = -g_{\mu\nu}k^2 + (1-\zeta)k_\mu k_\nu . \qquad (7.90)$$

To construct the inverse matrix we make a general symmetric ansatz

$$\mathcal{D}^{-1\,\nu\sigma}(k) = A(k^2)g^{\nu\sigma} + B(k^2)k^\nu k^\sigma . \qquad (7.91)$$

Requiring that

$$\mathcal{D}_{\mu\nu}(k)\mathcal{D}^{-1\,\nu\sigma}(k) = g_\mu{}^\sigma \qquad (7.92)$$

and comparing the coefficients, we get the conditions

$$-k^2 A(k^2) = 1 , \qquad (7.93a)$$
$$\zeta k^2 B(k^2) = (\zeta - 1) A(k^2) . \qquad (7.93b)$$

In the case $\zeta = 0$ these equations are not compatible. *Without the gauge-fixing term the matrix $\mathcal{D}_{\mu\nu}(k)$ cannot be inverted* (since the determinant vanishes) and the Feynman propagator cannot be constructed. If $\zeta \neq 0$, however, no problems arise and the system of equations (7.93) is solved by

$$A(k^2) = -\frac{1}{k^2} , \qquad B(k^2) = \frac{\zeta - 1}{\zeta}\frac{1}{(k^2)^2} , \qquad (7.94)$$

which leads to (7.69) after a suitable shifting of the poles.

EXERCISE

7.3 The Feynman Propagator for Arbitrary Values of the Gauge Parameter ζ

Problem. In Sect. 7.5 the special value $\zeta = 1$ (the Feynman gauge) was chosen for the gauge-fixing parameter from (7.2). Work out the generalization to the case of arbitrary ζ.

(a) The equal-time commutation relations (7.21) and (7.22) are modified as follows

$$[\hat{A}^\mu(\boldsymbol{x},t),\hat{A}^\nu(\boldsymbol{x}',t)] = 0 , \qquad (1a)$$
$$[\dot{\hat{A}}^\mu(\boldsymbol{x},t),\hat{A}^\nu(\boldsymbol{x}',t)] = i\left(g^{\mu\nu} - \frac{\zeta-1}{\zeta}g^{\mu 0}g^{\nu 0}\right)\delta^3(\boldsymbol{x}-\boldsymbol{x}') , \qquad (1b)$$
$$[\dot{\hat{A}}^\mu(\boldsymbol{x},t),\dot{\hat{A}}^\nu(\boldsymbol{x}',t)] = i\frac{\zeta-1}{\zeta}\left(g^{\mu 0}g^{\nu k} + g^{\nu 0}g^{\mu k}\right)\partial_k\delta^3(\boldsymbol{x}-\boldsymbol{x}') . \qquad (1c)$$

(b) The Feynman propagator solves the inhomogeneous wave equation for a unit point source:

$$\left(\Box g^\mu{}_\nu + (\zeta-1)\partial^\mu\partial_\nu\right)D_F^{\nu\sigma}(x-y) = g^{\mu\sigma}\delta^4(x-y) . \qquad (2)$$

Exercise 7.3

(c) The Feynman propagator in momentum space reads

$$D_F^{\mu\nu}(k) = \frac{-g^{\mu\nu}}{k^2 + i\epsilon} + \frac{\zeta - 1}{\zeta}\frac{k^\mu k^\nu}{(k^2 + i\epsilon)^2} \,. \tag{3}$$

Hint: The derivation of (3) from the properties of the field operators in coordinate space is quite involved. It is useful to split the field operator into two parts:[10]

$$\hat{A}_\mu(x) = \hat{A}_\mu^F(x) + \partial_\mu \hat{\chi}(x) \,, \tag{4a}$$

where

$$\hat{\chi}(x) = \frac{1-\zeta}{2}\frac{1}{\Delta}\left(x_0 \dot{\hat{\Lambda}}(x) - \frac{1}{2}\hat{\Lambda}(x)\right) \tag{4b}$$

with the abbreviation

$$\hat{\Lambda}(x) = \partial_\mu \hat{A}^\mu(x) \,. \tag{4c}$$

Solution. (a) If the gauge-fixing parameter is $\zeta \neq 1$ the Lagrangian (7.12) is changed to

$$\mathcal{L}'' = -\frac{1}{2}\partial_\mu A_\nu \partial^\mu A^\nu - \frac{\zeta - 1}{2}(\partial_\nu A^\nu)^2 \,. \tag{5}$$

The canonically conjugate field becomes

$$\pi_\mu = -\partial^0 A_\mu - (\zeta - 1)g_{\mu 0}\partial^\nu A_\nu \,. \tag{6}$$

This equation can be solved for the time derivative of the vector potential. The space and time components look quite different:

$$\dot{A}^i = -\pi^i \,, \qquad \dot{A}^0 = -\frac{1}{\zeta}\pi^0 - \frac{\zeta - 1}{\zeta}\partial_k A^k \,, \tag{7}$$

which can be combined into

$$\dot{A}^\mu = -\pi^\mu + g^{\mu 0}\frac{\zeta - 1}{\zeta}\left(\pi^0 - \partial_k A^k\right) \,. \tag{8}$$

With this result the time derivative is replaced by the canonically conjugate field and subsequently (after quantization) the ETCR (7.21) are used. Equation (1a) coincides with (7.22). For (1b) we find (the prime denotes the argument $(\boldsymbol{x}', t)$)

$$\begin{aligned}[\dot{\hat{A}}^\mu, \hat{A}'^\nu] &= -[\hat{\pi}^\mu, \hat{A}'^\nu] + g^{\mu 0}\frac{\zeta - 1}{\zeta}[\hat{\pi}^0, \hat{A}'^\nu] - g^{\mu 0}\frac{\zeta - 1}{\zeta}\partial_k[\hat{A}^k, \hat{A}'^\nu] \\
&= ig^{\mu\nu}\delta^3(\boldsymbol{x} - \boldsymbol{x}') - ig^{\mu 0}g^{\nu 0}\frac{\zeta - 1}{\zeta}\delta^3(\boldsymbol{x} - \boldsymbol{x}') \\
&= i\left(g^{\mu\nu} - \frac{\zeta - 1}{\zeta}g^{\mu 0}g^{\nu 0}\right)\delta^3(\boldsymbol{x} - \boldsymbol{x}') \,.\end{aligned} \tag{9}$$

Similarly the commutator (1c) reads

$$\begin{aligned}[\dot{\hat{A}}^\mu, \dot{\hat{A}}'^\nu] &= g^{\mu 0}\frac{\zeta - 1}{\zeta}\partial_l[\hat{A}^l, \hat{\pi}'^\nu] \\
&= i\frac{\zeta - 1}{\zeta}(g^{\mu 0}g^{\nu l} + g^{\nu 0}g^{\mu l})\partial_l\delta^3(\boldsymbol{x} - \boldsymbol{x}') \,.\end{aligned} \tag{10}$$

---

[10] B. Lautrup: Mat. Fys. Medd. Dan. Vid. Selsk. **35**, no. 11 (1967)

(b) The classical field solves the differential equation (7.7). The same equation also applies to the field operator, i.e.,

$$\left(\Box g^{\mu}{}_{\nu} + (\zeta - 1)\partial^{\mu}\partial_{\nu}\right)\hat{A}^{\nu} = 0 \,. \tag{11}$$

Exercise 7.3

To verify this we first derive the Hamilton density belonging to (5). Instead of (7.14) it reads

$$\begin{aligned}\mathcal{H}'' &= \pi^{\mu}\partial_0 A_{\mu} - \mathcal{L}'' \\ &= -\frac{1}{2}\pi^{\mu}\pi_{\mu} + \frac{1}{2}\partial_l A_{\nu}\partial^l A^l + \frac{\zeta-1}{2\zeta}\left(\pi^0 - \partial_l A^l\right)^2 \,. \end{aligned} \tag{12}$$

The Heisenberg equation of motion for the field operator $\hat{A}^0$ using the quantized version of the Hamiltonian (12) and the ETCR becomes

$$\begin{aligned}\dot{\hat{A}}^0 &= -\mathrm{i}\left[\hat{A}^0, \hat{H}\right] \\ &= \int \mathrm{d}^3 x' \left(\partial_l \hat{A}^0 \partial^{\prime l} - \frac{\zeta-1}{\zeta}\partial_l \hat{A}^0 \partial^{\prime k} + \frac{\zeta-1}{\zeta}\hat{\pi}_l \partial^l\right)\delta^3(\boldsymbol{x}-\boldsymbol{x}') \\ &= -\partial_l \partial^l \hat{A}^0 + \frac{\zeta-1}{\zeta}\partial_l\partial^l \hat{A}^0 - \frac{\zeta-1}{\zeta}\partial_l\partial^0 \hat{A}^l \,, \end{aligned} \tag{13}$$

which can be written as

$$\Box \hat{A}^0 + (\zeta-1)\partial^0 \partial_\nu \hat{A}^\nu = 0 \,. \tag{14}$$

The equation of motion for the space components can be derived similarly.

To obtain the differential equation for the Feynman propagator, the latter is rewritten as

$$\begin{aligned}\mathrm{i}D_F^{\nu\sigma}(x-y) &= \langle 0|T(\hat{A}^\nu(x)\hat{A}^\sigma(y))|0\rangle \\ &= \frac{1}{2}\langle 0|\{\hat{A}^\nu(x),\hat{A}^\sigma(y)\}|0\rangle \\ &\quad + \frac{1}{2}\epsilon(x_0-y_0)\langle 0|[\hat{A}^\nu(x),\hat{A}^\sigma(y)]|0\rangle \,. \end{aligned} \tag{15}$$

The first term in (15) vanishes if the differential operator (11) is applied. The second term leads to contributions that arise from the action of the time derivative on the sign function $\epsilon(x_0-y_0)$:

$$\frac{1}{2}\partial_0\partial^0 \epsilon(x_0-y_0)\langle 0|[\hat{A}^\mu(x),\hat{A}^\sigma(y)]|0\rangle$$

$$= \partial_0\Big(\underbrace{\delta(x_0-y_0)\langle 0|[\hat{A}^\mu,\hat{A}^\sigma]|0\rangle}_{=0} + \frac{1}{2}\epsilon(x_0-y_0)\langle 0|[\partial^0 \hat{A}^\mu,\hat{A}^\sigma]|0\rangle\Big)$$

$$= \delta(x_0-y_0)\langle 0|[\partial^0 \hat{A}^\mu,\hat{A}^\sigma]|0\rangle + \frac{1}{2}\epsilon(x_0-y_0)\langle 0|[\partial_0\partial^0 \hat{A}^\mu,\hat{A}^\sigma]|0\rangle \tag{16}$$

and further

$$\frac{1}{2}\partial^\mu\partial_\nu \epsilon(x_0-y_0)\langle 0|[\hat{A}^\nu,\hat{A}^\sigma]|0\rangle \tag{17}$$

$$= g^{\mu 0}\delta(x_0-y_0)\langle 0|[\partial_\nu \hat{A}^\nu,\hat{A}^\sigma]|0\rangle + \frac{1}{2}\epsilon(x_0-y_0)\langle 0|[\partial^\mu\partial_\nu \hat{A}^\nu,\hat{A}^\sigma]|0\rangle \,.$$

The combination of both extra terms with the help of (6) and (7.21) yields

Exercise 7.3

$$\delta(x_0 - y_0)\langle 0|[\partial_0 \hat{A}^\mu + g^{\mu 0}(\zeta - 1)\partial_\nu \hat{A}^\nu, \hat{A}^\sigma]|0\rangle$$
$$= \delta(x_0 - y_0)\langle 0| - [\hat{\pi}^\mu(x), \hat{A}^\sigma(y)]|0\rangle = \mathrm{i}\, g^{\mu\sigma}\, \delta^4(x-y)\,. \tag{18}$$

This proves (2).

(c) We have to construct the Feynman propagator

$$\mathrm{i}D_{\mathrm{F}}^{\mu\nu}(x' - x) = \Theta(x'_0 - x_0)\langle 0|\hat{A}^\mu(x')\hat{A}^\nu(x)|0\rangle$$
$$+ \Theta(x_0 - x'_0)\langle 0|\hat{A}^\nu(x)\hat{A}^\mu(x')|0\rangle\,. \tag{19}$$

An attempt to evaluate the vacuum expectation values in the accustomed way, as in (7.65), meets with problems. To wit, when the field operator is expanded into eigenmodes according to (7.26) it is found that the coefficients $\hat{a}_{k\lambda}$ and $\hat{a}_{k\lambda}^\dagger$ do not satisfy the usual commutation relations characteristic for creation and annihilation operators. In particular, the $\hat{a}_{k\lambda}$ do not all commute among each other (the troublemakers are the longitudinal and scalar photons $\lambda = 3$ and $\lambda = 0$). It is not possible to construct the Fock space in the usual way and (19) cannot be evaluted.

This problem can be circumvented if the field operator is split up as in (4a)–(4c). We will find that the gradient term in (4a) is constructed such that the remaining field operator $\hat{A}_\mu^{\mathrm{F}}(x)$ has the properties known from the Feynman gauge ($\zeta = 1$).

Let us first consider which equation of motion is satisfied by the auxiliary field $\hat{\chi}(x)$. The application of the d'Alembert operator leads to

$$\Box \hat{\chi} = (\partial_0{}^2 - \Delta)\hat{\chi}$$
$$= \frac{1-\zeta}{2}\frac{1}{\Delta}\left(x_0 \partial_0{}^3 \hat{\Lambda} + \frac{3}{2}\partial_0{}^2 \hat{\Lambda}\right) - \frac{1-\zeta}{2}\left(x_0 \partial_0 \hat{\Lambda} - \frac{1}{2}\hat{\Lambda}\right)\,. \tag{20}$$

Now the Lorentz gauge function $\hat{\Lambda}(x)$ according to (7.8) satisfies the homogeneous d'Alembert equation

$$\Box \hat{\Lambda} = 0 \quad \text{or} \quad \partial_0{}^2 \hat{\Lambda} = \Delta \hat{\Lambda}\,. \tag{21}$$

Thus equation (20) is simplified considerably:

$$\Box \hat{\chi} = (1-\zeta)\hat{\Lambda}\,. \tag{22}$$

The equation of motion of the original field $\hat{A}_\mu$ reads

$$\Box \hat{A}_\mu + (\zeta - 1)\partial_\mu \hat{\Lambda} = 0\,. \tag{23}$$

If the decomposition $\hat{A}_\mu = \hat{A}_\mu^{\mathrm{F}} + \partial_\mu \hat{\chi}$ is inserted the $\hat{\Lambda}$ term is cancelled because of (22). Thus the field $\hat{A}_\mu^{\mathrm{F}}$ satisfies the same equation of motion as in the case of the Feynman gauge:

$$\Box \hat{A}_\mu^{\mathrm{F}} = 0\,. \tag{24}$$

Furthermore, the ETCR have the same form as in the Feynman gauge. For example we consider (the index 0 designates equal time arguments)

$$[\dot{\hat{A}}_\mu^{\mathrm{F}}(x), \hat{A}_\nu^{\mathrm{F}}(x')]_0 = [\dot{\hat{A}}_\mu, \hat{A}'_\nu]_0 - [\partial_\mu \dot{\hat{\chi}}, \hat{A}'_\nu]_0$$
$$- [\dot{\hat{A}}_\mu, \partial'_\nu \hat{\chi}']_0 + [\partial_\mu \dot{\hat{\chi}}, \partial'_\nu \hat{\chi}']_0 \tag{25}$$

7.5 The Feynman Propagator for Photons

This can be reduced to the commutators between $\hat{A}_\mu$ and/or $\dot{\hat{A}}_\nu$. In an intermediate step the commutators of $\hat{\Lambda}$ and $\dot{\hat{\Lambda}}$ among each other and with $\hat{A}_\mu$ and $\dot{\hat{A}}_\nu$ are required. For example using (1a) and (1b) we find

Exercise 7.3

$$[\hat{\Lambda}, \hat{A}'_\mu]_0 = [\partial^0 \hat{A}_0, A'_\mu]_0 + \partial^k[\hat{A}_k, A'_\mu]_0$$
$$= i\left(g_{0\mu} - \frac{\zeta-1}{\zeta} g_{00}\, g_{0\mu}\right)\delta^3(\boldsymbol{x}-\boldsymbol{x}') = i\frac{1}{\zeta} g_{0\mu}\, \delta^3(\boldsymbol{x}-\boldsymbol{x}') \,. \quad (26a)$$

Similarly one finds

$$[\hat{\Lambda}, \dot{\hat{A}}'_\mu]_0 = -i\frac{1}{\zeta} g_{\mu k}\, \partial^k\, \delta^3(\boldsymbol{x}-\boldsymbol{x}') \tag{26b}$$

and

$$[\hat{\Lambda}, \hat{\Lambda}']_0 = 0 \,. \tag{26c}$$

To evaluate the commutators of $\dot{\hat{\Lambda}}$ we use the equation of motion (23) for $\mu = 0$ and (4c) to express this operator as

$$\dot{\hat{\Lambda}} = \frac{1}{\zeta}(\Delta \hat{A}_0 + \partial_k \dot{\hat{A}}^k) \,. \tag{27}$$

One finds the following commutators

$$[\dot{\hat{\Lambda}}, \hat{A}'_\mu]_0 = i\frac{1}{\zeta} g_{\mu k}\, \partial^k\, \delta^3(\boldsymbol{x}-\boldsymbol{x}') \,, \tag{28a}$$

$$[\dot{\hat{\Lambda}}, \dot{\hat{A}}'_\mu]_0 = i\frac{1}{\zeta} g_{\mu 0} \Delta\, \delta^3(\boldsymbol{x}-\boldsymbol{x}') \,, \tag{28b}$$

$$[\dot{\hat{\Lambda}}, \hat{\Lambda}']_0 = [\dot{\hat{\Lambda}}, \dot{\hat{\Lambda}}']_0 = 0 \,. \tag{28c}$$

Using (1), (26), and (28) it is not difficult to evaluate the commutators on the right-hand side of (25). As announced earlier all terms that depend on the gauge-fixing parameter ζ are found to cancel, leaving the result

$$\left[\hat{A}^F_\mu(x), \dot{\hat{A}}^F_\nu(x')\right]_0 = ig_{\mu\nu}\, \delta^3(\boldsymbol{x}-\boldsymbol{x}') \tag{29a}$$

as in (7.22). Similarly one finds

$$\left[\hat{A}^F_\mu(x), \hat{A}^F_\nu(x')\right]_0 = \left[\dot{\hat{A}}^F_\mu(x), \dot{\hat{A}}^F_\nu(x')\right]_0 = 0 \,. \tag{29b}$$

Because of (24) and (29) the field $\hat{A}^F_\mu$ has the usual Fourier decomposition (7.26), i.e., (with the abbreviation $N_k \equiv ((2\pi)^3 2\omega_k)^{-1/2}$)

$$\hat{A}^{F\mu}(x) = \int d^3x\, N_k \sum_{\lambda=0}^{3} \epsilon^\mu(\boldsymbol{k},\lambda)\left(\hat{a}_{\boldsymbol{k}\lambda}\, e^{-ik\cdot x} + \hat{a}^\dagger_{\boldsymbol{k}\lambda}\, e^{+ik\cdot x}\right), \tag{30}$$

with the creation and annihilation operators

$$[\hat{a}_{\boldsymbol{k}',\lambda'},\, \hat{a}^\dagger_{\boldsymbol{k}\lambda}] = -g_{\lambda'\lambda}\, \delta^3(\boldsymbol{k}'-\boldsymbol{k}) \,, \quad [\hat{a}_{\boldsymbol{k}',\lambda'},\, \hat{a}_{\boldsymbol{k}\lambda}] = [\hat{a}^\dagger_{\boldsymbol{k}',\lambda'},\, \hat{a}^\dagger_{\boldsymbol{k}\lambda}] = 0 \,. \tag{31}$$

For the Fourier decomposition of the gauge funtion $\hat{\Lambda}(x)$, we make the following ansatz:

$$\hat{\Lambda}(x) = \int d^3k\, N_k\, i\omega_k\, \frac{1}{\zeta}\left(\hat{\lambda}_{\boldsymbol{k}}\, e^{-ik\cdot x} - \hat{\lambda}^\dagger_{\boldsymbol{k}}\, e^{+ik\cdot x}\right). \tag{32}$$

Exercise 7.3

The operators $\hat{\lambda}_k$ and $\hat{\lambda}_k^\dagger$ obviously are constant in time (as were the $\hat{a}_{k\lambda}$, $\hat{a}_{k\lambda}^\dagger$) since $\hat{\Lambda}(x)$ as well as $\hat{A}_\mu^F(x)$ (but not $\hat{A}_\mu(x)$) satisfy the d'Alembert equation. There is a simple relation between the two sets of operators: using (22) we find

$$\hat{\Lambda} = \partial^\mu \hat{A}_\mu = \partial^\mu \hat{A}_\mu^F + \Box \hat{\chi} = \partial^\mu \hat{A}_\mu^F + (1-\zeta)\hat{\Lambda} \tag{33}$$

or

$$\hat{\Lambda} = \frac{1}{\zeta} \partial^\mu \hat{A}_\mu^F . \tag{34}$$

The four-divergences of $\hat{A}_\mu$ and $\hat{A}_\mu^F$ are thus seen to agree up to a numerical factor. If (34) is applied to (30) the Fourier decomposition of the gauge function becomes

$$\hat{\Lambda}(x) = \int d^3k\, N_k \frac{-i}{\zeta} \sum_{\lambda=0}^{3} k^\mu \epsilon_\mu(\boldsymbol{k},\lambda)\left(\hat{a}_{\boldsymbol{k}\lambda} e^{-ik\cdot x} - \hat{a}_{\boldsymbol{k}\lambda}^\dagger e^{+ik\cdot x}\right) . \tag{35}$$

According to (6.82) the polarization vectors for $\lambda = 1, 2$ satisfy the four-dimensional transversality condition and (35) is reduced to

$$\hat{\Lambda}(x) = \int d^3k\, N_k \frac{i}{\zeta}(k\cdot n)\left((\hat{a}_{\boldsymbol{k}3} - \hat{a}_{\boldsymbol{k}0})e^{-ik\cdot x} - (\hat{a}_{\boldsymbol{k}3}^\dagger - \hat{a}_{\boldsymbol{k}0}^\dagger)e^{+ik\cdot x}\right) . \tag{36}$$

If the auxiliary vector n^μ for simplicity is chosen as $n^\mu = (1, \mathbf{0})$ a comparison of (32) and (36) yields

$$\hat{\lambda}_{\boldsymbol{k}} = \hat{a}_{\boldsymbol{k}3} - \hat{a}_{\boldsymbol{k}0} , \qquad \hat{\lambda}_{\boldsymbol{k}}^\dagger = \hat{a}_{\boldsymbol{k}3}^\dagger - \hat{a}_{\boldsymbol{k}0}^\dagger . \tag{37}$$

This leads to the commutation relations

$$[\hat{\lambda}_{\boldsymbol{k}'}, \hat{\lambda}_{\boldsymbol{k}}] = [\hat{\lambda}_{\boldsymbol{k}'}^\dagger, \hat{\lambda}_{\boldsymbol{k}}^\dagger] = [\hat{\lambda}_{\boldsymbol{k}'}^\dagger, \hat{\lambda}_{\boldsymbol{k}}] = 0 , \tag{38a}$$

$$[\hat{a}_{\boldsymbol{k}'\lambda}^\dagger, \hat{\lambda}_{\boldsymbol{k}}] = [\hat{\lambda}_{\boldsymbol{k}'}^\dagger, \hat{a}_{\boldsymbol{k}\lambda}] = (g_{\lambda 3} - g_{\lambda 0})\delta^3(\boldsymbol{k}' - \boldsymbol{k}) , \tag{38b}$$

$$[\hat{a}_{\boldsymbol{k}'\lambda}, \hat{\lambda}_{\boldsymbol{k}}] = [\hat{a}_{\boldsymbol{k}'\lambda}^\dagger, \hat{\lambda}_{\boldsymbol{k}}^\dagger] = 0 . \tag{38c}$$

Using (32) it is now easy to write down the Fourier decompositon of the extra term present in the field operator (4a):

$$\partial_\mu \hat{\chi} = \partial_\mu \frac{1-\zeta}{2\zeta}\frac{1}{\Delta}\int d^3k\, N_k i\omega_k$$
$$\times \left[(-i\omega_k x_0 - \tfrac{1}{2})\hat{\lambda}_{\boldsymbol{k}} e^{-ik\cdot x} + (i\omega_k x_0 - \tfrac{1}{2})\hat{\lambda}_{\boldsymbol{k}}^\dagger e^{+ik\cdot x}\right] . \tag{39}$$

The inverse Laplacian $1/\Delta$ in momentum space reduces to a simple algebraic factor $1/(i\boldsymbol{k})^2 = -1/\omega_k^2$. Keeping in mind that in the case $\mu = 0$ the gradient ∂_μ also acts on the x_0 terms, we obtain

$$\partial_\mu \hat{\chi} = \frac{1-\zeta}{\zeta}\int d^3k\, N_k \frac{1}{2\omega_k}\Big[\left(-\omega_k g_{\mu 0} + (i\omega_k x_0 + \tfrac{1}{2})k_\mu\right)\hat{\lambda}_{\boldsymbol{k}} e^{-ik\cdot x}$$
$$+ \left(+\omega_k g_{\mu 0} + (i\omega_k x_0 - \tfrac{1}{2})k_\mu\right)\hat{\lambda}_{\boldsymbol{k}}^\dagger e^{+ik\cdot x}\Big] . \tag{40}$$

Now it is an easy task to calculate the Feynman propagator from its definition (19), expressing the field operators by the sum of (30) and (40). The vacuum is required to satisfy the conditions

$$\hat{a}_{k\lambda}|0\rangle = 0 \quad , \quad \hat{\lambda}_k|0\rangle = 0 \,. \tag{41}$$

The vacuum expectation value obtains nonvanishing contributions from the commutators (31a) and (38b), leading to the result

$$\langle 0|\hat{A}^\mu(x')\hat{A}^\nu(x)|0\rangle = \int d^3k'\, d^3k\, N_{k'} N_k\, \delta^3(\boldsymbol{k}' - \boldsymbol{k})\, e^{-ik\cdot(x'-x)}$$

$$\times \Bigg[\sum_{\lambda'\lambda} \epsilon^\mu(\boldsymbol{k}',\lambda')\epsilon^\nu(\boldsymbol{k},\lambda)(-g_{\lambda'\lambda})$$

$$+ \frac{1-\zeta}{\zeta}\frac{1}{2\omega_{k'}}\left(-\omega_{k'}g^{\mu 0} + (\mathrm{i}\omega_{k'}x_0' + \tfrac{1}{2})k'^\mu\right) \sum_\lambda \epsilon^\nu(\boldsymbol{k},\lambda)(-g_{\lambda 3} + g_{\lambda 0})$$

$$+ \sum_{\lambda'} \epsilon^\mu(\boldsymbol{k}',\lambda')\frac{1-\zeta}{\zeta}\frac{1}{2\omega_k}\left(\omega_k g^{\nu 0} + (\mathrm{i}\omega_k x_0 - \tfrac{1}{2})k^\nu\right)(g_{\lambda' 3} - g_{\lambda' 0})\Bigg]\,. \tag{42}$$

With our special choice for the vector n^μ the sum of longitudinal and scalar polarization vectors becomes

$$\epsilon^\mu(\boldsymbol{k},3) + \epsilon^\mu(\boldsymbol{k},0) = \big(0, \boldsymbol{k}/|\boldsymbol{k}|\big) + \big(1, \boldsymbol{0}\big) = \frac{1}{\omega_k}k^\mu \,. \tag{43}$$

This leads to the final result for the *Feynman propagator*:

$$D_F^{\mu\nu}(x'-x) = D_F^{(0)\mu\nu}(x'-x) + D_F^{(1)\mu\nu}(x'-x)\,, \tag{44}$$

with the familiar photon propagator in Feynman gauge

$$D_F^{(0)\mu\nu}(x'-x) = \mathrm{i}g^{\mu\nu}\int\frac{d^3k}{(2\pi)^3}\frac{1}{2\omega_k} \tag{45}$$

$$\times\left[\Theta(x_0'-x_0)\,e^{-ik\cdot(x'-x)} + \Theta(x_0-x_0')\,e^{+ik\cdot(x'-x)}\right]$$

and the *gauge-dependent extra term*

$$D_F^{(1)\mu\nu}(x'-x) = -\mathrm{i}\frac{\zeta-1}{\zeta}\int\frac{d^3k}{(2\pi)^3}\frac{1}{(2\omega_k)^2} \tag{46}$$

$$\times\Bigg[\Theta(x_0'-x_0)\,e^{-ik\cdot(x'-x)}\left(g^{\mu 0}k^\nu + g^{\nu 0}k^\mu - \frac{1}{\omega_k}k^\mu k^\nu\big(1 + \mathrm{i}\omega_k(x_0'-x_0)\big)\right)$$

$$+ \Theta(x_0-x_0')\,e^{+ik\cdot(x'-x)}\left(g^{\mu 0}k^\nu + g^{\nu 0}k^\mu - \frac{1}{\omega_k}k^\mu k^\nu\big(1 - \mathrm{i}\omega_k(x_0'-x_0)\big)\right)\Bigg]\,.$$

Thus the three-dimensional Fourier representation of the Feynman propagator is quite complicated. The *four-dimensional* Fourier representation, on the other hand, is very simple. The easiest way to verify this is by starting from the conjectured expression (3) and carrying out the k_0 integration, which will be found to produce (44)–(46).

Since (45) agrees with the usual Feynman propagator we are only interested in the extra term (46). We have to evaluate

$$D_F^{(1)\mu\nu}(x'-x) = \int\frac{d^4k}{(2\pi)^4}\, e^{-ik\cdot(x'-x)}\, D_F^{(1)\mu\nu}(k)$$

$$= \frac{\zeta-1}{\zeta}\int\frac{d^4k}{(2\pi)^4}\, e^{-ik\cdot(x'-x)}\,\frac{k^\mu k^\nu}{(k^2+\mathrm{i}\epsilon)^2}\,. \tag{47}$$

Exercise 7.3

Exercise 7.3

The k_0 integral of interest reads

$$I = \int_{-\infty}^{\infty} \frac{dk_0}{2\pi} e^{-ik_0(x'_0-x_0)} \frac{k^\mu k^\nu}{(k_0^2 - \omega_k^2 + i\epsilon)^2}$$

$$= \int_{-\infty}^{\infty} \frac{dk_0}{2\pi} e^{-ik_0(x'_0-x_0)} \frac{k^\mu k^\nu}{(k_0 - \omega_k + i\epsilon)^2 (k_0 + \omega_k - i\epsilon)^2} \,. \quad (48)$$

The integrand has *two double poles* at $k_0^\pm = \pm(\omega_k - i\epsilon)$. This integral can be solved with the help of the theorem of residues. Instead of the simple residue, now the derivative of the function under the integral enters according to

$$\oint dz \frac{f(z)}{(z-z_0)^2} = 2\pi i\, f'(z_0) \,. \quad (49)$$

This is the reason why (46) looks quite complicated. The product rule for the derivative of (48) leads to several terms.

Let us inspect the case $x'_0 - x_0 > 0$. Then the integration contour of (48) can be closed in the lower half of the complex plane and the double pole at $k_0^+ = \omega_k - i\epsilon$ contributes. The function under the integral reads

$$f_+(k_0) = \frac{k^\mu k^\nu}{(k_0 + \omega_k - i\epsilon)^2} e^{-ik_0(x'_0-x_0)} \,, \quad (50)$$

having the following derivative at $k_0 = k_0^+$:

$$f'_+(k_0^+) = \frac{g^{\mu 0} k^\nu + g^{\nu 0} k^\mu - ix_0 k^\mu k^\nu}{(2\omega_k)^2} e^{-i\omega_k(x'_0-x_0)} + k^\mu k^\nu \frac{-2}{(2\omega_k)^3} e^{-i\omega_k(x'_0-x_0)}$$

$$= \frac{1}{(2\omega_k)^2} e^{-i\omega_k(x'_0-x_0)} \left[g^{\mu 0} k^\nu + g^{\nu 0} k^\mu - \frac{1}{\omega_k} k^\mu k^\nu (1 + i\omega_k(x'_0 - x_0)) \right] \,. \quad (51)$$

This is just the part of (46) that is multiplied by $\Theta(x'_0 - x_0)$. The second contribution is obtained in the same manner.

7.7 Canonical Quantization in the Coulomb Gauge

If one is prepared to sacrifice manifest relativistic covariance of the quantization procedure, the use of the Coulomb gauge $\boldsymbol{\nabla} \cdot \boldsymbol{A} = 0$ suggests itself. As mentioned earlier, in this gauge the two "unphysical" degrees of freedom of the electromagnetic field are eliminated from the start and *only transverse photons are present*.

Let us first discuss the quantization of the free field A_μ, i.e., a system containing no electric charges or currents, $j_\mu = 0$. In this case the gauge can be chosen as

$$\boldsymbol{\nabla} \cdot \boldsymbol{A}(\boldsymbol{x},t) = 0 \quad \text{and} \quad A_0(\boldsymbol{x},t) = 0 \,, \quad (7.95)$$

i.e., the field A_0 is eliminated completely.[11] This choice also is known as the *radiation gauge*.

In the sense of the decomposition (6.21) the potential is purely transverse, $\boldsymbol{A} = \boldsymbol{A}_\perp$, $\boldsymbol{A}_\parallel = 0$. This of course will also hold for the electric field strength

$$\boldsymbol{E} = -\frac{\partial \boldsymbol{A}}{\partial t} - \boldsymbol{\nabla} A_0 = -\frac{\partial}{\partial t}\boldsymbol{A}_\perp = \boldsymbol{E}_\perp \qquad (7.96)$$

In the remainder of this section the symbol $\perp$ will be dropped for convenience. The canonically conjugate field associated with $\boldsymbol{A}$ reads

$$\boldsymbol{\pi}(\boldsymbol{x},t) = \boldsymbol{E}(\boldsymbol{x},t) \;, \qquad (7.97)$$

thus it is transverse too. One might be tempted to quantize the field by imposing the usual equal-time commutation relations (cf. (6.93) for the Proca field)

$$\left[\hat{A}^i(\boldsymbol{x},t),\, \hat{E}^j(\boldsymbol{x}',t)\right] \stackrel{?}{=} -\mathrm{i}\,\delta_{ij}\,\delta^3(\boldsymbol{x}-\boldsymbol{x}') \;, \qquad (7.98\mathrm{a})$$

$$\left[\hat{A}^i(\boldsymbol{x},t),\, \hat{A}^j(\boldsymbol{x}',t)\right] = \left[\hat{E}^i(\boldsymbol{x},t),\, \hat{E}^j(\boldsymbol{x}',t)\right] = 0 \;. \qquad (7.98\mathrm{b})$$

The first of these equations, however, must be *wrong*! Contrary to the case of the massive vector field the electromagnetic field also has to satisfy the transversality condition, thus the components of $\hat{A}^i$ or $\hat{E}^j$ are not independent. It is obvious that (7.98a) is not consistent with the transversality condition $\boldsymbol{\nabla} \cdot \hat{\boldsymbol{A}} = 0$ since[12]

$$\begin{aligned}\left[\boldsymbol{\nabla}\cdot\hat{\boldsymbol{A}}(\boldsymbol{x},t),\, \hat{E}^j(\boldsymbol{x}',t)\right] &= \partial_i\left[\hat{A}^i(\boldsymbol{x},t),\, \hat{E}^j(\boldsymbol{x}',t)\right] \\ &= -\mathrm{i}\,\partial_i\,\delta_{ij}\,\delta^3(\boldsymbol{x}-\boldsymbol{x}') \neq 0 \;.\end{aligned} \qquad (7.99)$$

The same problem is also encountered for the divergence of the field $\boldsymbol{E}$ since the Gauss law $\boldsymbol{\nabla} \cdot \hat{\boldsymbol{E}} = 0$ has to hold. It is not difficult, however, to modify equation (7.98a) in such a way that the ETCR become compatible with the transversality condition. For this purpose the transverse projection operator (6.23) can be employed,

$$(P_\perp)_{ij} = \delta_{ij} - \partial_i \frac{1}{\Delta} \partial_j \;, \qquad (7.100)$$

defining the *transverse delta function*

$$\delta^3_{\perp ij}(\boldsymbol{x}-\boldsymbol{x}') = \left(P_\perp\right)_{ij} \delta^3(\boldsymbol{x}-\boldsymbol{x}') \;. \qquad (7.101)$$

This function (or rather this distribution) by construction has a vanishing three-dimensional divergence:

$$\partial_i \left[\delta^3_{\perp ij}(\boldsymbol{x}-\boldsymbol{x}')\right] = 0 \;. \qquad (7.102)$$

The consistent quantization rule for the electromagnetic field in Coulomb gauge is obtained by starting from (7.98a) and projecting out the transverse part

---

[11] If charges are present the field component A_0 will not be nonvanishing. However, it will not be a dynamical variable of the theory (see Sect. 7.7.1).

[12] Here we use the summation convention for the repeated index i although the Kronecker delta symbol δ_{ij} is not a Lorentz tensor.

$$[\hat{A}^i(\boldsymbol{x},t),\,\hat{E}^j(\boldsymbol{x}',t)] = -\mathrm{i}\delta^3_{\perp ij}(\boldsymbol{x}-\boldsymbol{x}')\,. \tag{7.103}$$

In this way it is guaranteed that the quantization condition is compatible with the transversality condition. In coordinate space the transverse delta function is unfortunately a rather complicated object. However, in momentum space the projector (7.100) is a simple multiplicative factor. The transverse delta function can be represented by the Fourier integral

$$\delta^3_{\perp ij}(\boldsymbol{x}-\boldsymbol{x}') = \int\frac{\mathrm{d}^3k}{(2\pi)^3}\,\mathrm{e}^{\mathrm{i}\boldsymbol{k}\cdot(\boldsymbol{x}-\boldsymbol{x}')}\left(\delta_{ij} - \frac{k_i k_j}{\boldsymbol{k}^2}\right). \tag{7.104}$$

The properties of this distribution are discussed in Exercise 7.4 in some detail.

The plane-wave expansion of the field operator is particularly economical in the Coulomb gauge since here it contains only the two transverse field modes $\boldsymbol{\epsilon}(\boldsymbol{k},\lambda),\,\lambda=1,2$. Thus the vector potential can be written as

$$\hat{\boldsymbol{A}}(\boldsymbol{x},t) = \int\frac{\mathrm{d}^3k}{\sqrt{2\omega_k(2\pi)^3}}\sum_{\lambda=1}^{2}\boldsymbol{\epsilon}(\boldsymbol{k},\lambda)\left(\hat{a}_{\boldsymbol{k}\lambda}\,\mathrm{e}^{-\mathrm{i}k\cdot x} + \hat{a}^\dagger_{\boldsymbol{k}\lambda}\,\mathrm{e}^{\mathrm{i}k\cdot x}\right), \tag{7.105}$$

whereas the electric field strength becomes

$$\hat{\boldsymbol{E}}(\boldsymbol{x},t) = \int\frac{\mathrm{d}^3k}{\sqrt{2\omega_k(2\pi)^3}}\sum_{\lambda=1}^{2}\mathrm{i}\omega_k\,\boldsymbol{\epsilon}(\boldsymbol{k},\lambda)\left(\hat{a}_{\boldsymbol{k}\lambda}\,\mathrm{e}^{-\mathrm{i}k\cdot x} - \hat{a}^\dagger_{\boldsymbol{k}\lambda}\,\mathrm{e}^{\mathrm{i}k\cdot x}\right), \tag{7.106}$$

and the magnetic field strength is

$$\hat{\boldsymbol{B}}(\boldsymbol{x},t) = \int\frac{\mathrm{d}^3k}{\sqrt{2\omega_k(2\pi)^3}}\sum_{\lambda=1}^{2}\mathrm{i}\boldsymbol{k}\times\boldsymbol{\epsilon}(\boldsymbol{k},\lambda)\left(\hat{a}_{\boldsymbol{k}\lambda}\,\mathrm{e}^{-\mathrm{i}k\cdot x} - \hat{a}^\dagger_{\boldsymbol{k}\lambda}\,\mathrm{e}^{\mathrm{i}k\cdot x}\right). \tag{7.107}$$

Now one can expect that the operators $\hat{a}_{\boldsymbol{k}\lambda}$ and $\hat{a}^\dagger_{\boldsymbol{k}\lambda}$ will have the properties of creation and annihilation operators for transverse photons, satisfying the commutation relations

$$[\hat{a}_{\boldsymbol{k}'\lambda'},\,\hat{a}^\dagger_{\boldsymbol{k}\lambda}] = \delta^3(\boldsymbol{k}'-\boldsymbol{k})\,\delta_{\lambda\lambda'}\,, \tag{7.108a}$$

$$[\hat{a}_{\boldsymbol{k}'\lambda'},\,\hat{a}_{\boldsymbol{k}\lambda}] = [\hat{a}^\dagger_{\boldsymbol{k}'\lambda'},\,\hat{a}^\dagger_{\boldsymbol{k}\lambda}] = 0\,. \tag{7.108b}$$

As in the case of the massive vector field this can be verified by inverting the Fourier expansion (7.105) according to (13) in Example 6.3 and using the ETCR. Conversely, the validity of (7.103) can be confirmed by inserting the expansions for $\hat{\boldsymbol{A}}(\boldsymbol{x},t)$ and $\hat{\boldsymbol{E}}(\boldsymbol{x}',t)$ and using (7.108). The result is

$$[\hat{A}^i(\boldsymbol{x},t),\,\hat{E}^j(\boldsymbol{x}',t)]$$

$$= \int\frac{\mathrm{d}^3k}{\sqrt{2\omega_k(2\pi)^3}}\sum_{\lambda=1}^{2}\int\frac{\mathrm{d}^3k'}{\sqrt{2\omega_{k'}(2\pi)^3}}\sum_{\lambda'=1}^{2}\epsilon^i(\boldsymbol{k},\lambda)\epsilon^j(\boldsymbol{k}',\lambda')$$

$$\times\Big([\hat{a}_{\boldsymbol{k}\lambda},\,\hat{a}^\dagger_{\boldsymbol{k}'\lambda'}]\,\mathrm{e}^{-\mathrm{i}(k\cdot x - k'\cdot x')}(-\mathrm{i}\omega_{k'}) \tag{7.109}$$

$$+[\hat{a}^\dagger_{\boldsymbol{k}\lambda},\,\hat{a}_{\boldsymbol{k}'\lambda'}]\,\mathrm{e}^{\mathrm{i}(k\cdot x - k'\cdot x')}(+\mathrm{i}\omega_{k'})\Big)$$

$$= \int\frac{\mathrm{d}^3k}{2\omega_k(2\pi)^3}\,(-\mathrm{i}\omega_k)\left(\mathrm{e}^{\mathrm{i}\boldsymbol{k}\cdot(\boldsymbol{x}-\boldsymbol{x}')} + \mathrm{e}^{-\mathrm{i}\boldsymbol{k}\cdot(\boldsymbol{x}-\boldsymbol{x}')}\right)\sum_{\lambda=1}^{2}\epsilon^i(\boldsymbol{k},\lambda)\epsilon^j(\boldsymbol{k},\lambda)\,.$$

The transverse polarization vectors $\epsilon(\boldsymbol{k}, 1)$ and $\epsilon(\boldsymbol{k}, 2)$ together with the unit vector in the direction of momentum $\boldsymbol{k}/|\boldsymbol{k}|$ form an orthogonal basis of three-dimensional space, satisfying the completeness relation

$$\sum_{\lambda=1}^{2} \epsilon^i(\boldsymbol{k},\lambda)\epsilon^j(\boldsymbol{k},\lambda) + \frac{k^i k^j}{\boldsymbol{k}^2} = \delta_{ij} \ . \tag{7.110}$$

Since in the second term of (7.109) one may make the replacement $\boldsymbol{k} \to -\boldsymbol{k}$ the expected ETCR (7.103) is obtained:

$$\begin{aligned}
[\hat{A}^i(\boldsymbol{x},t), \hat{E}^j(\boldsymbol{x}',t)] &= -\mathrm{i} \int \frac{\mathrm{d}^3 k}{(2\pi)^3} \, \mathrm{e}^{\mathrm{i}\boldsymbol{k}\cdot(\boldsymbol{x}-\boldsymbol{x}')} \left(\delta_{ij} - \frac{k^i k^j}{\boldsymbol{k}^2} \right) \\
&= -\mathrm{i} \delta^3_{\perp ij}(\boldsymbol{x} - \boldsymbol{x}') \ .
\end{aligned} \tag{7.111}$$

The reason for the emergence of the transverse delta function in (7.103) becomes very lucid from this derivation.

The evaluation of the normal-ordered *Hamiltonian* proceeds as in Example 6.3. The use of (7.106) and (7.107) leads to the expected result

$$\begin{aligned}
\hat{H} &= \int \mathrm{d}^3 x \, \tfrac{1}{2} \mathbin{:}\!\left(\hat{\boldsymbol{E}}^2 + \hat{\boldsymbol{B}}^2 \right)\!\mathbin{:} \\
&= \frac{1}{2} \int \frac{\mathrm{d}^3 k}{2\omega_k} \sum_{\lambda\lambda'=1}^{2} \left[\omega_k^2 \, \epsilon(\boldsymbol{k},\lambda') \cdot \epsilon(\boldsymbol{k},\lambda) + (\boldsymbol{k} \times \epsilon(\boldsymbol{k},\lambda')) \cdot (\boldsymbol{k} \times \epsilon(\boldsymbol{k},\lambda)) \right] \\
&\quad \times \left(\hat{a}^\dagger_{\boldsymbol{k}\lambda'} \hat{a}_{\boldsymbol{k}\lambda} + \hat{a}^\dagger_{\boldsymbol{k}\lambda} \hat{a}_{\boldsymbol{k}\lambda'} \right) \\
&= \int \mathrm{d}^3 k \, \omega_k \sum_{\lambda=1}^{2} \hat{a}^\dagger_{\boldsymbol{k}\lambda} \hat{a}_{\boldsymbol{k}\lambda} \ .
\end{aligned} \tag{7.112}$$

Here $(\boldsymbol{k} \times \epsilon') \cdot (\boldsymbol{k} \times \epsilon) = \boldsymbol{k}^2 \epsilon' \cdot \epsilon - (\boldsymbol{k}\cdot\epsilon)(\boldsymbol{k}\cdot\epsilon')$ and the orthogonality conditions for the vectors ϵ and $\boldsymbol{k}$, and the energy-momentum relation $\omega_k^2 - \boldsymbol{k}^2 = 0$ have been used. Thus only the transverse photons contribute to the energy, as should be. Analogous results of course also hold for the *momentum*

$$\hat{\boldsymbol{P}} = \int \mathrm{d}^3 x \, \mathbin{:}\!\hat{\boldsymbol{E}} \times \hat{\boldsymbol{B}}\!\mathbin{:} \, = \int \mathrm{d}^3 k \sum_{\lambda=1}^{2} \boldsymbol{k} \, \hat{a}^\dagger_{\boldsymbol{k}\lambda} \hat{a}_{\boldsymbol{k}\lambda} \tag{7.113}$$

and for the *spin* of the photon field (cf. (30) of Example 6.3)

$$\hat{S}^{nl} = \mathrm{i} \int \mathrm{d}^3 k \sum_{\lambda,\lambda'=1}^{2} \epsilon^n(\boldsymbol{k},\lambda') \epsilon^l(\boldsymbol{k},\lambda) \left(\hat{a}^\dagger_{\boldsymbol{k}\lambda} \hat{a}_{\boldsymbol{k}\lambda'} - \hat{a}^\dagger_{\boldsymbol{k}\lambda'} \hat{a}_{\boldsymbol{k}\lambda} \right) \ . \tag{7.114}$$

Note that there is no longer a longitudinal contribution ($\lambda = 3$). The operator for the projection of spin in the direction of the momentum (helicity) again can be diagonalized with the use of transverse photons of circular polarization as in (36) of Example 6.3.

To obtain the *Feynman propagator in the Coulomb gauge* no new derivation is required. The result can be taken from Sect. 7.5, i.e., (here we reinstate the transversality index $\perp$)

$$iD_F^{\perp\mu\nu}(x-y) := \langle 0|T(\hat{A}_\perp^\mu(x)\hat{A}_\perp^\nu(y))|0\rangle \tag{7.115}$$

$$= \int \frac{d^3k}{2\omega_k(2\pi)^3} \left(\sum_{\lambda=1}^2 \epsilon^\mu(\boldsymbol{k},\lambda)\epsilon^\nu(\boldsymbol{k},\lambda)\right)$$

$$\times \left(\Theta(x_0-y_0)\,e^{-ik\cdot(x-y)} + \Theta(y_0-x_0)\,e^{+ik\cdot(x-y)}\right).$$

This agrees with the transverse part of the covariant Feynman propagator, which was introduced in (7.73):

$$D_F^{\perp\mu\nu} = D_{F\,(\text{trans})}^{\mu\nu}. \tag{7.116}$$

Expressed in terms of the covariant Feynman propagator $D_F^{\mu\nu}$, we can make the decomposition

$$D_F^{\perp\mu\nu} = D_F^{\mu\nu} - D_{F\,(\text{Coul})}^{\mu\nu} - D_{F\,(\text{resid})}^{\mu\nu}. \tag{7.117}$$

In the frame of reference where $n^\mu = (1,0,0,0)$ and the vectors $\epsilon(\boldsymbol{k},1)$ and $\epsilon(\boldsymbol{k},2)$ have only spatial compenents the subtracted terms ensure that $D_F^{\perp\mu\nu}$ vanishes if $\mu=0$ or $\nu=0$. This is to be expected since we have $\hat{A}_0(x)=0$.

7.7.1 The Coulomb Interaction

The Coulomb gauge allows only for transverse photons (both real and virtual). Where do we find the effects of the Coulomb interaction which cannot be described in this way? The answer rests in the potential A_0, which up to now we have simply put equal to zero. If there are electrical sources this is not justified; there will be a nonvanishing potential A_0 which describes the Coulomb interaction. This potential, however, *is not subject to quantization.*

The classical Lagrangian of the electromagnetic field being coupled to a current $j_\mu(\boldsymbol{x},t)$ reads (cf. (6.27))

$$\mathcal{L} = -\frac{1}{4}F_{\mu\nu}F^{\mu\nu} - j_\mu A^\mu = \mathcal{L}_{\text{e.m.}} + \mathcal{L}_{\text{int}}. \tag{7.118}$$

The vector field, because of the gauge condition $\boldsymbol{\nabla}\cdot\boldsymbol{A}=0$, *remains transverse*, $\boldsymbol{A}=\boldsymbol{A}^\perp$. The electric field strength, however, now has an additional *longitudinal part*

$$\boldsymbol{E} = \boldsymbol{E}^\perp + \boldsymbol{E}^\| \tag{7.119a}$$

with

$$\boldsymbol{E}^\perp = -\partial_0\boldsymbol{A}, \quad \boldsymbol{\nabla}\cdot\boldsymbol{E}^\perp = 0, \tag{7.119b}$$

$$\boldsymbol{E}^\| = -\boldsymbol{\nabla}A_0, \quad \boldsymbol{\nabla}\times\boldsymbol{E}^\| = 0. \tag{7.119c}$$

The potential A_0 satisfies the Poisson equation. It is determined simply from the instantaneous charge density of the sources:

$$A_0(\boldsymbol{x},t) = \frac{1}{4\pi}\int d^3x'\, \frac{j_0(\boldsymbol{x}',t)}{|\boldsymbol{x}-\boldsymbol{x}'|}. \tag{7.120}$$

Using the decompositon (7.119a), we find that the electromagnetic part of the Lagrangian (7.118) reads

7.7 Canonical Quantization in the Coulomb Gauge

$$\mathcal{L}_{\text{e.m.}} = \frac{1}{2}(\boldsymbol{E}^2 - \boldsymbol{B}^2)$$
$$= \frac{1}{2}(\boldsymbol{E}^{\|2} + \boldsymbol{E}^{\perp 2} - \boldsymbol{B}^2) + \boldsymbol{E}^{\|} \cdot \boldsymbol{E}^{\perp} . \tag{7.121}$$

The mixed longitudinal/transverse term can be written with the help of (7.119b) as a divergence

$$\boldsymbol{E}^{\|} \cdot \boldsymbol{E}^{\perp} = -(\boldsymbol{\nabla} A_0) \cdot \boldsymbol{E}^{\perp} = -\boldsymbol{\nabla} \cdot (A_0 \boldsymbol{E}^{\perp}) + A_0 \boldsymbol{\nabla} \cdot \boldsymbol{E}^{\perp}$$
$$= -\boldsymbol{\nabla} \cdot (A_0 \boldsymbol{E}^{\perp}) , \tag{7.122}$$

and thus can be discarded.

The transition to the Hamilton density makes use of the purely transverse canonically conjugate field (for the sign remember the footnote on page 158)

$$\boldsymbol{\pi}^{\perp} = -\frac{\partial \mathcal{L}}{\partial(\partial_0 \boldsymbol{A})} = \frac{\partial \mathcal{L}}{\partial \boldsymbol{E}^{\perp}} = \boldsymbol{E}^{\perp} , \tag{7.123}$$

leading to

$$\mathcal{H}_{\text{e.m.}} = \pi_\mu \dot{A}^\mu - \mathcal{L}_{\text{e.m.}} = -\boldsymbol{\pi}^{\perp} \cdot \dot{\boldsymbol{A}} - \mathcal{L}_{\text{e.m.}}$$
$$= -\boldsymbol{E}^{\perp} \cdot (-\boldsymbol{E}^{\perp}) - \frac{1}{2}(\boldsymbol{E}^{\|2} + \boldsymbol{E}^{\perp 2} - \boldsymbol{B}^2)$$
$$= \frac{1}{2}(\boldsymbol{E}^{\perp 2} + \boldsymbol{B}^2) - \frac{1}{2}\boldsymbol{E}^{\|2}$$
$$= \mathcal{H}_0 + \mathcal{H}^{\|} . \tag{7.124}$$

The longitudinal term $\mathcal{H}^{\|}$ describes the interaction between the charges. Thus it is useful to introduce a modified splitting between the free and the interacting part of the Hamilton density as follows

$$\mathcal{H} = (\mathcal{H}_0 + \mathcal{H}^{\|}) + \mathcal{H}_{\text{int}} = \mathcal{H}_0 + \mathcal{H}'_{\text{int}} \tag{7.125}$$

with

$$\mathcal{H}'_{\text{int}} = \mathcal{H}^{\|} + \mathcal{H}_{\text{int}} = \mathcal{H}^{\|} - \mathcal{L}_{\text{int}} . \tag{7.126}$$

After an integration by parts the interaction part of the Hamiltonian is found to be

$$H'_{\text{int}} = \int d^3x \left(-\frac{1}{2}\boldsymbol{\nabla} A_0 \cdot \boldsymbol{\nabla} A_0 + j_0 A^0 - \boldsymbol{j} \cdot \boldsymbol{A}\right)$$
$$= \int d^3x \left(\frac{1}{2} A_0 \boldsymbol{\nabla}^2 A_0 + j_0 A^0 - \boldsymbol{j} \cdot \boldsymbol{A}\right) . \tag{7.127}$$

With the help of the Poisson equation $\boldsymbol{\nabla}^2 A_0 = -j_0$ the first two terms can be combined and we arrive at

$$H'_{\text{int}} = H_{\text{Coul}} + H^{\perp}_{\text{int}} , \tag{7.128}$$

with

$$H_{\text{Coul}} = \int d^3x \, \frac{1}{2} j_0 A^0$$
$$= \frac{1}{2} \int d^3x \, d^3x' \, j_0(\boldsymbol{x},t) \frac{1}{4\pi|\boldsymbol{x}-\boldsymbol{x}'|} j_0(\boldsymbol{x}',t) , \tag{7.129a}$$

$$H^{\perp}_{\text{int}} = -\int d^3x \, \boldsymbol{j}(\boldsymbol{x},t) \cdot \boldsymbol{A}(\boldsymbol{x},t) . \tag{7.129b}$$

As expected, the *Coulomb interaction* makes its appearance. Equation (7.129a) consists of a double integral over the product of the charge density at two spatial points (the factor 1/2 serves to cancel the double counting of the interaction energy). The electromagnetic field here is not visible explicitly. Thus we have found that the Coulomb term H_{Coul} is not affected by field quantization. On the other hand, the remaining terms H_0 and $H_{\text{int}}^\perp$ contain only the transverse vector potential; here quantization can proceed as in the free-field case.

EXERCISE

7.4 The Transverse Delta Function

Problem. In (7.101) the transverse delta function was introduced, with use made of the projection operator $P_\perp$

$$\delta^3_{\perp ij}(\boldsymbol{x}) = \left(\delta_{ij} - \partial_i \frac{1}{\Delta}\partial_j\right) \delta^3(\boldsymbol{x}) \ . \tag{1}$$

Find a closed expression for this distribution in coordinate space. In particular, isolate the contribution proportional to $\delta_{ij}\,\delta^3(\boldsymbol{x})$.

Solution. The inverse Laplacian Δ^{-1} in (1) can be expressed by an integral over the Green's function of Δ as shown in (6.24). This leads to

$$\begin{aligned}
\delta^3_{\perp ij}(\boldsymbol{x}) &= \delta_{ij}\,\delta^3(\boldsymbol{x}) - \partial_i \int d^3x' \left(-\frac{1}{4\pi}\frac{1}{|\boldsymbol{x}-\boldsymbol{x}'|}\right) \partial'_j \delta^3(\boldsymbol{x}') \\
&= \delta_{ij}\,\delta^3(\boldsymbol{x}) - \frac{1}{4\pi}\int d^3x' \left(\partial'_j \frac{1}{|\boldsymbol{x}-\boldsymbol{x}'|}\right) \delta^3(\boldsymbol{x}') \\
&= \delta_{ij}\,\delta^3(\boldsymbol{x}) + \frac{1}{4\pi}\partial_i\partial_j \frac{1}{|\boldsymbol{x}|} \ .
\end{aligned} \tag{2}$$

Often it is sufficient simply to use this expression without further investigating its structure. One should be aware, however, that the second terms in (2) also has the character of a distribution since it contains second derivatives of the singular function $1/|\boldsymbol{x}|$. We know that

$$\Delta \frac{1}{|\boldsymbol{x}|} = -4\pi\,\delta^3(\boldsymbol{x}) \ , \tag{3}$$

so that the trace of the transverse delta function becomes

$$\begin{aligned}
\delta^3_{\perp ii}(\boldsymbol{x}) &= \delta_{ii}\,\delta^3(\boldsymbol{x}) + \frac{1}{4\pi}\Delta \frac{1}{|\boldsymbol{x}|} = (3-1)\,\delta^3(\boldsymbol{x}) = 2\,\delta^3(\boldsymbol{x}) \\
&= \frac{2}{3}\delta_{ii}\,\delta^3(\boldsymbol{x}) \ .
\end{aligned} \tag{4}$$

Therefore the second term in (2) contains a contribution of the type $-\frac{1}{3}\delta_{ij}\delta^3(\boldsymbol{x})$. To prove this and also to derive the remaining traceless part, we introduce a regularization of the divergent function $1/|\boldsymbol{x}|$, cutting off the singularity at the origin. The *regularized transverse delta function*[13] will be defined by

$$\delta^3_{\perp ij}(\boldsymbol{x}, \kappa) = (-\delta_{ij}\Delta + \partial_i\partial_j)\, g(\boldsymbol{x}, \kappa) \tag{5}$$

where (we write $r \equiv |\boldsymbol{x}|$)

$$g(\boldsymbol{x}, \kappa) = \frac{1}{4\pi r}\left(1 - e^{-\kappa r}\right). \tag{6}$$

Here κ is a cutoff parameter which ensures that the function g remains finite at small distances: $g(\boldsymbol{x}, \kappa) \to \kappa/4\pi$ for $r \to 0$. We are interested in the limit $\kappa \to \infty$, which will be taken at the end of the calculation.

Remark: The function (6) has a simple counterpart in momentum space. We have

$$g(\boldsymbol{x}, \kappa) = \int d^3k\,(2\pi)^3\, e^{i\boldsymbol{k}\cdot\boldsymbol{x}}\,\frac{1}{|\boldsymbol{k}|^2}\,\frac{\kappa^2}{|\boldsymbol{k}|^2+\kappa^2}. \tag{7}$$

The first part of the integral is the Coulomb potential in momentum space, which is cut off at large momenta $|\boldsymbol{k}| \gg \kappa$ by the regularizing factor $\kappa^2/(|\boldsymbol{k}|^2 + \kappa^2)$.

An elementary calculation leads to

$$\begin{aligned}\partial_i\partial_j g &= \frac{1}{4\pi}\left(\delta_{ij}\frac{-1+e^{-\kappa r}+\kappa r e^{-\kappa r}}{r^3}\right.\\ &\quad\left.+\frac{x_i x_j}{r^2}\frac{3-3e^{-\kappa r}-3\kappa r e^{-\kappa r}-\kappa^2 r^2 e^{-\kappa r}}{r^3}\right),\end{aligned} \tag{8}$$

and as a special case

$$\Delta g = \partial_i\partial_i g = \frac{1}{4\pi}\frac{-\kappa^2 e^{-\kappa r}}{r}. \tag{9}$$

Therefore (5) reads

$$\begin{aligned}\delta^3_{\perp ij}(\boldsymbol{x}, \kappa) &= \frac{1}{4\pi}\left(\delta_{ij}\frac{-1+e^{-\kappa r}+\kappa r e^{-\kappa r}+\kappa^2 r^2 e^{-\kappa r}}{r^3}\right.\\ &\quad\left.+\frac{x_i x_j}{r^2}\frac{3-3e^{-\kappa r}-3\kappa r e^{-\kappa r}-\kappa^2 r^2 e^{-\kappa r}}{r^3}\right).\end{aligned} \tag{10}$$

In order to isolate the delta-function contribution we first inspect how the limit $\kappa \to \infty$ works for the function $\Delta g(\boldsymbol{x}, \kappa)$ from (9). To show that (3) is obtained,

$$\Delta g(\boldsymbol{x}, \kappa) \to -\delta^3(\boldsymbol{x}), \tag{11}$$

we multiply Δg by a (sufficiently smooth) test function $f(\boldsymbol{x})$ and integrate over d^3x. The test function can be Taylor expanded at $\boldsymbol{x} = 0$:

$$f(\boldsymbol{x}) = f(\boldsymbol{0}) + x_i\partial_i f(\boldsymbol{0}) + \frac{1}{2}x_i x_j \partial_i\partial_j f(\boldsymbol{0}) + \ldots. \tag{12}$$

Then the integral reads

---
[13] See C. Cohen-Tannoudji, J. Dupont-Roc, G. Grynberg: *Photons and Atoms* (Wiley, New York 1989).

Exercise 7.4

Exercise 7.4

$$\int d^3x\, f(\boldsymbol{x})\Delta g(\boldsymbol{x},\kappa) = \int d^3x \left(f(\boldsymbol{0}) + x_i\partial_i f(\boldsymbol{0}) + \ldots\right)\frac{1}{4\pi}\frac{-\kappa^2 e^{-\kappa r}}{r}$$

$$= -\kappa^2 f(\boldsymbol{0}) \int_0^\infty dr\, r\, e^{-\kappa r} + \ldots$$

$$= -f(\boldsymbol{0}) + O\!\left(\frac{1}{\kappa^2}\right). \tag{13}$$

This derivation was based on the integral formula

$$\int_0^\infty dr\, r^n\, e^{-\kappa r} = \frac{n!}{\kappa^{n+1}}. \tag{14}$$

This formula implies that the higher terms of the Taylor expansion (13), after the integration over r, contain higher powers of $1/\kappa$ and thus can be discarded in the limit $\kappa \to \infty$. This confirms the limiting relation (11).

The expression (10) for the regularized transverse delta function can be cleverly rewritten:

$$\delta^3_{\perp ij}(\boldsymbol{x},\kappa) = p_{ij}(\boldsymbol{x},\kappa) + \frac{1}{4\pi r^3}\left(3\frac{x_i x_j}{r^2} - \delta_{ij}\right) q(\boldsymbol{x},\kappa) \tag{15}$$

with the function

$$p_{ij}(\boldsymbol{x},\kappa) = \frac{1}{4\pi}\frac{\kappa^2}{2r}\left(\delta_{ij} + \frac{x_i x_j}{r^2}\right)e^{-\kappa r}, \tag{16a}$$

$$q(\boldsymbol{x},\kappa) = 1 - \left(1 + \kappa r + \tfrac{1}{2}\kappa^2 r^2\right)e^{-\kappa r}. \tag{16b}$$

In the limit $\kappa \to \infty$ the first term in (16a) as in (11) simply leads to $\tfrac{1}{2}\delta_{ij}\delta^3(\boldsymbol{x})$. The integral over the second term vanishes if $i \neq j$ from symmetry considerations. (There might be finite contributions if the test function $f(\boldsymbol{x})$ is non-isotropic, which, however, would be of higher order in $1/\kappa$.) Thus because $\sum_i x_i x_i = r^2$ we immediately conclude

$$p_{ij}(\boldsymbol{x},\kappa \to \infty) = \frac{1}{2}\left(1 + \frac{1}{3}\right)\delta_{ij}\,\delta^3(\boldsymbol{x}) = \frac{2}{3}\delta_{ij}\,\delta^3(\boldsymbol{x}). \tag{17}$$

Although this result has already been deduced in (4), we now also know explicitly the traceless remainder. In (15) $q(\boldsymbol{x},\kappa)$ is an (isotropic) regularizing function which for $r > 0$ (or rather $r \gg 1/\kappa$) approaches unity, implying that the nonlocal residual term in this region behaves like $1/r^3$. The singularity at $r \to 0$ is cut off by the function $q(\boldsymbol{x},\kappa)$ since

$$\frac{q(\boldsymbol{x},\kappa)}{r^3} \to \frac{1}{6}\kappa^3 \quad \text{for} \quad r \ll \frac{1}{\kappa}, \tag{18}$$

which remains bounded for finite κ.

The integral of the residual term multiplied by a test function $f(\boldsymbol{x})$ then does not lead to a significant term proportional to $f(\boldsymbol{0})$ even in the limit $\kappa \to \infty$. To be more precise, the "suspicious" contribution to the integral over the residual term multiplied by a test function $f(\boldsymbol{x})$, which would characterize a delta function (or its derivatives), reads

$$\int_0^{1/\kappa} d^3x\, f(\boldsymbol{x})\frac{1}{4\pi}\frac{\kappa^3}{6}\left(3\frac{x_i x_j}{r^2} - \delta_{ij}\right). \tag{19}$$

The first term of the Taylor expansion of $f(\boldsymbol{x})$ could produce an integral of order unity but this vanishes because of symmetry reasons:

$$\frac{\kappa^3}{2\pi} f(0) \int_0^{1/\kappa} d^3x \left(3\frac{x_i x_j}{r^2} - \delta_{ij}\right) = 0. \tag{20}$$

For the next-higher order we find

$$\frac{\kappa^3}{2\pi} \partial_k f(0) \int_0^{1/\kappa} d^3x\, x_k \left(3\frac{x_i}{x_j}r^2 - \delta_{ij}\right)$$

$$\propto \partial_k f(0)\,\kappa^3 \int_0^{1/\kappa} dr\, r^3 = \partial_k f(0)\,\kappa^3 \frac{1}{\kappa^4} = O\left(\frac{1}{\kappa}\right), \tag{21}$$

where the value of the angular integral is unimportant.

EXERCISE

7.5 General Commutation Rules for the Electromagnetic Field

Problem. (a) Evaluate the commutator $[\hat{A}_\mu(x), \hat{A}_\nu(y)]$ of the free electromagnetic field in the Lorentz gauge for arbitrary space-time points x and y. Show that the commutator vanishes outside the light cone.
(b) Evaluate the commutators between the various components of the field-strength operators $\hat{E}(x)$ and $\hat{B}(x)$. Give an interpretation of the result.
(c) Repeat the previous calculations using the Coulomb gauge and compare the results.

Solution. (a) Insertion of the Fourier decomposition (7.26) of the potential operator $\hat{A}$ and use of the commutation relations for photon creation and annihilation operators (7.32) leads to (cf. the analogous treatment of the massive vector field in Exercise 6.4)

$$[\hat{A}_\mu(x), \hat{A}_\nu(y)] = \int \frac{d^3k}{(2\pi)^3} \frac{1}{2\omega_k} \left(\sum_{\lambda=0}^{3}(-g_{\lambda\lambda})\epsilon_\mu(\mathbf{k}, \lambda)\epsilon_\nu(\mathbf{k}, \lambda)\right)$$
$$\times \left(e^{-ik\cdot(x-y)} - e^{+ik\cdot(x-y)}\right). \tag{1}$$

Now the completeness relation (6.83) of the four polarization vectors can be used, leading to

$$[\hat{A}_\mu(x), \hat{A}_\nu(y)] = -g_{\mu\nu} \int \frac{d^3k}{(2\pi)^3} \frac{1}{2\omega_k} \left(e^{-ik\cdot(x-y)} - e^{+ik\cdot(x-y)}\right). \tag{2}$$

This integral is the invariant Pauli–Jordan–Schwinger function first introduced in Sect. 4.4. Since the field is now massless we use the notation $D(x-y)$ instead of $\Delta(x-y)$:

$$iD(x-y) = \int \frac{d^3k}{(2\pi)^3} \frac{1}{2\omega_k} \left(e^{-ik\cdot(x-y)} - e^{+ik\cdot(x-y)}\right)$$
$$= \int \frac{d^4k}{(2\pi)^3} \epsilon(k_0)\delta(k^2)\, e^{-ik\cdot(x-y)}, \tag{3}$$

where $k_0 = \omega_k = |\mathbf{k}|$. The commutator is thus given by

$$[\hat{A}_\mu(x), \hat{A}_\nu(y)] = -i g_{\mu\nu} D(x-y). \tag{4}$$

Exercise 7.5

The function $D(x)$ has a simple form both in momentum space (cf. (3)) and in coordinate space as we will show now. Equation (3) can be written as

$$D(x) = D^{(+)}(x) + D^{(-)}(x) = 2\operatorname{Re}\{D^{(+)}(x)\}\,. \tag{5}$$

The Fourier integral in $D^{(+)}(x)$ can be solved in polar coordinates:

$$\begin{aligned}D^{(+)}(x) &= -\mathrm{i}\int\frac{d^3k}{(2\pi)^3}\frac{1}{2\omega_k}\mathrm{e}^{-\mathrm{i}(\omega_k t - \boldsymbol{k}\cdot\boldsymbol{x})} \\ &= -\mathrm{i}\frac{1}{(2\pi)^2}\int_0^\infty dk\,\frac{k^2}{2\omega_k}\mathrm{e}^{-\mathrm{i}\omega_k t}\int_{-1}^{+1}d\cos\theta\,\mathrm{e}^{\mathrm{i}kr\cos\theta} \\ &= -\frac{1}{8\pi^2}\frac{1}{r}\int_0^\infty dk\,\left(\mathrm{e}^{\mathrm{i}k(r-t)} - \mathrm{e}^{-\mathrm{i}k(r+t)}\right)\,,\end{aligned} \tag{6}$$

with $r \equiv |\boldsymbol{x}|$, $t \equiv x_0$ and $k \equiv |\boldsymbol{k}| = \omega_k$. To ensure convergence of the k integration a cutoff factor $\mathrm{e}^{-\epsilon k}$ will be introduced, keeping in mind that at the end of the calculation the limit $\epsilon \to 0$ has to be taken. We have from (6) that

$$D^{(+)}(x) = -\mathrm{i}\frac{1}{8\pi^2}\frac{1}{r}\left(\frac{1}{r-t+\mathrm{i}\epsilon} + \frac{1}{r+t-\mathrm{i}\epsilon}\right)\,. \tag{7}$$

From this the commutation function

$$\begin{aligned}D(x) &= 2\operatorname{Re}\{D^{(+)}(x)\} = -\frac{1}{4\pi^2}\frac{1}{r}\operatorname{Re}\left\{\frac{\mathrm{i}}{r-t+\mathrm{i}\epsilon} + \frac{\mathrm{i}}{r+t-\mathrm{i}\epsilon}\right\} \\ &= -\frac{1}{4\pi^2}\frac{1}{r}\left(\frac{\epsilon}{(r-t)^2+\epsilon^2} - \frac{\epsilon}{(r+t)^2+\epsilon^2}\right)\end{aligned} \tag{8}$$

is obtained. In the limit $\epsilon \to 0$ we recognize the representation of the delta function

$$\delta(x) = \lim_{\epsilon \to 0}\frac{1}{\pi}\frac{\epsilon}{\epsilon^2+x^2}\,, \tag{9}$$

which implies

$$D(x) = -\frac{1}{4\pi}\frac{1}{r}\Big(\delta(r-t) - \delta(r+t)\Big)\,. \tag{10}$$

Using the identity $\delta(x^2) = [\delta(r-t)+\delta(r+t)]/2r$, we can simplify this to read

$$D(x) = -\frac{1}{2\pi}\epsilon(t)\,\delta(x^2)\,. \tag{11}$$

This agrees with the $m = 0$ limit of the massive Δ function given earlier in (4.168a) in Sect. 4.6.

We have found that the potential operators $\hat{A}_\mu(x)$ and $\hat{A}_\nu(y)$ commute nearly everywhere. Only on the light cone $(x-y)^2 = 0$ will there be a nonvanishing commutator, corresponding to the fact that the points can communicate by an electromagnetic signal transmission.[14]

(b) To evaluate the commutators between the field-strength operators

---

[14] Note that in the massive case, $m \neq 0$, the function $\Delta(x-y)$ is finite also within the light cone $(x-y)^2 > 0$, since massive particles can also travel at velocities smaller than the velocity of light.

$$\hat{E}^i = \partial^i \hat{A}^0 - \partial^0 \hat{A}^i , \qquad (12a)$$
$$\hat{B}^i = \epsilon^{ijk} \partial_j \hat{A}_k \qquad (12b)$$

we might start from scratch and proceed as in (a). It saves work, however, to make use of (4) and act with the derivative operators from (12) on this equation. In this way the commutator of the electric-field-strength operator is found to be

$$\begin{aligned}
\left[\hat{E}^i(x), \hat{E}^j(y)\right] &= \left[\partial^i \hat{A}^0(x) - \partial^0 \hat{A}^i(x), \partial^j \hat{A}^0(y) - \partial^0 \hat{A}^j(y)\right] \\
&= \partial_x^0 \partial_y^0 \left[\hat{A}^i(x), \hat{A}^j(y)\right] + \partial_x^i \partial_y^j \left[\hat{A}^0(x), \hat{A}^0(y)\right] \\
&= \left(\partial_x^0 \partial_y^0 g^{ij} + \partial_x^i \partial_y^j g^{00}\right)(-iD(x-y)) ,
\end{aligned} \qquad (13)$$

and thus, using $\partial_y^\mu D(x-y) = -\partial_x^\mu D(x-y) \equiv -\partial^\mu D(x-y)$,

$$\left[\hat{E}^i(x), \hat{E}^j(y)\right] = -i\left(\delta^{ij} \partial^0 \partial^0 - \partial^i \partial^j\right) D(x-y) . \qquad (14)$$

For the magnetic field strength the calculation proceeds as follows:

$$\begin{aligned}
\left[\hat{B}^i(x), \hat{B}^j(y)\right] &= \epsilon^{ikl} \epsilon^{jmn} \left[\partial_k \hat{A}_l(x), \partial_m \hat{A}_n(y)\right] \\
&= \epsilon^{ikl} \epsilon^{jmn} \partial_k^x \partial_m^y \left(-i(-\delta_{ln})D(x-y)\right) \\
&= i\left(\delta^{ij}\delta^{km} - \delta^{im}\delta^{jk}\right) \partial_k^x \partial_m^y D(x-y) \\
&= -i\left(\delta^{ij}\boldsymbol{\nabla}^2 - \partial^i \partial^j\right) D(x-y) .
\end{aligned} \qquad (15)$$

This result is similar to the electric-field-strength commutator (14). Both results indeed turn out to be identical since the function $D(x-y)$ satisfies d'Alembert's equation

$$\left(\boldsymbol{\nabla}^2 - \partial^0 \partial^0\right) D(x-y) = 0 , \qquad (16)$$

which implies

$$\left[\hat{B}^i(x), \hat{B}^j(y)\right] = -i\left(\delta^{ij} \partial^0 \partial^0 - \partial^i \partial^j\right) D(x-y) . \qquad (17)$$

The mixed electric/magnetic commutator takes on a somewhat different form

$$\begin{aligned}
\left[\hat{E}^i(x), \hat{B}^j(y)\right] &= \left[\partial^i \hat{A}^0(x) - \partial^0 \hat{A}^i(x), \epsilon^{jkl} \partial_k \hat{A}_l(y)\right] \\
&= \epsilon^{jkl}(-\partial_0^x \partial_k^y)\left(-ig^i{}_l D(x-y)\right) \\
&= i\epsilon^{ijk} \partial_0 \partial_k D(x-y) .
\end{aligned} \qquad (18)$$

The commutators (14), (17), and (18) all consist of second derivatives of the function $D(x-y)$. This guarantees that *all the field-strength operators commute outside of the light cone*. This is what we have to request if the electromagnetic field strength is to be interpreted as an observable quantity that has to satisfy the microcausality condition.[15]

---

[15] More information on the physical interpretation of the measuring process and on the uncertainty as applied to the field strength can be found in W. Heitler: *The Quantum Theory of Radiation* (Clarendon Press, Oxford 1954) and in the classical work of Bohr and Rosenfeld quoted in the footnote on page 103.

Exercise 7.5

We can obtain an explicit expression for the field-strength commutator (14) by twice differentiating (10):

$$
\begin{aligned}
\left[\hat{E}^i(x), \hat{E}^j(y)\right] &= \frac{i}{4\pi}\Bigg[\left(\delta_{ij} - \frac{r_i r_j}{r^2}\right)\frac{1}{r}\left(\delta''(r-t) - \delta''(r+t)\right) \\
&\quad + \left(3\frac{r_i r_j}{r^2} - \delta_{ij}\right)\left(\frac{1}{r^2}\left(\delta'(r-t) - \delta'(r+t)\right) - \frac{1}{r^3}\left(\delta(r-t) - \delta(r+t)\right)\right)\Bigg],
\end{aligned}
\tag{19}
$$

with the notation $r = |\boldsymbol{x}-\boldsymbol{y}|$, $r_i = x_i - y_i$, $t = x_0 - y_0$. At equal times, $t \to 0$, this expression is problematic since it contains badly defined products of the type $r^{-n}\delta(r)$. One has to go back to the regularized delta function (9) etc. to recognize that

$$
\left[\hat{E}^i(x), \hat{E}^j(y)\right]\Big|_{x_0 = y_0} = 0 \,.
\tag{20}
$$

This can also be deduced directly from (14) since the derivatives of D vanish at equal times $x_0 = y_0$.

(c) In the Coulomb gauge $\hat{A}^0 = 0$ only two polarization vector are present. Instead of (1) the commutator of the vector potential reads

$$
\begin{aligned}
\left[\hat{A}^i(x), \hat{A}^j(y)\right] &= \int \frac{d^3 k}{(2\pi)^3}\frac{1}{2\omega_k}\left(\sum_{\lambda=1}^{2}\epsilon^i(\boldsymbol{k},\lambda)\epsilon^j(\boldsymbol{k},\lambda)\right) \\
&\quad \times \left(e^{-ik\cdot(x-y)} - e^{+ik\cdot(x-y)}\right) \\
&= \int \frac{d^3 k}{(2\pi)^3}\frac{1}{2\omega_k}\left(\delta_{ij} - \frac{k^i k^j}{\boldsymbol{k}^2}\right)\left(e^{-ik\cdot(x-y)} - e^{+ik\cdot(x-y)}\right).
\end{aligned}
\tag{21}
$$

The polarization sum has generated a momentum-dependent factor which we identify as the *transverse projection operator* $P_\perp$, (7.100):

$$
\begin{aligned}
\left[\hat{A}^i(x), \hat{A}^j(y)\right] &= \left(\delta_{ij} - \frac{\partial^i \partial^j}{\Delta}\right)\int \frac{d^3 k}{(2\pi)^3}\frac{1}{2\omega_k}\left(e^{-ik\cdot(x-y)} - e^{+ik\cdot(x-y)}\right) \\
&= (P_\perp)_{ij}\, iD(x-y) \,.
\end{aligned}
\tag{22}
$$

$P_\perp$ in coordinate space is a *nonlocal* operator that has major consequences for the properties of the commutator: (22) is nonvanishing even for distances outside the light cone! To check this claim we write (22) as

$$
\left[\hat{A}^i(x), \hat{A}^j(y)\right] = i\delta_{ij}D(x-y) + \partial^i \partial^j H(x-y)
\tag{23}
$$

with the auxiliary function

$$
H(x) = \int \frac{d^3 k}{(2\pi)^3}\frac{1}{2\omega_k^3}\left(e^{-ik\cdot(x-y)} - e^{+ik\cdot(x-y)}\right).
\tag{24}
$$

After performing the angular integration this reads

$$
H(x) = \frac{-i}{8\pi^2}\frac{1}{r}\int_0^\infty dk\, \frac{1}{k^2}\left(e^{ikr} - e^{-ikr}\right)\left(e^{-ikt} - e^{ikt}\right).
\tag{25}
$$

This integral can be attacked by the theorem of residues after it has been reshaped a bit. To begin with, the integrand of (25) does not have a pole at the origin since the numerator also vanishes quadratically at $k \to 0$. Thus there

7.7 Canonical Quantization in the Coulomb Gauge

is nothing to keep us from replacing k^2 by $(k+i\epsilon)(k-i\epsilon)$ in the denominator. Subsequently the range of integration can be extended to the whole real axis by using the substitution $k \to -k$. The integral in (25) reads

$$
\begin{aligned}
I &= \int_0^\infty dk \frac{1}{(k-i\epsilon)(k+i\epsilon)} \left(e^{ik(r-t)} + e^{-ik(r-t)} - e^{ik(r+t)} - e^{-ik(r+t)} \right) \\
&= \int_{-\infty}^\infty dk \frac{1}{(k-i\epsilon)(k+i\epsilon)} e^{ikr} \left(e^{-ikt} - e^{+ikt} \right) .
\end{aligned}
\quad (26)
$$

The integrand now has two first-order poles at $k = \pm i\epsilon$. For $r > |t|$ the first exponential factor in (26) dominates and the integration contour can be closed in the lower half-plane. Using the residue at the pole $k = i\epsilon$, we find the integral becomes

$$ I = 2\pi i \frac{e^{\epsilon t} - e^{-\epsilon t}}{2i\epsilon} = 2\pi t . \quad (27) $$

To treat the opposite case, $r < |t|$, the integrals in (26) can be combined in a different way:

$$ I = \int_{-\infty}^\infty dk \frac{1}{(k-i\epsilon)(k+i\epsilon)} e^{-ikt} \left(e^{ikr} - e^{-ikr} \right) . \quad (28) $$

Now the integration contour can be closed in the upper (lower) half-plane if t is negative (positive), leading to

$$ I = 2\pi r \, \text{sgn}(t) . \quad (29) $$

The discontinuities can be expressed in terms of the sign function

$$ H(x) = -\frac{i}{8\pi} \frac{1}{r} \Big((t+r)\,\text{sgn}(t+r) - (t-r)\,\text{sgn}(t-r) \Big) \quad (30) $$

which can also be written as

$$ H(x) = -\frac{i}{4\pi} \begin{cases} \text{sgn}(t) & \text{for} \quad r < |t| \\ \dfrac{t}{r} & \text{for} \quad r > |t| \end{cases} . \quad (31) $$

This result is sketched in Fig. 7.3. An inspection of (23) reveals that the $\hat{A}$ commutator *does not vanish for space-like distances* since for $r > |t|$ the function $H(x)$ depends on the space position and its derivative is finite, $\partial^i \partial^j H(x) \neq 0$! This is a consequence of the instantaneous part of the potential in the Coulomb gauge which, viewed separately, appears to violate causality. Fortunately this presents no problem since the potential $A^\mu(x)$ is not an observable quantity. For the field strengths, on the other hand, no causality violation is found. Using $\hat{E}^i = -\partial^0 \hat{A}^i$ we find

$$
\begin{aligned}
\left[\hat{E}^i(x), \hat{E}^j(y) \right] &= \partial_x^0 \partial_y^0 \left[\hat{A}^i(x), \hat{A}^j(y) \right] = -i \partial^0 \partial^0 (P_\perp)_{ij} D(x-y) \\
&= -i \delta_{ij} \partial^0 \partial^0 D(x-y) \\
&\quad + \partial^i \partial^j \int \frac{d^3 k}{(2\pi)^3} \frac{1}{2\omega_k} \frac{\omega_k^2}{k^2} \left(e^{-ik\cdot(x-y)} - e^{+ik\cdot(x-y)} \right) \\
&= -i \left(\delta_{ij} \partial^0 \partial^0 - \partial^i \partial^j \right) D(x-y) .
\end{aligned}
\quad (32)
$$

Exercise 7.5

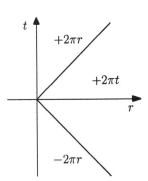

Fig. 7.3. The integral I in various regions of the r–t plane. The diagonal lines represent the light cone

Exercise 7.5

This is in agreement with the previous result (14), which was derived using the Lorentz gauge. As expected, expressions involving the field strength are gauge independent, and they do not violate causality.

8. Interacting Quantum Fields

8.1 Introduction

The previous chapters were dedicated to the quantization of *free* wave fields. The presentation of a detailed introduction to the relevant techniques is the main goal of this book. It cannot be overlooked, however, that the description of free fields on their own is a rather trivial task. The emergence of interesting physical processes can only be expected if there are *interactions* among the fields. From a mathematical point of view this happens if the Lagrangian contains more complicated parts that go beyond the bilinear terms encountered so far. Such terms may consist of higher powers of the field function (describing self-interactions) or of products involving fields of different kinds. Within the framework of classical theory this leads to a nonlinear field equation or coupled systems of field equations which are difficult to solve.

On the level of *interacting quantum fields* one is faced with even larger problems. Quantum fluctuations will induce complicated admixtures of particles of different kinds. A free-field operator has a very simple dynamics: the expansion into eigenmodes simply leads to constant creation and annihilation operators. The field operator of an interacting field theory, however, can no longer be described in this simple way: in general it will develop in time into a superposition of creation and annihilation operators for infinitely many virtual particles and antiparticles. As a consequence there is no way to find exact solutions for interacting quantum field theories (except for overly simplified mathematical models). Nevertheless, under favourable conditions, i.e., if the strength of the interaction is small, it is possible to obtain reliable approximate solutions. A systematic derivation of the *perturbation series for the scattering matrix* will be the main topic of this chapter. Nonperturbative quantum field theory, which is an actively pursued research topic, lies beyond the scope of this book.

8.2 The Interaction Picture

As we know from ordinary quantum mechanics the theory can be formulated in several different, but completely equivalent, "representations" or "pictures" that differ in the way the time dependence is treated. This ambiguity arises since all the physical obervables are expectation values or matrix elements of operators $\hat{O}$ evaluated with state vectors $|\alpha\rangle$. These matrix elements remain unchanged if the operators and the state vectors are jointly subjected to a unitary transformation.

The two extreme choices that can be constructed in this way are the *Schrödinger picture* (operators are constant, state vectors are time dependent) and the *Heisenberg picture* (state vectors are constant, operators are time dependent). In the context of quantum field theory one often prefers the Heisenberg picture. This is what we have done up to now: without explicitly stressing this point we have treated the field operators $\hat{\phi}(x,t)$ as time dependent.

In the *Heisenberg picture* the operators satisfy the Heisenberg equation of motion[1]

$$i\partial_t \hat{O}^{\mathrm{H}}(t) = [\hat{O}^{\mathrm{H}}(t), \hat{H}] \,, \tag{8.1}$$

if $\hat{H}$ is the (time independent) Hamiltonian of the system. We assume that the system is closed, otherwise it might experience an additional time dependence driven by external influences.

A formal solution for the time development of an operator $\hat{O}$ can be written down at once:

$$\hat{O}^{\mathrm{H}}(t) = e^{i\hat{H}t} \, \hat{O}^{\mathrm{H}}(0) \, e^{-i\hat{H}t} \,, \tag{8.2}$$

which is easily verified by forming the derivative. Instead of $t = 0$ of course we can chose any other time $t = t_0$ as a point of reference. This only amounts to the replacement of t by $t - t_0$ in the exponent. The state vectors in the Heisenberg picture are constant, i.e.,

$$|\alpha, t\rangle^{\mathrm{H}} = |\alpha, 0\rangle^{\mathrm{H}} \equiv |\alpha\rangle^{\mathrm{H}} \,. \tag{8.3}$$

The time dependence now can be transferred to the state vectors, which leads to the *Schrödinger picture*. This is achieved through the transformation

$$\hat{O}^{\mathrm{S}} = e^{-i\hat{H}t} \, \hat{O}^{\mathrm{H}}(t) \, e^{+i\hat{H}t} \tag{8.4a}$$

and

$$|\alpha, t\rangle^{\mathrm{S}} = e^{-i\hat{H}t} |\alpha\rangle^{\mathrm{H}} \,. \tag{8.4b}$$

Because of (8.2) and (8.4a) it is obvious that $\hat{O}^{\mathrm{S}}$ is time independent and the pictures agree at the (arbitrarily chosen) time $t = 0$:

$$\hat{O}^{\mathrm{S}} = \hat{O}^{\mathrm{H}}(0) \,, \tag{8.5a}$$

$$|\alpha, 0\rangle^{\mathrm{S}} = |\alpha\rangle^{\mathrm{H}} \,. \tag{8.5b}$$

The state vectors are constant and satisfy the Schrödinger equation

$$i\partial_t |\alpha, t\rangle^{\mathrm{S}} = \hat{H} |\alpha, t\rangle^{\mathrm{S}} \,. \tag{8.6}$$

The unitary transformation (8.4) of course leaves the matrix elements invariant, since

---

[1] Occasionally we have used the notation $\dot{\hat{O}}(t)$ for the time derivative of operators. In general such an object is meant to designate an operator with matrix elements that satisfy $\langle\beta|\dot{\hat{O}}|\alpha\rangle = \frac{\partial}{\partial t}\langle\beta|\hat{O}|\alpha\rangle$, irrespective of the picture. In the Heisenberg picture this definition of course agrees with the ordinary time derivative of the operator, $\dot{\hat{O}}^{\mathrm{H}} = \frac{\partial}{\partial t}\hat{O}^{\mathrm{H}}$.

$$\begin{aligned}{}^{\mathrm{S}}\langle\beta,t|\hat{O}^{\mathrm{S}}|\alpha,t\rangle^{\mathrm{S}} &= {}^{\mathrm{H}}\langle\beta|\mathrm{e}^{\mathrm{i}\hat{H}t}\,\mathrm{e}^{-\mathrm{i}\hat{H}t}\,\hat{O}^{\mathrm{H}}(t)\,\mathrm{e}^{\mathrm{i}\hat{H}t}\,\mathrm{e}^{-\mathrm{i}\hat{H}t}|\alpha\rangle^{\mathrm{H}}\\ &= {}^{\mathrm{H}}\langle\beta|\hat{O}^{\mathrm{H}}(t)|\alpha\rangle^{\mathrm{H}}\,,\end{aligned} \qquad (8.7)$$

where we used the fact that the Hamiltonian $\hat{H}$ is a hermitean operator, and thus $(\exp(-\mathrm{i}\hat{H}t))^{\dagger}=\exp(+\mathrm{i}\hat{H}t)$.

The Hamiltonian $\hat{H}$ does not have to be adorned with an index since it is the same in both pictures. This is seen immediately from (8.5) and the fact that $\hat{H}$ according to (8.1) is time independent also in the Heisenberg picture.[2]

It is important that the transition between the two pictures is mediated by a *canonical transformation* which leaves the commutation relations between operators invariant. If two operators in the Heisenberg picture satisfy the commutation relation

$$[\hat{A}^{\mathrm{H}},\hat{B}^{\mathrm{H}}]_{\pm}=\hat{C}^{\mathrm{H}} \qquad (8.8)$$

then the corresponding Schrödinger operators are immediately found to satisfy

$$\begin{aligned}{}[\hat{A}^{\mathrm{S}},\hat{B}^{\mathrm{S}}]_{\pm} &= \mathrm{e}^{-\mathrm{i}\hat{H}t}\left(\hat{A}^{\mathrm{H}}\,\mathrm{e}^{\mathrm{i}\hat{H}t}\,\mathrm{e}^{-\mathrm{i}\hat{H}t}\,\hat{B}^{\mathrm{H}}\pm\hat{B}^{\mathrm{H}}\,\mathrm{e}^{\mathrm{i}\hat{H}t}\,\mathrm{e}^{-\mathrm{i}\hat{H}t}\,\hat{A}^{\mathrm{H}}\right)\mathrm{e}^{\mathrm{i}\hat{H}t}\\ &= \mathrm{e}^{-\mathrm{i}\hat{H}t}\,[\hat{A}^{\mathrm{H}},\hat{B}^{\mathrm{H}}]_{\pm}\,\mathrm{e}^{\mathrm{i}\hat{H}t}=\mathrm{e}^{-\mathrm{i}\hat{H}t}\,\hat{C}^{\mathrm{H}}\,\mathrm{e}^{\mathrm{i}\hat{H}t}=\hat{C}^{\mathrm{S}}\,.\end{aligned} \qquad (8.9)$$

In principle it is possible to formulate the perturbation theory for interacting systems within the Schrödinger picture.[3] In such a formulation the field operators $\hat{\psi}^{\mathrm{S}}(\boldsymbol{x})$ will depend only on the spatial coordinate $\boldsymbol{x}$ and not on time t. This is a drawback if one is interested in studying relativistic theories since in this way the (manifest) Lorentz covariance is lost.

As a starting point for the development of a perturbation theory one assumes that the Hamiltonian of the system under study *can be split into two parts*:

$$\hat{H}=\hat{H}_0+\hat{H}_1\,. \qquad (8.10)$$

Here $\hat{H}_0$ is typically the Hamiltonian of a system of free fields. For this Hamiltonian alone an exact (albeit uninteresting) solution can be given. The strategy now is to put this "easy" part of the problem into the definition of the state vectors. Then only the "perturbation" operator $\hat{H}_1$ will be visible in the equations of motion. This is achieved by transforming to the *interaction picture* (which also is called the *Dirac picture* or *Tomonage picture*):

$$\hat{O}^{\mathrm{I}}(t) = \mathrm{e}^{\mathrm{i}\hat{H}_0^{\mathrm{S}}t}\,\hat{O}^{\mathrm{S}}\,\mathrm{e}^{-\mathrm{i}\hat{H}_0^{\mathrm{S}}t}\,, \qquad (8.11\mathrm{a})$$

$$|\alpha,t\rangle^{\mathrm{I}} = \mathrm{e}^{\mathrm{i}\hat{H}_0^{\mathrm{S}}t}\,|\alpha,t\rangle^{\mathrm{S}}\,, \qquad (8.11\mathrm{b})$$

with the free Hamiltonian $\hat{H}_0^{\mathrm{S}}=\hat{H}_0(t=0)$.

The connection with the Heisenberg picture is given by the following relations, using (8.4):

$$\hat{O}^{\mathrm{I}}(t) = \mathrm{e}^{\mathrm{i}\hat{H}_0^{\mathrm{S}}t}\,\mathrm{e}^{-\mathrm{i}\hat{H}t}\,\hat{O}^{\mathrm{H}}(t)\,\mathrm{e}^{\mathrm{i}\hat{H}t}\,\mathrm{e}^{-\mathrm{i}\hat{H}_0^{\mathrm{S}}t}\,, \qquad (8.12\mathrm{a})$$

$$|\alpha,t\rangle^{\mathrm{I}} = \mathrm{e}^{\mathrm{i}\hat{H}_0^{\mathrm{S}}t}\,\mathrm{e}^{-\mathrm{i}\hat{H}t}\,|\alpha\rangle^{\mathrm{H}}\,. \qquad (8.12\mathrm{b})$$

---

[2] As mentioned earlier we assume that there is no explicit time dependence enforced externally.

[3] For an early reference see, e.g., W. Heitler: *The Quantum Theory of Radiation* (Clarendon Press, Oxford 1954).

In the case $\hat{H}_1 = 0$ the interaction picture agrees with the Heisenberg picture. Furthermore, at $t = 0$ all three pictures agree:

$$|\alpha, 0\rangle^{\mathrm{I}} = |\alpha, 0\rangle^{\mathrm{S}} = |\alpha\rangle^{\mathrm{H}}, \qquad (8.13\mathrm{a})$$

$$\hat{O}^{\mathrm{I}}(0) = \hat{O}^{\mathrm{H}}(0) = \hat{O}^{\mathrm{S}}. \qquad (8.13\mathrm{b})$$

Since the connection between the three pictures is mediated by unitary transformations they all lead to the same matrix elements.

The Schrödinger equation for the time development of the state vector in the interaction picture assumes the form

$$\begin{aligned}
\mathrm{i}\partial_t |\alpha, t\rangle^{\mathrm{I}} &= -\hat{H}_0^{\mathrm{S}}\, \mathrm{e}^{\mathrm{i}\hat{H}_0^{\mathrm{S}} t}\, |\alpha, t\rangle^{\mathrm{S}} + \mathrm{e}^{\mathrm{i}\hat{H}_0^{\mathrm{S}} t}\, \mathrm{i}\partial_t |\alpha, t\rangle^{\mathrm{S}} \\
&= -\hat{H}_0^{\mathrm{S}} |\alpha, t\rangle^{\mathrm{I}} + \mathrm{e}^{\mathrm{i}\hat{H}_0^{\mathrm{S}} t}\, \hat{H}\, \mathrm{e}^{-\mathrm{i}\hat{H}_0^{\mathrm{S}} t}\, \mathrm{e}^{+\mathrm{i}\hat{H}_0^{\mathrm{S}} t} |\alpha, t\rangle^{\mathrm{S}} \\
&= \left(-\hat{H}_0^{\mathrm{S}} + \hat{H}^{\mathrm{I}}\right)|\alpha, t\rangle^{\mathrm{I}} = \left(-\hat{H}_0^{\mathrm{I}} + \hat{H}^{\mathrm{I}}\right)|\alpha, t\rangle^{\mathrm{I}},
\end{aligned} \qquad (8.14)$$

and thus

$$\mathrm{i}\partial_t |\alpha, t\rangle^{\mathrm{I}} = \hat{H}_1^{\mathrm{I}} |\alpha, t\rangle^{\mathrm{I}}, \qquad (8.15)$$

where we have used

$$\hat{H}_0^{\mathrm{I}} = \mathrm{e}^{\mathrm{i}\hat{H}_0^{\mathrm{S}} t}\, \hat{H}_0^{\mathrm{S}}\, \mathrm{e}^{-\mathrm{i}\hat{H}_0^{\mathrm{S}} t} = \hat{H}_0^{\mathrm{S}}. \qquad (8.16)$$

As intended the equation of motion (8.15) for the state vector contains only the "perturbation" operator $\hat{H}_1$. On the other hand, the time dependence of the operators is determined by the dynamics of the free field. The time derivative of the operator in the interaction picture (8.11a) yields

$$\mathrm{i}\partial_t \hat{O}^{\mathrm{I}}(t) = -\hat{H}_0^{\mathrm{S}}\, \hat{O}^{\mathrm{I}}(t) + \hat{O}^{\mathrm{I}}(t)\, \hat{H}_0^{\mathrm{S}} = [\hat{O}^{\mathrm{I}}(t), \hat{H}_0^{\mathrm{S}}]. \qquad (8.17)$$

This implies for the problem of interacting quantum fields that *the field operators retain the properties of the free fields*. For example, let us study the field operator $\hat{\phi}$, which in the Schrödinger picture can be expanded into basis functions $u_k(\boldsymbol{x})$ as

$$\hat{\phi}^{\mathrm{S}}(\boldsymbol{x}) = \sum_{k}\left(\hat{a}_k^{\mathrm{S}}\, u_k(\boldsymbol{x}) + \hat{a}_k^{\mathrm{S}\dagger}\, u_k^*(\boldsymbol{x})\right). \qquad (8.18)$$

The Hamiltonian of the free theory has the simple form

$$\hat{H}_0^{\mathrm{S}} = \sum_k \omega_k\, \hat{a}_k^{\mathrm{S}\dagger}\, \hat{a}_k^{\mathrm{S}}. \qquad (8.19)$$

Using (8.17) and (8.11a) we obtain the equation of motion for the annihilation operator $\hat{a}_k^{\mathrm{I}}$ in the interaction picture:

$$\begin{aligned}
\mathrm{i}\partial_t \hat{a}_k^{\mathrm{I}}(t) &= \left[\hat{a}_k^{\mathrm{I}}, \hat{H}_0^{\mathrm{S}}\right] = \mathrm{e}^{\mathrm{i}\hat{H}_0^{\mathrm{S}} t}\left[\hat{a}_k^{\mathrm{S}}, \hat{H}_0^{\mathrm{S}}\right]\mathrm{e}^{-\mathrm{i}\hat{H}_0^{\mathrm{S}} t} \\
&= \mathrm{e}^{\mathrm{i}\hat{H}_0^{\mathrm{S}} t} \sum_{k'} \omega_{k'} \left[\hat{a}_k^{\mathrm{S}}, \hat{a}_{k'}^{\mathrm{S}\dagger}\hat{a}_{k'}^{\mathrm{S}}\right] \mathrm{e}^{-\mathrm{i}\hat{H}_0^{\mathrm{S}} t} \\
&= \omega_k\, \mathrm{e}^{\mathrm{i}\hat{H}_0^{\mathrm{S}} t}\, \hat{a}_k^{\mathrm{S}}\, \mathrm{e}^{-\mathrm{i}\hat{H}_0^{\mathrm{S}} t} = \omega_k\, \hat{a}_k^{\mathrm{I}}(t),
\end{aligned} \qquad (8.20)$$

using the usual canonical commutation relations. The solution of this differential equation is simply given by

$$\hat{a}_{\boldsymbol{k}}^{\mathrm{I}}(t) = \mathrm{e}^{-\mathrm{i}\omega_k t}\,\hat{a}_{\boldsymbol{k}}^{\mathrm{I}}(0)\;. \tag{8.21}$$

In the interaction picture the creation and annihilation operators thus retain their character as single-particle operators in dependence on time. They are embellished with only an additional time-dependent phase factor. The *expansion of the field operator in the interaction picture* analogous to (8.18) thus will read

$$\hat{\phi}^{\mathrm{I}}(\boldsymbol{x},t) = \sum_{\boldsymbol{k}}\left(\hat{a}_{\boldsymbol{k}}\,\mathrm{e}^{-\mathrm{i}\omega_k t}\,u_{\boldsymbol{k}}(\boldsymbol{x}) + \hat{a}_{\boldsymbol{k}}^{\dagger}\,\mathrm{e}^{+\mathrm{i}\omega_k t}\,u_{\boldsymbol{k}}^{*}(\boldsymbol{x})\right) \tag{8.22}$$

with time-independent operators $\hat{a}_{\boldsymbol{k}} := \hat{a}_{\boldsymbol{k}}^{\mathrm{I}}(0)$ and $\hat{a}_{\boldsymbol{k}}^{\dagger} := \hat{a}_{\boldsymbol{k}}^{\mathrm{I}\dagger}(0)$. One also can arrive at (8.22) directly by transforming (8.18):

$$\hat{\phi}^{\mathrm{I}}(\boldsymbol{x},t) = \mathrm{e}^{\mathrm{i}\hat{H}_0^{\mathrm{S}} t}\,\hat{\phi}^{\mathrm{S}}(\boldsymbol{x})\,\mathrm{e}^{-\mathrm{i}\hat{H}_0^{\mathrm{S}} t}\;. \tag{8.23}$$

To prove this assertion one can employ the helpful operator identity (the Baker–Campbell–Hausdorff relation, see Exercise 1.3)

$$\mathrm{e}^{\mathrm{i}\hat{A}}\,\hat{B}\,\mathrm{e}^{-\mathrm{i}\hat{A}} = \hat{B} + \mathrm{i}[\hat{A},\hat{B}] + \frac{\mathrm{i}^2}{2!}\left[\hat{A},[\hat{A},\hat{B}]\right] + \cdots\;. \tag{8.24}$$

We obtain

$$\left[\hat{H}_0^{\mathrm{S}},\hat{a}_{\boldsymbol{k}}^{\mathrm{S}}\right] = \left[\sum_{\boldsymbol{k}'}\omega_{\boldsymbol{k}'}\hat{a}_{\boldsymbol{k}'}^{\mathrm{S}\dagger}\hat{a}_{\boldsymbol{k}'}^{\mathrm{S}},\hat{a}_{\boldsymbol{k}}^{\mathrm{S}}\right] = \omega_k \hat{a}_{\boldsymbol{k}}^{\mathrm{S}}\;,$$

$$\left[\hat{H}_0^{\mathrm{S}},\left[\hat{H}_0^{\mathrm{S}},\hat{a}_{\boldsymbol{k}}^{\mathrm{S}}\right]\right] = \omega_k^2 \hat{a}_{\boldsymbol{k}}^{\mathrm{S}}\;,\quad \ldots$$

and a similar result for $\hat{a}_{\boldsymbol{k}}^{\mathrm{S}\dagger}$. In this way (8.23) is transformed into (8.22).

The field operator in the interaction picture is constructed in such a way that it has *the same expansion (8.22) in terms of creation and annihilation operators as in the free theory* (when working in the Heisenberg picture in the latter case). As a consequence the canonical commutation relations are known for arbitrary times $t \neq t'$. For example, for scalar fields, according to Chap. 4.4, this takes the form

$$\left[\hat{\phi}^{\mathrm{I}}(\boldsymbol{x},t),\hat{\phi}^{\mathrm{I}\dagger}(\boldsymbol{x},t')\right] = \mathrm{i}\,\Delta(x-x')\;. \tag{8.25}$$

In the Heisenberg picture such a general expression for the field commutators is not known since the operators $\hat{\phi}^{\mathrm{H}}(\boldsymbol{x},t)$ have a complicated time dependence driven by the interaction. Note that the relation (8.9) for the transformation of commutators cannot be applied here since it is based on the assumption of equal time arguments.

8.3 The Time-Evolution Operator

Within the Dirac picture the effect of the interaction is of course hidden in the state vectors $|\alpha,t\rangle^{\mathrm{I}}$. Thus the Schrödinger equation (8.15) is the starting point for a perturbative treatment of the problem of interacting fields. For this we define the *time-evolution operator* (also known as the *Dyson operator*) $\hat{U}(t_1,t_0)$, which describes the connection between the state vectors at the times t_0 and t_1, i.e.,

$$|\alpha, t_1\rangle^{\mathrm{I}} = \hat{U}(t_1, t_0) |\alpha, t_0\rangle^{\mathrm{I}} . \tag{8.26}$$

With the help of (8.11b) and (8.4b) we can immediately write down a formal solution of this equation (with $\hat{H}_0 \equiv \hat{H}_0^{\mathrm{S}}$)

$$\begin{aligned}
|\alpha, t_1\rangle^{\mathrm{I}} &= \mathrm{e}^{\mathrm{i}\hat{H}_0 t_1} |\alpha, t_1\rangle^{\mathrm{S}} = \mathrm{e}^{\mathrm{i}\hat{H}_0 t_1} \mathrm{e}^{-\mathrm{i}\hat{H}(t_1-t_0)} |\alpha, t_0\rangle^{\mathrm{S}} \\
&= \mathrm{e}^{\mathrm{i}\hat{H}_0 t_1} \mathrm{e}^{-\mathrm{i}\hat{H}(t_1-t_0)} \mathrm{e}^{-\mathrm{i}\hat{H}_0 t_0} |\alpha, t_0\rangle^{\mathrm{I}} ,
\end{aligned} \tag{8.27}$$

and thus

$$\hat{U}(t_1, t_0) = \mathrm{e}^{\mathrm{i}\hat{H}_0 t_1} \mathrm{e}^{-\mathrm{i}\hat{H}(t_1-t_0)} \mathrm{e}^{-\mathrm{i}\hat{H}_0 t_0} . \tag{8.28}$$

Since the operators $\hat{H}_0$ and $\hat{H}$ in general do not commute, the ordering of the factors in (8.28) is very important.

The time-evolution operator satisfies a number of fundamental relations, which we will collect now. A trivial property is

$$\hat{U}(t_0, t_0) = 1\!\!1 . \tag{8.29}$$

Furthermore when two time translations are applied consecutively the following *group property* holds:

$$\hat{U}(t_2, t_1) \hat{U}(t_1, t_0) = \hat{U}(t_2, t_0) . \tag{8.30}$$

For the special case $t_2 = t_0$, this together with (8.29) gives an expression for the *inverse operator*

$$\hat{U}^{-1}(t_0, t_1) = \hat{U}(t_1, t_0) . \tag{8.31}$$

Finally $\hat{U}$ turns out to be a *unitary operator*

$$\hat{U}^\dagger(t_1, t_0) = \hat{U}^{-1}(t_1, t_0) , \tag{8.32}$$

which guarantees that the normalization of state vectors does not depend on time. Equation (8.32) can be derived from the hermitean property of the Hamiltonians $\hat{H}_0$ and $\hat{H}$, as we can prove with the help of the formal solution (8.28):

$$\begin{aligned}
\hat{U}^\dagger(t_1, t_0) &= \mathrm{e}^{\mathrm{i}\hat{H}_0^\dagger t_0} \mathrm{e}^{\mathrm{i}\hat{H}^\dagger(t_1-t_0)} \mathrm{e}^{-\mathrm{i}\hat{H}_0^\dagger t_1} = \left(\mathrm{e}^{\mathrm{i}\hat{H}_0 t_1} \mathrm{e}^{\mathrm{i}\hat{H}(t_0-t_1)} \mathrm{e}^{-\mathrm{i}\hat{H}_0 t_0} \right)^{-1} \\
&= \hat{U}^{-1}(t_1, t_0) .
\end{aligned} \tag{8.33}$$

We note that $\hat{U}(t,0) = \exp(\mathrm{i}\hat{H}_0^{\mathrm{S}} t)$ is just the transformation operator (8.12) linking the Heisenberg and the interaction pictures:

$$\hat{O}^{\mathrm{I}}(t) = \hat{U}(t,0) \hat{O}^{\mathrm{H}}(t) \hat{U}^{-1}(t,0) , \tag{8.34a}$$

$$|\alpha, t\rangle^{\mathrm{I}} = \hat{U}(t,0) |\alpha\rangle^{\mathrm{H}} . \tag{8.34b}$$

Since the operator $\hat{U}$ determines the time evolution of the state vector $|\alpha, t\rangle^{\mathrm{I}}$ it satisfies the differential equation (8.15),

$$\mathrm{i}\partial_t \hat{U}(t, t_0) = \hat{H}_1^{\mathrm{I}}(t) \hat{U}(t, t_0) , \tag{8.35}$$

with the boundary condition (8.29).

In quantum field theory as in ordinary quantum-mechanical scattering theory it proves advantageous to transform the differential equation (8.15) into an equivalent *integral equation*. Within the boundary condition (8.29), this integral equation reads

$$\hat{U}(t,t_0) = \mathbb{I} + (-\mathrm{i}) \int_{t_0}^{t} \mathrm{d}t' \, \hat{H}_1^I(t') \, \hat{U}(t',t_0) \,, \tag{8.36}$$

which is immediately verified by differentiating with respect to time.

Equation (8.36) is an integral equation of Volterra type (i.e., the independent variable enters as an integral boundary). Equations of this type can be solved by iteration under quite general conditions.

The process of successive re-insertion of the left-hand side of (8.36) leads to the *Neumann series*

$$\begin{aligned}
\hat{U}(t,t_0) &= \mathbb{I} + (-\mathrm{i}) \int_{t_0}^{t} \mathrm{d}t_1 \, \hat{H}_1^I(t_1) \\
&\quad + (-\mathrm{i})^2 \int_{t_0}^{t} \mathrm{d}t_1 \int_{t_0}^{t_1} \mathrm{d}t_2 \, \hat{H}_1^I(t_1)\hat{H}_1^I(t_2) \\
&\quad + \ldots \\
&\quad + (-\mathrm{i})^n \int_{t_0}^{t} \mathrm{d}t_1 \ldots \int_{t_0}^{t_{n-1}} \mathrm{d}t_n \, \hat{H}_1^I(t_1)\hat{H}_1^I(t_2)\cdots\hat{H}_1^I(t_n) \\
&\quad + \ldots \,.
\end{aligned} \tag{8.37}$$

From now on the index I will be dropped since all the results in this chapter will be based on the interaction picture.

The Neumann series consists of multiple integrals involving products of the interaction Hamiltonian $\hat{H}_1(t_i)$ taken at different times. We notice that the time arguments are sorted in descending order. The appearance of these mutually dependent upper boundaries makes these multiple integrals (8.37) quite difficult to handle. Fortunately, however, following an idea by Dyson[4] the integrations can be rewritten such that they all cover the full time interval $[t_0, t]$. To achieve this we need the *time-ordered product* (or Dyson product) as a tool. This construction has already been introduced (Sect. 4.5), and now we are in a position to understand its relevance. The operators in the product are put in the order of descending time argument:

$$T\big(\hat{H}_1(t_1)\,\hat{H}_1(t_2)\ldots\hat{H}_1(t_n)\big) = \hat{H}_1(t_{i_1})\,\hat{H}_1(t_{i_2})\ldots\hat{H}_1(t_{i_n}) \tag{8.38}$$
$$\text{where } t_{i_1} \geq t_{i_2} \geq \ldots \geq t_{i_n} \,.$$

The concept of a time ordering can be applied to all types of products of operators, in particular to field operators ($\hat{\phi}(x), \hat{\psi}(x), \hat{A}_\mu(x)$, etc,) and their canonically conjugate fields. There may arise problems if noncommuting operators taken at equal time arguments are involved. For a general definition one adds the rule that *a minus sign* arises for each interchange of fermion operators upon time ordering. Such sign factors to not arise in (8.38) since the Hamiltonian is of a scalar nature and therefore always consists of an even number of fermion operators (which involve half-integer spin).

Let us now investigate the second term in the series (8.37),

---

[4] F. Dyson, Phys. Rev. **75**, 486 and 1736 (1949).

$$\int_{t_0}^{t} dt_1 \int_{t_0}^{t_1} dt_2 \, \hat{H}_1(t_1) \hat{H}_1(t_2) = \int_{t_0}^{t} dt_2 \int_{t_2}^{t} dt_1 \, \hat{H}_1(t_1) \hat{H}_1(t_2) \quad (8.39a)$$

$$= \int_{t_0}^{t} dt_1 \int_{t_1}^{t} dt_2 \, \hat{H}_1(t_2) \hat{H}_1(t_1) \,. \quad (8.39b)$$

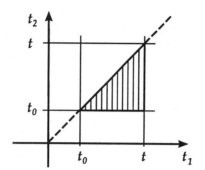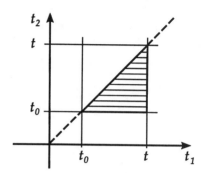

Fig. 8.1. The two equivalent ways to integrate over a triangular area, cf. (8.39a)

The integration extends over a triangular area in the $t_1 - t_2$ plane. As sketched in Fig. 8.1, when the boundaries are suitably chosen one may as well integrate first over the variable t_1 and then over t_2. In the second step of (8.39b), furthermore, the integration variables have been renamed, $t_1 \leftrightarrow t_2$. Adding up the two alternative but equivalent forms of integration, we arrive at

$$2 \int_{t_0}^{t} dt_1 \int_{t_0}^{t_1} dt_2 \, \hat{H}_1(t_1) \hat{H}_1(t_2)$$

$$= \int_{t_0}^{t} dt_1 \int_{t_0}^{t_1} dt_2 \, \hat{H}_1(t_1) \hat{H}_1(t_2) + \int_{t_0}^{t} dt_1 \int_{t_1}^{t} dt_2 \, \hat{H}_1(t_2) \hat{H}_1(t_1)$$

$$= \int_{t_0}^{t} dt_1 \int_{t_0}^{t} dt_2 \, T\!\left(\hat{H}_1(t_1) \hat{H}_1(t_2)\right) . \quad (8.40)$$

Thus the use of the time-ordering operator T has made it possible to extend both integrations to the full interval $[t_0, t]$. In fact, the same procedure can be extended to all the higher-order multiple integrals encountered in (8.37). In the nth order there are $n!$ possible permutations of the time arguments. Therefore (8.40) gets generalized into

$$n! \int_{t_0}^{t} dt_1 \ldots \int_{t_0}^{t_{n-1}} dt_n \, \hat{H}_1(t_1) \cdots \hat{H}_1(t_n)$$

$$= \int_{t_0}^{t} dt_1 \ldots \int_{t_0}^{t} dt_n \, T\!\left(\hat{H}_1(t_1) \cdots \hat{H}_1(t_n)\right) . \quad (8.41)$$

It is not difficult to prove this plausible result formally by employing mathematical induction.

Using (8.41) we find the *perturbation series for the time-evolution operator*

$$\hat{U}(t, t_0) = \sum_{n=0}^{\infty} \frac{1}{n!} (-\mathrm{i})^n \int_{t_0}^{t} dt_1 \ldots \int_{t_0}^{t} dt_n \, T\!\left(\hat{H}_1(t_1) \ldots \hat{H}_1(t_n)\right) . \quad (8.42)$$

We can easily verify that (8.42) solves the original differential equation (8.35). Taking the derivative with respect to t we get

$$i\partial_t \hat{U}(t,t_0)$$
$$= i\sum_{n=1}^{\infty} \frac{1}{n!}(-i)^n n \int_{t_0}^{t} dt_1 \ldots \int_{t_0}^{t} dt_{n-1} T\big(\hat{H}_1(t_1) \cdots \hat{H}_1(t_{n-1})\,\hat{H}_1(t)\big)$$
$$= \hat{H}_1(t) \sum_{n=1}^{\infty} \frac{1}{(n-1)!}(-i)^{n-1} \int_{t_0}^{t} dt_1 \ldots \int_{t_0}^{t} dt_{n-1} T\big(\hat{H}_1(t_1) \cdots \hat{H}_1(t_{n-1})\big)$$
$$= \hat{H}_1(t)\,\hat{U}(t,t_0)\,. \tag{8.43}$$

Here in the first step the symmetry of the integrand was used to obtain the factor n. Subsequently $\hat{H}_1(t)$ could be extracted from the T product since t is larger than all the other time arguments t_i.

We remark that it is possible to formally sum up the series (8.42), arriving at the *time-ordered exponential function*

$$\hat{U}(t,t_0) = T \exp\!\left(-i \int_{t_0}^{t} dt'\, \hat{H}_1(t')\right). \tag{8.44}$$

In essence, however, this is just a compact way of writing the series (8.42).

In any local field theory the Hamiltonian can be expressed as an integral over the Hamilton density $\hat{\mathcal{H}}_1(x)$ which consists of products of field operators and possibly their derivatives. Then (8.44) can be written as

$$\hat{U}(t,t_0) = T \exp\!\left(-i \int_{t_0}^{t} d^4x'\, \hat{\mathcal{H}}_1(x')\right). \tag{8.45}$$

The main use of the time-evolution operator lies in its application to scattering processes. Furthermore $\hat{U}(t,t_0)$ can be employed to calculate the *energy shift of a bound level* under the influence of an interaction. This is achieved by the Gell-Mann–Low theorem which we will discuss in Example 8.1.

8.4 The Scattering Matrix

The scattering matrix (S matrix) is a central concept in quantum field theory as well as in ordinary quantum mechanics. It describes the probability amplitude for a process in which the system makes a transition from an initial to a final state under the influence of an interaction. If one works in the interaction picture the time-evolution operator is the right tool to use to evaluate the scattering matrix. In constructing the S matrix we will first proceed in a naive way. Some problems with this approach will be revisited in the next chapter.

Let $|\Psi(t)\rangle$ denote the time-dependent state vector, which in the limit $t \to -\infty$ has evolved from the "free" initial state $|\Phi_i\rangle$:

$$\lim_{t\to-\infty} |\Psi(t)\rangle = |\Phi_i\rangle\,. \tag{8.46}$$

(Here we ignore the problem that because of vacuum fluctuations the interaction between quantum fields does not completely vanish in the asymptotical region). The S matrix element is defined as the projection of this state vector on the final state $\langle \Phi_f |$, which is defined by the quantum numbers f in the limit $t \to +\infty$:

$$\begin{aligned} S_{fi} &= \lim_{t \to +\infty} \langle \Phi_f | \Psi(t) \rangle \\ &= \langle \Phi_f | \hat{S} | \Phi_i \rangle \,. \end{aligned} \qquad (8.47)$$

Expressed in terms of the time-evolution operator this reads

$$S_{fi} = \lim_{t_2 \to +\infty} \lim_{t_1 \to -\infty} \langle \Phi_f | \hat{U}(t_2, t_1) | \Phi_i \rangle \,. \qquad (8.48)$$

Therefore the $\hat{S}$ operator defined in (8.47) is just given by

$$\hat{S} = \hat{U}(\infty, -\infty) \,. \qquad (8.49)$$

Therefore we already know the perturbation series for the S operator:

$$\begin{aligned} \hat{S} &= \sum_{n=0}^{\infty} \frac{1}{n!} (-\mathrm{i})^n \int_{-\infty}^{\infty} \mathrm{d}t_1 \ldots \int_{-\infty}^{\infty} \mathrm{d}t_n \, T\big(\hat{H}_1(t_1) \cdots \hat{H}_1(t_n)\big) & (8.50\mathrm{a}) \\ &= \sum_{n=0}^{\infty} \frac{1}{n!} (-\mathrm{i})^n \int \mathrm{d}^4 x_1 \ldots \mathrm{d}^4 x_n \, T\big(\hat{\mathcal{H}}_1(t_1) \cdots \hat{\mathcal{H}}_1(t_n)\big) \,. & (8.50\mathrm{b}) \end{aligned}$$

Being a special type of time-evolution operator, the S matrix is unitary according to (8.32):

$$\hat{S}^\dagger \hat{S} = \mathbb{1} \,. \qquad (8.51)$$

Knowing the S matrix element observable quantities, we can, for example, calculate scattering cross sections and decay rates by taking the square and performing some kinematical manipulations. The relevant transformation formula is quoted in Sect. 8.7.

EXAMPLE

8.1 The Gell-Mann–Low Theorem

In Sect. 8.3 we saw how the perturbation series for the time-evolution operator $\hat{U}(t,t_0)$ can be derived within the interaction picture. This provides us with a systematic method to construct the S matrix for scattering problems. In addition, the formalism also makes it possible to calculate the *energy shifts* of states in the discrete spectrum.

We start from a known discrete eigenstate $|\Phi\rangle$ of the unperturbed Hamiltonian

$$\hat{H}_0 |\Phi\rangle = E_0 |\Phi\rangle \,, \qquad (1)$$

and want to find the "corresponding" eigenstate $|\Psi\rangle$ of the full Hamiltonian

$$\hat{H} = \hat{H}_0 + \lambda \hat{H}_1 \,, \qquad (2)$$

which contains a perturbation operator $\hat{H}_1$:

$$\hat{H} |\Psi\rangle = E |\Psi\rangle \,. \qquad (3)$$

Example 8.1

The strength of the perturbation is parametrized by the *coupling constant* λ which will help us to keep track of the terms in the perturbation series.

If $\hat{H}_1$ is a *stationary perturbation*, for example an interaction potential in a many-particle system, the time-evolution operator at first sight might appear to be the wrong tool to use to describe the system. However, here the trick of *adiabatic switching* of the interaction comes to the rescue. The stationary problem is transformed into a time-dependent one by modifying the Hamiltonian according to

$$\hat{H}_\epsilon(t=0) = \hat{H} \quad , \quad \lim_{t\to\pm\infty} \hat{H}_\epsilon(t) = \hat{H}_0 . \tag{4}$$

An explicit prescription to achieve this may be formulated in terms of a switching function:

$$\hat{H}_\epsilon(t) = \hat{H}_0 + \lambda e^{-\epsilon|t|} \hat{H}_1 . \tag{5}$$

In this way the perturbation operator $\hat{H}_1$ at asymptotic times $t \to \pm\infty$ is dampened off exponentially. Of course physical observables should not depend on the details of this procedure since its implementation is arbitrary. Therefore it must be possible to take the limit $\epsilon \to 0$ at the end of the calculation, which turns out to be somewhat tricky.

In the *interaction picture* the equation of motion for the "adiabatic" state vector reads

$$i\partial_t \left|\Psi_\epsilon(t)\right\rangle^\mathrm{I} = \lambda e^{-\epsilon|t|}\, \hat{H}_1(t) \left|\Psi_\epsilon(t)\right\rangle^\mathrm{I} , \tag{6}$$

with the perturbation operator

$$\hat{H}_1(t) = e^{i\hat{H}_0 t}\, H_1(0)\, e^{-i\hat{H}_0 t} . \tag{7}$$

The corresponding *adiabatic time-evolution operator* satisfies the differential equation

$$i\partial_t \hat{U}_\epsilon(t,t_0) = \lambda e^{-\epsilon|t|}\, \hat{H}_1(t)\, \hat{U}_\epsilon(t,t_0) . \tag{8}$$

Because of (6), the state $\left|\Psi_\epsilon(t)\right\rangle^\mathrm{I}$ in the limit $t \to \infty$ obviously becomes constant. As an initial condition we demand that the state vector approaches the unperturbed state defined in (1):

$$\lim_{t\to-\infty}\left|\Psi_\epsilon(t)\right\rangle^\mathrm{I} = \left|\Phi\right\rangle . \tag{9}$$

At $t=0$, on the other hand, we know that $\hat{H}_\epsilon(0) = \hat{H}$. Thus one suspects that the state

$$\left|\Psi_\epsilon\right\rangle \equiv \left|\Psi_\epsilon(0)\right\rangle^\mathrm{I} = \hat{U}_\epsilon(0,-\infty)\left|\Phi\right\rangle \tag{10}$$

in some way is connected with the desired solution of the interacting Hamiltonian (3). Of course this can only hold true if the perturbation operator is switched on "sufficiently slowly", otherwise the switching-on will induce artificial dynamical excitations. Now it turns out that the limit of arbitrarily slow switching, $\lim_{\epsilon\to 0}\left|\Psi_\epsilon(0)\right\rangle^\mathrm{I}$, does not exist mathematically! The problem can be traced back to a *divergent phase factor* (see (26) below). Gell-Mann and

Example 8.1

Low[5] have shown how a meaningful result for the energy of the interacting system can be obtained despite this problem.

1. The state

$$|\Psi\rangle = \lim_{\epsilon \to 0} \frac{\hat{U}_\epsilon(0, -\infty)|\Phi\rangle}{\langle\Phi|\hat{U}_\epsilon(0, -\infty)|\Phi\rangle} \equiv \lim_{\epsilon \to 0} \frac{|\Psi_\epsilon\rangle}{\langle\Phi|\Psi_\epsilon\rangle} \qquad (11)$$

is an eigenstate of the full Hamiltonian, i.e.,

$$(H - E) \lim_{\epsilon \to 0} \frac{|\Psi_\epsilon\rangle}{\langle\Phi|\Psi_\epsilon\rangle} = 0 \,. \qquad (12)$$

This is true provided that the limit (11) exists and that the perturbation series of $|\Psi\rangle$ in powers of the coupling constant λ is well defined.

2. The *energy shift* of the state $|\Psi\rangle$ with respect to $|\Phi\rangle$ induced by the interaction can be calculated from the formula

$$\Delta E = E - E_0 = \lim_{\epsilon \to 0} i\epsilon\lambda \frac{\partial}{\partial \lambda} \ln \langle\Phi|\hat{U}_\epsilon(0, -\infty)|\Phi\rangle \,. \qquad (13)$$

The Gell-Mann–Low theorem does not guarantee the existence of $|\Psi\rangle$ defined in (11). However, *if* this state can be constructed then it solves the interacting problem.

To prove the assertions **1** and **2** we first apply the operator $\hat{H}_0 - E_0$ to the adiabatic state $|\Psi_\epsilon\rangle$. Using (1) we obtain

$$(\hat{H}_0 - E_0)|\Psi_\epsilon\rangle = (\hat{H}_0 - E_0)\hat{U}_\epsilon(0, -\infty)|\Phi\rangle = [\hat{H}_0, \hat{U}_\epsilon(0, -\infty)]|\Phi\rangle \,. \quad (14)$$

The commutator will be evaluated with the help of the perturbation series (8.37) for the time-evolution operator:

$$\begin{aligned}[\hat{H}_0, \hat{U}_\epsilon(0, -\infty)] &= \Bigg[\hat{H}_0, \sum_{n=0}^{\infty} \frac{(-i)^n}{n!}\lambda^n \int_{-\infty}^{0} dt_1 \ldots \int_{-\infty}^{0} dt_n \\ &\quad \times e^{+\epsilon(t_1+\ldots+t_n)} T(\hat{H}_1(t_1)\cdots\hat{H}_1(t_n))\Bigg] \,. \end{aligned} \qquad (15)$$

If the series is absolutely convergence $\hat{H}_0$ can be moved under the integral. For a fixed (but arbitrary) time ordering $t_{i_1} \geq t_{i_2} \geq \ldots \geq t_{i_n}$ we have

$$\begin{aligned} \left[\hat{H}_0, T(\hat{H}_1(t_1)\cdots\hat{H}_1(t_n))\right] &= \left[\hat{H}_0, \hat{H}_1(t_{i_1})\cdots\hat{H}_1(t_{i_n})\right] \\ &= \left[\hat{H}_0, \hat{H}_1(t_{i_1})\right]\hat{H}_1(t_{i_2})\cdots\hat{H}_1(t_{i_n}) + \ldots + \hat{H}_1(t_{i_1})\cdots\left[\hat{H}_0, \hat{H}_1(t_{i_n})\right] \\ &= (-i)\frac{\partial \hat{H}_1(t_i)}{\partial t_{i_1}}\hat{H}_1(t_{i_2})\cdots\hat{H}_1(t_{i_n}) + \ldots + \hat{H}_1(t_{i_1})\cdots(-i)\frac{\partial \hat{H}_1(t_{i_n})}{\partial t_{i_n}} \\ &= (-i)\sum_{i=1}^{n}\frac{\partial}{\partial t_i}\hat{H}_1(t_{i_1})\cdots\hat{H}_1(t_{i_n}) \\ &= (-i)\sum_{i=1}^{n}\frac{\partial}{\partial t_i}T(\hat{H}_1(t_1)\cdots\hat{H}_1(t_n)) \,. \end{aligned} \qquad (16)$$

---

[5] M. Gell-Mann and F. Low: Phys. Rev. **84**, 350 (1951).

Here the equation of motion for the perturbing Hamiltonian in the interaction picture was used:

$$i[\hat{H}_0, \hat{H}_1(t)] = \frac{\partial}{\partial t}\hat{H}_1(t) \,. \tag{17}$$

Equation (14) now reads

$$\begin{aligned}(\hat{H}_0 - E_0)|\Psi_\epsilon\rangle &= \sum_{n=0}^{\infty} \frac{(-\mathrm{i})^n}{n!} \lambda^n \int_{-\infty}^{0} dt_1 \cdots \int_{-\infty}^{0} dt_n \, \mathrm{e}^{+\epsilon(t_1+\ldots+t_n)} \\ &\quad \times (-\mathrm{i}) \sum_{i=1}^{n} \frac{\partial}{\partial t_i} T\big(\hat{H}_1(t_1)\cdots\hat{H}_1(t_n)\big)|\Phi\rangle \\ &= -\lambda \sum_{n=1}^{\infty} \frac{(-\mathrm{i})^{n-1}}{(n-1)!}\lambda^{n-1} \int_{-\infty}^{0} dt_1 \cdots \int_{-\infty}^{0} dt_n \, \mathrm{e}^{+\epsilon(t_1+\ldots+t_n)} \\ &\quad \times \frac{\partial}{\partial t_1} T\big(\hat{H}_1(t_1)\cdots\hat{H}_1(t_n)\big)|\Phi\rangle \end{aligned} \tag{18}$$

Because the integrand is symmetric under permutation of the time arguments all derivatives $\partial/\partial t_i$ are equivalent. Therefore it was admissible in (18) to replace the i summation by the factor n and to keep only a single differentiation $\partial/\partial t_1$. The remaining derivative can be eliminated through an integration by parts:

$$\begin{aligned}&\int_{-\infty}^{0} dt_1 \, \mathrm{e}^{\epsilon t_1} \frac{\partial}{\partial t_1} T\big(\hat{H}_1(t_1)\cdots\hat{H}_1(t_n)\big) \\ &= \mathrm{e}^{\epsilon t_1} T\big(\hat{H}_1(t_1)\cdots\hat{H}_1(t_n)\big)\Big|_{-\infty}^{0} - \int_{-\infty}^{0} dt_1 \left(\frac{\partial}{\partial t_1} \mathrm{e}^{\epsilon t_1}\right) T\big(\hat{H}_1(t_1)\cdots\hat{H}_1(t_n)\big) \\ &= \hat{H}_1(0) T\big(\hat{H}_1(t_2)\cdots\hat{H}_1(t_n)\big) - \epsilon \int_{-\infty}^{0} dt_1 \, \mathrm{e}^{\epsilon t_1} T\big(\hat{H}_1(t_1)\cdots\hat{H}_1(t_n)\big) \,. \end{aligned} \tag{19}$$

Therefore the right-hand side of (18) contains two contributions

$$\begin{aligned}&(\hat{H}_0 - E_0)|\Psi_\epsilon\rangle \\ &= -\lambda \hat{H}_1(0) \sum_{n=1}^{\infty} \frac{(-\mathrm{i})^{n-1}}{(n-1)!}\lambda^{n-1} \int_{-\infty}^{0} dt_2 \cdots \int_{-\infty}^{0} dt_n \, \mathrm{e}^{+\epsilon(t_2+\ldots+t_n)} \\ &\quad \times T\big(\hat{H}_1(t_2)\cdots\hat{H}_1(t_n)\big)|\Phi\rangle \\ &\quad + \epsilon\lambda \sum_{n=1}^{\infty} \frac{(-\mathrm{i})^{n-1}}{(n-1)!}\lambda^{n-1} \int_{-\infty}^{0} dt_1 \cdots \int_{-\infty}^{0} dt_n \, \mathrm{e}^{+\epsilon(t_1+\ldots+t_n)} T\big(\hat{H}_1(t_1)\cdots\hat{H}_1(t_n)\big)|\Phi\rangle \\ &= -\lambda \hat{H}_1(0)\, \hat{U}_\epsilon(0,-\infty)|\Phi\rangle \\ &\quad + \epsilon\lambda\mathrm{i}\frac{\partial}{\partial\lambda} \sum_{n=0}^{\infty} \frac{(-\mathrm{i})^n}{n!}\lambda^n \int_{-\infty}^{0} dt_1 \cdots \int_{-\infty}^{0} dt_n \, \mathrm{e}^{+\epsilon(t_1+\ldots+t_n)} T\big(\hat{H}_1(t_1)\cdots\hat{H}_1(t_n)\big)|\Phi\rangle \\ &\quad -\lambda\hat{H}_1(0)|\Psi_\epsilon\rangle + \mathrm{i}\epsilon\lambda\frac{\partial}{\partial\lambda}|\Psi_\epsilon\rangle \,. \end{aligned} \tag{20}$$

In the first term the summation index has been renamed, $n-1 \to n$. in the second term the series for the time-evolution operator $\hat{U}_\epsilon(0,-\infty)$ was regained

Example 8.1

by differentiating with respect to the parameter λ. Use was made of the fact that the interaction term is linear in the coupling constant λ.

Expressed in terms of the full Hamiltonian, (20) reads

$$(\hat{H} - E_0)|\Psi_\epsilon\rangle = i\epsilon\lambda \frac{\partial}{\partial \lambda}|\Psi_\epsilon\rangle . \tag{21}$$

If we assumed a regular limiting process $\lim_{\epsilon \to 0} |\Psi_\epsilon\rangle$ the right-hand side of (21) would vanish because of the extra factor ϵ. This would lead to the conclusion that the interacting state has the same energy E_0 as the "free" state, which is certainly wrong.

The correct way to proceed is given in (11). To arrive at the state vector employed in this equation we divide (21) by the scalar product $\langle\Phi|\Psi_\epsilon\rangle$ and perform a clever transformation on the right-hand side:

$$\begin{aligned}(\hat{H} - E_0)\frac{|\Psi_\epsilon\rangle}{\langle\Phi|\Psi_\epsilon\rangle} &= i\epsilon\lambda \frac{\frac{\partial}{\partial \lambda}|\Psi_\epsilon\rangle}{\langle\Phi|\Psi_\epsilon\rangle} \\ &= i\epsilon\lambda\left[\frac{\partial}{\partial \lambda}\frac{|\Psi_\epsilon\rangle}{\langle\Phi|\Psi_\epsilon\rangle} + \frac{\left(\frac{\partial}{\partial \lambda}\langle\Phi|\Psi_\epsilon\rangle\right)|\Psi_\epsilon\rangle}{\left(\langle\Phi|\Psi_\epsilon\rangle\right)^2}\right] \\ &= i\epsilon\lambda\left[\frac{\partial}{\partial \lambda}\frac{|\Psi_\epsilon\rangle}{\langle\Phi|\Psi_\epsilon\rangle} + \frac{|\Psi_\epsilon\rangle}{\langle\Phi|\Psi_\epsilon\rangle}\frac{\partial}{\partial \lambda}\ln\langle\Phi|\Psi_\epsilon\rangle\right]\end{aligned} \tag{22}$$

or

$$\left(\hat{H} - E_0 - i\epsilon\lambda\frac{\partial}{\partial \lambda}\ln\langle\Phi|\Psi_\epsilon\rangle\right)\frac{|\Psi_\epsilon\rangle}{\langle\Phi|\Psi_\epsilon\rangle} = i\epsilon\lambda\frac{\partial}{\partial \lambda}\frac{|\Psi_\epsilon\rangle}{\langle\Phi|\Psi_\epsilon\rangle} . \tag{23}$$

As a prerequisite for the Gell-Mann–Low theorem to be applicable, we have assumed that $\lim_{\epsilon \to 0} |\Psi_\epsilon\rangle/\langle\Phi|\Psi_\epsilon\rangle$ is well defined and finite in each order of the perturbation series. This property should also hold for the derivative with respect to the parameter λ. Then the right-hand side of (23) will vanish in the limit $\epsilon \to 0$ since it contains a factor ϵ. Thus

$$(\hat{H} - E)\lim_{\epsilon \to 0}\frac{|\Psi_\epsilon\rangle}{\langle\Phi|\Psi_\epsilon\rangle} = 0 , \tag{24}$$

and we have verified that the state vector $|\Psi\rangle$ (provided that it exists!) defined in (11) is indeed an eigenstate of the interacting system.

The *energy shift* ΔE between the perturbed and the unperturbed states can be read off from (23):

$$E = E_0 + \Delta E = E_0 + \lim_{\epsilon \to 0} i\epsilon\lambda\frac{\partial}{\partial \lambda}\ln\langle\Phi|\Psi_\epsilon\rangle . \tag{25}$$

This energy shift turns out to be the matrix element of the perturbation operator $\lambda\hat{H}_1$, modified by a normalization factor, as can be seen from the projection of (24) on $\langle\Phi|$ using $\langle\Phi|\hat{H}_0 = \langle\Phi|E_0$.

An inspection of the result (25) for the energy shift reveals the presence of a *divergent phase factor*, which was mentioned before. Integrating over λ for small values of ϵ leads to

Example 8.1

$$\frac{\partial}{\partial\lambda}\ln\langle\Phi|\Psi_\epsilon\rangle \simeq \frac{\Delta E(\lambda)}{i\epsilon\lambda},$$

$$\ln\langle\Phi|\Psi_\epsilon\rangle \simeq -i\frac{1}{\epsilon}\int^\lambda d\lambda'\,\frac{\Delta E(\lambda')}{\lambda'} \equiv -i\frac{1}{\epsilon}f(\lambda),$$

$$\langle\Phi|\Psi_\epsilon\rangle \simeq \exp\!\left(-i\frac{f(\lambda)}{\epsilon}\right). \tag{26}$$

The state vector $|\Psi_\epsilon\rangle$ thus has an essential singularity at $\epsilon = 0$, characterized by a phase factor that oscillates "infinitely fast".

We further remark that the derivation of $|\Psi\rangle$ and ΔE could equally well have been based on the time interval $0\ldots+\infty$. The formulae (11) and (13) in this case have to be replaced by

$$|\Psi'\rangle = \lim_{\epsilon\to 0}\frac{\hat{U}_\epsilon(0,\infty)|\Phi\rangle}{\langle\Phi|\hat{U}_\epsilon(0,\infty)|\Phi\rangle}, \tag{27}$$

$$\Delta E = \lim_{\epsilon\to 0}(-i\epsilon\lambda)\frac{\partial}{\partial\lambda}\ln\langle\Phi|\hat{U}_\epsilon(0,\infty)|\Phi\rangle. \tag{28}$$

The vectors $|\Psi\rangle$ and $|\Psi'\rangle$ describe the same physical state. In principle they could differ by a phase factor but this is ruled out by the normalization condition $\langle\Phi|\Psi\rangle = \langle\Phi|\Psi'\rangle = 1$ and we conclude that $|\Psi\rangle = |\Psi'\rangle$. This conclusion, however, can no longer be drawn if there is an energetic *degeneracy*. In particular it is not valid for states in the continuous spectrum (scattering states).

As a third alternative the energy shift ΔE can also be written in a symmetrical fashion:

$$\Delta E = \lim_{\epsilon\to 0}\frac{1}{2}i\epsilon\lambda\frac{\partial}{\partial\lambda}\ln\langle\Phi|\hat{U}_\epsilon(\infty,\infty)|\Phi\rangle. \tag{29}$$

8.5 Wick's Theorem

Based on the interaction picture we have deduced a series expansion of the $\hat{S}$ operator in powers of the interaction Hamiltonian $\hat{H}_1$. Thus in principle perturbation theory can be carried out at arbitrarily high orders. When attempting to do so one is confronted with the technical problem to evaluate time-ordered products of the type $T(\hat{\mathcal{H}}_1(x_1)\ldots\hat{\mathcal{H}}_1(x_n))$. The interaction Hamiltonian $\hat{\mathcal{H}}_1(x)$ will consist of products of field operators describing the interacting quantum fields. As one can imagine, the evaluation of higher-order T products can be quite laborious. In this section we will develop tools which will allow us to approach this task in an economical way.

For practical application one has to evaluate matrix elements of the $\hat{S}$ operator taken between initial and final states, which describe certain given multi-particle configurations. As an example let us take a look at the *scattering of two particles* described by the momentum balance $\hat{k}_1, \hat{k}_2 \to \hat{k}_3, \hat{k}_4$. Call $\hat{a}_k^\dagger$ and $\hat{a}_k$ the creation and annihilation operators of the corresponding field quanta, and the S matrix element reads

$$S_{fi} = \langle f|\hat{S}|i\rangle = \langle 0|\hat{a}_{k_4}\hat{a}_{k_3}\hat{S}\hat{a}^\dagger_{k_1}\hat{a}^\dagger_{k_2}|0\rangle. \tag{8.52}$$

The higher-order contributions to the operator $\hat{S}$ are complicated superpositions of the $\hat{a}_{\boldsymbol{k}}$ and $\hat{a}_{\boldsymbol{k}}^\dagger$ and possibly further creation and annihilation operators ($\hat{b}_{\boldsymbol{p}}, \hat{b}_{\boldsymbol{p}}^\dagger$, etc.) describing further quantum fields. To evaluate (8.52) one has to pick out those contributions that have the matching structure $f(\boldsymbol{k}_1, \boldsymbol{k}_2, \boldsymbol{k}_3, \boldsymbol{k}_4)\,\hat{a}_{\boldsymbol{k}_4}^\dagger \hat{a}_{\boldsymbol{k}_3}^\dagger \hat{a}_{\boldsymbol{k}_1} \hat{a}_{\boldsymbol{k}_2}$ which is the only way to produce a nonvanishing matrix element between the given initial and final states. In principle this is not a difficult calculation: by "manually" applying the canonical commutation relations one can reorder the operators (moving the $\hat{a}^\dagger$ to the left and the $\hat{a}$ to the right) until one is left with a simple result. This can be a tedious and rather messy task, especially at higher orders of perturbation theory. Fortunately there is a very elegant and efficient tool that can be used to systematically evaluate any complicated time-ordered product. This is *Wick's theorem*, which we will derive in the following.

Let us first recall the definition of the *normal product* first introduced Chap. 4. A field operator is split into a "positive-frequency" part (i.e., $\hat{\phi}^{(+)} \simeq \exp(-\mathrm{i}\omega_p t)$) and a "negative-frequency" part ($\hat{\phi}^{(-)} \simeq \exp(+\mathrm{i}\omega_p t)$). The first (second) term consists of annihilation (creation) operators. For a scalar field operator $\hat{\phi}(x)$ the decomposition reads

$$\hat{\phi}(x) = \hat{\phi}^{(+)}(x) + \hat{\phi}^{(-)}(x) \;. \tag{8.53}$$

The *normal product* of two *bosonic* field operators is defined as

$$\begin{aligned}:\hat{\phi}(x)\hat{\phi}(y): \;=\;& \hat{\phi}^{(-)}(x)\hat{\phi}^{(-)}(y) + \hat{\phi}^{(-)}(x)\hat{\phi}^{(+)}(y) \\ &+ \hat{\phi}^{(+)}(x)\hat{\phi}^{(+)}(y) + \hat{\phi}^{(-)}(y)\hat{\phi}^{(+)}(x) \;.\end{aligned} \tag{8.54}$$

In the last term the contribution of the creation operator $\hat{\phi}^{(-)}(y)$ has been moved so that it stand to the left of the annihilation part $\hat{\phi}^{(+)}(x)$. A similar prescription also applies to *fermionic* field operators, which we denote by $\hat{\psi}(x)$:

$$\begin{aligned}:\hat{\psi}(x)\hat{\psi}(y): \;=\;& \hat{\psi}^{(-)}(x)\hat{\psi}^{(-)}(y) + \hat{\psi}^{(-)}(x)\hat{\psi}^{(+)}(y) \\ &+ \hat{\psi}^{(+)}(x)\hat{\psi}^{(+)}(y) - \hat{\psi}^{(-)}(y)\hat{\psi}^{(+)}(x) \;.\end{aligned} \tag{8.55}$$

Here in the last term an *anticommutation* was performed, $\hat{\psi}^{(+)}(x)\hat{\psi}^{(-)}(y) \to -\hat{\psi}^{(-)}(y)\hat{\psi}^{(+)}(x)$.

It is obvious how these rules can be extended to normal products involving $n > 2$ operators: the terms obtained after a decomposition into creation and annihilation parts are reordered in such a way that all of the creators stand to the left of the annihilators. If fermionic operators are involved a sign factor arises, which is given by the parity of the permutation of the fermion operators. The sign is equal to -1 if an odd number of fermionic permutations is needed to achieve normal ordering and equal to $+1$ if an even number is needed. Exchanges involving bosonic operators according to this definition do not affect the sign.

In the following, the generic name $\hat{\Phi}_A(x)$ will be employed for both bosonic and fermionic operators. We note the following important property of normal products: except for a possible sign factor the order of the operators within the argument of a normal product does not matter. The factors can be permutated at will without changing more than the sign. This, of course, is an obvious consequence of the fact that the normal product enforces a certain ordering of the operators, regardless of their initial positions. As an example, for the normal product of two generic field operators we find.

$$\begin{aligned}
:\hat{\Phi}_A \hat{\Phi}_B: &= \hat{\Phi}_A^{(-)}\hat{\Phi}_B^{(-)} + \hat{\Phi}_A^{(-)}\hat{\Phi}_B^{(+)} + \hat{\Phi}_A^{(+)}\hat{\Phi}_B^{(+)} + \epsilon_{AB}\hat{\Phi}_B^{(-)}\hat{\Phi}_A^{(+)} \\
&= \epsilon_{AB}\hat{\Phi}_B^{(-)}\hat{\Phi}_A^{(-)} + \hat{\Phi}_A^{(-)}\hat{\Phi}_B^{(+)} + \epsilon_{AB}\hat{\Phi}_B^{(+)}\hat{\Phi}_A^{(+)} + \epsilon_{AB}\hat{\Phi}_B^{(-)}\hat{\Phi}_A^{(+)} \\
&= \epsilon_{AB}\left(\hat{\Phi}_B^{(-)}\hat{\Phi}_A^{(-)} + \epsilon_{AB}\hat{\Phi}_A^{(-)}\hat{\Phi}_B^{(+)} + \hat{\Phi}_B^{(+)}\hat{\Phi}_A^{(+)} + \hat{\Phi}_B^{(-)}\hat{\Phi}_A^{(+)}\right) \\
&= \epsilon_{AB}:\hat{\Phi}_B\hat{\Phi}_A:\ .
\end{aligned} \qquad (8.56)$$

where the sign is $\epsilon_{AB} = +1\,(-1)$ if $\hat{\Phi}_A$ and $\hat{\Phi}_B$ are bosonic (fermionic) operators. This is immediately generalized to normal products involving more than two factors. We note that the *time-ordered product* is also invariant under permutations, except for a sign factor. For example

$$T\bigl(\hat{\Phi}_A(x)\,\hat{\Phi}_B(y)\bigr) = \epsilon_{AB}\,T\bigl(\hat{\Phi}_B(y)\,\hat{\Phi}_A(x)\bigr)\ . \qquad (8.57)$$

This has the same reason as in the case of the normal product: the prescription of time ordering enforces a certain ordering of the operators.

Now we will derive a formula for the general evaluation of the time-ordered product of an arbitrary number of operators. Expressions of this kind are the central ingredient of the perturbation series. Our aim is to find a *decomposition of the T product into normal products*. Before investigating the general formula we will illustrate in detail the special cases of the T product of two and of three operators. From these special cases we will deduce the general rule.

For a start let us evaluate the time-ordered product $T\bigl(\hat{\Phi}_A(x_1)\,\hat{\Phi}_B(x_2)\bigr)$ of *two* field operators, expressed in terms of their normal product. First we assume that the time arguments are ordered such that $t_1 > t_2$, allowing us to write

$$\begin{aligned}
T\bigl(\hat{\Phi}_A(x_1)\,\hat{\Phi}_B(x_2)\bigr) &= \hat{\Phi}_A(x_1)\,\hat{\Phi}_B(x_2) \\
&= \hat{\Phi}_A^{(+)}(x_1)\hat{\Phi}_B^{(+)}(x_2) + \hat{\Phi}_A^{(-)}(x_1)\hat{\Phi}_B^{(-)}(x_2) \\
&\quad + \hat{\Phi}_A^{(-)}(x_1)\hat{\Phi}_B^{(+)}(x_2) + \hat{\Phi}_A^{(+)}(x_1)\hat{\Phi}_B^{(-)}(x_2)\ .
\end{aligned} \qquad (8.58)$$

Except for the last term the products are already normal ordered. The last term can be written as

$$\hat{\Phi}_A^{(+)}(x_1)\hat{\Phi}_B^{(-)}(x_2) = \epsilon_{AB}\hat{\Phi}_B^{(-)}(x_2)\hat{\Phi}_A^{(+)}(x_1) + \bigl[\hat{\Phi}_A^{(+)}(x_1),\hat{\Phi}_B^{(-)}(x_2)\bigr]_{\mp}\ . \qquad (8.59)$$

Now we make the essential observation that *(anti-)commutators of free-field operators are not operators themselves but are simple c numbers*. In Sect. 4.4 and in the Exercises 5.5, 6.4, and 7.5 this has been demonstrated explicitly for the various types of fields. The statement also remains true for the fields operators expressed in the interaction picture since their dynamics by construction is equal to that of free-field operators! For the problem at hand this implies that the (anti-)commutator in (8.59) can be replaced by its vacuum expectation value since it is just a c number. Taking into account $\hat{\Phi}_B^{(+)}|0\rangle = \langle 0|\hat{\Phi}_A^{(-)} = 0$, we can rewrite

$$\begin{aligned}
&\bigl[\hat{\Phi}_A^{(+)}(x_1),\hat{\Phi}_B^{(-)}(x_2)\bigr]_{\mp} \\
&\quad = \langle 0|\bigl[\hat{\Phi}_A^{(+)}(x_1),\hat{\Phi}_B^{(-)}(x_2)\bigr]_{\mp}|0\rangle = \langle 0|\hat{\Phi}_A^{(+)}(x_1)\hat{\Phi}_B^{(-)}(x_2)|0\rangle
\end{aligned}$$

$$= \langle 0|\hat{\Phi}_A(x_1)\hat{\Phi}_B(x_2)|0\rangle = \langle 0|T(\hat{\Phi}_A(x_1)\hat{\Phi}_B(x_2))|0\rangle \,. \tag{8.60}$$

The last step is based on the condition $t_1 > t_2$. Therefore (8.58) reads, with the help of (8.59) and (8.60),

$$T(\hat{\Phi}_A(x_1)\hat{\Phi}_B(x_2)) = :\hat{\Phi}_A(x_1)\hat{\Phi}_B(x_2): + \langle 0|T(\hat{\Phi}_A(x_1)\hat{\Phi}_B(x_2))|0\rangle \,. \tag{8.61}$$

This equation has been derived for $t_1 > t_2$. However, it also holds true in the opposite case $t_2 > t_1$ and thus is valid in general. This immediately follows from the previous observation that both the normal product and the T product are invariant under permutation of the factors, up to a sign factor which is the same in both cases.

To gain complete confidence in this result we explicitly repeat the calculation leading to (8.61) under the assumption $t_1 < t_2$:

$$\begin{aligned}
T(\hat{\Phi}_A(x_1)\hat{\Phi}_B(x_2)) &= \epsilon_{AB}\hat{\Phi}_B(x_2)\hat{\Phi}_A(x_1) = \dots \\
&= \epsilon_{AB}:\hat{\Phi}_B(x_2)\hat{\Phi}_A(x_1): + \epsilon_{AB}\langle 0|\hat{\Phi}_B(x_2)\hat{\Phi}_A(x_1)|0\rangle \\
&= :\hat{\Phi}_A(x_1)\hat{\Phi}_B(x_2): + \langle 0|T(\hat{\Phi}_A(x_1)\hat{\Phi}_B(x_2))|0\rangle \,,
\end{aligned} \tag{8.62}$$

as expected.

It is quite obvious that (8.61) must be true: both products $T(\cdots)$ and $:\cdots:$ according to their construction can differ only by a commutator which is a c number. Since the vacuum expectation value of a normal product vanishes the value of this number is given by $\langle 0|T(\hat{\Phi}_A(x_1)\hat{\Phi}_B(x_2))|0\rangle$, which is what (8.61) tells us. Since the *vacuum expectation value of a T product of two operators* occurs frequently it is given a name of its own and a compact notation. One defines the time-ordered contraction (or simply *the contraction*) of two operators as

$$\underline{\hat{\Phi}_A(x_1)\hat{\Phi}_B(x_2)} := \langle 0|T(\hat{\Phi}_A(x_1)\hat{\Phi}_B(x_2))|0\rangle \,. \tag{8.63}$$

Contractions can also appear in more complicated expressions involving operators, for example within the argument of a normal product. An example, written in shorthand notation, might read

$$:\hat{A}\hat{B}\underline{\hat{C}\hat{D}}\hat{E}\hat{F}\dots\hat{K}\hat{L}\hat{M}\dots: = \epsilon_P :\hat{A}\hat{B}\hat{F}\dots\hat{K}\hat{M}\dots:\underline{\hat{C}\hat{E}}\,\underline{\hat{D}\hat{L}} \,. \tag{8.64}$$

Here $\epsilon_P = \pm 1$ again is the parity of the permutation of fermionic operators. Of course nonvanishing contractions can exist only for operators of the same type since only in this case can one have a "balanced" combination of creators and annihilators in the vacuum expectation value. Later we will give explicit expressions for the contractions of various quantum fields.

Using the definition (8.63), we can write the result (8.61) as

$$T(\hat{\Phi}_A(x_1)\hat{\Phi}_B(x_2)) = :\hat{\Phi}_A(x_1)\hat{\Phi}_B(x_2): + :\underline{\hat{\Phi}_A(x_1)\hat{\Phi}_B(x_2)}: \,. \tag{8.65}$$

The normal product in the second term could be added without doing harm since this term is a c number.

The same procedure which lead us to (8.65) can be easily extended to the *product of three operators*. We first assume the time ordering $t_1, t_2 > t_3$. Then we find

$$T\big(\hat{\Phi}_A(x_1)\hat{\Phi}_B(x_2)\hat{\Phi}_C(x_3)\big) = T\big(\hat{\Phi}_A(x_1)\hat{\Phi}_B(x_2)\big)\hat{\Phi}_C(x_3)$$
$$= \;:\hat{\Phi}_A(x_1)\hat{\Phi}_B(x_2):\hat{\Phi}_C(x_3) + \underline{\hat{\Phi}_A(x_1)\hat{\Phi}_B(x_2)}\,\hat{\Phi}_C(x_3)\;. \qquad (8.66)$$

Only the factor $\hat{\Phi}_C^{(-)}(x_3)$ in the first term presents a problem since it violates normal ordering. We have (dropping the arguments x_i for simplicity)
$$:\hat{\Phi}_A\hat{\Phi}_B:\hat{\Phi}_C = \Big(\hat{\Phi}_A^{(-)}\hat{\Phi}_B^{(-)} + \hat{\Phi}_A^{(-)}\hat{\Phi}_B^{(+)} + \hat{\Phi}_A^{(+)}\hat{\Phi}_B^{(+)} + \epsilon_{AB}\hat{\Phi}_B^{(-)}\hat{\Phi}_A^{(+)}\Big)$$
$$\times \Big(\hat{\Phi}_C^{(-)} + \hat{\Phi}_C^{(+)}\Big)\;. \qquad (8.67)$$

In order to turn this into a normal product the factor $\hat{\Phi}_C^{(-)}$ has to be moved to the left of $\hat{\Phi}_A^{(+)}$ and $\hat{\Phi}_B^{(+)}$. The reordering leads to
$$\hat{\Phi}_A^{(-)}\hat{\Phi}_B^{(+)}\hat{\Phi}_C^{(-)} = \epsilon_{BC}\,\hat{\Phi}_A^{(-)}\hat{\Phi}_C^{(-)}\hat{\Phi}_B^{(+)} + \hat{\Phi}_A^{(-)}\big[\hat{\Phi}_B^{(+)},\hat{\Phi}_C^{(-)}\big]_{\mp}\;, \qquad (8.68)$$
and similarly
$$\hat{\Phi}_B^{(-)}\hat{\Phi}_A^{(+)}\hat{\Phi}_C^{(-)} = \epsilon_{AC}\,\hat{\Phi}_B^{(-)}\hat{\Phi}_C^{(-)}\hat{\Phi}_A^{(+)} + \hat{\Phi}_B^{(-)}\big[\hat{\Phi}_A^{(+)},\hat{\Phi}_C^{(-)}\big]_{\mp}\;, \qquad (8.69)$$
as well as
$$\hat{\Phi}_A^{(+)}\hat{\Phi}_B^{(+)}\hat{\Phi}_C^{(-)} = \hat{\Phi}_A^{(+)}\Big(\epsilon_{BC}\,\hat{\Phi}_C^{(-)}\hat{\Phi}_B^{(+)} + \big[\hat{\Phi}_B^{(+)},\hat{\Phi}_C^{(-)}\big]_{\mp}\Big)$$
$$= \epsilon_{BC}\,\epsilon_{AC}\,\hat{\Phi}_C^{(-)}\hat{\Phi}_A^{(+)}\hat{\Phi}_B^{(+)} + \epsilon_{BC}\big[\hat{\Phi}_A^{(+)},\hat{\Phi}_C^{(-)}\big]_{\mp}\hat{\Phi}_B^{(+)}$$
$$+ \hat{\Phi}_A^{(+)}\big[\hat{\Phi}_B^{(+)},\hat{\Phi}_C^{(-)}\big]_{\mp}\;. \qquad (8.70)$$

The four commutators again are pure c numbers. They can be combined in pairs with the result
$$:\hat{\Phi}_A\hat{\Phi}_B:\hat{\Phi}_C = \;:\hat{\Phi}_A\hat{\Phi}_B\hat{\Phi}_C: + \hat{\Phi}_A\big[\hat{\Phi}_B^{(+)},\hat{\Phi}_C^{(-)}\big]_{\mp} + \epsilon_{AB}\,\hat{\Phi}_B\big[\hat{\Phi}_A^{(+)},\hat{\Phi}_C^{(-)}\big]_{\mp}$$
$$= \;:\hat{\Phi}_A\hat{\Phi}_B\hat{\Phi}_C: + \hat{\Phi}_A\langle 0|T(\hat{\Phi}_B\hat{\Phi}_C)|0\rangle$$
$$+\epsilon_{AB}\,\hat{\Phi}_B\langle 0|T(\hat{\Phi}_A\hat{\Phi}_C)|0\rangle\;. \qquad (8.71)$$

In the first step ϵ_{BC} could be replaced by ϵ_{AB}; $\hat{\Phi}_A$ and $\hat{\Phi}_C$ have to be operators of the same type since otherwise their (anti-)commutator would vanish. For the transformation in (8.71), again (8.60) was employed. The result (8.71) has been derived under the assumption $t_1, t_2 > t_3$. However, since all terms in (8.71) are invarint under permutation of $\hat{\Phi}_A(x_1)$, $\hat{\Phi}_B(x_2)$, and $\hat{\Phi}_C(x_3)$ (up to a common sign factor) the result again is valid for arbitrary time orderings. Employing the contraction defined in (8.63) and using the rule (8.64), we write (8.66) as
$$T\big(\hat{\Phi}_A(x_1)\hat{\Phi}_B(x_2)\hat{\Phi}_C(x_3)\big) = \;:\hat{\Phi}_A(x_1)\hat{\Phi}_B(x_2)\hat{\Phi}_C(x_3):$$
$$+ :\underline{\hat{\Phi}_A(x_1)\hat{\Phi}_B(x_2)}\hat{\Phi}_C(x_3):$$
$$+ :\underline{\hat{\Phi}_A(x_1)}\hat{\Phi}_B(x_2)\underline{\hat{\Phi}_C(x_3)}:$$
$$+ :\hat{\Phi}_A(x_1)\underline{\hat{\Phi}_B(x_2)\hat{\Phi}_C(x_3)}:\;. \qquad (8.72)$$

As in the case of two operators, see (8.65), we have been able to express the time-ordered product of three operators in terms of a sum of normal products.

This result can be generalized to products of arbitrary complexity. This is the essence of *Wick's theorem*:[6]

- The *time-ordered product* of a set of operators can be decomposed into the sum of the corresponding *contracted normal products*. All contractions of pairs of operators that possibly can arise enter this sum.

Calling the operators $\hat{A}, \hat{B}, \ldots$ for simplicity, we can write Wick's theorem in condensed form as follows

$$T(\hat{A}\hat{B}\hat{C}\ldots\hat{X}\hat{Y}\hat{Z}) = \;:\hat{A}\hat{B}\hat{C}\ldots\hat{X}\hat{Y}\hat{Z}:$$
$$+ :\hat{A}\underline{\hat{B}}\hat{C}\ldots\hat{X}\hat{Y}\underline{\hat{Z}}: + :\hat{A}\underline{\hat{B}}\hat{C}\ldots\underline{\hat{X}}\hat{Y}\hat{Z}: + \ldots + :\hat{A}\hat{B}\hat{C}\ldots\hat{X}\underline{\hat{Y}\hat{Z}}:$$
$$+ :\underline{\hat{A}\hat{B}}\underline{\hat{C}\hat{D}}\ldots\hat{X}\hat{Y}\hat{Z}: + :\underline{\hat{A}\hat{B}\hat{C}\hat{D}}\ldots\hat{X}\hat{Y}\hat{Z}: + \ldots + :\hat{A}\hat{B}\hat{C}\ldots\underline{\hat{W}}\underline{\hat{X}}\underline{\hat{Y}}\underline{\hat{Z}}:$$

+ sum over triply contracted terms

+ higher contractions . $\hspace{5cm}$ (8.73)

We immediately recognize that our earlier results (8.65) and (8.72) are special cases of (8.73). The number of terms in the sum can become very large, although it will always be finite, of course. The general proof of Wick's theorem is the subject of Exercise 8.2.

Before applying (8.73) to special examples let us take a further look at the contractions of field operators, (8.63), which play such an important role in Wick's theorem. In anticipation of things to come we have already calculated the vacuum expectation values of time-ordered products of various free-field operator pairs. Since we are working in the interaction picture these results keep their validity. Let us recapitulate the results.

For a *scalar field* $\hat{\phi}(x)$ we found in Sect. 4.5 that

$$\overline{\hat{\phi}(x)\hat{\phi}^\dagger(y)} = \langle 0|T(\hat{\phi}(x)\hat{\phi}^\dagger(y))|0\rangle = \mathrm{i}\,\Delta_\mathrm{F}(x-y)\;. \tag{8.74}$$

Thus the contraction of $\hat{\phi}(x)$ and $\hat{\phi}^\dagger(y)$ is identical to the *Feynman propagator* for spin-0 particles. Its representation in momentum space is

$$\Delta_\mathrm{F}(x-y) = \int \frac{\mathrm{d}^4p}{(2\pi)^4} \frac{\mathrm{e}^{-\mathrm{i}p\cdot(x-y)}}{p^2 - m^2 + \mathrm{i}\epsilon}\;. \tag{8.75}$$

Similarly for a *Dirac field* (see Sect. 5.4) we obtained

$$\overline{\hat{\psi}_\alpha(x)\hat{\bar{\psi}}_\beta(y)} = \langle 0|T(\hat{\psi}_\alpha(x)\hat{\bar{\psi}}_\beta(y))|0\rangle = \mathrm{i}\,S_\mathrm{F}(x-y)\;, \tag{8.76}$$

where S_F is the Feynman propagator for spin-$\frac{1}{2}$ particles

$$S_\mathrm{F}(x-y) = \left(\mathrm{i}\gamma^\mu\partial_\mu + m\right)\Delta_\mathrm{F}(x-y) = \int \frac{\mathrm{d}^4p}{(2\pi)^4}\frac{\mathrm{e}^{-\mathrm{i}p\cdot(x-y)}}{\slashed{p} - m + \mathrm{i}\epsilon}\;. \tag{8.77}$$

Finally the contraction of two *photon* field operators is given by

$$\overline{\hat{A}_\mu(x)\hat{A}_\nu(y)} = \langle 0|T(\hat{A}_\mu(x)\hat{A}_\nu(y))|0\rangle = \mathrm{i}\,D_{\mathrm{F}\mu\nu}(x-y)\;. \tag{8.78}$$

---

[6] G.C. Wick: Phys. Rev. **80**, 268 (1950); F. Dyson: Phys. Rev. **82**, 428 (1951).

The covariant photon propagator was constructed in Sect. 7.5:

$$D_{F\mu\nu}(x-y) = -\int \frac{d^4k}{(2\pi)^4} e^{-ik\cdot(x-y)} \left(\frac{g_{\mu\nu}}{k^2 + i\epsilon} + \frac{1-\zeta}{\zeta}\frac{k_\mu k_\nu}{(k^2+i\epsilon)^2}\right). \quad (8.79)$$

Here ζ is the free parameter in the gauge-fixing term $\frac{1}{2}\zeta(\partial_\mu A^\mu)^2$ entering the Lagrange density of the free electromagnetic field. In the most simple case (the Feynman gauge, $\zeta = 1$) the extra term in (8.79) vanishes. Generally, physical observables must not depend on the value of ζ.

From a mathematical point of view the propagators $\Delta_F(x)$ etc. are *distributions*. To give them a well-defined mathematical meaning they have to be multiplied by test functions and integrated over d^4x. The propagataors are highly *singular* at small space-time distances. In the vicinity of the light cone, $x^2 = 0$, one finds (see section 4.6)

$$\Delta_F(x) \simeq -\frac{1}{4\pi}\delta(x) + \frac{i}{4\pi^2 x^2} - \frac{im^2}{8\pi^2}\ln(x^2)^{1/2}$$
$$+ \frac{m^2}{16\pi}\Theta(x^2) + O\left(|x^2|^{1/2}\ln|x^2|\right). \quad (8.80)$$

Thus the propagator exhibits several types of singularities. This leads to problems if products of such distributions are encountered: they can produce ill-defined divergent expressions. In momentum space the propagators have a much simpler structure. Infinities will arise nevertheless in the form of divergent momentum integrals. The handling of such divergencies using first regularization (which leads to finite but cutoff-dependent quantities) and then renormalization (to get rid of the cutoff) is a central problem of quantum field theory.

EXERCISE

8.2 Proof of Wick's Theorem

Problem. (a) Let $\hat{A}, \hat{B}, \ldots, \hat{Y}, \hat{Z}$ be time-independent linear operators. Assume that the time argument of $\hat{Z}$ is smaller than that of all the other operators, $t_Z < t_A, \ldots, t_Y$. Prove the following lemma, which describes the multiplication of a normal product by an operator,

$$:\hat{A}\hat{B}\ldots\hat{X}\hat{Y}:\hat{Z} = :\hat{A}\hat{B}\ldots\hat{X}\hat{Y}\hat{Z}: + :\underbracket{\hat{A}\hat{B}\ldots\hat{X}\hat{Y}\hat{Z}}: $$
$$+ :\hat{A}\underbracket{\hat{B}\ldots\hat{X}\hat{Y}\hat{Z}}: + \ldots + :\hat{A}\hat{B}\ldots\hat{X}\underbracket{\hat{Y}\hat{Z}}: \quad (1)$$

with the use of mathematical induction.
(b) Use the lemma from (a) to prove Wick's theorem (8.73). Can the assumption made for the time t_Z be lifted?

Solution. (a) The proof is much simplified if the following assumptions are made, which do not restrict generality.
  (i) $\hat{Z}$ is a creation operator.
  (ii) $\hat{A}, \ldots, \hat{Y}$ are annihilation operators.

Re (i): In general $\hat{Z}$ will be a sum of creation an annihilation parts,

Exercise 8.2

$$\hat{Z} = \hat{Z}^{(-)} + \hat{Z}^{(+)} \ . \tag{2}$$

For the annihilation part $\hat{Z}^{(+)}$ the claimed relation (1) is valid trivially because the product on the left-hand side of (1) is already normal ordered. On the right-hand side all contraction terms vanish because the condition $t_Z < t_A$ implies

$$\underbrace{\hat{A}(t_A)\hat{Z}^{(+)}}(t_Z) = \langle 0|T(\hat{A}(t_A)\hat{Z}^{(+)}(t_Z))|0\rangle = \langle 0|\hat{A}(t_A)\hat{Z}^{(+)}(t_Z)|0\rangle = 0 \ . \tag{3}$$

Therefore only annihilation operators $\hat{Z} = \hat{Z}^{(-)}$ have to be considered.

Re **(ii)**: Since the distribution law is valid for all products in (1) the operators can be split into creation and annihilation parts which then are treated separately. If (1) has been proven for annihilation operators then any additional creation operators $\hat{O}^{(-)}$ can be written in front of the operator products since they are in a normal-ordered position automatically. On the right-hand side of (1) no additional contraction terms arise in this way since according to **(i)**

$$\underbrace{\hat{O}^{(-)}\hat{Z}^{(-)}} = \langle 0|T(\hat{O}^{(-)}\hat{Z}^{(-)})|0\rangle = 0 \ . \tag{4}$$

To start with the proof of (1) by induction we study the simplest possible case involving two operators

$$:\hat{Y}:\hat{Z} = \hat{Y}\hat{Z} = T(\hat{Y}\hat{Z}) = :\hat{Y}\hat{Z}: + \underbrace{\hat{Y}\hat{Z}} \ . \tag{5}$$

Here in the second step the assumption $t_Z < t_Y$ was used, followed by (8.63), which defines the contraction. Now assume that (1) is valid for a product of n operators $\hat{B}\ldots\hat{Y}$. We multiply from the left by an additional annihilation operator $\hat{A}$ having a time argument, that also satisfies the condition $t_A > t_Z$:

$$\hat{A}:\hat{B}\hat{C}\ldots\hat{Y}:\hat{Z} = \hat{A}:\hat{B}\hat{C}\ldots\hat{Y}\hat{Z}: + :\hat{A}\underbrace{\hat{B}\hat{C}\ldots\hat{Y}\hat{Z}}: + \ldots + :\hat{A}\hat{B}\hat{C}\ldots\underbrace{\hat{Y}\hat{Z}}: \ . \tag{6}$$

We have to prove that the first term leads to a further contraction which involves $\hat{A}$ and $\hat{Z}$:

$$\hat{A}:\hat{B}\hat{C}\ldots\hat{X}\hat{Y}\hat{Z}: = :\hat{A}\hat{B}\hat{C}\ldots\hat{X}\hat{Y}\hat{Z}: + :\underbrace{\hat{A}\hat{B}\hat{C}\ldots\hat{X}\hat{Y}\hat{Z}}: \ . \tag{7}$$

According to the restrictions **(i)** and **(ii)** the normal product is trivial since $\hat{Z}$ is the sole creation operator:

$$\hat{A}:\hat{B}\hat{C}\ldots\hat{X}\hat{Y}\hat{Z}: = \epsilon\, \hat{A}\hat{Z}\, \hat{B}\hat{C}\ldots\hat{X}\hat{Y} \ . \tag{8}$$

The sign factor $\epsilon = \pm 1$ counts the number of commutations of fermion operators when $\hat{Z}$ is moved. The product of $\hat{A}\hat{Z}$, because of $t_A > t_Z$, satisfies

$$\hat{A}\hat{Z} = T(\hat{A}\hat{Z}) = :\hat{A}\hat{Z}: + \underbrace{\hat{A}\hat{Z}} \ , \tag{9}$$

and therefore

$$\hat{A}:\hat{B}\hat{C}\ldots\hat{Y}\hat{Z}: = \epsilon:\hat{A}\hat{Z}::\hat{B}\hat{C}\ldots\hat{Y}: + \epsilon\underbrace{\hat{A}\hat{Z}}:\hat{B}\hat{C}\ldots\hat{Y}: \ . \tag{10}$$

Now the operator $\hat{Z}$ has to be commuted back to the tail position of the product. This again leads to the factor ϵ, which cancels out. In the first term of (10) we again use the condition that $\hat{Z}$ is the only creation operator:

$$:\hat{A}\hat{Z}::\hat{B}\hat{C}\ldots\hat{Y}: = (\pm\hat{Z}\hat{A})(\hat{B}\hat{C}\ldots\hat{Y}) = \pm:\hat{Z}\hat{A}\hat{B}\hat{C}\ldots\hat{Y}:$$
$$= :\hat{A}\hat{Z}\hat{B}\hat{C}\ldots\hat{Y}: = \epsilon:\hat{A}\hat{B}\hat{C}\ldots\hat{Y}\hat{Z}: \ . \tag{11}$$

In view of the rule (8.64) for the normal product of a chain of operators containing a contraction we also have

$$\underline{\hat{A}\hat{Z}} : \hat{B}\hat{C}\ldots\hat{Y}: \, = \epsilon : \underline{\hat{A}}\hat{B}\hat{C}\ldots\hat{Y}\underline{\hat{Z}}: \,. \qquad (12)$$

Therefore (10) is transformed into the conjectured identity (7) which proves the lemma (1). We notice that (1) can be immediately generalized to the case where some of the operators in the normal product $:\hat{A}\hat{B}\ldots\hat{X}\hat{Y}:$ are contracted with each other.

Exercise 8.2

(b) The first step in the inductive chain for the proof of Wick's theorem,

$$T(\hat{A}\hat{B}) = :\hat{A}\hat{B}: + \hat{A}\hat{B}\,, \qquad (13)$$

has already been shown explicitly, see (8.65). Now let us assume that the theorem is valid for the time-ordered product of n operators

$$\begin{aligned}T(\hat{A}\hat{B}\ldots\hat{X}\hat{Y}) &= :\hat{A}\hat{B}\ldots\hat{X}\hat{Y}: + :\underline{\hat{A}\hat{B}}\ldots\hat{X}\hat{Y}: + \ldots \\ &\quad + :\underline{\hat{A}\hat{B}}\,\underline{\hat{C}\hat{D}}\ldots\hat{X}\hat{Y}: + \ldots \\ &\quad + \ldots \end{aligned} \qquad (14)$$

and multiply this identity from the right by a further operator $\hat{Z}$. If its time argument is earlier as that of the other operators, $t_Z < t_A,\ldots,t_Y$, then the left-hand side is simply given by

$$T(\hat{A}\hat{B}\ldots\hat{X}\hat{Y})\hat{Z} = T(\hat{A}\hat{B}\ldots\hat{X}\hat{Y}\hat{Z})\,. \qquad (15)$$

For each term on the right-hand side of (14) the lemma (1) can be applied. It is obvious that in this way all combinatorically possible contraction terms involving the operator $\hat{Z}$ are generated. This is what is claimed by Wick's theorem for the $n+1$ operators.

The condition $t_Z < t_A,\ldots,t_Y$ does not really restrict the generality of the proof. The operators in Wick's theorem can be reordered, $(\hat{A},\ldots\hat{Z}) \to (\hat{A}',\ldots\hat{Z}')$, in such a way that they are time ordered from the outset, $t_{A'} \geq t_{B'} \geq \ldots \geq t_{Z'}$. Within the T product this reordering at most can produce a sign factor. The same factor will also arise on the right-hand side when we permute the operators in the normal products. There are no additional terms and thus the equation is invariant under permutations. After reordering the proof given above will hold. At then end the inverse permutation $(\hat{A}',\ldots\hat{Z}') \to (\hat{A},\ldots\hat{Z})$ can be employed to restore the original ordering of the operators having no particular temporal sequence.

8.6 The Feynman Rules of Quantum Electrodynamics

Wick's theorem provides us with a powerfull tool for practical evaluations of the perturbation series for the S matrix. There are many theories involving coupled quantum fields which can be treated in this way. In the following examples the ϕ^4 theory and scalar electrodynamics will be studied. In this section we investigate quantum electrodynamics (QED), i.e., the theory of a charged spin-$\frac{1}{2}$ field being coupled to a massless spin-1 field (electrons and photons). This theory is ideally suited for a perturbative approach since the coupling constant $\alpha = e^2/4\pi\hbar c \simeq 1/137$ is very small. QED is studied in detail in the volume *Quantum Electrodynamics* in this series of books by using a heuristic approach based on the propagator formalism. In the following we will demonstrate that the same results, in particular the Feynman rules of QED, can be recovered in a systematic way from quantum field theory.

The classical Lagrangian of quantum electrodynamics was introduced in Exercise 6.2:

$$\mathcal{L} = \mathcal{L}_0^{\text{Dirac}} + \mathcal{L}_0^{\text{e.m.}} + \mathcal{L}_1 \,, \tag{8.81}$$

with

$$\mathcal{L}_0^{\text{Dirac}} = \bar{\psi}\left(\frac{\mathrm{i}}{2}\gamma_\mu \stackrel{\leftrightarrow}{\partial}{}^\mu - m\right)\psi \,, \tag{8.82a}$$

$$\mathcal{L}_0^{\text{e.m.}} = -\frac{1}{4}F_{\mu\nu}F^{\mu\nu} - \frac{1}{2}\zeta(\partial_\mu A^\mu)^2 \,, \tag{8.82b}$$

$$\mathcal{L}_1 = -e\bar{\psi}\gamma_\mu\psi A^\mu \,, \tag{8.82c}$$

where the electromagnetic field was coupled in by using the gauge invariant "minimal" substitution $\partial_\mu \to \partial_\mu + \mathrm{i}eA_\mu$. Quantizing this theory, we will employ the normal-ordering prescription so that the interaction term becomes

$$\hat{\mathcal{L}}_1 = -e\,{:}\hat{\bar{\psi}}\gamma_\mu\hat{\psi}\hat{A}^\mu{:} \,. \tag{8.83}$$

The Hamilton density describing the interaction is given simply by

$$\hat{\mathcal{H}}_1 = -\hat{\mathcal{L}}_1 = e\,{:}\hat{\bar{\psi}}\gamma_\mu\hat{\psi}\hat{A}^\mu{:} \,. \tag{8.84}$$

The canonical transformation leading from the Heisenberg to the interaction picture will not affect the form of this operator. (This can happen only if derivative operators are present, see Example 8.6). Therefore the perturbation series for the S operator reads

$$\begin{aligned}\hat{S} &= \mathbb{1} + \sum_{n=1}^\infty \hat{S}^{(n)} \\ &= \sum_{n=0}^\infty \frac{1}{n!}(-\mathrm{i}e)^n \int \mathrm{d}^4 x_1 \ldots \mathrm{d}^4 x_n \\ &\quad \times T\left[{:}\hat{\bar{\psi}}(x_1)\gamma_{\mu_1}\hat{\psi}(x_1)\hat{A}^{\mu_1}(x_1){:} \cdots {:}\hat{\bar{\psi}}(x_n)\gamma_{\mu_n}\hat{\psi}(x_n)\hat{A}^{\mu_n}(x_n){:}\right] \,.\end{aligned} \tag{8.85}$$

The terms in the sum have the structure

$$\hat{S}^{(n)} = \frac{1}{n!}\int \mathrm{d}^4 x_1 \ldots \mathrm{d}^4 x_n\, \hat{S}_n(x_1,\ldots,x_n) \,. \tag{8.86}$$

The function under the integral can be constructed with the help of Wick's theorem. It consists of a superposition of c-valued functions multiplied by normal products of the field operators $\hat{\bar{\psi}}$, $\hat{\psi}$, and $\hat{A}$, which we will write schematically as

$$\hat{S}_n(x_1,\ldots,x_n) = \sum_{\text{contractions}} K(x_1,\ldots,x_n)\,{:}\cdots\hat{\bar{\psi}}(x_i)\cdots\hat{\psi}(x_j)\cdots\hat{A}(x_k){:} \,. \tag{8.87}$$

The functions $K(x_1, \ldots, x_n)$ consist of contractions involving those field operators that are not contained in the normal product.

To explicitly evaluate a scattering process we need the matrix element of the $\hat{S}$ operator taken between states involving definite particle configurations:

$$S_{fi} = \langle k'_1 \lambda'_1, \ldots, \bar{p}'_1 \bar{s}'_1, \ldots p'_1 s'_1, \ldots | \hat{S} | p_1 s_1, \ldots, \bar{p}_1 \bar{s}_1, \ldots, k_1 \lambda_1, \ldots \rangle . \quad (8.88)$$

Here p_1, s_1 ($\bar{p}_1, \bar{s}_1$), etc. designate the momenta and spins of the electrons (positrons), and k_1, λ_1, etc. the momenta and polarizations of the photons. The many-particle states are constructed by applying the creation operators on the vacuum:

$$|p_1 s_1, \ldots, \bar{p}_1 \bar{s}_1, \ldots, k_1 \lambda_1, \ldots \rangle = \hat{b}^\dagger_{p_1 s_1} \cdots \hat{d}^\dagger_{\bar{p}_1 \bar{s}_1} \cdots \hat{a}^\dagger_{k_1 \lambda_1} \cdots |0\rangle , \quad (8.89a)$$

$$\langle k'_1 \lambda'_1, \ldots, \bar{p}'_1 \bar{s}'_1, \ldots, p'_1 s'_1, \ldots | = \langle 0 | \hat{a}_{k'_1 \lambda'_1} \cdots \hat{d}_{\bar{p}'_1 \bar{s}'_1} \cdots \hat{b}_{p'_1 s'_1} . \quad (8.89b)$$

(If a field mode is multiply occupied, which can happen only for the photons, an additional normalization factor will occur, which has been left out here).

The time-independent operators $\hat{b}$, $\hat{d}$, $\hat{a}$ are those entering the plane-wave expansion of the field operators:

$$\hat{\psi}(x) = \int \frac{d^3 p}{(2\pi)^{3/2}} \sqrt{\frac{m}{E_p}} \sum_s \left(\hat{b}_{ps} u(p,s) e^{-ip \cdot x} + \hat{d}^\dagger_{ps} v(p,s) e^{+ip \cdot x} \right) , \quad (8.90a)$$

$$\hat{\bar{\psi}}(x) = \int \frac{d^3 p}{(2\pi)^{3/2}} \sqrt{\frac{m}{E_p}} \sum_s \left(\hat{d}_{ps} \bar{v}(p,s) e^{-ip \cdot x} + \hat{b}^\dagger_{ps} \bar{u}(p,s) e^{+ip \cdot x} \right) , \quad (8.90b)$$

$$\hat{A}_\mu(x) = \int \frac{d^3 k}{\sqrt{(2\pi)^3 2\omega_k}} \sum_\lambda \left(\hat{a}_{k\lambda} \epsilon^\mu(k,\lambda) e^{-ik \cdot x} + \hat{a}^\dagger_{k\lambda} \epsilon^{*\mu}(k,\lambda) e^{+ik \cdot x} \right) . \quad (8.90c)$$

In the evaluation of matrix elements of the type (8.88), obviously the only terms of the expansion (8.87) that will contribute are those that contain a "matching" configuration of field operators in the normal product. To be more specific, for each electron/positron/photon in the *inital state* a matching annihilation operator is needed, i.e., a factor $\hat{\psi}^{(+)}/\hat{\bar{\psi}}^{(+)}/\hat{A}^{(+)}_\mu$. The plus sign is meant to designate the part of the field operator (8.90) having positive frequency (the annihilation part). Similarly for each electron/positron/photon in the *final state* a creation operator of the type $\hat{\psi}^{(-)}/\hat{\bar{\psi}}^{(-)}/\hat{A}^{(-)}_\mu$ must appear. If these conditions are not fulfilled the matrix element involving the states (8.89) will vanish since unpaired operators are present, which annihilate the vacuum state.

To make this perfectly clear we study the simplest possible case (which will turn out to be too simple for describing a physical process), i.e., the $\hat{S}$ operator in *first order* ($n = 1$):

$$\hat{S}^{(1)} = -ie \int d^4 x \, T\left[:\hat{\bar{\psi}}(x) \gamma_\mu \hat{\psi}(x) \hat{A}^\mu(x): \right]$$

$$= -ie \int d^4 x \, :\hat{\bar{\psi}}(x) \gamma_\mu \hat{\psi}(x) \hat{A}^\mu(x): . \quad (8.91)$$

The T symbol of course here can be omitted since there is only one time argument. Since each of the three field operators contains an annihilation and a creation part the expression (8.91) can be split into eight different parts, each corresponding to a specific physical situation. Since it is easy to be confused by lengthy algebraic expressions it is common and very helpful to use the graphical notation introduced by Feynman, i.e., the language of Feynman diagrams. We introduce the following graphical "translation rules" (see Fig. 8.2).

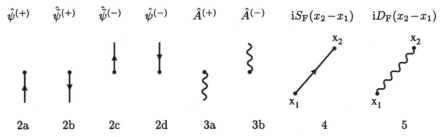

Fig. 8.2. The various building blocks of a Feynman graph in quantum electrodynamics. The time axis is directed upward.

1. Each point of interaction x is associated with a *vertex*. This corresponds to the algebraic factor $-ie\gamma_\mu$.
2. Each electron field operator is associated with an *external fermion line* glued to a vertex. These lines are fully drawn and carry an arrow. The following associations are made:

   (a) $\hat{\psi}^{(+)}(x)$ a line pointing upward ending at x (electron absorption),

   (b) $\hat{\bar{\psi}}^{(+)}(x)$ a line moving downward starting at x (positron absorption),

   (c) $\hat{\bar{\psi}}^{(-)}(x)$ a line pointing upward starting at x (electron emission),

   (d) $\hat{\psi}^{(-)}(x)$ a line pointing downward ending at x (positron emission).

3. Each photon field operator is associated with a wiggly *external photon line*:

   (a) $\hat{A}_\mu^{(+)}(x)$ a line starting at x pointing downward (photon absorption),

   (b) $\hat{A}_\mu^{(-)}(x)$ a line starting at x pointing upward (photon emission).

4. The contraction of two fermion operators,
   $$\underline{\hat{\psi}(x_2)\hat{\bar{\psi}}(x_1)} = iS_F(x_2 - x_1) ,$$
   is associated with a directed *internal fermion line* drawn from x_1 to x_2.

5. The contraction of two photon operators
   $$\underline{\hat{A}^\mu(x_2)\hat{A}^\mu(x_1)} = iD_F(x_2 - x_1) ,$$
   is associated with a wiggly *internal photon line* connecting x_1 and x_2.

Remark: The photon lines have no sense of direction (they bear no arrow) since the photon "is its own antiparticle". This is reflected in the symmetry

of the photon propagator
$$D_F^{\mu\nu}(x_2 - x_1) = D_F^{\mu\nu}(x_1 - x_2) \ .$$
The fermion propagator behaves differently since its field carries a charge. As discussed in Sect. 5.4 the Feynman propagator $S_F(x_2 - x_1)$ describes processes where a unit of charge is created at the point x_1 and annihilated at x_2. This can mean either that a virtual electron moves from x_1 to x_2 or that a virtual positron moves from x_2 to x_1.

The eight different contributions to the operator $\hat{S}^{(1)}$ are depicted in Fig. 8.3. All these graphs are *topologically equivalent* since they all have arisen from the same expression (8.91) and only differ by the direction of the external photon and fermion lines. The various graphs describe the following (hypothetical) processes:

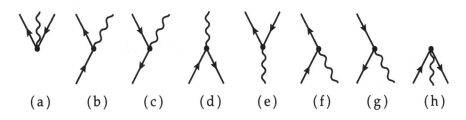

Fig. 8.3. Graphical representation of all first-order contributions to the $\hat{S}$ operator (8.91)

(a) Spontaneous creation of an electron–positron pair together with a photon,
(b) Emission of a photon by an electron,
(c) Emission of a photon by a positron,
(d) Annihilation of a pair producing a photon,
(e) Creation of a pair by a photon,
(f) Absorption of a photon by an electron,
(g) Absorption of a photon by a positron,
(h) Annihilation of a pair together with a photon.

One should note, however, that none of these processes is kinematically allowed: the corresponding S matrix elements all vanish since it is impossible to simultaneously satisfy the laws of energy and of momentum conservation and the dispersion relations for free particles. The system of equations
$$\pm p \pm p' \pm k = 0 \ , \quad p^2 = m^2 \ , \quad p'^2 = m^2 \ , \quad k^2 = 0 \qquad (8.92)$$
cannot be satisfied simultaneously for any of the eight sign combinations.

To obtain processes of physical interest one has to go to the *second-order* S operator ($n = 2$):
$$\begin{aligned}\hat{S}^{(2)} &= \frac{1}{2!}(-\mathrm{i}e)^2 \int \mathrm{d}^4x_1 \mathrm{d}^4x_2 \\ &\quad \times T\Big[{:}\hat{\bar{\psi}}(x_1)\gamma_\mu\hat{\psi}(x_1)\hat{A}^\mu(x_1){:} \ {:}\hat{\bar{\psi}}(x_2)\gamma_\nu\hat{\psi}(x_2)\hat{A}^\nu(x_2){:} \Big] \ . \end{aligned} \qquad (8.93)$$

This expression can be simplified with the help of Wick's theorem. First, however, we have to consider the effect of the *normal orderings* found inside the T product. It can be shown that "mixed" T products of this kind can also be decomposed according to Wick's theorem. In fact, the result even gets simplified since the following rule holds:

- Contractions between operators contained within the scope of a normal product do not enter in (8.73).

This means that the prescription of normal ordering of the interaction operator eliminates the interaction of a particle with itself at the same point x. Mathematically such a contribution would be described by the singular expression $\Delta_{\mathrm{F}}(0)$. Figure 8.4 shows a Feynman graph that is eliminated through normal ordering. We might as well have started from an interaction that is not normal ordered. Then the contribution of the "tadpole graph" shown in Fig. 8.4 would have to be kept, leading to an additional term (a mass shift of the electron) in the renormalization procedure, which has to be carried out anyhow. Since the mass is adjusted to its physical value the tadpole graph has no influence on observable quantities. If one adopts the normal-ordering prescription it is absent from the start.

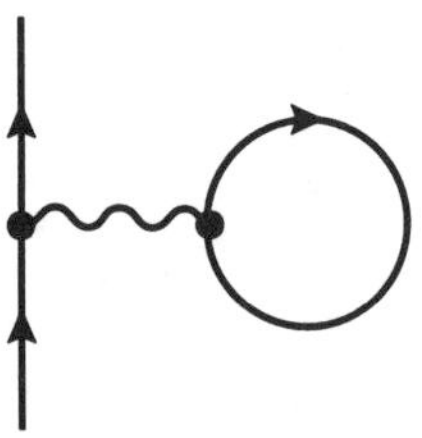

Fig. 8.4. The tadpole graph which gets eliminated if the normal-ordered interaction Lagrangian is adopted.

It is obvious that contributions of the type $\underbrace{\hat{\psi}(x_1)\hat{\psi}}(x_2)$ and $\underbrace{\hat{\bar{\psi}}(x_1)\hat{\bar{\psi}}}(x_2)$ do not arise since the creation and annihilation operators cannot be paired off in

$$\langle 0|T\big(\hat{\psi}(x_1)\hat{\psi}(x_2)\big)|0\rangle = \langle 0|T\big(\hat{\bar{\psi}}(x_1)\hat{\bar{\psi}}(x_2)\big)|0\rangle = 0\,. \tag{8.94}$$

After these preparations we are now able to write down the Wick expansion of the $\hat{S}$ *operator of QED in second order*, (8.93). There is one term without a contraction, three terms each involving one and two contractions, and one fully contracted term:

$$\begin{aligned}
\hat{S}^{(3)} &= \frac{(-\mathrm{i}e)^2}{2!}\int \mathrm{d}^4x_1\mathrm{d}^4x_2 :\hat{\bar{\psi}}(x_1)\gamma_\mu\hat{\psi}(x_1)\,\hat{\bar{\psi}}(x_2)\gamma_\nu\hat{\psi}(x_2)\,\hat{A}^\mu(x_1)\hat{A}^\nu(x_2): \quad \text{(a)}\\
&+ \frac{(-\mathrm{i}e)^2}{2!}\int \mathrm{d}^4x_1\mathrm{d}^4x_2 :\hat{\bar{\psi}}(x_1)\gamma_\mu\underbrace{\hat{\psi}(x_1)\,\hat{\bar{\psi}}(x_2)}\gamma_\nu\hat{\psi}(x_2)\,\hat{A}^\mu(x_1)\hat{A}^\nu(x_2): \quad \text{(b)}\\
&+ \frac{(-\mathrm{i}e)^2}{2!}\int \mathrm{d}^4x_1\mathrm{d}^4x_2 :\underbrace{\hat{\bar{\psi}}(x_1)\gamma_\mu\hat{\psi}(x_1)\,\hat{\bar{\psi}}(x_2)\gamma_\nu\hat{\psi}(x_2)}\,\hat{A}^\mu(x_1)\hat{A}^\nu(x_2): \quad \text{(c)}\\
&+ \frac{(-\mathrm{i}e)^2}{2!}\int \mathrm{d}^4x_1\mathrm{d}^4x_2 :\hat{\bar{\psi}}(x_1)\gamma_\mu\hat{\psi}(x_1)\,\hat{\bar{\psi}}(x_2)\gamma_\nu\hat{\psi}(x_2)\,\underbrace{\hat{A}^\mu(x_1)\hat{A}^\nu(x_2)}: \quad \text{(d)}\\
&+ \frac{(-\mathrm{i}e)^2}{2!}\int \mathrm{d}^4x_1\mathrm{d}^4x_2 :\hat{\bar{\psi}}(x_1)\gamma_\mu\underbrace{\hat{\psi}(x_1)\,\hat{\bar{\psi}}(x_2)}\gamma_\nu\hat{\psi}(x_2)\,\underbrace{\hat{A}^\mu(x_1)\hat{A}^\nu(x_2)}: \quad \text{(e)}\\
&+ \frac{(-\mathrm{i}e)^2}{2!}\int \mathrm{d}^4x_1\mathrm{d}^4x_2 :\underbrace{\hat{\bar{\psi}}(x_1)\gamma_\mu\hat{\psi}(x_1)\,\hat{\bar{\psi}}(x_2)\gamma_\nu\hat{\psi}(x_2)}\,\underbrace{\hat{A}^\mu(x_1)\hat{A}^\nu(x_2)}: \quad \text{(f)}\\
&+ \frac{(-\mathrm{i}e)^2}{2!}\int \mathrm{d}^4x_1\mathrm{d}^4x_2 :\underbrace{\hat{\bar{\psi}}(x_1)\gamma_\mu\underbrace{\hat{\psi}(x_1)\,\hat{\bar{\psi}}(x_2)}\gamma_\nu\hat{\psi}(x_2)}\,\hat{A}^\mu(x_1)\hat{A}^\nu(x_2): \quad \text{(g)}\\
&+ \frac{(-\mathrm{i}e)^2}{2!}\int \mathrm{d}^4x_1\mathrm{d}^4x_2 :\underbrace{\hat{\bar{\psi}}(x_1)\gamma_\mu\underbrace{\hat{\psi}(x_1)\,\hat{\bar{\psi}}(x_2)}\gamma_\nu\hat{\psi}(x_2)}\,\underbrace{\hat{A}^\mu(x_1)\hat{A}^\nu(x_2)}:\,. \quad \text{(h)}
\end{aligned} \tag{8.95}$$

8.6 The Feynman Rules of Quantum Electrodynamics

This is conveniently visualized with the help of the *Feynman diagrams* presented in Fig. 8.5.

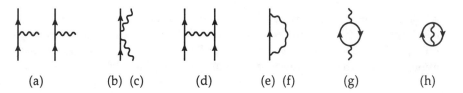

Fig. 8.5. The classes of Feynman diagrams corresponding to the eight terms of (8.95)

This figure shows a single diagram for each of the terms in (8.95), representing their topological structure. Of course we known that in each case there are various ways to associate the external lines with incoming or outgoing particles so that a given generic graph can descibe several distinct physical processes. Let us consider the terms in (8.95) separately.

(a) Two fermions (electron or positron) separately absorb or emit one photon each. Since the energy and momentum balance has to be satisfied separately for both partial graphs, $\hat{S}_{\rm a}^{(2)}$ does not contribute to the S matrix, as was the case for $\hat{S}^{(1)}$.

(b),(c) These two terms describe processes involving two free photons and two fermions. Depending on the association of the external lines with incoming and outgoing particles four different physical processes are covered, as shown in Fig. 8.6: (i) *Compton scattering* at an electron, (ii) Compton scattering at a positron, (iii) *electron–positron pair annihilation* into two photons, (iv) *electron–positron pair creation* by two photons. In each case there is a direct graph and an exchange graph which differ by the interchange of the two photon lines. Both contributions to the S matrix element have to be summed up coherently.

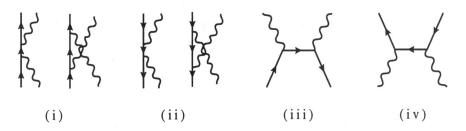

Fig. 8.6. Possible configurations of external lines in the generic Feynman diagram of Fig. 8.5b,c

The expressions (8.95b) and (8.95c) obtained from the decompositions of the $\hat{S}$ operator are identical. This becomes apparent if one renames $x_1 \leftrightarrow x_2$ and $\mu \leftrightarrow \nu$ and exchanges the ordering of the field operators (which is an admissible operation for the factors in a normal product):

$$\frac{(-\mathrm{i}e)^2}{2!} \int \mathrm{d}^4x_1 \mathrm{d}^4x_2 :\hat{\bar{\psi}}(x_1)\gamma_\mu\hat{\psi}(x_1)\,\hat{\bar{\psi}}(x_2)\gamma_\nu\hat{\psi}(x_2)\,\hat{A}^\mu(x_1)\hat{A}^\nu(x_2):$$

$$= \frac{(-\mathrm{i}e)^2}{2!} \int \mathrm{d}^4x_2 \mathrm{d}^4x_1 :\hat{\bar{\psi}}(x_2)\gamma_\nu\hat{\psi}(x_2)\,\hat{\bar{\psi}}(x_1)\gamma_\mu\hat{\psi}(x_1)\,\hat{A}^\nu(x_2)\hat{A}^\mu(x_1):$$

$$= \frac{(-ie)^2}{2!} \int d^4x_1 d^4x_2 \, :\hat{\bar{\psi}}(x_1)\gamma_\mu \hat{\psi}(x_1) \, \hat{\bar{\psi}}(x_2)\gamma_\nu \hat{\psi}(x_2) \, \hat{A}^\mu(x_1) \hat{A}^\nu(x_2): \, . \tag{8.96}$$

(d) The processes involving four external fermion lines described by this term are: (i) *electron–electron scattering*, (ii) *positron–positron scattering*, (iii) *electron–positron scattering*. The remaining possible configurations of the external lines again are forbidden for kinematic reasons.

Fig. 8.7. The same as Fig. 8.6 for the graph with four external fermion lines.

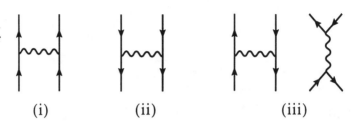

(e),(f) Here only two external fermion lines are present, therefore no interaction between particles separated asymptotically takes place. The graph describes the *self-energy* of the electron or positron.

(g) This is the *vacuum polarization* graph which originates from the creation and annihilation of a virtual electron–positron pair.

(h) This fully contracted *vacuum fluctuation term* does not contain a field operator and thus can only make a contribution to the S matrix if the initial and final states are identical. It can be shown that "closed bubbles" of this type can only produce phase factors which multiply the S matrix elements. Therefore they have no physical consequences and can be discarded (see Exercise 8.3).

In Example 8.4 the second-order S matrix elements derived from the Wick decomposition will be evaluated for a few QED processes, i.e., electron–electron and electron–photon scattering. For practical purposes one often takes the easier route. For a start all possible Feynman graphs (up to a chosen order) are drawn which describe the process to be investigated. Subsequently these graphs are translated into the corresponding algebraic expressions for the S matrix element. Let us collect the required "cooking recipes" in a comprehensive listing.

The Feynman Rules of QED in Coordinate Space

1. To evaluate an S matrix element of nth order all topologically distinct Feynman graphs with n vertices and within the desired configuration of external lines are drawn. Each vertex is assigned a coordinate variable x_i.

2. Vertex: $\qquad -ie(\gamma_\mu)_{\alpha\beta}$.

3. Internal photon line: $iD_F^{\mu\nu}(x_k - x_l)$.

4. Internal fermion line: $iS_{F\alpha\beta}(x_k - x_l)$.

5. External fermion line: $N_p\, u(p,s)\, e^{-ip\cdot x}$ (incoming electron),

$\qquad\qquad\qquad\qquad N_p\, \bar{v}(p,s)\, e^{+ip\cdot x}$ (incoming positron),

$\qquad\qquad\qquad\qquad N_p\, \bar{u}(p,s)\, e^{+ip\cdot x}$ (outgoing electron),

$\qquad\qquad\qquad\qquad N_p\, v(p,s)\, e^{+ip\cdot x}$ (outgoing positron).

6. External photon line: $N_k\, \epsilon^\mu(k,\lambda),\, e^{-ik\cdot x}$ (incoming photon),

$\qquad\qquad\qquad\qquad N_k\, \epsilon^{\mu*}(k,\lambda),\, e^{+ik\cdot x}$ (outgoing photon).

7. All coordinates x_i are integrated over.

8. Each closed fermion loop leads to a factor -1.

The normalization factors for the Dirac and the photon wave functions are

$$N_p = \sqrt{\frac{m}{(2\pi)^3 E_p}} \quad , \quad N_k = \sqrt{\frac{1}{(2\pi)^3 2\omega_k}} \,, \qquad (8.97)$$

if continuum normalization is applied. If instead one chooses the normalization to a box of finite volume $V = L^3$ the factor $(2\pi)^2$ is replaced by V.

The x integrations can be carried out (owing to the simple plane-wave factors) as shown in Example 8.4 for a few cases. As a result we arrive at an equivalent set of Feynman rules in momentum space.

The Feynman Rules of QED in Momentum Space

1. All the relevant topologically distinct Feynman graphs with n vertices are drawn. Each internal line is assigned a momentum variable p_i or k_i.

2. Vertex: $\quad -\mathrm{i}e(\gamma_\mu)_{\alpha\beta}$.

3. Internal photon line: $\mathrm{i}D_\mathrm{F}^{\mu\nu}(k_i)$.

4. Internal fermion line: $\mathrm{i}S_{\mathrm{F}\alpha\beta}(p_i)$.

5. External fermion line: $N_p\, u(p,s)\quad$ (incoming electron),

   $\qquad\qquad\qquad\qquad\ N_p\, \bar{v}(p,s)\quad$ (incoming positron),

   $\qquad\qquad\qquad\qquad\ N_p\, \bar{u}(p,s)\quad$ (outgoing electron),

   $\qquad\qquad\qquad\qquad\ N_p\, v(p,s)\quad$ (outgoing positron).

6. External photon line: $N_k\, \epsilon^\mu(k,\lambda)\quad$ (incoming photon),

   $\qquad\qquad\qquad\qquad\ N_k\, \epsilon^{\mu*}(k,\lambda)\quad$ (outgoing photon).

7. All momenta of the internal lines are integrated over: $\displaystyle\int \frac{\mathrm{d}^4 p}{(2\pi)^4}$.

8. Each closed fermion loop leads to a factor -1.

9. Each vertex is associated with a factor $(2\pi)^4\,\delta^4(p'-p\pm k)$.

A few explanatory remarks are in order.

- Two graphs that might appear to be equal are topologically distinct if the sequence of the vertices is different. For example, the box graphs of Fig. 8.8 (and several more graphs of similar appearance, see Exercise 8.3) both have to be considered and their contributions have to be added up coherently.
- The factor $1/n!$ resulting from the series expansion of the Dyson operator (8.50) is *not* present in the S matrix element. It is cancelled by a permutation factor $n!$ which arises from the ambiguous association of the vertices $x_1,\ldots,x_n$. This will be illustrated in Exercise 8.5.
- The minus sign from rule **8** results from the anticommutation of fermionic field operators. To demonstrate this let us consider the second-order vacuum polarization graph (8.95g). The corresponding $\hat{S}$ operator reads

$$\hat{S}_\mathrm{g}^{(2)} = \frac{(-\mathrm{i}e)^2}{2!}\int \mathrm{d}^4 x_1 \mathrm{d}^4 x_2 :\hat{\bar{\psi}}_\alpha(x_1)\gamma^\mu_{\alpha\beta}\hat{\psi}_\beta(x_1)\,\hat{\bar{\psi}}_\gamma(x_2)\gamma^\nu_{\gamma\delta}\hat{\psi}_\delta(x_2):$$

$$\times :\hat{A}_\mu(x_1)\hat{A}_\nu(x_2):\ . \qquad (8.98)$$

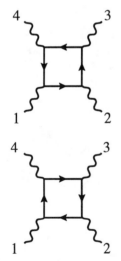

Fig. 8.8. Two box graphs that are not equivalent

One observes that the ordering of $\hat{\psi}$ and $\hat{\bar{\psi}}$ is different in the two contractions. This leads to a minus sign since

$$\hat{\psi}_\beta(x_1)\hat{\bar{\psi}}_\gamma(x_2) = \mathrm{i}S_{\mathrm{F}\beta\gamma}(x_1-x_2)\ , \qquad (8.99\mathrm{a})$$

$$\hat{\bar{\psi}}_\alpha(x_1)\hat{\psi}_\delta(x_2) = -\mathrm{i}S_{\mathrm{F}\delta\alpha}(x_2 - x_1)\,, \tag{8.99b}$$

and therefore

$$\begin{aligned}
\hat{S}_{\mathrm{g}}^{(2)} &= \frac{(-\mathrm{i}e)^2}{2!}\int \mathrm{d}^4x_1\mathrm{d}^4x_2\,\bigl(-\mathrm{i}S_{\mathrm{F}\delta\alpha}(x_2-x_1)\bigr)\gamma^\mu_{\alpha\beta}\bigl(\mathrm{i}S_{\mathrm{F}\beta\gamma}(x_1-x_2)\bigr)\gamma^\nu_{\gamma\delta}\\
&\quad \times\mathbin{:}\hat{A}_\mu(x_1)\hat{A}_\nu(x_2)\mathbin{:}\\
&= -\frac{(-\mathrm{i}e)^2}{2!}\int \mathrm{d}^4x_1\mathrm{d}^4x_2\,\mathrm{Sp}\bigl[\mathrm{i}S_{\mathrm{F}}(x_2-x_1)\gamma^\mu \mathrm{i}S_{\mathrm{F}}(x_1-x_2)\gamma^\nu\bigr]\\
&\quad \mathbin{:}\hat{A}_\mu(x_1)\hat{A}_\nu(x_2)\mathbin{:}\,. \tag{8.100}
\end{aligned}$$

It is quite obvious that for any process involving a closed fermion loop there always will be *one* contraction in which the ordering of the operators is inverted as in (8.98), leading to a minus sign.

EXERCISE

8.3 Disconnected Vacuum Graphs

Problem. Show that the $\hat{S}$ operator can be factorized according to

$$\hat{S} = \langle 0|\hat{S}|0\rangle \sum_{k=0}^\infty \frac{(-\mathrm{i})^k}{k!} \int \mathrm{d}^4x_1\cdots \mathrm{d}^4x_k\, T_{\mathrm{ext}}\bigl(\hat{\mathcal{H}}_1(x_1)\cdots \hat{\mathcal{H}}_1(x_k)\bigr)\,. \tag{1}$$

Here the modified time-ordered product T_{ext} is defined in such a way that only those terms are included in its expansion according to Wick's theorem that have a "connection to the outside". Completely closed vacuum bubbles are discarded. Use equation (1) to argue why disconnected Feynman diagrams need not be considered in the evaluation of S matrix elements.

Solution. We have to evaluate

$$\hat{S} = \sum_{n=0}^\infty \frac{(-\mathrm{i})^n}{n!}\int \mathrm{d}^4x_1\cdots \mathrm{d}^4x_n\, T\bigl(\hat{\mathcal{H}}_1(x_1)\cdots \hat{\mathcal{H}}_1(x_n)\bigr)\,, \tag{2}$$

where the T product can be expanded using Wick's theorem. The resulting terms can be combined into classes according to the number m of vertices whose field operators are fully contracted among each other. It is not predetermined which position in the contracted expression (or, equivalently, in the Feynman graph) is taken by the coordinates x_i. All permutations of the x_i are possible. The T product can now be written as

$$\begin{aligned}
T\bigl(\hat{\mathcal{H}}_1(x_1)\cdots \hat{\mathcal{H}}_1(x_n)\bigr) &= \sum_{m=0}^n \sum_{\mathrm{combin.}} T_{\mathrm{ext}}\bigl(\hat{\mathcal{H}}_1(x'_1)\cdots \hat{\mathcal{H}}_1(x'_{n-m})\bigr)\\
&\quad \times T_{\mathrm{vac}}\bigl(\hat{\mathcal{H}}_1(x'_{n-m+1})\cdots \hat{\mathcal{H}}_1(x'_n)\bigr)\,. \tag{3}
\end{aligned}$$

Here the sum over the combinations means the following. For the argument of the T_{ext} product take $n-m$ values out of the set of coordinates $x_1,\ldots,x_n$ and rename them as $x'_1,\ldots x'_{n-m}$. The remaining coordinates are the arguments of the T_{vac} product. The x_i in (2) are integration variables which can be given

Exercise 8.3

any name. Therefore in each term the combination sum can be transformed back, $x_i' \to x_i$. All terms in the sum having the same m thus are equal and the sum simply leads to a multiplicative factor which is given by the binomial coefficient $\binom{n}{m} = \frac{n!}{m!(n-m)!}$:

$$\begin{aligned}\hat{S} &= \sum_{n=0}^{\infty} \frac{(-i)^n}{n!} \int d^4x_1 \cdots d^4x_n \sum_{m=0}^{n} \frac{n!}{m!(n-m)!} \\ &\quad \times T_{\text{ext}}\big(\hat{\mathcal{H}}_1(x_1)\cdots\hat{\mathcal{H}}_1(x_{n-m})\big) T_{\text{vac}}\big(\hat{\mathcal{H}}_1(x_{n-m+1})\cdots\hat{\mathcal{H}}_1(x_n)\big) \\ &= \sum_{n=0}^{\infty}\sum_{m=0}^{n} \frac{(-i)^{n-m}}{(n-m)!}\frac{(-i)^m}{m!} \int d^4x_1\cdots d^4x_{n-m} \\ &\quad \times T_{\text{ext}}\big(\hat{\mathcal{H}}_1(x_1)\cdots\hat{\mathcal{H}}_1(x_{n-m})\big) \\ &\quad \times \int d^4x_{n-m+1}\cdots d^4x_n\, T_{\text{vac}}\big(\hat{\mathcal{H}}_1(x_{n-m+1})\cdots\hat{\mathcal{H}}_1(x_n)\big)\,. \quad (4)\end{aligned}$$

Upon transformation to the variables $n \to k = n - m$ the two sums get decoupled:

$$\begin{aligned}\hat{S} &= \sum_{k=0}^{\infty} \frac{(-i)^k}{k!} \int d^4x_1\cdots d^4x_k\, T_{\text{ext}}\big(\hat{\mathcal{H}}_1(x_1)\cdots\hat{\mathcal{H}}_1(x_k)\big) \\ &\quad \times \sum_{m=0}^{\infty}\frac{(-i)^m}{m!}\int d^4x_{k+1}\cdots d^4x_{k+m}\, T_{\text{vac}}\big(\hat{\mathcal{H}}_1(x_{k+1})\cdots\hat{\mathcal{H}}_1(x_{k+m})\big).\ (5)\end{aligned}$$

The second factor also can be written as

$$\begin{aligned}&\sum_{m=0}^{\infty}\frac{(-i)^m}{m!} \int d^4x_1\cdots d^4x_m\, T_{\text{vac}}\big(\hat{\mathcal{H}}_1(x_1)\cdots\hat{\mathcal{H}}_1(x_m)\big) \\ &= \Big\langle 0\Big|\sum_{m=0}^{\infty}\frac{(-i)^m}{m!}\int d^4x_1\cdots d^4x_m\, T\big(\hat{\mathcal{H}}_1(x_1)\cdots\hat{\mathcal{H}}_1(x_m)\big)\Big|0\Big\rangle \\ &= \langle 0|\hat{S}|0\rangle\,, \quad (6)\end{aligned}$$

since T_{vac} by definition is just the fully contracted part of the time-ordered product and the operator-valued part T_{ext} has a vanishing expectation value. This proves the assertion (1).

The factor $\langle 0|\hat{S}|0\rangle$ has no physical relevance. It is common to all S matrix elements and can be interpreted as the probability amplitude for a vacuum to vacuum transition. Since *the vacuum is required to be stable* (in the absence of external forces) this amplitude can only be a *phase factor* of modulus 1: $\langle 0|\hat{S}|0\rangle = e^{i\Phi}$ with real-valued Φ. Therefore it is sufficient to evaluate the matrix elements of the operator

$$\hat{S}' = \frac{\hat{S}}{\langle 0|\hat{S}|0\rangle}\,, \quad (7)$$

i.e., to replace T by T_{ext}.

The first terms of the perturbation series for the phase factor have the following form:

$$\langle 0|\hat{S}|0\rangle = 1 + \bigcirc\!\!\!\bigcirc + \bigcirc\!\!\!\sim\!\!\!\bigcirc + \cdots \quad . \tag{8}$$

Exercise 8.3

An attempt to actually evaluate these graphs would very soon tell us that they are represented by divergent expressions. However, as we have seen, we do not have to perform this calculation.

EXAMPLE

8.4 Møller Scattering and Compton Scattering

For some simple scattering processes we will explicitly evaluate the second-order S matrix element. We do not aim, however, at an exposition of the steps that lead from the S matrix elements to observable quantities such as scattering cross sections or decay rates. This has been presented in detail in the volume *Quantum Electrodynamics*. The main result is summarized in Sect. 8.7.

Electron–Electron Scattering

Let us study the process of Møller scattering in which two electrons with momenta and spins p_1, s_1 and p_2, s_2 in the initial state are scattered into the final state with p_1', s_1' and p_2', s_2'. According to (8.95d) the S matrix element is given by

$$\begin{aligned} S_{fi} &= \langle 0|\hat{b}_{p_2' s_2'} \hat{b}_{p_1' s_1'} \hat{S}^{(2)} \hat{b}^\dagger_{p_1 s_1} \hat{b}^\dagger_{p_2 s_2}|0\rangle \\ &= \frac{(-ie)^2}{2!}\int d^4x_1 d^4x_2 \, \langle 0|\hat{b}_{p_2' s_2'} \hat{b}_{p_1' s_1'} \\ &\quad \times :\hat{\bar{\psi}}^{(-)}(x_1)\gamma_\mu \hat{\psi}^{(+)}(x_1)\hat{\bar{\psi}}^{(-)}(x_2)\gamma_\nu \hat{\psi}^{(+)}(x_2): \hat{A}^\mu(x_1)\hat{A}^\nu(x_2) \\ &\quad \times \hat{b}^\dagger_{p_1 s_1} \hat{b}^\dagger_{p_2 s_2}|0\rangle \, . \end{aligned} \tag{1}$$

Here we have inserted the appropriate field operators $\hat{\psi}^{(+)}$ for the incoming and $\hat{\bar{\psi}}^{(-)}$ for the outgoing electrons according to Fig. 8.2. Using the plane-wave expansion (8.90) we arrive at

$$\begin{aligned} S_{fi} &= \frac{(-ie)^2}{2!}\int d^4x_1 d^4x_2 \sum_{\sigma_1 \sigma_2 \sigma_3 \sigma_4} \\ &\quad \times \int \frac{d^3q_1}{(2\pi)^{3/2}}\sqrt{\frac{m}{E_{q_1}}} \int \frac{d^3q_2}{(2\pi)^{3/2}}\sqrt{\frac{m}{E_{q_2}}} \int \frac{d^3q_3}{(2\pi)^{3/2}}\sqrt{\frac{m}{E_{q_3}}} \int \frac{d^3q_4}{(2\pi)^{3/2}}\sqrt{\frac{m}{E_{q_4}}} \\ &\quad \times \bar{u}(q_1,\sigma_1)e^{+iq_1\cdot x_1}\gamma_\mu u(q_2,\sigma_2)e^{-iq_2\cdot x_1}\bar{u}(q_3,\sigma_3)e^{+iq_3\cdot x_2}\gamma_\nu u(q_4,\sigma_4)e^{-iq_4\cdot x_2} \\ &\quad \times iD_F^{\mu\nu}(x_1-x_2)\,\langle 0|\hat{b}_{p_2' s_2'}\hat{b}_{p_1' s_1'}:\hat{b}^\dagger_{q_1\sigma_1}\hat{b}_{q_2\sigma_2}\hat{b}^\dagger_{q_3\sigma_3}\hat{b}_{q_4\sigma_4}:\hat{b}^\dagger_{p_1 s_1}\hat{b}^\dagger_{p_2 s_2}|0\rangle \, . \end{aligned} \tag{2}$$

The vacuum expectation value of the creation and annihilation operators as usual is evaluated by commuting the $\hat{b}^\dagger$ to the left and the $\hat{b}$ to the right:

$$-\langle 0|\hat{b}_{p_2' s_2'}\hat{b}_{p_1' s_1'}\hat{b}^\dagger_{q_1\sigma_1}\hat{b}^\dagger_{q_3\sigma_3}\hat{b}_{q_2\sigma_2}\hat{b}_{q_4\sigma_4}\hat{b}^\dagger_{p_1 s_1}\hat{b}^\dagger_{p_2 s_2}|0\rangle \tag{3}$$

$$= \Big(\delta_{s_1'\sigma_3}\delta^3(p_1'-q_3)\delta_{s_2'\sigma_1}\delta^3(p_2'-q_1) - \delta_{s_1'\sigma_1}\delta^3(p_1'-q_1)\delta_{s_2'\sigma_3}\delta^3(p_2'-q_3)\Big)$$

$$\times \Big(\delta_{s_1\sigma_4}\delta^3(p_1-q_4)\delta_{s_2\sigma_2}\delta^3(p_2-q_2) - \delta_{s_1\sigma_2}\delta^3(p_1-q_2)\delta_{s_2\sigma_4}\delta^3(p_2-q_4)\Big) \, .$$

Example 8.4

There are four possible ways to associate the external fermion lines of the Feynman graphs with the states of the scattering particles. However, as illustrated in Fig. 8.9, the integrands come in pairs which differ just by an interchange of the variables x_1 and x_2, thus leading to the same contribution. This is how the permutation factor 1/2! in (2) gets cancelled.

Fig. 8.9. Feynman graphs for Møller scattering. Only one direct and one exchange graph are topologically distinct

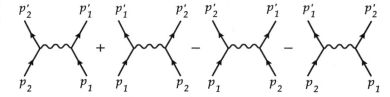

The final result for the S matrix element reads

$$S_{fi} = (-\mathrm{i}e)^2 \frac{1}{(2\pi)^6} \int \mathrm{d}^4 x_1 \mathrm{d}^4 x_2 \sqrt{\frac{m}{E_{p_1}}} \sqrt{\frac{m}{E_{p_2}}} \sqrt{\frac{m}{E_{p'_1}}} \sqrt{\frac{m}{E_{p'_2}}} \mathrm{i} D_F^{\mu\nu}(x_1 - x_2)$$

$$\times \Big[\mathrm{e}^{\mathrm{i}(p'_2 - p_2)\cdot x_2} \mathrm{e}^{\mathrm{i}(p'_1 - p_1)\cdot x_1} \bar{u}(p'_2, s'_2)\gamma_\mu u(p_2, s_2) \bar{u}(p'_1, s'_1)\gamma_\nu u(p_1, s_1)$$

$$- \mathrm{e}^{\mathrm{i}(p'_2 - p_1)\cdot x_2} \mathrm{e}^{\mathrm{i}(p'_1 - p_2)\cdot x_1} \bar{u}(p'_2, s'_2)\gamma_\mu u(p_1, s_1) \bar{u}(p'_1, s'_1)\gamma_\nu u(p_2, s_2) \Big]$$

$$:= S_{fi}^{\mathrm{dir}} + S_{fi}^{\mathrm{ex}} . \tag{4}$$

This expression consists of a *direct* and an *exchange graph* which are added with a relative minus sign. This is a consequence of the nondistinguishability of the electrons and of the Fermi–Dirac statistics.

The result (4) gets simplifed when it is transformed to *momentum space*. The Fourier integration in the direct term becomes

$$\int \mathrm{d}^4 x_1 \mathrm{d}^4 x_2 \, \mathrm{e}^{\mathrm{i}(p'_2 - p_2)\cdot x_2} \mathrm{e}^{\mathrm{i}(p'_1 - p_1)\cdot x_1} \int \frac{\mathrm{d}^4 q}{(2\pi)^4} \mathrm{e}^{-\mathrm{i}q\cdot(x_1 - x_2)} \mathrm{i} D_F^{\mu\nu}(q)$$

$$= \int \mathrm{d}^4 x_1 \int \frac{\mathrm{d}^4 q}{(2\pi)^4} \mathrm{e}^{\mathrm{i}(p'_1 - p_1 - q)\cdot x_1} \mathrm{i} D_F^{\mu\nu}(q) (2\pi)^4 \delta^4(p'_2 - p_2 + q)$$

$$= (2\pi)^4 \delta^4(p'_1 - p_1 + p'_2 - p_2) \mathrm{i} D_F^{\mu\nu}(p'_1 - p_1) \tag{5}$$

and similarly for the exchange term

$$\int \mathrm{d}^4 x_1 \mathrm{d}^4 x_2 \, \mathrm{e}^{\mathrm{i}(p'_2 - p_1)\cdot x_2} \mathrm{e}^{\mathrm{i}(p'_1 - p_2)\cdot x_1} \int \frac{\mathrm{d}^4 q}{(2\pi)^4} \mathrm{e}^{-\mathrm{i}q\cdot(x_1 - x_2)} \mathrm{i} D_F^{\mu\nu}(q)$$

$$= (2\pi)^4 \delta^4(p'_1 - p_2 + p'_2 - p_1) \mathrm{i} D_F^{\mu\nu}(p'_1 - p_2) . \tag{6}$$

In both cases the energy-momentum conservation of the scattering process is satisfied. The S matrix element finally reads

$$S_{fi} = (-\mathrm{i}e)^2 \frac{1}{(2\pi)^6} \sqrt{\frac{m}{E_{p_1}}} \sqrt{\frac{m}{E_{p_2}}} \sqrt{\frac{m}{E_{p'_1}}} \sqrt{\frac{m}{E_{p'_2}}} (2\pi)^4 \delta^4(p'_1 + p'_2 - p_1 - p_2)$$

$$\times \Big[\bar{u}(p'_2, s'_2)\gamma_\mu u(p_2, s_2) \, \mathrm{i} D_F^{\mu\nu}(p'_1 - p_1) \, \bar{u}(p'_1, s'_1)\gamma_\nu u(p_1, s_1)$$

$$- \bar{u}(p'_2, s'_2)\gamma_\mu u(p_1, s_1) \, \mathrm{i} D_F^{\mu\nu}(p'_1 - p_2) \, \bar{u}(p'_1, s'_1)\gamma_\nu u(p_2, s_2) \Big] . \tag{7}$$

8.6 The Feynman Rules of Quantum Electrodynamics

Example 8.4

The photon propagator in momentum space in the covariant gauge has the general form

$$D_{\mathrm{F}}^{\mu\nu}(q) = -\frac{g^{\mu\nu}}{q^2 + i\epsilon} + \frac{\zeta - 1}{\zeta}\frac{q^\mu q^\nu}{(q^2 + i\epsilon)^2} \, . \tag{8}$$

The gauge-dependent term, however, does not contribute to the S matrix since it leads to spinor products of the form

$$\bar{u}(p_1', s_1')\left(\slashed{p}_1' - \slashed{p}_1\right) u(p_1, s_1) = 0 \, , \tag{9}$$

because $(\slashed{p}_1 - m)u(p_1, s_1) = 0$ and $\bar{u}(p_1', s_1')(\slashed{p}_1' - m) = 0$. Of course the reason the gauge-dependent term vanishes is the conservation law for the electromagnetic transition current which multiplies the propagator.

The result (7) has been obtained in the volume *Quantum Electrodynamics* by using the heuristic propagator formalism. Field quantization provides a solid theoretical framework for this and for similar results.

Compton Scattering

The initial state for Compton scattering contains an electron p, s and a photon k, λ, whereas the same particles in the final states are characterized by p', s' and k', λ'. Because (8.95c) and (8.95d) are identical, the S matrix element reads

$$\begin{aligned}
S_{fi} &= \langle 0 | \hat{b}_{p's'} \hat{a}_{k'\lambda'} \hat{S}^{(2)} \hat{a}_{k,\lambda}^\dagger \hat{b}_{ps}^\dagger | 0 \rangle \\
&= 2\frac{(-\mathrm{i}e)^2}{2!} \int \mathrm{d}^4 x_1 \mathrm{d}^4 x_2 \, \langle 0 | \hat{b}_{p's'} \hat{a}_{k'\lambda'} : \hat{\bar{\psi}}^{(-)}(x_1)\gamma^\mu \underbrace{\hat{\psi}(x_1)\hat{\bar{\psi}}(x_2)}\gamma^\nu \hat{\psi}^{(+)}(x_2) : \\
&\quad \times : \hat{A}_\mu^{(-)}(x_1)\hat{A}_\nu^{(+)}(x_2) + \hat{A}_\mu^{(+)}(x_1)\hat{A}_\nu^{(-)}(x_2) : \hat{a}_{k\lambda}^\dagger \hat{b}_{ps}^\dagger | 0 \rangle \\
&= 2\frac{(-\mathrm{i}e)^2}{2!} \int \mathrm{d}^4 x_1 \mathrm{d}^4 x_2 \sum_{\sigma_1 \sigma_2} \sum_{\lambda_1 \lambda_2} \\
&\quad \times \int \frac{\mathrm{d}^3 q_1}{(2\pi)^{3/2}}\sqrt{\frac{m}{E_{q_1}}} \int \frac{\mathrm{d}^3 q_2}{(2\pi)^{3/2}}\sqrt{\frac{m}{E_{q_2}}} \int \frac{\mathrm{d}^3 k_1}{\sqrt{(2\pi)^3 2\omega_1}} \int \frac{\mathrm{d}^3 k_2}{\sqrt{(2\pi)^3 2\omega_2}} \\
&\quad \times \bar{u}(q_1, \sigma_1) \, \mathrm{e}^{+\mathrm{i}q_1 \cdot x_1} \gamma^\mu \, \mathrm{i}S_{\mathrm{F}}(x_1 - x_2) \, \gamma^\nu u(q_2, \sigma_2) \, \mathrm{e}^{-\mathrm{i}q_2 \cdot x_2} \\
&\quad \times \Big[\, \epsilon_\mu^*(k_1, \lambda_1) \, \mathrm{e}^{+\mathrm{i}k_1 \cdot x_1} \epsilon_\nu(k_2, \lambda_2) \, \mathrm{e}^{-\mathrm{i}k_2 \cdot x_2} \\
&\quad \times \langle 0 | \hat{b}_{p's'} \hat{a}_{k'\lambda'} : \hat{b}_{q_1\sigma_1}^\dagger \hat{b}_{q_2\sigma_2} \hat{a}_{k_1\lambda_1}^\dagger \hat{a}_{k_2\lambda_2} : \hat{a}_{k\lambda}^\dagger \hat{b}_{ps}^\dagger | 0 \rangle \\
&\quad + \epsilon_\mu(k_1, \lambda_1) \, \mathrm{e}^{-\mathrm{i}k_1 \cdot x_1} \epsilon_\nu^*(k_2, \lambda_2) \, \mathrm{e}^{+\mathrm{i}k_2 \cdot x_2} \\
&\quad \times \langle 0 | \hat{b}_{p's'} \hat{a}_{k'\lambda'} : \hat{b}_{q_1\sigma_1}^\dagger \hat{b}_{q_2\sigma_2} \hat{a}_{k_1\lambda_1} \hat{a}_{k_2\lambda_2}^\dagger : \hat{a}_{k\lambda}^\dagger \hat{b}_{ps}^\dagger | 0 \rangle \Big] \, .
\end{aligned} \tag{10}$$

The vacuum expectation values of the electron and photon operators reduce to a single term each and we obtain for the S matrix element of Compton scattering in coordinate space.

$$\begin{aligned}
S_{fi} &= (-\mathrm{i}e)^2 \frac{1}{(2\pi)^6} \int \mathrm{d}^4 x_1 \mathrm{d}^4 x_2 \sqrt{\frac{m}{E_p}}\sqrt{\frac{m}{E_{p'}}} \frac{1}{\sqrt{2\omega_k}} \frac{1}{\sqrt{2\omega_{k'}}} \\
&\quad \times \bar{u}(p', s')\gamma^\mu \, \mathrm{i}S_{\mathrm{F}}(x_1 - x_2) \, \gamma^\nu u(p, s) \, \mathrm{e}^{+\mathrm{i}p' \cdot x_1} \, \mathrm{e}^{-\mathrm{i}p \cdot x_2} \\
&\quad \times \Big[\epsilon_\mu^*(k', \lambda')\epsilon_\nu(k, \lambda) \, \mathrm{e}^{+\mathrm{i}k' \cdot x_1} \, \mathrm{e}^{-\mathrm{i}k \cdot x_2} + \epsilon_\mu(k, \lambda)\epsilon_\nu^*(k', \lambda') \, \mathrm{e}^{-\mathrm{i}k \cdot x_1} \, \mathrm{e}^{+\mathrm{i}k' \cdot x_2} \Big] \, .
\end{aligned} \tag{11}$$

Example 8.4

Again the x integrations can be carried out using the Feynman propagator of the electron in momentum space

$$S_F(q) = \frac{1}{\not{q} - m + i\epsilon} , \qquad (12)$$

leading to

$$\begin{aligned}S_{fi} &= (-ie)^2 \frac{1}{(2\pi)^6}\sqrt{\frac{m}{E_p}}\sqrt{\frac{m}{E_{p'}}}\frac{1}{\sqrt{2\omega_k}}\frac{1}{\sqrt{2\omega_{k'}}}(2\pi)^4\delta^4(p'+k'-p-k)\\ &\quad\times\Big[\bar{u}(p',s')\gamma^\mu\, iS_F(p+k)\,\gamma^\nu u(p,s)\,\epsilon^*_\mu(k',\lambda')\epsilon_\nu(k,\lambda)\\ &\qquad+\bar{u}(p',s')\gamma^\mu\, iS_F(p-k')\,\gamma^\nu u(p,s)\,\epsilon_\mu(k,\lambda)\epsilon^*_\nu(k',\lambda')\Big].\end{aligned}\qquad(13)$$

Again a direct and an exchange term are present, which this time are added with a positive relative sign since the photons satisfy Bose statistics. The corresponding Feynman graphs[7] are shown in Fig. 8.10. In contrast to Møller scattering the two terms do not arise from the (anti-)symmetry of the initial or final states under the exchange of identical particles. Rather they originate from the product of photon field operators $\hat{A}_\mu(x_1)\hat{A}_\nu(x_2)$ which can be decomposed in two ways into a product of a creation and an annihilation operator.

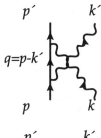

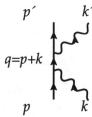

Fig. 8.10. Direct and exchange graph describing Compton scattering of an electron.

Compton Scattering of a Positron

To conclude this section we will investigate the changes encountered when the electron in Compton scattering is replaced by a positron. The lowest-order S matrix element in this case reads

$$\begin{aligned}S_{fi} &= 2\frac{(-ie)^2}{2!}\int d^4x_1 d^4x_2\,\langle 0|\hat{d}_{p's'}\hat{a}_{k'\lambda'}\\ &\quad\times :\hat{\bar{\psi}}^{(+)}(x_1)\gamma^\mu\underbrace{\hat{\psi}(x_1)\hat{\bar{\psi}}(x_2)}\gamma^\nu\hat{\psi}^{(-)}(x_2):\, :\hat{A}_\mu(x_1)\hat{A}_\nu(x_2):\,\hat{a}^\dagger_{k\lambda}\hat{d}^\dagger_{ps}|0\rangle.\end{aligned}\qquad(14)$$

This expression can be evaluated in the same way as (10). The vacuum expectation value of the Dirac creation and annihilation operators for the case of electron scattering,

$$\langle 0|\hat{b}_{p's'}:\hat{b}^\dagger_{q_1\sigma_1}\hat{b}_{q_2\sigma_2}:\hat{b}^\dagger_{ps}|0\rangle = \delta_{s'\sigma_1}\,\delta^3(p'-q_1)\,\delta_{s\sigma_2}\,\delta^3(p-q_2)\qquad(15)$$

is now replaced by

$$\langle 0|\hat{d}_{p's'}:\hat{d}_{q_1\sigma_1}\hat{d}^\dagger_{q_2\sigma_2}:\hat{d}^\dagger_{ps}|0\rangle = -\delta_{s'\sigma_2}\,\delta^3(p'-q_2)\,\delta_{s\sigma_1}\,\delta^3(p-q_1).\qquad(16)$$

In addition to an extra minus sign we note that the role of the initial and final states has been interchanged. As a consequence (11) is replaced by

$$\begin{aligned}S_{fi} &= -(-ie)^2\frac{1}{(2\pi)^6}\int d^4x_1 d^4x_2\sqrt{\frac{m}{E_p}}\sqrt{\frac{m}{E_{p'}}}\frac{1}{\sqrt{2\omega_k}}\frac{1}{\sqrt{2\omega_{k'}}}\\ &\quad\times \bar{v}(p,s)\gamma^\mu\, iS_F(x_1-x_2)\,\gamma^\nu v(p',s')\,e^{-ip\cdot x_1}\,e^{+ip'\cdot x_2}\end{aligned}\qquad(17)$$

---

[7] In contrast to the Feynman graphs in coordinate space, here the photon lines have also been drawn with an arrow. This serves to keep track of the direction of the momentum flow.

$$\times \Big[\epsilon_\mu^*(k',\lambda')\epsilon_\nu(k,\lambda)\,\mathrm{e}^{+ik'\cdot x_1}\,\mathrm{e}^{-ik\cdot x_2}$$
$$+ \epsilon_\mu(k,\lambda)\epsilon_\nu^*(k',\lambda')\,\mathrm{e}^{-ik\cdot x_1}\,\mathrm{e}^{+ik'\cdot x_2} \Big] .$$

In momentum space (13) is replaced by

$$S_{fi} = -(-\mathrm{i}e)^2 \frac{1}{(2\pi)^6} \sqrt{\frac{m}{E_p}}\sqrt{\frac{m}{E_{p'}}} \frac{1}{\sqrt{2\omega_k}} \frac{1}{\sqrt{2\omega_{k'}}} (2\pi)^4\,\delta^4(p'+k'-p-k)$$
$$\times \Big[\bar{v}(p,s)\gamma^\mu\,\mathrm{i}S_F(-p+k')\,\gamma^\nu v(p',s')\,\epsilon_\mu^*(k',\lambda')\epsilon_\nu(k,\lambda)$$
$$+ \bar{v}(p,s)\gamma^\mu\,\mathrm{i}S_F(-p-k)\,\gamma^\nu v(p',s')\,\epsilon_\mu(k,\lambda)\epsilon_\nu^*(k',\lambda') \Big] . \quad (18)$$

The corresponding Feynman diagrams in momentum space are shown in Fig. 8.11. The direction of the arrow describing the propagation of the antiparticle is reversed and at the same time it is assigned a negative energy and momentum, $-p$, leaving the overall balance unchanged. Also, the ordering of the spinors is reversed: $\bar{u}(p',s')\ldots u(p,s) \to \bar{v}(p,s)\ldots v(p',s')$. These changes reflect the concept of Stückelberg and Feynman that *antiparticles behave like negative- energy particles that move backward in time.*

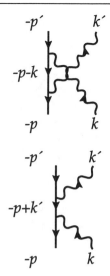

Fig. 8.11. Direct and exchange graph describing Compton scattering of a positron.

EXERCISE

8.5 The Feynman Graphs of Photon–Photon Scattering

Problem. Construct the S matrix element for the QED process of photon–photon scattering (Delbrück scattering) in fourth-order perturbation theory. Show that the factor $1/n!$ in the perturbation series gets cancelled.

Solution. The general expression for the fourth-order S matrix element is

$$S_{fi} = \frac{(-\mathrm{i}e)^4}{4!}\int \mathrm{d}^4 x_1\,\mathrm{d}^4 x_2\,\mathrm{d}^4 x_3\,\mathrm{d}^4 x_4$$
$$\times \langle f | T[:\hat{\bar\psi}(x_1)\hat{A}(x_1)\hat\psi(x_1):\;:\hat{\bar\psi}(x_2)\hat{A}(x_2)\hat\psi(x_2):$$
$$\times :\hat{\bar\psi}(x_3)\hat{A}(x_3)\hat\psi(x_3):\;:\hat{\bar\psi}(x_4)\hat{A}(x_4)\hat\psi(x_4):]|i\rangle \quad (1)$$

having two photons each in the initial and final state.

$$|i\rangle = \hat{a}_{k_2\lambda_2}^\dagger \hat{a}_{k_1\lambda_1}^\dagger |0\rangle \quad , \quad \langle f| = \langle 0|\hat{a}_{k_1'\lambda_1'}\hat{a}_{k_2'\lambda_2'} . \quad (2)$$

If the T product in (1) is decomposed according to Wick's theorem only those terms contribute in which the fermion operators are completely contracted. In all there are six different choices for the four contractions, which can be schematically labeled by an ordered 4-tupel of the coordinate indices (1234), (1243), (1324), (1342), (1423), (1432). The first term, for example, is

$$(1234) = :\hat{\bar\psi}_1\hat{A}_1\hat\psi_1\,\hat{\bar\psi}_2\hat{A}_2\hat\psi_2\,\hat{\bar\psi}_3\hat{A}_3\hat\psi_3\,\hat{\bar\psi}_4\hat{A}_4\hat\psi_4: . \quad (3)$$

Exercise 8.5

Since the x_i are just integration variables they can be freely renamed. Therefore all the terms are equal and the contraction (3) is simply multiplied by the factor 6.

The evaluation of the Fock-space matrix element

$$\langle 0| \hat{a}_{k'_1 \lambda'_1} \hat{a}_{k'_2 \lambda'_2} :\hat{A}(x_1)\hat{A}(x_2)\hat{A}(x_3)\hat{A}(x_4): \hat{a}^\dagger_{k_2 \lambda_2} \hat{a}^\dagger_{k_1 \lambda_1} |0\rangle \tag{4}$$

poses a combinatorical problem. Each of the $\hat{A}(x_i)$ is expanded into plane waves and will contribute either a creation or an annihilation operator to (4). This leads to matrix elements of the type

$$\langle 0| \hat{a}_{k'_1 \lambda'_1} \hat{a}_{k'_2 \lambda'_2} \hat{a}^\dagger_i \hat{a}^\dagger_j \hat{a}_k \hat{a}_l \hat{a}^\dagger_{k_2 \lambda_2} \hat{a}^\dagger_{k_1 \lambda_1} |0\rangle \,, \tag{5}$$

where the creation and annihilation operators have to be paired off. In this way (4) gives us $4! = 24$ contributions (each of the coordinates can be associated with each of the four photons). However, four terms can each be grouped together since they differ only by a cyclical permutation of the coordinates. Combined with the earlier factor 6 we arrive at a multiplicity factor of 24 which just serves to cancel the denominator $1/4!$ in (1). We are left with six different contributions, which are distinguished by the sequence of photon vertices. Thus Delbrück scattering is described by the coherent superposition of the six amplitudes depicted in Fig. 8.12.

Fig. 8.12. The six topologically different graphs for photon–photon scattering

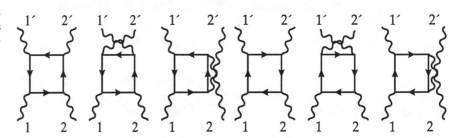

The S matrix element in momentum space is given by the following analytical expression (with the abbreviation $N_k = \big((2\pi)^3 2\omega_k\big)^{-1/2}$):

$$\begin{aligned}
S_{fi} = &-(-\mathrm{i}e)^4 \Big(\prod_{i=1}^{4} N_{k_i}\Big) \int \mathrm{d}^4 x_1 \, \mathrm{d}^4 x_2 \, \mathrm{d}^4 x_3 \, \mathrm{d}^4 x_4 \; \mathrm{e}^{-\mathrm{i}(k_1 \cdot x_1 + k_2 \cdot x_2 - k'_1 \cdot x_3 - k'_2 \cdot x_4)} \\
&\times \mathrm{Sp}\big[\slashed{\epsilon}(\boldsymbol{k}_1, \lambda_1) \,\mathrm{i}S_\mathrm{F}(x_2 - x_1) \,\slashed{\epsilon}(\boldsymbol{k}_2, \lambda_2) \,\mathrm{i}S_\mathrm{F}(x_3 - x_2) \\
&\qquad \slashed{\epsilon}(\boldsymbol{k}'_2, \lambda'_2) \,\mathrm{i}S_\mathrm{F}(x_4 - x_3) \,\slashed{\epsilon}(\boldsymbol{k}'_1, \lambda'_1) \,\mathrm{i}S_\mathrm{F}(x_1 - x_4)\big] \\
&+ 5 \text{ permutations} \,,
\end{aligned} \tag{6}$$

or in momentum space

$$\begin{aligned}
S_{fi} = &-(-\mathrm{i}e)^4 \Big(\prod_{i=1}^{4} N_{k_i}\Big) \delta^4(k'_1 + k'_2 - k_1 - k_2) \\
&\times \int \mathrm{d}^4 k \, \mathrm{Sp}\big[\slashed{\epsilon}(\boldsymbol{k}_1, \lambda_1) \,\mathrm{i}S_\mathrm{F}(k) \,\slashed{\epsilon}(\boldsymbol{k}_2, \lambda_2) \,\mathrm{i}S_\mathrm{F}(k + k_2) \\
&\qquad \slashed{\epsilon}(\boldsymbol{k}'_2, \lambda'_2) \,\mathrm{i}S_\mathrm{F}(k + k_2 - k'_2) \,\slashed{\epsilon}(\boldsymbol{k}'_1, \lambda'_1) \,\mathrm{i}S_\mathrm{F}(k - k_1)\big] \\
&+ 5 \text{ permutations} \,.
\end{aligned} \tag{7}$$

EXAMPLE ▬▬▬▬▬▬▬▬▬▬▬▬▬▬▬

8.6 Scalar Electrodynamics

Quantum electrodynamics (QED) describes the interaction of Dirac particles with the electromagnetic field. Since nature provides elementary spin-$\frac{1}{2}$ particles (the leptons e, μ, τ), which interact predominantly through their electric charges, this theory can be easily checked and has great practical relevance. Owing to the lack of "elementary" electrically charged spin-0 particles the same cannot be said for scalar electrodynamics. Pions (and scalar mesons), for example, experience the strong interaction and their electromagnetic charge plays only a minor role. Nevertheless it is instructive to work out the theory of interacting Klein–Gordon and Maxwell fields.

Since spin as an internal degree of freedom is absent in scalar electrodynamics, this theory in principle should be simpler than spinor QED. We will be confronted, however, with a new problem related to the fact that the interaction term now contains *derivatives of the fields*. Such "gradient couplings" complicate the transition from the Lagrangian to the Hamiltonian and even will produce an apparent violation of Lorentz invariance. Upon closer inspection fortunately this problem will turn out to be fictitious.

The Lagrangian of scalar electrodynamics,

$$\mathcal{L} = \mathcal{L}_0^{\text{KG}} + \mathcal{L}_0^{\text{em}} + \mathcal{L}_1 \; , \tag{1}$$

is composed of the free Lagrangians $\mathcal{L}_0$ of the Klein–Gordon field (see (4.54)) and of the Maxwell field (see Chap. 7.2). The interaction term $\mathcal{L}_1$ is obtained as usual through the gauge-invariant minimal-coupling prescription, i.e., one makes the replacement $\partial_\mu \to (\partial_\mu + ieA_\mu)\phi$ and $\phi^* \overleftarrow{\partial}_\mu \to \phi^*(\overleftarrow{\partial}_\mu - ieA_\mu)$. This leads to

$$\mathcal{L}_1 = -ie\phi^* \overleftrightarrow{\partial}_\mu \phi A^\mu + e^2 \phi^*\phi A_\mu A^\mu \; . \tag{2}$$

To implement canonical quantization, the transition to Hamilton's formalism is made. For this the canonically conjugate fields

$$\pi = \frac{\partial \mathcal{L}}{\partial(\partial_0 \phi)} = \partial^0 \phi^* - ieA^0 \phi^* \; , \tag{3a}$$

$$\pi^* = \frac{\partial \mathcal{L}}{\partial(\partial_0 \phi^*)} = \partial^0 \phi + ieA^0 \phi \tag{3b}$$

are needed. Since $\mathcal{L}_1$ contains a *time derivative*, $-ie\phi^* \overleftrightarrow{\partial}_0 \phi A^0$, (3) features an additional term.

The Hamilton density $\mathcal{H}$ as usual can be obtained from the Lagrange density by

$$\begin{aligned} \mathcal{H} &= \pi\dot\phi + \pi^*\dot\phi^* - \mathcal{L}_0^{\text{KG}} - \mathcal{L}_1 + \mathcal{H}_0^{\text{em}} \\ &= \pi\dot\phi + \pi^*\dot\phi^* - (\dot\phi^*\dot\phi - \nabla\phi^* \cdot \nabla\phi - m^2\phi^*\phi) - \mathcal{L}_1 + \mathcal{H}_0^{\text{em}} \; . \end{aligned} \tag{4}$$

Now the "velocities" $\dot\phi, \dot\phi^*$ are to be replaced by the "momenta" π, π^* making use of (3a). The contribution of the free photon field can be copied from Chap. 7. In summary we find

Example 8.6

$$\begin{aligned}
\mathcal{H} &= \pi(\pi^* - \mathrm{i}eA^0\phi) + \pi^*(\pi + \mathrm{i}eA^0\phi^*) \\
&\quad - \Big((\pi + \mathrm{i}eA^0\phi^*)(\pi^* - \mathrm{i}eA^0\phi) - \boldsymbol{\nabla}\phi^* \cdot \boldsymbol{\nabla}\phi + \mathcal{H}_0^{\mathrm{em}} - m^2\phi^*\phi\Big) - \mathcal{L}_1 \\
&= (\pi^*\pi + \boldsymbol{\nabla}\phi^* \cdot \boldsymbol{\nabla}\phi + m^2\phi^*\phi) + \mathcal{H}_0^{\mathrm{em}} - e^2\phi^*\phi(A^0)^2 - \mathcal{L}_1 \\
&= \mathcal{H}_0^{\mathrm{KG}} + \mathcal{H}_0^{\mathrm{em}} + \mathcal{H}_1 ,
\end{aligned} \quad (5)$$

with the interaction Hamilton density

$$\begin{aligned}
\mathcal{H}_1 &= -\mathcal{L}_1 - e^2\phi^*\phi(A^0)^2 \\
&= \mathrm{i}e\phi^* \overset{\leftrightarrow}{\partial}_\mu \phi A^\mu - e^2\phi^*\phi A_\mu A^\mu - e^2\phi^*\phi(A^0)^2 .
\end{aligned} \quad (6)$$

The elimination of $\dot{\phi}$ and $\dot{\phi}^*$ from the first term leads to

$$\begin{aligned}
\mathcal{H}_1 &= \mathrm{i}e\phi^* \overset{\leftrightarrow}{\partial}_k \phi A^k + \mathrm{i}e\phi^*(\pi^* - \mathrm{i}eA^0\phi)A^0 - \mathrm{i}e(\pi + \mathrm{i}e\phi^* A^0)\phi A^0 \\
&\quad - e^2\phi^*\phi A_\mu A^\mu - e^2\phi^*\phi(A^0)^2 \\
&= \mathrm{i}e\phi^* \overset{\leftrightarrow}{\partial}_k \phi A^k + \mathrm{i}e(\pi^*\phi^* - \pi\phi)A^0 - e^2\phi^*\phi A_\mu A^\mu + e^2\phi^*\phi(A^0)^2 .
\end{aligned} \quad (7)$$

The last term in $\mathcal{H}_1$ is known as a *normal-dependent term*. The origin of this name derives from the fact that the quantization scheme can be formulated in a more general form starting from an arbitrary space-like hyper surface $\sigma(x)$ instead of the equal-time condition $x_0 = x_0'$. In this case the extra term becomes $e^2\phi^*(x)\phi(x)\big(n_\mu(x)A^\mu(x)\big)^2$, where $n_\mu(x)$ is the normal vector defined on the surface $\sigma(x)$. Since we always employ the special hypersurface $x_0 = \mathrm{const}$, in our case the normal direction is $n_\mu = (1,0,0,0)$. In this way a particular direction in space-time is singled out, which appears to destroy the Lorentz covariance of the theory. When working out the perturbation expansion, however, one observes that the boson propagator also contains a noncovariant part that exactly cancels the normal-dependent term! We will explicitly verify this in first-order perturbation theory.

As described in Chap. 4 the quantization is performed in terms of the fields ϕ, ϕ^* and π, π^*. We make the replacement $\phi \to \hat{\phi}$, $\pi \to \hat{\pi}$, $A_\mu \to \hat{A}_\mu$ and for the scalar field impose the condition

$$[\hat{\phi}(\boldsymbol{x},t),\hat{\pi}(\boldsymbol{x}',t)] = [\hat{\phi}^\dagger(\boldsymbol{x},t),\hat{\pi}^\dagger(\boldsymbol{x}',t)] = \mathrm{i}\delta^3(\boldsymbol{x}-\boldsymbol{x}') , \quad (8)$$

while the other equal-time commutators are required to vanish. To eliminate vacuum contributions we will impose normal ordering on the products of operators.

The discussion so far is based on the *Heisenberg picture* where the dynamics of the system is fully described by the field operators. To carry out the perturbation expansion we need the operators to be in the *interaction picture*, however. As explained in Sect. 8.2 the transition is accomplished by the canonical transformation

$$\hat{O}^{\mathrm{I}}(t) = \hat{U}(t)\,\hat{O}^{\mathrm{H}}(t)\,\hat{U}^{-1}(t) . \quad (9)$$

According to (8.12) the transformation operator is explicitly given by

$$\hat{U}(t) = \mathrm{e}^{\mathrm{i}\hat{H}_0^{\mathrm{S}}t}\,\mathrm{e}^{-\mathrm{i}\hat{H}t} . \quad (10)$$

The canonically conjugate fields in the interaction picture are given by

$$\hat{\pi}^{\mathrm{I}\dagger}(x) = \partial^0 \hat{\phi}^{\mathrm{I}}(x) \quad , \quad \hat{\pi}^{\mathrm{I}}(x) = \partial^0 \hat{\phi}^{\mathrm{I}\dagger}(x) \; . \tag{11}$$

Of course the connection between $\hat{\phi}$ and $\hat{\pi}$ is the same as in the noninteracting case, since the interaction picture was introduced in order to have field operators satisfying the dynamics of free fields.

Proof of (11): The time derivative of $\hat{\phi}^{\mathrm{I}}$ according to the product rule is

$$\partial^0 \hat{\phi}^{\mathrm{I}} = (\partial^0 \hat{U})\hat{\phi}^{\mathrm{H}}\hat{U}^{-1} + \hat{U}(\partial^0 \hat{\phi}^{\mathrm{H}})\hat{U}^{-1} + \hat{U}\hat{\phi}^{\mathrm{H}}(\partial^0 \hat{U}^{-1}) \; . \tag{12}$$

The time derivative of the time-evolution operator can be written as

$$\begin{aligned}\partial^0 \hat{U} &= \mathrm{e}^{\mathrm{i}\hat{H}_0^{\mathrm{S}}t}\bigl(\mathrm{i}\hat{H}_0^{\mathrm{S}} - \mathrm{i}\hat{H}\bigr)\mathrm{e}^{-\mathrm{i}\hat{H}t} = -\mathrm{i}\mathrm{e}^{\mathrm{i}\hat{H}_0^{\mathrm{S}}t}\mathrm{e}^{-\mathrm{i}\hat{H}t}\bigl(\mathrm{e}^{\mathrm{i}\hat{H}t}\hat{H}_1^{\mathrm{S}}\mathrm{e}^{-\mathrm{i}\hat{H}t}\bigr)\\ &= -\mathrm{i}\hat{U}\hat{H}_1^{\mathrm{H}} \; , \end{aligned} \tag{13}$$

and

$$\partial^0 \hat{U}^{-1} = \mathrm{i}\hat{H}_1^{\mathrm{H}}\hat{U}^{-1} \; . \tag{14}$$

With the equation of motion for the Heisenberg field operator

$$\mathrm{i}\partial^0 \hat{\phi}^{\mathrm{H}} = [\hat{\phi}^{\mathrm{H}}, \hat{H}] \tag{15}$$

(12) becomes

$$\partial^0 \hat{\phi}^{\mathrm{I}} = -\mathrm{i}\hat{U}\Bigl([\hat{H}_1^{\mathrm{H}}, \hat{\phi}^{\mathrm{H}}] + [\hat{\phi}^{\mathrm{H}}, \hat{H}]\Bigr)\hat{U}^{-1} = -\mathrm{i}\hat{U}[\hat{\phi}^{\mathrm{H}}, \hat{H}_0^{\mathrm{H}}]\hat{U}^{-1} \; . \tag{16}$$

The required commutators follow from (5) and the ETCR (8):

$$[\hat{\phi}^{\mathrm{H}}(x), \hat{H}_0^{\mathrm{H}}] = \int \mathrm{d}^3x' \, [\hat{\phi}^{\mathrm{H}}(x), \hat{\pi}^{\mathrm{H}\dagger}(x')\hat{\pi}^{\mathrm{H}}(x')] = \mathrm{i}\hat{\pi}^{\mathrm{H}\dagger}(x) \; . \tag{17}$$

This proves the first part of (11); the second assertion can be confirmed in the same way.

The interaction Lagrangian $\hat{\mathcal{H}}_1$ of (7) is transformed according to (9). The spatial (but not the temporal!) gradient coupling ∂_k does not act on the transformation operator $\hat{U}(t)$, therefore a multiple insertion of $\mathbb{1} = \hat{U}^{-1}(t)\hat{U}(t)$ between the factors in (7) simply yields

$$\begin{aligned}\hat{\mathcal{H}}_1^{\mathrm{I}} &= \mathrm{i}e\hat{\phi}^{\mathrm{I}\dagger} \overset{\leftrightarrow}{\partial_k} \hat{\phi}^{\mathrm{I}} \hat{A}^{\mathrm{I}k} + \mathrm{i}e\bigl(\hat{\pi}^{\mathrm{I}\dagger}\hat{\phi}^{\mathrm{I}\dagger} - \hat{\pi}^{\mathrm{I}}\hat{\phi}^{\mathrm{I}}\bigr)A^{\mathrm{I}0}\\ &\quad - e^2 \hat{\phi}^{\mathrm{I}\dagger}\hat{\phi}^{\mathrm{I}} \hat{A}^{\mathrm{I}\mu}\hat{A}^{\mathrm{I}\mu} + e^2 \hat{\phi}^{\mathrm{I}\dagger}\hat{\phi}^{\mathrm{I}} (\hat{A}^{\mathrm{I}0})^2 \; .\end{aligned} \tag{18}$$

This result looks like the corresponding expression in the Heisenberg picture (although the nature of the field operators of course is quite different). This changes, however, when the canonically conjugate field operator $\hat{\pi}$ is expressed in terms of the time derivative of $\hat{\phi}$. According to (11) this results in

$$\begin{aligned}\hat{\mathcal{H}}_1^{\mathrm{I}} &= \mathrm{i}e\hat{\phi}^{\mathrm{I}\dagger} \overset{\leftrightarrow}{\partial_\mu} \hat{\phi}^{\mathrm{I}} \hat{A}^{\mathrm{I}\mu} - e^2 \hat{\phi}^{\mathrm{I}\dagger}\hat{\phi}_{\mathrm{I}} \hat{A}^{\mathrm{I}}_\mu \hat{A}^{\mathrm{I}\mu} + e^2 \hat{\phi}^{\mathrm{I}\dagger}\hat{\phi}^{\mathrm{I}} (\hat{A}^{\mathrm{I}0})^2 \; .\\ &= -\hat{\mathcal{L}}_1^{\mathrm{I}} + e^2 \hat{\phi}^{\mathrm{I}\dagger}\hat{\phi}^{\mathrm{I}} (\hat{A}^{\mathrm{I}0})^2 \; .\end{aligned} \tag{19}$$

In comparison with (6) the "normal-dependent" term in the quantized theory thus has the opposite sign within the interaction picture. Of course this does not change the nature of the problem posed by this term.

As mentioned earlier, the meaning of operator products will be specified by the *normal-ordering* prescription:

Example 8.6

$$\hat{\mathcal{H}}_1'^I = \mathrm{i} e :\hat{\phi}^{I\dagger} \stackrel{\leftrightarrow}{\partial}_\mu \hat{\phi}^I \hat{A}^{I\mu}: - e^2 :\hat{\phi}^{I\dagger} \hat{\phi}^I \hat{A}^I_\mu \hat{A}^{I\mu}: + e^2 :\hat{\phi}^{I\dagger} \hat{\phi}^I (\hat{A}^{I0})^2: . \qquad (20)$$

The spatial integral over (20) is the interaction Hamiltonian $\hat{H}_1^I$, which is the basis of perturbation theory as derived in Sect. 8.3. In the following the index I will be dropped for brevity's sake.

Inspecting (20) we note that the interaction Hamiltonian $\hat{\mathcal{H}}_1$ contains a linear (e) as well as a quadratic (e^2) dependence on the coupling constant. Therefore the nth term of the perturbation series of the $\hat{S}$ operator will be a superposition of contributions ranging between the orders e^n and e^{2n}. To obtain a power series in e the terms have to be reshuffled. For the case $n = 1$ we find

$$\begin{aligned}
\hat{S}^{(1)} &= -\mathrm{i} \int \mathrm{d}^4 x \; T\Big(\mathcal{H}_1(x)\Big) \\
&= -\mathrm{i} \int \mathrm{d}^4 x \; \Big(\mathrm{i} e :\hat{\phi}^\dagger(x) \stackrel{\leftrightarrow}{\partial}_\mu \hat{\phi}(x) \hat{A}^\mu(x): \\
&\quad -e^2 :\hat{\phi}^\dagger(x)\hat{\phi}(x) \hat{A}_\mu(x)\hat{A}^\mu(x): + e^2 :\hat{\phi}^\dagger(x)\hat{\phi}(x) \hat{A}_0(x)\hat{A}^0(x):\Big) .
\end{aligned} \qquad (21)$$

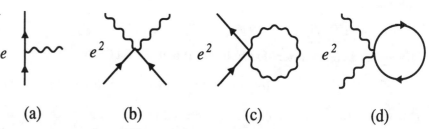

Fig. 8.13. Lowest-order Feynman diagrams of scalar electrodynamics

The corresponding Feynman diagrams are shown in Fig. 8.13a,b. The quadratic term leads to a four-legged vertex where two boson lines and two photon lines meet. This is also known by the imaginative name "seagull graph". Depending on the association of the external lines, diagram 8.13a could describe the emission/absorption of a single photon or the creation or annihilation of a fermion pair. We know, however, that such processes are kinematically forbidden. The diagram 8.13b, on the other hand, describes the allowed processes of Compton scattering, or the creation or annihilation of a fermion pair by two photons. Furthermore there are graphs involving two external lines only (Fig. 8.13c,d). The closed loops would make a contribution to the "self-energy" of the spin-0 boson or the photon, respectively. As we have already observed in QED the normal ordering of $\mathcal{H}_1$ eliminates these unobservable processes from the theory since the expectation value of the normal product with respect to the boson or the photon vacuum vanishes.

As mentioned earlier, there is a further contribution to the S matrix element of the same order e^2, which has to be added coherently. This contribution comes from $\hat{S}^{(2)}$, which in full is given by

$$\hat{S}^{(2)} = \frac{(-\mathrm{i})^2}{2!} \int \mathrm{d}^4 x \, \mathrm{d}^4 y \; T\Big(\hat{\mathcal{H}}_1(x)\hat{\mathcal{H}}_1(y)\Big)$$

8.6 The Feynman Rules of Quantum Electrodynamics

Example 8.6

$$\begin{aligned}= \frac{(-\mathrm{i})^2}{2!}\int \mathrm{d}^4x\,\mathrm{d}^4y\,T\Big\{&\Big(\mathrm{i}e:\hat\phi^\dagger(x)\stackrel{\leftrightarrow}{\partial}_\mu\hat\phi(x)\hat A^\mu(x):\\ &-e^2:\hat\phi^\dagger(x)\hat\phi(x)\,\hat A_\mu(x)\hat A^\mu(x):+e^2:\hat\phi^\dagger(x)\hat\phi(x)\,\hat A_0(x)\hat A^0(x):\Big)\\ \times&\Big(\mathrm{i}e:\hat\phi^\dagger(y)\stackrel{\leftrightarrow}{\partial}_\nu\hat\phi(y)\hat A^\nu(y):\\ &-e^2:\hat\phi^\dagger(y)\hat\phi(y)\,\hat A_\nu(y)\hat A^\nu(y):+e^2:\hat\phi^\dagger(y)\hat\phi(y)\,\hat A_0(y)\hat A^0(y):\Big)\Big\}\,.\end{aligned}\quad(22)$$

The decomposition of the T product into normal products according to Wick's theorem results in quite a number of terms. Looking only at the operator structure, we can classify the arguments of the T product into three different groups: $e^2(\hat\phi^\dagger\hat\phi\hat A)(\hat\phi^\dagger\hat\phi\hat A)$, $e^3(\hat\phi^\dagger\hat\phi\hat A)(\hat\phi^\dagger\hat\phi\hat A\hat A)$, and $e^4(\hat\phi^\dagger\hat\phi\hat A\hat A)(\hat\phi^\dagger\hat\phi\hat A\hat A)$. Each of these terms can be decomposed into normal products and contractions in several ways.

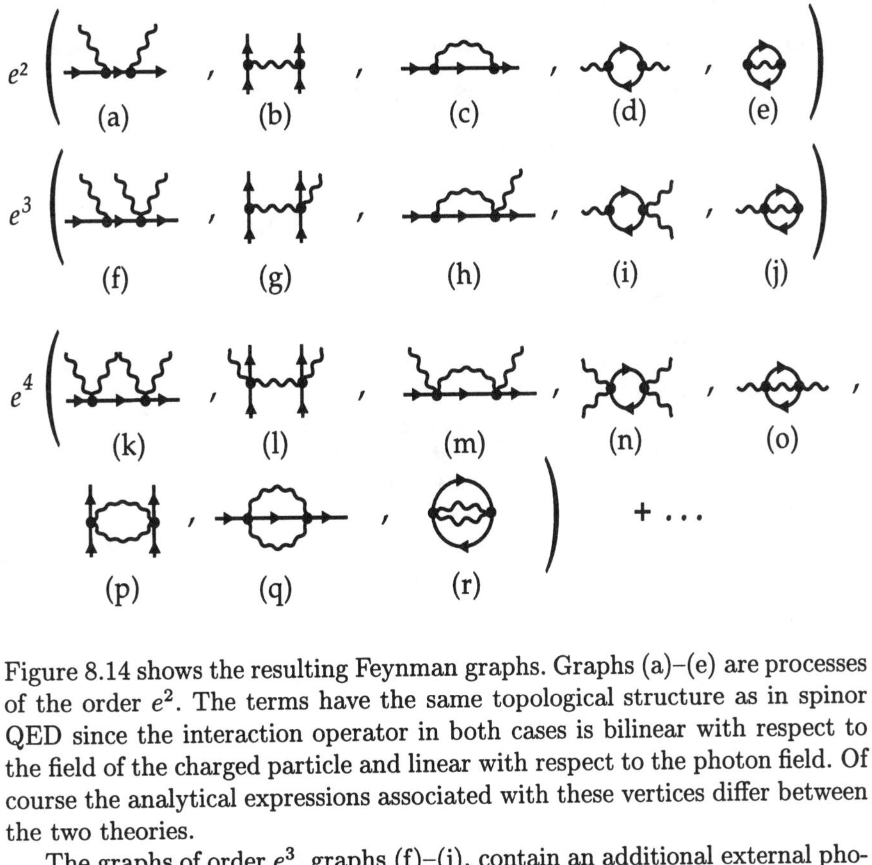

Fig. 8.14. A collection of Feynman graphs resulting from the expansion of the S matrix in second order

Figure 8.14 shows the resulting Feynman graphs. Graphs (a)–(e) are processes of the order e^2. The terms have the same topological structure as in spinor QED since the interaction operator in both cases is bilinear with respect to the field of the charged particle and linear with respect to the photon field. Of course the analytical expressions associated with these vertices differ between the two theories.

The graphs of order e^3, graphs (f)–(j), contain an additional external photon line glued to one of the two vertices. The graphs of order e^4 have two vertices with two photon lines each. The photon lines can correspond to free photons (graphs (k)–(o)) or they may be contracted to internal photon lines (graphs (p)–(r)).

The trivial nonconnected graphs have not been drawn in the figure since they simply correspond to the product of two first-order graphs. Clearly some

Example 8.6

of the processes of Fig. 8.14 that violate the joint energy and momentum conservation are kinematically forbidden. This is the case for graphs (h)–(j).

Let us reiterate that Fig. 8.14 contains only the graphs involving two vertices. The orders e^3 and e^4 of the power series expansion will also receive contributions from graphs with three and four vertices. Only for the order e^2 is it sufficient to restrict attention to the graphs from Figs. 8.13 and 8.14.

It would be tedious and not very enlightening if we set out to write down the algebraic expressions for the contributions to $\hat{S}^{(2)}$ made by each of the graphs in Fig. 8.14. We will do this only for the representative example of the two-boson two-photon interaction in order e^2 corresponding to the graph 8.14(a). The contribution to the $\hat{S}$ operator is obtained from a contraction $\hat{\phi}(x)\hat{\phi}^\dagger(y)$:

$$\hat{S}^{(2)}_{2B2\gamma} = 2\frac{(-i)^2}{2!}(ie)^2 \int d^4x\, d^4y : \hat{\phi}^\dagger(x)\stackrel{\leftrightarrow}{\partial_\mu}\hat{\phi}(x)\,\hat{\phi}^\dagger(y)\stackrel{\leftrightarrow}{\partial_\nu}\hat{\phi}(y)\,\hat{A}^\mu(x)\hat{A}^\nu(y): . \quad(23)$$

The factor 2 arises since there is a second contraction $\hat{\phi}^\dagger(x)\ldots\hat{\phi}(y)$ which leads to the same expression after we rename the variables $x \leftrightarrow y$

The contraction of the boson field operators $\hat{\phi}(x)$ and $\hat{\phi}^\dagger(y)$ usually gives rise to the Feynman propagator $\Delta_F(x-y)$:

$$\begin{aligned} i\Delta_F(x-y) &= \hat{\phi}(x)\hat{\phi}^\dagger(y) = \langle 0|T\big(\hat{\phi}(x)\hat{\phi}^\dagger(y)\big)|0\rangle \\ &= i\int \frac{d^4p}{(2\pi)^4}\frac{e^{-ip\cdot(x-y)}}{p^2 - m^2 + i\epsilon}. \end{aligned} \quad(24)$$

This function appears in (23) if the first gradient operator acts to the left and the second one acts to the right. In addition, however, there are contributions where the gradient of the field operator gets contracted. This has to be done carefully since the operator ∂_μ does not commute with the T product. Rather, the time derivative ∂_0 also acts on the step function $\Theta(x_0 - y_0)$ which implicitly is contained in the time-ordered product.

Let us first evaluate the action of a single gradient operator on the bosonic Feynman propagator. Using the product rule we find

$$\begin{aligned} i\partial^y_\nu \Delta_F(x-y) &= \partial^y_\nu \langle 0|T\big(\hat{\phi}(x)\hat{\phi}^\dagger(y)\big)|0\rangle \\ &= \partial^y_\nu\Big(\langle 0|\hat{\phi}(x)\hat{\phi}^\dagger(y)|0\rangle \Theta(x_0 - y_0) + \langle 0|\hat{\phi}^\dagger(y)\hat{\phi}(x)|0\rangle \Theta(y_0 - x_0)\Big) \\ &= \langle 0|\hat{\phi}(x)\partial_\nu\hat{\phi}^\dagger(y)|0\rangle \Theta(x_0 - y_0) + \langle 0|\partial_\nu\hat{\phi}^\dagger(y)\hat{\phi}(x)|0\rangle \Theta(y_0 - x_0) \\ &\quad + (-)g_{\nu 0}\delta(x_0 - y_0)\langle 0|\hat{\phi}(x)\hat{\phi}^\dagger(y)|0\rangle + g_{\nu 0}\delta(x_0 - y_0)\langle 0|\hat{\phi}^\dagger(y)\hat{\phi}(x)|0\rangle \\ &= \langle 0|T\big(\hat{\phi}(x)\partial_\nu\hat{\phi}^\dagger(y)\big)|0\rangle - g_{\nu 0}\delta(x_0 - y_0)\langle 0|\,[\hat{\phi}(x),\hat{\phi}^\dagger(y)]\,|0\rangle. \end{aligned} \quad(25)$$

The last term does not contribute: The commutator is just the Pauli–Jordan function discussed in Sect. 4.4

$$\begin{aligned} i\Delta(x-y) &= [\hat{\phi}(x),\hat{\phi}^\dagger(y)] \\ &= \int \frac{d^3p}{(2\pi)^3 2\omega_p}\left(e^{-ip\cdot(x-y)} - e^{+ip\cdot(x-y)}\right). \end{aligned} \quad(26)$$

This function enters (25) for equal time arguments $x_0 = y_0$. We know from (4.104) that

$$\Delta(0, \boldsymbol{x} - \boldsymbol{y}) = 0, \tag{27}$$

which also can be read from the integral representation (26). Thus we have shown

$$\underbrace{\hat{\phi}(x)\partial_\nu^y \hat{\phi}^\dagger(y)}_{} = \mathrm{i}\,\partial_\nu^y \Delta_\mathrm{F}(x-y), \tag{28}$$

as one would expect naively. However, if a second gradient operator acts, the situation changes:

$$\mathrm{i}\partial_\mu^x \partial_\nu^y \Delta_\mathrm{F}(x-y) = \partial_\mu^x \partial_\nu^y \langle 0|T(\hat{\phi}(x)\hat{\phi}^\dagger(y))|0\rangle$$
$$= \partial_\mu^x \langle 0|T(\hat{\phi}(x)\partial_\nu \hat{\phi}^\dagger(y))|0\rangle$$
$$= \partial_\mu^x \Big(\langle 0|\hat{\phi}(x)\partial_\nu \hat{\phi}^\dagger(y)|0\rangle\,\Theta(x_0-y_0) + \langle 0|\partial_\nu \hat{\phi}^\dagger(y)\hat{\phi}(x)|0\rangle\,\Theta(y_0-x_0)\Big)$$
$$= \langle 0|T(\partial_\mu\hat{\phi}(x)\partial_\nu\hat{\phi}^\dagger(y))|0\rangle + g_{\mu 0}\delta(x_0-y_0)\langle 0|[\hat{\phi}(x),\partial_\nu\hat{\phi}^\dagger(y)]|0\rangle . \tag{29}$$

The extra term in this case does not vanish! Using (26) we find

$$g_{\mu 0}\delta(x_0-y_0)\langle 0|[\hat{\phi}(x),\partial_\nu\hat{\phi}^\dagger(y)]|0\rangle = g_{\mu 0}\delta(x_0-y_0)\,\mathrm{i}\partial_\nu^y \Delta(x-y)$$
$$= g_{\mu 0}\delta(x_0-y_0) \int \frac{\mathrm{d}^3 p}{(2\pi)^3 2\omega_p} \Big(\mathrm{i}p_\nu\,\mathrm{e}^{+\mathrm{i}p\cdot(x-y)} - (-\mathrm{i}p_\nu)\,\mathrm{e}^{-\mathrm{i}p\cdot(x-y)}\Big)$$
$$= \mathrm{i}g_{\mu 0}\delta(x_0-y_0) \int \frac{\mathrm{d}^3 p}{(2\pi)^3 2\omega_p} p_\nu \Big(\mathrm{e}^{+\mathrm{i}p\cdot(x-y)} + \mathrm{e}^{-\mathrm{i}p\cdot(x-y)}\Big). \tag{30}$$

The integral vanishes for spatial indices $\nu \neq 0$ since both terms cancel if one makes the replacement $\boldsymbol{p} \to -\boldsymbol{p}$ in the second integral. For $\nu = 0$, however, both terms have the same sign (remember $p_0 \equiv +\omega_p$) leading to

$$\mathrm{i}g_{\mu 0}\delta(x_0-y_0)g_{\nu 0} \int \frac{\mathrm{d}^3 p}{(2\pi)^3 2\omega_p} 2\omega_p\,\mathrm{e}^{\mathrm{i}p\cdot(x-y)} = \mathrm{i}g_{\mu 0}\,g_{\nu 0}\,\delta^4(x-y). \tag{31}$$

Thus the *contraction of the gradients of two field operators* reads

$$\underbrace{\partial_\mu \hat{\phi}(x)\partial_\nu \hat{\phi}^\dagger(y)}_{} = \mathrm{i}\partial_\mu^x \partial_\nu^y \Delta_\mathrm{F}(x-y) - \mathrm{i}\,g_{\mu 0}\,g_{\nu 0}\,\delta^4(x-y). \tag{32}$$

Thus we have found an extra term that singles out the time-like component. Taken on its own this term would destroy the Lorentz covariance of the theory. It is important to keep in mind that the operations that led to (24), (28), and (32) were based on the fact that $\hat{\phi}(x)$ satisfies the free-field commutation relations. The derivations are valid in the interaction picture, not in the Heisenberg picture.

Now we are able to evaluate (23):

$$\hat{S}^{(2)}_{2B2\gamma} = 2\frac{(-\mathrm{i})^2}{2!}(\mathrm{i}e)^2 \int \mathrm{d}^4x\,\mathrm{d}^4y : \Big(\hat{\phi}^\dagger(x)\partial_\mu\underbrace{\hat{\phi}(x)\hat{\phi}^\dagger(y)}_{}\partial_\nu\hat{\phi}(y)$$
$$+ \partial_\mu\hat{\phi}^\dagger(x)\underbrace{\hat{\phi}(x)\partial_\nu\hat{\phi}^\dagger(y)}_{}\hat{\phi}(y) - \partial_\mu\hat{\phi}^\dagger(x)\underbrace{\hat{\phi}(x)\hat{\phi}^\dagger(y)}_{}\partial_\nu\hat{\phi}(y)$$

Example 8.6

$$-\hat{\phi}^\dagger(x)\partial_\mu\hat{\phi}(x)\partial_\nu\hat{\phi}^\dagger(y)\hat{\phi}(y)\Big)\Big):\!:\hat{A}^\mu(x)\hat{A}^\nu(y)\!:$$

$$= (-i)^2(ie)^2\int d^4x\,d^4y : \Big(\hat{\phi}^\dagger(x)\partial_\nu\hat{\phi}(y)\,i\partial^x_\mu\Delta_F(x-y)$$

$$+\partial_\mu\hat{\phi}^\dagger(x)\hat{\phi}(y)\,i\partial^y_\nu\Delta_F(x-y) - \partial_\mu\hat{\phi}^\dagger(x)\partial_\nu\hat{\phi}(y)\,i\Delta_F(x-y)$$

$$-\hat{\phi}^\dagger(x)\hat{\phi}(y)\Big(i\partial^x_\mu\partial^y_\nu\,\Delta_F(x-y) - i\,g_{\mu 0}g_{\nu 0}\,\delta^4(x-y)\Big)\Big):$$

$$\times :\hat{A}^\mu(x)\hat{A}^\nu(y):$$

$$= (-i)^2(ie)^2\int d^4x\,d^4y :\hat{\phi}^\dagger(x)\overset{\leftrightarrow}{\partial^x_\mu}\,i\Delta_F(x-y)\overset{\leftrightarrow}{\partial^y_\nu}\hat{\phi}(y):\,:\hat{A}^\mu(x)\hat{A}^\nu(y):$$

$$+ \text{extra term}\,. \tag{33}$$

The noncovariant extra term reads

$$i(-i)^2(ie)^2\int d^4x\,d^4y\,:\hat{\phi}^\dagger(x)\hat{\phi}(y):\,g_{\mu 0}\,g_{\nu 0}\,\delta^4(x-y)\,:\hat{A}^\mu(x)\hat{A}^\nu(y):$$

$$= ie^2\int d^4x\,:\hat{\phi}^\dagger(x)\hat{\phi}(x):\,:\hat{A}_0(x)\hat{A}_0(x):\,. \tag{34}$$

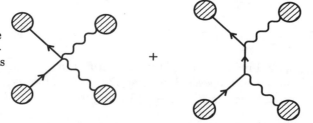

Fig. 8.15. Two subgraphs of a complicated Feynman graph which are of the same order and always appear together. If their contributions are combined the normal-dependent terms get cancelled.

We realize that (34) just *cancels the normal-dependent term* that had shown up in $S^{(1)}$ (see (21)). Here we have shown this to happen for the case of the two-photon two-boson $\hat{S}$ operator in lowest order (e^2). However, the conclusion holds true in general. As depicted in Fig. 8.15, a two-photon vertex may arise as a part of a more complicated higher-order process. Inescapably it will be accompanied automatically by another graph of equal order containing two one-photon vertices and an intermediate boson line. It is quite obvious that the normal-dependent terms will always be cancelled, as in our example.[8]

Thus it is sufficient to replace the Hamiltonian (20) by the simpler covariant expression

$$\hat{\mathcal{H}}_1 = ie\,:\hat{\phi}^\dagger\overset{\leftrightarrow}{\partial_\mu}\hat{\phi}\,\hat{A}^\mu:\,-\,e^2\,:\hat{\phi}^\dagger\hat{\phi}\,\hat{A}_\mu\hat{A}^\mu:\,. \tag{35}$$

This leads to the correct result for the S operator if at the same time a *modified time-ordered product* T^* is introduced that has the property

$$\partial^x_\mu\partial^y_\nu T^*\big(\hat{\phi}(x)\hat{\phi}^\dagger(y)\big) = T\big(\partial_\mu\hat{\phi}(x)\partial_\nu\hat{\phi}^\dagger(y)\big)\,. \tag{36}$$

---

[8] A general proof of this was given in: F. Rohrlich: Phys. Rev. **80**, 666 (1950).

8.6 The Feynman Rules of Quantum Electrodynamics

Example 8.6

When contracting two boson field operators, the modified T^* is to be used, which implies that (32) is replaced by the naive expression

$$\overline{\partial_\mu \hat{\phi}(x) \partial_\nu \hat{\phi}^\dagger(y)} = \mathrm{i}\partial_\mu^x \partial_\nu^y \Delta_\mathrm{F}(x-y) \ . \tag{37}$$

To summarize: *the normal-dependent terms can simply be ignored in the perturbation expansion.*

To arrive at the Feynman rules of scalar electrodynamics we evaluate the S matrix element for *spin-0 Compton scattering* as an example:

$$S_{fi} = \langle 0 | \hat{a}_{k'\lambda'} \hat{c}_{p'} \hat{S} \hat{c}_p^\dagger \hat{a}_{k\lambda}^\dagger | 0 \rangle \ . \tag{38}$$

Here $\hat{c}_p^\dagger$ and $\hat{a}_{k\lambda}^\dagger$ are creation operators for a boson with momentum p and for a photon with momentum k and polarisation λ. The *first-order* contribution to the $\hat{S}$ operator (21) reads

$$S_{fi}^{(1)} = \mathrm{i}e^2 \int \mathrm{d}^4x \, \langle 0 | \hat{a}_{k'\lambda'} \hat{c}_{p'} : \hat{\phi}^\dagger(x) \hat{\phi}(x) \, \hat{A}_\mu(x) \hat{A}^\mu(x) : \hat{c}_p^\dagger \hat{a}_{k\lambda}^\dagger | 0 \rangle \ . \tag{39}$$

Now the expansions of the field operators with respect to free states (in the interaction picture) are inserted. The matrix element will contribute only if creation and annihilation operators are "paired off" and we easily find

$$\begin{aligned}
S_{fi}^{(1)} &= \mathrm{i}e^2 \int \mathrm{d}^4x \, N_{p'} N_{k'} N_p N_k \, \mathrm{e}^{\mathrm{i}p'\cdot x} \, \mathrm{e}^{-\mathrm{i}p\cdot x} \cdot \\
&\quad \times \left(\epsilon_\mu(k',\lambda') \mathrm{e}^{\mathrm{i}k'\cdot x} \epsilon^\mu(k,\lambda) \mathrm{e}^{-\mathrm{i}k\cdot x} + \epsilon_\mu(k,\lambda) \mathrm{e}^{-\mathrm{i}k\cdot x} \epsilon^\mu(k',\lambda') \mathrm{e}^{\mathrm{i}k'\cdot x} \right) \\
&= \mathrm{i}e^2 \, 2(2\pi)^4 \delta^4(p'+k'-p-k) \, N_{p'} N_{k'} N_p N_k \ ,
\end{aligned} \tag{40}$$

with the normalization factors $N_p = \left((2\pi)^3 2\omega_p\right)^{-1/2}$, etc. The factor 2 originates from the two possible associations of the field operators in the product $:\hat{A}_\mu(x)\hat{A}^\mu(x):$.

The evaluation of the matrix element (33) of the *second-order* $\hat{S}$ operator is a bit more involved:

$$\begin{aligned}
S_{fi}^{(2)} &= (-\mathrm{i})^2(\mathrm{i}e)^2 \int \mathrm{d}^4x \, \mathrm{d}^4y \, \langle 0 | \hat{a}_{k'\lambda'} \hat{c}_{p'} : \hat{\phi}^\dagger(x) \overset{\leftrightarrow}{\partial}_\mu^x \mathrm{i}\Delta_\mathrm{F}(x-y) \overset{\leftrightarrow}{\partial}_\nu^y \hat{\phi}(y) : \\
&\quad \times : \hat{A}^\mu(x) \hat{A}^\nu(y) : \hat{c}_p^\dagger \hat{a}_{k\lambda}^\dagger | 0 \rangle \ .
\end{aligned} \tag{41}$$

The matrix element is evaluated as in (40). After inserting the Fourier-transformed Feynman propagator

$$\Delta_\mathrm{F}(x-y) = \int \frac{\mathrm{d}^4q}{(2\pi)^4} \, \mathrm{e}^{-\mathrm{i}q\cdot(x-y)} \Delta_\mathrm{F}(q) \ , \tag{42}$$

we find

$$\begin{aligned}
S_{fi}^{(2)} &= (-\mathrm{i})^2 (\mathrm{i}e^2) N_{p'} N_{k'} N_p N_k \int \mathrm{d}^4x \, \mathrm{d}^4y \int \frac{\mathrm{d}^4q}{(2\pi)^4} \, \mathrm{i}\Delta_\mathrm{F}(q) \\
&\quad \times \mathrm{e}^{\mathrm{i}p'\cdot x} \overset{\leftrightarrow}{\partial}_\mu^x \mathrm{e}^{-\mathrm{i}q\cdot(x-y)} \overset{\leftrightarrow}{\partial}_\nu^y \mathrm{e}^{-\mathrm{i}p\cdot y} \\
&\quad \times \left(\epsilon^\mu(k',\lambda')\epsilon^\nu(k,\lambda) \mathrm{e}^{\mathrm{i}k'\cdot x} \mathrm{e}^{-\mathrm{i}k\cdot y} + \epsilon^\mu(k,\lambda)\epsilon^\nu(k',\lambda') \mathrm{e}^{-\mathrm{i}k\cdot x} \mathrm{e}^{\mathrm{i}k'\cdot y} \right) .
\end{aligned} \tag{43}$$

Example 8.6

The gradients in (43) in momentum space reduce to simple multiplication:

$$e^{ip'\cdot x} \overleftrightarrow{\partial}_\mu^x e^{-iq\cdot(x-y)} \overleftrightarrow{\partial}_\nu^y e^{-ip\cdot y}$$
$$= (-i)(q_\mu + p'_\mu)(-i)(q_\nu + p_\nu)\, e^{i(p'-q)\cdot x - i(p-q)\cdot y} \; . \tag{44}$$

All the integrations in (43) break down and we are left with

$$S^{(2)}_{fi} = (-i)^2(ie)^2 N_{p'} N_{k'} N_p N_k \, (2\pi)^4 \delta^4(p' + k' - p - k) \tag{45}$$

$$\times \Big((-i)(p+k+p')_\mu (-i)(p+k+p)_\nu \, i\Delta_F(p+k)\, \epsilon^\mu(k',\lambda')\epsilon^\nu(k,\lambda)$$

$$+ (-i)(p-k'+p')_\mu (-i)(p-k'+p)_\nu \, i\Delta_F(p-k')\, \epsilon^\mu(k,\lambda)\epsilon^\nu(k',\lambda') \Big) \; .$$

As expected, there is a direct and an exchange contribution, as in spinor QED. In addition, however, the two-photon vertex from equation (40) also contributes to Compton scattering. The three graphs to be added up coherently are shown in Fig. 8.16. We will not pursue the further steps that lead from (45) to the cross section for scalar Compton scattering. This can be found in Chap. 8 of the volume *Quantum Electrodynamics*.

Fig. 8.16. The three graphs describing Compton scattering of the order e^2 in scalar electrodynamics

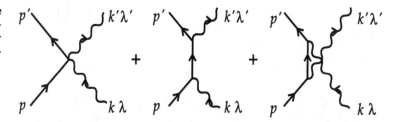

Our purpose here is to systematically derive the Feynman rules of scalar electrodynamics within the formalism of canonical field quantization. As usual all the topologically distinct Feynman graphs with the intended configuration of external lines have to be drawn. Both one-photon and two-photons vertices have to be taken into account. In addition to the general rules for treating external and internal lines given in Sect. 8.6 the following *Feynman rules of scalar electrodynamics* in momentum space apply.

1. Each one-photon vertex is associated with a factor $e(p_\mu + p'_\mu)$, multiplied by the delta function $(2\pi)^4\delta^4(p' - p \pm k)$.

2. Each two-photon vertex is associated with a factor $-2e^2 g_{\mu\nu}$, multiplied by the delta function $(2\pi)^4\delta^4(p' - p \pm k \pm k')$.

3. For each pair of two-photon vertices that are mutually connecced by two photon lines a factor $\frac{1}{2}$ has to be taken into account.

Rules 1 and 2 are obvious from the example of scalar Compton scattering, which we have worked out in detail. With some practice the vertex factor can be read directly from the structure of the interaction Hamilton density (20) or Lagrange density (2)

The *symmetry factor* $\frac{1}{2}$ contained in rule 3 needs some explanation. The "double seagull graph" in Fig. 8.17a arises from the contraction of the $\hat{A}$ field operators of a pair of two-photon vertices. The Wick decomposition of the time-ordered product $T(\hat{\mathcal{H}}_1(x)\hat{\mathcal{H}}_1(y))$ leads to two contractions of this type:

$$\underbracket{\hat{A}_\mu(x)\underbracket{\hat{A}_\nu(x)\,\hat{A}^\mu(y)}\hat{A}^\nu(y)} + \underbracket{\hat{A}_\mu(x)\underbracket{\hat{A}_\nu(x)\,\hat{A}^\mu(y)\hat{A}^\nu(y)}} . \tag{46}$$

According to rule 2 *each* of the vertices was associated with a factor 2 to account for the fact that the photon lines can be interchanged. However, here this obviously leads to double counting and has to be compensated for by the factor $\frac{1}{2}$. Note that the analog of Fig. 8.17a represented by the double boson line, cf. Fig. 8.17b, does not lead to a correction factor. In this case the photon lines (or rather the association of the photon lines with the field operators $\hat{A}_\mu(x_i)$) can be interchanged independently at both vertices, whereas the contraction of the two boson lines is unique.

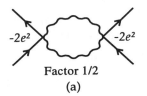
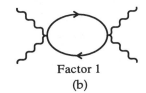

Fig. 8.17. A symmetry factor $\frac{1}{2}$ arises in the case of double photon lines

EXAMPLE

8.7 ϕ^4 Theory

The simplest interacting field theory describes one type of particles that can interact among each other. For the case of a neutral spin-0 field the Lagrange density can be split up as follows

$$\mathcal{L} = \mathcal{L}_0^{\mathrm{KG}} + \mathcal{L}_1 , \tag{1}$$

with the free Lagrange density

$$\mathcal{L}_0^{\mathrm{KG}} = \frac{1}{2}\partial_\mu\phi\partial^\mu\phi - \frac{1}{2}m^2\phi^2 . \tag{2}$$

The interaction term $\mathcal{L}_1$ is usually expressed as a power series expansion in ϕ. The simplest case applies if this series contains only a single term. The "most popular" model of this type is the ϕ^4 theory:

$$\mathcal{L}_1 := -gV(\phi) = -\frac{g}{4!}\phi^4 \quad , \quad \mathcal{H}_1 = -\mathcal{L}_1 . \tag{3}$$

Here g is a (dimensionless) coupling constant. The factor $1/4!$ was included in order to simplify later results as it will cancel combinatorial factors that arise

Example 8.7

in the perturbation expansion. The "potential" $V(\phi)$ of the self-interaction ϕ^4 is quartic, rising fast at large values of $|\phi| \to \infty$. From a practical point of view the ϕ^3 theory would be even more convenient. However, the corresponding potential $V(\phi) = \frac{1}{3!}\phi^3$ clearly is not bounded from below, allowing for an unhindered drop in energy. This renders the pure ϕ^3 theory pathological since it will not possess a stable vacuum state (state of lowest energy). This would not be noticeable within a formal evaluation of the terms of perturbation; nevertheless we prefer to study the more "realistic" ϕ^4 theory.[9]

After quantization of the ϕ field and the introduction of the normal-ordering prescription in the interaction operator (3) the series expansion of the $\hat{S}$ operator becomes

$$\begin{aligned}\hat{S} &= \mathbb{1} + \frac{-ig}{4!}\int d^4x\, {:}\hat{\phi}^4(x){:} \\ &+ \frac{1}{2!}\left(\frac{-ig}{4!}\right)^2 \int d^4x_1\, d^4x_2\, T\left({:}\hat{\phi}^4(x_1){:}\,{:}\hat{\phi}^4(x_2){:}\right) \\ &+ \ldots . \end{aligned} \quad (4)$$

The Feynman rules of the ϕ^4 theory are best found from a simple example such as *two-particle scattering* $p_1 + p_2 \to p'_1 + p'_2$. In contrast to the cases treated up to now, in the ϕ^4 theory the quartic structure of the interaction terms allows such scattering to show up already in first-order perturbation theory. The corresponding S matrix element reads

$$\begin{aligned}S_{fi}^{(1)} &= \langle 0|\hat{a}_{p'_1}\hat{a}_{p'_2}\, \hat{S}^{(1)}\, \hat{a}^\dagger_{p_2}\hat{a}^\dagger_{p_1}|0\rangle \\ &= \frac{-ig}{4!}\int d^4x\, \langle 0|\hat{a}_{p'_1}\hat{a}_{p'_2}\, {:}\hat{\phi}^4(x){:}\, \hat{a}^\dagger_{p_2}\hat{a}^\dagger_{p_1}|0\rangle . \end{aligned} \quad (5)$$

Splitting the products of field operators into creation and annihilation parts leads to sixteen terms, which can be grouped together as follows:

$${:}\hat{\phi}^4{:} = \hat{\phi}^{(-)4} + 4\hat{\phi}^{(-)3}\hat{\phi}^{(+)} + 6\hat{\phi}^{(-)2}\hat{\phi}^{(+)2} + 4\hat{\phi}^{(-)}\hat{\phi}^{(+)3} + \hat{\phi}^{(+)4} . \quad (6)$$

Equation (5) picks out the part containing two creation and two annihilation operators. Use of the plane-wave expansion of $\hat{\phi}$ leads to

$$\begin{aligned}S_{fi}^{(1)} &= \frac{-ig}{4!}6\int d^4x \int d^3q_1\, d^3q_2\, d^3q_3\, d^3q_4 \left(\prod_{i=1}^{4} N_{q_i}\right) e^{i(q_1+q_2-q_3-q_4)\cdot x} \\ &\times \langle 0|\hat{a}_{p'_1}\hat{a}_{p'_2}\, \hat{a}^\dagger_{q_1}\hat{a}^\dagger_{q_2}\hat{a}_{q_3}\hat{a}_{q_4}\, \hat{a}^\dagger_{p_2}\hat{a}^\dagger_{p_1}|0\rangle . \end{aligned} \quad (7)$$

The Fock-space matrix element as usual is evaluated by commuting $\hat{a}$ and $\hat{a}^\dagger$. The creation and annihilation operators must be exactly "paired off", resulting in

---

[9] In fact the ϕ^4 theory in four dimensions most probably is also not a realistic model. Detailed investigations have shown that through renormalization effects, i.e., through quantum fluctuations, the interaction completely shields itself, leading to an effective coupling constant $g_{\text{ren}} = 0$ after renormalization. Thus ϕ^4 theory according to present-day knowledge is "trivial", i.e., noninteracting. For more information see M. Lüscher and P. Weisz: Nucl. Phys. **B290**, 25 (1987); J. Fröhlich, Nucl. Phys. **B200**, 281(1982).

$$\langle 0|\hat{a}_{p'_1}\hat{a}_{p'_2}\hat{a}^\dagger_{q_1}\hat{a}^\dagger_{q_2}\hat{a}_{q_3}\hat{a}_{q_4}\hat{a}^\dagger_{p_2}\hat{a}^\dagger_{p_1}|0\rangle$$
$$= \left(\delta_{q_1 p'_1}\delta_{q_2 p'_2} + \delta_{q_1 p'_2}\delta_{q_2 p'_1}\right)\left(\delta_{q_3 p_1}\delta_{q_4 p_2} + \delta_{q_3 p_2}\delta_{q_4 p_1}\right). \quad (8)$$

The two terms within the brackets produce exactly the same integrand in (7), which leads to a factor of $2 \times 2 = 4$. As a result

$$S^{(1)}_{fi} = \frac{-ig}{4!} 6 \cdot 4 \left(\prod_{i=1}^{4} N_{p_i}\right) \int d^4 x \, e^{i(p'_1 + p'_2 - p_1 - p_2)\cdot x}$$
$$= -ig\left(\prod_{i=1}^{4} N_{p_i}\right) (2\pi)^4 \delta^4(p'_1 + p'_2 - p_1 - p_2). \quad (9)$$

Thus a combinatorical factor 4! has shown up, which cancels the normalization factor introduced when the coupling constant g in (3) was defined. That this will always happen is easy to understand since there always are 4! equivalent alternative to "link the four field operators in $\hat{\phi}^4(x)$ to the outside world". In the language of Feynman graphs the ϕ^4 interaction is described by a *four-boson vertex*. This vertex is associated with the following:

- Four-boson vertex: factor $-ig$.

Now we evaluate the *second-order* S matrix element.

$$S^{(2)}_{fi} = \frac{1}{2!}\left(\frac{-ig}{4!}\right)^2 \int d^4x_1 d^4x_2 \, \langle 0|\hat{a}_{p'_1}\hat{a}_{p'_2} T(:\hat{\phi}^4(x_1)::\hat{\phi}^4(x_2):)\hat{a}^\dagger_{p_2}\hat{a}^\dagger_{p_1}|0\rangle. \quad (10)$$

This of course is again evaluated via Wick's theorem. In our case this reads

$$T(:\hat{\phi}^4(x_1)::\hat{\phi}^4(x_2):) = :\hat{\phi}^4(x_1)\hat{\phi}^4(x_2): \quad (11a)$$
$$+ 4^2 \, \underbracket{\hat{\phi}(x_1)\hat{\phi}(x_2)} :\hat{\phi}^3(x_1)\hat{\phi}^3(x_2): \quad (11b)$$
$$+ \frac{4^2 \cdot 3^2}{2!}\left(\underbracket{\hat{\phi}(x_1)\hat{\phi}(x_2)}\right)^2 :\hat{\phi}^2(x_1)\hat{\phi}^2(x_2): \quad (11c)$$
$$+ \frac{4^2 \cdot 3^2 \cdot 2^2}{3!}\left(\underbracket{\hat{\phi}(x_1)\hat{\phi}(x_2)}\right)^3 :\hat{\phi}(x_1)\hat{\phi}(x_2): \quad (11d)$$
$$+ \frac{4^2 \cdot 3^2 \cdot 2^2}{4!}\left(\underbracket{\hat{\phi}(x_1)\hat{\phi}(x_2)}\right)^4. \quad (11e)$$

The combinatorial prefactors take into account the number of different ways the (multiple) contractions between the two groups of four field operators each can be formed. The denominators in (11c)–(11e) are important. It is their task to ensure that the same configuration of contractions is not counted twice. The various terms in (11) can be translated into the Feynman graphs of Fig. 8.18.

Without normal ordering being imposed on $\hat{\mathcal{H}}_1$, additional closed-loop graphs would have arisen:

in $\hat{S}^{(1)}$: ⊶ ,

Example 8.7

in $\hat{S}^{(3)}$: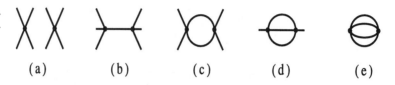

They produce constant (i.e., momentum-independent) self-energy contributions which are eliminated when the theory is renormalized. No harm is done if they are discarded from the start.

Fig. 8.18. The second-order diagrams of ϕ^4 theory

(a)   (b)   (c)   (d)   (e)

Using (11c) we can evaluate the second-order S matrix element (10):

$$S_{fi}^{(2)} = \frac{1}{2!}\left(\frac{-\mathrm{i}g}{4!}\right)^2 \frac{4^2 \cdot 3^2}{2!} \int \mathrm{d}^4x_1\,\mathrm{d}^4x_2\,\bigl(\mathrm{i}\Delta_\mathrm{F}(x_1-x_2)\bigr)^2$$
$$\times \langle 0|\hat{a}_{\boldsymbol{p}_1'}\hat{a}_{\boldsymbol{p}_2'}:\hat{\phi}^2(x_1)\hat{\phi}^2(x_2):\hat{a}_{\boldsymbol{p}_2}^\dagger \hat{a}_{\boldsymbol{p}_1}^\dagger |0\rangle\;. \tag{12}$$

The vacuum expectation value differs from that in (5) only by the fact that two different coordinates x_1 and x_2 are involved. As in (6) there are six terms containing two creation and two annihilation operators:

$$\begin{aligned}:\hat{\phi}^2(x_1)\hat{\phi}^2(x_2): \;&=\; \hat{\phi}^{(-)2}(x_1)\hat{\phi}^{(+)2}(x_2) \\ &\quad + 4\hat{\phi}^{(-)}(x_1)\hat{\phi}^{(-)}(x_2)\hat{\phi}^{(+)}(x_1)\hat{\phi}^{(+)}(x_2) \\ &\quad + \hat{\phi}^{(-)2}(x_2)\hat{\phi}^{(+)2}(x_1) \\ &\quad + \ldots\;. \end{aligned} \tag{13}$$

Using this (12) becomes

$$S_{fi}^{(2)} = \frac{1}{2!}\left(\frac{-\mathrm{i}g}{4!}\right)^2 \frac{4^2 \cdot 3^2}{2!} \int \mathrm{d}^4x_1\,\mathrm{d}^4x_2\,\bigl(\mathrm{i}\Delta_\mathrm{F}(x_1-x_2)\bigr)^2$$
$$\times \int \mathrm{d}^3q_1\mathrm{d}^3q_2\mathrm{d}^3q_3\mathrm{d}^3q_4 \prod_{i=1}^{4} N_{q_i} \tag{14}$$
$$\times \Bigl(\mathrm{e}^{\mathrm{i}(q_1\cdot x_1+q_2\cdot x_1-q_3\cdot x_2-q_4\cdot x_2)}\langle 0|\hat{a}_{\boldsymbol{p}_1'}\hat{a}_{\boldsymbol{p}_2'}\hat{a}_{\boldsymbol{q}_1}^\dagger \hat{a}_{\boldsymbol{q}_2}^\dagger \hat{a}_{\boldsymbol{q}_3}\hat{a}_{\boldsymbol{q}_4}\hat{a}_{\boldsymbol{p}_2}\hat{a}_{\boldsymbol{p}_1}|0\rangle$$
$$+\,4\,\mathrm{e}^{\mathrm{i}(q_1\cdot x_1-q_2\cdot x_1+q_3\cdot x_2-q_4\cdot x_2)}\langle 0|\hat{a}_{\boldsymbol{p}_1'}\hat{a}_{\boldsymbol{p}_2'}\hat{a}_{\boldsymbol{q}_1}^\dagger \hat{a}_{\boldsymbol{q}_3}^\dagger \hat{a}_{\boldsymbol{q}_2}\hat{a}_{\boldsymbol{q}_4}\hat{a}_{\boldsymbol{p}_2}\hat{a}_{\boldsymbol{p}_1}|0\rangle$$
$$+\,\mathrm{e}^{\mathrm{i}(-q_1\cdot x_1-q_2\cdot x_1+q_3\cdot x_2+q_4\cdot x_2)}\langle 0|\hat{a}_{\boldsymbol{p}_1'}\hat{a}_{\boldsymbol{p}_2'}\hat{a}_{\boldsymbol{q}_3}^\dagger \hat{a}_{\boldsymbol{q}_4}^\dagger \hat{a}_{\boldsymbol{q}_1}\hat{a}_{\boldsymbol{q}_2}\hat{a}_{\boldsymbol{p}_2}\hat{a}_{\boldsymbol{p}_1}|0\rangle\Bigr)\;.$$

The vacuum expectation values all have the structure (8). As a consequence the q_i integrations collapse and after identical terms are combined the following result is obtained:

$$S_{fi}^{(2)} = \frac{1}{2!}\left(\frac{-\mathrm{i}g}{4!}\right)^2 \frac{4^2 \cdot 3^2}{2!}\cdot 4\cdot 2 \prod_{i=1}^{4} N_{p_i} \int \mathrm{d}^4x_1\,\mathrm{d}^4x_2\,\bigl(\mathrm{i}\Delta_\mathrm{F}(x_1-x_2)\bigr)^2$$
$$\times \Bigl(\mathrm{e}^{\mathrm{i}(p_1'+p_2')\cdot x_1}\,\mathrm{e}^{-\mathrm{i}(p_1+p_2)\cdot x_2} + \mathrm{e}^{\mathrm{i}(p_1'-p_1)\cdot x_1}\,\mathrm{e}^{\mathrm{i}(p_2'-p_2)\cdot x_2}$$
$$+\,\mathrm{e}^{\mathrm{i}(p_2'-p_1)\cdot x_1}\,\mathrm{e}^{\mathrm{i}(p_1'-p_2)\cdot x_2}\Bigr)\;. \tag{15}$$

8.6 The Feynman Rules of Quantum Electrodynamics

The prefactor amounts to $\frac{1}{2}(-ig)^2$. Equation (15) can also be represented graphically. It consists of the coherent summation of the three Feynman graphs depicted in Fig. 8.19.

Example 8.7

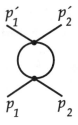 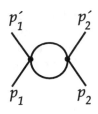 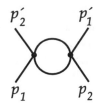

Fig. 8.19. The three topologically distinct boson–boson scattering graphs

Each of these graphs contains a *loop* formed by the two boson propagators connecting x_1 and x_2. Both associations of the coordinates x_i with the vertices make equal contributions, which leads to the factor 2 in (15). We already know that a graph of order n carries a permutation factor $n!$ which cancels the $1/n!$ contained in Dyson's perturbation series.

The result (15) is easily transformed to *momentum space*:

$$S_{fi}^{(2)} = \frac{1}{2}(-ig)^2 \prod_{i=1}^{4} N_{p_i} \int \frac{d^4q_1}{(2\pi)^4} \frac{d^4q_2}{(2\pi)^4} \; i\Delta_F(q_1) \, i\Delta_F(q_2)$$

$$\times \int d^4x_1 \, d^4x_2 \; e^{-iq_1\cdot(x_1-x_2)} \, e^{-iq_2\cdot(x_1-x_2)}$$

$$\times \Big(e^{i(p_1'+p_2')\cdot x_1} e^{-i(p_1+p_2)\cdot x_2} + e^{i(p_1'-p_1)\cdot x_1} e^{i(p_2'-p_2)\cdot x_2}$$

$$+ e^{i(p_2'-p_1)\cdot x_1} e^{i(p_1'-p_2)\cdot x_2} \Big)$$

$$= \frac{1}{2}(-ig)^2 \, \delta^4(p_1'+p_2'-p_1-p_2) \prod_{i=1}^{4} N_{p_i}$$

$$\times \Big(J(p_1+p_2) + J(p_1'-p_1) + J(p_2'-p_1) \Big) \; . \tag{16}$$

Here the *loop integral* $J(p)$ is defined as follows:

$$J(p) = \int d^4q \; \frac{i}{q^2-m^2+i\epsilon} \; \frac{i}{(q-p)^2-m^2+i\epsilon} \; . \tag{17}$$

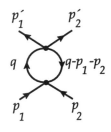 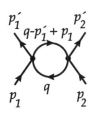 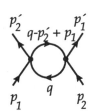

Fig. 8.20. Same as Fig. 8.19 indicating the flow of momentum

Example 8.7

The integration variable q denotes the momentum that "runs through the loop". Figure 8.20 depicts the Feynman graphs in momentum space where the lines have now been endowed with arrows that represent the direction of momentum flow.

Remark: Although we will not deal with the topic of renormalization in this book it should be noted that results derived in perturbation theory, such as (16), have to be treated with care. By simply counting the powers of momentum, one observes that the integral (17) is *logarithmically divergent* at large values of q:

$$J(p) \simeq \int d^4q \, \frac{1}{q^4} \simeq \ln \Lambda \qquad (18)$$

where Λ is a cutoff momentum introduced to "regularize" the integral. It can be shown that this cutoff dependence does not influence physical observables. It only affects the *renormalization of the coupling constant*, leading from its ill-defined "bare" value g to the renormalized coupling constant g_ren, which is to be identified with the observable value.

The result (16) could have been given without intermediate derivations directly by using the Feynman rules in momentum space. Only the factor $\frac{1}{2}$ needs some consideration. Its origin is the same as in Example 8.6 (cf. Fig. 8.17).

If more complicated graphs are involved it may become quite tedious to find the correct prefactor. Fortunately there is a supplementary Feynman rule that deals with this problem.

- Each Feynman graph carries a *symmetry factor* $1/S$ where $S = \nu \prod_{n=2,3,\ldots} (n!)^{\pi_n}$.

Here π_n is the number of pairs of vertices connected by n identical lines. Lines having an opposite sense of direction (like electrons and positrons) do not count as identical. The factor ν counts the number of graphs that are topologically equivalent, keeping the configuration of external lines fixed. The latter definition may sound a bit obscure, so we will illustrate it by an example. In ϕ^4 theory the following fourth-order "cat's-eye" graphs can be drawn:

These two diagrams, however, are completely equivalent; they represent identically the same mathematical expression, only depicted in two different ways. Therefore there is *only one* such contribution to the boson scattering amplitude and the $1/n!$ prefactor will not be cancelled. To take this into account the graph has to be multiplied by $\frac{1}{2}$. The overall symmetry factor of the cat's eye graph is $1/S = 1/4$ since $\nu = 2$ and $\pi_2 = 1$.

Example 8.7

As a counterexample look at the self-energy graph containing a three-boson loop:

 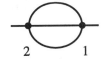

The two versions of this graph look very similar but here the sequence of the vertices is different. Therefore both graphs contribute and no compensating factor $1/\nu$ is required. There is still a symmetry factor, of course, since the two vertices are connected by identical lines: $\nu = 1$, $\pi_2 = 0$, $\pi_3 = 1$, which results in $1/S = 1/6$.

We finally note that in some theories the obnoxious symmetry factors do not emerge at all. This is true in particular for quantum electrodynamics. Because there is only a single simply structured vertex and because the fermion lines carry a sense of direction, QED graphs with internal lines are always topologically distinct. This is illustrated by the comparison of the following corresponding graphs of QED (left) and ϕ^3 theory (right). The right-hand graph carries a symmetry factor $1/S = 1/2$.

8.7 Appendix: The Scattering Cross Section

For easy reference in this appendix we present the connection between the S matrix element and the physical scattering cross section. This topic is discussed in more detail in the volume *Quantum Electrodynamics*.

Starting from the scattering matrix element S_{fi}, we define the *invariant amplitude* M_{fi} by extracting some kinematical factors. For a scattering process with two particles in the initial state and n particles in the final state (momentum balance $p_1 + p_2 \to p'_1 + p'_2 + \ldots + p'_n$) the connection is

$$S_{fi} = \mathrm{i}(2\pi)^4\, \delta^4\!\left(p_1 + p_2 - \sum_{i=1}^{n} p'_i\right) M_{fi} \prod_{i=1}^{2} \sqrt{\frac{N_i}{2E_i(2\pi)^3}} \prod_{i=1}^{n} \sqrt{\frac{N'_i}{2E'_i(2\pi)^3}} \tag{8.101}$$

Here the factors N_i depend on the normalization conventions used for the different types of particles involved: $N_i = 1$ for spin-0 bosons and photons,[10] $N_i = 2m$ for spin-$\frac{1}{2}$ particles.

---

[10] If Gaussian units are used the normalization factor for photons is $N_i = 4\pi$.

The *differential scattering cross section* is given by

$$\sigma = \frac{S}{4\sqrt{(p_1 \cdot p_2)^2 - m_1^2 m_2^2}} N_1 N_2 (2\pi)^4 \, \delta^4\left(p_1 + p_2 - \sum_{i=1}^n p_i'\right) |M_{fi}|^2$$
$$\times \prod_{i=1}^n \frac{N_i' d^3 p_i'}{2E_i'(2\pi)^3} \,. \tag{8.102}$$

The square-root factor is the invariant particle flux, which is essentially given by the relative velocity of the incoming particles. Furthermore (8.102) contains a delta function that accounts for the energy and momentum balance and a phase space factor for each of the outgoing particles. The degeneracy factor S has to be included if the final states contains indistinguishable particles. It is defined by $S = \prod_k 1/g_k!$ where g_k counts the multiplicity of particles of species k. If polarizations are not measured one has to average over the inital spins and to sum over the final spins.

9. The Reduction Formalism

9.1 Introduction

The description of interacting quantum fields is beset with both conceptual and technical problems. In particular, the field operators do not have a simple interpretation as creation and annihilation operators of isolated particles with well-defined properties. On the other hand, one often encounters experimental conditions characterized by the interaction of individual particles. *Scattering experiments* are defined in such a way that asymptotically, at times well before or well after the collision, specific configurations of free particles are found, having well-defined momentum, spin, etc. In the last section we developed a method, based on the interaction picture, that allowed the construction of the transition amplitude of a scattering process, i.e., the S matrix.

In the following we will take a closer look at this subject and add some conceptual and mathematical details.[1] The treatment will we based on the *Heisenberg picture*, with the use of time-independent state vectors and field operators that carry the full dynamics. The main result will be the asymptotic condition for the field operators and the so-called Lehmann–Symanzik–Zimmermann (LSZ) formalism, which provides a "reduction formula" for calculating S matrix elements. When evaluated in perturbation theory the latter will turn out to be equivalent to our earlier results. Our presentation is intended to supplement and elaborate the results obtained in the last chapter, not as an exhaustive or rigorous study of interacting quantum field theory. Therefore most of the discussion will be limited to the simplest case of scalar fields.

As the starting point for field quantization we postulate that the *canonical commutation relations* are valid also for the interacting field operators. In the case of a neutral spin-0 field this amounts to (cf. (4.5))

$$\left[\hat{\phi}(x), \dot{\hat{\phi}}(y)\right]_{x_0=y_0} = i\delta^3(\boldsymbol{x}-\boldsymbol{y}), \tag{9.1a}$$

$$\left[\hat{\phi}(x), \hat{\phi}(y)\right]_{x_0=y_0} = \left[\dot{\hat{\phi}}(x), \dot{\hat{\phi}}(y)\right]_{x_0=y_0} = 0, \tag{9.1b}$$

---

[1] We do not intend to present a truely rigorous formulation, however. This is done in *axiomatic* or *constructive quantum field theory* which tries to set up a mathematically consistent framework starting from a small number of clearly stated basic postulates. A standard introduction to this field is: R.F. Streater, A.S. Wightman: *PCT, Spin and Statistics, and All That* (Benjamin, New York 1964).

assuming that there are no gradient couplings that modify the canonically conjugate field. In contrast to the noninteracting case, however, these commutation relations *cannot be extended to arbitrary coordinates involving unequal times* $x_0 \neq y_0$.

The commutator between the field operator and the momentum operator, which is the generator of infinitesimal space and time translations, keeps the form

$$[\hat{P}^\mu, \hat{\phi}(x)] = -i\partial^\mu \hat{\phi}(x) , \qquad (9.2)$$

which has the formal solution

$$\hat{\phi}(x) = e^{i\hat{P}\cdot x} \hat{\phi}(0) e^{-i\hat{P}\cdot x} . \qquad (9.3)$$

9.2 In and Out Fields

In the following, we study a system that asymptotically consists of a collection of well separated particles. For simplicity the discussion will be restricted, for the time being, to the case of a single neutral scalar field.

Let us assume that *in the asymptotic region* the familiar construction of the Fock space can be carried out. This implies in particular the *existence of a vacuum state* $|0\rangle$. This state will be normalized,

$$\langle 0|0 \rangle = 1 , \qquad (9.4)$$

and will satisfy the condition of Poincaré invariance. The energy and angular momentum of the vacuum by convention are defined to be zero:

$$\hat{P}_\mu |0\rangle = 0 \quad , \quad \hat{M}_{\mu\nu}|0\rangle = |0\rangle . \qquad (9.5)$$

The *asymptotically* $(x_0 \to \pm\infty)$ *free quantum field* is decribed by Heisenberg field operators marked by the index "in" or "out". These operators are required to satisfy the *noninteracting field equation*, which for spinless bosons is just the Klein–Gordon equation

$$(\Box + m^2)\hat{\phi}_{\rm in}(x) = 0 \quad , \quad (\Box + m^2)\hat{\phi}_{\rm out}(x) = 0 . \qquad (9.6)$$

where the physical mass m of the particle enters.[2]

The field operators $\hat{\phi}_{\rm in/out}$ can be treated in the usual way. An expansion with respect to the plane-wave basis $u_p(x)$ (cf. (4.27)),

$$\hat{\phi}_{\rm in}(\boldsymbol{x}, t) = \int d^3p \left(\hat{a}_{\boldsymbol{p},{\rm in}} u_{\boldsymbol{p}}(\boldsymbol{x}, t) + \hat{a}^\dagger_{\boldsymbol{p},{\rm in}} u^*_{\boldsymbol{p}}(\boldsymbol{x}, t) \right) , \qquad (9.7)$$

leads to a set of (constant) creation and annihilation operators that satisfy the usual bosonic commutation relations. In this way the n-particle Fock-space states,

$$|\boldsymbol{p}_1, \ldots, \boldsymbol{p}_n; {\rm in}\rangle = \hat{a}^\dagger_{\boldsymbol{p}_1,{\rm in}} \cdots \hat{a}^\dagger_{\boldsymbol{p}_n,{\rm in}} |0\rangle , \qquad (9.8)$$

---

[2] In this chapter a few remarks will be made on mass renormalization. Therefore we have to distinguish between the *physical mass* m, seen in experiments, and the *bare mass* m_0, present in the original free Lagrangian of the theory. The latter will be affected by the interaction.

are constructed. (For simplicity we assume that all the momenta are different, otherwise a normalization factor has to be included as in (3.34).) The out operators satisfy the same equations (9.7) and (9.8) as the in operators. We postulate that *the vacuum state is stable and unique:*[3]

$$|0\rangle = |0;\text{in}\rangle = |0;\text{out}\rangle . \qquad (9.9)$$

The quantities of interest are the *S matrix elements*, i.e., the quantum amplitudes describing the transition from an n-particle in state to an m-particle out state:

$$S_{fi} = \langle q_1,\ldots,q_m;\text{out}|p_1,\ldots,p_n;\text{in}\rangle . \qquad (9.10)$$

This can be written as an expectation value of the *S operator* which connects the in and out operators:

$$\hat{\phi}_{\text{out}}(x) = \hat{S}^{-1}\hat{\phi}_{\text{in}}(x)\hat{S} . \qquad (9.11)$$

This operator should be *unitary*:

$$\hat{S}^\dagger \hat{S} = \mathbb{I} . \qquad (9.12)$$

To ensure the mathematical validity of this condition the *postulate of asymptotic completeness* is imposed, according to which the in and the out states span the same Hilbert space, which is also assumed to agree with the Hilbert space of the interacting theory. This is not a trivial postulate: if particles can be permanently combined into bound states the structure of the Hilbert space is modified. In the following, we will assume that problems from stable bound states do not arise.

Since the basis functions $u_p(x)$ are linearly independent, and since they are not affected by $\hat{S}$, the transformation relation (9.11) also holds for the creation and annihilation operators:

$$\hat{a}_{p,\text{out}} = \hat{S}^{-1}\hat{a}_{p,\text{in}}\hat{S} \quad , \quad \hat{a}^\dagger_{p,\text{out}} = \hat{S}^{-1}\hat{a}^\dagger_{p,\text{in}}\hat{S} . \qquad (9.13)$$

The $\hat{S}$ operator also describes the transformation between the in and out state vectors since

$$\begin{aligned}
|p_1,\ldots,p_n;\text{in}\rangle &= \hat{a}^\dagger_{p_1,\text{in}}\cdots\hat{a}^\dagger_{p_n,\text{in}}|0\rangle \\
&= \hat{S}\hat{a}^\dagger_{p_1,\text{out}}\hat{S}^{-1}\hat{S}\cdots\hat{S}^{-1}\hat{S}\hat{a}^\dagger_{p_n,\text{out}}\hat{S}^{-1}|0\rangle \\
&= \hat{S}\hat{a}^\dagger_{p_1,\text{out}}\cdots\hat{a}^\dagger_{p_n,\text{out}}|0\rangle \\
&= \hat{S}|p_1,\ldots,p_n;\text{out}\rangle .
\end{aligned} \qquad (9.14)$$

Here the stability and uniqueness of the vacuum state were used (a possible unmeasurable phase factor was put to unity):

$$\hat{S}^{-1}|0\rangle = \hat{S}|0\rangle = |0\rangle . \qquad (9.15)$$

In analogy with (9.14) we also have

$$\langle q_1,\ldots,q_m;\text{out}| = \langle q_1,\ldots,q_m;\text{in}|\hat{S} , \qquad (9.16)$$

---

[3] This will not be true if an open system is studied. Under the influence of an external field the in and out vacua can differ since the external field can produce particles.

so that the S matrix element can be expressed in two ways:

$$\begin{aligned} S_{fi} &= \langle q_1, \ldots, q_m; \text{in} | \hat{S} | p_1, \ldots, p_n; \text{in} \rangle \\ &= \langle q_1, \ldots, q_m; \text{out} | \hat{S} | p_1, \ldots, p_n; \text{out} \rangle . \end{aligned} \qquad (9.17)$$

The $\hat{S}$ operator should also be consistent with the *symmetry properties* of the theory. Therefore $\hat{S}$ has to commute with the generators $\hat{Q}$ of the symmetry transformations:

$$[\hat{S}, \hat{Q}] = 0 . \qquad (9.18)$$

In particular this is required for the translations and Lorentz transformations

$$\hat{\mathcal{U}} \hat{S} \hat{\mathcal{U}}^{-1} = \hat{S} , \qquad (9.19)$$

where $\hat{\mathcal{U}}$ is the unitary operator of Poincaré transformations (cf. Sect. 4.3) which depends on ten parameters. We will come back to the transformation properties of the S matrix in Sect. 10.5.

We mention that the $\hat{S}$ operator acts like the unit operator if it is restricted to the subspace of *single-particle states*:

$$\hat{S} | p; \text{in} \rangle = | p; \text{in} \rangle . \qquad (9.20)$$

Thus there is only one type of single-particle state to which no label in or out has to be attached:

$$| p; \text{in} \rangle = | p; \text{out} \rangle = | p \rangle . \qquad (9.21)$$

This is hardly surprising since a single particle is necessarily free and does not experience an interaction. To be more precise: there is always an unavoidable interaction with a "cloud" of virtual field quanta surrounding the particle. This cloud, however, has already been taken into account in the construction of the asymptotic states since (9.6) contains the physical instead of the bare mass.

To evaluate any physically relevant quantity some knowledge of the full Heisenberg field operator $\hat{\phi}(x)$ is required. The operator $\hat{\phi}(x)$ in the context of axiomatic quantum field theory is known as the *interpolating field*. There is a (formal) connection between the interpolating field and the asymptotic field operators $\hat{\phi}_{\text{in/out}}(x)$ through the *Yang-Feldman equations* which we will sketch in the following. A modified version of the Yang–Feldman equations will be derived in Exercise 9.1.

As a starting point we consider the field equation for the interacting field operator, given by

$$(\Box + m^2) \hat{\phi}(x) = \hat{j}(x) . \qquad (9.22)$$

Here the "source" $\hat{j}(x)$ is an operator that in general depends on the field operator. Provided that the interaction term in the Lagrange density $\mathcal{L} = \mathcal{L}_0 + \mathcal{L}_1$ does not contain derivative couplings we simply have

$$\hat{j}(x) = \frac{\partial \hat{\mathcal{L}}_1}{\partial \hat{\phi}(x)} . \qquad (9.23)$$

In addition, $\hat{j}(x)$ will contain a term that accounts for the use of the *physical mass* m instead of the bare mass m_0 in the Klein–Gordon operator. This term reads $\hat{j}_{\delta m}(x) = (m^2 - m_0^2)\hat{\phi}^2(x) := -\delta m^2 \hat{\phi}^2(x)$. The differential equation (9.22) can be formally solved by using the Green's function technique. This leads to the *Yang–Feldman equations*[4]

$$\hat{\phi}(x) = \hat{\phi}_{\text{in}}(x) - \int d^4x' \, \Delta_R(x - x') \hat{j}(x') \,, \tag{9.24a}$$

$$\hat{\phi}(x) = \hat{\phi}_{\text{out}}(x) - \int d^4x' \, \Delta_A(x - x') \hat{j}(x') \,. \tag{9.24b}$$

The first term on the right-hand side is a solution of the homogeneous field equation. In (9.24) the Klein–Gordon propagator that matches the boundary condition is chosen. The incoming free solution $\hat{\phi}_{\text{in}}$ is associated with the *retarded propagator* Δ_R (see Chap. 4.6), which moves a "perturbation" forward in time. The outgoing solution $\hat{\phi}_{\text{out}}$ is associated with the *advanced propagator* Δ_A.

It is obvious that both versions of (9.24) solve the field equation (9.22), the difference lies in the asymptotics. The asymptotic behavior of the field operator involves rather subtle mathematical problems. If the source acts only for a finite time interval, i.e.,

$$\hat{j}(x) = 0 \quad \text{for} \quad |t| > T \,, \tag{9.25}$$

we can immediately deduce that

$$\lim_{x_0 \to -\infty} \hat{\phi}(x) = \hat{\phi}_{\text{in}}(x) \,, \tag{9.26a}$$

$$\lim_{x_0 \to +\infty} \hat{\phi}(x) = \hat{\phi}_{\text{out}}(x) \,. \tag{9.26b}$$

This simply follows from the causality features of the propagators: $\Delta_R(x - x') = 0$ for $x_0 < x_0'$ and $\Delta_A(x - x') = 0$ for $x_0 > x_0'$. Unfortunately in general the condition (9.25) is not satisfied. The self-interaction, which in perturbation theory is described in terms of the emission and reabsorption of virtual field quanta, cannot be switched off, so the contribution of the second term in (9.24) never will completely vanish.

In an attempt to find a better description for the asymptotics of the field operator $\hat{\phi}(x)$ one may weaken the condition of equality and only postulate proportionality with the free field operator. Instead of (9.26),

$$\lim_{x_0 \to -\infty} \hat{\phi}(x) = \sqrt{Z} \, \hat{\phi}_{\text{in}}(x) \,, \tag{9.27a}$$

$$\lim_{x_0 \to +\infty} \hat{\phi}(x) = \sqrt{Z} \, \hat{\phi}_{\text{out}}(x) \tag{9.27b}$$

will be postulated. Here Z is a *renormalization constant* that has to be specified later. In the absence of interaction the constant is trivial, $Z = 1$. The quantity $\sqrt{Z}$ has a simple intuitive meaning: it describes the probability amplitude that the operator $\hat{\phi}(x)$ creates a single-particle state $|1\rangle$ when applied to the vacuum since

$$\langle 1|\hat{\phi}(x)|0\rangle = \sqrt{Z} \, \langle 1|\hat{\phi}_{\text{in}}(x)|0\rangle \,. \tag{9.28}$$

---

[4] C.N. Yang, D. Feldman: Phys. Rev. **79**, 972 (1950).

Because of the interaction, however, the operator $\hat{\phi}(x)$ can also create complicated many-particle states, therefore the value of the constant Z should be reduced compared to $Z = 1$:

$$0 \leq Z < 1 . \tag{9.29}$$

This question will be studied in more detail in Sect. 9.3.

Equations (9.27) still do not form a satisfactory description of the asymptotical behavior of the field operator. To wit, if indeed the condition $Z \neq 1$ holds this immediately contradicts the equal-time commutation relations (9.1). We would expect

$$\begin{aligned} \mathrm{i}\,\delta^3(\boldsymbol{x}-\boldsymbol{y}) &= \lim_{x_0\to\infty}\big[\hat{\phi}(x),\dot{\hat{\phi}}(y)\big]_{x_0=y_0} = Z\big[\hat{\phi}(x)_{\mathrm{in}},\dot{\hat{\phi}}_{\mathrm{in}}(y)\big]_{x_0=y_0} \\ &= Z\,\mathrm{i}\,\delta^3(\boldsymbol{x}-\boldsymbol{y}) , \end{aligned} \tag{9.30}$$

since of course also the in-field operators have to satisfy the ETCR. From this one would conclude $Z = 1$, in contradiction to the asserted condition (9.29). To escape from this dilemma one has to take a closer look at the mathematical notion of convergence when applied to operators. The theory of functional analysis warns us to be careful when vector spaces of *infinite dimension* are involved and it distinguishes between different types of convergence. Let us look at a sequence of operators $\hat{O}_i, i=1,\ldots,\infty$, which in some sense approaches the operator $\hat{O}$. We indicate this naively by writing

$$\hat{O}_i \to \hat{O} \quad \text{for} \quad i \to \infty . \tag{9.31}$$

Strictly speaking, however, the arrow here can have different meanings. The *strong operator convergence* or norm convergence holds if $\hat{O}_i|\Psi\rangle$ approaches $\hat{O}|\Psi\rangle$ for every vector $|\Psi\rangle$ in the Hilbert space $\mathcal{H}$, i.e.,

$$\big\|\hat{O}_i|\Psi\rangle - \hat{O}|\Psi\rangle\big\| \to 0 \quad \text{for all} \quad |\Psi\rangle \in \mathcal{H} . \tag{9.32}$$

Here $\|\ldots\|$ is the norm of the state. In our case this will be the norm induced by the scalar product. For many purposes, in particular in the realm of scattering theory, the imposition of (9.32) will be too restrictive. Here one defines the *weak operator convergence* as follows

$$\langle\chi|\hat{O}_i|\Psi\rangle - \langle\chi|\hat{O}|\Psi\rangle \to 0 \quad \text{for all} \quad |\chi\rangle \quad \text{and} \quad |\Psi\rangle \in \mathcal{H} . \tag{9.33}$$

Thus only the *matrix elements* of the sequence of operator $\hat{O}_i$ have to approach the matrix element of the operator $\hat{O}$. For the envisaged aim this is a sufficient condition since physical observables are described by (squared) matrix elements. From a mathematical point of view, however, the condition (9.33) is weaker than (9.32) if an *infinite-dimensional* Hilbert space is involved. This is easily demonstrated by writing down the square of the norm (9.32) and inserting a complete set of basis states $|\chi\rangle$:

$$\begin{aligned} \big\|(\hat{O}_i-\hat{O})|\Psi\rangle\big\|^2 &= \langle\Psi|(\hat{O}_i^\dagger-\hat{O}^\dagger)(\hat{O}_i-\hat{O})|\Psi\rangle \\ &= \sum_\chi \langle\Psi|\hat{O}_i^\dagger-\hat{O}^\dagger|\chi\rangle\langle\chi|\hat{O}_i-\hat{O}|\Psi\rangle \\ &= \sum_\chi \big|\langle\Psi|\hat{O}_i-\hat{O}|\chi\rangle\big|^2 . \end{aligned} \tag{9.34}$$

Provided that the strong convergence criterion is fulfilled, this sum approaches zero, which will then also hold for each of the (positive-definite) terms in the sum. Thus (9.32) implies (9.33). The reverse conclusion cannot be drawn, however. Since the range of the sum in (9.34) is infinite the vanishing of the individual terms does *not* imply that the sum itself vanishes if the convergence is nonuniform.[5] Thus (9.33) indeed is a weaker condition than (9.32).

If scattering states are involved that describe particles "escaping to infinity" the asymptotic behavior has to be described by using weak convergence. This applies to ordinary quantum scattering theory as well as to quantum field theory. The *asymptotic condition for the field operator* is obtained if (9.27) is interpreted in the sense of weak operator convergence:[6]

$$\lim_{x_0 \to -\infty} \langle b|\hat{\phi}(x)|a\rangle = \sqrt{Z}\, \langle b|\hat{\phi}_{\text{in}}(x)|a\rangle \,, \tag{9.35a}$$

$$\lim_{x_0 \to +\infty} \langle b|\hat{\phi}(x)|a\rangle = \sqrt{Z}\, \langle b|\hat{\phi}_{\text{out}}(x)|a\rangle \,. \tag{9.35b}$$

These equations are to hold for any pair of Heisenberg state vectors $|a\rangle$ and $|b\rangle$. In this way the apparent contradiction we encountered when calculating the commutator of field operators is alleviated: if weak operator convergence is invoked, the limits $\hat{A}_i \to \hat{A}$ and $\hat{B}_i \to \hat{B}$ do *not* imply that the product of the operators approaches $\hat{A}_i \hat{B}_i \to \hat{A}\hat{B}$, therefore (9.30) is not applicable.

Remark: Even the condition (9.35) is mathematically incorrect since the matrix elements are oscillatory functions in time, which do not posses a well-defined limit. Strictly speaking the theory should be formulated using *smeared field operators*. These are defined by projecting the field operators onto *spatially localized wave packets*:

$$\hat{\phi}^\alpha(t) = i\int d^3x\, u_\alpha^*(\boldsymbol{x},t) \stackrel{\leftrightarrow}{\partial_0} \hat{\phi}(\boldsymbol{x},t) \,, \tag{9.36a}$$

$$\hat{\phi}^\alpha_{\text{in/out}} = i\int d^3x\, u_\alpha^*(\boldsymbol{x},t) \stackrel{\leftrightarrow}{\partial_0} \hat{\phi}_{\text{in/out}}(\boldsymbol{x},t) \,. \tag{9.36b}$$

Here the $u_\alpha(\boldsymbol{x},t)$ stand for a complete set of *localized* solutions of the Klein–Gordon equation:

$$\left(\Box + m^2\right) u_\alpha(x) = 0 \,. \tag{9.37}$$

In (9.36) the interacting or the free field operator is projected onto this set of states by using the Klein–Gordon scalar product (cf. (4.25)). The notation in (9.36b) implies that the smeared asymptotic field operators are time independent, which is seen as follows. Using (9.37) and (9.6) and integrating by parts we find

$$\begin{aligned}
\partial_0 \hat{\phi}^\alpha_{\text{in}} &= i\int d^3x\, \left(u_\alpha^*(\partial_0^2 \hat{\phi}_{\text{in}}) - (\partial_0^2 u_\alpha^*)\phi_{\text{in}} \right) \\
&= i\int d^3x\, \left(u_\alpha^*(\partial_0^2 \hat{\phi}_{\text{in}}) - (\nabla^2 - m^2)u_\alpha^* \phi_{\text{in}} \right) \\
&= i\int d^3x\, u_\alpha^* (\Box + m^2)\hat{\phi}_{\text{in}} = 0 \,.
\end{aligned} \tag{9.38}$$

---

[5] As an example think of the Fourier series $\frac{\sin x}{1} + \frac{\sin 3x}{3} + \frac{\sin 5x}{5} + \ldots = \frac{\pi}{4}$ in which each of the terms vanishes in the limit $x \to 0$, whereas the sum remains finite.

[6] H. Lehmann, K. Symanzik, W. Zimmermann: Nuovo Cim. **1**, 205 (1955).

The last step will not be true, of course, for the interacting field operator $\hat{\phi}^\alpha$. The corresponding matrix elements, however, asymptotically approach a constant value since

$$\lim_{x_0 \to \mp\infty} \langle b|\hat{\phi}^\alpha(x_0)|a\rangle = \sqrt{Z}\,\langle b|\hat{\phi}^\alpha_{\text{in/out}}|a\rangle \,. \tag{9.39}$$

In practical work one often avoids explicitly using the smearing prescription, i.e., one uses (9.35) instead of (9.39). If mathematical problems are encountered, however, one should revert to the wave-packet formalism.

If plane waves $u_p(x)$ instead of the wave packets $u_\alpha(x)$ are used when forming the smeared field operator then $\hat{\phi}^\alpha$ reduces to

$$\hat{a}_p(x_0) = \mathrm{i}\int \mathrm{d}^3 x\, u_p^*(x)\, \overset{\leftrightarrow}{\partial_0}\, \hat{\phi}(x) \,, \tag{9.40a}$$

$$\hat{a}_{p;\text{in/out}} = \mathrm{i}\int \mathrm{d}^3 x\, u_p^*(x)\, \overset{\leftrightarrow}{\partial_0}\, \hat{\phi}_{\text{in/out}}(x) \,. \tag{9.40b}$$

Here $\hat{a}_{p;\text{in/out}}$ is the usual annihilation operator for free particles but the time-dependent operator $\hat{a}_p(x_0)$ does not possess a simple interpretation in terms of particles because of the interaction. The asymptotic condition (9.39) in this less-rigorous form reads

$$\lim_{x_0 \to \mp\infty} \langle b|\hat{a}_p(x_0)|a\rangle = \sqrt{Z}\,\langle b|\hat{a}_p^{\text{in/out}}|a\rangle \,. \tag{9.41}$$

EXERCISE

9.1 Derivation of the Yang–Feldman Equation

Problem. Derive the Yang–Feldman equation using the asymptotic condition (9.41) for the field operators.

Solution. With (9.41) and (9.40a) the matrix element of the in annihilation operator becomes

$$\sqrt{Z}\,\langle b|\hat{a}_{p;\text{in}}|a\rangle = \lim_{x_0' \to -\infty} \mathrm{i}\int \mathrm{d}^3 x'\, u_p^*(x')\, \overset{\leftrightarrow}{\partial_0'}\, \langle b|\hat{\phi}(x')|a\rangle \,. \tag{1}$$

The right-hand side can be transformed into a four-dimensional integral since for any point of time x_0

$$\int \mathrm{d}^3 x'\, F(\boldsymbol{x}',-\infty) = \int \mathrm{d}^3 x'\, F(\boldsymbol{x}',x_0) - \int \mathrm{d}^3 x' \int_{-\infty}^{x_0} \mathrm{d}x_0'\, \partial_0' F(\boldsymbol{x}',x_0') \,, \tag{2}$$

and thus

$$\sqrt{Z}\,\langle b|\hat{a}_{p;\text{in}}|a\rangle$$
$$= \mathrm{i}\int \mathrm{d}^3 x'\, u_p^*(x')\, \overset{\leftrightarrow}{\partial_0'}\, \langle b|\hat{\phi}(x')|a\rangle\Big|_{x_0'=x_0}$$
$$- \mathrm{i}\int \mathrm{d}^3 x' \int_{-\infty}^{x_0} \mathrm{d}x_0'\, \partial_0'\Big[u_p^*(x')\, \overset{\leftrightarrow}{\partial_0'}\, \langle b|\hat{\phi}(x')|a\rangle\Big]$$
$$= \langle b|\hat{a}_p(x_0)|a\rangle$$

$$-\mathrm{i}\int\mathrm{d}^3x'\int_{-\infty}^{x_0}\mathrm{d}x'_0\left[u_p^*(x')\partial_0'^2\langle b|\hat{\phi}(x')|a\rangle - \partial_0'^2 u_p^*(x')\langle b|\hat{\phi}(x')|a\rangle\right]. \quad (3)$$

Using the Klein–Gordon equation for the plane wave

$$\partial_0^2 u_p(x) = (\boldsymbol{\nabla}^2 - m^2)u_p(x), \quad (4)$$

twice integrating by parts, and employing the field equation (9.22), we arrive at

$$\sqrt{Z}\,\langle b|\hat{a}_{p;\mathrm{in}}|a\rangle = \langle b|\hat{a}_p(x_0)|a\rangle - \mathrm{i}\int\mathrm{d}^3x'\int_{-\infty}^{x_0}\mathrm{d}x'_0\,u_p^*(x')(\Box'^2 + m^2)\langle b|\hat{\phi}(x')|a\rangle$$

$$= \langle b|\hat{a}_p(x_0)|a\rangle - \mathrm{i}\int\mathrm{d}^3x'\int_{-\infty}^{x_0}\mathrm{d}x'_0\,u_p^*(x')\langle b|\hat{j}(x')|a\rangle. \quad (5)$$

The same derivation can be repeated for the creation operators, leading to

$$\sqrt{Z}\,\langle b|\hat{a}_{p;\mathrm{in}}^\dagger|a\rangle = \langle b|\hat{a}_p^\dagger(x_0)|a\rangle + \mathrm{i}\int\mathrm{d}^3x'\int_{-\infty}^{x_0}\mathrm{d}x'_0\,u_p(x')\langle b|\hat{j}(x')|a\rangle. \quad (6)$$

The field operator can now be constructed by making use of the creation and annihilation operators:

$$\hat{\phi}(x) = \int\mathrm{d}^3p\,\left(\hat{a}_p(x_0)u_p(x) + \hat{a}_p^\dagger(x_0)u_p^*(x)\right), \quad (7)$$

and similarly for $\hat{\phi}_{\mathrm{in}}(x)$. Multiplying (5) and (6) by $u_p(x)$ and $u_p^*(x)$, respectively, and adding them we obtain

$$\sqrt{Z}\,\langle b|\hat{\phi}_{\mathrm{in}}(x)|a\rangle \quad (8)$$

$$= \langle b|\hat{\phi}(x)|a\rangle - \mathrm{i}\int\mathrm{d}^3x'\int_{-\infty}^{x_0}\mathrm{d}x'_0\int\mathrm{d}^3p\,\left(u_p(x)u_p^*(x') - u_p^*(x)u_p(x')\right)\langle b|\hat{j}(x')|a\rangle.$$

The momentum integral is just the Pauli–Jordan function introduced in Sect. 4.4:

$$\int\mathrm{d}^3p\,\left(u_p(x)u_p^*(x') - u_p^*(x)u_p(x')\right) = \int\mathrm{d}^3p\,\frac{1}{2\omega_p(2\pi)^3}\left(\mathrm{e}^{-\mathrm{i}p\cdot(x-x')} - \mathrm{e}^{\mathrm{i}p\cdot(x-x')}\right)$$

$$= \mathrm{i}\,\Delta(x-x'), \quad (9)$$

which leads to

$$\sqrt{Z}\,\langle b|\hat{\phi}_{\mathrm{in}}(x)|a\rangle = \langle b|\hat{\phi}(x)|a\rangle + \int\mathrm{d}^4x'\,\Theta(x_0 - x'_0)\Delta(x-x')\langle b|\hat{j}(x')|a\rangle. \quad (10)$$

Multiplication of the Pauli–Jordan function with the step function produces the retarded propagator (see (4.158)):

$$\Delta_{\mathrm{R}}(x-x') = \Theta(x_0 - x'_0)\Delta(x-x'). \quad (11)$$

This concludes the proof of the Yang–Feldman equation

$$\langle b|\hat{\phi}(x)|a\rangle = \sqrt{Z}\,\langle b|\hat{\phi}_{\mathrm{in}}(x)|a\rangle - \int\mathrm{d}^4x'\,\Delta_{\mathrm{R}}(x-x')\langle b|\hat{j}(x')|a\rangle. \quad (12)$$

Exercise 9.1

A similar calculation for the limit $x_0 \to +\infty$ leads to

$$\langle b|\hat{\phi}(x)|a\rangle = \sqrt{Z}\,\langle b|\hat{\phi}_{\text{out}}(x)|a\rangle - \int d^4x'\,\Delta_A(x-x')\,\langle b|\hat{j}(x')|a\rangle\,. \tag{13}$$

Comparing these results with (9.24) we observe two differences: in the spirit of the weak operator convergence, instead of the operators now their matrix elements are invoked, furthermore the asymptotic fields are multiplied by the renormalization constant $\sqrt{Z}$.

9.3 The Lehmann–Källen Spectral Representation

The invariant functions, in particular the Feynman propagator $\Delta_F(x-y)$ and the Pauli–Jordan function $\Delta(x-y)$, play an important role in quantum field theory. They are defined in terms of products of the field operators $\hat{\phi}(x)$ and $\hat{\phi}(y)$ and for free fields can be calculated in closed form (cf. Sects. 4.4 and 4.5). For interacting fields such an exact solution is no longer available. However, there is a very useful formalism that represents the exact functions as superpositions of the corresponding free functions taken at different mass values, multiplied by a "spectral density distribution". This spectal density in general too cannot be evaluated exactly. It is subject to constraints, however, which allows us to draw certain general conclusions, in particular with regard to the renormalization constant. This will give us a better understanding of (9.29).

All the invariant functions can be derived from the vacuum expectation value of a product of field operators, which we will denote by

$$W(x-y) = \langle 0|\hat{\phi}(x)\hat{\phi}(y)|0\rangle\,. \tag{9.42}$$

This important object is known as the (2-point) *Wightman function*.[7] Equation (9.42) can be reduced to a combination of matrix elements of the field operator by *inserting a complete set of states*

$$\sum_\alpha |\alpha\rangle\langle\alpha| = \mathbb{1}\,. \tag{9.43}$$

This basis contains single-particle momentum eigenstates $|p\rangle$ where $p^2 = m^2$, as well as more complicated many-particle states, the detailed structure of which is of no concern for us. In any case we know that the $|\alpha\rangle$ will be *eigenstates of the momentum operator*:

$$\hat{P}_\mu|\alpha\rangle = p_\mu^{(\alpha)}|\alpha\rangle\,. \tag{9.44}$$

These states should have a physically attainable space-like momentum and positive energy:

$$\left(p_\mu^{(\alpha)}\right)^2 \geq 0 \quad \text{and} \quad p_0^{(\alpha)} \geq 0\,. \tag{9.45}$$

---

[7] A.S. Wightman: Phys. Rev. **101**, 860 (1956) used this function as the basis for developing his axiomatic formulation of quantum field theory.

9.3 The Lehmann–Källen Spectral Representation

Using the transformation law of the field operator under translations (9.3) and the translation invariance of the vacuum (9.5), we can write the Wightman function as

$$W(x-y) = \sum_\alpha \langle 0| e^{i\hat{P}\cdot x} \hat{\phi}(0) e^{-i\hat{P}\cdot x} |\alpha\rangle \langle\alpha| e^{i\hat{P}\cdot y} \hat{\phi}(0) e^{-i\hat{P}\cdot y} |0\rangle$$

$$= \sum_\alpha e^{-ip^{(\alpha)}\cdot(x-y)} |\langle\alpha|\hat{\phi}(0)|0\rangle|^2 . \qquad (9.46)$$

Now the *spectral density* $\rho(p^2)$ is defined as

$$\tilde{\rho}(p) = (2\pi)^3 \sum_\alpha \delta^4(p - p^{(\alpha)}) |\langle\alpha|\hat{\phi}(0)|0\rangle|^2$$

$$\equiv \Theta(p_0)\rho(p^2) . \qquad (9.47)$$

In the second step the condition (9.45) has been used, accoding to which all momenta a located within the forward light cone. The function $\rho(p^2)$ is real valued, positive definite, and Lorentz invariant. It depends on the scalar variable p^2. Using the definition (9.47) a four-dimensional momentum integration can be introduced in (9.46):

$$W(x-y) = \int \frac{d^4p}{(2\pi)^3} \Theta(p_0) \rho(p^2) e^{-ip\cdot(x-y)} . \qquad (9.48)$$

As an additional artifice now an auxiliary integration over a variable s is introduced,

$$\rho(p^2) = \int_0^\infty ds\, \delta(p^2 - s)\rho(s) , \qquad (9.49)$$

which allows us to remove the spectral density from the momentum integral:

$$W(x-y) = \int_0^\infty ds\, \rho(s) \int \frac{d^4p}{(2\pi)^3} \Theta(p_0) \delta(p^2 - s) e^{-ip\cdot(x-y)} . \qquad (9.50)$$

In this way the momentum integral has been fully decoupled from the properties of the interaction, it only depends on the parameter s. In fact, we have already encountered the integral in connection with the invariant functions, see (4.150). We are dealing with the function $i\Delta^{(+)}$,

$$i\Delta^{(+)}(x; m^2) = \int \frac{d^4p}{(2\pi)^3} \Theta(p_0)\, \delta(p^2 - m^2) e^{-ip\cdot x} , \qquad (9.51)$$

which can be viewed as the Wightman function of a *free* scalar field of mass $m = \sqrt{s}$. The Wightman function of the interacting field theory thus can be represented as an integral over its free counterpart, multiplied by the spectral density. The integral extends over all possible invariant masses, spanning the range from zero to infinity:

$$W(x-y) = \int_0^\infty ds\, \rho(s)\, i\Delta^{(+)}(x - y; s) . \qquad (9.52)$$

A relation of this kind is called *spectral representation* or *Lehmann–Källén representation*.[8] It is obvious how the various invariant functions of interest can be derived from the Wightman function, thus producing their spectral representations. For the interacting scalar *Feynman propagator* we find

$$\begin{aligned}
i\Delta'_F(x-y) &= \langle 0|T(\hat\phi(x)\hat\phi(y))|0\rangle \\
&= \Theta(x_0-y_0)W(x-y) + \Theta(y_0-x_0)W(y-x) \\
&= \int_0^\infty ds\,\rho(s)\Big[\Theta(x_0-y_0)\,i\Delta^{(+)}(x-y;s) \\
&\qquad + \Theta(y_0-x_0)\,i\Delta^{(+)}(y-x;s)\Big] \\
&= \int_0^\infty ds\,\rho(s)\,i\Delta_F(x-y;s)\,,
\end{aligned} \qquad (9.53)$$

where (4.154) and (4.157) have been used. This relation of course also can be transformed to momentum space:

$$\Delta'_F(p) = \int_0^\infty ds\,\rho(s)\,\frac{1}{p^2 - s + i\epsilon}\,. \qquad (9.54)$$

The *Pauli–Jordan function* for the free theory was defined in terms of the commutator

$$i\Delta(x-y) = [\hat\phi(x),\hat\phi(y)]\,. \qquad (9.55)$$

If interactions are present this commutator in general is a complicated operator-valued expression. Only at equal times $x_0 = y_0$ is it reduced to a c number since the time derivative leads to the delta function (9.1a). However, (9.55) can be extended to the interacting case if the definition is based on the *vacuum expectation value of the commutator*:

$$i\Delta'(x-y) := \langle 0|[\hat\phi(x),\hat\phi(y)]|0\rangle = W(x-y) - W(y-x)\,. \qquad (9.56)$$

Clearly then, (9.52) leads to the following spectral representation of this function:

$$\Delta'(x-y) = \int_0^\infty ds\,\rho(s)\Delta(x-y;s)\,. \qquad (9.57)$$

The spectral density has the following intuitive meaning. Given an interacting single-particle state $\hat\phi(0)|0\rangle$ it gives us the probability of finding a free state with squared invariant mass s, where s can take on any physically accessible values. Indeed $\rho(s)$ is normalized to 1, as a probability distribution should be. This follows from the equal-time commutation relation postulated in (9.1) for the interacting field operator. Taking the derivative of (9.56) with respect to the time argument y_0 and equating $x_0 = y_0$ leads to

---

[8] G. Källén: Helv. Phys. Acta **25**, 417 (1952); H. Lehmann: Nuovo Cim. **11**, 342 (1954).

9.3 The Lehmann-Källen Spectral Representation

$$\begin{aligned}
\mathrm{i}\delta^3(\boldsymbol{x}-\boldsymbol{y}) &= \left[\hat{\phi}(x),\dot{\hat{\phi}}(y)\right]_{x_0=y_0} = \partial_{y_0}\langle 0|[\hat{\phi}(x),\hat{\phi}(y)]|0\rangle_{x_0=y_0} \\
&= \mathrm{i}\partial_{y_0}\Delta'(x-y)\big|_{x_0=y_0} = \int_0^\infty \mathrm{d}s\,\rho(s)\,\mathrm{i}\partial_{y_0}\Delta(x-y;s)\big|_{x_0=y_0} \\
&= \int_0^\infty \mathrm{d}s\,\rho(s)\,\mathrm{i}\delta^3(\boldsymbol{x}-\boldsymbol{y})
\end{aligned} \qquad (9.58)$$

where (4.105) was used for the derivative of the free commutator function. This leads to the following "sum rule" for the spectral density

$$\int_0^\infty \mathrm{d}s\,\rho(s) = 1\,. \qquad (9.59)$$

We have mentioned already that all accessible many-particle states contribute to the spectral density. The state with the lowest energy will be the single-particle state which has a *discrete* value of the invariant mass $s = m^2$. If there are no stable bound states (as we have assumed in connection with the asymptotic condition in the last section) there will be a threshold at $s = m_1^2 = 4m^2$ where the *continuum of two-particle states* begins. [9] Higher up there will follow other thresholds for more complicated many-particle states. Figure 9.1 gives a schematic representation of the spectral density $\rho(s)$. Besides the discrete contribution of the single-particle states at $s = m^2$ the threshold of the two-particle continuum at m_1^2 is shown. This continuum displays a resonance at m_r^2.

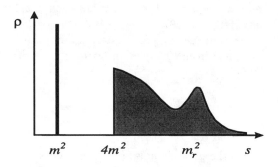

Fig. 9.1. Schematic representation of the spectral function of a scalar quantum field which describes particles of mass m. The threshold for pair creation lies at $s = m_1^2 = 4m^2$

As announced earlier the sum rule (9.59) leads to a constraint for the renormalization constant Z. We split (9.47) into the contribution of the single-particle states $|q\rangle$ and a remainder

$$\begin{aligned}
\Theta(p_0)\rho(p^2) &= (2\pi)^3\int \mathrm{d}^3q\,\delta^4(p-q)\,|\langle q|\hat{\phi}(0)|0\rangle|^2 \\
&\quad + (2\pi)^3\sum_{"\alpha>1"}\delta^4(p-p^{(\alpha)})\,|\langle\alpha|\hat{\phi}(0)|0\rangle|^2\,.
\end{aligned} \qquad (9.60)$$

With (9.28) and (9.7) the matrix element of the field operator reads

---

[9] If particular selection rules prevail that forbid the appearence of two-particle states the threshold may be higher.

$$\langle q|\hat{\phi}(0)|0\rangle = \sqrt{Z}\langle q|\hat{\phi}_{\rm in}(0)|0\rangle = \sqrt{Z}u_q^*(0) = \sqrt{Z}\frac{1}{\sqrt{(2\pi)^3\,2\omega_q}}\;. \tag{9.61}$$

Then (9.60) becomes

$$\Theta(p_0)\rho(p^2) = Z\int d^3q\,\frac{1}{2\omega_q}\delta^4(p-q) + {\rm Rem.}\;. \tag{9.62}$$

The q integral can be extended to four dimensions by using the familiar identity (see (4.101))

$$\frac{1}{2\omega_q} = \int_{-\infty}^{\infty}dq_0\,\Theta(q_0)\,\delta(q^2-m^2)\;, \tag{9.63}$$

which leads to

$$\begin{aligned}\Theta(p_0)\rho(p^2) &= Z\int d^4q\,\Theta(q_0)\,\delta(q^2-m^2)\,\delta^4(p-q) + {\rm Rem.}\\ &= Z\,\Theta(p_0)\,\delta(p^2-m^2) + {\rm Rem.}\;.\end{aligned} \tag{9.64}$$

Then the sum rule (9.59) becomes

$$1 = Z + \int_{m_1^2}^{\infty}ds\,\rho(s)\;. \tag{9.65}$$

In an interacting field theory certainly more complicated states will be encountered, besides the single-particle states, which have nonvanishing matrix elements $\langle\alpha|\hat{\phi}(0)|0\rangle$. Thus the second term in (9.65) will make a contribution and consequently the renormalization constant Z will be smaller than 1, as asserted in (9.29).

A convenient equation determining the exact Feynman propagator can be obtained from (9.53) by inserting the decomposition of the spectral function (9.64):

$$\Delta'_{\rm F}(p) = \int_0^{\infty}ds\,\rho(s)\Delta_{\rm F}(p,s) = Z\,\Delta_{\rm F}(p,m^2) + \int_{m_1^2}^{\infty}ds\,\rho(s)\Delta_{\rm F}(p,s)\;. \tag{9.66}$$

Since the possible poles (or branch cuts) of the second term will lie at higher energies, the renormalization constant Z can be extracted from the residue of the single-particle pole of the exact Feynman propagator as follows:

$$Z = \lim_{p^2\to m^2}(p^2-m^2)\Delta'_{\rm F}(p)\;. \tag{9.67}$$

9.4 The LSZ Reduction Formula

It is one of the main goals of quantum field theory to provide the means for calculating S matrix elements. The latter have been introduced in Sect. 9.2 as the scalar product between in and out state vectors. This is an abstract definition which does not immediately lend itself for practical calculations.

For this purpose in the following we will derive a "reduction formula", based on the asymptotic condition for the field operator, which will express the S matrix element in terms of the vacuum expectation value of a time-ordered product of field operators. The fundamental object of interest will be the *n-particle Green's function*, or simply the *n-point function*, defined as

$$G^{(n)}(x_1,\ldots,x_n) = \langle 0|T(\hat{\phi}(x_1)\cdots\hat{\phi}(x_n))|0\rangle \, . \tag{9.68}$$

The connection between this function and the S matrix element was elucidated by Lehmann, Symanzik, and Zimmermann (see the footnote on page 275) and is known as the *LSZ reduction formula*. In the following, we will present its derivation for the case of a neutral scalar field.

The goal is to express the S matrix element (9.10) for a scattering process of $n \to m$ particles,

$$S_{fi} = \langle q_1,\ldots,q_m; \text{out}|p_1,\ldots,p_n; \text{in}\rangle \, , \tag{9.69}$$

as a vacuum expectation value of a product of field operators $\hat{\phi}(x)$. This connection is made in three steps, which can be roughly sketched as follows: (1) a creation operator $\hat{a}_{p_i;\text{in}}$ is extracted from the inital state; (2) it is expressed in terms of the asymptotic field operators $\hat{\phi}_\text{in}(x)$ and $\hat{\phi}_\text{out}(x)$; (3) by using the asymptotic condition the transition to the interacting field operator $\hat{\phi}(x)$ is made. This procedure is repeated successively for all particles present in the initial state and similarly also for the particles in the final state, so that finally a vacuum expectation value of field-operator products is obtained.

In the first step we address a particle in the initial state, which has momentum p_1. The S matrix element can be written as

$$\begin{aligned}S_{fi} &= \langle q_1,\ldots,q_m; \text{out}|\hat{a}^\dagger_{p_1;\text{in}}|p_2,\ldots,p_n;\text{in}\rangle \\ &= \langle q_1,\ldots,q_m; \text{out}|\hat{a}^\dagger_{p_1;\text{out}}|p_2,\ldots,p_n;\text{in}\rangle \\ &\quad + \langle q_1,\ldots,q_m; \text{out}|\hat{a}^\dagger_{p_1;\text{in}} - \hat{a}^\dagger_{p_1;\text{out}}|p_2,\ldots,p_n;\text{in}\rangle \, . \end{aligned} \tag{9.70}$$

The operator $\hat{a}^\dagger_{p_1;\text{out}}$ acts as an annihilation operator on the final states, so that the first term in (9.70) will only contribute if one of the momenta $q_1,\ldots,q_m$ agrees with p_1. If this is the case case we get

$$S^{(1)}_{fi} = \sum_{k=1}^m \langle q_1,\ldots,q_{k-1},q_{k+1},\ldots,q_m;\text{out}|p_2,\ldots,p_n;\text{in}\rangle\, \delta^3(q_k - p_1) \, . \tag{9.71}$$

To keep the following calculation simple we will assume that all momenta of the incoming and outgoing particles are different. In this way we get rid of processes in which some of the particles simply "run through" without participating in the scattering. Under this condition only the second term in (9.70) has to be taken into account. It would not be difficult, however, to include also the terms contained in (9.71) since they simply consist of S matrix elements like (9.69) with a reduced number of particles.

By using (9.40b) the creation operators can be transformed into projections of the asymptotic field operators onto the basis of plane waves (or preferrably wave packets):

$$\hat{a}^\dagger_{p_1;\text{in}} - \hat{a}^\dagger_{p_1;\text{out}} = \mathrm{i} \int \mathrm{d}^3 x \left(\hat{\phi}_{\text{in}}(x) - \hat{\phi}_{\text{out}}(x) \right) \overleftrightarrow{\partial}_{x_0} u_{p_1}(x) \,. \tag{9.72}$$

Since this expression is constant it can be evaluated at any chosen point of time x_0. In particular we can take the limit $x_0 \to \infty$ in the first term and $x_0 \to -\infty$ in the second term. Under these conditions it is possible to replace the asymptotic field operator by its interacting counterpart operator according to (9.35):

$$\begin{aligned}
\hat{a}^\dagger_{p_1;\text{in}} - \hat{a}^\dagger_{p_1;\text{out}} &= \mathrm{i} \int \mathrm{d}^3 x \left(\lim_{x_0 \to -\infty} \hat{\phi}_{\text{in}}(x) \overleftrightarrow{\partial}_{x_0} u_{p_1}(x) \right. \\
&\qquad \left. - \lim_{x_0 \to +\infty} \hat{\phi}_{\text{out}}(x) \overleftrightarrow{\partial}_{x_0} u_{p_1}(x) \right) \\
&= -\frac{\mathrm{i}}{\sqrt{Z}} \left(\lim_{x_0 \to +\infty} - \lim_{x_0 \to -\infty} \right) \int \mathrm{d}^3 x \, \hat{\phi}(x) \overleftrightarrow{\partial}_{x_0} u_{p_1}(x) \,,
\end{aligned} \tag{9.73}$$

where the latter step, strictly speaking, is valid only in the sense of weak convergence, i.e., it should be written in terms of the matrix elements instead of the operator itself. After switching to a four-dimensional integral with the help of the identity

$$\left(\lim_{x_0 \to +\infty} - \lim_{x_0 \to -\infty} \right) F(x) = \int_{-\infty}^{\infty} \mathrm{d}x_0 \, \partial_{x_0} F(x) \,, \tag{9.74}$$

we can transform (9.73) into

$$\begin{aligned}
\hat{a}^\dagger_{p_1;\text{in}} - \hat{a}^\dagger_{p_1;\text{out}} &= -\frac{\mathrm{i}}{\sqrt{Z}} \int \mathrm{d}^4 x \, \partial_{x_0} \left(\hat{\phi}(x) \overleftrightarrow{\partial}_{x_0} u_{p_1}(x) \right) \\
&= -\frac{\mathrm{i}}{\sqrt{Z}} \int \mathrm{d}^4 x \left(\hat{\phi}(x) \overrightarrow{\partial}^2_{x_0} u_{p_1}(x) - \hat{\phi}(x) \overleftarrow{\partial}^2_{x_0} u_{p_1}(x) \right) \\
&= -\frac{\mathrm{i}}{\sqrt{Z}} \int \mathrm{d}^4 x \left(\hat{\phi}(x)(\nabla^2 - m^2) u_{p_1}(x) - \hat{\phi}(x) \overleftarrow{\partial}^2_{x_0} u_{p_1}(x) \right) \\
&= \frac{\mathrm{i}}{\sqrt{Z}} \int \mathrm{d}^4 x \, \hat{\phi}(x) \overleftarrow{(\Box + m^2)} u_{p_1}(x) \,.
\end{aligned} \tag{9.75}$$

Here we have used that $u_{p_1}(x)$ satisfies the Klein–Gordon equation, subsequently an integration by parts was performed.

The achievement of our calculation has been to reduce the number of particles in the initial state from n to $n-1$:

$$S_{fi} = \frac{\mathrm{i}}{\sqrt{Z}} \int \mathrm{d}^4 x \, \langle q_1, \ldots, q_m; \text{out} | \hat{\phi}(x) | p_2; \ldots, p_n; \text{in} \rangle \overleftarrow{(\Box_x + m^2)} u_{p_1}(x) \,. \tag{9.76}$$

It is obvious how to proceed further: By repeating the above procedure step by step all particles can be extracted from the state vectors. We illustrate how the second particle having momentum p_2 is extracted from the inital state. We need

$$\begin{aligned}
\hat{\phi}(x) \hat{a}^\dagger_{p_2;\text{in}} &= \mathrm{i} \lim_{y_0 \to -\infty} \int \mathrm{d}^3 y \, \hat{\phi}(x) \hat{\phi}_{\text{in}}(y) \overleftrightarrow{\partial}_{y_0} u_{p_2}(y) \\
&= \frac{\mathrm{i}}{\sqrt{Z}} \lim_{y_0 \to -\infty} \int \mathrm{d}^3 y \, \hat{\phi}(x) \hat{\phi}(y) \overleftrightarrow{\partial}_{y_0} u_{p_2}(y) \,.
\end{aligned} \tag{9.77}$$

9.4 The LSZ Reduction Formula

Now comes the tricky point: because of the limit $y_0 \to -\infty$ it does no harm to introduce the *time-ordered product* in (9.77):

$$\hat{\phi}(x)\hat{a}^\dagger_{\boldsymbol{p}_2;\text{in}} = \frac{i}{\sqrt{Z}} \lim_{y_0 \to -\infty} \int d^3y\, T(\hat{\phi}(x)\hat{\phi}(y))\, \overleftrightarrow{\partial_{y_0}}\, u_{\boldsymbol{p}_2}(y)\,. \tag{9.78}$$

Introducing a four-dimensional integration as in (9.74), we can change the asymptotic limit from the past to the future:

$$\begin{aligned}\hat{\phi}(x)\hat{a}^\dagger_{\boldsymbol{p}_2;\text{in}} &= \frac{i}{\sqrt{Z}} \lim_{y_0 \to +\infty} \int d^3y\, T(\hat{\phi}(x)\hat{\phi}(y))\, \overleftrightarrow{\partial_{y_0}}\, u_{\boldsymbol{p}_2}(y) \\ &\quad - \frac{i}{\sqrt{Z}} \int d^4y\, \partial_{y_0}\left(T(\hat{\phi}(x)\hat{\phi}(y))\, \overleftrightarrow{\partial_{y_0}}\, u_{\boldsymbol{p}_2}(y)\right). \end{aligned} \tag{9.79}$$

When we evaluate the *first* term the field operators are interchanged because of the time ordering:

$$\begin{aligned}&\frac{i}{\sqrt{Z}} \lim_{y_0 \to +\infty} \int d^3y\, \hat{\phi}(y)\hat{\phi}(x)\, \overleftrightarrow{\partial_{y_0}}\, u_{\boldsymbol{p}_2}(y) \\ &= i \lim_{y_0 \to +\infty} \int d^3y\, \hat{\phi}_{\text{out}}(y)\hat{\phi}(x)\, \overleftrightarrow{\partial_{y_0}}\, u_{\boldsymbol{p}_2}(y) \\ &= \hat{a}^\dagger_{\boldsymbol{p}_2;\text{out}}\hat{\phi}(x)\,. \end{aligned} \tag{9.80}$$

Since all final momenta q_i are assumed to be different from p_2, (9.80) makes no contribution. The *second* term in (9.79) can be treated as that in (9.75) which leads to

$$\hat{\phi}(x)\hat{a}^\dagger_{\boldsymbol{p}_2;\text{in}} = \frac{i}{\sqrt{Z}} \int d^4y\, T(\hat{\phi}(x)\hat{\phi}(y))\, \overleftarrow{(\Box_y + m^2)}\, u_{\boldsymbol{p}_2}(y)\,. \tag{9.81}$$

Therefore the S matrix element at the second stage of the reduction process has become

$$\begin{aligned}S_{fi} &= \left(\frac{i}{\sqrt{Z}}\right)^2 \int d^4x\, d^4y\, \langle q_1,\ldots,q_m;\text{out}|T(\hat{\phi}(x)\hat{\phi}(y))|p_3,\ldots,p_n;\text{in}\rangle \\ &\quad \times \overleftarrow{(\Box_x + m^2)}\overleftarrow{(\Box_y + m^2)}\, u_{\boldsymbol{p}_1}(x)u_{\boldsymbol{p}_2}(y)\,. \end{aligned} \tag{9.82}$$

It is obvious how this formula is generalized for the n particles in the initial states. To complete the argument we will now take a look at the corresponding derivation for the particles in the final state. As an example we take the second reduction step where the following matrix element has to be calculated:

$$\begin{aligned}&\langle q_1,\ldots,q_m;\text{out}|\hat{\phi}(x)|p_2,\ldots,p_n;\text{in}\rangle \\ &= \langle q_2,\ldots,q_m;\text{out}|\hat{a}_{q_1;\text{out}}\hat{\phi}(x)|p_2,\ldots,p_n;\text{in}\rangle\,. \end{aligned} \tag{9.83}$$

The steps leading from (9.77) to (9.81) can be repeated with hardly any change:

$$\begin{aligned}\hat{a}_{q_1;\text{out}}\hat{\phi}(x) &= \frac{i}{\sqrt{Z}} \lim_{y_0 \to \infty} \int d^3y\, u^*_{q_1}(y)\, \overleftrightarrow{\partial_{y_0}}\, \hat{\phi}(y)\hat{\phi}(x) \\ &= \frac{i}{\sqrt{Z}} \lim_{y_0 \to \infty} \int d^3y\, u^*_{q_1}(y)\, \overleftrightarrow{\partial_{y_0}}\, T(\hat{\phi}(y)\hat{\phi}(x)) \\ &= \hat{\phi}(x)\hat{a}_{q_1;\text{in}} + \frac{i}{\sqrt{Z}} \int d^4y\, \partial_{y_0}\left(u^*_{q_1}(y)\, \overleftrightarrow{\partial_{y_0}}\, T(\hat{\phi}(y)\hat{\phi}(x))\right) \end{aligned}$$

$$= \hat{\phi}(x)\hat{a}_{q_1;\text{in}} + \frac{\mathrm{i}}{\sqrt{Z}} \int \mathrm{d}^4 y\, u^*_{q_1}(y)\, \overrightarrow{(\Box_y + m^2)}\, T(\hat{\phi}(x)\hat{\phi}(y))\,, \tag{9.84}$$

where the first term again will drop out if the momenta are different. The further steps are obvious and we arrive at the *LSZ reduction formula for spin-0 particles*:

$$\begin{aligned}
S_{fi} &= \langle q_1,\ldots,q_m;\text{in}|\hat{S}|p_1,\ldots,p_n;\text{in}\rangle \\
&= \left(\frac{\mathrm{i}}{\sqrt{Z}}\right)^{n+m} \int \mathrm{d}^4 y_1 \ldots \mathrm{d}^4 y_m \mathrm{d}^4 x_1 \ldots \mathrm{d}^4 x_n \\
&\quad \times u^*_{q_1}(y_1)\cdots u^*_{q_m}(y_m)\, \overrightarrow{(\Box_{y_1} + m^2)}\cdots \overrightarrow{(\Box_{y_m} + m^2)} \\
&\quad \times \langle 0|T(\hat{\phi}(x_1)\cdots\hat{\phi}(y_m))|0\rangle \\
&\quad \times \overleftarrow{(\Box_{x_1} + m^2)}\cdots \overleftarrow{(\Box_{x_n} + m^2)}\, u_{p_1}(x_1)\cdots u_{p_n}(x_n)\,.
\end{aligned} \tag{9.85}$$

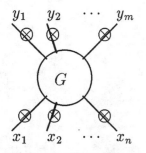

Fig. 9.2. The multi-particle Green's function with amputated external legs

The S matrix element for a process involving n bosons in the initial and m bosons in the final state essentially is determined by the $n+m$-particle Green's function $G^{(n+m)}(x_1,\ldots,y_m)$. We note that all the fields are treated on an equal footing. The Green's function does not "know" about the incoming and the outgoing particles and about their momenta. This connection is made by applying the Klein–Gordon operators and by projecting on the plane waves. Symbolically, (9.85) can be represented as in Fig. 9.2. Here the central circle represents the $(n+m)$-particle Green's function and the crosses appended to the external boson lines stand for the Klein–Gordon operators. Since the differential operator $(\Box + m^2)$ is the inverse of the boson propagator its action in (9.85) is called the *amputation of the external boson lines*.

It is useful to transform the reduction formula (9.85) to *momentum space*. The Fourier-transformed n-particle Green's function is defined as

$$\begin{aligned}
G^{(n)}(k_1,\ldots,k_n) &= \int \mathrm{d}^4 x_1 \ldots \mathrm{d}^4 x_n\, \mathrm{e}^{\mathrm{i}(k_1\cdot x_1 + \ldots + k_n\cdot x_n)} \\
&\quad \times \langle 0|T(\hat{\phi}(x_1)\cdots\hat{\phi}(x_n))|0\rangle\,.
\end{aligned} \tag{9.86}$$

Since the wave functions $u_p(x)$ are plane waves the integrations in (9.85) are Fourier transformations and we obtain, after integrating by parts, the *LSZ reduction formula in momentum space*:

$$\begin{aligned}
S_{fi} &= \left(\frac{-\mathrm{i}}{\sqrt{Z}}\right)^{n+m} N_{q_1}\cdots N_{q_m} N_{p_1}\cdots N_{p_n} \\
&\quad \times (q_1^2 - m^2)\cdots(q_m^2 - m^2)(p_1^2 - m^2)\cdots(p_n^2 - m^2) \\
&\quad \times G^{(n+m)}(q_1,\ldots,q_m,-p_1,\ldots,-p_n)\,,
\end{aligned} \tag{9.87}$$

with the normalization factors $N_p = ((2\pi)^3 2\omega_p)^{-1/2}$. Thus the S matrix element is directly linked to the many-particle Green's function in momentum space. The momenta of the incoming and outgoing particles are counted with opposite signs. The function $G^{(n+m)}$ is multipled by factors of the type $(p^2 - m^2)$. Since the external particles by definition are free they should satisfy the mass shell condition $p^2 = m^2$, i.e., the multiplying factors are zero! Thus to end up with a nonvanishing scattering matrix element S_{fi}, the Green's function $G^{(n+m)}$ must posses *first-order pole on the mass shell* of the particle momenta. Equation (9.87) then tells us that *the S matrix element is the (multiple) residue of the corresponding Fourier-transformed multi-particle Green's function*.

Remark on the pole structure of the Green's function. A closer inspection of the n-point function confirms that indeed it has the required poles at $p_i^2 = m^2$ (for all $i = 1, \ldots, n$). This divergent behavior originates from the time integrations in (9.86) which extend to infinity. Let us look, for example, at the coordinate $i = 1$. We evaluate the matrix element (9.86) by inserting a complete set of intermediate states $|\alpha\rangle\langle\alpha|$, as was done in (9.43):

$$G^{(n)}(k_1, \ldots, k_n) = \sum_\alpha \int d^4x_1 \ldots d^4x_n \, e^{i(k_1 \cdot x_1 + \ldots + k_n \cdot x_n)}$$
$$\times \langle 0|\hat{\phi}(x_1)|\alpha\rangle\langle\alpha|T(\hat{\phi}(x_2)\cdots\hat{\phi}(x_n))|0\rangle$$
$$\times \Theta(t_1 - \max(t_2, \ldots, t_n)) + \ldots . \quad (9.88)$$

Here that region of the integration variable was isolated for which the operator $\hat{\phi}(x_1)$ stands at the leftmost position in the time-ordered product. Since $|\alpha\rangle$ is a momentum eigenstate we have, according to (9.3),

$$\langle 0|\hat{\phi}(x_1)|\alpha\rangle = e^{-ip_\alpha \cdot x_1}\langle 0|\hat{\phi}(0)|\alpha\rangle, \quad (9.89)$$

and the x_1 integral in (9.88) becomes

$$I_1 = \int d^4x_1 \, e^{i(k_1-p_\alpha)\cdot x_1}\Theta(t_1-\tau) = \int d^3x \, e^{-i(\mathbf{k}_1-\mathbf{p}_\alpha)\cdot \mathbf{x}_1}\int_\tau^\infty dt_1 \, e^{i(k_1^0-p_\alpha^0)t}$$
$$= (2\pi)^3\delta^3(\mathbf{k}_1-\mathbf{p}_\alpha)\int_\tau^\infty dt_1 \, e^{i(k_1^0-p_\alpha^0)t}, \quad (9.90)$$

with the abbreviation $\tau = \max(t_2, \ldots, t_n)$. The time integral has to be interpreted as a singular distribution. Introducing an exponential damping factor we can write it as

$$\int_\tau^\infty dt \, e^{i\omega t} \to \int_\tau^\infty dt \, e^{(i\omega-\epsilon)t} = \frac{1}{i(\omega+i\epsilon)} e^{i(\omega+i\epsilon)t}\Big|_\tau^\infty = \frac{i}{\omega+i\epsilon} e^{-i\omega\tau} \quad (9.91)$$

Using this prescription, (9.90) becomes

$$I_1 = (2\pi)^3\delta^3(\mathbf{k}_1-\mathbf{p}_\alpha)\frac{i}{k_1^0-p_\alpha^0+i\epsilon} e^{-i(k_1^0-p_\alpha^0)\tau}$$
$$= i(2\pi)^3\delta^3(\mathbf{k}_1-\mathbf{p}_\alpha)\frac{k_1^0+p_\alpha^0}{k_1^2-p_\alpha^2+i\epsilon'} e^{-i(k_1^0-p_\alpha^0)\tau}. \quad (9.92)$$

As discussed in Sect. 9.3, the spectrum of the states $|\alpha\rangle$ will have a *discrete eigenvalue* for the single-particle states $|p\rangle$ at $p^2 = m^2$ (in addition there will be continua at higher values of the invariant mass p_α^2). Thus we learn from (9.92) that the n-particle Green's function indeed has poles "on the mass shell" of the free particles.

The use of the reduction technique summarized in the formula (9.85) of course is not limited to neutral spin-0 fields. Similar reduction formulae can also be derived for other quantum fields. Our discussion, however, will concentrate on Klein–Gordon fields, with the exception of Exercise 9.2 where the LSZ formula for Dirac fields is derived.

EXERCISE

9.2 The Reduction Formula for Spin-$\frac{1}{2}$ Particles

Problem. Derive the LSZ reduction formula for S matrix elements involving spin-$\frac{1}{2}$ particles, leaning on the treatment of the Klein–Gordon field in Sect. 9.4.

Solution. The formalism for scalar fields can be easily transferred to the case of Dirac fields. Free asymptotic in and out field operators are introduced which satisfy the weak limiting condition

$$\lim_{x_0 \to -\infty} \langle b|\hat{\psi}(x)|a\rangle = \sqrt{Z_2}\langle b|\hat{\psi}_{\rm in}|a\rangle , \tag{1a}$$

$$\lim_{x_0 \to +\infty} \langle b|\hat{\psi}(x)|a\rangle = \sqrt{Z_2}\langle b|\hat{\psi}_{\rm out}|a\rangle . \tag{1b}$$

Here Z_2 is the customary name for the fermion renormalization constant. The asymptotic in field operator can be expanded into plane waves as in Sect. 5.3:

$$\hat{\psi}_{\rm in}(x) = \int d^3p \sum_s \left(\hat{b}_{\rm in}(p,s) u_{ps}(x) + \hat{d}^\dagger_{\rm in}(p,s) v_{ps}(x) \right) \tag{2}$$

and similarly for the out field. We have used the following abbreviated notation for the normalized Dirac plane waves of particles and antiparticles:

$$u_{ps}(x) = \frac{1}{(2\pi)^{3/2}} \sqrt{\frac{m}{\omega_p}}\, u(p,s)\, e^{-ip\cdot x} , \tag{3a}$$

$$v_{ps}(x) = \frac{1}{(2\pi)^{3/2}} \sqrt{\frac{m}{\omega_p}}\, v(p,s)\, e^{+ip\cdot x} . \tag{3b}$$

These solutions satisfy the free Dirac equations

$$(i\slashed{\nabla} - m) u_{ps}(x) = (\slashed{p} - m) u_{ps}(x) = 0 , \tag{4a}$$

$$(i\slashed{\nabla} - m) v_{ps}(x) = (-\slashed{p} - m) v_{ps}(x) = 0 . \tag{4b}$$

To derive the reduction formula, we must invert (2) to obtain the creation operators for in particles and antiparticles

$$\hat{b}^\dagger_{\rm in}(p,s) = \int d^3x\, \hat{\psi}^\dagger_{\rm in}(x) u_{ps}(x) , \tag{5a}$$

$$\hat{d}^\dagger_{\rm in}(p,s) = \int d^3x\, v^\dagger_{ps}(x) \hat{\psi}_{\rm out}(x) . \tag{5b}$$

Similarly the annihilation operators for out particles and antiparticles are

$$\hat{b}_{\rm out}(p,s) = \int d^3x\, u^\dagger_{ps}(x) \hat{\psi}_{\rm out}(x) , \tag{6a}$$

$$\hat{d}_{\rm out}(p,s) = \int d^3x\, \hat{\psi}^\dagger_{\rm out}(x) v_{ps}(x) . \tag{6b}$$

Exercise 9.2

The general S matrix element for a scattering involving n particles and $\bar{n}$ antiparticles in the initial state and m particles and $\bar{m}$ particles in the final state reads

$$S_{fi} = \langle f; \text{out} | i; \text{in} \rangle \tag{7}$$
$$= \langle q_1 r_1, \ldots, q_m r_m; \bar{q}_1 \bar{r}_1, \ldots, \bar{q}_{\bar{m}} \bar{r}_{\bar{m}}; \text{out} | p_1 s_1, \ldots p_n s_n; \bar{p}_1 \bar{s}_1, \ldots, \bar{p}_{\bar{n}} \bar{s}_{\bar{n}}; \text{in} \rangle .$$

At the first stage four different reductions are possible. For example, the extraction of particle $(p_1 s_1)$ from the initial state proceeds as follows. Abbreviate $|i; \text{in}\rangle = \hat{b}^\dagger_{\text{in}}(p_1, s_1) |i - (p_1 s_1); \text{in}\rangle$ and the S matrix element reads

$$\begin{aligned}
S_{fi} &= \langle f; \text{out} | \hat{b}^\dagger_{\text{in}}(p_1, s_1) | i - (p_1 s_1); \text{in} \rangle \\
&= \lim_{x_0 \to -\infty} \int d^3 x \, \langle f; \text{out} | \hat{\psi}^\dagger_{\text{in}}(x) | i - (p_1 s_1); \text{in} \rangle u_{p_1 s_1}(x) \\
&= \frac{1}{\sqrt{Z_2}} \lim_{x_0 \to -\infty} \int d^3 x \, \langle f; \text{out} | \hat{\psi}^\dagger(x) | i - (p_1 s_1); \text{in} \rangle u_{p_1 s_1}(x) \\
&= \frac{1}{\sqrt{Z_2}} \lim_{x_0 \to +\infty} \int d^3 x \, \langle f; \text{out} | \hat{\psi}^\dagger(x) | i - (p_1 s_1); \text{in} \rangle u_{p_1 s_1}(x) \\
&\quad - \frac{1}{\sqrt{Z_2}} \left(\lim_{x_0 \to -\infty} - \lim_{x_0 \to +\infty} \right) \int d^3 x \, \langle f; \text{out} | \hat{\psi}^\dagger(x) | i - (p_1 s_1); \text{in} \rangle u_{p_1 s_1}(x) .
\end{aligned} \tag{8}$$

The first term can be written as

$$\lim_{x_0 \to +\infty} \int d^3 x \, \langle f; \text{out} | \hat{\psi}^\dagger_{\text{out}}(x) | i - (p_1 s_1); \text{in} \rangle u_{p_1 s_1}(x)$$
$$= \langle f; \text{out} | \hat{b}^\dagger_{\text{out}}(p_1, s_1) | i - (p_1 s_1); \text{in} \rangle \tag{9}$$

and vanishes under the assumption that all initial and final momenta are different. The second contribution in (8) is written as a four-dimensional integral. The operator entering the matrix element is

$$\begin{aligned}
&-\frac{1}{\sqrt{Z_2}} \left(\lim_{x_0 \to -\infty} - \lim_{x_0 \to +\infty} \right) \int d^3 x \, \hat{\psi}^\dagger(x) u_{p_1 s_1}(x) \\
&= -\frac{1}{\sqrt{Z_2}} \int d^4 x \, \partial_0 \left[\hat{\psi}^\dagger(x) u_{p_1 s_1}(x) \right] \\
&= -\frac{1}{\sqrt{Z_2}} \int d^4 x \, \hat{\psi}^\dagger(x) \left(\overleftarrow{\partial_0} + \overrightarrow{\partial_0} \right) u_{p_1 s_1}(x) .
\end{aligned} \tag{10}$$

The free Dirac equation (4a) (multiplied by γ^0) reads

$$\partial_0 u_{p_1 s_1} = i\gamma_0 \left(i\gamma_k \partial^k - m \right) u_{p_1 s_1} , \tag{11}$$

so that (10) becomes

$$\begin{aligned}
&-\frac{1}{\sqrt{Z_2}} \int d^4 x \, \hat{\psi}^\dagger(x) \left[\overleftarrow{\partial_0} + i\gamma_0 \left(i\gamma_k \overrightarrow{\partial^k} - m \right) \right] u_{p_1 s_1}(x) \\
&= -\frac{i}{\sqrt{Z_2}} \int d^4 x \, \hat{\psi}^\dagger(x) \gamma^0 \left[-i\gamma_0 \overleftarrow{\partial^0} - i\gamma_k \overleftarrow{\partial^k} - m \right] u_{p_1 s_1}(x) \\
&= -\frac{i}{\sqrt{Z_2}} \int d^4 x \, \hat{\bar{\psi}}(x) \left(-i\overleftarrow{\nabla}_x - m \right) u_{p_1 s_1}(x)
\end{aligned} \tag{12}$$

after an integration by parts. The reduced S matrix element then becomes

Exercise 9.2

$$S_{fi} = \frac{-i}{\sqrt{Z_2}} \int d^4x \, \langle f; \text{out}|\hat{\bar{\psi}}(x)|i-(p_1s_1); \text{in}\rangle \, \overleftarrow{(-i\overleftarrow{\slashed{\partial}}_x - m)} \, u_{p_1s_1}(x) \, . \quad (13)$$

The other three types of first-stage reductions are treated in the same way. The results are listed below.

- An antiparticle $(\bar{p}_1\bar{s}_1)$ in the initial state:

$$S_{fi} = \frac{i}{\sqrt{Z_2}} \int d^4x \, \bar{v}_{\bar{p}_1\bar{s}_1} \, \overrightarrow{(i\overrightarrow{\slashed{\partial}}_x - m)} \, \langle f; \text{out}|\hat{\psi}(x)|i-(\bar{p}_1\bar{s}_1); \text{in}\rangle \, . \quad (14)$$

- A particle (q_1r_1) in the final state:

$$S_{fi} = \frac{-i}{\sqrt{Z_2}} \int d^4x \, \bar{u}_{q_1r_1} \, \overrightarrow{(i\overrightarrow{\slashed{\partial}}_x - m)} \, \langle f-(q_1r_1); \text{out}|\hat{\psi}(x)|i; \text{in}\rangle \, . \quad (15)$$

- An antiparticle $(\bar{q}_1\bar{r}_1)$ in the final state:

$$S_{fi} = \frac{i}{\sqrt{Z_2}} \int d^4x \, \langle f-(\bar{q}_1\bar{r}_1); \text{out}|\hat{\bar{\psi}}(x)|i; \text{in}\rangle \, \overleftarrow{(-i\overleftarrow{\slashed{\partial}}_x - m)} \, v_{\bar{q}_1\bar{r}_1}(x) \, . \quad (16)$$

The reductions of higher stages can be derived by using the trick of introducing the time-ordered product as in (9.78). This results in the general *LSZ reduction formula for spin-$\frac{1}{2}$ particles*:

$$\begin{aligned}
S_{fi} &= \left(\frac{-i}{\sqrt{Z_2}}\right)^{n+m} \left(\frac{i}{\sqrt{Z_2}}\right)^{\bar{n}+\bar{m}} \\
&\times \int d^4x_1 \ldots d^4x_n d^4\bar{x}_1 \ldots d^4\bar{x}_{\bar{n}} d^4y_1 \ldots d^4y_m d^4\bar{y}_1 \ldots d^4\bar{y}_{\bar{m}} \\
&\times \bar{u}_{q_mr_m}(y_m) \, \overrightarrow{(i\overrightarrow{\slashed{\partial}}_{y_m} - m)} \cdots \bar{u}_{q_1r_1}(y_1) \, \overrightarrow{(i\overrightarrow{\slashed{\partial}}_{y_1} - m)} \\
&\times \bar{v}_{\bar{p}_{\bar{n}}\bar{s}_{\bar{n}}}(\bar{x}_{\bar{n}}) \, \overrightarrow{(i\overrightarrow{\slashed{\partial}}_{\bar{x}_{\bar{n}}} - m)} \cdots \bar{v}_{\bar{p}_1\bar{s}_1}(\bar{x}_1) \, \overrightarrow{(i\overrightarrow{\slashed{\partial}}_{\bar{x}_1} - m)} \\
&\times \langle 0|T(\hat{\bar{\psi}}(\bar{y}_{\bar{m}}) \cdots \hat{\bar{\psi}}(\bar{y}_1) \hat{\psi}(y_m) \cdots \hat{\psi}(y_1) \hat{\bar{\psi}}(x_1) \cdots \hat{\bar{\psi}}(x_n) \hat{\psi}(\bar{x}_1) \cdots \hat{\psi}(\bar{x}_{\bar{n}}))|0\rangle \\
&\times \overleftarrow{(-i\overleftarrow{\slashed{\partial}}_{x_1} - m)} \, u_{p_1s_1}(x_1) \cdots \overleftarrow{(-i\overleftarrow{\slashed{\partial}}_{x_n} - m)} \, u_{p_ns_n}(x_n) \\
&\times \overleftarrow{(-i\overleftarrow{\slashed{\partial}}_{\bar{y}_1} - m)} \, v_{\bar{q}_1\bar{r}_1}(\bar{y}_1) \cdots \overleftarrow{(-i\overleftarrow{\slashed{\partial}}_{\bar{y}_{\bar{m}}} - m)} \, v_{\bar{q}_{\bar{m}}\bar{r}_{\bar{m}}}(\bar{y}_{\bar{m}}) \, . \quad (17)
\end{aligned}$$

The resemblance between (17) and the scalar reduction formula (9.85) is evident. The S matrix element again consists of the many-particle Green's function, which gets multiplied by inverse propagators associated with the external lines and gets projected on plane-wave spinors. Note that the Green's function is an object that carries $(n+m+\bar{n}+\bar{m})$ four-dimensional Dirac indices which have to be paired off with the associated operators $(\pm i\slashed{\partial} - m)$ that amputate the external lines. Incoming antiparticles play the same role as outgoing particles, and vice versa.

9.5 Perturbation Theory for the n-Point Function

The LSZ reduction formula provides an elegant way to connect the S matrix element with the multi-particle Green's function. Of course, for any practical purpose this does not solve the problem but only shifts it to the task of evaluating the function $G^{(n)}(x_1,\ldots,x_n)$. As discussed extensively in the last chapter this can be achieved by using perturbation theory, which is the only available analytical tool for this purpose. All the results obtained in Chap. 8, in particular the Feynman rules for constructing the terms of the perturbation series, can be repeated using the language of Green's functions.

Here we will derive the perturbation theory for the many-particle Green's function applied to a scalar field. Subsequently Example 9.4 will illustrate, for some simple Feynman graphs of the ϕ^4 theory, that the LSZ reduction formalism produces the same results for the S matrix elements as does our earlier treatment, which was based on the interaction picture.

Obviously one encounters an essential problem when trying to evaluate the vacuum expectation value (9.68): the operators $\hat{\phi}(x)$ belong to the *interacting* "interpolating" fields about which we do not have detailed knowledge. It would be much more convenient to deal with the *asymptotic* fields instead, for example, with the in field $\hat{\phi}_{\text{in}}(x)$. The latter has a simple dynamics and any expectation values can be easily evaluated by using the decomposition into creation and annihilation operators for free particles. Therefore we will strive for a way to express $\hat{\phi}(x)$ in terms of $\hat{\phi}_{\text{in}}(x)$, and similarly for the canonically conjugate field operator $\hat{\pi}(x)$. The connection is provided by a (time-dependent) unitary transformation operator[10] $\hat{U}(t)$:

$$\hat{\phi}(\boldsymbol{x},t) = \hat{U}^\dagger(t)\,\hat{\phi}_{\text{in}}(\boldsymbol{x},t)\,\hat{U}(t) \quad , \quad \hat{\pi}(\boldsymbol{x},t) = \hat{U}^\dagger(t)\,\hat{\pi}_{\text{in}}(\boldsymbol{x},t)\,\hat{U}(t) \;. \qquad (9.93)$$

As we shall see soon this transformation is closely related to Dyson's time-evolution operator (cf. Sect. 8.3) which formed the basic building block for developing the perturbation series.

Let us study a system described by a Hamiltonian that can be split into a free and an interacting part:

$$\hat{H}(\hat{\phi},\hat{\pi}) = \hat{H}_0(\hat{\phi},\hat{\pi}) + \hat{H}_1(\hat{\phi},\hat{\pi}) \;. \qquad (9.94)$$

H_0 and H_1 are polynomials in the field operators $\hat{\phi}$ and $\hat{\pi}$.

The transformation operator $\hat{U}(t)$ defined in (9.93) satisfies a simple equation of motion, which can be obtained from the two different Heisenberg equations for the asymptotic and the interacting field operator:

$$\begin{aligned}
\mathrm{i}\partial_0 \hat{\phi}_{\text{in}}(x) &= \left[\hat{\phi}_{\text{in}}(x)\,,\,\hat{H}_{\text{in}}(\hat{\phi}_{\text{in}},\hat{\pi}_{\text{in}})\right] \;, & (9.95\text{a})\\
\mathrm{i}\partial_0 \hat{\phi}(x) &= \left[\hat{\phi}(x)\,,\,\hat{H}(\hat{\phi},\hat{\pi})\right] \;. & (9.95\text{b})
\end{aligned}$$

Since the in-field operator by construction is noninteracting, (9.95a) contains only the asymptotic Hamiltonian $\hat{H}_{\text{in}}$, expressed in terms of the asymptotic field.[11] Equation (9.95b), on the other hand, describes the full dynamics of the interacting quantum system.

---

[10] The existence of this operator is not guaranteed mathematically if the system allows for stable bound state. Here we will ignore this problem and will take the existence of the operator $\hat{U}(t)$ for granted.

[11] Note that $\hat{H}_{\text{in}}(\hat{\phi}_{\text{in}},\hat{\pi}_{\text{in}})$ differs from $\hat{H}_0(\hat{\phi}_{\text{in}},\hat{\pi}_{\text{in}})$ because the former contains the physical mass m and the latter the bare mass m_0. See (1) and (2) in Example 9.4.

Taking the time derivative of (9.93) and using (9.95) one arrives, after some effort (see Exercise 9.3), at the *equation of motion for $\hat{U}(t)$*:

$$i\partial_0 \hat{U}(t) = \hat{H}_1^c(t)\,\hat{U}(t) \,. \tag{9.96}$$

Here $\hat{H}_1^c(t)$ is the interaction Hamiltonian, expressed in terms of the in-fields, to which an undetermined c number function of time, $c(t)$, is added:

$$\hat{H}_1^c(t) = \hat{H}(\hat{\phi}_{\text{in}}, \hat{\pi}_{\text{in}}) - \hat{H}_{\text{in}}(\hat{\phi}_{\text{in}}, \hat{\pi}_{\text{in}}) + c(t) \,. \tag{9.97}$$

The function $c(t)$ need not to be specified since it only produces an irrelevant phase factor.

Now we observe that (9.96) looks exactly like the equation of motion (8.35) for Dyson's time-evolution operator. Thus the construction of the perturbation series for the operator $\hat{U}(t)$ can be simply copied from Chap. 8. Again the differential equation (9.96) is rewritten as an integral equation

$$\hat{U}(t) = \hat{U}(t_0) - i\int_{t_0}^t d\tau\, \hat{H}_1^c(\tau)\,\hat{U}(\tau) \,, \tag{9.98}$$

which is solved iteratively in terms of the Neumann series. When we introduce the time-ordered product, the *perturbation series for the operator $\hat{U}(t)$* takes the same form as (8.42):

$$\hat{U}(t) = \sum_{n=0}^\infty \frac{1}{n!}(-i)^n \int_{t_0}^t dt_1 \ldots \int_{t_0}^t dt_n\, T\bigl(\hat{H}_1^c(t_1)\ldots \hat{H}_1^c(t_n)\bigr)\,\hat{U}(t_0) \,. \tag{9.99}$$

This can be formally summed to yield the time-ordered exponential function (8.44):

$$\hat{U}(t) = T \exp\!\Bigl(-i \int_{t_0}^t d\tau\, \hat{H}_1^c(\tau)\Bigr)\,\hat{U}(t_0) \,. \tag{9.100}$$

The analogy to the Dyson operator is complete if we define

$$\hat{U}(t, t_0) = \hat{U}(t)\,\hat{U}^{-1}(t_0) \,, \tag{9.101}$$

so that the operator satisfies the condition $\hat{U}(t, t) = \mathbb{I}$. The result (9.100) for the operator $\hat{U}(t)$ can be used to express the many-particle Green's function,

$$G^{(n)}(x_1, \ldots, x_n) = \langle 0 | T\bigl(\hat{\phi}(x_1)\cdots\hat{\phi}(x_n)\bigr) | 0 \rangle \,, \tag{9.102}$$

in terms of a vacuum expectation value which solely depends on the in fields:

$$\begin{aligned} G^{(n)}(x_1, \ldots, x_n) =\ & \langle 0 | T\bigl(\hat{U}^\dagger(t_1)\hat{\phi}_{\text{in}}(x_1)\hat{U}(t_1)\,\hat{U}^\dagger(t_2)\hat{\phi}_{\text{in}}(x_2)\hat{U}(t_2) \\ & \times \cdots \hat{U}^\dagger(t_n)\hat{\phi}_{\text{in}}(x_n)\hat{U}(t_n)\bigr) | 0 \rangle \,. \end{aligned} \tag{9.103}$$

Now the unit operator is inserted twice, $\hat{U}^\dagger(t)\hat{U}(t) = \mathbb{I}$ on the left and $\hat{U}^\dagger(t')\hat{U}(t') = \mathbb{I}$ on the right, and (9.101) is used:

$$\begin{aligned} G^{(n)}(x_1, \ldots, x_n) =\ & \langle 0 | T\bigl(\hat{U}^\dagger(t)\hat{U}^\dagger(t, t_1)\hat{\phi}_{\text{in}}(x_1)\hat{U}(t_1, t_2)\hat{\phi}_{\text{in}}(x_2) \\ & \times \cdots \hat{U}(t_{n-1}, t_n)\hat{\phi}_{\text{in}}(x_n)\hat{U}(t_n, t')\hat{U}(t')\bigr) | 0 \rangle \,. \end{aligned} \tag{9.104}$$

The time arguments t and t' are chosen to be so large that all the other times fall into the interval $t > t_1, \ldots, t_n > t'$, so that the factors $\hat{U}^\dagger(t)$ and $\hat{U}(t')$ can be removed from the time-ordered product. Equation (9.104) can be considerably simplified if one recalls that the factors in the argument of a T product can be placed at arbitrary positions. The time-ordering prescription will take care of the correct ordering anyway. Therefore all the factors $\hat{U}(t_i, t_{i+1})$ in (9.104) can be combined as follows:

$$\hat{U}(t,t_1)\hat{U}(t_1,t_2)\cdots \hat{U}(t_n,t') = \hat{U}(t,t') \,. \tag{9.105}$$

Thus (9.104) becomes

$$G^{(n)}(x_1,\ldots,x_n) = \langle 0|\hat{U}^\dagger(t) T\big(\hat{\phi}_{\rm in}(x_1)\hat{\phi}_{\rm in}(x_2)\cdots \hat{\phi}_{\rm in}(x_n)\hat{U}(t,t')\big)\hat{U}(t')|0\rangle \,. \tag{9.106}$$

To evaluate this matrix element we have to know the action of the operator $\hat{U}(\pm\infty)$ on the vacuum. We make the plausible assumption (which can be backed by a formal proof) that $\hat{U}$ in the asymptotic limit $t \to \pm\infty$ will not influence the particle contents of the vacuum:

$$\hat{U}(-\infty)|0\rangle = \alpha|0\rangle \quad , \quad \hat{U}(+\infty)|0\rangle = \beta|0\rangle \,, \tag{9.107}$$

where the values of the c numbers α and β do not have physical relevance. To fix these numbers we look at the n-point function of "zeroth order", which is simply the overlap

$$\begin{aligned}\langle 0\,|\,0\rangle &= \langle 0|\hat{U}^\dagger(+\infty)\, T\big(\hat{U}(+\infty,-\infty)\big)\hat{U}(-\infty)|0\rangle \\ &= \beta^*\alpha\langle 0|T\big(\hat{U}(+\infty,-\infty)\big)|0\rangle \,. \end{aligned} \tag{9.108}$$

Postulating that the vacuum state is normalized according to $\langle 0|0\rangle = 1$, we find

$$\mathcal{N} \equiv \beta^*\alpha = \frac{1}{\langle 0|T\exp\big(-{\rm i}\int_{-\infty}^{+\infty}{\rm d}\tau\,\hat{H}_1^c(\tau)\big)|0\rangle} \,. \tag{9.109}$$

Using this normalization factor the final result for the *n-point function* reads

$$G^{(n)}(x_1,\ldots,x_n) \tag{9.110}$$
$$= \mathcal{N}\langle 0|T\Big[\hat{\phi}_{\rm in}(x_1)\hat{\phi}_{\rm in}(x_2)\cdots \hat{\phi}_{\rm in}(x_n)\exp\Big(-{\rm i}\int_{-\infty}^{+\infty}{\rm d}\tau\,\hat{H}_1^c(\tau)\Big)\Big]|0\rangle \,.$$

Clearly we are allowed to make the replacement $\hat{H}_1^c \to \hat{H}_1$ in this formula since the same factor $\exp\big(-{\rm i}\int_{-\infty}^{+\infty}{\rm d}\tau\,c(\tau)\big)$ enters in the numerator and denominator and thus is cancelled.

The *perturbation series of the n-point function* is obtained from (9.110) simply by inserting the Taylor expansion of the exponential function:

$$\begin{aligned}G^{(n)}(x_1,\ldots,x_n) &= \mathcal{N}\sum_{k=1}^\infty \frac{(-{\rm i})^k}{k!}\int {\rm d}^4y_1\ldots {\rm d}^4y_k \tag{9.111} \\ &\quad \times \langle 0|T\big(\hat{\phi}_{\rm in}(x_1)\hat{\phi}_{\rm in}(x_2)\cdots \hat{\phi}_{\rm in}(x_n)\hat{\mathcal{H}}_1(y_1)\cdots \hat{\mathcal{H}}_1(y_k)\big)|0\rangle \,.\end{aligned}$$

The vacuum expectation value can be evaluated by using *Wick's theorem*, leading to a summation over all fully contracted operator products.

The normalization factor $\mathcal{N}$ plays the same role as in Exercise 8.3. It is identical with the phase factor $\langle 0|\hat{S}|0\rangle$ which describes unobservable vacuum fluctuations. Therefore the factor $\mathcal{N}$ *can simply be ignored provided that all graphs with disconnected vacuum bubbles are left out* in the evaluation of (9.111). A graph is disconnected if it contains a group of operators $\hat{\mathcal{H}}_1(y_i)$ that are only contracted among each other, and to not have a contraction to a field operator $\hat{\phi}_{\text{in}}(x_j)$.

EXERCISE

9.3 The Equation of Motion for the Operator $\hat{U}(t)$

Problem. Show that the transformation operator $\hat{U}(t)$ defined in (9.93) satisfies the following commutation relations:

$$[i\partial_0 \hat{U} \hat{U}^\dagger - \hat{H}_1(\hat{\phi}_{\text{in}}, \hat{\pi}_{\text{in}}), \hat{\phi}_{\text{in}}] = 0, \quad (1a)$$

$$[i\partial_0 \hat{U} \hat{U}^\dagger - \hat{H}_1(\hat{\phi}_{\text{in}}, \hat{\pi}_{\text{in}}), \hat{\pi}_{\text{in}}] = 0, \quad (1b)$$

and obtain the equations of motion for $\hat{U}(t)$.

Solution. Taking the time derivative of (9.93) leads to

$$i\partial_0 \hat{\phi} = i\partial_0 \hat{U}^\dagger \hat{\phi}_{\text{in}} \hat{U} + \hat{U}^\dagger i\partial_0 \hat{\phi}_{\text{in}} \hat{U} + \hat{U}^\dagger \hat{\phi}_{\text{in}} \partial_0 \hat{U}. \quad (2)$$

The time derivatives of $\hat{U}^\dagger$ and $\hat{U}$ are related because of the condition of unitarity $\hat{U}^\dagger \hat{U} = \mathbb{I}$:

$$\partial_0 \hat{U}^\dagger = -\hat{U}^\dagger \partial_0 \hat{U} \hat{U}^\dagger. \quad (3)$$

Furthermore, using the Heisenberg equations (9.95), we replace the derivatives of the field operators in (2) by the commutators with the free or the interacting Hamiltonian. This leads to

$$[\hat{\phi}, \hat{H}(\hat{\phi}, \hat{\pi})] = -\hat{U}^\dagger i\partial_0 \hat{U} \hat{U}^\dagger \hat{\phi}_{\text{in}} \hat{U} + \hat{U}^\dagger [\hat{\phi}_{\text{in}}, \hat{H}_{\text{in}}(\hat{\phi}_{\text{in}}, \hat{\pi}_{\text{in}})] \hat{U} + \hat{U}^\dagger \hat{\phi}_{\text{in}} i\partial_0 \hat{U}, (4)$$

and after multiplying with $\hat{U}$ from the left and $\hat{U}^\dagger$ from the right:

$$\hat{U}[\hat{\phi}, \hat{H}(\hat{\phi}, \hat{\pi})]\hat{U}^\dagger = -i\partial_0 \hat{U} \hat{U}^\dagger \hat{\phi}_{\text{in}} + [\hat{\phi}_{\text{in}}, \hat{H}_{\text{in}}(\hat{\phi}_{\text{in}}, \hat{\pi}_{\text{in}})] + \hat{\phi}_{\text{in}} i\partial_0 \hat{U} \hat{U}^\dagger. \quad (5)$$

According to (9.93) the operator $\hat{U}$ transforms $\hat{\phi}$ into $\hat{\phi}_{\text{in}}$:

$$\hat{U} \hat{\phi} \hat{U}^\dagger = \hat{\phi}_{\text{in}}, \qquad \hat{U} \hat{\pi} \hat{U}^\dagger = \hat{\pi}_{\text{in}}. \quad (6)$$

This remains true for any product of field operators since one can always insert $\mathbb{I} = \hat{U}^\dagger \hat{U}$ between the factors. Since the Hamiltonian $\hat{H}(\hat{\phi}, \hat{\pi})$ is a polynomial in $\hat{\phi}$ and $\hat{\pi}$, (6) implies

$$\hat{U} \hat{H}(\hat{\phi}, \hat{\pi}) \hat{U}^\dagger = \hat{H}(\hat{\phi}_{\text{in}}, \hat{\pi}_{\text{in}}). \quad (7)$$

Therefore (5) can be written as

$$[\hat{\phi}_{\text{in}}, \hat{H}(\hat{\phi}_{\text{in}}, \hat{\pi}_{\text{in}})] = -i[\partial_0 \hat{U} \hat{U}^\dagger, \hat{\phi}_{\text{in}}] + [\hat{\phi}_{\text{in}}, \hat{H}_{\text{in}}(\hat{\phi}_{\text{in}}, \hat{\pi}_{\text{in}})]. \quad (8)$$

This agrees with the assertion (1a). The corresponding relation (1b) for the canonically conjugate field operator is derived in the same way.

9.5 Perturbation Theory for the n-Point Function

The combination $\hat{O} = i\partial_0 \hat{U}\hat{U}^\dagger - \hat{H}_1(\hat{\phi}_{\text{in}}, \hat{\pi}_{\text{in}})$ therefore is seen to commutate with any of the field operators $\hat{\phi}_{\text{in}}$ and $\hat{\pi}_{\text{in}}$. Since these are assumed to form a complete set of operators we conclude that $\hat{O}$ is a multiple of unity:[12]

$$i\partial_0 \hat{U}\hat{U}^\dagger - \hat{H}_1(\hat{\phi}_{\text{in}}, \hat{\pi}_{\text{in}}) = c(t)\mathbb{1} \ . \tag{9}$$

This leads to the equation of motion (9.96)

$$i\partial_0 \hat{U} = \big(\hat{H}_1(\hat{\phi}_{\text{in}}, \hat{\pi}_{\text{in}}) + c(t)\big)\hat{U}(t) \equiv \hat{H}_1^c\,\hat{U}(t) \ . \tag{10}$$

Exercise 9.3

EXAMPLE

9.4 Green's Functions and the S Matrix of ϕ^4 Theory

We come back to the ϕ^4 theory introduced in Example 8.7, which represents an interacting field theory with a particularly simple structure. We will show that the same results of perturbation theory obtained earlier can also be deduced with the help of the n-point function and the LSZ reduction formula.

The normal-ordered Hamiltonian of the ϕ^4 theory, expressed in terms of the interacting field operators $\hat{\phi}(x)$ and $\hat{\pi}(x)$, reads

$$\begin{aligned}\hat{H}(\hat{\phi}, \hat{\pi}) &= \int d^3x \, \frac{1}{2} \!:\!\big(\hat{\pi}^2 + (\boldsymbol{\nabla}\hat{\phi})^2 + m_0^2 \hat{\phi}^2\big)\!: + \int d^3x \, \frac{g}{4!} \!:\!\hat{\phi}^4\!: \\ &= \hat{H}_0(\hat{\phi}, \hat{\pi}) + \hat{H}_1(\hat{\phi}, \hat{\pi}) \ . \end{aligned} \tag{1}$$

Here g is the coupling constant and m_0 is the *bare mass* introduced in the original free Lagrangian. The *asymptotic Hamiltonian*, on the other hand, describes *free particles* that posses the *observable mass* m:

$$\hat{H}_{\text{in}}(\hat{\phi}_{\text{in}}, \hat{\pi}_{\text{in}}) = \int d^3x \, \frac{1}{2} \!:\!\big(\hat{\pi}_{\text{in}}^2 + (\boldsymbol{\nabla}\hat{\phi}_{\text{in}})^2 + m^2 \hat{\phi}_{\text{in}}^2\big)\!: \ . \tag{2}$$

Accordingly the interaction Hamiltonian defined in (9.97) reads

$$\begin{aligned}\hat{H}_1(\hat{\phi}_{\text{in}}, \hat{\pi}_{\text{in}}) &= \hat{H}(\hat{\phi}_{\text{in}}, \hat{\pi}_{\text{in}}) - \hat{H}_{\text{in}}(\hat{\phi}_{\text{in}}, \hat{\pi}_{\text{in}}) \\ &= \int d^3x \left(\frac{1}{2}(m_0^2 - m^2)\!:\!\hat{\phi}_{\text{in}}^2\!: + \frac{g}{4!}\!:\!\hat{\phi}_{\text{in}}^4\!:\right) \\ &\equiv \hat{H}_1^{\delta m^2} + \hat{H}_1^{\phi^4} \ . \end{aligned} \tag{3}$$

Besides the "real" interaction, which is of quartic type, this Hamiltonian also contains a quadratic *mass-renormalization term*. This term is responsible for shifting the mass by an amount $\delta m^2 = m_0^2 - m^2$. As we will see this will guarantee that the particles obtain the correct physical value of the mass.

The first terms of the perturbation series for the *two-point Green's function* are given by

$$\begin{aligned}G^{(2)}(x_1, x_2) &= \langle 0|T\big(\hat{\phi}_{\text{in}}(x_1)\hat{\phi}_{\text{in}}(x_2) \exp(-i\int d^4x\, \hat{\mathcal{H}}_1(x))\big)|0\rangle \\ &= \langle 0|T\big(\hat{\phi}_{\text{in}}(x_1)\hat{\phi}_{\text{in}}(x_2)\big)|0\rangle\end{aligned}$$

---

[12] A more precise mathematical formulation would be: the $\hat{\phi}_{\text{in}}$ and $\hat{\pi}_{\text{in}}$ form an irreducible ring of operators.

Example 9.4

$$+ (-\mathrm{i})\frac{1}{2}\delta m^2 \int \mathrm{d}^4x \, \langle 0|T(\hat{\phi}_{\mathrm{in}}(x_1)\hat{\phi}_{\mathrm{in}}(x_2){:}\hat{\phi}_{\mathrm{in}}^2(x){:})|0\rangle$$
$$+ (-\mathrm{i})\frac{g}{4!} \int \mathrm{d}^4x \, \langle 0|T(\hat{\phi}_{\mathrm{in}}(x_1)\hat{\phi}_{\mathrm{in}}(x_2){:}\hat{\phi}_{\mathrm{in}}^4(x){:})|0\rangle$$
$$+ \ldots \, . \qquad (4)$$

This can be evaluated via Wick's theorem. The ϕ^4 interaction term does not contribute in the first order since no complete contraction is possible; remember that the normal-ordering prescription excludes self contractions of the operators $\hat{\phi}_{\mathrm{in}}(x)$. The factor $\frac{1}{2}$ in the δm^2 term cancels because there are two contractions leading to equal results. Accordingly, (4) leads to

$$\begin{aligned} G^{(2)}(x_1, x_2) &= \mathrm{i}\Delta_{\mathrm{F}}(x_1 - x_2) \\ &+ \int \mathrm{d}^4x \, \mathrm{i}\Delta_{\mathrm{F}}(x_1 - x)(-\mathrm{i}\delta m^2)\mathrm{i}\Delta_{\mathrm{F}}(x - x_2) + O(g^2) \, . \end{aligned} \qquad (5)$$

The graphical representation of the perturbation series for the two-point function is

$$\begin{array}{c} \bigotimes_G = \Big| + \Big\ast + \bigcirc + \Big\ast\Big\ast + \ldots \end{array} \qquad (6)$$

The mass renormalization term leads to a new type of Feynman graph which is represented by a cross instead of a vertex. An extra Feynman rule is established to deal with such graphs. In coordinate space it reads:

- Each mass renormalization term (cross) is associated with the factor $-\mathrm{i}\delta m^2 \int \mathrm{d}^4x$.

Note that the Feynman propagator Δ_{F} contains the "correct" physical mass m since this, according to (2), is the mass of the quanta belonging to the in field $\hat{\phi}_{\mathrm{in}}$. The interaction $H_1^{\phi^4}$ leads to self-energy corrections that would change the mass. The terms involving $\hat{H}_1^{\delta m^2}$ serve to *cancel this mass shift* in such a way that the exact Green's function $G^{(2)}(x_1, x_2)$ carries the physical mass. The expression $\hat{H}_1^{\delta m^2}$ is known as the *mass counter term*.

Let us study the effect the mass counter term would have in the perturbation series when taken on its own. The two-point function is represented by an infinite series of the "barbed-wire" graphs shown in (6). The corresponding analytical expression is particularly simple in *momentum space*. Fourier transforming (5), we obtain (cf. Sect. 12.6 concerning the notation of n-point functions in momentum space)

$$\begin{aligned} G^{(2)}_{\delta m^2}(p, -p) &= \int \mathrm{d}^4p \, \mathrm{e}^{\mathrm{i}p \cdot (x_1 - x_2)} G^{(2)}(x_1 - x_2) \\ &= \mathrm{i}\Delta_{\mathrm{F}}(p) + \mathrm{i}\Delta_{\mathrm{F}}(-\mathrm{i}\delta m^2)\mathrm{i}\Delta_{\mathrm{F}}(p) \\ &\quad + \mathrm{i}\Delta_{\mathrm{F}}(-\mathrm{i}\delta m^2)\mathrm{i}\Delta_{\mathrm{F}}(p)(-\mathrm{i}\delta m^2)\mathrm{i}\Delta_{\mathrm{F}}(p) + \ldots \, . \end{aligned} \qquad (7)$$

This is just the geometric series written in the form

$$\frac{1}{X} + \frac{1}{X}Y\frac{1}{X} + \frac{1}{X}Y\frac{1}{X}Y\frac{1}{X} + \ldots = \frac{1}{X - Y} \, , \qquad (8)$$

where $X = 1/\mathrm{i}\Delta_F(p) = -\mathrm{i}(p^2 - m^2 + \mathrm{i}\epsilon)$ and $Y = -\mathrm{i}\delta m^2$. As a consequence the two-point functions gives

$$G^{(2)}_{\delta m^2}(p, -p) = \frac{\mathrm{i}}{p^2 - m^2 - \delta m^2 + \mathrm{i}\epsilon} \,. \tag{9}$$

Thus we see that the counter term just has the effect of shifting the pole of the Green's function, i.e., of changing the value of the mass. Obviously one will chose the value of δm^2 such that it compensates the mass shift generated by the "true" interaction (i.e., those graphs of (6) that are not contained in (7)). This guarantees that the pole of the exact Green's function $G^{(2)}$ is located at the correct physical mass m.

An ugly feature of this procedure should not be hidden: the value of δm^2 cannot be calculated in perturbation theory since the Feynman graphs involving closed loops are *divergent*. By introducing the counter term δm^2, however, at least we have succeeded in eliminating the infinite (or, to phrase it differently, cutoff-dependent) part of the self-energy from the theory.

The *four-point Green's function* reads

$$G^{(4)}(x_1, x_2, x_3, x_4) \tag{10}$$
$$= \langle 0|T(\hat{\phi}_{\mathrm{in}}(x_1)\hat{\phi}_{\mathrm{in}}(x_2)\hat{\phi}_{\mathrm{in}}(x_3)\hat{\phi}_{\mathrm{in}}(x_4))|0\rangle$$
$$+ (-\mathrm{i})\frac{1}{2}\delta m^2 \int \mathrm{d}^4 x \, \langle 0|T(\hat{\phi}_{\mathrm{in}}(x_1)\hat{\phi}_{\mathrm{in}}(x_2)\hat{\phi}_{\mathrm{in}}(x_3)\hat{\phi}_{\mathrm{in}}(x_4){:}\hat{\phi}^2_{\mathrm{in}}(x){:})|0\rangle$$
$$+ (-\mathrm{i})\frac{g}{4!} \int \mathrm{d}^4 x \, \langle 0|T(\hat{\phi}_{\mathrm{in}}(x_1)\hat{\phi}_{\mathrm{in}}(x_2)\hat{\phi}_{\mathrm{in}}(x_3)\hat{\phi}_{\mathrm{in}}(x_4){:}\hat{\phi}^4_{\mathrm{in}}(x){:})|0\rangle + O(g^2)$$
$$\equiv G^{(4)}_0(x_1, x_2, x_3, x_4) + G^{(4)}_{\delta m^2}(x_1, x_2, x_3, x_4) + G^{(4)}_{\phi^4}(x_1, x_2, x_3, x_4) + O(g^2) \,.$$

For the zeroth-order of perturbation theory Wick's theorem produces three different fully contracted terms:

$$\begin{aligned} G^{(4)}_0(x_1, x_2, x_3, x_4) &= \mathrm{i}\Delta_F(x_1 - x_2)\,\mathrm{i}\Delta_F(x_3 - x_4) \\ &\quad + \mathrm{i}\Delta_F(x_1 - x_3)\,\mathrm{i}\Delta_F(x_2 - x_4) \\ &\quad + \mathrm{i}\Delta_F(x_1 - x_4)\,\mathrm{i}\Delta_F(x_2 - x_3) \,. \end{aligned} \tag{11}$$

This has the following graphical representation:

$$G^{(4)}_0(x_1, x_2, x_3, x_4) = \begin{array}{c} x_3 \text{---} x_4 \\ x_1 \text{---} x_2 \end{array} + \begin{array}{c} x_2 \text{---} x_4 \\ x_1 \text{---} x_3 \end{array} + \begin{array}{c} x_2 \text{---} x_3 \\ x_1 \text{---} x_4 \end{array} \tag{12}$$

Obviously these graphs describe the *independent propagation of two free particles*. Since the particles are nondistinguishable all combinations of the coordinates enter on equal footing.

Mass renormalization leads to six terms in all:

$$G^{(4)}_{\delta m^2}(x_1, x_2, x_3, x_4) = \mathrm{i}\Delta_F(x_1 - x_2) \int \mathrm{d}^4 x \, \mathrm{i}\Delta_F(x_3 - x)\,\Delta_F(x - x_4)$$
$$+ \text{5 permutations} \,. \tag{13}$$

The various terms arise because for each of the free Feynman propagators in (11) a mass correction can be made. The corresponding graphs are

Example 9.4

$$x_3 \text{———} x_4 \atop x_1 \text{—×—} x_2 \quad + \quad {x_2 \text{———} x_4 \atop x_1 \text{—×—} x_3} \quad + \quad {x_2 \text{———} x_3 \atop x_1 \text{—×—} x_4} \quad + \quad {x_3 \text{—×—} x_4 \atop x_1 \text{———} x_2} \quad + \quad {x_2 \text{—×—} x_4 \atop x_1 \text{———} x_3} \quad + \quad {x_2 \text{—×—} x_3 \atop x_1 \text{———} x_4}$$

The interaction term proper, $G_{\phi^4}^{(4)}(x_1, x_2, x_3, x_4)$, reads

$$G_{\phi^4}^{(4)}(x_1, x_2, x_3, x_4)$$
$$= -\mathrm{i}\frac{g}{4!} \int \mathrm{d}^4x \, 4! \, \underbrace{\hat{\phi}_{\mathrm{in}}(x_1)\hat{\phi}_{\mathrm{in}}(x_2)\hat{\phi}_{\mathrm{in}}(x_3)\hat{\phi}_{\mathrm{in}}(x_4) \, \hat{\phi}_{\mathrm{in}}(x)\hat{\phi}_{\mathrm{in}}(x)\hat{\phi}_{\mathrm{in}}(x)\hat{\phi}_{\mathrm{in}}(x)}$$

$$= -\mathrm{i}g \int \mathrm{d}^4 x \, \mathrm{i}\Delta_{\mathrm{F}}(x_1 - x)\mathrm{i}\Delta_{\mathrm{F}}(x_2 - x) \, \mathrm{i}\Delta_{\mathrm{F}}(x_3 - x) \, \mathrm{i}\Delta_{\mathrm{F}}(x_4 - x)$$

$$= {x_4 \atop x_1} \!\!\times\!\! {x_3 \atop x_2} \, . \tag{14}$$

From this four-point function the *S matrix element* can be calculated by using the LSZ reduction formula (9.85):

$$S_{fi} = \langle p_1', p_2'; \mathrm{in}| \hat{S} |p_1, p_2; \mathrm{in}\rangle \tag{15}$$
$$= \left(\frac{\mathrm{i}}{\sqrt{Z}}\right)^4 \prod_{i=1}^{4} N_{p_i} \int \mathrm{d}^4 y_1 \mathrm{d}^4 y_2 \mathrm{d}^4 x_1 \mathrm{d}^4 x_2 \, \mathrm{e}^{\mathrm{i}(p_1' \cdot y_1 + p_2' \cdot y_2)} \, \mathrm{e}^{-\mathrm{i}(p_1 \cdot x_1 + p_2 \cdot x_2)}$$
$$\times \, (\overrightarrow{\Box_{y_1} + m^2})(\overrightarrow{\Box_{y_2} + m^2}) \, G^{(4)}(y_1, y_2, x_1, x_2) \, (\overleftarrow{\Box_{x_1} + m^2})(\overleftarrow{\Box_{x_2} + m^2}) \, .$$

Because

$$(\Box + m^2)\Delta_{\mathrm{F}}(x) = -\delta^4(x) \, , \tag{16}$$

the Klein–Gordon differential operators cancel the four Feynman propagators and four integrations in (14), leaving us with

$$S_{fi} = \left(\frac{\mathrm{i}}{\sqrt{Z}}\right)^4 \prod_{i=1}^{4} N_{p_i} \int \mathrm{d}^4 y_1 \mathrm{d}^4 y_2 \mathrm{d}^4 x_1 \mathrm{d}^4 x_2 \, \mathrm{e}^{\mathrm{i}(p_1' \cdot y_1 + p_2' \cdot y_2 - p_1 \cdot x_1 - p_2 \cdot x_2)}$$
$$\times (-\mathrm{i}g) \int \mathrm{d}^4 x \, \left(-\mathrm{i}\delta^4(y_1 - x)\right)\left(-\mathrm{i}\delta^4(y_2 - x)\right)$$
$$\times \left(-\mathrm{i}\delta^4(x_1 - x)\right)\left(-\mathrm{i}\delta^4(x_2 - x)\right)$$
$$= \left(\frac{1}{\sqrt{Z}}\right)^4 (-\mathrm{i}g) \prod_{i=1}^{4} N_{p_i} \int \mathrm{d}^4 x \, \mathrm{e}^{\mathrm{i}(p_1' + p_2' - p_1 - p_2) \cdot x}$$
$$= \frac{1}{Z^2}(-\mathrm{i}g) \prod_{i=1}^{4} N_{p_i} (2\pi)^4 \, \delta^4(p_1' + p_2' - p_1 - p_2) \, . \tag{17}$$

This essentially agrees with the result of (9) in Example 8.7.

Extending the calculation to higher orders of the perturbation series is quite simple. The Green's function $G^{(4)}(x_1, x_2, x_3, x_4)$ will always contain four external propagators $\mathrm{i}\Delta_{\mathrm{F}}(x_k - y_k)$, $k = 1, \ldots, 4$, which are connected to a more or less complicated web of internal lines. These external propagators are "amputated" when we switch to the S matrix. For example, for the nontrivial second-order contribution to the S matrix element of boson–boson scattering we find the graph

+ permutations ,

in full agreement with Example 8.7.

The only difference between this and our earlier treatment lies in the presence of the renormalization constant $(\sqrt{Z})^{-4}$. In lowest-order perturbation theory this factor can be ignored, however. The renormalization constant Z itself is a function of the coupling constant g, as can be seen from (9.67):

$$Z(g) = 1 + O(g^2) \ . \tag{18}$$

Therefore it is consistent to make the replacement $Z \to 1$ in (17). For higher orders of perturbation theory this argument no longer applies. In this case, however, additional graphs involving loops will also contribute and a consistent renormalization of the ϕ^4 theory has to be carried out. It turns out that the Z factors lead to a *renormalization of the coupling constant*, which results in the replacement of the ill-defined value g by the physical coupling constant g_{ren}.

As a practical rule of thumb we summarize:

- In the lowest nonvanishing order of perturbation theory (i.e., in the *tree approximation*) one may simply put $Z = 1$ and $\delta m^2 = 0$. This leads to the correct S matrix element if the physical values of the mass m and coupling constant g are used.

10. Discrete Symmetry Transformations

10.1 Introduction

In Sect. 4.3 we have studied the transformation properties of quantum fields. The discussion was devoted to *continuous* transformations that can be constructed by starting from infinitesimal transformations "close to unity". If a theory is invariant under such a transformation it will possess a Noether current and thus there will be a conservation law. In addition, however, there is the class of *discrete* symmetries, which have to be described differently. Discrete symmetries can be employed to relate the behavior of different physical systems, for example, those that differ by an interchange of particles and antiparticles. New conserved quantities (e.g., parity, charge parity) and selection rules can be generated by discrete symmetries. We will study three types of discrete transformations which are of fundamental importance: *space inversion* $\hat{\mathcal{P}}$, *charge conjugation* $\hat{\mathcal{C}}$, and *time reversal* $\hat{\mathcal{T}}$. All noninteracting field theories are invariant under these transformations but this will change if interactions are present that break the symmetry. A prominent example for this is the famous maximal violation of space inversion symmetry (parity) in weak interactions. Also the time-reversal symmetry is (very slightly) violated in nature as witnessed by the decay of the $K^0/\bar{K}^0$ mesons.

In the following, we will one by one investigate the transformations $\hat{\mathcal{P}}$, $\hat{\mathcal{C}}$, and $\hat{\mathcal{T}}$. This will culminate in a discussion of the combination of these three transformations, which leads to one of the most general and fundamental results of quantum field theory: according to the *CPT theorem* any "reasonable" theory has to be invariant under the combination of the three transformations. As a consequence the masses and lifetimes of particles and antiparticles are exactly equal.

10.2 Scalar Fields

10.2.1 Space Inversion

Let us investigate the connection between two physical systems that are identical except for a reflection of the space coordinates:

$$x \to x' \equiv (\boldsymbol{x}', t') = (-\boldsymbol{x}, t) . \tag{10.1}$$

In the classical theory of a free scalar field the equation of motion (which only contains second-order derivatives) and the Lagrangian (containing squares of first-order derivatives) are not affected by a space reflection. Therefore the theory is invariant under (10.1) provided that the field function $\phi(x)$ is unchanged or, at most, is multiplied by a phase factor η_P of modulus $|\eta_P| = 1$ to preserve normalization:

$$\phi(x) \to \phi'(x') = \eta_P \phi(x) \ . \tag{10.2}$$

Remembering Sect. 4.3, we demand that the quantized theory satisfies (10.2) applied to the matrix elements of the field operator. This leads to the *transformation law for the field operator* (compare (4.75)):

$$\hat{\mathcal{P}}\hat{\phi}(\boldsymbol{x},t)\hat{\mathcal{P}}^{-1} = \eta_P \hat{\phi}(-\boldsymbol{x},t) \quad \text{and} \quad \hat{\mathcal{P}}\hat{\phi}^\dagger(\boldsymbol{x},t)\hat{\mathcal{P}}^{-1} = \eta_P^* \hat{\phi}^\dagger(-\boldsymbol{x},t) \ , \tag{10.3}$$

where $\hat{\mathcal{P}}$ is the unitary parity operator. The value of η_P has yet to be fixed. As we will show at the end of this section, in the case of a charged field one can choose an arbitrary complex phase factor. If the particles are neutral (i.e., the field operator is hermitean, $\hat{\phi} = \hat{\phi}^\dagger$) according to (10.3) the factor η_P has to be real, which implies

$$\eta_P = \pm 1 \ . \tag{10.4}$$

The particles described by these two alternatives are said to have positive or negative *intrinsic parity*. Examples for both cases, i.e. *scalar* and *pseudoscalar* particles, can be found in nature. The intrinsic parity of a given particle has to be determined by experiment. (Note that this is only possible if there are interactions. The intrinsic parity of a noninteracting field has no physical meaning).

To learn more about the operator $\hat{\mathcal{P}}$ we study its action for the case of a free charged Klein–Gordon field. We insert the plane-wave decompositon (4.58) of the field operator into (10.3)

$$\int d^3 p \left[\hat{\mathcal{P}} \hat{a}_{\boldsymbol{p}} \hat{\mathcal{P}}^{-1} u_{\boldsymbol{p}}(\boldsymbol{x},t) + \hat{\mathcal{P}} \hat{b}_{\boldsymbol{p}}^\dagger \hat{\mathcal{P}}^{-1} u_{\boldsymbol{p}}^*(\boldsymbol{x},t) \right]$$
$$= \eta_P \int d^3 p \left[\hat{a}_{\boldsymbol{p}} u_{\boldsymbol{p}}(-\boldsymbol{x},t) + \hat{b}_{\boldsymbol{p}}^\dagger u_{\boldsymbol{p}}^*(-\boldsymbol{x},t) \right] \ . \tag{10.5}$$

The operator $\hat{\mathcal{P}}$ acts in the Hilbert space of second quantization, and thus can be moved past the wave function u. Upon space inversion the plane waves $u_{\boldsymbol{p}}(\boldsymbol{x},t) = N_p \exp[\mathrm{i}(\omega_p t - \boldsymbol{p} \cdot \boldsymbol{x})]$ satisfy $u_{\boldsymbol{p}}(-\boldsymbol{x},t) = u_{-\boldsymbol{p}}(\boldsymbol{x},t)$. Therefore by comparing the coefficients we deduce from (10.5) that

$$\hat{\mathcal{P}} \hat{a}_{\boldsymbol{p}} \hat{\mathcal{P}}^{-1} = \eta_P \hat{a}_{-\boldsymbol{p}} \quad , \quad \hat{\mathcal{P}} \hat{b}_{\boldsymbol{p}}^\dagger \hat{\mathcal{P}}^{-1} = \eta_P \hat{b}_{-\boldsymbol{p}}^\dagger \ , \tag{10.6a}$$

and because of the unitarity of $\hat{\mathcal{P}}$

$$\hat{\mathcal{P}} \hat{a}_{\boldsymbol{p}}^\dagger \hat{\mathcal{P}}^{-1} = \eta_P^* \hat{a}_{-\boldsymbol{p}}^\dagger \quad , \quad \hat{\mathcal{P}} \hat{b}_{\boldsymbol{p}} \hat{\mathcal{P}}^{-1} = \eta_P^* \hat{b}_{-\boldsymbol{p}} \ . \tag{10.6b}$$

As one might expect, the inversion of the space coordinates results in the reversal of the (three-dimensional) momentum vectors.

It will be shown in Exercise 10.1 how an explicit expression for the operator $\hat{\mathcal{P}}$ can be given that satisfies the condition (10.6). For real values of η_P the operator takes the somewhat involved form

$$\hat{\mathcal{P}} = \exp\left[\mathrm{i}\frac{\pi}{2} \int d^3 p \left((\hat{a}_{\boldsymbol{p}}^\dagger \hat{a}_{-\boldsymbol{p}} + \hat{b}_{\boldsymbol{p}}^\dagger \hat{b}_{-\boldsymbol{p}}) - \eta_P (\hat{a}_{\boldsymbol{p}}^\dagger \hat{a}_{\boldsymbol{p}} + \hat{b}_{\boldsymbol{p}}^\dagger \hat{b}_{\boldsymbol{p}}) \right) \right] \ . \tag{10.7}$$

This can be verified by using the operator identity

$$e^{\hat{A}}\hat{B}e^{-\hat{A}} = \hat{B} + [\hat{A},\hat{B}] + \frac{1}{2!}[\hat{A},[\hat{A},\hat{B}]] + \ldots \qquad (10.8)$$

from Exercise 1.3. With the help of (10.7) it is easy to study the action of the parity operator $\hat{\mathcal{P}}$ in detail. For a start it is immediately obvious that the vacuum $|0\rangle$ is an eigenstate of $\hat{\mathcal{P}}$ having the eigenvalue $+1$:

$$\hat{\mathcal{P}}|0\rangle = |0\rangle . \qquad (10.9)$$

This must be true since all higher-order terms of the Taylor expansion of (10.7) always contain an annihilation operator at the right-most positon and thus make no contribution to (10.9). Thus the vacuum has positive parity. (Note, however, that this has no absolute meaning. If the opposite parity were assigned simultaneously to all physical states, including the vacuum, the observables would remain unchanged.) The transformation property of Fock states follows from (10.6) and (10.9). For the one-particle state we get

$$\hat{\mathcal{P}}|\boldsymbol{p}\rangle = \hat{\mathcal{P}}\hat{a}_{\boldsymbol{p}}^{\dagger}|0\rangle = \hat{\mathcal{P}}\,\hat{a}_{\boldsymbol{p}}^{\dagger}\,\hat{\mathcal{P}}^{-1}\,\hat{\mathcal{P}}|0\rangle = \eta_{\mathrm{P}}|-\boldsymbol{p}\rangle , \qquad (10.10)$$

etc. Also the transformation property of operators related to the "kinematic" conserved quantities can be obtained easily if they are expressed in terms of creation and annihilation operators and then (10.6) is applied. For the Hamiltonian (4.60) we find

$$\begin{aligned}\hat{\mathcal{P}}\hat{H}\hat{\mathcal{P}}^{-1} &= \int \mathrm{d}^3p\,\omega_p \hat{\mathcal{P}}(\hat{a}_{\boldsymbol{p}}^{\dagger}\hat{a}_{\boldsymbol{p}} + \hat{b}_{\boldsymbol{p}}^{\dagger}\hat{b}_{\boldsymbol{p}})\hat{\mathcal{P}}^{-1} \\ &= \eta_{\mathrm{P}}^2 \int \mathrm{d}^3p\,\omega_p (\hat{a}_{-\boldsymbol{p}}^{\dagger}\hat{a}_{-\boldsymbol{p}} + \hat{b}_{-\boldsymbol{p}}^{\dagger}\hat{b}_{-\boldsymbol{p}}) \\ &= \hat{H} . \qquad (10.11)\end{aligned}$$

Thus the parity operator commutes with the Hamiltonian

$$[\hat{\mathcal{P}},\hat{H}] = 0 \qquad (10.12)$$

and thus is independent of time. Therefore the parity of a state is a conserved quantity. Similarly we find the following transformation properties for the momentum operator (4.61) and the angular-momentum operator (4.62):

$$\hat{\mathcal{P}}\,\hat{\boldsymbol{P}}\,\hat{\mathcal{P}}^{-1} = -\hat{\boldsymbol{P}} , \qquad (10.13a)$$
$$\hat{\mathcal{P}}\,\hat{\boldsymbol{L}}\,\hat{\mathcal{P}}^{-1} = +\hat{\boldsymbol{L}} . \qquad (10.13b)$$

This is the appropriate result for a polar and an axial vector. Equations (10.11) and (10.13a) can be combined into a four-dimensional expression

$$\hat{\mathcal{P}}\,\hat{P}^{\mu}\,\hat{\mathcal{P}}^{-1} = \hat{P}_{\mu} . \qquad (10.14)$$

The *current-density operator* of the Klein–Gordon field transforms as follows

$$\begin{aligned}\hat{\mathcal{P}}\hat{j}^{\mu}(\boldsymbol{x},t)\hat{\mathcal{P}}^{-1} &= \mathrm{i}\hat{\mathcal{P}}\,\hat{\phi}^{\dagger}(\boldsymbol{x},t)\hat{\mathcal{P}}^{-1}\,\overleftrightarrow{\partial}^{\mu}\,\hat{\mathcal{P}}\hat{\phi}(\boldsymbol{x},t)\hat{\mathcal{P}}^{-1} \\ &= \mathrm{i}\,\eta_{\mathrm{P}}^{*}\eta_{\mathrm{P}}\,\hat{\phi}^{\dagger}(-\boldsymbol{x},t)\,\overleftrightarrow{\partial}^{\mu}\,\hat{\phi}(-\boldsymbol{x},t) \\ &= \hat{j}_{\mu}(-\boldsymbol{x},t) , \qquad (10.15)\end{aligned}$$

i.e., the sign of the spatial components is inverted. The result (10.15) implies that the operator of total charge $\hat{Q}$ is invariant under parity transformations

$$[\hat{\mathcal{P}}, \hat{Q}] = 0 \,. \tag{10.16}$$

Also the *canonical commutation relations* are invariant under space reflection. According to (10.6) and (4.59) the creation and annihilation operators obey

$$[\hat{\mathcal{P}}\,\hat{a}_{\boldsymbol{p}}\,\hat{\mathcal{P}}^{-1}, \hat{\mathcal{P}}\,\hat{a}^\dagger_{\boldsymbol{p}'}\,\hat{\mathcal{P}}^{-1}] = \eta_{\mathrm{P}}^2\,[\hat{a}_{-\boldsymbol{p}}, \hat{a}^\dagger_{-\boldsymbol{p}'}] = \delta^3(\boldsymbol{p} - \boldsymbol{p}') \,, \tag{10.17}$$

etc. Accordingly the transformed field operators satisfy the customary equal-time commutation relations

$$[\hat{\mathcal{P}}\hat{\phi}(\boldsymbol{x},t)\hat{\mathcal{P}}^{-1}, \hat{\mathcal{P}}\dot{\hat{\phi}}^\dagger(\boldsymbol{x},t)\hat{\mathcal{P}}^{-1}] = \mathrm{i}\,\delta^3(\boldsymbol{x} - \boldsymbol{x}') \,, \tag{10.18}$$

etc.

The considerations of this section so far have been restricted to the case of a free boson field. This applies in particular to the explicit expression for the parity operator (10.7). If there are *interactions* the field operator in the Heisenberg picture has a complicated dynamical evolution and also the creation and annihilation operators become time dependent. However, we can choose a point in time, such as $t = 0$, for which the interacting field operator coincides with its free counterpart, so that $\hat{a}^\dagger_{\boldsymbol{p}}(0)$ creates a single field quantum with momentum $\boldsymbol{p}$. Starting from this point, we can transform the parity operator to arbitrary times by

$$\hat{\mathcal{P}}(t) = \mathrm{e}^{\mathrm{i}\hat{H}t}\,\hat{\mathcal{P}}(0)\,\mathrm{e}^{-\mathrm{i}\hat{H}t} \,. \tag{10.19}$$

If the interacting system is parity invariant, i.e., if (10.12) is valid, then $\hat{\mathcal{P}}$ will be a constant operator. In the general case (10.19) has to be employed to define the parity operator. Here $\hat{\mathcal{P}}(0)$ is given by (10.7) where the creation and annihilation operators at time $t = 0$ have to be used.

These considerations on the treatment of interacting systems of course are not restricted to the parity operator. The same also holds for the other discrete symmetries, which will be treated in the next sections. We will not bother to spell this out seperately in each case.

A Remark on the Phase Factor η_{P}. As mentioned already it can be shown that the phase factor η_{P} in (10.3) can be chosen arbitrarily in the case of charged fields. This has its roots in the existence of the charge operator $\hat{Q}$ which has the property (see Exercise 4.4)

$$[\hat{Q}, \hat{\phi}(x)] = -\hat{\phi}(x) \quad,\quad [\hat{Q}, \hat{\phi}^\dagger(x)] = \hat{\phi}^\dagger(x) \,. \tag{10.20}$$

The operator $\hat{Q}$ can be employed to construct a unitary transformation which modifies the phase of the field operator:

$$\begin{aligned}
\mathrm{e}^{\mathrm{i}\alpha\hat{Q}}\hat{\phi}\,\mathrm{e}^{-\mathrm{i}\alpha\hat{Q}} &= \hat{\phi} + [\mathrm{i}\alpha\hat{Q}, \hat{\phi}] + \frac{1}{2!}[\mathrm{i}\alpha\hat{Q}, [\mathrm{i}\alpha\hat{Q}, \hat{\phi}]] + \ldots \\
&= \hat{\phi} - \mathrm{i}\alpha\hat{\phi} + \frac{1}{2!}(-\mathrm{i}\alpha)^2\hat{\phi} + \ldots \\
&= \hat{\phi}\,\mathrm{e}^{-\mathrm{i}\alpha} \,. \tag{10.21}
\end{aligned}$$

Then one can define a modified parity operator $\hat{\mathcal{P}}'$

$$\hat{\mathcal{P}}' = \hat{\mathcal{P}}\,\mathrm{e}^{\mathrm{i}\alpha\hat{Q}} \,, \tag{10.22}$$

which is unitary as well and which leads to the transformation law

$$\begin{aligned}
\hat{\mathcal{P}}' \hat{\phi}(\boldsymbol{x},t) \hat{\mathcal{P}}'^{-1} &= \hat{\mathcal{P}} \, \mathrm{e}^{\mathrm{i}\alpha \hat{Q}} \, \hat{\phi}(\boldsymbol{x},t) \, \mathrm{e}^{-\mathrm{i}\alpha \hat{Q}} \, \hat{\mathcal{P}}^{-1} \\
&= \mathrm{e}^{-\mathrm{i}\alpha} \, \hat{\mathcal{P}} \hat{\phi}(\boldsymbol{x},t) \hat{\mathcal{P}}^{-1} \\
&= \mathrm{e}^{-\mathrm{i}\alpha} \, \eta_\mathrm{P} \, \hat{\phi}(-\boldsymbol{x},t) \, .
\end{aligned} \qquad (10.23)$$

Therefore the phase $\eta_\mathrm{P} \to \mathrm{e}^{-\mathrm{i}\alpha}\eta_\mathrm{P}$ can be re-gauged at will. Since $\hat{Q}$ is a conserved quantity $\hat{\mathcal{P}}'$ has the same properties as the original operator $\hat{\mathcal{P}}$ and can be used as the parity operator with equal justification.

This argument has been quite general. It will apply also to the phase factors encountered for the other discrete transformation and to other types of charged fields.

10.2.2 Charge Conjugation

This important discrete transformation has nothing to do with the space or time coordinates, instead it is related to the particle–antiparticle degree of freedom of a theory. The charge-conjugation transformation replaces a particle by the corresponding antiparticle and vice versa. For the field operator of a charged Klein–Gordon field this amounts to the following operation. A unitary operator $\hat{\mathcal{C}}$ is constructed which has the property

$$\hat{\mathcal{C}} \, \hat{\phi}(x) \, \hat{\mathcal{C}}^{-1} = \eta_\mathrm{C} \hat{\phi}^\dagger(x) \quad , \quad \hat{\mathcal{C}} \, \hat{\phi}^\dagger(x) \, \hat{\mathcal{C}}^{-1} = \eta_\mathrm{C}^* \, \hat{\phi}(x) \, , \qquad (10.24)$$

where $|\eta_\mathrm{C}| = 1$. As in (10.5) the plane-wave decomposition immediately yields

$$\begin{aligned}
\hat{\mathcal{C}} \, \hat{a}_{\boldsymbol{p}} \, \hat{\mathcal{C}}^{-1} &= \eta_\mathrm{C} \hat{b}_{\boldsymbol{p}} \quad , \quad & \hat{\mathcal{C}} \, \hat{b}_{\boldsymbol{p}} \, \hat{\mathcal{C}}^{-1} &= \eta_\mathrm{C}^* \, \hat{a}_{\boldsymbol{p}} \, , \\
\hat{\mathcal{C}} \, \hat{a}_{\boldsymbol{p}}^\dagger \, \hat{\mathcal{C}}^{-1} &= \eta_\mathrm{C}^* \, \hat{b}_{\boldsymbol{p}}^\dagger \quad , \quad & \hat{\mathcal{C}} \, \hat{b}_{\boldsymbol{p}}^\dagger \, \hat{\mathcal{C}}^{-1} &= \eta_\mathrm{C} \hat{a}_{\boldsymbol{p}}^\dagger \, .
\end{aligned} \qquad (10.25)$$

Thus the particle and antiparticle operators are simply interchanged. As in (10.7) the operator $\hat{\mathcal{C}}$ can be constructed explicitly (cf. Exercise 10.1), with the result (assuming a real-valued η_C)

$$\hat{\mathcal{C}} = \exp\left[\mathrm{i}\frac{\pi}{2} \int \mathrm{d}^3 p \, \left(\hat{b}_{\boldsymbol{p}}^\dagger \hat{a}_{\boldsymbol{p}} + \hat{a}_{\boldsymbol{p}}^\dagger \hat{b}_{\boldsymbol{p}} - \eta_\mathrm{C}(\hat{a}_{\boldsymbol{p}}^\dagger \hat{a}_{\boldsymbol{p}} + \hat{b}_{\boldsymbol{p}}^\dagger \hat{b}_{\boldsymbol{p}}) \right) \right] \, . \qquad (10.26)$$

The vacuum is an eigenstate of $\hat{\mathcal{C}}$ having the eigenvalue $+1$:

$$\hat{\mathcal{C}}|0\rangle = |0\rangle \, . \qquad (10.27)$$

In general, because $\hat{\mathcal{C}}^2 = 1$, a $\hat{\mathcal{C}}$ eigenstate has either positive or negative "charge parity". The operators of energy, momentum, and angular momentum are invariant under $\hat{\mathcal{C}}$ transformations:

$$\begin{aligned}
\hat{\mathcal{C}} \, \hat{P}^\mu \, \hat{\mathcal{C}}^{-1} &= \hat{P}^\mu \, , & (10.28\mathrm{a}) \\
\hat{\mathcal{C}} \, \hat{\boldsymbol{L}} \, \hat{\mathcal{C}}^{-1} &= \hat{\boldsymbol{L}} \, , & (10.28\mathrm{b})
\end{aligned}$$

since here the particles and antiparticles enter in a symmetric fashion. On the other hand all components of the *current-density* operator $\hat{j}^\mu(x)$ change their sign:

$$\begin{aligned}
\hat{\mathcal{C}} \, \hat{j}^\mu(x) \, \hat{\mathcal{C}}^{-1} &= \mathrm{i}\hat{\mathcal{C}} \, \hat{\phi}^\dagger \overset{\leftrightarrow}{\partial^\mu} \hat{\phi} \, \hat{\mathcal{C}}^{-1} = \mathrm{i}\hat{\mathcal{C}} \, \hat{\phi}^\dagger \, \hat{\mathcal{C}}^{-1} \overset{\leftrightarrow}{\partial^\mu} \hat{\mathcal{C}} \, \hat{\phi} \, \hat{\mathcal{C}}^{-1} = \mathrm{i}\hat{\phi} \overset{\leftrightarrow}{\partial^\mu} \hat{\phi}^\dagger \\
&= -\hat{j}^\mu(x) \, .
\end{aligned} \qquad (10.29)$$

After integrating $\hat{j}^0$ over d^3x we find that the charge operator anticommutes with $\hat{C}$, $\hat{C}\hat{Q} = -\hat{Q}\hat{C}$. This tells us that states with a well-defined nonzero charge cannot be eigenstates of the operator $\hat{C}$. Only certain special systems having vanishing total charge allow the definition of a *charge parity*.

Again the *canonical commutation relations* are invariant under charge conjugation, since

$$[\hat{C}\,\hat{b}_{\boldsymbol{p}}\,\hat{C}^{-1}, \hat{C}\,\hat{b}^\dagger_{\boldsymbol{p}'}\,\hat{C}^{-1}] = [\hat{b}_{\boldsymbol{p}}, \hat{b}^\dagger_{\boldsymbol{p}'}] = \delta^3(\boldsymbol{p} - \boldsymbol{p}') \,, \qquad (10.30)$$

etc.

10.2.3 Time Reversal

The third discrete transformation keeps the spatial coordinates fixed while the sign of the time coordinate is inverted

$$x \to x' \equiv (\boldsymbol{x}', t') = (\boldsymbol{x}, -t) \,. \qquad (10.31)$$

Although this looks very similar to the space inversion (10.1), the time reversal is something special, both from the conceptual and form the mathematical point of view. In classical mechanics the transition to the time-mirrored system corresponds to "running the movie backwards", i.e., all particles follow their trajectories in the opposite direction. The momenta and angular momenta of all particles point in the opposite direction and *the roles of the initial and final configurations are interchanged*. The latter property is peculiar to the time-reversal transformation

Let us first consider time reversal prior to second quantization, i.e., applied to a "classical" Klein–Gordon field $\phi(x)$. The general ansatz (4.71) reads

$$\phi'(x') = \Lambda \phi(x) \,. \qquad (10.32)$$

For spatial reflections of the spin-0 field Λ has simply been a number. This will not suffice to describe time reversal. For example, let us consider a Klein–Gordon wave function having a well-defined energy E_n:

$$\phi_n(\boldsymbol{x}, t) = e^{-iE_n t} u_n(\boldsymbol{x}) \,, \qquad (10.33)$$

where the time evolution $0 \ldots t$ is described by the phase factor. Now the time-reversed wave function can be obtained in two different ways. Application of (10.32) leads to

$$\phi'_n(\boldsymbol{x}, -t) = \Lambda \, e^{-iE_n t} u_n(\boldsymbol{x}) \,. \qquad (10.34)$$

On the other hand, it should be possible to obtain the same result by switching to the time-reversed wave function at $t = 0$ and subsequently propagating the resulting wave function in time, going into the negative direction $0 \ldots -t$:

$$\phi'_n(\boldsymbol{x}, -t) = e^{+iE'_n t} \Lambda u_n(\boldsymbol{x}) \,. \qquad (10.35)$$

Equations (10.34) and (10.35) agree provided that

$$\Lambda \, e^{-iE_n t} = e^{+iE'_n t} \Lambda \,. \qquad (10.36)$$

If a c number were chosen for Λ then $E'_n = -E_n$ would have to hold. Clearly this condition has no meaningful physical interpretation since it would imply that the energy spectrum of the time-reversed system is unbounded from below. As an alternative choice we can choose

$$\Lambda = \eta_T K \tag{10.37}$$

where η_T is a phase factor and K is the *operator of complex conjugation*, defined by[1]

$$K c K^{-1} = c^* \tag{10.38}$$

for complex numbers c. Then the transformation law (10.32) reads

$$\phi'(\boldsymbol{x}, -t) = \eta_T K \phi(\boldsymbol{x}, t) = \eta_T \phi^*(\boldsymbol{x}, t) \, . \tag{10.39}$$

K is an *antilinear operator* characterized by the property

$$K(c_1 \phi_1 + c_2 \phi_2) = c_1^* K \phi_1 + c_2^* K \phi_2 \, . \tag{10.40}$$

When we apply time-reversal twice, the original wave function is regained provided that η_T is a phase factor:

$$|\eta_T|^2 = 1 \, . \tag{10.41}$$

Now the transformation law (10.39) for the wave function is to be extended to the field operator $\hat{\phi}(x)$. As in (4.72) we define transformed state vectors

$$|\alpha'\rangle = \hat{T}|\alpha\rangle \quad , \quad |\beta'\rangle = \hat{T}|\beta\rangle \, , \tag{10.42}$$

and require that (10.39) holds for the matrix elements of the field operator. Here one has to take into account that the initial and final states are to be exchanged in the transformed system. Thus instead of (4.74) we now require

$$\langle \alpha'|\hat{\phi}(\boldsymbol{x}, -t)|\beta'\rangle = \eta_T \langle \beta|\hat{\phi}^\dagger(\boldsymbol{x}, t)|\alpha\rangle \, . \tag{10.43}$$

Because of the interchanged states this does not allow us to immediately read off an equation for the field operator. For this the operator $\hat{T}$ has to be constructed in such a way that it reverts the interchange. As demonstrated in the following paragraph, this is the characteristic property of an *antiunitary transformation*.

Antiunitary Operators. Let $\hat{V}$ be an antilinear operator in Hilbert space which satisfies

$$\hat{V}(c_1|\alpha_1\rangle + c_2|\alpha_2\rangle) = c_1^* \hat{V}|\alpha_1\rangle + c_2^* \hat{V}|\alpha_2\rangle \, . \tag{10.44}$$

The *hermitean conjugate* operator $\hat{V}^\dagger$ is defined through

$$\langle \alpha|\hat{V}^\dagger|\beta\rangle = \langle \beta|\hat{V}|\alpha\rangle \tag{10.45}$$

and is antilinear as well. The operator $\hat{V}$ is *antiunitary* if it satisfies

$$\hat{V}\hat{V}^\dagger = \hat{V}^\dagger \hat{V} = 1 \, . \tag{10.46}$$

An *antiunitary transformation* in Hilbert space,

$$|\alpha'\rangle = \hat{V}|\alpha\rangle \quad , \quad |\beta'\rangle = \hat{V}|\beta\rangle \, , \tag{10.47}$$

---

[1] The antilinear operator of time-reversal transformations was first encountered in nonrelativistic quantum mechanics where it was introduced by E. Wigner in 1932.

preserves the norm of a state in the same way as a unitary transformation does. However, when applied to scalar products it interchanges the "bra" and "ket" vectors since

$$\langle \beta'|\alpha'\rangle = \langle \beta'|\hat{V}|\alpha\rangle = \langle \alpha|\hat{V}^\dagger|\beta'\rangle = \langle \alpha|\hat{V}^\dagger\hat{V}|\beta\rangle = \langle \alpha|\beta\rangle = \langle \beta|\alpha\rangle^* \ . \qquad (10.48)$$

The matrix elements of a linear operator $\hat{A}$ taken between transformed states can be rewritten as

$$\begin{aligned}\langle \beta'|\hat{A}|\alpha'\rangle &= \langle \hat{A}^\dagger\beta'|\alpha'\rangle = \langle \hat{A}^\dagger\beta'|\hat{V}|\alpha\rangle = \langle \alpha|\hat{V}^\dagger|\hat{A}^\dagger\beta'\rangle \\ &= \langle \alpha|\hat{V}^\dagger\hat{A}^\dagger\hat{V}|\beta\rangle \\ &= \langle \alpha|(\hat{V}^{-1}\hat{A}\hat{V})^\dagger|\beta\rangle \ . \end{aligned} \qquad (10.49)$$

It is possible to show that every antiunitary operator can be factored into the product of a unitary operator $\hat{U}$ and the operator of complex conjugation K.

$$\hat{V} = \hat{U}K \ . \qquad (10.50)$$

If the time reversal operator is chosen to be antiunitary then (10.49) holds and the interchange of initial and final states is reverted. Thus (10.43) leads to the operator equation

$$\left(\hat{T}^{-1}\hat{\phi}(\boldsymbol{x},-t)\hat{T}\right)^\dagger = \eta_\mathrm{T}\hat{\phi}^\dagger(\boldsymbol{x},t) \ , \qquad (10.51)$$

which can also be written as

$$\hat{T}^{-1}\hat{\phi}(\boldsymbol{x},-t)\hat{T} = \eta_\mathrm{T}^*\hat{\phi}(\boldsymbol{x},t) \qquad (10.52)$$

or, using (10.41), as

$$\hat{T}\hat{\phi}(\boldsymbol{x},t)\hat{T}^{-1} = \eta_\mathrm{T}\hat{\phi}(\boldsymbol{x},-t) \ . \qquad (10.53)$$

Though the analogy to the space inversion (10.3) is obvious, one has to keep in mind the antiunitary character of $\hat{T}$.

If we insert the plane-wave expansion of the field operator, (10.53) becomes

$$\begin{aligned}\int\mathrm{d}^3 p\, N_p\Big(\hat{T}\,\hat{a}_p\,\hat{T}^{-1}u_p^*(\boldsymbol{x},t) + \hat{T}\,\hat{b}_p^\dagger\,\hat{T}^{-1}u_p(\boldsymbol{x},t)\Big) \\ = \eta_\mathrm{T}\int\mathrm{d}^3 p\, N_p\big(\hat{a}_p\, u_p(\boldsymbol{x},-t) + \hat{b}_p^\dagger\, u_p^*(\boldsymbol{x},-t)\big) \ , \end{aligned} \qquad (10.54)$$

where the antilinearity of the transformation is responsible for the complex conjugation of the basis wave functions u_p on the left-hand side. Now the plane waves have the property $u_p(\boldsymbol{x},-t) = u_{-p}^*(\boldsymbol{x},t)$ so that we can deduce the transformation law for the creation and annihilation operators under time reversal:

$$\begin{aligned}\hat{T}\,\hat{a}_p\,\hat{T}^{-1} &= \eta_\mathrm{T}\,\hat{a}_{-p} \ , & \hat{T}\,\hat{b}_p\,\hat{T}^{-1} &= \eta_\mathrm{T}^*\,\hat{b}_{-p} \ , \\ \hat{T}\,\hat{a}_p^\dagger\,\hat{T}^{-1} &= \eta_\mathrm{T}^*\,\hat{a}_{-p}^\dagger \ , & \hat{T}\,\hat{b}_p^\dagger\,\hat{T}^{-1} &= \eta_\mathrm{T}\,\hat{b}_{-p}^\dagger \ . \end{aligned} \qquad (10.55)$$

We note that (10.55) agrees with (10.6). Therefore the action of the operators $\hat{\mathcal{P}}$ and $\hat{T}$ in Hilbert space is the same. However, $\hat{T}$ is associated with an additional complex conjugation. The energy and momentum operators of the free Klein–Gordon field are transformed as in the case of spatial inversion:

$$\hat{T}\,\hat{H}\,\hat{T}^{-1} = \hat{H} \quad , \quad \hat{T}\,\hat{\boldsymbol{P}}\,\hat{T}^{-1} = -\hat{\boldsymbol{P}} \ . \qquad (10.56)$$

The angular-momentum operator has the property

$$\hat{\mathcal{T}}\,\hat{\boldsymbol{L}}\,\hat{\mathcal{T}}^{-1} = -\hat{\boldsymbol{L}}\,, \tag{10.57}$$

which agrees with the classical notion of a trajectory being traced in the opposite direction. The sign of (10.57) differs from that encountered for the parity transformation (10.13b), because the factor i in (4.62) is complex conjugated.

The *current-density* operator is transformed as follows:

$$\begin{aligned}
\hat{\mathcal{T}}\hat{j}^\mu(\boldsymbol{x},t)\hat{\mathcal{T}}^{-1} &= \hat{\mathcal{T}}\,\mathrm{i}\hat{\phi}^\dagger(\boldsymbol{x},t)\hat{\mathcal{T}}^{-1}\stackrel{\leftrightarrow}{\partial}{}^\mu \hat{\mathcal{T}}\hat{\phi}(\boldsymbol{x},t)\hat{\mathcal{T}}^{-1} \\
&= -\mathrm{i}\hat{\phi}^\dagger(\boldsymbol{x},-t)\stackrel{\leftrightarrow}{\partial}{}^\mu\hat{\phi}(\boldsymbol{x},-t) \\
&= \hat{j}_\mu(\boldsymbol{x},-t)\,.
\end{aligned} \tag{10.58}$$

As expected the spatial current $\hat{\boldsymbol{j}}$ reverses its sign, whereas the charge density $\hat{j}^0$ remains unchanged.

Finally we are interested in the influence of time reversal on the *canonical commutation relations*

$$[\hat{\phi}(\boldsymbol{x},t),\dot{\hat{\phi}}^\dagger(\boldsymbol{x}',t)] = \mathrm{i}\,\delta^3(\boldsymbol{x}-\boldsymbol{x}')\,. \tag{10.59}$$

Application of $\hat{\mathcal{T}}\ldots\hat{\mathcal{T}}^{-1}$ on the left-hand side leads to

$$\begin{aligned}
&\hat{\mathcal{T}}\hat{\phi}(\boldsymbol{x},t)\hat{\mathcal{T}}^{-1}\hat{\mathcal{T}}\dot{\hat{\phi}}^\dagger(\boldsymbol{x}',t)\hat{\mathcal{T}}^{-1} - \hat{\mathcal{T}}\dot{\hat{\phi}}^\dagger(\boldsymbol{x}',t)\hat{\mathcal{T}}^{-1}\hat{\mathcal{T}}\hat{\phi}(\boldsymbol{x},t)\hat{\mathcal{T}}^{-1} \\
&= [\hat{\phi}(\boldsymbol{x},-t),\dot{\hat{\phi}}^\dagger(\boldsymbol{x}',-t)]\,.
\end{aligned} \tag{10.60}$$

On the right-hand side the sign gets inverted since $\hat{\mathcal{T}}\mathrm{i}\hat{\mathcal{T}}^{-1} = -\mathrm{i}$. Therefore the transformed canonical commutation relation becomes

$$[\hat{\phi}(\boldsymbol{x},-t),\dot{\hat{\phi}}^\dagger(\boldsymbol{x}',-t)] = -\mathrm{i}\,\delta^3(\boldsymbol{x}-\boldsymbol{x}')\,. \tag{10.61}$$

This is consistent with the result one obtains through the replacement $t \to -t$ in the original ETCR since the time derivative leads to a reversal of the sign.

EXERCISE

10.1 The Operators $\hat{\mathcal{P}}$ and $\hat{\mathcal{C}}$ for Scalar Fields

Problem. (a) Construct an operator $\hat{\mathcal{P}}$ that satisfies (10.6), i.e., inverts the momentum argument of the creation and annihilation operators
(b) Do the same for the charge conjugation operator $\hat{\mathcal{C}}$ that satisfies (10.25).

Solution. (a) Clearly it suffices to satisfy the relations (10.6) for the operators $\hat{a}_{\boldsymbol{p}}$ and $\hat{b}_{\boldsymbol{p}}$. As a first attempt we make the ansatz

$$\hat{\mathcal{P}}_1 = \mathrm{e}^{\mathrm{i}\alpha\hat{A}}\,, \tag{1}$$

where the hermitean operator $\hat{A}$ and the c number coefficient α have to be determined. The operator (1) should transform a $-\boldsymbol{p}$ particle into a $+\boldsymbol{p}$ particle, for arbitrary values of $\boldsymbol{p}$. The following ansatz suggests itself:

$$\hat{A} = \int\mathrm{d}^3p\,\left(\hat{a}^\dagger_{\boldsymbol{p}}\hat{a}_{-\boldsymbol{p}} + \hat{b}^\dagger_{\boldsymbol{p}}\hat{b}_{-\boldsymbol{p}}\right)\,. \tag{2}$$

This operator clearly is hermitean. Using (4.59) we find it satisfies the commutation relations

Exercise 10.1

$$[\hat{A}, \hat{a}_{\bm p}] = \int d^3p' [\hat{a}^\dagger_{\bm p'} \hat{a}_{-\bm p'}, \hat{a}_{\bm p}] = -\hat{a}_{-\bm p} , \tag{3a}$$

$$[\hat{A}, \hat{b}_{\bm p}] = \int d^3p' [\hat{b}^\dagger_{\bm p'} \hat{b}_{-\bm p'}, \hat{b}_{\bm p}] = -\hat{b}_{-\bm p} . \tag{3b}$$

The double commutators become

$$[\hat{A}, [\hat{A}, \hat{a}_{\bm p}]] = +\hat{a}_{\bm p} , \quad [\hat{A}, [\hat{A}, \hat{b}_{\bm p}]] = +\hat{b}_{\bm p} , \tag{4}$$

and the formation law for higher-order commutators is obvious. With the help of the operator identity (10.8) the action of the unitary transformation $\hat{\mathcal{P}}_1$ on the annihilation operator $\hat{a}_{\bm p}$ can be derived:

$$\begin{aligned}\hat{\mathcal{P}}_1 \hat{a}_{\bm p} \hat{\mathcal{P}}_1^{-1} &= \hat{a}_{\bm p} - i\alpha \hat{a}_{-\bm p} + \frac{(i\alpha)^2}{2!} \hat{a}_{\bm p} - \frac{(i\alpha)^3}{3!} \hat{a}_{-\bm p} \pm \ldots \\ &= \hat{a}_{\bm p}\Big(1 - \frac{\alpha^2}{2!} \pm \ldots\Big) - i\hat{a}_{-\bm p}\Big(\alpha - \frac{\alpha^3}{3!} \pm \ldots\Big) \\ &= \hat{a}_{\bm p} \cos\alpha - i\hat{a}_{-\bm p} \sin\alpha . \end{aligned} \tag{5}$$

The same relation also holds for the antiparticle annihilation operator $\hat{b}_{\bm p}$. With the choice $\alpha = \pi/2$ the result simplifies to

$$\hat{\mathcal{P}}_1 \hat{a}_{\bm p} \hat{\mathcal{P}}_1^{-1} = -i\hat{a}_{-\bm p} , \quad \hat{\mathcal{P}}_1 \hat{b}_{\bm p} \hat{\mathcal{P}}_1^{-1} = -i\hat{b}_{-\bm p} . \tag{6}$$

Except for a factor of $i\eta_P$ or $i\eta_P^*$, respectively, these agree with the desired result (10.6). To implement this factor a second unitary transformation is required:

$$\hat{\mathcal{P}}_2 = e^{i\hat{B}} . \tag{7}$$

Since now the momentum should be preserved, we try the hermitean ansatz

$$\hat{B} = \int d^3p \left(\beta \hat{a}^\dagger_{\bm p} \hat{a}_{\bm p} + \gamma \hat{b}^\dagger_{\bm p} \hat{b}_{\bm p} \right) . \tag{8}$$

The commutators now read

$$[\hat{B}, \hat{a}_{\bm p}] = -\beta \hat{a}_{\bm p} , \quad [\hat{B}, \hat{b}_{\bm p}] = -\gamma \hat{b}_{\bm p} , \tag{9a}$$

$$[\hat{B}, [\hat{B}, \hat{a}_{\bm p}]] = +\beta^2 \hat{a}_{\bm p} , \quad [\hat{B}, [\hat{B}, \hat{b}_{\bm p}]] = +\gamma^2 \hat{b}_{\bm p} , \tag{9b}$$

etc. Correspondingly (5) is replaced by

$$\hat{\mathcal{P}}_2 \hat{a}_{\bm p} \hat{\mathcal{P}}_2^{-1} = \hat{a}_{\bm p} - i\beta \hat{a}_{\bm p} + \frac{(i\beta)^2}{2!} \hat{a}_{\bm p} - \frac{(i\beta)^3}{3!} \hat{a}_{\bm p} \pm \ldots = \hat{a}_{\bm p} e^{-i\beta} , \tag{10}$$

and similarly for $\hat{b}_{\bm p}$. With the choice $\beta = -\pi/2 - \varphi$ and $\gamma = -\pi/2 + \varphi$ the transformation $\hat{\mathcal{P}}_2$ ensures that the annihilation operators are multiplied by the desired factor. The phase φ is defined as

$$\varphi = \arg \eta_P \quad \text{or} \quad \eta_P = e^{i\varphi} . \tag{11}$$

It is easy to verify that the operators $\hat{A}$ and $\hat{B}$ commute, $[\hat{A}, \hat{B}] = 0$. Therefore $\exp \hat{A} \exp \hat{B} = \exp(\hat{A} + \hat{B})$ holds and the combined transformation can be written as

$$\begin{aligned}\hat{\mathcal{P}} = \hat{\mathcal{P}}_1 \hat{\mathcal{P}}_2 &= e^{i\hat{B}} e^{i\hat{A}} \\ &= \exp\left[i \int d^3p \left(\frac{\pi}{2}(\hat{a}^\dagger_{\bm p} \hat{a}_{-\bm p} + \hat{b}^\dagger_{\bm p} \hat{b}_{-\bm p}) - \Big(\frac{\pi}{2} + \varphi\Big) \hat{a}^\dagger_{\bm p} \hat{a}_{\bm p} - \Big(\frac{\pi}{2} - \varphi\Big) \hat{b}^\dagger_{\bm p} \hat{b}_{\bm p} \right)\right] . \end{aligned} \tag{12}$$

For real values of $\eta_P = \pm 1$ this result is simplified to (10.7). This is obvious for $\eta_P = +1$. In the case $\eta_P = -1$ one can choose $\varphi = -\pi$ in the first and $\varphi = +\pi$ in the second term since φ is only determined modulo 2π.

Exercise 10.1

(b) We define

$$\hat{C}_1 = e^{i\delta \hat{C}} \tag{13}$$

with the rather obvious ansatz

$$\hat{C} = \int d^3 p \left(\hat{b}^\dagger_{\boldsymbol{p}} \hat{a}_{\boldsymbol{p}} + \hat{a}^\dagger_{\boldsymbol{p}} \hat{b}_{\boldsymbol{p}} \right) . \tag{14}$$

The commutators of this operator are

$$[\hat{C}, \hat{a}_{\boldsymbol{p}}] = -\hat{b}_{\boldsymbol{p}} \quad , \quad [\hat{C}, \hat{b}_{\boldsymbol{p}}] = -\hat{a}_{\boldsymbol{p}} , \tag{15a}$$

$$[\hat{C}, [\hat{C}, \hat{a}_{\boldsymbol{p}}]] = +\hat{a}_{\boldsymbol{p}} \quad , \quad [\hat{C}, [\hat{C}, \hat{b}_{\boldsymbol{p}}]] = +\hat{b}_{\boldsymbol{p}} , \tag{15b}$$

etc. As in (5) this leads to

$$e^{i\delta \hat{C}} \hat{a}_{\boldsymbol{p}} \, e^{-i\delta \hat{C}} = \hat{a}_{\boldsymbol{p}} \cos\delta - i\hat{b}_{\boldsymbol{p}} \sin\delta . \tag{16}$$

Choosing $\delta = \pi/2$, we get

$$\hat{C}_1 \hat{a}_{\boldsymbol{p}} \hat{C}_1^{-1} = -i\hat{b}_{\boldsymbol{p}} \quad , \quad \hat{C}_1 \hat{b}_{\boldsymbol{p}} \hat{C}_1^{-1} = -i\hat{a}_{\boldsymbol{p}} . \tag{17}$$

As in (6) the factors $i\eta_C$ or $i\eta_C^*$ are missing, which can be furnished by applying the operator $\hat{\mathcal{P}}_2$. Since $\hat{C}$ and $\hat{B}$ commute the charge-conjugation operator becomes

$$\hat{C} = e^{i\hat{B}} e^{i\frac{\pi}{2}\hat{C}} \tag{18}$$

$$= \exp\left[i \int d^3 p \left(\frac{\pi}{2}(\hat{b}^\dagger_{\boldsymbol{p}} \hat{a}_{\boldsymbol{p}} + \hat{a}^\dagger_{\boldsymbol{p}} \hat{b}_{\boldsymbol{p}}) - \left(\frac{\pi}{2} - \varphi\right) \hat{a}^\dagger_{\boldsymbol{p}} \hat{a}_{\boldsymbol{p}} - \left(\frac{\pi}{2} + \varphi\right) \hat{b}^\dagger_{\boldsymbol{p}} \hat{b}_{\boldsymbol{p}} \right) \right] ,$$

with the phase $\varphi = \arg \eta_C$.

10.3 Dirac Fields

The application of discrete symmetry transformations to spinor fields is somewhat more involved than the corresponding treatment of scalar fields since the factor Λ will now be a matrix in spinor space. Fortunately we can rely on the results presented in Chaps. 4 and 12 of the companion volume *Relativistic Quantum Mechanics (RQM)* in this series of books. We only have to translate the results presented there for classical Dirac wave functions to the language of second quantization.

For simplicity in this section we will omit the phase factors, i.e., we make the special choice of phases $\eta_P = \eta_C = \eta_T = 1$. Note that this is justified since the remark on page 305 also applies to spinor fields. If there is a charge operator that satisfies (10.20) (this may refer to the electric charge or to other types of conserved quantities such as the lepton charge) then the phases can be freely chosen.

10.3.1 Space Inversion

The transformation law for Dirac wave functions under space inversion is derived in Chap. 4 of *RQM*:

$$\psi(\boldsymbol{x},t) \to \psi'(-\boldsymbol{x},t) = \gamma_0 \psi(\boldsymbol{x},t) \ . \tag{10.62}$$

This relation translates to equations for the field operator:

$$\hat{\mathcal{P}}\,\hat{\psi}(\boldsymbol{x},t)\hat{\mathcal{P}}^{-1} = \gamma_0 \hat{\psi}(-\boldsymbol{x},t) \tag{10.63}$$

and

$$\hat{\mathcal{P}}\,\hat{\bar{\psi}}(\boldsymbol{x},t)\hat{\mathcal{P}}^{-1} = \hat{\bar{\psi}}(-\boldsymbol{x},t)\gamma^0 \ . \tag{10.64}$$

Inserting the expansion (5.62) we obtain the transformation law for the creation and annihilation operators:

$$\int \mathrm{d}^3 p\, N_p \sum_s \Big(\hat{\mathcal{P}}\,\hat{b}(p,s)\hat{\mathcal{P}}^{-1}\, u(p,s)\, \mathrm{e}^{-\mathrm{i}p\cdot x} + \hat{\mathcal{P}}\,\hat{d}^\dagger(p,s)\hat{\mathcal{P}}^{-1}\, v(p,s)\, \mathrm{e}^{+\mathrm{i}p\cdot x}\Big)$$

$$= \int \mathrm{d}^3 p\, N_p \sum_s \Big(\hat{b}(p,s)\gamma_0\, u(p,s)\, \mathrm{e}^{-\mathrm{i}\tilde{p}\cdot x} + \hat{d}^\dagger(p,s)\gamma_0\, v(p,s)\, \mathrm{e}^{+\mathrm{i}\tilde{p}\cdot x}\Big) \ . \tag{10.65}$$

Here the abbreviation $\tilde{p} = (p_0, -\boldsymbol{p})$ denotes the four-momentum with inverted spatial components. The matrix γ_0 on the right-hand side has the effect of flipping the direction of the three-momentum and introduces an additional minus sign for the antiparticle component:

$$\gamma_0 u(p,s) = u(\tilde{p},s) \quad , \quad \gamma_0 v(p,s) = -v(\tilde{p},s) \ . \tag{10.66}$$

Proof: Using (5.63) and $\gamma_0 \slashed{p} = \slashed{\tilde{p}}\gamma_0$ as well as $\gamma_0 u(0,s) = u(0,s)$ one finds

$$\gamma_0 u(p,s) = \gamma_0 \frac{\slashed{p}+m}{\sqrt{2m(p_0+m)}} u(0,s) = \frac{\slashed{\tilde{p}}+m}{\sqrt{2m(p_0+m)}}\gamma_0\, u(0,s)$$

$$= u(\tilde{p},s) \ , \tag{10.67}$$

and similarly for $v(p,s)$ with $\gamma_0 v(0,s) = -v(0,s)$. Thus we find

$$\hat{\mathcal{P}}\,\hat{b}(p,s)\hat{\mathcal{P}}^{-1} = \hat{b}(\tilde{p},s) \quad , \quad \hat{\mathcal{P}}\,\hat{d}(p,s)\hat{\mathcal{P}}^{-1} = -\hat{d}(\tilde{p},s) \ ,$$

$$\hat{\mathcal{P}}\,\hat{b}^\dagger(p,s)\hat{\mathcal{P}}^{-1} = \hat{b}^\dagger(\tilde{p},s) \quad , \quad \hat{\mathcal{P}}\,\hat{d}^\dagger(p,s)\hat{\mathcal{P}}^{-1} = -\hat{d}^\dagger(\tilde{p},s) \ . \tag{10.68}$$

Comparing this result with (10.6) we note the minus sign for the antiparticle operators. We have found an important physical fact: *particles and antiparticles have opposite intrinsic parity*. This is not just a mathematical curiosity, rather it has immediate consequences, for example for the decay of positronium (the bound e^+e^- system) for which the ground state has negative parity, contrary to what one might expect (cf. Example 6.3).

Again it is possible to find an explicit expression for the parity operator:

$$\hat{\mathcal{P}} = \exp\Big[\mathrm{i}\frac{\pi}{2}\int \mathrm{d}^3 p \sum_s \big(\hat{b}^\dagger(p,s)\hat{b}(\tilde{p},s) + \hat{d}^\dagger(p,s)\hat{d}(\tilde{p},s)$$

$$-\hat{b}^\dagger(p,s)\hat{b}(p,s) + \hat{d}^\dagger(p,s)\hat{d}(p,s)\big)\Big] \ . \tag{10.69}$$

This agrees with the spin-0 result, see (10.7), if one adopts $\eta_P = +1$ for particles and $\eta_P = -1$ for antiparticles. The derivation proceeds as in Exercise 10.1.

The transformation law for space inversion of the *momentum operator* agrees with (10.14):

$$\hat{\mathcal{P}}\,\hat{P}^\mu\,\hat{\mathcal{P}}^{-1} = \hat{P}_\mu \,. \tag{10.70}$$

The extra minus sign in (10.68) is unimportant since only bilinear terms of the type $\hat{b}^\dagger \hat{b}$ and $\hat{d}^\dagger \hat{d}$ are present.

The *angular-momentum operator* transforms as a pseudo vector, which is easily confirmed by using $\gamma_0 \boldsymbol{\Sigma} \gamma_0 = \gamma_0(\gamma_5 \gamma_0 \boldsymbol{\gamma})\gamma_0 = \boldsymbol{\Sigma}$:

$$\begin{aligned}
\hat{\mathcal{P}}\,\hat{\boldsymbol{J}}\,\hat{\mathcal{P}}^{-1} &= \int \mathrm{d}^3 x \, \hat{\psi}^\dagger(-\boldsymbol{x},t)\gamma_0\big(-\mathrm{i}\boldsymbol{x}\times\boldsymbol{\nabla} + \tfrac{1}{2}\boldsymbol{\Sigma}\big)\gamma_0 \hat{\psi}(-\boldsymbol{x},t) \\
&= \int \mathrm{d}^3 x \, \hat{\psi}^\dagger(\boldsymbol{x},t)\big(-\mathrm{i}\boldsymbol{x}\times\boldsymbol{\nabla} + \tfrac{1}{2}\boldsymbol{\Sigma}\big)\hat{\psi}(-\boldsymbol{x},t) = +\hat{\boldsymbol{J}} \,.
\end{aligned} \tag{10.71}$$

Accordingly, the *helicity* of a particle changes its sign upon space inversion.

The *current-density operator* (we use the simplest version since normal ordering and antisymmetrization do not matter here) is transformed as

$$\begin{aligned}
\hat{\mathcal{P}}\hat{j}^\mu(\boldsymbol{x},t)\hat{\mathcal{P}}^{-1} &= \hat{\mathcal{P}}\hat{\bar{\psi}}(\boldsymbol{x},t)\hat{\mathcal{P}}^{-1}\gamma^\mu \hat{\mathcal{P}}\hat{\psi}(\boldsymbol{x},t)\hat{\mathcal{P}}^{-1} \\
&= \hat{\bar{\psi}}(-\boldsymbol{x},t)\gamma_0 \gamma^\mu \gamma_0 \hat{\psi}(-\boldsymbol{x},t) = \hat{\bar{\psi}}(-\boldsymbol{x},t)\gamma_\mu \hat{\psi}(-\boldsymbol{x},t) \\
&= \hat{j}_\mu(-\boldsymbol{x},t) \,.
\end{aligned} \tag{10.72}$$

The time (space) components of the current density are even (odd) under space inversion.

10.3.2 Charge Conjugation

Charge conjugation exchanges the roles of particle and antiparticle spinors. According to (12.15) in *RQM* this is effected by the transformation

$$\psi(x) \to \psi_c(x) = C\gamma_0 K \psi(x) = C\gamma_0 \psi^*(x) = C\bar{\psi}^\mathrm{T}(x) \,. \tag{10.73}$$

Both wave functions $\psi_c(x)$ and $\psi(x)$ satisfy the Dirac equation but their coupling to the electromagetic field involves oppposite signs of the charge e. The charge conjugation matrix C is constructed such that it produces transposed gamma matrices:

$$C\,\gamma^\mu\,C^{-1} = -\gamma^{\mu\mathrm{T}} \,. \tag{10.74}$$

C has the following properties:

$$C^{-1} = C^\dagger = C^\mathrm{T} = -C \,. \tag{10.75}$$

Within the standard representation of the Dirac matrices it is given by $C = \mathrm{i}\gamma^2\gamma^0$. Because of (10.73) the transformation law for the field operator reads

$$\hat{\mathcal{C}}\hat{\psi}(x)\hat{\mathcal{C}}^{-1} = C\hat{\bar{\psi}}^\mathrm{T}(x) \tag{10.76}$$

and similarly for the adjoint field operator

$$\hat{\mathcal{C}}\hat{\bar{\psi}}(x)\hat{\mathcal{C}}^{-1} = -\hat{\psi}^\mathrm{T}(x)\,C^\dagger \tag{10.77}$$

or
$$\hat{\mathcal{C}}\,\hat{\psi}^\dagger(x)\,\hat{\mathcal{C}}^{-1} = \hat{\psi}^T(x)\,\gamma^0\,C^\dagger \ . \tag{10.78}$$

Insertion of the plane-wave expansion in (10.76) leads to the condition

$$\int d^3 p\, N_p \sum_s \left(\hat{\mathcal{C}}\,\hat{b}(p,s)\hat{\mathcal{C}}^{-1} u(p,s)\,e^{-ip\cdot x} + \hat{\mathcal{C}}\,\hat{d}^\dagger(p,s)\hat{\mathcal{C}}^{-1} v(p,s)\,e^{+ip\cdot x}\right)$$
$$= \int d^3 p\, N_p \sum_s \left(\hat{b}^\dagger(p,s)\,C\,\bar{u}^T(p,s)\,e^{+ip\cdot x} + \hat{d}(p,s)\,C\,\bar{v}^T(p,s)\,e^{-ip\cdot x}\right) . \tag{10.79}$$

Under charge conjugation the unit spinors $u \leftrightarrow v$ are interchanged:

$$C\,\bar{u}^T(p,s) = v(p,s) \quad , \quad C\,\bar{v}^T(p,s) = u(p,s) \ . \tag{10.80}$$

Proof: (10.80) may be verified by using explicit expressions for the unit spinors. Also a more intuitive argument can be given. It is easy to verify that $\bar{u}^T(p,s)$ satisfies the Dirac equation for antiparticle spinors $(\not{p}+m)v(p,s)=0$ since we find, according to (10.74) and (5.64), that

$$(\not{p}+m)C\bar{u}^T(p,s) = C(-\not{p}^T+m)\bar{u}^T(p,s) = C\left[\bar{u}(p,s)(\not{p}-m)\right]^T = 0 \ . \tag{10.81}$$

The use of the same spin variable s on both sides of (10.80) is justified: a brief calculation shows that $C\bar{u}^T(p,s)$ is an eigenstate of the spin projection operator

$$\Sigma(s) = \frac{1}{2}(1+\gamma_5\not{s}) \tag{10.82}$$

satisfying

$$\Sigma(s)C\bar{u}^T(p,s) = C\bar{u}^T(p,s) \quad \text{and} \quad \Sigma(-s)C\bar{u}^T(p,s) = 0 \ . \tag{10.83}$$

The same reasoning can be applied to $C\bar{v}^T(p,s)$.

The transformation law for the creation and annihilation operators is obtained from (10.79) and (10.80):

$$\hat{\mathcal{C}}\,\hat{b}(p,s)\hat{\mathcal{C}}^{-1} = \hat{d}(p,s) \quad , \quad \hat{\mathcal{C}}\,\hat{d}(p,s)\hat{\mathcal{C}}^{-1} = \hat{b}(p,s) \ ,$$
$$\hat{\mathcal{C}}\,\hat{b}^\dagger(p,s)\hat{\mathcal{C}}^{-1} = \hat{d}^\dagger(p,s) \quad , \quad \hat{\mathcal{C}}\,\hat{d}^\dagger(p,s)\hat{\mathcal{C}}^{-1} = \hat{b}^\dagger(p,s) \ . \tag{10.84}$$

This again allows an explicit construction of $\hat{\mathcal{C}}$. Since the relations (10.84) are identical with the spin-0 result (10.25) we can simply copy the previous result:

$$\hat{\mathcal{C}} = \exp\left[i\frac{\pi}{2}\int d^3 p \sum_s (\hat{d}^\dagger(p,s)\hat{b}(p,s) + \hat{b}^\dagger(p,s)\hat{d}(p,s)\right.$$
$$\left. - \hat{b}^\dagger(p,s)\hat{b}(p,s) - \hat{d}^\dagger(p,s)\hat{d}(p,s))\right] \ . \tag{10.85}$$

Again the energy and momentum operators are invariant under charge conjugation,

$$\hat{\mathcal{C}}\,\hat{P}^\mu\,\hat{\mathcal{C}} = \hat{P}^\mu \ , \tag{10.86}$$

and the *charge operator* transforms as

$$\begin{aligned}
\hat{C}\hat{j}^\mu \hat{C}^{-1} &= \hat{C}\frac{1}{2}[\hat{\bar{\psi}}, \gamma^\mu \hat{\psi}]\hat{C}^{-1} = \frac{1}{2}[-\hat{\psi}^T C^\dagger, \gamma^\mu C \hat{\bar{\psi}}^T] \\
&= -\frac{1}{2} C^\dagger_{\alpha\beta} \gamma^\mu_{\beta\gamma} C_{\gamma\delta} [\hat{\psi}_\alpha, \hat{\bar{\psi}}_\delta] = +\frac{1}{2}\gamma^\mu_{\delta\alpha}[\hat{\psi}_\alpha, \hat{\bar{\psi}}_\delta] \\
&= -\frac{1}{2}[\hat{\bar{\psi}}, \gamma^\mu \hat{\psi}] = -\hat{j}^\mu\,, \quad (10.87)
\end{aligned}$$

where (10.74), (10.76), and (10.77) have been used. Here it was important to employ the antisymmetrized version of the current operator. Otherwise we would have obtained

$$\hat{C}\hat{\bar{\psi}}\gamma^\mu \hat{\psi}\hat{C}^{-1} = \hat{\psi}^T \gamma^{\mu T} \hat{\bar{\psi}}^T \neq -\hat{\bar{\psi}}\gamma^\mu \hat{\psi}\,. \quad (10.88)$$

The two expressions differ by an undetermined (divergent) term which has its origin in the anticommutator of $\hat{\psi}(x)$ and $\hat{\bar{\psi}}(x)$. An alternative possibility to eliminate this vacuum term makes use of the normal-ordering prescription as demonstrated in Exercise 5.2. This is expressed by

$$\hat{C}:\hat{\bar{\psi}}\gamma^\mu \hat{\psi}:\hat{C}^{-1} = -:\hat{\bar{\psi}}\gamma^\mu \hat{\psi}:\,. \quad (10.89)$$

The *canonical anticommutation relations* again are invariant under charge conjugation:

$$\begin{aligned}
\{\hat{C}\hat{\psi}_\alpha(x)\hat{C}^{-1}, \hat{C}\hat{\bar{\psi}}_\beta(x')\hat{C}^{-1}\}_{x_0=x'_0} &= \{(C\hat{\bar{\psi}}^T(x))_\alpha, (\hat{\psi}^T(x)\gamma^0 C^\dagger)_\beta\}_{x_0=x'_0} \\
&= C_{\alpha\mu}\gamma^0_{\nu\mu}\{\hat{\psi}^\dagger_\nu(x), \hat{\psi}_\sigma(x')\}_{x_0=x'_0} \gamma^0_{\sigma\tau} C^\dagger_{\tau\beta} \\
&= \delta^3(\boldsymbol{x}-\boldsymbol{x}')(C\gamma^0\gamma^0 C^\dagger)_{\alpha\beta} = \delta^3(\boldsymbol{x}-\boldsymbol{x}')(CC^{-1})_{\alpha\beta} \\
&= \delta^3(\boldsymbol{x}-\boldsymbol{x}')\delta_{\alpha\beta}\,. \quad (10.90)
\end{aligned}$$

10.3.3 Time Reversal

The definition of the time reversal applied to Dirac wave functions can be found in Chap. 6 of *RQM*:

$$\psi(\boldsymbol{x}, t) \to \psi'(\boldsymbol{x}, -t) = T\psi(\boldsymbol{x}, t) = T_0 K \psi(\boldsymbol{x}, t) = T_0 \psi^*(\boldsymbol{x}, t)\,. \quad (10.91)$$

Here $T = T_0 K$ consists of a matrix T_0 acting in spinor space and the operator of complex conjugation K. The antilinear operator T satisfies the defining relation

$$T^{-1}\gamma^\mu T = \gamma_\mu \quad (10.92)$$

or

$$T_0^{-1} \gamma^\mu T_0 = \gamma^*_\mu = \gamma^{\mu T}\,. \quad (10.93)$$

In the standard representation of the Dirac matrices

$$T_0 = i\gamma^1\gamma^3 = -i\gamma^5 C \quad (10.94)$$

satisfies this condition. This matrix has the properties

$$T_0 = T_0^{-1} = T_0^\dagger = -T_0^* = -T_0^T\,. \quad (10.95)$$

The free Dirac equation

$$(\mathrm{i}\gamma^0\partial_0 + \mathrm{i}\gamma^k\partial_k - m)\psi(\boldsymbol{x},t) = 0 \tag{10.96}$$

is form invariant under the transformation (10.91). This is verified by multiplying by T from the left and using (10.92):

$$\begin{aligned}
(T\mathrm{i}T^{-1}T\gamma^0T^{-1}\partial_0 + T\mathrm{i}T^{-1}T\gamma^kT^{-1}\partial_k - m)T\psi(\boldsymbol{x},t) &= 0, \\
(-\mathrm{i}\gamma^0\partial_0 - \mathrm{i}\gamma_k\partial_k - m)T\psi(\boldsymbol{x},t) &= 0, \\
(\mathrm{i}\gamma^0\partial_0 + \mathrm{i}\gamma^k\partial_k - m)T\psi(\boldsymbol{x},-t) &= 0.
\end{aligned} \tag{10.97}$$

For the matrix elements of the field operator $\hat{\psi}(x)$, (10.91) implies

$$\langle\alpha'|\hat{\psi}(\boldsymbol{x},-t)|\beta'\rangle = \langle\beta|T_0\hat{\psi}^\dagger(\boldsymbol{x},t)|\alpha\rangle \tag{10.98}$$

or, similar to (10.51),

$$\left(\hat{T}^{-1}\hat{\psi}(\boldsymbol{x},-t)\hat{T}\right)^\dagger = T_0\,\hat{\psi}^\dagger(\boldsymbol{x},t)\,. \tag{10.99}$$

Here a phase factor might emerge for which we have simply chosen the value $\eta_\mathrm{T} = 1$. The transformation law for the field operator can also be written as

$$\hat{T}\,\hat{\psi}(\boldsymbol{x},t)\,\hat{T}^{-1} = T_0\,\hat{\psi}(\boldsymbol{x},-t) \tag{10.100}$$

and

$$\hat{T}\,\hat{\bar{\psi}}(\boldsymbol{x},t)\,\hat{T}^{-1} = \hat{\bar{\psi}}(\boldsymbol{x},-t)\,T_0^\dagger \tag{10.101}$$

We note in passing that twice applying time inversion leads back to the original spinor, multiplied by a phase factor -1:

$$\hat{T}\hat{T}\,\hat{\psi}(\boldsymbol{x},t)\hat{T}^{-1}\hat{T}^{-1} = \hat{T}\,T_0\,\hat{\psi}(\boldsymbol{x},-t)\hat{T}^{-1} = T_0^*T_0\hat{\psi}(\boldsymbol{x},t) = -\hat{\psi}(\boldsymbol{x},t)\,. \tag{10.102}$$

This is reminiscent of the characteristic property of spinors that have to be rotated twice (i.e., by an angle of 4π) to get reproduced.

The transformation law for the creation and annihilation operators this time takes a somewhat more complicated form. Equation (10.100) leads to the condition

$$\int \mathrm{d}^3p\, N_p \sum_s \left(\hat{T}\hat{b}(p,s)\hat{T}^{-1}u^*(p,s)\,\mathrm{e}^{+\mathrm{i}p\cdot x} + \hat{T}\hat{d}^\dagger(p,s)\hat{T}^{-1}v^*(p,s)\,\mathrm{e}^{-\mathrm{i}p\cdot x}\right)$$

$$= \int \mathrm{d}^3p\, N_p \sum_s \left(\hat{b}(p,s)\,T_0 u(p,s)\,\mathrm{e}^{+\mathrm{i}\tilde{p}\cdot x} + \hat{d}^\dagger(p,s)\gamma_0\,T_0 v(p,s)\,\mathrm{e}^{-\mathrm{i}\tilde{p}\cdot x}\right), \tag{10.103}$$

where again the abbreviation $\tilde{p} = (p_0, -\boldsymbol{p})$ was employed. The action of the matrix T_0 on the unit spinor leads to the reversal of momentum and spin and to an additional phase factor. To derive the details we can employ the explicit form of the unit spinors (5.63) and use the anticommutation property of the gamma matrices $\gamma^\mu\gamma^\nu = -\gamma^\nu\gamma^\mu + 2g^{\mu\nu}$. The application of (10.94) produces

$$\begin{aligned}
T_0 u(p,s) &= \mathrm{i}\gamma^1\gamma^3\frac{\slashed{p}+m}{\sqrt{2m(p_0+m)}}u(0,s) \\
&= \mathrm{i}\frac{p_0\gamma^0 - p_1\gamma^1 + p_2\gamma^2 - p_3\gamma^3 + m}{\sqrt{2m(p_0+m)}}\gamma^1\gamma^3\,u(0,s)\,. \tag{10.104}
\end{aligned}$$

In the standard representation we have

$$\gamma^1\gamma^3 = \begin{pmatrix} 0 & \sigma_1 \\ -\sigma_1 & 0 \end{pmatrix}\begin{pmatrix} 0 & \sigma_3 \\ -\sigma_3 & 0 \end{pmatrix} = \begin{pmatrix} i\sigma_2 & 0 \\ 0 & i\sigma_2 \end{pmatrix}. \tag{10.105}$$

The Pauli matrix

$$i\sigma_2 = \begin{pmatrix} 0 & 1 \\ -1 & 0 \end{pmatrix} \tag{10.106}$$

flips the spin:

$$i\sigma_2 \chi_{\pm 1/2} = \mp \chi_{\mp 1/2} \quad , \text{ thus } \quad \gamma^1\gamma^3 u(0,s) = (-1)^{s+1/2} u(0,\tilde{s}), \tag{10.107}$$

where $\tilde{s} = (s_0, -\boldsymbol{s})$ denotes the flipped spin. The annoying sign of the $p_2\gamma^2$ term in (10.104) can be eliminated by complex conjugation of the whole expression since (in the standard representation) $u(0,s)$ and all gamma matrices except γ^2 are real valued:

$$\begin{aligned} T_0\, u(p,s) &= i\left[\frac{\slashed{p}+m}{\sqrt{2m(p_0+m)}}(-1)^{s+1/2}u(0,\tilde{s})\right]^* \\ &= i(-1)^{s+1/2} u^*(\tilde{p},\tilde{s}). \end{aligned} \tag{10.108}$$

The antiparticle spinors can be treated in the same way:

$$T_0\, v(p,s) = i(-1)^{s+1/2} v^*(\tilde{p},\tilde{s}). \tag{10.109}$$

Now we are in a position to deduce the transformation law for the creation and annihilation operators from (10.103):

$$\begin{aligned} \hat{T}\hat{b}(p,s)\hat{T}^{-1} &= i(-1)^{s-1/2}\hat{b}(\tilde{p},\tilde{s}) \quad,\quad \hat{T}\hat{d}(p,s)\hat{T}^{-1} = -i(-1)^{s-1/2}\hat{d}(\tilde{p},\tilde{s}), \\ \hat{T}\hat{b}^\dagger(p,s)\hat{T}^{-1} &= -i(-1)^{s-1/2}\hat{b}^\dagger(\tilde{p},\tilde{s}) \quad,\quad \hat{T}\hat{d}^\dagger(p,s)\hat{T}^{-1} = i(-1)^{s-1/2}\hat{d}^\dagger(\tilde{p},\tilde{s}). \end{aligned} \tag{10.110}$$

If we had chosen a different basis for the polarization states (e.g., eigenstates of the helicity operator) we would have obtained other phase factors. Apart from these sign factors, (10.110) tells us that time reversal leads to a flip of the momentum vector and the spin direction.

The energy and momentum operators transform as

$$\hat{T}\hat{P}^\mu\hat{T}^{-1} = \hat{P}_\mu \tag{10.111}$$

and the angular-momentum operator (5.23) is found to change its sign upon time reversal:

$$\begin{aligned} \hat{T}\hat{\boldsymbol{J}}(t)\hat{T}^{-1} &= \hat{T}\int d^3x\, \hat{\psi}^\dagger(\boldsymbol{x},t)\left(-i\boldsymbol{x}\times\boldsymbol{\nabla} + \tfrac{1}{2}\boldsymbol{\Sigma}\right)\hat{\psi}(\boldsymbol{x},t)\hat{T}^{-1} \\ &= \int d^3x\, \hat{T}\hat{\psi}^\dagger(\boldsymbol{x},t)\hat{T}^{-1}\left(-i\boldsymbol{x}\times\boldsymbol{\nabla} + \tfrac{1}{2}\boldsymbol{\Sigma}\right)^* \hat{T}\hat{\psi}(\boldsymbol{x},t)\hat{T}^{-1} \\ &= \int d^3x\, \hat{\psi}^\dagger(\boldsymbol{x},-t)\left(+i\boldsymbol{x}\times\boldsymbol{\nabla} + \tfrac{1}{2}T_0^\dagger \boldsymbol{\Sigma}^* T_0\right)\hat{\psi}(\boldsymbol{x},-t) \\ &= -\hat{\boldsymbol{J}}(-t). \end{aligned} \tag{10.112}$$

Here we used (10.93) and various properties of the gamma matrices:

$$\begin{aligned} T_0^\dagger \Sigma^* T_0 &= T_0^{-1}(\gamma_5\gamma^0\gamma)^* T_0 = T_0^{-1}\gamma_5\gamma^0\gamma^* T_0 = \gamma_5\gamma^0 T_0^{-1}\gamma^* T_0 \\ &= \gamma_5\gamma^0(-\gamma) = -\Sigma \; . \end{aligned} \qquad (10.113)$$

When deriving the transformation law for the current operator we have to take into account the antilinear character of the time-reversal transformation $\hat{T}$: $\hat{T}\gamma^\mu = \gamma^{\mu*}\hat{T}$. Using (10.93) we obtain

$$\begin{aligned} \hat{T}\hat{j}^\mu(x)\hat{T}^{-1} &= \frac{1}{2}\hat{T}\left[\hat{\bar{\psi}}(x),\gamma^\mu\hat{\psi}(x)\right]\hat{T}^{-1} \\ &= \frac{1}{2}\hat{T}\hat{\bar{\psi}}(x)\hat{T}^{-1}\gamma^{\mu*}\,\hat{T}\hat{\psi}(x)\hat{T}^{-1} - \frac{1}{2}\hat{T}\hat{\psi}^{\mathrm{T}}(x)\hat{T}^{-1}\gamma^{\mu\mathrm{T}*}\,\hat{T}\hat{\bar{\psi}}^{\mathrm{T}}(x)\hat{T}^{-1} \\ &= \frac{1}{2}\hat{\bar{\psi}}(\boldsymbol{x},-t)T_0^\dagger\gamma^{\mu*}\,T_0\hat{\psi}(\boldsymbol{x},-t) - \frac{1}{2}\hat{\psi}^{\mathrm{T}}(\boldsymbol{x},-t)T_0^{\mathrm{T}}\gamma^{\mu\mathrm{T}*}T_0^{\dagger\mathrm{T}}\hat{\bar{\psi}}^{\mathrm{T}}(\boldsymbol{x},-t) \\ &= \frac{1}{2}\hat{\bar{\psi}}(\boldsymbol{x},-t)\gamma_\mu\hat{\psi}(\boldsymbol{x},-t) - \frac{1}{2}\hat{\psi}^{\mathrm{T}}(\boldsymbol{x},-t)\gamma_\mu^{\mathrm{T}}\,\hat{\bar{\psi}}^{\mathrm{T}}(\boldsymbol{x},-t) \\ &= \frac{1}{2}\left[\hat{\bar{\psi}}(\boldsymbol{x},-t),\gamma_\mu\hat{\psi}(\boldsymbol{x},-t)\right] \\ &= \hat{j}_\mu(\boldsymbol{x},-t) \; , \end{aligned} \qquad (10.114)$$

which agrees with the scalar result (10.58).

The invariance of the canonical anticommutation relations under time reversal is verified immediately:

$$\begin{aligned} \left\{\hat{T}\hat{\psi}_\alpha(x)\hat{T}^{-1},\hat{T}\hat{\psi}_\beta^\dagger(x')\hat{T}^{-1}\right\}_{x_0=x_0'} & \\ = \left\{\left(T_0\hat{\psi}(x)\right)_\alpha,\left(\hat{\psi}^\dagger(x')T_0^\dagger\right)_\beta\right\}_{x_0=x_0'} &= T_{0\alpha\delta}T_{0\sigma\beta}^\dagger\left\{\hat{\psi}_\delta(x),\hat{\psi}_\sigma^\dagger(x')\right\}_{x_0=x_0'} \\ = \left(T_0 T_0^\dagger\right)_{\alpha\beta}\delta^3(\boldsymbol{x}-\boldsymbol{x}') &= \delta_{\alpha\beta}\,\delta^3(\boldsymbol{x}-\boldsymbol{x}') \; . \end{aligned} \qquad (10.115)$$

10.4 The Electromagnetic Field

Following the detailed exposition of the discrete transformations applied to spin-0 and spin-$\frac{1}{2}$ fields we will now briefly address the spin-1 case. The discussion will be restricted to the electromagnetic field $A_\mu(x)$. There is a simple argument that allows us to deduce the transformation properties of the electromagnetic field from the corresponding results for the matter fields we have already derived. The photon field satisfies Maxwell's equations, which contain the electromagnetic current as a source term. In quantum electrodynamics the Heisenberg field operator $\hat{A}^\mu(x)$ satisfies the equation of motion (assuming Feynman gauge)

$$\Box \hat{A}^\mu(x) = \frac{e}{2}\left[\hat{\bar{\psi}}(x),\gamma^\mu\hat{\psi}(x)\right] \; . \qquad (10.116)$$

This equation allows us to infer the transformation properties of the operator $\hat{A}^\mu(x)$ from our knowledge of the current. We find the following transformation laws:

space inversion, cf. (10.72),

$$\hat{\mathcal{P}}\,\hat{A}^\mu(\boldsymbol{x},t)\,\hat{\mathcal{P}}^{-1} = \hat{A}_\mu(-\boldsymbol{x},t) \; ; \qquad (10.117)$$

charge conjugation, cf. (10.87),

$$\hat{\mathcal{C}}\,\hat{A}^\mu(\boldsymbol{x},t)\,\hat{\mathcal{C}}^{-1} = -\hat{A}^\mu(\boldsymbol{x},t)\;; \tag{10.118}$$

time reversal, cf. (10.114),

$$\hat{\mathcal{T}}\,\hat{A}^\mu(\boldsymbol{x},t)\,\hat{\mathcal{T}}^{-1} = \hat{A}_\mu(\boldsymbol{x},-t)\,. \tag{10.119}$$

As a result of these transformation laws the interaction term $j_\mu A^\mu$ and thus the Hamiltonian of quantum electrodynamics is invariant under all the three discrete transformations:

$$\hat{\mathcal{U}}\,\hat{H}_{\mathrm{QED}}\,\hat{\mathcal{U}}^{-1} = \hat{H}_{\mathrm{QED}}\,, \tag{10.120}$$

where $\hat{\mathcal{U}}$ stands for $\hat{\mathcal{P}}$, $\hat{\mathcal{C}}$, and $\hat{\mathcal{T}}$. This invariance is by no means a trivial property of the Hamiltonian, rather it depends on the specific form of the electromagnetic interaction.

The transformation properties of the creation and annihilation operators are most conveniently studied in the *Coulomb gauge* (cf. Sect. 7.7). This gauge has the advantage that only the physically relevant degrees of freedom with transverse polarization are present, since $\boldsymbol{\nabla}\cdot\hat{\boldsymbol{A}}=0$. The plane-wave decomposition of the field operator (7.105) reads

$$\hat{\boldsymbol{A}}(\boldsymbol{x},t) = \int \mathrm{d}^3 k\, N_k \sum_{\lambda=1}^{2}\Big[\boldsymbol{\epsilon}(\boldsymbol{k},\lambda)\hat{a}_{\boldsymbol{k}\lambda}\,\mathrm{e}^{-\mathrm{i}k\cdot x} + \boldsymbol{\epsilon}^*(\boldsymbol{k},\lambda)\hat{a}^\dagger_{\boldsymbol{k}\lambda}\,\mathrm{e}^{+\mathrm{i}k\cdot x}\Big]\,, \tag{10.121}$$

with the transverse polarization vectors $\boldsymbol{\epsilon}(\boldsymbol{k},\lambda)\cdot\boldsymbol{k}=0$.

The equation (10.117) for space inversion becomes

$$\int \mathrm{d}^3 k\, N_k \sum_{\lambda=1}^{2}\Big[\hat{\mathcal{P}}\hat{a}_{\boldsymbol{k}\lambda}\hat{\mathcal{P}}^{-1}\boldsymbol{\epsilon}(\boldsymbol{k},\lambda)\,\mathrm{e}^{-\mathrm{i}k\cdot x} + \hat{\mathcal{P}}\hat{a}^\dagger_{\boldsymbol{k}\lambda}\hat{\mathcal{P}}^{-1}\boldsymbol{\epsilon}^*(\boldsymbol{k},\lambda)\,\mathrm{e}^{+\mathrm{i}k\cdot x}\Big] \tag{10.122}$$

$$= -\int \mathrm{d}^3 k\, N_k \sum_{\lambda=1}^{2}\Big[\hat{a}_{-\boldsymbol{k}\lambda}\boldsymbol{\epsilon}(-\boldsymbol{k},\lambda)\,\mathrm{e}^{-\mathrm{i}k\cdot x} + \hat{a}^\dagger_{-\boldsymbol{k}\lambda}\boldsymbol{\epsilon}^*(-\boldsymbol{k},\lambda)\,\mathrm{e}^{+\mathrm{i}k\cdot x}\Big]\,.$$

The vectors $\boldsymbol{\epsilon}(\boldsymbol{k},1)$, $\boldsymbol{\epsilon}(\boldsymbol{k},2)$, and $\boldsymbol{k}$ form a right-handed orthogonal dreibein. Of course this also holds true for the inverted momentum argument $-\boldsymbol{k}$. We can choose the vectors $\boldsymbol{\epsilon}$ as follows:

$$\boldsymbol{\epsilon}(-\boldsymbol{k},1) = +\boldsymbol{\epsilon}(\boldsymbol{k},1) \quad,\quad \boldsymbol{\epsilon}(-\boldsymbol{k},2) = -\boldsymbol{\epsilon}(\boldsymbol{k},2)\,. \tag{10.123}$$

Then (10.122) leads to

$$\hat{\mathcal{P}}\,\hat{a}_{\boldsymbol{k}\lambda}\,\hat{\mathcal{P}}^{-1} = (-1)^\lambda\,\hat{a}_{-\boldsymbol{k}\lambda}\,. \tag{10.124}$$

Often it is preferable to use circular polarization (i.e., helicity states of the photon) which are described by the spherical basis vectors (cf. Example 6.3)

$$\boldsymbol{\epsilon}(\boldsymbol{k},\pm) = \frac{1}{\sqrt{2}}\big(\boldsymbol{\epsilon}(\boldsymbol{k},1)\pm\mathrm{i}\,\boldsymbol{\epsilon}(\boldsymbol{k},2)\big)\,. \tag{10.125}$$

In this case (10.123) is replaced by

$$\boldsymbol{\epsilon}(-\boldsymbol{k},\sigma) = \boldsymbol{\epsilon}(-\boldsymbol{k},-\sigma)\,, \tag{10.126}$$

which implies

$$\hat{\mathcal{P}}\,\hat{a}_{\boldsymbol{k}\sigma}\,\hat{\mathcal{P}}^{-1} = -\hat{a}_{-\boldsymbol{k}-\sigma}\,. \tag{10.127}$$

Thus space inversion leads to a sign change of the helicity.

The corresponding results for *charge conjugation* follow from (10.118):

$$\hat{\mathcal{C}}\,\hat{a}_{k\lambda}\,\hat{\mathcal{C}}^{-1} = -\hat{a}_{k\lambda} \quad \text{or} \quad \hat{\mathcal{C}}\,\hat{a}_{k\sigma}\,\hat{\mathcal{C}}^{-1} = -\hat{a}_{k\sigma}\,, \qquad (10.128)$$

and for *time reversal* we find

$$\hat{\mathcal{T}}\,\hat{a}_{k\lambda}\,\hat{\mathcal{T}}^{-1} = (-1)^{\lambda}\hat{a}_{-k\lambda} \quad \text{or} \quad \hat{\mathcal{T}}\,\hat{a}_{k\sigma}\,\hat{\mathcal{T}}^{-1} = -\hat{a}_{-k\sigma}\,. \qquad (10.129)$$

The result (10.128) has immediate physical consequence since it implies that any *multi-photon state* is an eigenstate of the charge conjugation operator. The *charge parity* will be either even or odd, depending on the number of photons:

$$\begin{aligned}\hat{\mathcal{C}}|n\gamma\rangle &= \hat{\mathcal{C}}\,\hat{a}^{\dagger}_{k_1\lambda_1}\hat{\mathcal{C}}^{-1}\cdots\hat{\mathcal{C}}\,\hat{a}^{\dagger}_{k_n\lambda_n}\hat{\mathcal{C}}^{-1}\hat{\mathcal{C}}|0\rangle = (-1)^n \hat{a}^{\dagger}_{k_1\lambda_1}\cdots\hat{a}^{\dagger}_{k_n\lambda_n}|0\rangle \\ &= (-1)^n|n\gamma\rangle\,, \end{aligned} \qquad (10.130)$$

where it was assumed that the vacuum has even charge parity, $\hat{\mathcal{C}}|0\rangle = |0\rangle$. If a theory preserves C parity (which is true, e.g., for quantum electrodynamics) then (10.130) can lead to a selection rule for the decay of a system into photons (see Example 10.2).

As a further application of (10.130) we can deduce a version of *Furry's theorem*: QED does not allow the transition between states consisting of an even and an odd number of photons. The charge conjugation invariance of the $\hat{S}$ operator (cf. Sect. 10.5) implies that the S matrix element

$$\langle n'\gamma|\hat{S}|n\gamma\rangle = \langle n'\gamma|\hat{\mathcal{C}}^{-1}\hat{S}\hat{\mathcal{C}}|n\gamma\rangle = (-1)^{n+n'}\langle n'\gamma|\hat{S}|n\gamma\rangle \qquad (10.131)$$

vanishes if $n + n'$ is odd. In its general form (see *Quantum Electrodynamics*, Chap. 4) Furry's theorem holds also for virtual photons: the perturbation series of QED does not contain contributions from Feynman graphs with fermion loops having an odd number of vertices.

EXAMPLE

10.2 The Classification of Positronium States

The positronium system can serve as a showcase to illustrate the use of discrete symmetries for the classification of quantum states and for deducing selection rules. The bound system of an electron and a positron has energy levels that essentially agree with the hydrogen atom (scaled down by a factor of $1/2$ because of the reduced mass $m_{\text{red}} = m/2$). Contrary to the case for hydrogen, however, since positronium is composed of a particle and its antiparticle its states can be classified by the eigenvalues of charge parity. As a consequence certain decays of positronium into photons are strictly forbidden, as we will derive now.

It is very difficult to find an exact description of the bound states of the e^+e^- system since this would amount to solving the relativistic two-body problem. An equation can be formulated that in principle is exact (the Bethe–Salpeter equation; see *Quantum Electrodynamics*, Chap. 6) but this equation

10.4 The Electromagnetic Field

Example 10.2

cannot be solved exactly. Fortunately, a classification of the states can be achieved without the need for an exact solution.

According to the general principle of Lorentz invariance the state vector of positronium $|\text{Ps}\rangle$ can be classified by the eigenvalues of the operators $\hat{P}_\mu$, $\hat{J}^2$, and $\hat{J}_z$. In addition we have the operators of space inversion $\hat{\mathcal{P}}$ and charge conjugation $\hat{\mathcal{C}}$, which commute with the set of kinematic operators. The corresponding eigenvalues are

$$\hat{\mathcal{P}}|\text{Ps}\rangle = \pi_\text{P}|\text{Ps}\rangle \quad , \quad \hat{\mathcal{C}}|\text{Ps}\rangle = \pi_\text{C}|\text{Ps}\rangle \qquad (1)$$

where $\pi_\text{P} = \pm 1$ denotes space parity and $\pi_\text{C} = \pm 1$ denotes charge parity. Now we make the following ansatz for the state vector of positronium in the center-of-mass system ($\boldsymbol{P} = 0$):

$$|\text{Ps}\rangle = \int d^3p \sum_{ss'} R(\boldsymbol{p};s,s') \hat{b}^\dagger_{\boldsymbol{p},s} \hat{d}^\dagger_{-\boldsymbol{p},s'} |0\rangle , \qquad (2)$$

where $R(\boldsymbol{p};s,s')$ is the wave function in momentum space. Here s and s' are the projection of the electron and positron spins onto the z axis. The state (2) contains *one* electron–positron pair with a combined momentum of zero and is an approximation to the true bound state since the particle number in QED is not a conserved quantiy. In general higher multi-pair configurations with total charge zero, such as $\hat{b}^\dagger\hat{b}^\dagger\hat{d}^\dagger\hat{d}^\dagger|0\rangle$ etc., can contribute to the state vector. In positronium, however, such admixtures are very small, owing to the essentially nonrelativistic nature of this system. In any case, for the purpose of classifying the bound states it is sufficient to use the ansatz (2); any complicated higher-order admixtures would have the same symmetry properties.

Let us study the action of the parity operator on the state (2). Using (10.68) we find

$$\begin{aligned}
\hat{\mathcal{P}}|\text{Ps}\rangle &= \int d^3p \sum_{ss'} R(\boldsymbol{p};s,s') \hat{\mathcal{P}}\hat{b}_{\boldsymbol{p},s}\hat{\mathcal{P}}^{-1} \hat{\mathcal{P}}\hat{d}^\dagger_{-\boldsymbol{p},s'}\hat{\mathcal{P}}^{-1} \hat{\mathcal{P}}|0\rangle \\
&= \int d^3p \sum_{ss'} R(\boldsymbol{p};s,s') \eta_\text{P}^* \hat{b}^\dagger_{-\boldsymbol{p},s} (-\eta_\text{P})\hat{d}^\dagger_{\boldsymbol{p},s'} |0\rangle \\
&= \int d^3p \sum_{ss'} R_\text{P}(\boldsymbol{p};s,s') \hat{b}^\dagger_{\boldsymbol{p},s} \hat{d}^\dagger_{-\boldsymbol{p},s'} |0\rangle ,
\end{aligned} \qquad (3)$$

with the parity-transformed wave function

$$R_\text{P}(\boldsymbol{p};s,s') = -R(-\boldsymbol{p};s,s') . \qquad (4)$$

The minus sign has it origin in the negative intrinsic parity of the antiparticle. Similarly for the charge conjugate state we find, using (10.84),

$$\begin{aligned}
\hat{\mathcal{C}}|\text{Ps}\rangle &= \int d^3p \sum_{ss'} R(\boldsymbol{p};s,s') \hat{\mathcal{C}}\hat{b}^\dagger_{\boldsymbol{p},s}\hat{\mathcal{C}}^{-1} \hat{\mathcal{C}}\hat{d}^\dagger_{-\boldsymbol{p},s'}\hat{\mathcal{C}}^{-1} \hat{\mathcal{C}}|0\rangle \\
&= \int d^3p \sum_{ss'} R(\boldsymbol{p};s,s') \hat{d}^\dagger_{\boldsymbol{p},s} \hat{b}^\dagger_{-\boldsymbol{p},s'} |0\rangle \\
&= \int d^3p \sum_{ss'} R_\text{C}(\boldsymbol{p};s,s') \hat{b}^\dagger_{\boldsymbol{p},s} \hat{d}^\dagger_{-\boldsymbol{p},s'} |0\rangle .
\end{aligned} \qquad (5)$$

Example 10.2

Taking into account the anticommutation property of the fermion operators, we find the charge-conjugate wave function to be

$$R_C(\boldsymbol{p}; s, s') = -R(-\boldsymbol{p}; s', s) \ . \tag{6}$$

The spins of electrons and positron are exchanged and the direction of the momentum is inverted. To understand the consequences of (4) and (6) we employ the *nonrelativistic approximation* to the wave function for simplicity. This allows us to factorize the wave function into its orbital and spin parts:

$$R(\boldsymbol{p}; s, s') = R(p) Y_{Lm_L}(\Omega_{\boldsymbol{p}}) \chi^S_{m_S}(s, s') \ , \tag{7}$$

where $\chi^S_{m_S}(s, s')$ denotes the spin eigenfunction, which is either a triplet ($S = 0$) or a singlet ($S = 1$). If we denote the (nonrelativistic) unit spinors for electrons and positrons by χ^- and χ^+ the eigenfunctions with good total spin are given by

$$\chi_0^{S=0} = \frac{1}{\sqrt{2}}(\chi_\uparrow^- \chi_\downarrow^+ - \chi_\downarrow^- \chi_\uparrow^+) \ , \tag{8a}$$

$$\chi^{S=1}_{m_S} = \begin{cases} \chi_\uparrow^- \chi_\uparrow^+ & m_S = +1 \ , \\ \frac{1}{\sqrt{2}}(\chi_\uparrow^- \chi_\downarrow^+ + \chi_\downarrow^- \chi_\uparrow^+) & \text{for} \quad m_S = 0 \ , \\ \chi_\downarrow^- \chi_\downarrow^+ & m_S = -1 \ . \end{cases} \tag{8b}$$

Since space inversion for the spherical harmonics implies $Y_{LM}(\Omega_{-\boldsymbol{p}}) = (-1)^L Y_{LM}(\Omega_{\boldsymbol{p}})$, so the parity-transformed wave function (4) is given by

$$R_P(\boldsymbol{p}; s, s') = -(-1)^L R(\boldsymbol{p}; s, s') \ . \tag{9}$$

Thus the positronium states indeed are parity eigenstates characterized by the eigenvalue

$$\pi_P = -(-1)^L \ . \tag{10}$$

The spin wave function (8) is either even or odd under spin exchange:

$$\chi^S_{m_S}(s, s') = (-1)^{S+1} \chi^S_{m_S}(s, s') \ . \tag{11}$$

Therefore (6) reads

$$R_C(\boldsymbol{p}; s, s') = -(-1)^{S+1}(-1)^L R(\boldsymbol{p}; s, s') \ , \tag{12}$$

and the *charge parity* of positronium is given by

$$\pi_C = (-1)^{L+S} \ . \tag{13}$$

We can also define the combined space and charge (PC) parity, which has the eigenvalue

$$\pi_{PC} = -(-1)^S \ . \tag{14}$$

Since quantum electrodynamics is PC invariant the states can be classified according to π_{PC} instead of via the eigenvalues of $\hat{S}^2$. Therefore the distinction between singlet and triplet states remains strictly valid also in the relativistic theory, although spin and angular momentum here are intermingled.

Table 10.1 displays the eigenvalues of angular momentum and parity for various positronium states. The first column contains the traditional spectroscopic notation $^{2S+1}L_J$.

Table 10.1. Classification of the lowest states of positronium. Left part: singlet states, right part: triplet states.

Example 10.2

Parapositronium					Orthopositronium				
$^{2S+1}L_J$	J	π_P	π_C	π_{PC}	$^{2S+1}L_J$	J	π_P	π_C	π_{PC}
$^1S_0$	0	-1	$+1$	-1	$^3P_0$	0	$+1$	$+1$	$+1$
$^1P_1$	1	$+1$	-1	-1	$^3S_1 + {}^3D_1$	1	-1	-1	$+1$
$^1D_2$	2	-1	$+1$	-1	$^3P_1$	1	$+1$	$+1$	$+1$
$^1F_3$	3	$+1$	-1	-1	$^3P_2 + {}^3F_2$	2	$+1$	$+1$	$+1$
$^1G_4$	4	-1	$+1$	-1	$^3D_2$	2	-1	-1	$+1$

The left half of the table refers to the singlet states (parapositronium, $S = 0$), the right half to the triplet states (orthopositronium, $S = 1$). Inspecting the orthopositronium states we note that some of the entries, such as $^3S_1$ and $^3D_1$, share the same set of quantum numbers J, π_P, π_C. Since the angular momentum L is not a strictly conserved quantity these states will be mixed.

The C invariance of the electromagnetic interaction leads to a *selection rule for the n-photon decay* of positronium. We know from (10.130) that the multi-photon system has charge parity

$$\hat{C} |n\gamma\rangle = (-1)^n |n\gamma\rangle . \tag{15}$$

This value has to match with the positronium charge parity (13):

$$(-1)^{L+S} = (-1)^n . \tag{16}$$

Therefore the singlet states can decay into an even (odd) number of photons only if L is even (odd), whereas the triplet state shows the opposite behavior. In particular the ground state of positronium (1^1S_0) decays into two photons, and the metastable lowest orthopositronium state (1^3S_1) decays into three photons, a process which is strongly suppressed by an additional factor of the coupling constant α. We quote the corresponding lifetime values:

$$T_{1^1S_0 \to 2\gamma} = 1.25 \times 10^{-10} \text{ s} \quad , \quad T_{1^3S_1 \to 3\gamma} = 1.4 \times 10^{-7} \text{ s} . \tag{17}$$

Remark: The two-photon decay of the state $^3S_1$ would be forbidden even if charge parity were not conserved. The *Landau–Yang rule*[2] tells us that two photons cannot form a state with a total angular momentum $J = 1$. The absence of $^3D_2 \not\to 2\gamma$ transitions, however, is founded solely on the C-parity argument.

---

[2] L.D. Landau: Dokl. Akad. Nauk. **60**, 207 (1948); C.N. Yang: Phys. Rev. **77**, 242 (1950). Basic idea of the proof: a vector state (spin 1) constructed out of two photons can be characterized by the cross product of the polarization vectors $\epsilon(k_1, \lambda_1) \times \epsilon(k_2, \lambda_2)$. However, this is inconsistent with the fact that photons satisfy Bose statistics which demands that the state has to be *symmetric* under the exchange $k_1, \lambda_1 \leftrightarrow k_2, \lambda_2$.

10.5 Invariance of the S Matrix

If a theory is invariant under a symmetry transformation $\hat{\mathcal{U}}$ this will be reflected in the properties the scattering matrix. An invariance is present if the asymptotic field operators of the interacting theory $\hat{\phi}_{\text{in/out}}(x)$, which were introduced in Chap. 9 follow the same transformation law as do the free fields:

$$\hat{\mathcal{U}}\,\hat{\phi}_{\text{in/out}}(x)\,\hat{\mathcal{U}}^{-1} = \Lambda\,\hat{\phi}_{\text{in/out}}(x') \;. \tag{10.132}$$

Here the generic notation $\hat{\mathcal{U}}$ is meant to encompass Poincaré transformations, space reflection $\hat{\mathcal{P}}$, charge conjugation $\hat{\mathcal{C}}$, etc. x' denotes the transformed coordinate, Λ is a c-number-valued matrix acting in the space of the field components. Now let us study the implications of (10.132) for the $\hat{S}$ operator, which according to (9.11) connects the in and out states:

$$\hat{\phi}_{\text{out}}(x) = \hat{S}^{-1}\hat{\phi}_{\text{in}}(x)\hat{S} \;. \tag{10.133}$$

To deduce the transformation properties of $\hat{S}$ we apply $\hat{\mathcal{U}}$ to (10.133):

$$\hat{\mathcal{U}}\,\hat{\phi}_{\text{out}}(x)\,\hat{\mathcal{U}}^{-1} = \hat{\mathcal{U}}\,\hat{S}^{-1}\hat{\phi}_{\text{in}}(x)\,\hat{S}\,\hat{\mathcal{U}}^{-1} \;. \tag{10.134}$$

The left-hand side can be written as

$$\hat{\mathcal{U}}\,\hat{\phi}_{\text{out}}(x)\,\hat{\mathcal{U}}^{-1} = \Lambda\hat{\phi}_{\text{out}}(x') = \Lambda\hat{S}^{-1}\hat{\phi}_{\text{in}}(x')\hat{S} \;, \tag{10.135}$$

and the right-hand side, with the use of (10.132), becomes

$$\begin{aligned}\hat{\mathcal{U}}\,\hat{S}^{-1}\hat{\phi}_{\text{in}}(x)\,\hat{S}\,\hat{\mathcal{U}}^{-1} &= (\hat{\mathcal{U}}\hat{S}^{-1}\hat{\mathcal{U}}^{-1})\,\hat{\mathcal{U}}\hat{\phi}_{\text{in}}(x)\hat{\mathcal{U}}^{-1}\,(\hat{\mathcal{U}}\hat{S}\hat{\mathcal{U}}^{-1}) \\ &= (\hat{\mathcal{U}}\hat{S}^{-1}\hat{\mathcal{U}}^{-1})\,\Lambda\hat{\phi}_{\text{in}}(x')\,(\hat{\mathcal{U}}\hat{S}\hat{\mathcal{U}}^{-1}) \;. \end{aligned} \tag{10.136}$$

The matrix Λ can be factored out because it is not a Hilbert-space operator. Therefore the transformed version of (10.133) leads to the condition

$$\hat{S}^{-1}\hat{\phi}_{\text{in}}(x')\hat{S} = (\hat{\mathcal{U}}\hat{S}^{-1}\hat{\mathcal{U}}^{-1})\,\hat{\phi}_{\text{in}}(x')\,(\hat{\mathcal{U}}\hat{S}\hat{\mathcal{U}}^{-1}) \;. \tag{10.137}$$

This equation is satisfied if the operator $\hat{S}$ transforms according to

$$\hat{S} = \hat{\mathcal{U}}\,\hat{S}\,\hat{\mathcal{U}}^{-1} \;. \tag{10.138}$$

Phrased differently, the operator $\hat{S}$ has to commute with the transformation operator:

$$[\hat{S},\hat{\mathcal{U}}] = 0 \;. \tag{10.139}$$

This has consequences for the matrix elements of $\hat{S}$. Using (10.138) in the equivalent form $\hat{S} = \hat{\mathcal{U}}^\dagger \hat{S}\hat{\mathcal{U}}$ we get the condition

$$\begin{aligned} S_{\beta\alpha} &= \langle\beta;\text{in}|\hat{S}|\alpha;\text{in}\rangle = \langle\beta;\text{in}|\hat{\mathcal{U}}^\dagger\hat{S}\hat{\mathcal{U}}|\alpha;\text{in}\rangle = \langle\beta';\text{in}|\hat{S}|\alpha';\text{in}\rangle \\ &= S_{\beta'\alpha'} \equiv S_{U\beta,U\alpha} \;. \end{aligned} \tag{10.140}$$

The S matrix element (and thus the transition probability) has the same value for the transitions $\alpha \to \beta$ and $\alpha' \to \beta'$ where the primed states are obtained by applying the transformation $\hat{\mathcal{U}}$ to the unprimed states. If the operator $\hat{\mathcal{U}}$

is not only unitary but also hermitean,[3] $\hat{\mathcal{U}}^\dagger = \hat{\mathcal{U}}$, (10.139) leads to a *selection rule* with respect to the U quantum number. If α and β are eigenstates of $\hat{\mathcal{U}}$ with real-valued (because of the postulated hermitecity) eigenvalues λ_α and λ_β,

$$\hat{\mathcal{U}}\,|\alpha;\text{in}\rangle = \lambda_\alpha\,|\alpha;\text{in}\rangle \quad , \quad \hat{\mathcal{U}}\,|\beta;\text{in}\rangle = \lambda_\beta\,|\beta;\text{in}\rangle, \qquad (10.141)$$

then the following condition holds:

$$\begin{aligned}\langle\beta;\text{in}|\hat{S}\hat{\mathcal{U}}|\alpha;\text{in}\rangle &= \lambda_\alpha S_{\beta\alpha} \\ &= \langle\beta;\text{in}|\hat{\mathcal{U}}\hat{S}|\alpha;\text{in}\rangle = \langle\beta;\text{in}|\hat{\mathcal{U}}^\dagger \hat{S}|\alpha;\text{in}\rangle \\ &= \lambda_\beta S_{\beta\alpha}\,. \end{aligned} \qquad (10.142)$$

From this we conclude that S matrix elements between states of different symmetry ($\lambda_\alpha \neq \lambda_\beta$) will vanish.

The *time-reversal transformation* $\hat{\mathcal{U}} = \hat{\mathcal{T}}$ calls for a separate treatment, owing to its antiunitarity. When the operator $\hat{\mathcal{T}}$ is applied, the in and out states are exchanged:

$$|\alpha';\text{out}\rangle = \hat{\mathcal{T}}|\alpha;\text{in}\rangle \quad , \quad |\beta';\text{out}\rangle = \hat{\mathcal{T}}|\beta;\text{in}\rangle. \qquad (10.143)$$

Then the condition for the invariance of the operator $\hat{S}$ under the antiunitary transformation $\hat{\mathcal{T}}$ instead of (10.138) becomes (cf. (10.49))

$$\hat{S} = \left(\hat{\mathcal{T}}\,\hat{S}\,\hat{\mathcal{T}}^{-1}\right)^\dagger. \qquad (10.144)$$

This leads to

$$\begin{aligned}\langle\beta;\text{in}|\hat{S}|\alpha;\text{in}\rangle &= \langle\beta;\text{in}|\hat{\mathcal{T}}^{-1}\left(\hat{\mathcal{T}}\hat{S}\hat{\mathcal{T}}^{-1}\right)\hat{\mathcal{T}}|\alpha;\text{in}\rangle = \langle\beta;\text{in}|\hat{\mathcal{T}}^{-1}\hat{S}^\dagger \hat{\mathcal{T}}|\alpha;\text{in}\rangle \\ &= \langle\beta';\text{out}|\hat{S}^\dagger|\alpha';\text{out}\rangle = \langle\alpha';\text{out}|\hat{S}|\beta';\text{out}\rangle \\ &= \langle\alpha';\text{in}|\hat{S}|\beta';\text{in}\rangle,\end{aligned} \qquad (10.145)$$

and finally

$$S_{\beta\alpha} = S_{\alpha'\beta'} \equiv S_{T\alpha,T\beta}. \qquad (10.146)$$

This condition is known as the *reciprocity theorem*, which relates the transiton $\alpha \to \beta$ to the reverse transition between the time-mirrored states $T\beta \to T\alpha$. A time-mirrored stated is obtained by inverting the momenta and spins of all particles. Also a phase factor depending on the values of the angular momentum may be encountered here.

The discussion concering selection rules (see (10.141) and (10.142)) does not apply to time-reversal transformations. An immediate obstacle is the observation that the antiunitary operator $\hat{\mathcal{T}}$ cannot possess eigenstates as postulated in (10.141). This is so because if we assume

$$\hat{\mathcal{T}}\,|\alpha;\text{in}\rangle = \lambda_\alpha\,|\alpha;\text{in}\rangle, \qquad (10.147)$$

then also each state $c|\alpha;\text{in}\rangle$ with an arbitrary complex phase factor c would be an admissible eigenstate. However, this leads to a contradiction since

---

[3] For the operators of space inversion and charge conjugation the phases η_P, η_C can be chosen such that the conditions $\hat{\mathcal{P}}^2 = 1$, $\hat{\mathcal{C}}^2 = 1$ are satisfied. For the case of continuous symmetries an analogous reasoning applies if instead of $\hat{\mathcal{U}}$ the hermitean generator $\hat{G}$ defined in (4.77) is employed.

$$\hat{T}\,c\,|\alpha;\text{in}\rangle = c^*\hat{T}\,|\alpha;\text{in}\rangle = c^*\lambda_\alpha\,|\alpha;\text{in}\rangle \neq c\lambda_\alpha\,|\alpha;\text{in}\rangle\,. \tag{10.148}$$

Therefore the concept of a "temporal parity", which might lead to new selection rules does not exist.

To illustrate the types of processes that are related to each other through the discrete symmetries let us use an example taken from strong-interaction physics, namely the charge-exchange reaction of a proton and a negative pion,

$$p + \pi^- \to n + \pi^0$$

Table 10.2 shows how the original momenta, the z components of angular momentum, and the particle types are affected by the application of the transformations P, C, T and the combined transformation CPT.

Table 10.2. Transformation properties of spin and momentum in the reaction $p + \pi^- \to n + \pi^0$.

	$p(\boldsymbol{p}, s_3)$	+	$\pi^-(\boldsymbol{k})$	$\to$	$n(\boldsymbol{p}', s_3')$	+	$\pi^0(\boldsymbol{k}')$
P	$p(-\boldsymbol{p}, s_3)$	+	$\pi^-(-\boldsymbol{k})$	$\to$	$n(-\boldsymbol{p}', s_3')$	+	$\pi^0(-\boldsymbol{k}')$
C	$\bar{p}(\boldsymbol{p}, s_3)$	+	$\pi^+(\boldsymbol{k})$	$\to$	$\bar{n}(\boldsymbol{p}', s_3')$	+	$\pi^0(\boldsymbol{k}')$
T	$n(-\boldsymbol{p}', -s_3')$	+	$\pi^0(-\boldsymbol{k}')$	$\to$	$p(-\boldsymbol{p}, -s_3)$	+	$\pi^-(-\boldsymbol{k})$
CPT	$\bar{n}(\boldsymbol{p}', -s_3')$	+	$\pi^0(\boldsymbol{k}')$	$\to$	$\bar{p}(\boldsymbol{p}, -s_3)$	+	$\pi^+(\boldsymbol{k})$

Since the three discrete symmetries are conserved by the strong interaction, the S matrix elements and thus the scattering cross sections (except for a possible modification of the kinematic factor) agree for the various processes shown in the table. Note that in reality there will be small differences caused by the symmetry-breaking weak interaction. For the CPT mirrored reaction, however, the agreement is exact, as we will learn in the next section.

10.6 The CPT Theorem

The three discrete transformations studied in this chapter, i.e., space inversion, time reversal, and charge conjugation, are exact symmetries for *free* quantum fields. Whether this remains true in the presence of interactions is not known a priori; this will depend on the detailed structure of the interaction term. Naively one might expect that the discrete symmetries are as fundamental as the condition of Lorentz invariance and that they therefore should hold in every physical theory. The extremely well studied case of quantum electrodynamics appears to support this view. Thus the "fall of parity" in 1957 came as a great surprise: experiments revealed that the electrons emitted in beta decay predominantly have left-handed helicity. Today we know that all of the three symmetries (P, C, T) are broken in nature. We have to accept the fact that the distinction between left and right, between particle and antiparticle, and between motion forward and backward in time has an absolute meaning on a microsopic level. Here we will not be concerned with detailed examples of consequences of this symmetry breaking. Any book on elementary particle physics or on gauge theory can be consulted.

Whether one of the discrete symmetries is conserved or broken has to be determined by experiment; field theory can incorporate both alternatives equally well. There is one symmetry, however, that is deeply entrenched in the formalism and constitutes one of the most fundamental predictions of quantum field theory. To formulate this statement we introduce the combined transformation consisting of time reversal, charge conjugation, and space inversion:

$$\hat{\Theta} = \hat{\mathcal{C}}\hat{\mathcal{P}}\hat{\mathcal{T}} \ . \tag{10.149}$$

The operator $\hat{\Theta}$ is antiunitary (because of the complex conjugation contained in $\hat{\mathcal{T}}$). It turns out that $\hat{\Theta}$ describes a symmetry for any "acceptable" field theory.

The CPT Theorem

For any *local* quantum field theory that can be described by a *hermitean-* and *Lorentz-invariant* Lagrangian $\hat{\mathcal{L}}(x)$ and whose field operators satisfy the *spin-statistics theorem* the following relation holds true:

$$\hat{\Theta}\,\hat{\mathcal{L}}(x)\,\hat{\Theta}^{-1} = \hat{\mathcal{L}}(-x) \ . \tag{10.150}$$

The action integral, the field equations, and also the canonical commutation relations are invariant under the transformation $\hat{\Theta}$. The condition (10.150) implies that also the Hamiltonian is invariant:

$$\hat{\Theta}\,\hat{H}\,\hat{\Theta}^{-1} = \hat{H} \ . \tag{10.151}$$

Thus $\hat{\Theta}$ is a symmetry transformation.

The CPT theorem (note that the ordering of the operations is not important, the names TCP or PCT theorem are also in use) first was formulated by J. Schwinger and B. Zumino. The proof dates back to G. Lüders and W. Pauli[4]. Later the CPT theorem also was studied carefully within the framework of axiomatic field theory[5] and still weaker conditions have been found. But even in the form stated above the conditions are already weak enough. One would hardly like to give up any of them when constructing a theory. Therefore the CPT symmetry must be of general validity. To avoid any ambiguity let us point out that the required Lorentz invariance of course refers to *proper* Lorentz transformation which do not contain space and time reversals.

We will not present a rigorous proof of the CPT theorem (see the quoted references) but we will illustrate its principle. Let us first collect and combine the information on the behavior of the field operators under $\hat{\mathcal{T}}$, $\hat{\mathcal{P}}$, and $\hat{\mathcal{C}}$ transformations which were derived in the previous sections.

Scalar field:

$$\begin{aligned}
\hat{\Theta}\hat{\phi}(x)\hat{\Theta}^{-1} &= \hat{\mathcal{C}}\hat{\mathcal{P}}\hat{\mathcal{T}}\,\hat{\phi}(\boldsymbol{x},t)\,\hat{\mathcal{T}}^{-1}\hat{\mathcal{P}}^{-1}\hat{\mathcal{C}}^{-1} \\
&= \eta_\mathrm{T}\,\hat{\mathcal{C}}\hat{\mathcal{P}}\,\hat{\phi}(\boldsymbol{x},-t)\,\hat{\mathcal{P}}^{-1}\hat{\mathcal{C}}^{-1} = \eta_\mathrm{P}\eta_\mathrm{T}\,\hat{\mathcal{C}}\,\hat{\phi}(-\boldsymbol{x},-t)\,\hat{\mathcal{C}}^{-1} \\
&= \eta_\mathrm{C}\eta_\mathrm{P}\eta_\mathrm{T}\,\hat{\phi}^\dagger(-x) \ .
\end{aligned} \tag{10.152}$$

---

[4] W. Pauli in: *Niels Bohr and the Development of Physics*, (Pergamon Press, London 1955) p. 30; G. Lüders: Ann. Phys. (NY) **2**, 1 (1957).

[5] R. Jost: Helv. Phys. Acta **30**, 409 (1957);
see also the book by R.F. Streater and A.S. Wightman, quoted on page 269.

The freedom of choice for the phase factors allows us to use

$$\eta_C \eta_P \eta_T = 1 \,. \tag{10.153}$$

Electromagnetic field:

$$\begin{aligned}
\hat{\Theta} \hat{A}^\mu(x) \hat{\Theta}^{-1} &= \hat{C}\hat{P} \hat{A}_\mu(\boldsymbol{x},-t) \hat{P}^{-1}\hat{C}^{-1} = \hat{C} \hat{A}^\mu(-\boldsymbol{x},-t)\hat{C}^{-1} \\
&= -\hat{A}^\mu(-x) \,.
\end{aligned} \tag{10.154}$$

Other vector fields are treated in the same way.

Spin-$\frac{1}{2}$ field:

$$\begin{aligned}
\hat{\Theta} \hat{\psi}(x) \hat{\Theta}^{-1} &= \hat{C}\hat{P} T_0 \hat{\psi}(\boldsymbol{x},-t) \hat{P}^{-1}\hat{C}^{-1} = \hat{C} T_0 \gamma^0 \hat{\psi}(-\boldsymbol{x},-t)\hat{C}^{-1} \\
&= T_0 \gamma^0 C \hat{\bar{\psi}}^{\mathrm{T}}(-x) = -\mathrm{i}\gamma^5 \gamma^0 \hat{\bar{\psi}}^{\mathrm{T}}(-x) \,,
\end{aligned} \tag{10.155}$$

where the last step has made use of the explicit expressions for the matrices T_0 and C in the standard representation. As a corollary to (10.155) we have

$$\hat{\Theta} \hat{\bar{\psi}}(x) \hat{\Theta}^{-1} = \hat{\psi}^{\mathrm{T}}(-x) \, \mathrm{i}\gamma^5 \gamma^0 \,. \tag{10.156}$$

In studies of the spin-$\frac{1}{2}$ field it is essential that a Lorentz-invariant theory can depend only on *bilinear covariant* expressions (see *RQM*, Chap. 5). The Lagrangian of the theory will be built up from these quantities, which are Lorentz tensors of rank 0, 1, or 2 (scalar, vector, or tensor). It is a rather simple task to derive the transformation properties of the bilinear covariants. Let us define

$$\text{scalar:} \quad \hat{S}_{ba}(x) = \frac{1}{2}\big[\hat{\bar{\psi}}_b(x), \hat{\psi}_a(x)\big] \,, \tag{10.157a}$$

$$\text{pseudo scalar:} \quad \hat{P}_{ba}(x) = \frac{1}{2}\big[\hat{\bar{\psi}}_b(x), \mathrm{i}\gamma^5 \hat{\psi}_a(x)\big] \,, \tag{10.157b}$$

$$\text{vector:} \quad \hat{V}^\mu_{ba}(x) = \frac{1}{2}\big[\hat{\bar{\psi}}_b(x), \gamma^\mu \hat{\psi}_a(x)\big] \,, \tag{10.157c}$$

$$\text{axial vector:} \quad \hat{P}^\mu_{ba}(x) = \frac{1}{2}\big[\hat{\bar{\psi}}_b(x), \gamma^5 \gamma^\mu \hat{\psi}_a(x)\big] \,, \tag{10.157d}$$

$$\text{tensor:} \quad \hat{T}^{\mu\nu}_{ba}(x) = \frac{1}{2}\big[\hat{\bar{\psi}}_b(x), \sigma^{\mu\nu} \hat{\psi}_a(x)\big] \,, \tag{10.157e}$$

The indices a, b denote the fermion type, such as e, μ, ν, quark flavors, etc. This will be important for interactions that allow transmutation among different particle types. Within the second-quantized theory the ordering of operators has to be fixed. In (10.157) we have employed the *antisymmetrization* prescription of the field operators for the purpose of eliminating contributions from the "Dirac sea". The same effect can also be achieved by the prescription of normal ordering. We have shown this equivalence in connection with (10.87) for the example of the vector $\hat{V}^\mu_{aa}(x)$ (which is just the electromagnetic current operator $\hat{j}^\mu(x)$).

The transformation properties of the bilinear covariants are displayed in Table 10.3; for a proof see Exercise 10.3. The results for the single discrete transformations $\hat{C}, \hat{P}, \hat{T}$ and for the combined transformation $\hat{\Theta}$ are given. The abbreviation $\tilde{x}$ stands for $\tilde{x} = (x_0, -\boldsymbol{x})$.

The last column in this table points to a very favorable fact: the transformation $\hat{\Theta}$ has the ability to reproduce the original operator except for the

Table 10.3. Transformation properties of the bilinear covariants under charge conjugation, space inversion, time reversal, and the combined CPT operation

	$\hat{\mathcal{C}}$	$\hat{\mathcal{P}}$	$\hat{\mathcal{T}}$	$\hat{\Theta} = \hat{\mathcal{C}}\hat{\mathcal{P}}\hat{\mathcal{T}}$
$\hat{S}_{ba}(x)$	$+\hat{S}_{ab}(x)$	$+S_{ba}(\tilde{x})$	$+\hat{S}_{ba}(-\tilde{x})$	$+\hat{S}_{ab}(-x)$
$\hat{P}_{ba}(x)$	$+\hat{P}_{ab}(x)$	$-P_{ba}(\tilde{x})$	$-\hat{P}_{ba}(-\tilde{x})$	$+\hat{P}_{ab}(-x)$
$\hat{V}^\mu_{ba}(x)$	$-\hat{V}^\mu_{ab}(x)$	$+V^{ba}_\mu(\tilde{x})$	$+\hat{V}^{ba}_\mu(-\tilde{x})$	$-\hat{V}^\mu_{ab}(-x)$
$\hat{P}^\mu_{ba}(x)$	$+\hat{P}^\mu_{ab}(x)$	$-P^{ba}_\mu(\tilde{x})$	$+\hat{P}^{ba}_\mu(-\tilde{x})$	$-\hat{P}^\mu_{ab}(-x)$
$\hat{T}^{\mu\nu}_{ba}(x)$	$-\hat{T}^{\mu\nu}_{ab}(x)$	$+T^{ba}_{\mu\nu}(\tilde{x})$	$-\hat{T}^{ba}_{\mu\nu}(-\tilde{x})$	$+\hat{T}^{\mu\nu}_{ab}(-x)$

interchanges $a \leftrightarrow b$ and $x \leftrightarrow -x$. In addition there is a sign factor: $+1$ *for tensors of even rank* and -1 *for tensors of odd rank*. This is the key to the proof of the CPT theorem. The Lagrangian of any field theory has to be invariant under (proper) Lorentz transformations, i.e., it is a Lorentz scalar. Therefore it will consist of contractions of Lorentz tensors such that the indices μ etc. will always enter twice and the minus signs associated with odd tensors will cancel each other.

Although up to now we have considered only the bilinear covariants derived from Dirac fields the argument also is valid if $\hat{\mathcal{L}}(x)$ contains other ingredients. Because of the assumed locality of the theory, $\hat{\mathcal{L}}(x)$ will be a finite polynomial of the field operators, containing c numbers (the coupling constants) and possibly gradient operators ∂_μ as coefficients. Upon $\hat{\Theta}$ transformations the gradient produces a minus sign, $\partial_\mu \leftrightarrow -\partial_\mu$, which is just what we need to cancel the minus sign of the tensor that it is contracted. Also the vector field $\hat{A}_\mu(x)$ according to (10.154) complies with this rule.

These arguments essentially prove the CPT theorem (10.156). Let us add some further remarks. The interchange of the indices $a \leftrightarrow b$ does not affect the invariance property since $\hat{\mathcal{L}}$ is symmetrical with respect to these "flavor" indices, owing to the assumed hermitecity. For example the hermitean form of a "transition tensor" of the type (10.157) reads

$$\begin{aligned} :\hat{\bar{\psi}}_a M \hat{\psi}_b: + \text{h.c.} &= :\hat{\bar{\psi}}_a M \hat{\psi}_b: + :\hat{\bar{\psi}}_b \gamma^0 M^\dagger \gamma^0 \hat{\psi}_a: \\ &= :\hat{\bar{\psi}}_a M \hat{\psi}_b: + :\hat{\bar{\psi}}_b M \hat{\psi}_a: . \end{aligned} \quad (10.158)$$

All five bilinear covariants have the property $\gamma^0 M^\dagger \gamma^0 = M$.

Let us once more stress that the derivation of the CPT theorem relies on the correct statistics, i.e., the commutation of bosonic and anticommutation of fermionic field operators. For the Dirac fields this was used in the construction of the Table 10.3. To see that Bose fields have to commute we study the example of a bosonic transition vector current $\hat{W}^\mu(x) = :\hat{\phi}^\dagger_a(x)\partial^\mu \hat{\phi}_b(x): + \text{h.c.}$. Application of $\hat{\Theta}$ leads to

$$\begin{aligned} \hat{\Theta} \hat{W}^\mu(x) \hat{\Theta}^{-1} &= -:\hat{\phi}_a(-x)\partial^\mu \hat{\phi}^\dagger_b(-x): - :(\partial^\mu \hat{\phi}_b(-x))\hat{\phi}^\dagger_a(-x): \\ &= -:\partial^\mu \hat{\phi}^\dagger_b(-x)\hat{\phi}_a(-x): - :\hat{\phi}^\dagger_a(-x)\partial^\mu \hat{\phi}_b(-x): \\ &= -\hat{W}^\mu(-x) , \end{aligned} \quad (10.159)$$

provided that Bose statistics can be applied, i.e., that the exchange of the field operators under the normal product produces no sign change.

Exercise 10.3

EXERCISE

10.3 Transformation Rules for the Bilinear Covariants

Problem. Confirm the results of Table 10.2 for the transformation properties of the bilinear covariants under charge conjugation $\hat{\mathcal{C}}$, space inversion $\hat{\mathcal{P}}$, time reversal $\hat{\mathcal{T}}$, and the combined transformation $\hat{\Theta} = \hat{\mathcal{C}}\hat{\mathcal{P}}\hat{\mathcal{T}}$.

Solution. We have to transform the antisymmetrized bilinear forms

$$\hat{M}_{ba}(x) = \left[\hat{\bar{\psi}}_b(x),\, M\hat{\psi}_a(x)\right] = \hat{\bar{\psi}}_b(x) M \hat{\psi}_a(x) - \hat{\psi}_a^{\mathrm{T}}(x) M^{\mathrm{T}} \hat{\bar{\psi}}_b^{\mathrm{T}}(x) \tag{1}$$

for the five Dirac matrices $M = \mathbb{I}, i\gamma^5, \gamma^\mu, \gamma^5\gamma^\mu, \sigma^{\mu\nu} = \frac{i}{2}\left(\gamma^\mu\gamma^\nu - \gamma^\nu\gamma^\mu\right)$.

1. Charge Conjugation. Equations (10.76) and (10.77) lead to

$$\begin{aligned}
\hat{\mathcal{C}} \hat{M}_{ba}(x) \hat{\mathcal{C}}^{-1} &= -\hat{\psi}_b^{\mathrm{T}} C^\dagger M C \hat{\bar{\psi}}_a^{\mathrm{T}} + (C\hat{\bar{\psi}}_a^{\mathrm{T}})^{\mathrm{T}} M^{\mathrm{T}} (\hat{\psi}_b^{\mathrm{T}} C^\dagger)^{\mathrm{T}} \\
&= -\hat{\psi}_b^{\mathrm{T}} C^\dagger M C \hat{\bar{\psi}}_a^{\mathrm{T}} + \hat{\bar{\psi}}_a (C^\dagger M C)^{\mathrm{T}} \hat{\psi}_b^{\mathrm{T}} \\
&= \pm \hat{M}_{ab}(x),
\end{aligned} \tag{2}$$

assuming that

$$C^\dagger M C = C M C^{-1} = \pm M^{\mathrm{T}}. \tag{3}$$

That this relation is correct can be confirmed for each of the five cases. Using $C\gamma^5 = \gamma^5 C$, $\gamma^{5\mathrm{T}} = \gamma^5$ and (10.74) we find

$$\begin{aligned}
\text{scalar:} \quad & C \mathbb{I} C^{-1} = \mathbb{I} = +\mathbb{I}^{\mathrm{T}}, & (4\mathrm{a}) \\
\text{pseudo scalar:} \quad & C i\gamma^5 C^{-1} = i\gamma^5 = +(i\gamma^5)^{\mathrm{T}}, & (4\mathrm{b}) \\
\text{vector:} \quad & C \gamma^\mu C^{-1} = -\gamma^{\mu\mathrm{T}}, & (4\mathrm{c}) \\
\text{axial vector:} \quad & C \gamma^5 \gamma^\mu C^{-1} = -\gamma^5 \gamma^{\mu\mathrm{T}} = -(\gamma^\mu \gamma^5)^{\mathrm{T}} = +(\gamma^5 \gamma^\mu)^{\mathrm{T}}, & (4\mathrm{d}) \\
\text{tensor:} \quad & C \sigma^{\mu\nu} C^{-1} = \tfrac{i}{2}\left(\gamma^{\mu\mathrm{T}} \gamma^{\nu\mathrm{T}} - \gamma^{\nu\mathrm{T}} \gamma^{\mu\mathrm{T}}\right) \\
& \phantom{C \sigma^{\mu\nu} C^{-1}} = -\tfrac{i}{2}\left(\gamma^\mu \gamma^\nu - \gamma^\nu \gamma^\mu\right)^{\mathrm{T}}. & (4\mathrm{e})
\end{aligned}$$

This confirms the first column of the table.

2. Space Inversion. Using (10.63) and (10.64) we find

$$\begin{aligned}
\hat{\mathcal{P}} \hat{M}_{ba}(x) \hat{\mathcal{P}}^{-1} &= \hat{\bar{\psi}}_b(\tilde{x}) \gamma^0 M \gamma^0 \hat{\psi}_a(\tilde{x}) - \hat{\psi}_a^{\mathrm{T}}(\tilde{x}) (\gamma^0 M \gamma^0)^{\mathrm{T}} \hat{\bar{\psi}}_b^{\mathrm{T}}(\tilde{x}) \\
&= \pm \hat{M}_{ba}(\tilde{x}),
\end{aligned} \tag{5}$$

assuming that

$$\gamma^0 M \gamma^0 = \pm M, \tag{6}$$

which is confirmed by

$$\begin{aligned}
\text{scalar:} \quad & \gamma^0 \mathbb{I} \gamma^0 = +\mathbb{I}^{\mathrm{T}}, & (7\mathrm{a}) \\
\text{pseudo scalar:} \quad & \gamma^0 i\gamma^5 \gamma^0 = -i\gamma^5, & (7\mathrm{b}) \\
\text{vector:} \quad & \gamma^0 \gamma^\mu \gamma^0 = +\gamma_\mu, & (7\mathrm{c}) \\
\text{axial vector:} \quad & \gamma^0 \gamma^5 \gamma^\mu \gamma^0 = -\gamma^5 \gamma^0 \gamma^\mu \gamma^0 = -\gamma^5 \gamma_\mu, & (7\mathrm{d}) \\
\text{tensor:} \quad & \gamma^0 \sigma^{\mu\nu} \gamma^0 = \tfrac{i}{2}\left(\gamma_\mu \gamma_\nu - \gamma_\nu \gamma_\mu\right) = +\sigma_{\mu\nu}. & (7\mathrm{e})
\end{aligned}$$

3. *Time Reversal.* The relations (10.100) and (10.101) lead to

$$\hat{T}\hat{M}_{ba}(x)\hat{T}^{-1} = \hat{\bar{\psi}}_b(-\tilde{x})\, T_0^\dagger M^* T_0\, \hat{\psi}_a(-\tilde{x}) - \hat{\psi}_a^{\mathrm{T}}(-\tilde{x})(T_0^\dagger M^* T_0)^{\mathrm{T}}\hat{\bar{\psi}}_b^{\mathrm{T}}(-\tilde{x})$$
$$= \pm \hat{M}_{ba}(-\tilde{x})\,, \tag{8}$$

assuming that

$$T_0^\dagger M^* T_0 = T_0 M^* T_0^{-1} = \pm M\,. \tag{9}$$

The T_0 matrix satisfies $T_0 \gamma^{\mu *} T_0^{-1} = \gamma_\mu$, cf. (10.93), and $T_0 \gamma^5 = \gamma^5 T_0$. This leads to

$$\begin{aligned}
\text{scalar:} &\quad T_0 \mathbb{1}^* T_0^{-1} = +\mathbb{1}\,, &&(10\text{a}) \\
\text{pseudo scalar:} &\quad T_0 (\mathrm{i}\gamma^5)^* T_0^{-1} = -\mathrm{i}T_0 \gamma^5 T_0^{-1} = -\mathrm{i}\gamma^5\,, &&(10\text{b}) \\
\text{vector:} &\quad T_0 \gamma^{\mu *} T_0^{-1} = +\gamma_\mu\,, &&(10\text{c}) \\
\text{axial vector:} &\quad T_0 (\gamma^5 \gamma^\mu)^* T_0^{-1} = \gamma^5 T_0 \gamma^{\mu *} T_0^{-1} = +\gamma^5 \gamma_\mu\,, &&(10\text{d}) \\
\text{tensor:} &\quad T_0 \sigma^{\mu\nu *} T_0^{-1} = -\tfrac{\mathrm{i}}{2} T_0 (\gamma^\mu \gamma^\nu - \gamma^\nu \gamma^\mu)^* T_0^{-1} = -\sigma_{\mu\nu}\,. &&(10\text{e})
\end{aligned}$$

4. CPT Transformation. The results could be worked out explicitely using (10.155) and (10.156):

$$\hat{\Theta}\hat{\psi}(x)\hat{\Theta}^{-1} = -\mathrm{i}\gamma^5\gamma^0 \hat{\bar{\psi}}^{\mathrm{T}}(-x)\,,\quad \hat{\Theta}\hat{\bar{\psi}}(x)\hat{\Theta}^{-1} = \hat{\psi}^{\mathrm{T}}(-x)\,\mathrm{i}\gamma^5\gamma^0\,. \tag{11}$$

It is more convenient, however, simply to consecutively apply the operations of the first three columns of the Table 10.2.

EXAMPLE

10.4 The Relation Between Particles and Antiparticles

An important insight can be derived from the *CPT* theorem: the masses and lifetimes of particles and antiparticles are exactly equal. This statement is not as trivial as it may sound: the whole point of the discussion was to allow the violation of charge conjugation symmetry! In other aspects, such as their scattering processes, particles and antiparticles may well exhibit different behavior.

1. Mass Equality

A particle a in its rest frame is described as an eigenstate of the Hamiltonian $\hat{H}$ (including all interactions) and the angular-momentum operators $\hat{J}^2$ and $\hat{J}_z$ having the eigenvalues M_a, $j(j+1)$, and m. The mass M_a is determined by the relation

$$\begin{aligned}
M_a &= \langle a,m|\hat{H}|a,m\rangle = \langle a,m|\hat{\Theta}^{-1}\hat{\Theta}\hat{H}\hat{\Theta}^{-1}\hat{\Theta}|a,m\rangle \\
&= \langle a,m|\hat{\Theta}^{-1}\hat{H}\hat{\Theta}|a,m\rangle\,,
\end{aligned} \tag{1}$$

since the Hamiltonian is *CPT* invariant. Taking into account the antiunitary character of $\hat{\Theta}$ we can write this as (cf. (10.49))

$$M_a = \langle a_\theta, m|\hat{H}|a_\theta, m\rangle^* = \langle a_\theta, m|\hat{H}|a_\theta, m\rangle\,. \tag{2}$$

Example 10.4

When applied to an angular-momentum eigenstate the CPT operator leads to the interchange of particles and antiparticles and flips the angular-momentum projection $m \leftrightarrow -m$ (we have shown this only for the spins of the particles),

$$|a_\theta, m\rangle \equiv \hat{\Theta} |a, m\rangle = e^{i\theta} |\bar{a}, -m\rangle \,, \tag{3}$$

where the phase factor $e^{i\theta}$ need not to be specified in detail. Equation (2) then becomes

$$M_a = \langle \bar{a}, -m | \hat{H} | \bar{a}, -m \rangle = M_{\bar{a}} \,, \tag{4}$$

since owing to the isotropy of space the mass of a particle of course cannot depend on the sign of the angular-momentum projection.

2. Equality of Lifetimes

The treatment of lifetimes is more involved than the mass calculation; therefore the following discussion will be restricted to *first-order perturbation theory*. The Hamiltonian is split into two parts, $\hat{H} = \hat{H}_0 + \hat{H}_1$, where $\hat{H}_0$ generates a stable particle state which decays under the action of the small perturbation $\hat{H}_1$. The operator $\hat{H}_1$ typically might represent effects of the electromagnetic or the weak interaction.

The decay width (i.e., the inverse lifetime of the particle) in perturbation theory is given by "Fermi's golden rule", which contains the squared transition matrix element of the perturbation operator $\hat{H}_1$:

$$\Gamma_a = 2\pi \sum_b \delta(E_b - E_a) |\langle b; \text{out} | \hat{H}_1 | a, m \rangle|^2 \,. \tag{5}$$

Here $|a, m\rangle$ denotes the state vector of the decaying particle having magnetic quantum number m. $|b; \text{out}\rangle$ denotes a many-particle state satisfying the boundary condition of outgoing waves. The subsequent argument depends heavily on the fact that Γ_a is the *total decay width*, i.e., that (5) contains a summation (or integration) over all possible final states $|b, \text{out}\rangle$. The delta function ensures that energy is conserved in the decay process.

Now we assert that (5) agrees with

$$\Gamma_{\bar{a}} = 2\pi \sum_{\bar{b}} \delta(E_{\bar{b}} - E_{\bar{a}}) |\langle \bar{b}; \text{out} | \hat{H}_1 | \bar{a}, m \rangle|^2 \,. \tag{6}$$

To prove this assertion we insert the CPT operator into (5) and take into account that $\hat{\Theta}$ switches the in and out states:

$$\hat{\Theta} |b; \text{out}\rangle = |b_\theta; \text{in}\rangle \,, \tag{7}$$

where the state b_θ is obtained from b according to (3), by exchanging particles and antiparticles and flipping the spins.

$$\begin{aligned} \Gamma_a &= 2\pi \sum_b \delta(E_b - E_a) |\langle b; \text{out} | \hat{\Theta}^{-1} \hat{\Theta} \hat{H}_1 \hat{\Theta}^{-1} \hat{\Theta} | a, m \rangle|^2 \\ &= 2\pi \sum_b \delta(E_b - E_a) |\langle b_\theta; \text{in} | \hat{\Theta} \hat{H}_1 \hat{\Theta}^{-1} | \bar{a}, -m \rangle|^2 \\ &= 2\pi \sum_b \delta(E_b - E_a) |\langle b_\theta; \text{out} | \hat{S}^\dagger \hat{H}_1 | \bar{a}, -m \rangle|^2 \,. \end{aligned} \tag{8}$$

In the last step the $\hat{\Theta}$ invariance of $\hat{H}_1$ has been used (it is obvious from the derivation of the CPT theorem that each term in the Hamiltonian is invariant) and the fact that the $\hat{S}$ operator satisfies $\langle b'; \text{in}| = \langle b'; \text{out}|\hat{S}^\dagger$. The $\hat{S}$ operator generates a unitary transformation so that the states $\hat{S}|b_\theta; \text{out}\rangle$ span the same space as the original states $|b_\theta; \text{out}\rangle$. As a consequence, since a summation over all (energetically accessible) states is performed we simply may omit the factor $\hat{S}$ in (8). This is formally confirmed by twice inserting complete sets of out states:

Example 10.4

$$\Gamma_a = 2\pi \sum_b \delta(E_b - E_a) \sum_{b'b''} \langle a, -m|\hat{H}_1|b'; \text{out}\rangle \langle b', \text{out}|\hat{S}|b_\theta; \text{out}\rangle$$
$$\times \langle b_\theta; \text{out}|\hat{S}^\dagger|b''; \text{out}\rangle \langle b'', \text{out}|\hat{H}_1|a, -m\rangle \,. \tag{9}$$

Now the completeness relation for the b_θ states can be employed,[6]

$$\sum_b \delta(E_b - E_a)\langle b', \text{out}|\hat{S}|b_\theta; \text{out}\rangle \langle b_\theta; \text{out}|\hat{S}^\dagger|b''; \text{out}\rangle$$
$$= \delta(E_{b'} - E_a)\langle b', \text{out}|\hat{S}\hat{S}^\dagger|b''; \text{out}\rangle = \delta(E_{b'} - E_a)\,\delta_{b'b''} \,, \tag{10}$$

leading to

$$\Gamma_a = 2\pi \sum_{b'} \delta(E_b - E_a)|\langle b', \text{out}|\hat{H}_1|\bar{a}, -m\rangle|^2 \,. \tag{11}$$

A comparison with (6) proves the equality of particle and antiparticle lifetimes,[7]

$$\Gamma_a = \Gamma_{\bar{a}} \,, \tag{12}$$

as the energies are equal and the lifetime does not depend on the sign of m.

Note: The proof has worked only because there is a summation over all final states b. If a particle can decay into several final states the CPT theory does not rule out that the *branching ratios* in the particle and the antiparticle channel may be different! For example, if there are two decay channels

$$a \to b + c \quad \text{and} \quad a \to d + e \tag{13}$$

the values of the partial width may be different:

$$\frac{\Gamma_{a \to b+c}}{\Gamma_{a \to d+e}} \neq \frac{\Gamma_{\bar{a} \to \bar{b}+\bar{c}}}{\Gamma_{\bar{a} \to \bar{d}+\bar{e}}} \,. \tag{14}$$

The mechanism responsible for such deviations is the interaction between the particles in the final state, which may lead to interference effects that are different for particles and antiparticles.

The laws $M_a = M_{\bar{a}}$ and $\Gamma_a = \Gamma_{\bar{a}}$ and thus the validity of the CPT theorem have been checked experimentally with high precision. By far the most sensitive probe for this purpose is the neutral K meson. Since the decays of K^0

---

[6] This has become possible since the operator $\hat{S}$ – in contrast to $\hat{H}_1$ – only connects states of equal energy so that the delta function can be pulled out of the sum.
[7] T.D. Lee, R. Oehme, C.N. Yang: Phys. Rev. **106**, 340 (1957);
G. Lüders, B. Zumino: ibid. p. 385.

Example 10.4

and its antiparticle $\bar{K}^0$ interfere, an upper bound for the mass difference of these particles can be determined with exceedingly high precision:

$$\left|\frac{M_{\bar{K}^0} - M_{K^0}}{M_{K^0}}\right| < 4 \times 10^{-18} \ . \tag{15}$$

For other particles the masses have to be measured separately. The comparison of these values typically leads to $|\Delta M/M| < 10^{-5}$. A typical result for the agreement of particle and antiparticle lifetimes has been obtained for muons:

$$\left|\frac{\Gamma_{\mu^+} - \Gamma_{\mu^-}}{\Gamma_{\mu^+}}\right| < 10^{-3} \ . \tag{16}$$

Part III

Quantization with Path Integrals

11. The Path-Integral Method

11.1 Introduction

Up to now this book has exclusively dealt with the canonical method of field quantization. Field operators have been introduced which act in the Hilbert space of "second quantization" and canonical commutation rules have been postulated.

As it turns out, this is not the only way to proceed. There is an alternative approach to the problem of field quantization, which is phrased in a very different language. The method of *path integrals* does not speak about operators at all; instead a special kind of highly dimensional integral over *classical* fields is employed. The quantum properties of a system come into play because the motion of a particle between two points can proceed on a large (infinite) variety of classical trajectories and each of these alternatives makes its coherent contribution to the transition amplitude. This results in a path integral from which, at least in principle, all properties of a system can be deduced by using functional techniques. In the remainder of this book we will learn the basic properties and uses of the path-integral method, first addressing ordinary quantum mechanics and subsequently – in the next chapter – quantum field theory.[1]

The path-integral formalism leads to the same results as obtained by canonical quantization; both formulations of quantum field theory are equivalent. For some systems, however, the canonical method is quite awkward to formulate and to use. This holds true for the quantization of fields with constraints, a problem one regularly encounters with gauge theories. Because (nonabelian) gauge fields are the backbone of modern theoretical physics, and also because of its formal elegance and appeal, the path-integration formalism nowadays is the preferred route to field quantization.

11.2 Path Integrals in Nonrelativistic Quantum Mechanics

To understand the concept of a path integral and its basic properties we will start with the simplest possible case. Let us study the dynamics of a *nonrel-*

---

[1] Path integration has made its entrance in a variety of domains of theoretical physics, its use not being restricted to quantum theory. A comprehensive introduction is given by H. Kleinert: *Path Integrals in Quantum Mechanics, Statistics, and Polymer Physics* (World Scientific, Singapore 1990).

ativistic system with one degree of freedom, i.e., of a particle moving freely or under the influence of a potential in one space dimension. Within Hamilton's formulation of classical mechanics the system is described by a space coordinate q and the canonically conjugate momentum coordinate p. Upon canonical quantization the coordinates are represented by linear operators that satisfy the commutation relation

$$[\hat{p}, \hat{q}] = -i\hbar . \tag{11.1}$$

The eigenstates of these operators span the Hilbert space. In the Schrödinger picture we have

$$\begin{aligned} \hat{q}_S |q\rangle &= q |q\rangle , \\ \hat{p}_S |p\rangle &= p |p\rangle . \end{aligned} \tag{11.2}$$

The state vectors are normalized "to delta functions"

$$\langle q'|q\rangle = \delta(q' - q) , \tag{11.3a}$$

$$\langle p'|p\rangle = 2\pi\hbar \delta(p' - p) , \tag{11.3b}$$

and obey the completeness relations

$$\int dq \, |q\rangle\langle q| = 1 , \tag{11.4a}$$

$$\int \frac{dp}{2\pi\hbar} \, |p\rangle\langle p| = 1 . \tag{11.4b}$$

The coordinate-space representation of the momentum eigenstate is a plane wave:

$$\langle q|p\rangle = e^{ipq/\hbar} . \tag{11.5}$$

The factors $2\pi\hbar$ in (11.3) and (11.4) have been chosen such that (11.5) does not contain a normalization factor.

The quantum state of the system is described by a vector $|\Psi\rangle$ in Hilbert space. Within the *Schödinger picture* this state depends on time and satisfies the equation of motion

$$i\hbar \frac{\partial}{\partial t} |\Psi(t)\rangle_S = \hat{H}(\hat{p}, \hat{q}) |\Psi(t)\rangle_S . \tag{11.6}$$

The formal solution of this equation is

$$|\Psi(t)\rangle_S = e^{-i\hat{H}t/\hbar} |\Psi(0)\rangle_S . \tag{11.7}$$

The corresponding wave function in coordinate representation reads

$$\psi(q, t) = \langle q|\Psi(t)\rangle_S . \tag{11.8}$$

Within the *Heisenberg picture* the operators are time dependent and satisfy the Heisenberg equation

$$-i\hbar \dot{\hat{q}}_H = [\hat{H}, \hat{q}_H] . \tag{11.9}$$

Therefore the time evolution of the coordinate operator is given by

$$\hat{q}_H(t) = e^{i\hat{H}t/\hbar} \hat{q}_S e^{-i\hat{H}t/\hbar} . \tag{11.10}$$

11.2 Path Integrals in Nonrelativistic Quantum Mechanics

This time-dependent operator has a complete set of eigenstates

$$\hat{q}_{\rm H}(t)|q,t\rangle = q|q,t\rangle, \tag{11.11}$$

which evolve in time according to

$$|q,t\rangle = {\rm e}^{{\rm i}\hat{H}t/\hbar}|q\rangle, \; . \tag{11.12}$$

Note that the states $|q,t\rangle$ constructed in this way play the role of a "moving basis" in Hilbert space. They do not correspond to the true state vectors of the system, which in the Heisenberg picture have to be constant in time!

Since (11.12) is a unitary transformation, the orthonormality and completeness relations (11.3) and (11.4) remain valid for the time-dependent states $|q,t\rangle$:

$$\langle q',t|q,t\rangle = \langle q'|{\rm e}^{-{\rm i}\hat{H}t/\hbar}{\rm e}^{{\rm i}\hat{H}t/\hbar}|q\rangle = \langle q'|q\rangle = \delta(q'-q), \tag{11.13a}$$

$$\int {\rm d}q\,|q,t\rangle\langle q,t| = \int {\rm d}q\,{\rm e}^{{\rm i}\hat{H}t/\hbar}|q\rangle\langle q|{\rm e}^{-{\rm i}\hat{H}t/\hbar} = {\rm e}^{{\rm i}\hat{H}t/\hbar}{\rm e}^{-{\rm i}\hat{H}t/\hbar} = 1. \tag{11.13b}$$

The wave function (11.8) can be interpreted as the amplitude in the expansion of the Heisenberg state vector $|\Psi\rangle_{\rm H} = |\Psi(0)\rangle_{\rm S}$ with respect to the "moving basis":

$$\psi(q,t) = \langle q,t|\Psi\rangle_{\rm H}. \tag{11.14}$$

The starting point of the path-integral method is the idea that instead of working with Hilbert space operators the transition amplitude

$$\langle q',t'|q,t\rangle = \langle q'|{\rm e}^{-{\rm i}\hat{H}(t'-t)/\hbar}|q\rangle \tag{11.15}$$

can be viewed as the fundamental entity. The object $\langle q',t'|q,t\rangle$ is called the *Feynman kernel*. Knowledge of the Feynman kernel is equivalent to having solved the Schrödinger equation (11.6): the time development of the wave function for arbitrary t' can be obtained by a simple integration

$$\begin{aligned}\psi(q',t') = \langle q',t'|\Psi\rangle_{\rm H} &= \int {\rm d}q\,\langle q',t'|q,t\rangle\langle q,t|\Psi\rangle_{\rm H} \\ &= \int {\rm d}q\,\langle q',t'|q,t\rangle\,\psi(q,t). \end{aligned} \tag{11.16}$$

The path-integral formalism provides a means to construct the transition amplitude (11.15) from the classical Hamiltonian of the system alone, without any explicit reference to noncommuting operators or Hilbert space vectors.[2] For this purpose the time interval (t,t') is split up into many small slices, which for simplicity will be taken to be of equal length:

$$t_n = t + n\epsilon \quad \text{where} \quad t'-t = N\epsilon. \tag{11.17}$$

At each of the grid points $n = 1,\ldots,N-1$ a complete set of basis states $|q_n,t_n\rangle$ can be inserted:

---

[2] The basic idea of path integration can be traced back to the paper P.A.M. Dirac: Phys. Zeitschr. der Sowjetunion **3**, 64 (1933). The method was worked out and made popular by Feynman, see e.g. the monograph R.P. Feynman, A.R. Hibbs: *Quantum Mechanics and Path Integration* (MacGraw-Hill, New York 1965).

$$\langle q',t'|q,t\rangle = \int dq_{N-1}\ldots\int dq_2\int dq_1\ \langle q',t'|q_{N-1},t_{N-1}\rangle\cdots$$
$$\times\langle q_2,t_2|q_1,t_1\rangle\langle q_1,t_1|q,t\rangle\ . \tag{11.18}$$

According to (11.12) each of the matrix elements under the integral can be written as

$$\langle q_{n+1},t_{n+1}|q_n,t_n\rangle = \langle q_{n+1}|e^{-i\hat{H}(\hat{p},\hat{q})\epsilon/\hbar}|q_n\rangle\ . \tag{11.19}$$

This object is also known as the "transfer matrix" $T(q_{n+1},q_n)$.

Equation (11.18) can be interpreted as follows. A particle that propagates from point q at time t to point q' at time t' can take an arbitrary intermediate trajectory. Such a path is characterized by the coordinate values q_i at the intermediate grid points in the time interval (t,t'). One such path is shown in Figure 11.1 as a zigzag curve. It is essential that all conceivable paths are taken into account. According to the superposition principle of quantum mechanics they all contribute to the transition amplitude (11.18) in a coherent fashion. Of course some regions in the space of trajectories will turn out to be more important than others, as dictated by the behavior of the product of matrix elements under the integral.

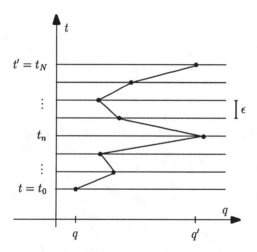

Fig. 11.1. A typical path connecting the points q,t and q',t'

The decomposition (11.18) of course is only helpful if the amplitudes $\langle q_{n+1},t_{n+1}|q_n,t_n\rangle$ are evaluated more easily than the full expression $\langle q',t'|q,t\rangle$. This indeed is the case, provided that the time interval $t'-t$ is chopped into sufficiently short pieces, i.e., if ϵ is small. Then the time-evolution operator can be approximated by the first two terms of its Taylor series

$$\langle q_{n+1},t_{n+1}|q_n,t_n\rangle = \langle q_{n+1}|1-\frac{i\epsilon}{\hbar}\hat{H}(\hat{p},\hat{q})|q_n\rangle + O(\epsilon^2)\ . \tag{11.20}$$

The Hamiltonian depends on $\hat{p}$ and $\hat{q}$. To evaluate its matrix elements we insert a complete set of momentum states according to (11.4b):

$$\langle q_{n+1}|\hat{H}(\hat{p},\hat{q})|q_n\rangle = \int\frac{dp_n}{2\pi\hbar}\ \langle q_{n+1}|p_n\rangle\langle p_n|\hat{H}(\hat{p},\hat{q})|q_n\rangle\ . \tag{11.21}$$

Since the operators $\hat{p}$ and $\hat{q}$ can act to the left or to the right on their respective eigenstates the matrix elements of the Hamilton operator are reduced to the classical Hamilton function:

$$\langle p_n|\hat{H}(\hat{p},\hat{q})|q_n\rangle = \langle p_n|q_n\rangle H(p_n,q_n) \,. \tag{11.22}$$

This step in general has its problems if $\hat{H}$ contains mixed products of the anticommuting operators $\hat{p}$ and $\hat{q}$. Then (11.22) tacitly assumes a certain ordering of the factors: the momentum operators $\hat{p}$ stand to the left of the coordinate operators $\hat{q}$. This is rather arbitrary and one might prefer a more symmetric prescription, called *Weyl's operator ordering*, which averages over all possible orderings with equal weight (see Example 11.2). It is possible to show that this choice for the Hamilton operator implies the replacement

$$\langle p_n|\hat{H}(\hat{p},\hat{q})|q_n\rangle \to \langle p_n|q_n\rangle H\left(p_n, \tfrac{1}{2}(q_{n+1}+q_n)\right) \tag{11.23}$$

instead of (11.22). As we do not want to commit ourselves we will write $H(p_n, \bar{q}_n)$ in the following, where $\bar{q}_n = q_n$ or $\bar{q}_n = \tfrac{1}{2}(q_{n+1}+q_n)$ can be chosen. Using (11.2) and (11.5) we find that the infinitesimal transition amplitude (11.20) becomes

$$\langle q_{n+1}, t_{n+1}|q_n, t_n\rangle = \int \frac{dp_n}{2\pi\hbar} \exp\left[ip_n(q_{n+1}-q_n)/\hbar\right] \left(1 - iH(p_n,\bar{q}_n)\frac{\epsilon}{\hbar}\right)$$
$$+ O(\epsilon^2) \,. \tag{11.24}$$

Now the essential (and mathematically nontrivial) step has to be ventured: making the partition finer and finer and at the end, going to the limit $\epsilon \to 0$ or $N \to \infty$. The Feynman kernel (11.18) then becomes

$$\langle q', t'|q, t\rangle = \lim_{N\to\infty} \int \prod_{n=1}^{N-1} dq_n \prod_{n=0}^{N-1} \frac{dp_n}{2\pi\hbar} \exp\left(\frac{i\epsilon}{\hbar}\sum_{n=0}^{N-1} p_n \frac{q_{n+1}-q_n}{\epsilon}\right)$$
$$\times \prod_{n=0}^{N-1}\left(1 - \frac{i\epsilon}{\hbar} H(p_n,\bar{q}_n)\right) \,. \tag{11.25}$$

The integrand in this equation can be written more elegantly. The well-known representation of the exponential function

$$\lim_{N\to\infty}\left(1 + \frac{x}{N}\right)^N = e^x \tag{11.26}$$

has the following generalization:

$$\lim_{N\to\infty}\prod_{n=0}^{N-1}\left(1 + \frac{x_n}{N}\right) = \exp\left(\lim_{N\to\infty}\frac{1}{N}\sum_{n=0}^{N-1} x_n\right) \,. \tag{11.27}$$

If one identifies $x_n = -\left[i(t'-t)/\hbar\right]H(p_n,\bar{q}_n)$, this identity can be used to rewrite (11.25):

$$\langle q', t'|q, t\rangle = \lim_{N\to\infty} \int \prod_{n=1}^{N-1} dq_n \prod_{n=0}^{N-1} \frac{dp_n}{2\pi\hbar}$$
$$\times \exp\left[\frac{i\epsilon}{\hbar}\sum_{n=0}^{N-1}\left(p_n \frac{q_{n+1}-q_n}{\epsilon} - H(p_n,\bar{q}_n)\right)\right] \,. \tag{11.28}$$

In the limit $N \to \infty$ the grid points q_n and p_n are arbitrarily close. They can be viewed as the sampled values of continuously defined functions $q(t)$ and $p(t)$, i.e., $q_n = q(t_n)$ and $p_n = p(t_n)$. The construction we have sketched is known as a functional integral or, more specifically, as a *path integral*. It is customary to introduce a special notation for path integrals:

$$\int \prod_{n=1}^{N-1} dq_n \to \int \mathcal{D}q \quad \text{and} \quad \int \prod_{n=0}^{N-1} \frac{dp_n}{2\pi\hbar} \to \int \mathcal{D}p \;. \tag{11.29}$$

If we replace the divided difference in the exponent of (11.28) by a time derivative and the sum by an integral,

$$\frac{q_{n+1} - q_n}{\epsilon} \to \dot{q}(t_n) \;, \quad \epsilon \sum_{n=0}^{N-1} f(t_n) \to \int_t^{t'} d\tau\, f(\tau) \;, \tag{11.30}$$

then the propagation amplitude (the Feynman kernel) is expressed by the following *path integral in phase space*:

$$\langle q', t' | q, t \rangle = \int \mathcal{D}q \int \mathcal{D}p \, \exp\left[\frac{i}{\hbar} \int_t^{t'} d\tau\, (p\dot{q} - H(p,q))\right] \;. \tag{11.31}$$

The path integral extends over all functions $p(t)$ in momentum space and over those functions $q(t)$ in coordinate space that satisfy the boundary conditions

$$q(t) = q \quad \text{and} \quad q(t') = q' \;. \tag{11.32}$$

We do not attempt to present a strict mathematical derivation of the path integral and its formal properties.[3] The path integral, although intuitively appealing, is beset with several problems. The contributing paths belong to a very general class of functions, they need not be smooth or even continuous. A severe problem lies in the behavior of the integrand, which is a pure phase factor and thus has the same modulus for all trajectories! Under this condition one can hope to achieve convergence only if the fast oscillating contributions of neighboring trajectories are "averaged to zero". The problem results from the superposition principle of quantum mechanics according to which probability amplitudes have to be added coherently.

Functional integrals are encountered also in the field of statistical physics. In this context convergence of the integrals is less of a problem since the integrand falls off asymptotically. As early as 1925, N. Wiener introduced path integrals to describe *diffusion* processes. The propagation of a particle undergoing diffusion is described as a "random walk". The probabilities for all paths $\boldsymbol{x}(t)$ leading from $(\boldsymbol{x}, t)$ to $(\boldsymbol{x}', t')$ are added up incoherently. To see the connection with quantum mechanics we remember that the Schrödinger equation turns into the diffusion equation if imaginary time arguments are considered. This has the consequence that the argument of the exponential function in Wiener's path integral has a (negative) real value, thus ensuring convergence of the integral. The quantum-mechanical path integral unfortunately does not have this reassuring property. However, one can use the artifice of making an analytical continuation of the transition amplitude to imaginary times (a Wick

---

[3] Mathematical aspects are treated, e.g., in B. Simon: *Functional Integration and Quantum Physics* (Academic Press, New York 1979).

rotation; see Sect. 11.5) and in this way ensuring the exponential fall-off of the integrand.

11.3 Feynman's Path Integral

Very often the Hamiltonian of a system has the standard form

$$H(p,q) = \frac{1}{2m}p^2 + V(q) . \tag{11.33}$$

In this case, i.e., if the momentum dependence is confined to a quadratic term (the kinetic energy), the integration over p can be carried out explicitly. The transition amplitude (11.24) for neighboring grid points in first order is given by

$$\langle q_{n+1}, t_{n+1} | q_n, t_n \rangle \simeq \int \frac{\mathrm{d}p_n}{2\pi\hbar} \exp\left[\frac{i\epsilon}{\hbar}\left(p_n \frac{q_{n+1} - q_n}{\epsilon} - H(p_n, \bar{q}_n)\right)\right]$$

$$= \int \frac{\mathrm{d}p_n}{2\pi\hbar} \exp\left[\frac{i\epsilon}{\hbar}\left(p_n \dot{q}_n - \frac{1}{2m}p_n^2 - V(\bar{q}_n)\right)\right] . \tag{11.34}$$

The function in the exponent of (11.34) just looks like the Lagrange function. There is a subtle difference, however, since the momentum p_n has been introduced as an independent variable when the intermediate states $|p_n\rangle\langle p_n|$ were inserted. The velocity $\dot{q}_n$, on the other hand, is an abbreviation for the divided difference in (11.30). Therefore the object found in the argument of the exponential function

$$\tilde{L}(q, \dot{q}, p) := p\dot{q} - H(p, q) \tag{11.35}$$

in this sense is not identical with the ordinary Lagrange function $L(q, \dot{q})$.

The exponent in (11.34) contains a quadratic form and thus we are confronted with an ordinary *Gaussian integral*. By completing the squares

$$p_n^2 - 2m\dot{q}_n p_n = (p_n - m\dot{q}_n)^2 - m^2 \dot{q}_n^2 \tag{11.36}$$

and shifting the integration variable $p'_n = p_n - m\dot{q}_n$ we obtain a standard integral:

$$\langle q_{n+1}, t_{n+1} | q_n, t_n \rangle \simeq \exp\left[\frac{i\epsilon}{\hbar}\left(\frac{m}{2}\dot{q}_n^2 - V(\bar{q}_n)\right)\right] \int_{-\infty}^{\infty} \frac{\mathrm{d}p'_n}{2\pi\hbar} \exp\left(\frac{-i\epsilon}{\hbar}\frac{p'^2_n}{2m}\right) .$$

$$\tag{11.37}$$

This can be solved by the Gaussian integral formula

$$\int_{-\infty}^{\infty} \mathrm{d}x\, e^{-zx^2} = \sqrt{\frac{\pi}{z}} . \tag{11.38}$$

This integral is well defined and absolutely convergent if the parameter z has a positive real part, $\mathrm{Re}\, z > 0$. In our problem z is purely imaginary and the integrand shows undamped oscillations. The use of (11.38) for arbitrary values of z in the complex plane corresponds to an analytic continuation. The resulting function has a branch cut at the singular point $z = 0$. With some care the Gaussian integral formula may even be applied in the region $\mathrm{Re}\, z < 0$, where the integrand grows exponentially.

Using the identity (11.38) we obtain

$$\langle q_{n+1}, t_{n+1} | q_n, t_n \rangle = \left(\frac{2\pi i \hbar \epsilon}{m} \right)^{-\frac{1}{2}} \exp\left[\frac{i\epsilon}{\hbar} \left(\frac{m}{2} \dot{q}_n^2 - V(\bar{q}_n) \right) \right] . \tag{11.39}$$

Inserting this into (11.28) we obtain the Feynman kernel as a *path integral in coordinate space*:

$$\begin{aligned}
\langle q', t' | q, t \rangle &= \lim_{N \to \infty} \left(\frac{2\pi i \hbar \epsilon}{m} \right)^{-N/2} \int \prod_{n=1}^{N-1} dq_n \, \exp\left[\frac{i\epsilon}{\hbar} \sum_{n=0}^{N-1} \left(\frac{m}{2} \dot{q}_n^2 - V(\bar{q}_n) \right) \right] \\
&= \mathcal{N} \int \mathcal{D}q \, \exp\left[\frac{i}{\hbar} \int_t^{t'} d\tau \left(\frac{m}{2} \dot{q}^2 - V(q) \right) \right] \\
&= \mathcal{N} \int \mathcal{D}q \, \exp\left[\frac{i}{\hbar} W(q, \dot{q}) \right] .
\end{aligned} \tag{11.40}$$

Here $\mathcal{N}$ is a yet undetermined normalization constant. Again the integral extends over all possible paths $q(t)$ satisfying the boundary conditions $q(t) = q$, $q(t') = q'$. The contribution of each path to the integral is of equal modulus and its phase is given by the *action functional*

$$W(q, \dot{q}) = \int_t^{t'} d\tau \left(\frac{m}{2} \dot{q}^2 - V(q) \right) = \int_t^{t'} d\tau L(q, \dot{q}) . \tag{11.41}$$

This observation can be used as the starting point to derive the *classical limit* of quantum mechanics. If $W \gg \hbar$ holds then almost everywhere in function space the strongly oscillating contributions to (11.40) made by neighboring paths will cancel each other. This cancellation can be avoided only if $W(q, \dot{q})$ does not change much in the vicinity of a path, i.e., *if the action is stationary* with respect to variations $\delta q(t)$ and $\delta \dot{q}(t)$. This is just the condition that determines the classically allowed trajectory $q_{\mathrm{cl}}(t)$ according to Hamilton's principle of least action:

$$\delta \int_t^{t'} d\tau \, L(q, \dot{q}) = 0 . \tag{11.42}$$

Thus the classical description of a system can be derived from quantum theory if the path integral is solved using the *stationary phase approximation*. Quantum effects may still be visible if there are several distinct classical trajectories q_{cl}^n which can interfere, each carrying its own phase factor $\exp\left(\frac{i}{\hbar} W(q_{\mathrm{cl}}^n, \dot{q}_{\mathrm{cl}}^n) \right)$.

Equation (11.40) contains the normalization constant $\mathcal{N}$ that we left open. As it turns out, the limit $N \to \infty$ in (11.40) is not well defined mathematically and the constant $\mathcal{N}$ is given by a divergent expression. Fortunately this does not preclude us from gaining useful information from the path integral since it is the functional dependence and not the absolute value that matters. Therefore one can use (11.40) without particularly caring about the normalization.

EXERCISE

11.1 The Path Integral for the Propagation of a Free Particle

Problem. Evaluate the Feynman kernel $\langle q',t'|q,t\rangle$ for the free propagation of a particle in one dimension. Show that the resulting phase corresponds to propagation along the classical path. Show that $\langle q',t'|q,t\rangle$ solves the time-dependent Schrödinger equation and satisfies the boundary condition $\lim_{t'\to t}\langle q',t'|q,t\rangle = \delta(q'-q)$.

Solution. To evaluate the path integral in coordinate-space form,

$$\langle q',t'|q,t\rangle = \lim_{N\to\infty}\left(\frac{2\pi i\hbar\epsilon}{m}\right)^{-N/2}\int\prod_{n=1}^{N-1}dq_n \exp\left[\frac{im}{2\hbar\epsilon}\sum_{n=0}^{N-1}(q_{n+1}-q_n)^2\right], \quad (1)$$

we successively integrate over the variables $q_1,\ldots,q_{N-1}$. This requires the knowledge of the integral

$$\int_{-\infty}^{\infty}dq_n \exp\left[-\alpha(q_{n+1}-q_n)^2\right]\exp\left[-\beta(q_n-q_{n-1})^2\right]$$

$$= \int_{-\infty}^{\infty}dq_n \exp\left[-(\alpha+\beta)\left(q_n - \frac{\alpha q_{n+1}+\beta q_{n-1}}{\alpha+\beta}\right)^2 - \frac{\alpha\beta}{\alpha+\beta}(q_{n+1}-q_{n-1})^2\right]$$

$$= \exp\left[-\frac{\alpha\beta}{\alpha+\beta}(q_{n+1}-q_{n-1})^2\right]\sqrt{\frac{\pi}{\alpha+\beta}}. \quad (2)$$

Completing the square and shifting the integration variable we have transformed this into a standard Gaussian integral. For the first integration, $n=1$, we have to choose $\alpha=\beta=-im/2\epsilon$. This leads to

$$\left(\frac{2\pi i\hbar\epsilon}{m}\right)^{-1}\int_{-\infty}^{\infty}dq_1 \exp\left[\frac{im}{2\hbar\epsilon}(q_1-q_0)^2\right]\exp\left[\frac{im}{2\hbar\epsilon}(q_2-q_1)^2\right]$$

$$= \sqrt{\frac{m}{2\pi i\hbar\, 2\epsilon}}\exp\left[\frac{1}{2\epsilon}\frac{im}{2\hbar}(q_2-q_0)^2\right], \quad (3)$$

where two square-root factors from (1) have been "used up". The expression (3) has the same structure as the exponential factors in (1), except for the replacement $\epsilon \to 2\epsilon$. The procedure can be repeated for the $n=2$ integration:

$$\sqrt{\frac{m}{2\pi i\hbar\, 2\epsilon}}\sqrt{\frac{m}{2\pi i\hbar\epsilon}}\int_{-\infty}^{\infty}dq_2 \exp\left[\frac{1}{2\epsilon}\frac{im}{2\hbar}(q_2-q_0)^2\right]\exp\left[\frac{1}{\epsilon}\frac{im}{2\hbar}(q_3-q_2)^2\right]$$

$$= \sqrt{\frac{m}{2\pi i\hbar\, 3\epsilon}}\exp\left[\frac{1}{3\epsilon}\frac{im}{2\hbar}(q_3-q_0)^2\right]. \quad (4)$$

In this way, after $N-1$ integrations the Feynman kernel becomes

$$\langle q',t'|q,t\rangle = \lim_{N\to\infty}\sqrt{\frac{m}{2\pi i\hbar N\epsilon}}\exp\left[\frac{1}{N\epsilon}\frac{im}{2\hbar}(q_N-q_0)^2\right]$$

$$= \sqrt{\frac{m}{2\pi i\hbar(t'-t)}}\exp\left[\frac{im}{2\hbar}\frac{(q'-q)^2}{t'-t}\right]. \quad (5)$$

Exercise 11.1

In fact this expression does not depend on the number N of partition points and the limiting process in (5) turns out to be trivial. Such a simplification, of course, occurs only for the special case of free motion.

The *classical path* between the points (q,t) and (q',t') is given by the straight line

$$q(\tau) = q + \frac{q'-q}{t'-t}\tau , \qquad (6)$$

leading to the classical action integral

$$\begin{aligned} W(t',t) &= \int_t^{t'} d\tau\, L(q,\dot{q}) = \int_t^{t'} d\tau\, \frac{1}{2}m\dot{q}^2 = \frac{1}{2}m\frac{(q'-q)^2}{(t'-t)^2}(t'-t) \\ &= \frac{m}{2}\frac{(q'-q)^2}{t'-t} . \end{aligned} \qquad (7)$$

This agrees with the phase factor in (5). The result (5) is closely related to the *free Green's function of the Schrödinger equation* as discussed, for example, in Chap. 1 of the volume *Quantum Electrodynamics*. The free retarded Green's function $G_0^+(q',t';q,t)$ is obtained if (5) is multiplied by the factor $-i\Theta(t'-t)$. By differentiating (5) one immediately confirms that the Feynman kernel satisfies the time-dependent Schrödinger equation. We find

$$\begin{aligned} i\hbar\frac{\partial}{\partial t'}\langle q',t'|q,t\rangle &= \left(\frac{m}{2}\frac{(q'-q)^2}{(t'-t)^2} - \frac{i}{2}\frac{\hbar}{t'-t}\right)\sqrt{\frac{m}{2\pi i\hbar(t'-t)}} \\ &\quad \times \exp\left[\frac{im}{2\hbar}\frac{(q'-q)^2}{t'-t}\right] , \end{aligned} \qquad (8)$$

and the same result is obtained for the kinetic energy $-\frac{\hbar^2}{2m}(\partial^2/\partial q'^2)\langle q',t'|q,t\rangle$, thus

$$\left(i\hbar\frac{\partial}{\partial t'} + \frac{\hbar^2}{2m}\frac{\partial^2}{\partial q'^2}\right)\langle q',t'|q,t\rangle = 0 . \qquad (9)$$

The boundary condition

$$\lim_{t'\to t}\langle q',t'|q,t\rangle = \delta(q'-q) \qquad (10)$$

is verified by multiplication with a test function $f(q')$ and integration over q'. By assumption, the test function can be expanded into a Taylor series at $q' = q$

$$f(q') = f(q) + \sum_{n=1}^{\infty}\frac{1}{n!}\left.\frac{d^n}{dq'^n}f(q')\right|_{q'=q}(q'-q)^n . \qquad (11)$$

Abbreviating $\tilde{q} := (q'-q)/\sqrt{\delta t}$ we find

$$\begin{aligned} &\lim_{\delta t\to 0}\int_{-\infty}^{\infty}dq'\, f(q')\sqrt{\frac{m}{2\pi i\hbar\,\delta t}}\exp\left[\frac{im}{2\hbar}\frac{(q'-q)^2}{\delta t}\right] \qquad (12) \\ &= f(q) + \lim_{\delta t\to 0}\sqrt{\frac{m}{2\pi i\hbar}}\sum_{n=1}^{\infty}\delta t^{n/2}\frac{1}{n!}f^{(n)}(q)\int_{-\infty}^{\infty}d\tilde{q}\,\tilde{q}^n\exp\left[\frac{im}{2\hbar}\tilde{q}^2\right] \\ &= f(q) , \end{aligned} \qquad (13)$$

which is the defining relation for the delta function.

EXAMPLE

11.2 Weyl Ordering of Operators

The transition from classical to quantum mechanics is accompanied by the problem of operator ordering if the Hamiltonian contains mixed terms, which simultaneously depends on the space and momentum coordinates. Since the opearators $\hat{p}$ and $\hat{q}$ do not commute, a classically well-defined product of the form $p^n q^m$ can be translated into a multitude of quantum operators that differ in the ordering of the factors $\hat{q}$ and $\hat{p}$. In the path-integral approach to quantization this problem does not seem to occur on first sight: everything is expressed in terms of classical quantities, and noncommuting operators are nowhere to be seen. This is not the whole truth, however. Also the path-integral formulation is not unique since the evaluation of the path integral can proceed in different ways.

In Sect. 11.2 we have made the assumption that the momentum operators always stand to the left of the coordinate operators; cf. (11.20) and (11.21). Of course this is an arbitrary choice that prefers one possible ordering over the others. The most unbiased choice would treat all orderings on an equal footing by taking the average value. This prescription is called *Weyl ordering*. For a product of operators $\hat{O}_1, \ldots, \hat{O}_n$ the Weyl ordering is defined as the sum over all permutations $P(i_1, \ldots, i_n)$ of the indices:

$$\left(\hat{O}_1 \hat{O}_2 \cdots \hat{O}_n\right)_W = \frac{1}{\text{number of permutations}} \sum_P \hat{O}_{i_1} \hat{O}_{i_2} \cdots \hat{O}_{i_n} . \tag{1}$$

Let us write down some examples:

$$\left(\hat{p}\,\hat{q}\right)_W = \frac{1}{2}(\hat{p}\hat{q} + \hat{q}\hat{p}) , \tag{2a}$$

$$\left(\hat{p}\,\hat{q}^2\right)_W = \frac{1}{3}(\hat{p}\hat{q}^2 + \hat{q}\hat{p}\hat{q} + \hat{q}^2\hat{p}) , \tag{2b}$$

$$\left(\hat{p}^2\,\hat{q}^2\right)_W = \frac{1}{6}(\hat{p}^2\hat{q}^2 + \hat{q}^2\hat{p}^2 + \hat{p}\hat{q}^2\hat{p} + \hat{q}\hat{p}^2\hat{q} + \hat{p}\hat{q}\hat{p}\hat{q} + \hat{q}\hat{p}\hat{q}\hat{p}) , \tag{2c}$$

and so on. If one of the factors enters linearly the Weyl product is

$$(\hat{p}\,\hat{q}^n)_W = \frac{1}{n+1} \sum_{k=0}^{n} \hat{q}^{n-k}\,\hat{p}\,\hat{q}^k . \tag{3}$$

For a quadratic factor we obtain a more complicated expression:

$$\left(\hat{p}^2\hat{q}^n\right)_W = \frac{2}{(n+1)(n+2)} \sum_{k=0}^{n} \sum_{l=0}^{n-k} \hat{q}^{n-k-l}\,\hat{p}\,\hat{q}^k\,\hat{p}\,\hat{q}^l . \tag{4}$$

The prefactor is the inverse number of possible combinations of the indices k and l.

The expressions (1)–(3) can be written in a seemingly different form by employing the commutation relation $[\hat{p}, \hat{q}] = -i\hbar$. For example, (2) also can be written as

$$\left(\hat{p}\,\hat{q}^2\right)_W = \frac{1}{4}(\hat{p}\,\hat{q}^2 + 2\hat{q}\hat{p}\hat{q} + \hat{q}^2\hat{p}) , \tag{5}$$

Example 11.2

or in a still different way

$$(\hat{p}\hat{q}^2)_{\mathrm{W}} = \frac{1}{2}(\hat{p}\hat{q}^2 + \hat{q}^2\hat{p}) \,. \tag{6}$$

There is an elegant method to construct the Weyl product, which is based on the following identity:

$$\left(\alpha\hat{p} + \beta\hat{q}\right)^N = \sum_{k=0}^{N} \binom{N}{k} \alpha^{N-k}\beta^k \left(\hat{p}^{N-k}\hat{q}^k\right)_{\mathrm{W}} \,. \tag{7}$$

For ordinary commuting numbers this would be just the familiar binomial formula. Since $\hat{p}$ and $\hat{q}$ are noncommuting operators the mixed products are different from each other. However, when multiplying out the binomial expression on the left-hand side of (7), we encounter all possible orderings of the operators which is just the definition of the Weyl product. Therefore the $k + l$th power of $\alpha\hat{p} + \beta\hat{q}$ can serve as the generating function for the Weyl product $(\hat{p}^k\hat{q}^l)_{\mathrm{W}}$ if one differentiates with respect to the formal parameters α and β:

$$(\hat{p}^k\hat{q}^l)_{\mathrm{W}} = \frac{1}{(k+l)!} \left(\frac{\partial}{\partial \alpha}\right)^k \left(\frac{\partial}{\partial \beta}\right)^l (\alpha\hat{p} + \beta\hat{q})^{k+l} \,. \tag{8}$$

Now we will prove the following conjecture: the Feynman kernel of a quantum system described by a Weyl-ordered Hamiltonian, $\hat{H}(\hat{p}, \hat{q}) = \hat{H}_{\mathrm{W}}(\hat{p}, \hat{q})$, is given by the following path integral:

$$\langle q', t' | q, t \rangle = \lim_{N \to \infty} \int \prod_{n=1}^{N-1} \mathrm{d}q_n \int \prod_{n=0}^{N-1} \frac{\mathrm{d}p_n}{2\pi\hbar}$$
$$\times \exp\left[\frac{\mathrm{i}\epsilon}{\hbar} \sum_{n=0}^{N-1} \left(p_n \frac{q_{n+1} - q_n}{\epsilon} - H(p_n, \bar{q}_n)\right)\right] \,, \tag{9}$$

with the abbreviation

$$\bar{q}_n := \frac{1}{2}(q_{n+1} + q_n) \,. \tag{10}$$

This agrees with the definition (11.28) of the path integral if the classical Hamiltonian is expressed as a function of the *averaged coordinate* $\bar{q}_n$. This is quite plausible since the symmetric treatment of the integration interval corresponds to the equal roles of the operators $\hat{p}$ and $\hat{q}$ in the Weyl-ordered Hamilton operator $\hat{H}(\hat{p}, \hat{q})$.

To prove (9) we can repeat the earlier calculation; only (11.21) has to be replaced by

$$\langle q_{n+1} | \hat{H}_{\mathrm{W}}(\hat{p}, \hat{q}) | q_n \rangle = \int \frac{\mathrm{d}p_n}{2\pi\hbar} \, \exp\left[\frac{\mathrm{i}}{\hbar} p_n (q_{n+1} - q_n)\right] H\left(p_n, \frac{q_{n+1} - q_n}{2}\right) \,. \tag{11}$$

This identity has to be proven for general Hamilton operators $\hat{H}_{\mathrm{W}}(\hat{p}, \hat{q})$. We make the assumption that $\hat{H}_{\mathrm{W}}$ can be expanded into a power series:

$$\hat{H}_{\mathrm{W}}(\hat{p}, \hat{q}) = \sum_{k,l} c_{kl} \left(\hat{p}^k \hat{q}^l\right)_{\mathrm{W}} \,. \tag{12}$$

In view of (8) it is sufficient to prove the assertion for a general polynomial $(\alpha\hat{p} + \beta\hat{q})^N$, i.e.,

$$\langle q'|(\alpha\hat{p} + \beta\hat{q})^N|q\rangle = \int \frac{dp}{2\pi\hbar} e^{ip(q'-q)/\hbar} \left(\alpha p + \beta \frac{q'+q}{2}\right)^N. \qquad (13)$$

According to (8) the result will also hold for each term of the series expansion (12) and thus for the complete Hamiltonian $\hat{H}_W$ itself. For brevity we will omit the indices, writing $q \equiv q_n$, $q' \equiv q_{n+1}$. The proof of (13) will be based on mathematical induction. For $N = 1$ the assertion is satisfied:

$$\begin{aligned}\langle q'|\alpha\hat{p} + \beta\hat{q}|q\rangle &= \int \frac{dp}{2\pi\hbar} \frac{1}{2}\Big(\langle q'|p\rangle\langle p|\alpha\hat{p}+\beta\hat{q}|q\rangle + \langle q'|\alpha\hat{p}+\beta\hat{q}|p\rangle\langle p|q\rangle\Big) \\ &= \int \frac{dp}{2\pi\hbar} \langle q'|p\rangle\langle p|q\rangle \Big(\frac{1}{2}(\alpha p + \beta q) + \frac{1}{2}(\alpha p + \beta q')\Big) \\ &= \int \frac{dp}{2\pi\hbar} e^{ip(q'-q)/\hbar}\left(\alpha p + \beta\frac{q+q'}{2}\right), \qquad (14)\end{aligned}$$

Assume that (13) holds for a certain value of N. Using $\hat{p} = (\hbar/i)\partial/\partial q$ we can write the $(N+1)$th order as

$$\begin{aligned}\langle q'|(\alpha\hat{p}+\beta\hat{q})^{N+1}|q\rangle &= \langle q'|\left(\alpha\frac{\hbar}{i}\frac{\partial}{\partial\hat{q}} + \beta\hat{q}\right)(\alpha\hat{p}+\beta\hat{q})^N|q\rangle \\ &= \left(\alpha\frac{\hbar}{i}\frac{\partial}{\partial q'} + \beta q'\right)\langle q'|(\alpha\hat{p}+\beta\hat{q})^N|q\rangle. \qquad (15)\end{aligned}$$

Since the Nth-order assertion is assumed to hold this can be rewritten as follows:

$$\begin{aligned}\langle q'|(\alpha\hat{p}+\beta\hat{q})^{N+1}|q\rangle &= \left(\alpha\frac{\hbar}{i}\frac{\partial}{\partial q'} + \beta q'\right)\int \frac{dp}{2\pi\hbar} e^{ip(q'-q)/\hbar}\left(\alpha p + \beta\frac{q+q'}{2}\right)^N \\ &= \int \frac{dp}{2\pi\hbar} e^{ip(q'-q)/\hbar}\left(\alpha p + \alpha\frac{\hbar}{i}\frac{\partial}{\partial q'} + \beta q'\right)\left(\alpha p + \beta\frac{q+q'}{2}\right)^N \\ &= \int \frac{dp}{2\pi\hbar} e^{ip(q'-q)/\hbar}\left(\alpha p + \beta\frac{q+q'}{2} + \alpha\frac{\hbar}{i}\frac{\partial}{\partial q'} + \beta\frac{q'-q}{2}\right) \\ &\quad \times \left(\alpha p + \beta\frac{q+q'}{2}\right)^N \\ &= \int \frac{dp}{2\pi\hbar} e^{ip(q'-q)/\hbar}\left(\alpha p + \beta\frac{q+q'}{2}\right)^{N+1}. \qquad (16)\end{aligned}$$

In the last step two terms have cancelled each other. This is verified by the following calculation (the arrows indicate to which function the differential operators is applied):

$$\begin{aligned}&\int \frac{dp}{2\pi\hbar} e^{ip(q'-q)/\hbar}\left(\alpha\frac{\hbar}{i}\frac{\partial}{\partial q'} + \beta\frac{q'-q}{2}\right)\left(\alpha p + \beta\frac{q+q'}{2}\right)^N \\ &= \int \frac{dp}{2\pi\hbar} e^{ip(q'-q)/\hbar}\left(\alpha\frac{\hbar}{i}\frac{\overrightarrow{\partial}}{\partial q'} + \beta\frac{\hbar}{i}\frac{\overleftarrow{\partial}}{\partial p}\right)\left(\alpha p + \beta\frac{q+q'}{2}\right)^N \\ &= \int \frac{dp}{2\pi\hbar} e^{ip(q'-q)/\hbar}\frac{\hbar}{i}\left(\alpha\frac{\overrightarrow{\partial}}{\partial q'} - \beta\frac{\overrightarrow{\partial}}{\partial p}\right)\left(\alpha p + \beta\frac{q+q'}{2}\right)^N, \qquad (17)\end{aligned}$$

Example 11.2

where the last step involved an integration by parts. The integrand in (17) vanishes because of

$$\left(\alpha\frac{\partial}{\partial q'} - \beta\frac{\partial}{\partial p}\right) f\left(\alpha p + \beta\frac{q+q'}{2}\right) = \left(\frac{1}{2}\alpha\beta - \frac{1}{2}\beta\alpha\right)f' = 0. \tag{18}$$

This completes the proof of (13) and (11) which in turn confirms (9). Thus we have shown that the "midpoint prescription" for the evaluation of the path integral corresponds to the Weyl ordering of the operators in the quantum Hamiltonian.

11.4 The Multi-Dimensional Path Integral

The extension of the path integral to systems of several degrees of freedom $(q_1, \ldots, q_D; p_1, \ldots, p_D) := (q; p)$ does not present any problems. (Note: we do not use a special notation to mark D-dimensional vectors.) The phase-space path integral for the Feynman kernel from (11.31) now reads

$$\langle q_1', \ldots, q_D', t' | q_1, \ldots, q_D, t \rangle$$
$$= \int \prod_{\alpha=1}^{D} \mathcal{D}q_\alpha \prod_{\beta=1}^{D} \mathcal{D}q_\beta \, \exp\left[\frac{i}{\hbar}\int_{t}^{t'} d\tau \left(\sum_{\alpha=1}^{D} p_\alpha \dot{q}_\alpha - H(p,q)\right)\right] \tag{11.43}$$

where the individual path integrations again are defined by the limiting process specified in (11.29). The integration extends over all paths $p(t)$ in momentum space and over those paths $q(t)$ in coordinate space which satisfy the boundary conditions $q(t) = q$ and $q(t') = q'$. The D-dimensional path integral in Feynman's form reads

$$\langle q_1', \ldots, q_D', t' | q_1, \ldots, q_D, t \rangle = \mathcal{N} \int \prod_{\alpha=1}^{D} \mathcal{D}q_\alpha \, e^{\frac{i}{\hbar}\int_{t}^{t'} d\tau \, L(q,\dot{q})}. \tag{11.44}$$

We remark that Feynman's coordinate-space path integral in principle is less general than the phase-space path integral. The transition from (11.43) to (11.44) was based on the assumption that the momentum dependence of the Hamiltonian is given by a quadratic function with constant coefficients, cf. (11.33). This restriction can be relaxed a little, admitting a kinetic energy term which consists of a *general quadratic form* with q-dependent coefficients. The Lagrangian in D dimensions is taken as

$$\begin{aligned} L(q,\dot{q}) &= \frac{1}{2}\sum_{\alpha\beta} \dot{q}_\alpha M_{\alpha\beta}(q)\dot{q}_\beta + \sum_\alpha b_\alpha(q)\dot{q}_\alpha - V(q) \\ &\equiv \frac{1}{2}\dot{q}^\mathrm{T} M(q)\dot{q} + b^\mathrm{T}(q)\dot{q} - V(q) \, . \end{aligned} \tag{11.45}$$

Here M is a nonsingular symmetric real $D \times D$ matrix and b is a D dimensional vector, both can be functions of q. The following canonically conjugate momentum is obtained from (11.45):

$$p = \frac{\partial L}{\partial \dot{q}^\mathrm{T}} = M(q)\dot{q} + b \, . \tag{11.46}$$

Introducing the shifted momentum $p' = p - b$ we can solve (11.46) for the velocity

$$\dot{q} = M^{-1}(q)(p - b) \equiv M^{-1}(q) p' . \qquad (11.47)$$

The Hamiltonian becomes

$$\begin{aligned} H(q,p) &= p^T \dot{q} - L(q, \dot{q}(p)) \\ &= (p'+b)^T M^{-1} p' - \tfrac{1}{2} p'^T M^{-1} M M^{-1} p' - b^T M^{-1} p' + V(q) \\ &= \tfrac{1}{2} p'^T M^{-1} p' + V(q) . \end{aligned} \qquad (11.48)$$

The phase-space path integral can be built up from the amplitudes $\langle q_{n+1}, t_{n+1} | q_n, t_n \rangle$ according to

$$\langle q', t' | q, t \rangle = \lim_{N \to \infty} \int \prod_{\alpha=1}^{D} \prod_{n=1}^{N-1} \mathrm{d} q_{\alpha n} \, \langle q', t' | q_{N-1}, t_{N-1} \rangle \ldots \langle q_1, t_1 | q, t \rangle . \qquad (11.49)$$

Inserting momentum eigenstates $|p_n\rangle \langle p_n|$ as in (11.21), we obtain an expression analogous to (11.28)

$$\langle q_{n+1}, t_{n+1} | q_n, t_n \rangle = \int \prod_{\alpha=1}^{D} \frac{\mathrm{d} p_{\alpha n}}{2\pi\hbar} \exp\left[\frac{i}{\hbar}\epsilon(p_n \dot{q}_n - H(p_n, q_n))\right] \qquad (11.50)$$

$$= \int \prod_{\alpha=1}^{D} \frac{\mathrm{d} p'_{\alpha n}}{2\pi\hbar} \exp\left[\frac{i}{\hbar}\epsilon\left(-\tfrac{1}{2} {p'_n}^T M^{-1} p'_n + {p'_n}^T \dot{q}_n + b^T \dot{q}_n - V(q_n)\right)\right] .$$

Note that the "velocity" is defined by $\dot{q}_n := (q_{n+1} - q_n)/\epsilon$ (compare (11.30)) and is not identical with the variable $\dot{q}$ contained in the Lagrangian (11.45). Therefore it cannot be expressed in terms of the integration variable p_n according to (11.48). Being of Gaussian type, the integral in (11.50) can be solved. According to Exercise 11.3 the following formula holds

$$\int \mathrm{d}^D v \, \exp\left[-\tfrac{1}{2} v^T A v + \rho^T v\right]$$
$$= (2\pi)^{D/2} \exp\left[-\tfrac{1}{2} \mathrm{Tr} \ln A\right] \exp\left[\tfrac{1}{2}\rho^T A^{-1} \rho\right] . \qquad (11.51)$$

Here $\exp\left(-\tfrac{1}{2}\mathrm{Tr}\ln A\right)$ is another way of writing $(\det A)^{-\tfrac{1}{2}}$. Functions of a matrix like $\ln A$ can be defined via their series expansion. The Gaussian integral formula can be applied to (11.50) leading to

$$\langle q_{n+1}, t_{n+1} | q_n, t_n \rangle \qquad (11.52)$$
$$= (2\pi\hbar)^{-D}(2\pi)^{D/2} \exp\left[-\tfrac{1}{2}\mathrm{Tr} \ln\left(\frac{i\epsilon}{\hbar} M^{-1}\right)\right]$$
$$\times \exp\left[\tfrac{1}{2}\left(\frac{i\epsilon}{\hbar}\dot{q}_n^T\right)\left(\frac{i\epsilon}{\hbar} M^{-1}\right)^{-1}\left(\frac{i\epsilon}{\hbar}\dot{q}_n\right)\right] \exp\left[\frac{i\epsilon}{\hbar}\left(b^T \dot{q}_n - V(q_n)\right)\right]$$
$$= (2\pi\hbar)^{-D}(2\pi)^{D/2} \exp\left[-\tfrac{1}{2}\mathrm{Tr} \ln\left(\frac{i\epsilon}{\hbar}\mathbb{1}\right)\right] \exp\left[-\tfrac{1}{2}\mathrm{Tr} \ln M^{-1}\right]$$

$$\times \exp\left[\frac{1}{2}\frac{i\epsilon}{\hbar}\dot{q}_n^T M \dot{q}_n\right] \exp\left[\frac{i\epsilon}{\hbar}\left(b^T \dot{q}_n - V(q_n)\right)\right]$$

$$= (2\pi i\hbar\epsilon)^{-D/2} \exp\left[\frac{1}{2}\text{Tr}\ln M\right] \exp\left[\frac{i\epsilon}{\hbar}\left(\frac{1}{2}\dot{q}_n^T M \dot{q}_n + b^T \dot{q}_n - V(q_n)\right)\right].$$

In the last step $\ln M^{-1} = -\ln M$ was used which follows from $M^{-1}M = \mathbb{I}$. Let us compare (11.52) with the result we obtained in one dimension and for constant mass, (11.39). Setting $D = 1$, $M = m$, and $b = 0$, we reduce (11.52) in this special case to the old result because $\exp\left(\frac{1}{2}\text{Tr}\ln m\right) = m^{1/2}$.

In the general case given by the Lagrange density (11.45) the Feynman kernel (11.49) becomes, via (11.52),

$$\langle q', t'|q, t\rangle = \lim_{N\to\infty} (2\pi i\hbar\epsilon)^{-ND/2} \int \prod_{n=1}^{N-1}\prod_{\alpha=1}^{D} dq_{\alpha n} \exp\left[\frac{1}{2}\text{Tr}\ln M(q_n)\right]$$

$$\times \exp\left[\frac{i\epsilon}{\hbar}\sum_{n=0}^{N-1} L(q_n, \dot{q}_n)\right]. \quad (11.53)$$

If the "mass matrix" $M(q)$ does *not depend on the coordinates* then the extra factor $\exp\left(\frac{1}{2}\text{Tr}\ln M\right)$ can be extracted from the integral and is absorbed by the normalization constant $\mathcal{N}$. Then the coordinate-space path integral again becomes

$$\langle q', t'|q, t\rangle = \mathcal{N}' \int \mathcal{D}q \exp\left[\frac{i\epsilon}{\hbar}\int_t^{t'} d\tau\, L(q, \dot{q})\right], \quad (11.54)$$

and agrees with the familiar expression (11.44). The new normalization factor is obtained from $\left((2\pi i\hbar\epsilon)^D/\det M\right)^{N/2}$. The limit $N \to \infty$ is not well defined but this does not preclude practical applications of the path integral, as mentioned earlier.

Remark: If *the mass matrix $M(q)$ depends on the coordinates* Feynman's formula (11.54) has to be modified since the extra factor

$$\prod_{n=1}^{N-1} \exp\left[\frac{1}{2}\text{Tr}\ln M(q_n)\right] = \exp\left[\frac{1}{2\epsilon}\sum_{n=1}^{N-1}\epsilon\,\text{Tr}\ln M(q_n)\right] \quad (11.55)$$

cannot be drawn before the integral. In the limit $\epsilon \to 0$, $N \to \infty$ the sum in the exponent becomes an integral over time. If we write symbolically $1/\epsilon \to \delta(0)$ the path integral becomes

$$\langle q', t'|q, t\rangle = \mathcal{N}' \int \mathcal{D}q \exp\left[\frac{i}{\hbar}\int_t^{t'} d\tau\left(L(q,\dot{q}) - \frac{i\hbar}{2}\delta(0)\,\text{Tr}\ln M(q)\right)\right]. \quad (11.56)$$

In certain cases the extra term in the exponent of (11.56) is required to cancel a singular contribution that arises from another source.[4]

The ansatz (11.45) for the Lagrangian quadratic in the velocities $\dot{q}$ is general enough to describe most of the systems usually encountered. However,

---

[4] T.D. Lee, C.N. Yang: Phys. Rev. **128**, 885 (1962).

a possible complication arises if the matrix M does not have an inverse. As an important example where this happens consider the case that some of the velocities $\dot{q}_\alpha$ do not contribute to the kinetic energy in (11.45), such that the corresponding rows (and columns) in M vanish. Then the above derivation does not apply and one has to go back to the original phase-space path integral (11.43), which has more general validity.

EXERCISE

11.3 Gaussian Integrals in D Dimensions

Problem. Let v and ρ be D-dimensional vectors and A a real symmetric and positive-definite $D \times D$ matrix. Prove the following identities for the generalized Gaussian integral (the integrations extend over the whole space $\mathbb{R}^D$):

(a) $\displaystyle\int d^D v\, e^{-\frac{1}{2} v^T A v} = (2\pi)^{D/2} e^{-\frac{1}{2}\operatorname{Tr}\ln A}$, (1)

(b) $\displaystyle\int d^D v\, e^{-\frac{1}{2} v^T A v + \rho^T v} = (2\pi)^{D/2} e^{-\frac{1}{2}\operatorname{Tr}\ln A}\, e^{\frac{1}{2}\rho^T A^{-1} \rho}$, (2)

(c) $\displaystyle\int d^D v\, v_{k_1} \cdots v_{k_n} e^{-\frac{1}{2} v^T A v}$
$= (2\pi)^{D/2} e^{-\frac{1}{2}\operatorname{Tr}\ln A} \left(A^{-1}_{k_1 k_2} \cdots A^{-1}_{k_{n-1} k_n} + \text{permut.}\right)$ (3)

if n is even. For odd n the integral vanishes.

(d) If A is a positive-definite hermitean matrix and z is a complex vector:

$$\int dz_1 dz_1^* \ldots dz_D dz_D^*\, e^{-z^\dagger A z} = (2\pi)^D\, e^{-\operatorname{Tr}\ln A}\,.$$ (4)

The integration extends over the complex plane according to

$$\int dz\, dz^*\, f(z, z^*) = 2 \int_{-\infty}^{\infty} d\operatorname{Re} z \int_{-\infty}^{\infty} d\operatorname{Im} z\, f(z, z^*)\,.$$ (5)

Solution. (a) In one dimension the ordinary Gaussian integral reads

$$\int_{-\infty}^{\infty} dv\, e^{-\frac{1}{2} a v^2} = \left(\frac{2\pi}{a}\right)^{1/2},$$ (6)

where a is a positive constant. By assumption, A is a real symmetric matrix which can be diagonalized according to

$$\begin{aligned}\tilde{v} &= Uv\,,\\ \tilde{A} &= UAU^{-1} = UAU^T = \operatorname{diag}(\alpha_1, \ldots, \alpha_D)\,,\end{aligned}$$ (7)

where U is an orthogonal rotation matrix. The numbers α_i are the eigenvalues of the matrix A. The transformation (7) by construction leaves the bilinear form $v^T A v$ invariant. Furthermore we have $UU^T = 1$ and thus the determinant is $\det U = 1$. This implies that the coordinate transformation described

Exercise 11.3

by U does not change the volume element. Then the integral in (1) can be split into a product of simple one-dimensional integrals:

$$\int d^D v \, e^{-\frac{1}{2}v^T A v} = \int d^D \tilde{v} \, e^{-\frac{1}{2}\tilde{v}^T \tilde{A} \tilde{v}} = \prod_{i=1}^{D} \int_{-\infty}^{\infty} d\tilde{v}_i \, e^{-\frac{1}{2}\alpha_i \tilde{v}_i^2}$$

$$= \prod_{i=1}^{D} \left(\frac{2\pi}{\alpha_i} \right)^{1/2} = (2\pi)^{D/2} \, (\det A)^{-1/2} \,, \qquad (8)$$

since the determinant of a matrix is the product of all of its eigenvalues. Clearly the integral will only exist if all the eigenvalues of the matrix are positive, which means that the matrix A has to be positive definite. The final result in (1) follows from the identity

$$\ln \det A = \ln \prod_{i=1}^{D} \alpha_i = \sum_{i=1}^{D} \ln \alpha_i = \mathrm{Tr} \ln A \,. \qquad (9)$$

(b) By completing the square the integrand can be reduced to case **(a)**. In one dimension the transformation

$$av^2 - 2rv = a\left(v - \frac{r}{a}\right)^2 - \frac{1}{a}r^2 \qquad (10)$$

would be used. This can be generalized to the following identity for the bilinear form in D dimensions:

$$v^T A v - 2\rho^T v = (v - A^{-1}\rho)^T A (v - A^{-1}\rho) - \rho^T A^{-1}\rho \,, \qquad (11)$$

which can be verified by using $(A^{-1})^T = A^{-1}$ and $\rho^T v = v^T \rho$. After shifting the integration variable $v' = v - A^{-1}\rho$, (2) is obtained immediately.

(c) Gaussian integrals containing an arbitrary polynomial in the integral besides the exponential function can be solved in closed form. The derivation can be based on an often-used trick: an auxiliary vector ρ is introduced which enters as a linear term in the exponent. The corresponding integral (2) is differentiated with respect to the elements ρ_i of the auxiliary vector, each time producing a factor of v_i. At the end of the calculation the auxiliary variable is removed by setting $\rho = 0$. This results in

$$\int d^D v \, v_{k_1} \cdots v_{k_n} \, e^{-\frac{1}{2}v^T A v}$$

$$= \frac{\partial}{\partial \rho_{k_1}} \cdots \frac{\partial}{\partial \rho_{k_n}} \int d^D v \, e^{-\frac{1}{2}v^T A v + \rho^T v} \bigg|_{\rho=0}$$

$$= (2\pi)^{D/2} e^{-\frac{1}{2}\mathrm{Tr} \ln A} \, \frac{\partial}{\partial \rho_{k_1}} \cdots \frac{\partial}{\partial \rho_{k_n}} e^{\frac{1}{2}\rho^T A^{-1} \rho} \bigg|_{\rho=0} . \qquad (12)$$

From

$$\frac{\partial}{\partial \rho_k} \left(\frac{1}{2} \rho^T A^{-1} \rho \right) = \frac{1}{2} \sum_{j=1}^{D} \left(A_{kj}^{-1} \rho_j + \rho_j A_{jk}^{-1} \right) = (A^{-1}\rho)_k \qquad (13)$$

it becomes obvious that the sought for integral vanishes if $n = 1$. In the case of two factors, v_{k_1} and v_{k_2}, one finds

$$\frac{\partial}{\partial \rho_{k_2}} \frac{\partial}{\partial \rho_{k_1}} e^{\frac{1}{2}\rho^T A^{-1}\rho} = \frac{\partial}{\partial \rho_{k_2}} \left((A^{-1}\rho)_{k_1} e^{\frac{1}{2}\rho^T A^{-1}\rho} \right)$$

$$= \left(A^{-1}_{k_1 k_2} + (A^{-1}\rho)_{k_1}(A^{-1}\rho)_{k_2} \right) e^{\frac{1}{2}\rho^T A^{-1}\rho}. \quad (14)$$

Exercise 11.3

This implies

$$\int d^D v \, v_{k_1} v_{k_2} \, e^{-\frac{1}{2}v^T A v} = (2\pi)^{D/2} e^{-\frac{1}{2}\mathrm{Tr}\,\ln A} \, A^{-1}_{k_1 k_2}. \quad (15)$$

For increasingly higher values of n, the application of the product rule of differentiation produces a rapidly growing number of terms. It is obvious that the exponential function is multiplied by a polynomial in ρ of the order n which consists solely of either even or odd powers. Therefore a zeroth-order term surviving the limit $\rho \to 0$ will exist only if n is even. It is not difficult to confirm that this constant term consists of products of $\frac{n}{2}$ factors of the type $A^{-1}_{k_1 k_2} \ldots A^{-1}_{k_{n-1} k_n}$, where all possible permutations of the indices k_i will occur. The only exception is the interchange of the two indices in a matrix A^{-1}: in order to avoid double counting only one of the two combinations is to be included.

(d) The proof from **(a)** can be easily extended to the complex domain. The matrix A must then be hermitean and it is diagonalized by a unitary matrix U. Equation (8) becomes

$$\int dz_1 dz_1^* \ldots dz_D dz_D^* \, e^{-z^\dagger A z} = \int d\tilde{z}_1 d\tilde{z}_1^* \ldots d\tilde{z}_D d\tilde{z}_D^* \, e^{-\sum_i \alpha_i |\tilde{z}_i|^2}$$

$$= \prod_{i=1}^{D} \int d\tilde{z}_i d\tilde{z}_i^* \, e^{-\alpha_i |\tilde{z}_i|^2}. \quad (16)$$

The integration extends over the whole complex plane. To show that the integration measure is invariant under the unitary transformation we split the matrix U into two real matrices V and W according to $U = V + iW$. The unitarity condition $UU^\dagger = \mathbb{I}$ then becomes

$$VV^T + WW^T = \mathbb{I} \quad , \quad WV^T - VW^T = 0. \quad (17)$$

The real and imaginary components of $z = x + iy$ transform as $\tilde{x} + i\tilde{y} = (V + iW)(x + iy)$ or

$$\begin{pmatrix} \tilde{x} \\ \tilde{y} \end{pmatrix} = \begin{pmatrix} V & -W \\ W & V \end{pmatrix} \begin{pmatrix} x \\ y \end{pmatrix} \equiv M \begin{pmatrix} x \\ y \end{pmatrix}. \quad (18)$$

The $2D \times 2D$ matrix M is real and orthogonal since (17) implies

$$MM^T = \begin{pmatrix} V & -W \\ W & V \end{pmatrix} \begin{pmatrix} V^T & W^T \\ -W^T & V^T \end{pmatrix}$$

$$= \begin{pmatrix} VV^T + WW^T & VW^T - WV^T \\ WV^T - VW^T & WW^T + VV^T \end{pmatrix} = \mathbb{I}, \quad (19)$$

which proves the invariance of the volume element of the D-dimensional integration. Using (5) we have for each factor in (16)

Exercise 11.3

$$\int d\tilde{z}\, d\tilde{z}^* \, e^{-\alpha|\tilde{z}|^2} = 2\int d\operatorname{Re}\tilde{z}\, d\operatorname{Im}\tilde{z}\, e^{-\alpha(\operatorname{Re}\tilde{z})^2}\, e^{-\alpha(\operatorname{Im}\tilde{z})^2}$$

$$= 2\left(\int_{-\infty}^{\infty} dx\, e^{-\alpha x^2}\right)^2 = 2\frac{\pi}{\alpha} \tag{20}$$

and therefore

$$\int dz_1 dz_1^* \ldots dz_D dz_D^*\, e^{-z^\dagger A z} = \prod_{i=1}^{D} \frac{2\pi}{\alpha_i} = \frac{(2\pi)^D}{\det A} = (2\pi)^D\, e^{-\operatorname{Tr}\ln A}\ . \tag{21}$$

The factor 2 in the integration measure (5) originates in the Jacobian of the transformation according to $z = x + iy$, $z^* = x - iy$, $|dz\, dz^*| = 2\, dx\, dy$.

11.5 The Time-Ordered Product and n-Point Functions

The path-integral representation of the Feynman kernel $\langle q', t'|q, t\rangle$ can be employed to evaluate various transition matrix elements and expectation values. We are particulary interested in the matrix elements of the coordinate operator $\hat{q}(t)$ (in the Heisenberg picture) and of products of $\hat{q}$ operators. These matrix elements take a simple form in the path-integral representation, as we will show for the example of a one-dimensional quantum system. Generalizing the result to several dimensions and ultimately to quantum fields is straightforward.

Assume that t_i is a time coordinate within the inverval $t \leq t_i \leq t'$. If one wants to evaluate the matrix element of the operator $\hat{q}(t_i)$ taken between the states $\langle q', t'|$ and $|q, t\rangle$ it will be advisable to choose a partition of the time coordinate such that t_i coincides with one of the grid points. Then the operator $\hat{q}(t_i)$ will act on its eigenstate $|q_i, t_i\rangle$ and will be replaced by the eigenvalue $q(t_i)$, introducing a simple c number into the integrand:

$$\langle q', t'|\hat{q}(t_i)|q, t\rangle = \int dq_{N-1} \ldots \int dq_1\, \langle q', t'|q_{N-1}, t_{N-1}\rangle \cdots$$
$$\times \langle q_{i+1}, t_{i+1}|\hat{q}(t_i)|q_i, t_i\rangle \cdots \langle q_1, t_1|q, t\rangle$$
$$= \int \mathcal{D}q\, \mathcal{D}p\, q(t_i)\, \exp\left[\frac{i}{\hbar}\int_t^{t'} d\tau(p\dot{q} - H)\right], \tag{11.57}$$

a result which is hardly surprising. How can (11.57) be generalized to products of more than one $\hat{q}$ operator? Let us take two intermediate time coordinates t_i and t_j, assuming at first that $t_i > t_j$. Then we have

$$\langle q', t'|\hat{q}(t_i)\hat{q}(t_j)|q, t\rangle = \int dq_{N-1} \ldots \int dq_1\, \langle q', t'|q_{N-1}, t_{N-1}\rangle \cdots$$
$$\times \langle q_{i+1}, t_{i+1}|\hat{q}(t_i)|q_i, t_i\rangle \cdots \langle q_{j+1}, t_{j+1}|\hat{q}(t_j)|q_j, t_j\rangle \cdots \langle q_1, t_1|q, t\rangle$$
$$= \int \mathcal{D}q\, \mathcal{D}p\, q(t_i) q(t_j)\, \exp\left[\frac{i}{\hbar}\int_t^{t'} d\tau(p\dot{q} - H)\right]. \tag{11.58}$$

11.5 The Time-Ordered Product and n-Point Functions

Under the opposite condition, $t_j > t_i$, this calulation is not valid since the path integral assumes an increasing (from right to left) ordering of the grid points in time. Instead it is clear that we obtain exactly the same integral if the ordering of the operators is inverted, $\hat{q}(t_i)\hat{q}(t_j) \to \hat{q}(t_j)\hat{q}(t_i)$. Both cases can be combined by using the *time-ordered product* $T(\cdots)$ which we have already encountered several times:

$$\langle q',t'|T[\hat{q}(t_i)\hat{q}(t_j)]|q,t\rangle = \int \mathcal{D}q\,\mathcal{D}p\, q(t_i)q(t_j) \exp\left[\frac{\mathrm{i}}{\hbar}\int_t^{t'} \mathrm{d}\tau\,(p\dot{q}-H)\right]. \tag{11.59}$$

It is most interesting that the path-integral formalism has led us in a natural way to the same concept of time-ordered products that played such an important role for the perturbation theory developed in Chap. 8.

Of course (11.59) is immediately generalized to time-ordered products involving more than two factors:

$$\langle q',t'|T[\hat{q}(t_1)\hat{q}(t_2)\cdots\hat{q}(t_n)]|q,t\rangle$$
$$= \int \mathcal{D}q\,\mathcal{D}p\ \ q(t_1)\cdots q(t_N) \exp\left[\frac{\mathrm{i}}{\hbar}\int_t^{t'} \mathrm{d}\tau\,(p\dot{q}-H)\right]. \tag{11.60}$$

For a system with quadratic velocity dependence in the Lagrangian, Feynman's coordinate-space path integral can be employed:

$$\langle q',t'|T[\hat{q}(t_1)\cdots\hat{q}(t_n)]|q,t\rangle = \mathcal{N}\int \mathcal{D}q\, q(t_1)\ldots q(t_n) \exp\left(\frac{\mathrm{i}}{\hbar}\int_t^{t'} \mathrm{d}\tau\, L\right). \tag{11.61}$$

A quantity of particular interest is the *ground-state expectation value of the time-ordered product of $\hat{q}$ operators*. We have studied the analogous object in the framework of quantum field theory, where it was identified with the field propagator and could be used to construct the S matrix. We are interested in the *two-point function*, which for the simple case of a one-dimensional system depends only on the two time coordinates t_1 and t_2:

$$G(t_1,t_2) = \langle 0|T[\hat{q}(t_1)\hat{q}(t_2)]|0\rangle. \tag{11.62}$$

Using a clever trick we can extract this function from the transition matrix element (11.61). For this purpose we insert two complete sets of states on the left and on the right of (11.61). This time we employ the eigenstates of the Hamiltonian in the Schrödinger picture:

$$\hat{H}|n\rangle_\mathrm{S} = E_n|n\rangle_\mathrm{S} \tag{11.63}$$

which leads to

$$\begin{aligned}|q,t\rangle &= \mathrm{e}^{\mathrm{i}\hat{H}t/\hbar}\sum_n |n\rangle_\mathrm{S}\,_\mathrm{S}\langle n|q\rangle_\mathrm{S} = \sum_n \mathrm{e}^{\mathrm{i}E_n t/\hbar}|n\rangle_\mathrm{S}\,_\mathrm{S}\langle n|q\rangle_\mathrm{S} \\ &= \sum_n \mathrm{e}^{\mathrm{i}E_n t/\hbar}\psi_n^*(q)|n\rangle_\mathrm{S}.\end{aligned} \tag{11.64}$$

Here $\psi_n(q)$ denotes the coordinate-space wave function belonging to the state $|n\rangle$. Then the matrix element in (11.61) can be rewritten as

$$\langle q',t'|T[\hat{q}(t_1)\hat{q}(t_2)]|q,t\rangle \qquad (11.65)$$

$$= \sum_{n,n'} \langle q',t'|n'\rangle_S {}_S\langle n'|T[\hat{q}(t_1)\hat{q}(t_2)]|n\rangle_S {}_S\langle n|q,t\rangle$$

$$= \sum_{n,n'} e^{-i(E_{n'}t'-E_n t)/\hbar} \psi_{n'}(q')\psi_n^*(q) \quad {}_S\langle n'|T[\hat{q}(t_1)\hat{q}(t_2)]|n\rangle_S .$$

One of the many contributing terms on the right-hand side (the one having $n' = n = 0$) contains the two-point function $G(t_1, t_2)$ we are interested in.

How can this contribution be extracted form the sum? Taking the limit $t' \to +\infty$, $t \to -\infty$ at first does not seem to help much since the exponential functions at infinity exhibit undamped oscillatory behavior. At this point, however, one may take recourse to a mathematical artifice by which the oscillations are turnend into an exponential fall-off: we change the limiting prescription and *attach an imaginary part to the time coordinate*. To be more precise, a rotation by an angle $0 < \delta < \pi$ in the complex plane in the mathematically negative direction is performed (see Fig. 11.2), leading to the limits $t' \to \infty e^{-i\delta}$ and $t \to -\infty e^{-i\delta}$. This has the desired consequence: the factor $\exp[-i(E_{n'}t' - E_n t)]$ drops off exponentially (assuming that all energies E_n can be chosen positive).

Fig. 11.2. The complex rotation of the time axis

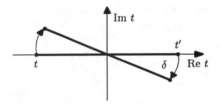

Expressed in terms of the new rotated time coordinates $\tau = e^{i\delta}t$ and $\tau' = e^{i\delta}t'$ the limit of the matrix element reads (11.65)

$$\lim_{\substack{t'\to\infty \\ t\to-\infty}} \langle q',t'|T[\hat{q}(t_1)\hat{q}(t_2)]|q,t\rangle = \lim_{\substack{\tau'\to e^{i\delta}\infty \\ \tau\to-e^{i\delta}\infty}} \langle q',e^{-i\delta}\tau'|T[\hat{q}(t_1)\hat{q}(t_2)]|q,e^{-i\delta}\tau\rangle$$

$$\leftrightarrow \lim_{\substack{\tau'\to\infty \\ \tau\to-\infty}} \langle q',e^{-i\delta}\tau'|T[\hat{q}(t_1)\hat{q}(t_2)]|q,e^{-i\delta}\tau\rangle .$$

(11.66)

The decisive step is taken in the second line by going to real values of the rotated time coordinate τ. This will only be admissible mathematically if the matrix element is an analytic function in the variables t and t'; then the transition in (11.66) amounts to an *analytical continuation*. To phrase it differently: by using (11.66) we attribute to the matrix element a value obtained by starting from the well-defined quantity on the right-hand side and making an analytical continuation to $\delta \to 0$. The particular value chosen for the angle δ is not important and the result must not depend on it. The most elegant formulation is achieved if an angle of $\delta = \pi/2$ is chosen, rotating the time axis into the purely imaginary direction, $t' \to -i\infty$, $t \to +i\infty$. Such a rotation of the time-like component of a Minkowski vector by 90^0 in the

11.5 The Time-Ordered Product and n-Point Functions

complex plane is called a *Wick rotation* or the *extension to the euclidian region*. The Wick rotation is a very popular tool in field theory, and we will come back to it in Sect. 12.2.

When applied to the two-point function the Wick rotation calls for the calculation of (11.65) for imaginary time arguments $t = -i\tau$, τ being real:

$$\langle q', -i\tau' | T[\hat{q}(t_1)\hat{q}(t_2)] | q, -i\tau \rangle$$
$$= \sum_{n,n'} e^{-(E_{n'}\tau' - E_n \tau)} \psi_{n'}(q') \psi_n^*(q) \, {}_S\langle n' | T[\hat{q}(t_1)\hat{q}(t_2)] | n \rangle_S . \quad (11.67)$$

Compared to (11.65) this expression has a decisive advantage: for imaginary times the exponential functions no longer oscillate, rather they fall off rapidly. In the limit of large τ the sum will be dominated by the term with the slowest fall-off rate, which happens to be the *ground state* $n = 0$ having the lowest energy (we assume that the ground state is not degenerate):

$$\langle q', -i\tau' | T[\hat{q}(t_1)\hat{q}(t_2)] | q, -i\tau \rangle$$
$$\to e^{-E_0(\tau' - \tau)} \psi_0(q') \psi_0^*(q) \langle 0 | T[\hat{q}(t_1)\hat{q}(t_2)] | 0 \rangle . \quad (11.68)$$

We remark that this reasoning remains valid also for other choices of the rotation angle δ taken from the interval $0 < \delta < \pi$. Then the exponent would contain an additional factor $i e^{-i\delta}$ with a positive real part.

The prefactor on the right-hand side of (11.67) obviously is the limit of the Feynman kernel in the euclidian region:

$$\langle q', -i\tau' | q, -i\tau \rangle = \sum_n e^{-E_n(\tau' - \tau)} \psi_n(q') \psi_n^*(q) \quad (11.69)$$
$$\to e^{-E_0(\tau' - \tau)} \psi_0(q') \psi_0^*(q) = \psi_0(q', -i\tau') \psi_0^*(q, -i\tau) .$$

Thus we obtain the two-point function as the limit of the ratio between (11.68) and (11.69):

$$\langle 0 | T[\hat{q}(t_1)\hat{q}(t_2)] | 0 \rangle = \lim_{\substack{\tau' \to \infty \\ \tau \to -\infty}} \frac{\langle q', -i\tau' | T[\hat{q}(t_1)\hat{q}(t_2)] | q, -i\tau \rangle}{\langle q', -i\tau' | q, -i\tau \rangle} . \quad (11.70)$$

Now the analytical continuation back to real time values can be taken (i.e., we rotate by the angle $+90^0$ in the complex plane). Expressed in terms of the path-integral representations (11.61) and (11.40) for the numerator and denominator of (11.70), the final result for the two-point function is

$$\langle 0 | T[\hat{q}(t_1)\hat{q}(t_2)] | 0 \rangle = \lim_{\substack{t' \to \infty \\ t \to -\infty}} \frac{\langle q', t' | T[\hat{q}(t_1)\hat{q}(t_2)] | q, t \rangle}{\langle q', t' | q, t \rangle} \quad (11.71)$$

$$= \lim_{\substack{t' \to \infty \\ t \to -\infty}} \frac{\int \mathcal{D}q \int \mathcal{D}p \; q(t_1) q(t_2) \exp\left[\frac{i}{\hbar} \int_{-\infty}^{\infty} dt \, (p\dot{q} - H)\right]}{\int \mathcal{D}q \int \mathcal{D}p \; \exp\left[\frac{i}{\hbar} \int_{-\infty}^{\infty} dt \, (p\dot{q} - H)\right]} .$$

This relation can immediately extended to the case of arbitrary *n-point functions*:

$$\langle 0|T[\hat{q}(t_1)\cdots\hat{q}(t_n)]|0\rangle$$
$$= \lim_{\substack{t'\to\infty \\ t\to-\infty}} \frac{\int \mathcal{D}q \int \mathcal{D}p \; q(t_1)\cdots q(t_n) \exp\left[\frac{i}{\hbar}\int_{-\infty}^{\infty} dt\,(p\dot{q}-H)\right]}{\int \mathcal{D}q \int \mathcal{D}p \; \exp\left[\frac{i}{\hbar}\int_{-\infty}^{\infty} dt\,(p\dot{q}-H)\right]}. \quad (11.72)$$

If the conditions described in Sect. 11.3 are fulfilled, (11.72) can also be written as a Feynman-type path integral:

$$\langle 0|T[\hat{q}(t_1)\cdots\hat{q}(t_n)]|0\rangle$$
$$= \lim_{\substack{t'\to\infty \\ t\to-\infty}} \frac{\int \mathcal{D}q \; q(t_1)\cdots q(t_n) \exp\left[\frac{i}{\hbar}\int_{-\infty}^{\infty} dt\, L(q,\dot{q})\right]}{\int \mathcal{D}q \; \exp\left[\frac{i}{\hbar}\int_{-\infty}^{\infty} dt\, L(q,\dot{q})\right]}. \quad (11.73)$$

If convergence problems are encountered when dealing with the path integral one should remember that the limiting process can be taken in the complex plane, i.e., $t' \to e^{-i\delta}\infty$ and $t \to -e^{-i\delta}\infty$.

11.6 The Vacuum Persistence Amplitude $W[J]$

A useful approach to learn about the dynamical properties of a quantum system would be to study how it reacts to a "kick". In other words the system is subjected to an externally applied perturbation and the influence of this perturbation on transition amplitudes, for example, on the by now familiar amplitude $\langle q',t'|q,t\rangle$, is evaluated. The perturbation will be described by a *classical source term* $J(t)$ which one may visualize as an external macroscopic force field. To ensure that $J(t)$ enters the equation of motion as an inhomogeneous source a term $J(t)q$ has to be included in the Lagrangian of the system.

Since the validity of the path-integral formalism is not affected by such a modification we already know what the Feynman kernel (11.31) looks like in the presence of the perturbation:

$$\langle q',t'|q,t\rangle_J = \int \mathcal{D}q \int \mathcal{D}p \; \exp\left[\frac{i}{\hbar}(p\dot{q}-H(p,q)+J(t)q)\right], \quad (11.74)$$

where the index J indicates the presence of the perturbation. The object $\langle q',t'|q,t\rangle_J$ is a *functional* depending on the function $J(t)$. Let us study the properties of (11.74) in the limit of large times, $t' \to +\infty$, $t \to -\infty$. To obtain a meaningful limit we assume that the source is switched off adiabatically at large times (see Fig. 11.3)

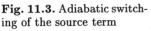

Fig. 11.3. Adiabatic switching of the source term

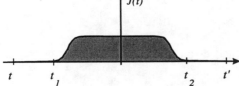

$$\lim_{t \to \pm\infty} J(t) = 0 \,. \tag{11.75}$$

The time values t_1 and t_2 with $t < t_1 < 0$, $t' > t_2 > 0$ are to be chosen such that the function $J(t)$ vanishes outside of the interval $[t_1, t_2]$. After inserting two complete sets of eigenstates of the coordinate operator $\hat{q}$ (in the "moving basis" of the Heisenberg picture, cf. (11.12)) we obtain ← time dependent but no integration over t

$$\langle q', t' | q, t \rangle_J = \int dq_2 \int dq_1 \, \langle q', t' | q_2, t_2 \rangle \langle q_2, t_2 | q_1, t_1 \rangle_J \langle q_1, t_1 | q, t \rangle \,. \tag{11.76}$$

Since in the outer region the perturbation J is switched off, the first and third matrix element in (11.76) can be expanded with respect to the eigenstates of the unperturbed Hamiltonian $\hat{H}$. As in (11.64) we obtain

$$\begin{aligned}
\langle q', t' | q_2, t_2 \rangle &= \langle q' | \exp\left[-\frac{i}{\hbar}\hat{H}(t' - t_2)\right] | q_2 \rangle \\
&= \sum_n \langle q' | \exp\left[-\frac{i}{\hbar}\hat{H}t'\right] | n \rangle \langle n | \exp\left[\frac{i}{\hbar}\hat{H}t_2\right] | q_2 \rangle \\
&= \sum_n \langle q' | n \rangle \langle n | q_2 \rangle \exp\left[-\frac{i}{\hbar}E_n(t' - t_2)\right] \\
&= \sum_n \psi_n(q') \psi_n^*(q_2) \exp\left[-\frac{i}{\hbar}E_n(t' - t_2)\right] \,, \tag{11.77}
\end{aligned}$$

and correspondingly

$$\langle q_1, t_1 | q, t \rangle = \sum_n \psi_n(q_1) \psi_n^*(q) \exp\left[-\frac{i}{\hbar}E_n(t_1 - t)\right] \,. \tag{11.78}$$

Then (11.76) reads

$$\begin{aligned}
\langle q', t' | q, t \rangle_J &= \sum_{n, n'} \psi_{n'}(q') \psi_n^*(q) \exp\left[-\frac{i}{\hbar}(E_{n'}t' - E_n t)\right] \\
&\quad \times \int dq_2 \int dq_1 \, \psi_{n'}^*(q_2, t_2) \langle q_2, t_2 | q_1, t_1 \rangle_J \psi_n(q_1, t_1) \,.
\end{aligned} \tag{11.79}$$

We are interested in the Feynman kernel (11.74) for large times, $t' \to \infty$, $t \to -\infty$. Because of the oscillating phase factors this limit is not well defined when taken literally. Here the same artifice of going to complex times can be applied, which we introduced in the last section for the evaluation of the n-point function. As in (11.66) we make an analytic continuation to the euclidian region (i.e., to imaginary times $\tau = it$):

$$\lim_{\substack{t' \to \infty \\ t \to -\infty}} \langle q', t' | q, t \rangle_J = \lim_{\substack{\tau' \to i\infty \\ \tau \to -i\infty}} \langle q', -i\tau' | q, -i\tau \rangle_J \leftrightarrow \lim_{\substack{\tau' \to \infty \\ \tau \to -\infty}} \langle q', -i\tau' | q, -i\tau \rangle_J \,. \tag{11.80}$$

Because of the exponential decay only the contribution of the ground state $n = n' = 0$ survives, since it shows the slowest fall-off. Thus we have (for $\tau' \to \infty$, $\tau \to -\infty$)

$$\langle q', -\mathrm{i}\tau' | q, -\mathrm{i}\tau \rangle_J$$

$$\to \psi_0(q')\,\mathrm{e}^{-\frac{1}{\hbar}E_0\tau'}\,\psi_0^*(q)\,\mathrm{e}^{+\frac{1}{\hbar}E_0\tau} \tag{11.81}$$

$$\times \int \mathrm{d}q_2 \int \mathrm{d}q_1\, \psi_0^*(q_2, t_2)\, \langle q_2, t_2 | q_1, t_1 \rangle_J\, \psi_0(q_1, t_1)$$

$$= \psi_0(q', -\mathrm{i}\tau')\psi_0^*(q, -\mathrm{i}\tau) \int \mathrm{d}q_2 \int \mathrm{d}q_1\, \langle 0 | q_2, t_2 \rangle \langle q_2, t_2 | t_1, t_1 \rangle_J \langle q_1, t_1 | 0 \rangle\,.$$

The integral expression on the right-hand side obviously can be interpreted as the probability amplitude for the ground state $|0\rangle$ remaining unchanged under the action of the perturbation $J(t)$ in the time interval $t_1 \ldots t_2$, i.e., for the process in which the perturbation does not induce excitations. Anticipating the field-theoretic vocabulary we will call it the *vacuum–vacuum amplitude* or *vacuum persistence amplitude*, designated by the functional $W[J]$:

$$W[J] = \langle 0|0 \rangle_J = \int \mathrm{d}q_2 \int \mathrm{d}q_1\, \langle 0|q_2, t_2 \rangle \langle q_2, t_2 | q_1, t_1 \rangle_J \langle q_1, t_1 | 0 \rangle\,. \tag{11.82}$$

The vacuum functional defined in this way can be evaluated by using (11.81):

$$W[J] = \lim_{\substack{t' \to -\mathrm{i}\infty \\ t \to \mathrm{i}\infty}} \frac{\langle q', t' | q, t \rangle_J}{\psi_0(q', t')\psi_0^*(q, t)} = \lim_{\substack{t' \to -\mathrm{i}\infty \\ t \to \mathrm{i}\infty}} \frac{\langle q', t' | q, t \rangle_J}{\langle q', t' | q, t \rangle}\,. \tag{11.83}$$

Here the time interval during which the perturbation acts can be taken to be arbitrarily large, $t_2 \to \infty$, $t_1 \to -\infty$. Except for the normalization factor the amplitude in (11.83) agrees with the Feynman kernel for the propagation from (q, t) to (q', t') in the presence of the perturbation $J(t)q$. Therefore $W[J]$ can be represented by the path integral

$$W[J] = \mathcal{N} \int \mathcal{D}q \int \mathcal{D}p\, \exp\left[\frac{\mathrm{i}}{\hbar} \int_{-\infty}^{+\infty} \mathrm{d}t\, \big(p\dot{q} - H(p,q) + Jq\big)\right]\,, \tag{11.84}$$

where we have switched back to the real time coordinate. The normalization constant $\mathcal{N}$ can be fixed by postulating

$$W[0] = 1\,, \tag{11.85}$$

since the ground state has to be stable if there is no perturbation, $J = 0$.

In the calculation of the path integral over $\mathcal{D}q$ in principle boundary conditions have to be imposed, $\lim_{t \to \infty} q(t) = q'$ and $\lim_{t \to -\infty} q(t) = q$. The derivation of (11.84) indicates, however, that the value of $W[J]$ will actually not depend on the choice of these boundary values.

For systems characterized by a quadratic velocity dependence of the kinetic energy, (11.84) can again be expressed in Feynman's form:

$$W[J] = \mathcal{N}' \int \mathcal{D}q\, \exp\left[\frac{\mathrm{i}}{\hbar} \int_{-\infty}^{+\infty} \mathrm{d}t\, \big(L(q, \dot{q}) + Jq\big)\right]\,. \tag{11.86}$$

Generalizing (11.84) or (11.86) from one dimension to quantum systems having D dimensions presents no problem. The vacuum–vacuum amplitude W is a functional of $\alpha = 1, \ldots, D$ independent source functions $J_\alpha(t)$, i.e.,

11.6 The Vacuum Persistence Amplitude $W[J]$

$$W[J_1,\ldots,J_D] = \mathcal{N}\int \prod_{\alpha=1}^{D} \mathcal{D}q_\alpha \prod_{\alpha=1}^{D} \mathcal{D}p_\alpha \qquad (11.87)$$

$$\times \exp\left[\frac{i}{\hbar}\int_{-\infty}^{+\infty}\!dt\left(\sum_{\alpha=1}^{D} p_\alpha \dot{q}_\alpha - H\right) + \sum_{\alpha=1}^{D} J_\alpha q_\alpha\right]$$

and similarly for (11.86)

The vacuum–vacuum amplitude $W[J]$ is an object of fundamental importance within quantum field theory. From the knowledge of this functional the quantities can be derived that are needed in perturbation theory, namely the n-particle Green's functions. Furthermore the functional $W[J]$ can serve as a starting point for the nonperturbative treatment of the theory.

To see the use of the vacuum functional we form the n-fold functional derivative of $W[J]$ with respect to the function $J(t)$:

$$\frac{\delta^n W[J]}{\delta J(t_1)\ldots\delta J(t_n)} = \left(\frac{i}{\hbar}\right)^n \mathcal{N}\int\mathcal{D}q\int\mathcal{D}p\; q(t_1)\cdots q(t_n)$$

$$\times \exp\left[\frac{i}{\hbar}\int_{-\infty}^{+\infty}\!dt\left(p\dot{q} - H + Jq\right)\right]. \qquad (11.88)$$

This expression agrees with the path integral in (11.73) except for the presence of J in the exponent. Thus *the n-point function is obtained as the functional derivative of $W[J]$*, taken at the vanishing $J=0$:

$$\langle 0|T[\hat{q}(t_1)\cdots\hat{q}(t_n)]|0\rangle = \left(\frac{\hbar}{i}\right)^n \left.\frac{\delta^n W[J]}{\delta J(t_1)\ldots\delta J(t_n)}\right|_{J=0}. \qquad (11.89)$$

Note that the normalization factor $\mathcal{N}$ in (11.88) just agrees with the denominator in (11.73).

Remark. In our final expressions we have simply written $t'\to\infty$, $t\to-\infty$ for the limits of the time coordinates in the path integrals. As mentioned already several times one should keep in mind that the underlying limiting process was justified only when rotating the integration contour in the complex time plane by a negative angle (e.g., $\delta = \frac{\pi}{2}$). There is an alternative technical device that can be employed if one wants to avoid the introduction of complex times. The Hamiltonian can be endowed with a "damping term" that enforces asymptotic convergence. A possible choice for such a term is $H' = -i\epsilon q^2$ where ϵ is a small positive number. In this way the path integral obtains an additional factor $\exp\left(-\frac{1}{\hbar}\int_{-\infty}^{+\infty}\!dt\,\epsilon q^2\right)$, which has a dampening effect at large times $|t|$. The choice of the quadratic dependence on q was quite arbitrary but this choice turns out to be very convenient (allowing for Gaussian integration). Using this prescription we find that the expression (11.86), taken as an example, becomes

$$W[J] = \mathcal{N}'\int\mathcal{D}q\,\exp\left[\frac{i}{\hbar}\int_{-\infty}^{+\infty}\!dt\left(L(q,\dot{q}) + J(t)q + i\epsilon q^2\right)\right]. \qquad (11.90)$$

In the following, however, we will not make use of this prescription. Rather, we prefer the euclidian formulation, which can claim greater generality.

12. Path Integrals in Field Theory

12.1 The Path Integral for Scalar Quantum Fields

Up to now we have investigated systems possessing one (q) or several $(q_\alpha, \alpha = 1, \ldots, D)$ degrees of freedom and studied their quantization using path integrals. In this chapter we will pass over to (relativistic) field theories, starting with the case of a scalar neutral field $\phi(x)$. When discussing canonical quantization we already noticed that the field function $\phi(\boldsymbol{x}, t)$ can be viewed as a generalized coordinate vector $q_i(t)$ depending on the "continuous index" $\boldsymbol{x}$ in the place of the discrete index i. Thus a field is a system with an infinite number of degrees of freedom. Its dynamics is described by a Lagrange function which in local field theories can be written as an integral over the Lagrange density:

$$L = \int d^3x \, \mathcal{L}(\phi(x), \partial_\mu \phi(x)) \,. \tag{12.1}$$

Introducing the canonically conjugate field

$$\pi(x) = \frac{\partial \mathcal{L}}{\partial(\partial_0 \phi)} \tag{12.2}$$

the corresponding Hamiltonian is given by

$$H = \int d^3x \, \mathcal{H}(\pi, \phi) = \int d^3x \left(\pi \, \partial_0 \phi - \mathcal{L} \right) \,. \tag{12.3}$$

Within the formalism of canonical quantization the field functions $\phi(x)$ and $\pi(x)$ are converted into operators $\hat{\phi}(x)$ and $\hat{\pi}(x)$ for which equal-time canonical commutation relations are postulated

$$[\hat{\phi}(\boldsymbol{x}, t), \hat{\pi}(\boldsymbol{x}', t)] = i\hbar \delta^3(\boldsymbol{x} - \boldsymbol{x}'), \tag{12.4a}$$
$$[\hat{\phi}(\boldsymbol{x}, t), \hat{\phi}(\boldsymbol{x}', t)] = [\hat{\pi}(\boldsymbol{x}, t), \hat{\pi}(\boldsymbol{x}', t)] = 0 \,. \tag{12.4b}$$

In complete analogy with the treatment of a one-dimensional quantum system presented in the last chapter we study the field operator in the Heisenberg picture $\hat{\phi}(\boldsymbol{x}, t)$. It satisfies the equation of motion

$$-i\hbar \dot{\hat{\phi}}(\boldsymbol{x}, t) = \left[\hat{H}, \hat{\phi}(\boldsymbol{x}, t) \right] , \tag{12.5}$$

which is formally solved by

$$\hat{\phi}(\boldsymbol{x}, t) = e^{i\hat{H}t/\hbar} \, \hat{\phi}(\boldsymbol{x}, 0) \, e^{-i\hat{H}t/\hbar} \,. \tag{12.6}$$

The Heisenberg field operator has a set of time-dependent eigenstates satisfying

$$\hat{\phi}(\boldsymbol{x},t)\,|\phi,t\rangle = \phi(\boldsymbol{x})\,|\phi,t\rangle\,. \tag{12.7}$$

This definition corresponds to the "moving basis" introduced in (11.11). Note that we did not write the time coordinate as an argument of the state vector $|\phi,t\rangle$. This vector has a functional dependence on $\phi(\boldsymbol{x})$, it is not related to the local value of the field at the point $\boldsymbol{x}$.

The time dependence of the basis vector $|\phi,t\rangle$, expressed in terms of the constant Heisenberg vector $|\phi\rangle$, is determined by

$$|\phi,t\rangle = \mathrm{e}^{\mathrm{i}\hat{H}t/\hbar}|\phi,0\rangle \equiv \mathrm{e}^{\mathrm{i}\hat{H}t/\hbar}|\phi\rangle\,. \tag{12.8}$$

Now we want to determine the transition amplitude between two state vectors taken at different times. This amplitude, which was named the *Feynman kernel* in Chap. 11 reads

$$\langle \phi',t'|\phi,t\rangle = \langle \phi'|\,\mathrm{e}^{-\mathrm{i}(t'-t)\hat{H}/\hbar}\,|\phi\rangle\,. \tag{12.9}$$

This is the probability amplitude for making a transition from the field configuration $\phi(\boldsymbol{x})$ at time t leading to the field configuration $\phi'(\boldsymbol{x})$ at time t'. The knowledge of (12.9) is sufficient to answer any question on the dynamics of the quantum field by forming suitable projections.

It is quite obvious how the path integral (11.31) of ordinary quantum mechanics should be generalized to describe quantum field theory:

$$\langle \phi',t'|\phi,t\rangle = \int \mathcal{D}\phi \int \mathcal{D}\pi \, \exp\left[\frac{\mathrm{i}}{\hbar}\int_t^{t'}d\tau\int d^3x\,\bigl(\pi\partial_0\phi - \mathcal{H}(\pi,\phi)\bigr)\right]\,. \tag{12.10}$$

The corresponding boundary conditions are

$$\phi(\boldsymbol{x},t') = \phi'(\boldsymbol{x}) \quad \text{and} \quad \phi(\boldsymbol{x},t) = \phi(\boldsymbol{x})\,, \tag{12.11}$$

where we have chosen particular functions $\phi'(\boldsymbol{x})$ and $\phi(\boldsymbol{x})$.

Equation (12.10) only makes sense if an operative prescription is given for evaluating the path integral over a field $\int \mathcal{D}\phi(\boldsymbol{x},t)$, thus generalizing the path integral over a function $\int \mathcal{D}q(t)$. A way to achieve this comes to mind immediately: the space coordinates can be restricted to a finite volume V and discretized. One may slice space into a set of M "elementary cells" of volume ΔV, which for simplicity will be taken of equal size, centered at the coordinates $\boldsymbol{x}_l, l = 1,\ldots,M$. In this way the continuous field function $\phi(\boldsymbol{x},t)$ is made into a finite-dimensional vector $\phi_l(t) := \phi(\boldsymbol{x}_l,t)$ with discrete index l. Now the construction of the path integral, based on the splitting of the time interval $t,\ldots,t'$ into N segments, can be simply copied from Chap. 11. Equation (12.10) then reads explicitly

$$\langle \phi',t'|\phi,t\rangle = \lim \prod_{l=1}^{M}\left[\prod_{n=1}^{N-1} d\phi_{l\,n} \prod_{n=0}^{N-1}\frac{\Delta V\,d\pi_{l\,n}}{2\pi\hbar}\right], \tag{12.12}$$

$$\times \exp\left[\frac{\mathrm{i}}{\hbar}\sum_{n=0}^{N-1}\epsilon\sum_{l=1}^{M}\Delta V\left(\pi_{l\,n}\frac{\phi_{l\,n+1}-\phi_{l\,n}}{\epsilon} - \mathcal{H}_{l\,n}\right)\right],$$

with the field variables $\phi_{l\,n} := \phi(\boldsymbol{x}_l, t_n)$ and $\pi_{l\,n} := \pi(\boldsymbol{x}_l, t_n)$ defined on the spatial and temporal lattice points.

The limiting process implied in (12.12) consists of three distinct steps:

- $\lim\limits_{N \to \infty}$ under the condition $N\epsilon = t' - t$ fixed,
- $\lim\limits_{M \to \infty}$ under the condition $M\Delta V = V$ fixed,
- $\lim\limits_{V \to \infty}$.

In Exercise 12.1 the steps leading to (12.12) will be worked out in more detail.

If the Lagrange function has a simple quadratic "velocity" dependence, once again it is possible to go from the phase-space path integral (12.10) to Feynman's coordinate space form. Note that the condition is satisfied for ordinary field theories. Let us illustrate this for the example of a scalar field with self-interation $V(\phi)$. Here the Lagrange density reads ($c = 1$)

$$\mathcal{L} = \frac{\hbar^2}{2} \partial_\mu \phi \, \partial^\mu \phi - \frac{1}{2} m^2 \phi^2 - V(\phi) \,, \tag{12.13}$$

and the Hamilton density is given by

$$\mathcal{H} = \frac{1}{2\hbar^2} \pi^2 + \frac{1}{2} \hbar^2 (\boldsymbol{\nabla}\phi)^2 + \frac{1}{2} m^2 \phi^2 + V(\phi) \,, \tag{12.14}$$

where the canonically conjugate field is defined as

$$\pi = \hbar^2 \partial_0 \phi \,. \tag{12.15}$$

In this case the *Feynman path integral* equivalent to (12.12) becomes

$$\langle \phi', t' | \phi, t \rangle = \mathcal{N} \int \mathcal{D}\phi \, \exp\left[\frac{\mathrm{i}}{\hbar} \int_t^{t'} \mathrm{d}\tau \int \mathrm{d}^3 x \, \mathcal{L}(\phi, \dot\phi)\right] \,, \tag{12.16}$$

as derived in Exercise 12.1.

Also the remaining results of the last chapter can be transferred from finite-dimensional systems to field theory without difficulties (at least at the formal level). As an example the *vacuum expectation value of the time-ordered product of field operators* $\hat\phi(\boldsymbol{x}, t) \equiv \hat\phi(x)$, i.e., the *n*-point function or many-body Green's function is given by

$$\langle 0 | T(\hat\phi(x_1) \cdots \hat\phi(x_n)) | 0 \rangle = \mathcal{N} \int \mathcal{D}\phi \, \phi(x_1) \cdots \phi(x_n) \, \exp\left[\frac{\mathrm{i}}{\hbar} \int \mathrm{d}^4 x \, \mathcal{L}(\phi, \dot\phi)\right] \,. \tag{12.17}$$

The time limits for the action integral in the exponent extends to infinity. To achieve convergence and to filter out the contribution of the ground state one has to perform the rotation in the complex time plane $t \to \pm\infty \, \mathrm{e}^{-\mathrm{i}\delta}$ discussed in Sect. 11.5.

Also the artifice of introducing an *external source* can be exploited in field theory. The source now is a function $J(x)$ depending on space and time, which gets multiplied by the local field $\phi(x)$. Generalizing (12.16), this leads to the transition amplitude

$$\langle \phi', t' | \phi, t \rangle_J = \mathcal{N} \int_\phi^{\phi'} \mathcal{D}\phi \, \exp\left[\frac{\mathrm{i}}{\hbar} \int_t^{t'} \mathrm{d}^4 x \, \left(\mathcal{L}(\phi, \dot\phi) + J\phi\right)\right] \,. \tag{12.18}$$

The *vacuum–vacuum transition functional* reads (cf. (11.86)):

$$W[J] = \langle 0 | 0 \rangle_J = \mathcal{N} \int \mathcal{D}\phi \, \exp\left[\frac{\mathrm{i}}{\hbar} \int \mathrm{d}^4 x \, \left(\mathcal{L}(\phi, \dot{\phi}) + J\phi\right)\right] . \qquad (12.19)$$

The vacuum functional $W[J]$ plays a central role in field theory since it can be used to evaluate the Green's functions, i.e., the vacuum expectation values of T products. From (12.17) and (12.19) we conclude

$$\langle 0 | T(\hat{\phi}(x_1) \cdots \hat{\phi}(x_n)) | 0 \rangle = \left(\frac{\hbar}{\mathrm{i}}\right)^n \left. \frac{\delta^n W[J]}{\delta J(x_1) \ldots \delta J(x_n)} \right|_{J=0} . \qquad (12.20)$$

EXERCISE

12.1 Construction of the Field-Theoretical Path Integral

Problem. Repeat the steps involved in the construction of the quantum-mechanical path integral for the case of a scaler field $\phi(\boldsymbol{x}, t)$.

Solution. We have to evaluate the transition matrix element $\langle \phi', t' | \phi, t \rangle$. Configuration space is discretized into M cells at $\boldsymbol{x}_l$, $l = 1, \ldots, M$, and the time interval is split into N segments at $t_n, n = 0, \ldots, N$ (so that $t_0 = t$, $t_1 = t + \epsilon, \ldots, t_N = t'$). Then for each of the cells a basis of eigenstates of the Heisenberg field operator $\hat{\phi}_{ln} := \hat{\phi}(\boldsymbol{x}_l, t_n)$ can be constructed that satisfies the completeness relation

$$\int \mathrm{d}\phi_{ln} \, | \phi_l, t_n \rangle \langle \phi_l, t_n | = 1 . \qquad (1)$$

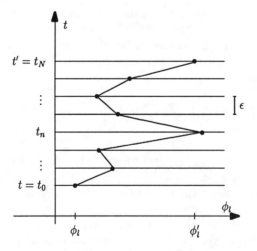

Fig. 12.1. A typical path for the time development of the field function in a given cell l

The transition amplitude can be split into a product of the amplitudes in the M cells. Subsequently the time evolution is discretized as in Sect. 11.2

$$\langle \phi', t' | \phi, t \rangle \simeq \langle \phi'_1, \phi'_2, \ldots, \phi'_M, t' | \phi_1, \phi_2, \ldots, \phi_M, t \rangle$$

$$= \prod_{l=1}^{M} \langle \phi'_l, t' | \phi_l, t \rangle$$

$$= \prod_{l=1}^{M} \int d\phi_{l\,N-1} \ldots \int d\phi_{l\,2} \int d\phi_{l\,1} \langle \phi'_l, t' | \phi_{l\,N-1}, t_{N-1} \rangle \cdots$$
$$\times \langle \phi_{l\,2}, t_2 | \phi_{l\,1}, t_1 \rangle \langle \phi_l, t_1 | \phi_1, t \rangle \,. \tag{2}$$

Exercise 12.1

If N is very large it is sufficient to know the transition amplitudes for infinitesimal time differences:

$$\langle \phi_{l\,n+1}, t_{n+1} | \phi_{l\,n}, t_n \rangle = \langle \phi_{l\,n+1} | e^{-\frac{i}{\hbar} \hat{H} \epsilon} | \phi_{l\,n} \rangle$$
$$= \langle \phi_{l\,n+1} | 1 - \frac{i}{\hbar} \hat{H} \epsilon | \phi_{l\,n} \rangle + O(\epsilon^2) \,. \tag{3}$$

The Hamilton operator of the discretized theory (finite M) reads

$$\hat{H}(\hat{\pi}_1, \ldots, \hat{\pi}_M; \hat{\phi}_1, \ldots, \hat{\phi}_M) = \sum_{l=1}^{M} \Delta V \, \hat{\mathcal{H}}(\hat{\pi}_l, \hat{\phi}_l) \,. \tag{4}$$

It depends on the values of the discretized conjugate field operator

$$\hat{\pi}_l = \frac{\partial \hat{\mathcal{L}}}{\partial \dot{\hat{\phi}}_l} \,. \tag{5}$$

To evaluate (3) a complete set of eigenstates

$$\hat{\pi}_l | \pi_{l\,n} \rangle = \pi_{l\,n} | \pi_{l\,n} \rangle \tag{6}$$

can be inserted which satisfies

$$\int \frac{\Delta V \, d\pi_{l\,n}}{2\pi\hbar} | \pi_{l\,n} \rangle \langle \pi_{l\,n} | = 1 \,. \tag{7}$$

Then the matrix element of the Hamiltonian reads

$$\langle \phi_{l\,n+1} | \hat{H}(\hat{\pi}, \hat{\phi}) | \phi_{l\,n} \rangle$$
$$= \int \frac{\Delta V \, d\pi_{l\,n}}{2\pi\hbar} \langle \phi_{l\,n+1} | \pi_{l\,n} \rangle \langle \pi_{l\,n} | \hat{H} | \phi_{l\,n} \rangle$$
$$= \int \frac{\Delta V \, d\pi_{l\,n}}{2\pi\hbar} \langle \phi_{l\,n+1} | \pi_{l\,n} \rangle \langle \pi_{l\,n} | \phi_{l\,n} \rangle \Delta V \mathcal{H}(\pi_{l\,n}, \phi_{l\,n}) \,. \tag{8}$$

Now the generalized equation (11.5) can be used, which reads

$$\langle \phi_{l\,n} | \pi_{l\,n} \rangle = \exp\left(\frac{i}{\hbar} \Delta V \pi_{l\,n} \phi_{l\,n} \right) \,. \tag{9}$$

In the continuum limit this equation becomes

$$\langle \phi | \pi \rangle = \exp\left[\frac{i}{\hbar} \int d^3x \, \pi(\boldsymbol{x}) \phi(\boldsymbol{x}) \right] \,. \tag{10}$$

One immediately verifies that (10) is necessary in order to satisfy the operator relation

$$\hat{\pi} = -i\hbar \frac{\delta}{\delta \hat{\phi}} \,. \tag{11}$$

Exercise 12.1

Using (8) and (9) we find the overlap matrix element (3) becomes

$$\langle \phi_{l\,n+1}, t_{n+1} | \phi_{l\,n}, t_n \rangle = \int \frac{\Delta V \mathrm{d}\pi_{l\,n}}{2\pi\hbar} \exp\left[\frac{i}{\hbar} \Delta V \pi_{l\,n}(\phi_{l\,n+1} - \phi_{l\,n})\right]$$
$$\times \left(1 - \frac{i}{\hbar}\epsilon \Delta V \mathcal{H}_{l\,n}\right) + O(\epsilon^2), \quad (12)$$

with the abbreviation $\mathcal{H}_{l\,n} := \mathcal{H}(\pi_{l\,n}, \phi_{l\,n})$. Linking these factors together (2) finally leads to

$$\langle \phi', t' | \phi, t \rangle \simeq \prod_{l=1}^{M}\left(\prod_{n=1}^{N-1} \int \mathrm{d}\phi_{l\,n} \prod_{n=0}^{N-1} \int \frac{\Delta V \mathrm{d}\pi_{l\,n}}{2\pi\hbar}\right) \quad (13)$$
$$\times \exp\left(\frac{i}{\hbar} \sum_{n=0}^{N-1} \sum_{l=1}^{M} \Delta V \pi_{l\,n}(\phi_{l\,n+1} - \phi_{l\,n})\right) \exp\left(-\frac{i}{\hbar} \sum_{n=0}^{N-1} \epsilon \sum_{l=1}^{M} \Delta V \mathcal{H}_{l\,n}\right).$$

In the continuum limit this becomes

$$\langle \phi', t' | \phi, t \rangle = \lim_{V \to \infty} \lim_{M \to \infty} \lim_{N \to \infty} \prod_{l=1}^{M}\left(\prod_{n=1}^{N-1} \int \mathrm{d}\phi_{l\,n} \int \frac{\Delta V \mathrm{d}\pi_{l\,n}}{2\pi\hbar}\right)$$
$$\times \exp\left[\frac{i}{\hbar} \sum_{n=0}^{N-1} \epsilon \sum_{l=1}^{M} \Delta V \left(\pi_{l\,n} \frac{\phi_{l\,n+1} - \phi_{l\,n}}{\epsilon} - \mathcal{H}_{l\,n}\right)\right]$$
$$\equiv \int \mathcal{D}\phi \int \mathcal{D}\pi \exp\left[\frac{i}{\hbar} \int_{t}^{t'} \mathrm{d}\tau \int \mathrm{d}^3 x \left(\pi \partial_0 \phi - \mathcal{H}(\pi, \phi)\right)\right]. \quad (14)$$

The phase-space path integral (14) can be transformed into Feynman's form if the Hamiltonian depends quadratically on the conjugate field π. Let us assume that $\mathcal{H}$ has the standard form (12.14)

$$\mathcal{H}(\pi, \phi) = \frac{1}{2\hbar^2}\pi^2 + \frac{1}{2}\hbar^2(\boldsymbol{\nabla}\phi)^2 + \frac{1}{2}m^2\phi^2 + V(\phi). \quad (15)$$

Then the transition amplitude (12) for neighboring points in time in the first order is given by

$$\langle \phi_{l\,n+1}, t_{n+1} | \phi_{l\,n}, t_n \rangle$$
$$\simeq \int_{-\infty}^{\infty} \frac{\Delta V \mathrm{d}\pi_{l\,n}}{2\pi\hbar} \exp\left[\frac{i}{\hbar}\epsilon \Delta V \left(\pi_{l\,n}\dot\phi_{l\,n} - \frac{1}{2\hbar^2}\pi_{l\,n}^2 - \mathcal{H}'_{l\,n}\right)\right]. \quad (16)$$

The function

$$\mathcal{H}'_{l\,n} = \frac{1}{2}\hbar^2(\boldsymbol{\nabla}\phi)^2 + \frac{1}{2}m^2\phi^2 + V(\phi) = \mathcal{H}_{l\,n} - \frac{1}{2\hbar^2}\pi_{l\,n}^2 = \mathcal{H}_{l\,n} - \frac{\hbar^2}{2}\dot\phi_{l\,n}^2 \quad (17)$$

collects the π-independent terms of the Hamiltonian. Equation (16) is a Gaussian integral which can be brought into the standard form by completing the square:

$$\langle \phi_{l\,n+1}, t_{n+1} | \phi_{l\,n}, t_n \rangle \simeq \frac{\Delta V}{2\pi\hbar} \exp\left[\frac{i}{\hbar}\epsilon\Delta V\left(\frac{1}{2}\hbar^2\dot\phi_{l\,n}^2 - \mathcal{H}'_{l\,n}\right)\right]$$
$$\times \int_{-\infty}^{\infty} \mathrm{d}\pi' \exp\left[-\frac{i\epsilon\Delta V}{2\hbar^3}\pi'^2\right]$$

$$= \left(\frac{2\pi i\epsilon}{\Delta V \hbar}\right)^{-1/2} \exp\left[\frac{i}{\hbar}\epsilon\Delta V\left(\hbar^2 \dot\phi_{ln}^2 - \mathcal{H}_{ln}\right)\right] . \quad (18)$$

This leads to the following expression for the Feynman kernel

$$\langle \phi', t'|\phi, t\rangle = \lim_{V\to\infty}\lim_{M\to\infty}\lim_{N\to\infty} \prod_{l=1}^{M}\prod_{n=1}^{N-1}\int d\phi_{ln} \left(\frac{2\pi i\epsilon}{\Delta V \hbar}\right)^{-N/2}$$

$$\times \exp\left[\frac{i}{\hbar}\sum_{n=0}^{N-1}\epsilon \sum_{l=1}^{M}\Delta V \mathcal{L}_{ln}\right]$$

$$\equiv \mathcal{N}\int \mathcal{D}\phi \, \exp\left[\frac{i}{\hbar}\int d^4x\, \mathcal{L}(\phi, \dot\phi)\right] . \quad (19)$$

Exercise 12.1

12.2 Euclidian Field Theory

Equations (12.16)–(12.19) are of central importance for the application of the path-integral formalism to problems of quantum field theory. We already have mentioned several times, however, that the convergence of the integrals is not guaranteed because of the oscillating exponential function. To circumvent this problem, in Sect. 11.5 we introduced complex time values, rotating the integration contour in the complex time plane by a negative angle, $t = e^{-i\delta}\tau$ with τ real. In this way the exponent obtains a negative real part, so that the integrand falls off, thus guaranteeing convergence. This procedure is based on the assumption that there are no singularities in the covered region of the complex t plane. Under this condition an analytic continuation in t is possible and the results do not depend on the chosen value of the angle δ. In some simple cases, for example, for the noninteracting Feynman propagator, the validity of this assumption can be verified by an explicit calculation. Fortunately the analytic continuation is justified also under much more general conditions.[1]

The complex-rotation presciption becomes particularly elegant if the angle $\delta = \pi/2$ is chosen, i.e., if purely *imaginary times* are introduced. We will formulate this in a relativistically covariant fashion.

The ordinary coordinates $x_\mu = (x_0, x_1, x_2, x_3) \equiv (x_0, \boldsymbol{x})$ form a fourvector in Minkowski space. Now a new vector of coordinates $x_{\text{E}\mu} = (x_{\text{E}1}, x_{\text{E}2}, x_{\text{E}3}, x_{\text{E}4}) \equiv (\boldsymbol{x}_\text{E}, x_4)$ is introduced, identifying

$$x_4 = ix_0 \quad , \quad \boldsymbol{x}_\text{E} = \boldsymbol{x} . \quad (12.21)$$

Thus the time coordinate is rotated by an angle $\pi/2$ while the space coordinates are kept unchanged. If x_4 is postulated to be a real number then the new vectors form a *four-dimensional euclidian space*. The factor i just has the effect of cancelling the difference in sign between space and time components in the Minkowski metric. The squared norm of an euclidian vector is

---

[1] K. Osterwalder, E. Schrader: Comm. Math. Phys. **42**, 440 (1975);
J. Schwinger: Proc. Natl. Acad. Sci. **44**, 956 (1958).

$$x_E^2 = x_{E\mu}x_{E\mu} = x_4^2 + x_1^2 + x_2^2 + x_3^2 = -x_0^2 + \boldsymbol{x}^2 = -x_\mu x^\mu = -x^2 \ . \quad (12.22)$$

Note that it is no longer necessary to distinguish between the covariant and contravariant components of a vector, $x_{E\mu} = x_E^\mu$.

The connection between Minkowski and euclidian space in the complex time plane is illustrated in Figure 12.2. The prescription introduced in Chap. 11 now reads: in the evaluation of the path integral the time contour is to be taken along the *real x_4 axis* (the dashed line in Fig. 12.2b). This leads to the "euclidian transition amplitude". Subsequently an analytic continuation has to be made which leads back to the ordinary Minkowski time. This artifice of going to imaginary times is called a *Wick rotation*. It is an important technical tool used very often in quantum field theory. We emphasize that the Wick rotation (or alternatively the introduction of a damping term as in (11.90)) is necessary to ensure that quantities like the vacuum amplitude $W[J]$ are mathematically well defined.

Fig. 12.2. Wick rotation: The integration along the real euclidian x_4 axis in Minkowski space corresponds to an integration along the imaginary x_0 axis in negative direction

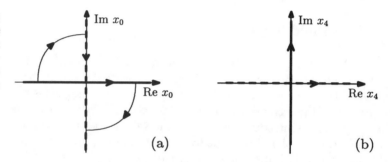

Let us add some remarks on the technical aspect of calculating in euclidian space. In view of the importance of plane waves and Fourier transforms we introduce the *euclidian momentum space*. The euclidian *momentum vector* is defined as $p_{E\mu} = (p_{E1}, p_{E2}, p_{E3}, p_{E4}) \equiv (\boldsymbol{p}_E, p_4)$ with

$$p_4 = -ip_0 \quad , \quad \boldsymbol{p}_E = \boldsymbol{p} \ , \quad (12.23)$$

where p_4 is a real variable. The norm of the momentum vector is

$$p_E^2 = p_4^2 + p_1^2 + p_2^2 + p_3^2 = -p_0^2 + \boldsymbol{p}^2 = -p^2 \ . \quad (12.24)$$

Unlike in (12.21), which holds for the space coordinate, the momentum variable p_0 (the energy) is rotated in the mathematically positive direction, as shown in Fig. 12.3. The sign in (12.23) was chosen such that the product of time and energy (or frequency) is invariant, i.e., $p_4 x_4 = (-ip_0)(ix_0) = p_0 x_0$. This is an arbitrary choice which ensures, however, that the travelling direction of a plane wave remains unchanged when the coordinates are transformed, since

$$e^{ip_\mu x^\mu} = e^{i(p_0 x_0 - \boldsymbol{p} \cdot \boldsymbol{x})} = e^{i(p_4 x_4 - \boldsymbol{p}_E \cdot \boldsymbol{x}_E)} \ . \quad (12.25)$$

For the scalar product of the euclidian four-vectors we find

$$p_E \cdot x_E = p_4 x_4 + \boldsymbol{p}_E \cdot \boldsymbol{x}_E = p_0 x_0 + \boldsymbol{p} \cdot \boldsymbol{x} \ , \quad (12.26)$$

which does not agree with $p \cdot x = p_0 x_0 - \boldsymbol{p} \cdot \boldsymbol{x}$.

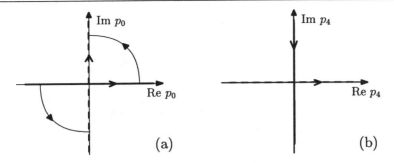

Fig. 12.3. The Wick rotation in momentum space is defined with an opposite direction

Let us consider the four-dimensional Fourier transform. The volume element transforms as

$$d^4 x = -i\, d^4 x_E \quad , \quad d^4 p = i\, d^4 p_E \ . \tag{12.27}$$

The Fourier transform of a function $f(p^2)$ that depends on the square of the four-momentum in euclidian coordinates becomes

$$\begin{aligned}
\tilde{f}(x) &= \int \frac{d^4 p}{(2\pi)^4}\, e^{-ip\cdot x} f(p^2) \\
&= i \int \frac{d^4 p_E}{(2\pi)^4}\, e^{-i(p_4 x_4 - \boldsymbol{p}_E \cdot \boldsymbol{x}_E)} f(-p_E^2) \\
&= i \int \frac{d^4 p'_E}{(2\pi)^4}\, e^{-i(p'_4 x_4 + \boldsymbol{p}'_E \cdot \boldsymbol{x}_E)} f(-p_E'^2) \\
&= i \int \frac{d^4 p_E}{(2\pi)^4}\, e^{-i p_E \cdot x_E} f(-p_E^2) \ .
\end{aligned} \tag{12.28}$$

In the second step we inverted the spatial momentum vector $\boldsymbol{p}'_E = -\boldsymbol{p}_E$, and susequently renamed the integration variable back to p_E.

If vector fields $A^\mu(x)$ are involved in a theory, their behavior under Wick rotations has to be specified too. The following rule is used: a Lorentz vector field is transformed into a euclidian vector field $A_E^\mu(x_E)$ according to

$$A_4(x_E) = -i\, A_0(x) \quad , \quad \boldsymbol{A}_E(x_E) = \boldsymbol{A}(x) \ , \tag{12.29}$$

where $A_{E\mu}$ is assumed to have real-valued components. The prescription (12.29) is inverse to the transformation of the coordinate vector (12.21), $x_4 = +ix_0$. The prescription is suggested by the principle of gauge invariance according to which the A^μ field has the transformation properties of a gradient (remember the gauge-covariant derivative $D^\mu = \partial^\mu + ieA^\mu$). In the formation of the derivative, the factor i gets into the denominator, which leads to an inverted sign. For the time component we have $D_0 = \partial/\partial x_0 + ieA_0 = \partial/\partial(-ix_4) + ieA_0 = i\,(\partial/\partial x_4 + ieA_4)$, by using (12.29).

To conclude these remarks we give the expression for the euclidian vacuum functional of a scalar field theory (12.19):

$$W_E[J] = \mathcal{N}_E \int \mathcal{D}\phi \, \exp\left[\frac{i}{\hbar}(-i) \int d^4 x_E \left(\mathcal{L}\left(\phi, i\frac{\partial \phi}{\partial x_4}\right) + J\phi\right)\right] \ , \tag{12.30}$$

where the time integration according to Fig. 12.2 extends along the real x_4 axis (instead of $-i\infty \ldots +i\infty$). In the case of a scalar field with self-interaction $V(\phi)$ the Lagrange density according to (12.13) reads

$$\begin{aligned}
\mathcal{L} &= \frac{\hbar^2}{2}(\partial_0\phi\,\partial_0\phi - \boldsymbol{\nabla}\phi\cdot\boldsymbol{\nabla}\phi) - \frac{1}{2}m^2\phi^2 - V(\phi)\\
&= -\left[\frac{\hbar^2}{2}(\partial_4\phi\,\partial_4\phi + \boldsymbol{\nabla}\phi\cdot\boldsymbol{\nabla}\phi) + \frac{1}{2}m^2\phi^2 + V(\phi)\right]\\
&= -\left[\frac{\hbar^2}{2}\partial_{\mathrm{E}\mu}\phi\,\partial_{\mathrm{E}\mu}\phi + \frac{1}{2}m^2\phi^2 + V(\phi)\right]\,. \quad (12.31)
\end{aligned}$$

The euclidian vacuum functional of the scalar field theory thus is given by

$$W_{\mathrm{E}}[J] = \mathcal{N}_{\mathrm{E}}\int\mathcal{D}\phi\,\exp\left[-\frac{1}{\hbar}\int\mathrm{d}^4 x_{\mathrm{E}}\left(\frac{\hbar^2}{2}\partial_{\mathrm{E}\mu}\phi\,\partial_{\mathrm{E}\mu}\phi + \frac{1}{2}m^2\phi^2 + V(\phi) - J\phi\right)\right]. \quad (12.32)$$

This result has the intended property: the weight factor for the paths is a real-valued function which falls off exponentially since the exponent in (12.32) is always positive (at least in the asymptotic region where J is switched off). This is ensured povided that the potential of the self-interaction $V(\phi)$ satisfies the stability condition, i.e., provided that it is bounded from below. By shifting the energy scale, $V(\phi) \geq 0$ can be chosen.

Let us note that the relation (12.32) in the form

$$W_{\mathrm{E}} \propto \int\mathcal{D}\phi\,\mathrm{e}^{-\frac{1}{\hbar}S_{\mathrm{E}}[\phi]}\,, \quad (12.33)$$

with a real euclidian action $S_{\mathrm{E}} \geq 0$, serves as the starting point for constructing a far-reaching *analogy between euclidian quantum field theory and statistical mechanics*.[2] The path integral over all field configurations corresponds to the canonical sum of states (the partition function)

$$Z = \sum_n \mathrm{e}^{-\frac{1}{kT}E_n}\,, \quad (12.34)$$

which extends over all possible configurations of a classical mechanical system having energies of E_n. Strictly speaking this provides only a formal mathematical connection between problems that have very different underlying physics. Nevertheless it becomes possible in this way to apply the elaborate tools of statistical mechanics to the realm of quantum field theory. Finding connections between apparently unrelated fields can be very rewarding.

The euclidian formulation of field theory also has important consequences for practical calculations. To evaluate path integrals numerically the field variables have to be discetized on a space-time lattice.[3] This leads to multiple integrals in a very highly dimensional space (there are typically 10^4 or more integration variables). The only chance to do such an integration lies in sampling the integrand on a rather small number of cleverly chosen statistical grid points ("Monte Carlo integration"). If the integrand were to exhibit

---

[2] An extensive presentation of functional methods in quantum field thory and in statistical mechanics can be found, e.g., in J. Zinn-Justin: *Quantum Field Theory and Critical Phenomena* (Oxford University Press, Oxford 1989); M. LeBellac: *Quantum and Statistical Field Theory* (Oxford University Press, Oxford 1991).

[3] See W. Greiner, A. Schäfer: *Quantum Chromodynamics* (Springer, Berlin, Heidelberg 1994); M. Creutz: *Quarks, Gluons and Lattices* (Cambridge University Press, Cambridge 1983).

oscillations, as it does for the original path integral before the Wick rotation is applied, there would be no chance whatsoever of achieving numerical convergence. With the use of the euclidian formulation, however, numerical simulations of lattice field theory have become an important tool in modern theoretical physics.

12.3 The Feynman Propagator

Let us explicitly evaluate the vacuum functional $W_0[J]$ for the case of a *free* uncharged (i.e., real-valued) scalar field $\phi(x)$ described by the Lagrange density

$$\mathcal{L}_0 = \frac{\hbar^2}{2} \partial_\mu \phi \, \partial^\mu \phi - \frac{1}{2} m^2 \phi^2 \ . \tag{12.35}$$

We have to solve the path integral

$$W_0[J] = \mathcal{N} \int \mathcal{D}\phi \, \exp\left[\frac{i}{\hbar} \int d^4x \left(\frac{\hbar^2}{2} \partial_\mu \phi \, \partial^\mu \phi - \frac{1}{2} m^2 \phi^2 + J\phi\right)\right] \ . \tag{12.36}$$

Expressed in terms of euclidian coordinates we find

$$\mathcal{L}_0^{\mathrm{E}} = -\left(\frac{\hbar^2}{2} \partial_\mu^{\mathrm{E}} \phi \, \partial_\mu^{\mathrm{E}} \phi + \frac{1}{2} m \phi^2\right) \ , \tag{12.37}$$

and the corresponding *euclidian vacuum functional* reads

$$W_0^{\mathrm{E}}[J] = \mathcal{N}_{\mathrm{E}} \int \mathcal{D}\phi \, \exp\left[-\frac{1}{\hbar} \int d^4x_{\mathrm{E}} \left(\frac{\hbar^2}{2} \partial_\mu^{\mathrm{E}} \phi \, \partial_\mu^{\mathrm{E}} \phi + \frac{1}{2} m^2 \phi^2 - J\phi\right)\right] \ . \tag{12.38}$$

This path integral has to be evaluated and subsequently to be continued analytically back to Minkowski coordinates. The exponents of (12.38) contains the square of the field variable ϕ. Therefore (12.38) is a *Gaussian integral* that can be solved analytically. It is, however, a Gaussian integral on a somewhat higher level of abstraction, i.e., a *functional integral*. Fortunately we will be able to find the solution by using the methods we already have at hand, although the derivation will not be rigorous from a mathematical point of view.

Equation (12.38) contains an integration over the elements of a function space of infinite dimension, i.e., over the wave functions ϕ in Hilbert space. Instead of this demanding task we will integrate over the vectors of a finite-dimensional vector space (of dimension N). In the limit $N \to \infty$ in this way the whole function space should be covered. Strictly speaking, of course, we would have to prove that the result obtained in this way does not depend on the details of the limiting procedure, but we will not address mathematical questions of this type. Equation (12.38) can be written in the following general form:

$$W_0^{\mathrm{E}}[J] = \mathcal{N}_{\mathrm{E}} \int \mathcal{D}\phi \, e^{-\frac{1}{2}(\phi, A\phi) + (\rho, \phi)} \ . \tag{12.39}$$

Here we have introduced the following scalar product

$$(\psi, \phi) = \int d^4 x_E \, \psi(x_E) \phi(x_E) \,, \tag{12.40}$$

which is a generalization of

$$(v, w) = v^T w = \sum_{i=1}^{N} v_i w_i \tag{12.41}$$

for a finite-dimensional (real) vector space. The coordinate x_E thus plays the role of a "continuous index". Correspondingly the "matrix" A has become a function of two continuous coordinates and we have

$$(\phi, A\phi) = \int d^4 x'_E \, d^4 x_E \, \phi(x'_E) A(x'_E, x_E) \, \phi(x_E) \tag{12.42}$$

in analogy with the finite-dimensional expression

$$(v, Aw) = v^T A w = \sum_{ij=1}^{N} v_i A_{ij} w_j \,. \tag{12.43}$$

The action integral in (12.38) contains

$$\begin{aligned} A(x'_E, x_E) &= \frac{1}{\hbar}\left(\hbar^2 \partial'^E_\mu \partial^E_\mu + m^2\right) \delta^4\left(x'_E - x_E\right) \\ &= \frac{1}{\hbar}\left(-\hbar^2 \Box_E + m^2\right) \delta^4\left(x'_E - x_E\right) \,, \end{aligned} \tag{12.44a}$$

$$\rho(x_E) = \frac{1}{\hbar} J(x_E) \,. \tag{12.44b}$$

Now we make use of formula (2) from Exercise 11.3 for the general Gaussian integral in N dimensions. Except for the factor $(2\pi)^{N/2}$ which can be combined with the integration measure this formula contains no reference to the space dimension. Therefore we will be bold and use the finite-dimensional Gaussian formula also for the functional integral in (12.39):

$$\int \mathcal{D}\phi \, e^{-\frac{1}{2}(\phi, A\phi) + (\rho, \phi)} = (\det A)^{-\frac{1}{2}} e^{\frac{1}{2}(\rho, A^{-1}\rho)} \,. \tag{12.45}$$

For the euclidian vacuum functional this implies

$$W_0^E[J] = \mathcal{N}_E (\det A)^{-\frac{1}{2}} \exp\left[\frac{1}{2}\frac{1}{\hbar^2} \int d^4 x'_E \, d^4 x_E \, J(x'_E) A^{-1}(x'_E, x_E) J(x_E)\right] \,. \tag{12.46}$$

We need the inverse of the "continuous matrix" $A(x', x)$ and (apparently) also the determinant $\det A$. Fortunately the latter does not need to be calculated after all. Instead we can rely on the normalization condition for the vacuum amplitude in the absence of the external perturbation J

$$\begin{aligned} 1 \stackrel{!}{=} W_0^E[0] &= \mathcal{N}_E \int \mathcal{D}\phi \, \exp\left[-\frac{1}{\hbar} \int d^4 x_E \left(\frac{\hbar^2}{2} \partial^E_\mu \phi \, \partial^E_\mu \phi + \frac{1}{2} m^2 \phi^2\right)\right] \\ &= \mathcal{N}_E (\det A)^{-\frac{1}{2}} \,. \end{aligned} \tag{12.47}$$

Therefore $\mathcal{N}_E = (\det A)^{-\frac{1}{2}}$ so that the factor in front of the exponential function in (12.46) drops out. This is very favourable since an attempt to evaluate $\det A$ would have led to a divergent result.

12.3 The Feynman Propagator

This can be shown with the help of a Fourier transformation. We have to evaluate

$$\det A = e^{\operatorname{Tr} \ln A} \tag{12.48}$$

where $A(x', x)$ now is a continuous matrix. Let $f(A)$ be an arbitrary function of such a matrix that can be expressed in terms of a power series:

$$f(A) = \sum_n c_n A^n . \tag{12.49}$$

Written out in components (for simplicity we consider only one dimension) this is to be understood as

$$f(A)(x', x) = c_0 + c_1 A(x', x) + c_2 \int dx_1 \, A(x', x_1) A(x_1, x) + \ldots \tag{12.50}$$

$$+ c_n \int dx_1 \ldots dx_{n-1} A(x', x_1) A(x_1, x_2) \cdots A(x_{n-1}, x) + \ldots .$$

If the matrix $A(x', x)$ depends only on the difference of coordinates $x' - x$, as is the case for (12.44), the Fourier transform is given by

$$A(x', x) = \int \frac{dk}{2\pi} e^{-ik(x'-x)} A(k) . \tag{12.51}$$

As an example, the second-order term in this expansion becomes in momentum space (12.50)

$$\int dx_1 \, A(x', x_1) A(x_1, x) = \int dx_1 \int \frac{dk'}{2\pi} \int \frac{dk}{2\pi} e^{-ik'(x'-x_1)} e^{-ik(x_1-x)} A(k') A(k)$$

$$= \int \frac{dk}{2\pi} e^{-ik(x'-x)} \big(A(k)\big)^2 . \tag{12.52}$$

Clearly the general term of nth order contains delta functions from the $n-1$ integrations over x, each of which cancels a k integration, leading to

$$\int dx_1 \ldots dx_{n-1} \, A(x', x_1) \cdots A(x_{n-1}, x) = \int \frac{dk}{2\pi} e^{-ik(x'-x)} (A(k))^n . \tag{12.53}$$

Therefore any function $f(A)$ satisfies

$$f(A)(x', x) = \int \frac{dk}{2\pi} e^{-ik(x'-x)} f(A(k)) . \tag{12.54}$$

The notation $f(A)(x', x)$ has been intentionally chosen, in order to emphasize that first the function of the matrix A has to be formed for which subsequently the "components" at x' and x have to be taken. Keeping this in mind, of course, we could also have used the more conventional notation $f(A(x', x))$.

In our case (using euclidian four vectors) the Fourier transform is given by

$$A(p_E) = \frac{1}{\hbar} \left(p_E^2 + m^2 \right) , \tag{12.55}$$

which is immediately verified by applying the differential operator (12.44) to the Fourier representation of the delta function:

$$A(x'_E, x_E) = \frac{1}{\hbar} \left(\hbar^2 \partial_\mu^{'E} \partial_\mu^E + m^2 \right) \int \frac{d^4 p_E}{(2\pi\hbar)^4} \exp\left[-i p_E \cdot (x'_E - x_E)/\hbar\right]$$

$$= \int \frac{d^4 p_E}{(2\pi\hbar)^4} \exp\left[-i p_E \cdot (x'_E - x_E)/\hbar\right] \frac{1}{\hbar} (p_E^2 + m^2) . \tag{12.56}$$

The trace from (12.48) then reads, with the help of (12.54)

$$\text{Tr }\ln A = \int d^4 x_E\,(\ln A)(x_E, x_E) = \int d^4 x_E \int \frac{d^4 p_E}{(2\pi\hbar)^4}\,\ln\!\left(\frac{1}{\hbar}(p_E^2 + m^2)\right). \quad (12.57)$$

As announced this is a divergent result.

The same conclusion can be reached even faster, albeit less rigorously, by considering the determinant as the product of all the eigenvalues of a matrix. The differential operator $\frac{1}{\hbar}\left(-\hbar^2\,\Box_E + m^2\right)$ from (12.44) has the eigenvalues $\chi_{p_E} = \frac{1}{\hbar}\left(p_E^2 + m^2\right)$, the eigenfunctions being the plane waves. Since the spectrum forms a continuum, the determinant is clearly given by a divergent infinite product

$$\det A = \prod_{p_E} \frac{1}{\hbar}\left(p_E^2 + m^2\right). \quad (12.58)$$

Fortunately the pathology of $\det A$ has no negative consequences as it is a constant factor, i.e., it does not depend on J, and gets eliminated by the normalization condition (12.47). We are left with

$$W_0^E[J] = \exp\left[\frac{1}{2}\frac{1}{\hbar^2}\int d^4 x_E'\,d^4 x_E\,J(x_E')A^{-1}(x_E', x_E)J(x_E)\right]. \quad (12.59)$$

The inverse of the matrix $A(x_E', x_E)$ from (12.44) is defined by

$$\int d^4 x_E\,A^{-1}(x_E', x_E)A(x_E, x_E'') = \delta^4(x_E' - x_E''). \quad (12.60)$$

In momentum space $A(p_E)$ is a multiplication operator and its inverse is given by

$$A^{-1}(p_E) = \frac{\hbar}{p_E^2 + m^2}, \quad (12.61)$$

which in coordinate space amounts to

$$A^{-1}(x_E', x_E) = \int \frac{d^4 p_E}{(2\pi\hbar)^4}\,\exp\left[-ip_E\cdot(x_E' - x_E)/\hbar\right]\frac{\hbar}{p_E^2 + m^2}. \quad (12.62)$$

This clearly satisfies (12.60). Because of the translational invariance of the theory, A and A^{-1} depend only on the difference of coordinates $x_E' - x_E$.

Now we define the function

$$\begin{aligned}\Delta_F^E(x_E' - x_E) &:= \frac{1}{\hbar}A^{-1}(x_E', x_E) \\ &= \int \frac{d^4 p_E}{(2\pi\hbar)^4}\,\exp\left[-ip_E\cdot(x_E' - x_E)/\hbar\right]\frac{1}{p_E^2 + m^2}. \quad (12.63)\end{aligned}$$

This is the *euclidian Feynman propagator of a massive spin-0 field*. The function $\Delta_F^E(x_E' - x_E)$ solves the euclidian version of the Klein–Gordon equation with a delta-function inhomogeneity,

$$(-\hbar^2\,\Box_E + m^2)\Delta_F^E(x_E' - x_E) = \delta^4(x_E' - x_E), \quad (12.64)$$

as a propagator should do. The function is uniquely defined by the Fourier integral: because of the euclidian metric p_E^2 is a positive definite quadratic form on the real axis. The integrand in (12.62) has poles at imaginary frequencies $p_4^E = \pm i\sqrt{p^2 + m^2}$, which are not approached when integrating along the real p_4^E axis.

This problem emerges, however, if we perform the analytic continuation going back to the Minkowski time coordinate. According to the Wick-rotation prescription $x_4^E = ix_0$, the analytically continued Feynman propagator is given by

$$\Delta_F^E\left(i(x_0' - x_0), \boldsymbol{x}' - \boldsymbol{x}\right) = \int \frac{d^4 p_E}{(2\pi\hbar)^4} \exp\left[-\frac{i}{\hbar}\left(p_4^E i(x_0' - x_0) + \boldsymbol{p}_E \cdot (\boldsymbol{x}' - \boldsymbol{x})\right)\right]$$
$$\times \frac{1}{p_{E4}^2 + \boldsymbol{p}_E^2 + m^2} .$$
(12.65)

Putting $p_0 = ip_{E4}$ and substituting $\boldsymbol{p}_E \to -\boldsymbol{p}$ we obtain (cf. (12.28))

$$\Delta_F^E\left(i(x_0' - x_0), \boldsymbol{x}' - \boldsymbol{x}\right) = i \int \frac{d^4 p}{(2\pi\hbar)^4} \exp\left[\frac{i}{\hbar}\left(p_0(x_0' - x_0) - \boldsymbol{p} \cdot (\boldsymbol{x}' - \boldsymbol{x})\right)\right]$$
$$\times \frac{1}{p_0^2 - \boldsymbol{p}^2 - m^2}$$
$$= i \int \frac{d^4 p}{(2\pi\hbar)^4} \exp\left(-ip \cdot x/\hbar\right) \frac{1}{p^2 - m^2}$$
$$= i\Delta_F(x' - x) .$$
(12.66)

Thus we have recovered the *Feynman propagator* Δ_F, by analytic continuation of Δ_F^E.

The integration contour for p_0 is uniquely fixed, going from $-i\infty$ to $+i\infty$. According to the theorem of residues it can be deformed to the real axis if the poles $p_0 = \pm\sqrt{\boldsymbol{p}^2 + m^2}$ are circumvented using the contour C_F depicted in Fig. 12.4. As we know, an equivalent prescription shifts the poles by attaching a small negative imaginary part to the mass (cf. Sect. 4.5):

$$\Delta_F(x' - x) = \int \frac{d^4 p}{(2\pi\hbar)^4} \exp\left[-ip \cdot (x' - x)/\hbar\right] \frac{1}{p^2 - m^2 + i\epsilon} .$$
(12.67)

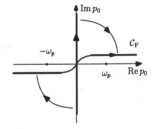

Fig. 12.4. Transition from the euclidian contour to the Feynman contour for the Feynman propagator Δ_F

Thus the euclidian formulation automatically has lead us to the propagator that satisfies the correct boundary conditions! Working in Minkowski coordinates throughout, we would not have known without further deliberation which of the various Green's functions of the Klein–Gordon operator is to be used to construct the inverse $A^{-1}(x', x)$.

We finally note that the correct Feynman propagator (12.67) can also be obtained avoiding the Wick rotation if a convergence factor is built into the path integral, as described at the end of the last chapter:

$$W_0[J] = \mathcal{N} \int \mathcal{D}\phi \, \exp\left[\frac{i}{\hbar} \int d^4 x \left(\frac{\hbar^2}{2} \partial_\mu \phi \, \partial^\mu \phi - \frac{1}{2} m^2 \phi^2 + J\phi + \frac{1}{2} i\epsilon \phi^2 \right)\right] .$$
(12.68)

This is just what happens if the mass m is given a small negative imaginary part.

We also should stress that the above calculation has confirmed that the Wick rotation is admissible at least for noninteracting scalar fields. No singularities in the complex time plane were encountered and the analytic continuation could be performed without problems.

The final result of our calculation is an explicit expression for the *vacuum functional of the noninteracting Klein–Gordon field* obtained from (12.59):

$$W_0[J] = \exp\left[-\frac{1}{2}\frac{i}{\hbar}\int d^4x' d^4x \; J(x')\Delta_F(x'-x)J(x)\right] \; . \tag{12.69}$$

12.4 Generating Functional and Green's Function

The vacuum functional $W[J]$ constitutes a very important instrument for the study of quantum field theories. Indeed, the knowledge of the functional $W[J]$ allows one to calculate all the Green's functions of a theory, i.e., the vacuum expectation values of the time-ordered products of field operators, in the language of canonical quantization. As in Sect. 9.4 we define the *n-point Green's function*

$$G^{(n)}(x_1, \ldots, x_n) = \langle 0 | T(\hat{\phi}(x_1) \cdots \hat{\phi}(x_n)) | 0 \rangle \; . \tag{12.70}$$

As in (11.89) the functions $G^{(n)}$ can be obtained by forming functional derivatives of the vacuum functional with respect to the source, taken at $J=0$:

$$G^{(n)}(x_1, \ldots, x_n) = \left(\frac{\hbar}{i}\right)^n \frac{\delta^n W[J]}{\delta J(x_1) \ldots \delta J(x_n)}\bigg|_{J=0} \; . \tag{12.71}$$

Therefore $W[J]$ is the *generating functional* for the Green's functions. This name generalizes the notion of a generating function, which one encounters in several branches of mathematics, for example, in connection with the orthogonal polynomials. For a function $F(y_1, \ldots, y_N)$ of N variables the Taylor expansion at the point $y=0$ reads

$$F(y_1, \ldots, y_N) = \sum_{n=0}^{\infty} \frac{1}{n!} \sum_{k_1=1}^{N} \cdots \sum_{k_n=1}^{N} P_n(k_1, \ldots, k_n) \, y_{k_1} \cdots y_{k_n} \; , \tag{12.72}$$

with

$$P_n(k_1, \ldots, k_n) = \frac{\partial F(y)}{\partial y_{k_1} \ldots \partial y_{k_n}}\bigg|_{y=0} \; . \tag{12.73}$$

Here $F(y)$ is the generating function for the infinite set of coefficients of the Taylor series $P_n(k_1, \ldots, k_n)$. Generalizing this to the continuous case $k_i \to x$, $\sum_{k_i} \to \int d^4x$ we are lead to

$$W[J] = \sum_{n=0}^{\infty} \int d^4x_1 \ldots \int d^4x_n \frac{1}{n!} \left(\frac{i}{\hbar}\right)^n G^{(n)}(x_1, \ldots, x_n) \, J(x_1) \cdots J(x_n) \; , \tag{12.74}$$

in keeping with (12.71). Clearly, if the vacuum functional $W[J]$ is known exactly, the functions $G^{(n)}(x_1, \ldots, x_n)$ can be obtained by forming derivatives. The Green's functions are the coefficients of the *Volterra series* of the functional $W[J]$. Of course this is helpful only if the functional $W[J]$ is known, for which usually only approximations are available. In particular, for theories involving a very weak or very strong coupling, a perturbation expansion with respect to the coupling constant or its inverse makes sense.

12.4 Generating Functional and Green's Function

A closed expression for the vacuum functional can be given only for the *noninteracting theory* (see (12.69)). From this, the *free n-point Green's functions* are obtained immediately:

$$G_0^{(n)}(x_1,\ldots,x_n) = \left(\frac{\hbar}{i}\right)^n \frac{\delta^n W_0[J]}{\delta J(x_1)\cdots\delta J(x_n)}\bigg|_{J=0}, \tag{12.75}$$

where the index 0 signals the absence of interactions. Since we have at our disposal an explicit expression for $W_0[J]$, (12.69), the n-point functions $G_0^{(n)}$ are easily obtained by successively forming functional derivatives. The first differentiation of (12.69) leads to

$$\frac{\hbar}{i}\frac{\delta W_0[J]}{\delta J(x_1)} = \frac{\hbar}{i}\left(\frac{-i}{2\hbar}\right)\left[2\int d^4x\, J(x)\Delta_F(x-x_1)\right]\exp\left(-\frac{i}{2\hbar}(J,\Delta_F J)\right), \tag{12.76}$$

where the symmetry property $\Delta_F(x-y) = \Delta_F(y-x)$ was used, leading to a factor of two. In the exponent we used the abbreviation

$$(J,\Delta_F J) := \int d^4x' d^4x\, J(x')\Delta_F(x'-x)J(x). \tag{12.77}$$

The result (12.76) shows that there is no such thing as a "one-point Green's function" $G_0^{(1)}(x_1)$ since the factor in the square brackets in (12.76) vanishes at $J=0$:

$$G_0^{(1)}(x_1) = \frac{\hbar}{i}\frac{\delta W_0[J]}{\delta J(x_1)}\bigg|_{J=0} = 0. \tag{12.78}$$

A further differentiation of (12.76) leads to

$$\left(\frac{\hbar}{i}\right)^2 \frac{\delta^2 W_0[J]}{\delta J(x_1)\delta J(x_2)}$$
$$= -\frac{\hbar}{i}\left[\Delta_F(x_1-x_2) + \frac{-i}{\hbar}\int d^4x\, J(x)\Delta_F(x-x_1)\int d^4x\, J(x)\Delta_F(x-x_2)\right]$$
$$\times \exp\left(-\frac{i}{2\hbar}(J,\Delta_F J)\right). \tag{12.79}$$

Since the first term does not depend on J we obtain a nonvanishing result for the *free two-point Green's function*:

$$G_0^{(2)}(x_1,x_2) = \left(\frac{\hbar}{i}\right)^2 \frac{\delta^2 W_0[J]}{\delta J(x_1)\delta J(x_2)}\bigg|_{J=0} = i\hbar\,\Delta_F(x_1-x_2). \tag{12.80}$$

Thus up to a constant factor the two-point function is identical to the Feynman propagator.

All the free Green's functions of higher order can be obtained by elementary, albeit increasingly tedious, forming of derivatives. Without working this out in detail we can make a general observation: the result vanishes for odd n values. In this case, because of the quadratic J dependence in the exponent, one obtains an odd polynomial in J, which vanishes when we take the limit $J=0$ (see also Exercise 11.3c).

Now we perform the next two steps in the calculation. Functional differentiation of (12.79) with respect to $J(x_1)$ produces (we replace $x_1 \to x_2$, $x_2 \to x_3$)

$$\left(\frac{\hbar}{i}\right)^3 \frac{\delta W_0[J]}{\delta J(x_1)\delta J(x_2)\delta J(x_3)} = \left[\frac{\hbar}{i}\Delta_F(x_1-x_2)\int d^4x\, J(x)\Delta_F(x-x_3)\right.$$

$$+ \frac{\hbar}{i}\Delta_F(x_1-x_3)\int d^4x\, J(x)\Delta_F(x-x_2)$$

$$+ \frac{\hbar}{i}\Delta_F(x_2-x_3)\int d^4x\, J(x)\Delta_F(x-x_1)$$

$$\left. - \int d^4x\, J(x)\Delta_F(x-x_1)\int d^4x\, J(x)\Delta_F(x-x_2)\int d^4x\, J(x)\Delta_F(x-x_3)\right]$$

$$\times \exp\left(-\frac{i}{2\hbar}(J,\Delta_F J)\right). \tag{12.81}$$

Since (12.81) vanishes for $J=0$ there is no free three-point Green's function, in agreement with the general rule:

$$G_0^{(3)}(x_1,x_2,x_3) = \left(\frac{\hbar}{i}\right)^3 \frac{\delta^3 W_0[J]}{\delta J(x_1)\delta J(x_2)\delta J(x_3)}\bigg|_{J=0} = 0. \tag{12.82}$$

The next nontrivial Green's function is the *four-point function*, which is obtained by differentiating (12.81). We give only the result obtained in the limit $J=0$:

$$G_0^{(4)}(x_1,x_2,x_3,x_4) = (i\hbar)^2\Big[\Delta_F(x_1-x_2)\Delta_F(x_3-x_4)$$

$$+ \Delta_F(x_1-x_3)\Delta_F(x_2-x_4)$$

$$+ \Delta_F(x_1-x_4)\Delta_F(x_2-x_3)\Big]. \tag{12.83}$$

Results like (12.80) and (12.83) are conveniently represented in the language of *Feynman graphs*. Each of the popagator factors $i\hbar\Delta_F(x_i-x_j)$ is translated into a line having end points x_i and x_j. Equations (12.80) and (12.83) then have the following graphical representations:

$$G_0^{(2)}(x_1,x_2) = \overline{x_1 \quad x_2}, \tag{12.84a}$$

$$G_0^{(4)}(x_1,x_2,x_3,x_4) = \begin{matrix}x_1 \text{---} x_2 \\ x_3 \text{---} x_4\end{matrix} + \begin{matrix}x_1 & x_2 \\ | & | \\ x_3 & x_4\end{matrix} + \begin{matrix}x_1 & x_2 \\ \times \\ x_3 & x_4\end{matrix}. \tag{12.84b}$$

Intuitively this corresponds to the free propagation of one (12.84a) or two (12.84b) free particles between the given space-time points x_i. The presence of three terms for the four-point function clearly has its origin in the fact that the particles are indistinguishable, leading to exchange graphs.

By inspection of (12.84b) the general building principle for higher-order Green's functions becomes clear: the free $2n$-point function $G_0^{(2n)}$ consists of products of n disjoint two-point Green's functions (Feynman progagators) $G_0^{(2)}(x_i,x_j)$:

$$G_0^{(2n)}(x_1,\ldots,x_{2n}) = \sum_{\text{Permut.}} G_0^{(2)}(x_{k_1},x_{k_2})\cdots G_0^{(2)}(x_{k_{2n-1}},x_{k_{2n}}). \tag{12.85}$$

The sum over the permutations of the end points x_k is to be formed in such a way that each pair (x_i, x_j) enters only once and the pairs are ordered, $i > j$, to avoid double counting. When we change from $G_0^{(2n-2)}$ to $G_0^{(2n)}$ the number of possible decompositions into pairs grows by a factor of $2n - 1$ (because x_1 can be combined with $(2n - 1)$ other coordinates, which leaves $(2n - 2)$ coordinates that are to be combined). Therefore the number of terms in (12.85) is $(2n - 1)!! = 1 \times 3 \times 5 \cdots \times (2n - 1)$: for example, $G_0^{(6)}$ contains $5!! = 15$ terms.

We note that the result will depend only on the distances $x_i - x_j$, which is a consequence of the translation invariance of the theory.

The Green's functions of higher order, which are defined by (12.70) or (12.71) for free fields are rather redundant. Since the particles pass each other without interacting everything can be expressed by simple combinations of single-particle propagators $G_0^{(2)}(x, x')$. There is a convenient way to eliminate trivial contributions of this type by introducing a modified *generating functional* $Z[J]$ *for irreducible Green's functions*. It is defined as the logarithm of the functional $W[J]$ according to

$$W[J] = e^{\frac{i}{\hbar} Z[J]} \ . \tag{12.86}$$

Because of $W[0] = 1$ the new functional satisfies the normalization condition $Z[0] = 0$. This new functional is used to define a new type of Green's function $G_c^{(n)}(x_1, \ldots, x_n)$:

$$\frac{i}{\hbar} Z[J] = \sum_{n=0}^{\infty} \int d^4x_1 \ldots \int d^4x_n \frac{1}{n!} \left(\frac{i}{\hbar}\right)^n G_c^{(n)}(x_1, \ldots, x_n) J(x_1) \cdots J(x_n) \tag{12.87}$$

as in (12.74). The coefficient functions according to (12.71) are given by

$$G_c^{(n)}(x_1, \ldots, x_n) = \left(\frac{\hbar}{i}\right)^{n-1} \frac{\delta^n Z[J]}{\delta J(x_1) \ldots \delta J(x_n)} \bigg|_{J=0} \ . \tag{12.88}$$

The $G_c^{(n)}$ are called *connected n-point functions*.

We will later confirm by an explicit calculation that the functions G_c^n are represented by Feynman graphs for which all of the n external legs are connected among each other in a more or less complicated manner. There are no "reducible" contributions that can be split into independent factors as in (12.84b).

For free fields this property of course can be seen immediately. Because of (12.69), the irreducible generating functional is simply given by

$$Z_0[J] = -\frac{1}{2} \int d^4x' d^4x \ J(x') \Delta_F(x' - x) J(x) \tag{12.89}$$

so that

$$G_{0c}^{(2)}(x_1, x_2) = \left(\frac{\hbar}{i}\right) \frac{\delta^2 Z_0[J]}{\delta J(x_1) \, \delta J(x_2)} = i\hbar \Delta_F(x_1 - x_2) = G_0^{(2)}(x_1 - x_2) \ . \tag{12.90}$$

Since (12.84a) is a connected graph the two Green's functions $G_0^{(2)}$ and $G_{0c}^{(2)}$ coincide in this case. It turns out that this remains true also in the interacting case. The free n-point functions $G_{0c}^{(n)}$ of higher order, however, differ drastically from $G_0^{(2)}$: they all vanish. Since $Z[J]$ is bilinear in J there simply are no derivatives of order $n > 2$. This corresponds to the elementary fact that (barring interactions!) it is not possible to draw connected graphs with more than two external legs.

12.5 Generating Functional for Interacting Fields

The calculations presented in the last section were clean and easy, but they did not produce insights of much consequence since no interactions were admitted. Now the discussion will be extended to interacting systems, starting with the example of a self-interacting scalar field ϕ. The interaction will be described by a local potential function $V(\phi)$ multiplied by a coupling constant g which serves for bookkeeping purposes. The full Lagrangian of the system is given by

$$\mathcal{L} = \mathcal{L}_0 + \mathcal{L}_{\text{int}} = \frac{\hbar^2}{2} \partial_\mu \phi \, \partial^\mu \phi - \frac{1}{2} m^2 \phi^2 - g V(\phi) \, . \tag{12.91}$$

The *interacting generating functional* according to (12.19) is given by

$$\begin{aligned}
W[J] &= \mathcal{N} \int \mathcal{D}\phi \, \exp\left[\frac{i}{\hbar} \int d^4x \, (\mathcal{L}_0 + \mathcal{L}_{\text{int}} + J\phi)\right] \\
&= \mathcal{N} \int \mathcal{D}\phi \, \exp\left[-\frac{i}{\hbar} \int d^4x \, g V(\phi)\right] \exp\left[\frac{i}{\hbar} \int d^4x \, (\mathcal{L}_0 + J\phi)\right] \, .
\end{aligned} \tag{12.92}$$

In the second line the interaction term was split off as a separate factor. If it were not for this factor the expression (12.92) would agree with the generating functional for free fields $W_0[J]$, the explicit form of which is known from (12.69). The offending part is the ϕ dependence contained in the interaction term, which prevents us from putting the factor $\exp\left[-\frac{i}{\hbar} \int d^4x \, g V(\phi)\right]$ before the integral. Using a familiar trick, however, it is possible to do this nevertheless! Proceeding as we did when extracting the n-point function from the generating functional we replace the field ϕ by a (functional) differential operator with respect to the source $J(x)$:

$$\frac{\delta}{\delta J(x)} \exp\left[\frac{i}{\hbar} \int d^4x' \, (\mathcal{L}_0 + J\phi)\right] = \frac{i}{\hbar} \phi(x) \exp\left[\frac{i}{\hbar} \int d^4x' \, (\mathcal{L}_0 + J\phi)\right] \, . \tag{12.93}$$

The auxiliary variable J had been introduced just to make transformations of this kind possible. Although the ϕ dependence of the first factor in (12.92) is quite involved, nothing prevents us from replacing ϕ by the functional-derivative operator

$$\phi(x) \to \frac{\hbar}{i} \frac{\delta}{\delta J(x)} \tag{12.94}$$

wherever it enters. Now we assume that the differentiation with respect to the parameter J can be interchanged with the functional integration and pull the whole interaction term before the integral. Then the *interacting generating functional* becomes

$$W[J] = \mathcal{N} \exp\left[-\frac{i}{\hbar}\int d^4x\, gV\left(\frac{\hbar}{i}\frac{\delta}{\delta J(x)}\right)\right] W_0[J]. \tag{12.95}$$

The normalization factor, which does not depend on J, is determined as in (11.85) such that $W[0] = 1$ holds:

$$\mathcal{N}^{-1} = \exp\left[-\frac{i}{\hbar}\int d^4x\, gV\left(\frac{\hbar}{i}\frac{\delta}{\delta J(x)}\right)\right] W_0[J]\bigg|_{J=0}. \tag{12.96}$$

Equation (12.95) saves us from the need to evaluate the full path integral (12.92) of the interacting theory, which we could not have achieved anyway since we know only how to do Gaussian integrals. As compensation, (12.95) contains a complicated combination of functional derivatives. Fortunately there is a systematic way to find approximate solutions. If the exponential function is expanded into its power series the *perturbation series for the generating functional* is obtained. This leads to an expansion with respect to powers of the interaction strength g. Let us write this series in the form

$$W[J] = W_0[J]\left(1 + gw_1[J] + g^2 w_2[J] + \ldots\right). \tag{12.97}$$

The relation of the expansion coefficients $w_1[J], w_2[J], \ldots$ to the terms of the Taylor expansion in (12.95) is slightly involved. One has to take into account that because of the the normalization condition $W[0] = 1$, the normalization factor (12.96) also depends on the coupling constant g (but not on J). We define a set of functionals $u_k[J]$ through

$$\exp\left[-\frac{i}{\hbar}\int d^4x\, gV\left(\frac{\hbar}{i}\frac{\delta}{\delta J}\right)\right] W_0[J]$$
$$\equiv W_0[J]\left(1 + gu_1[J] + g^2 u_2[J] + g^3 u_3[J] + \ldots\right). \tag{12.98}$$

From the Taylor expansion of the exponential function one learns that

$$u_1[J] = W_0^{-1}[J]\left[\frac{-i}{\hbar}\int d^4x\, V\left(\frac{\hbar}{i}\frac{\delta}{\delta J}\right)\right] W_0[J], \tag{12.99a}$$

$$u_2[J] = W_0^{-1}[J]\frac{1}{2!}\left[\frac{-i}{\hbar}\int d^4x\, V\left(\frac{\hbar}{i}\frac{\delta}{\delta J}\right)\right]\left[\frac{-i}{\hbar}\int d^4y\, V\left(\frac{\hbar}{i}\frac{\delta}{\delta J}\right)\right] W_0[J], \tag{12.99b}$$

etc. To obtain the power-series representation of the generating functional $W[J]$ the division of two series has to be performed:

$$W[J] = W_0[J]\,\frac{1 + gu_1[J] + g^2 u_2[J] + g^3 u_3[J] + \ldots}{1 + gu_1[0] + g^2 u_2[0] + g^3 u_3[0] + \ldots}. \tag{12.100}$$

This will result in the coefficient functionals $w_k[J]$ defined in the series expansion (12.97). In higher orders the connection between the two sets of coefficient functionals $u_k[J]$ and $w_k[J]$ is quite involved (see Exercise 12.2). Going up to the second order we have

$$w_1[J] = u_1[J] - u_1[0], \tag{12.101a}$$
$$w_2[J] = \bigl(u_2[J] - u_2[0]\bigr) - \bigl(u_1[J] - u_1[0]\bigr)u_1[0]. \tag{12.101b}$$

The power series in the denominator of (12.100) thus appears to lead to annoying complications in the perturbation series. As we will see in Example 12.4, however, these extra terms have the beneficial consequence of cancelling the unphysical (and divergent!) contributions of unconnected "vacuum bubbles".

EXERCISE

12.2 Series Expansion of the Generating Functional

Problem. (a) Express the coefficients $w_k[J]$ in the series expansion (12.97) of the generating functional $W[J]$ in terms of the coefficients $u_k[J]$ in (12.98), and give explicit expressions up to the third order.
(b) Find the corresponding series expansion for the generating functional $Z[J]$ of the connected Green's functions.

Solution. (a) We have to perform a division of two power series, cf. (12.100). Written in a general way the coefficients c_k have to be determined for which

$$\frac{\sum_{k=0}^{\infty} a_k x^k}{\sum_{k=0}^{\infty} b_k x^k} = \sum_{k=0}^{\infty} c_k x^k . \tag{1}$$

Moving the denominator to the right-hand side and performing the multiplication, we see that the coefficients satisfy

$$a_k = \sum_{i=0}^{k} b_{k-i} c_i . \tag{2}$$

This is a recursion relation which allows for the successive evaluation of the c_i. Starting with $a_0 = b_0 = 1$ we obtain

$$\begin{aligned}
c_0 &= 1 , \\
c_1 &= a_1 - b_1 , \\
c_2 &= a_2 - b_2 - (a_1 - b_1) b_1 , \\
c_3 &= a_3 - b_3 - (a_2 - b_2) b_1 + (a_1 - b_1)(b_1^2 - b_2) , \\
&\vdots \quad .
\end{aligned} \tag{3}$$

We remark that it is also possible to give a closed expression for the coefficients in terms of an $n \times n$ determinant.[4]

$$c_n = (-1)^n \begin{vmatrix} b_1 - a_1 & 1 & 0 & \cdots & 0 \\ b_2 - a_2 & b_1 & 1 & \cdots & 0 \\ b_3 - a_3 & b_2 & b_1 & \cdots & 0 \\ \vdots & \vdots & \vdots & \ddots & \vdots \\ b_n - a_n & b_{n-1} & b_{n-2} & \cdots & b_1 \end{vmatrix} . \tag{4}$$

For the series expansion of the generating functional

$$W[J] = W_0[J] \left(1 + g w_1[J] + g^2 w_2[J] + \ldots \right) , \tag{5}$$

---
[4] See e.g. I.S. Gradshteyn, I.M. Ryzhik: *Table of Integrals, Series, and Products* (Academic Press, New York 1965).

equation (3) implies

Exercise 12.2

$$w_1[J] = u_1[J] - u_1[0], \tag{6a}$$

$$w_2[J] = (u_2[J] - u_2[0]) - (u_1[J] - u_1[0])u_1[0], \tag{6b}$$

$$w_3[J] = (u_3[J] - u_3[0]) - (u_2[J] - u_2[0])u_1[0]$$
$$+ (u_1[J] - u_1[0])(u_1^2[0] - u_2[0]), \tag{6c}$$

$$\vdots$$

We note that all terms contain factors of the type $u_k[J] - u_k[0]$. This implies $w_k[0] = 0$, thus ensuring that the normalization condition $W[0] = W_0[0] = 1$ is met.

(b) The generating functional for *connected* Green's functions

$$\frac{i}{\hbar}Z[J] = \ln W[J] \tag{7}$$

can be expanded into a power series according to

$$Z[J] = Z_0[J] + g z_1[J] + g^2 z_2[J] + \ldots . \tag{8}$$

The expansion coefficients $z_k[J]$ can be obtained from (5) and (7) by employing the Taylor expansion of the logarithm:

$$\frac{i}{\hbar}Z[J] = \ln W[J] = \ln W_0[J] + \ln\big(1 + g\, w_1[J] + g^2 w_2[J] + g^3 w_3[J] + \ldots\big)$$
$$= \ln W_0[J] + \big(g\, w_1[J] + g^2 w_2[J] + g^3 w_3[J] + \ldots\big) \tag{9}$$
$$- \frac{1}{2}\big(g w_1[J] + g^2 w_2[J] + \ldots\big)^2 + \frac{1}{3}\big(g\, w_1[J] + \ldots\big)^3 + \ldots .$$

For the first three coefficients of the power series of $\frac{i}{\hbar}Z[J]$ we find

$$\frac{i}{\hbar}z_1[J] = w_1[J], \tag{10a}$$

$$\frac{i}{\hbar}z_2[J] = w_2[J] - \frac{1}{2}w_1^2[J], \tag{10b}$$

$$\frac{i}{\hbar}z_3[J] = w_3[J] - w_2[J]w_1[J] + \frac{1}{3}w_1^3[J], \tag{10c}$$

$$\vdots$$

Everything can be expressed in terms of $u_k[J]$ if the relations (6) are inserted.

In Example 12.4 we demonstrate, for $z_2[J]$ calculated within the ϕ^4 theory, how the subtraction of $-\frac{1}{2}w_1^2[J]$ in (10b) serves to cancel the contributions of nonconnected graphs.

EXERCISE

12.3 A Differential Equation for $W[J]$

Problem. The generating functional $W[J]$ of the scalar field theory with self-interaction satisfies the following linear functional differential equation:

Exercise 12.3

$$\frac{\hbar}{i}\left(\hbar^2 \Box + m^2\right)\frac{\delta W[J]}{\delta J(x)} - \mathcal{L}'_{\text{int}}\left(\frac{\hbar}{i}\frac{\delta}{\delta J(x)}\right)W[J] = J(x)\,W[J]\,. \tag{1}$$

Here $\mathcal{L}'_{\text{int}}$ denotes the derivative of the interaction Lagrange density with respect to its argument. This differential equation can be used to deduce properties of $W[J]$ without taking recourse to the path-integral representation.[5]

(a) Prove the validity of (1) with the help of the path-integral representation of $W[J]$.

(b) Show that (1) leads to the well-known free generating functional $W_0[J]$ in the limit of vanishing interaction $\mathcal{L}_{\text{int}} \equiv 0$.

(c) Show that the differential equation (1) in the general case is solved by

$$W[J] = \mathcal{N}\exp\left[\frac{i}{\hbar}\int d^4x\,\mathcal{L}_{\text{int}}\left(\frac{\hbar}{i}\frac{\delta}{\delta J(x)}\right)\right]W_0[J]\,. \tag{2}$$

Solution. (a) We start from the path-integral representation of the generating functional

$$W[J] = \mathcal{N}\int \mathcal{D}\phi\,e^{\frac{i}{\hbar}A[\phi,J]} \tag{3}$$

with the derivative

$$\frac{\hbar}{i}\frac{\delta}{\delta J(x)}W[J] = \mathcal{N}\int\mathcal{D}\phi\,\phi(x)\,e^{\frac{i}{\hbar}A[\phi,J]}\,. \tag{4}$$

Here $A[\phi,J]$ denotes the action integral (including the source term)

$$A[\phi,J] = \int d^4x\left(\mathcal{L}_0 + \mathcal{L}_{\text{int}} + J\phi\right) \tag{5}$$

with the free action

$$\int d^4x\,\mathcal{L}_0(\phi) = \int d^4x\,\frac{1}{2}\left(\hbar^2\partial_\mu\phi\,\partial^\mu\phi - m^2\phi^2\right) = -\int d^4x\,\frac{1}{2}\phi\left(\hbar^2\Box + m^2\right)\phi\,. \tag{6}$$

Now the Klein–Gordon operator is applied to (4):

$$\left(\hbar^2\Box + m^2\right)\frac{\hbar}{i}\frac{\delta W[J]}{\delta J(x)} = \mathcal{N}\int \mathcal{D}\phi\,\left(\hbar^2\Box + m^2\right)\phi\,e^{\frac{i}{\hbar}A[\phi,J]}\,. \tag{7}$$

We notice that according to (6) the Klein–Gordon operator also stands in the exponent of (7). Therefore the factor $(\hbar^2\Box+m^2)\phi(x)$ can be replaced by a differentiation $\delta/\delta\phi(x)$ according to the identity

$$\frac{\hbar}{i}\frac{\delta}{\delta\phi(x)}e^{\frac{i}{\hbar}A[\phi,J]} = \frac{\delta A[\phi,J]}{\delta\phi(x)}e^{\frac{i}{\hbar}A[\phi,J]} \tag{8}$$

$$= \left(-(\hbar^2\Box+m^2)\phi(x) + \mathcal{L}'_{\text{int}}(\phi(x)) + J(x)\right)e^{\frac{i}{\hbar}A[\phi,J]}\,.$$

Therefore (7) can be written as

$$(\hbar^2\Box+m^2)\frac{\hbar}{i}\frac{\delta W[J]}{\delta J(x)} = \mathcal{N}\int\mathcal{D}\phi\left(-\frac{\hbar}{i}\frac{\delta}{\delta\phi(x)} + \mathcal{L}'_{\text{int}}(\phi(x)) + J(x)\right)e^{\frac{i}{\hbar}A[\phi,J]}\,. \tag{9}$$

---

[5] K. Symanzik: Z. Naturforsch. **9a**, 809 (1954).

The first term on the right-hand side represents a total derivative. Upon integration over $\mathcal{D}\phi$ this leads only to "surface terms", which we shall disregard as usual:

$$(\hbar^2 \Box + m^2) \frac{\hbar}{i} \frac{\delta W[J]}{\delta J(x)} = \mathcal{N} \int \mathcal{D}\phi \left(\mathcal{L}'_{\text{int}}(\phi) + J(x) \right) e^{\frac{i}{\hbar} A[\phi, J]} . \tag{10}$$

The factor in the brackets can be pulled before the integral provided that $\phi(x)$ is replaced by the functional derivative operator $(\hbar/i)\delta/\delta J(x)$ which acts on the exponential function. The remain integral is just the original generating functional $W[J]$. This proves the differential equation (1).

(b) The free generating functional satisfies the differential equation

$$\frac{\hbar}{i}(\hbar^2 \Box + m^2) \frac{\delta W_0[J]}{\delta J(x)} = J(x) W_0[J] . \tag{11}$$

The Klein–Gordon operator can be inverted with the help of the propagator Δ_F:

$$\begin{aligned}
\frac{\delta W_0[J]}{\delta J(x)} &= \frac{i}{\hbar}(\hbar^2 \Box + m^2)^{-1} J(x) W_0[J] \\
&= -\frac{i}{\hbar} \int d^4 x' \, \Delta_F(x - x') J(x') W_0[J] .
\end{aligned} \tag{12}$$

This functional differential equation is solved by an exponential function,

$$W_0[J] = \exp\left(-\frac{i}{\hbar} \frac{1}{2} \int d^4 x \, d^4 x' \, J(x) \Delta_F(x - x') J(x') \right) , \tag{13}$$

where the integration constant was fixed from the normalization condition $W_0[0] = 1$. This result agrees with the solution found in Sect. 12.3.

(c) To prove the assertion, (2) has to be inserted into the differential equation (1). On the left-hand side the factor

$$\exp\left(\frac{i}{\hbar} \int d^4 x \, \mathcal{L}_{\text{int}}\left(\frac{\hbar}{i} \frac{\delta}{\delta J(x)} \right) \right) \equiv e^{\hat{O}} \tag{14}$$

can be moved to the front. $e^{\hat{O}}$ can be written as a power series of the derivative operator $\delta/\delta J(x)$. Because

$$\left[\frac{\delta}{\delta J(x)}, \frac{\delta}{\delta J(y)} \right] = 0 , \tag{15}$$

each term in the series commutes with the derivative operators that stand in front of $W[J]$. Similarly $e^{\hat{O}}$ also commutes with the Klein–Gordon operator (which contains only ordinary instead of functional derivatives). Therefore the left-hand side of (1) reads

$$\begin{aligned}
e^{\hat{O}} &\left(\frac{\hbar}{i}(\hbar^2 \Box + m^2) \frac{\delta}{\delta J(x)} - \mathcal{L}'_{\text{int}}\left(\frac{\hbar}{i} \frac{\delta}{\delta J(x)} \right) \right) W_0[J] \\
&= e^{\hat{O}} \left(J(x) - \mathcal{L}'_{\text{int}}\left(\frac{\hbar}{i} \frac{\delta}{\delta J(x)} \right) \right) W_0[J] .
\end{aligned} \tag{16}$$

Here the differential equation (11) for the free generating functional $W_0[J]$ has been used.

Exercise 12.3

Exercise 12.3

The treatment of the right-hand side of (1), $J(x)\,\mathrm{e}^{\hat{O}} W_0[J]$, is a bit trickier since the operator $\hat{O}$ contains derivatives that do not commute with $J(x)$. We will show that the ensuing commutator produces the second term in (16). For this we first write $\mathcal{L}_{\text{int}}$ as a power series:

$$\mathcal{L}_{\text{int}}\left(\frac{\hbar}{\mathrm{i}}\frac{\delta}{\delta J(y)}\right) = \sum_n c_n \left(\frac{\hbar}{\mathrm{i}}\frac{\delta}{\delta J(y)}\right)^n . \qquad (17)$$

The commutator of each term of this series with $J(x)$ can be easily found with the help of the identity (see, e.g., (2.29))

$$\left[\frac{\delta}{\delta J(y)},\, J(x)\right] = \delta^4(x-y) . \qquad (18)$$

For example, we obtain for $n=2$

$$\begin{aligned}\left[\left(\frac{\delta}{\delta J(y)}\right)^2,\, J(x)\right] &= \frac{\delta}{\delta J(y)}\left[\frac{\delta}{\delta J(y)},\, J(x)\right] + \left[\frac{\delta}{\delta J(y)},\, J(x)\right]\frac{\delta}{\delta J(y)} \\ &= 2\,\delta^4(x-y)\frac{\delta}{\delta J(y)} . \end{aligned} \qquad (19)$$

The general result reads,

$$\left[\left(\frac{\delta}{\delta J(y)}\right)^n,\, J(x)\right] = n\,\delta^4(x-y)\left(\frac{\delta}{\delta J(y)}\right)^{n-1} , \qquad (20)$$

as one can easily prove using complete induction. With the series expansion (17) the following commutator is obtained:

$$\begin{aligned}\left[\hat{O},\, J(x)\right] &= \left[\int \mathrm{d}^4 y\,\mathcal{L}_{\text{int}}\left(\frac{\hbar}{\mathrm{i}}\frac{\delta}{\delta J(y)}\right),\, J(x)\right] \\ &= \int \mathrm{d}^4 y\,\sum_n c_n \left[\left(\frac{\hbar}{\mathrm{i}}\frac{\delta}{\delta J(y)}\right)^n,\, J(x)\right] \\ &= \int \mathrm{d}^4 y\,\frac{\hbar}{\mathrm{i}}\delta^4(x-y)\sum_n c_n\, n\left(\frac{\hbar}{\mathrm{i}}\frac{\delta}{\delta J(y)}\right)^{n-1} \\ &= \frac{\hbar}{\mathrm{i}}\mathcal{L}'_{\text{int}}\left(\frac{\hbar}{\mathrm{i}}\frac{\delta}{\delta J(x)}\right) . \end{aligned} \qquad (21)$$

What we need is not (21) but rather the commutator of the exponential function of $\hat{O}$ with $J(x)$. Here the following operator identity is helpful:

$$\hat{A}\,\mathrm{e}^{\hat{B}} = \mathrm{e}^{\hat{B}}\hat{A} + \mathrm{e}^{\hat{B}}[\hat{A},\,\hat{B}] + \frac{1}{2!}\mathrm{e}^{\hat{B}}[[\hat{A},\,\hat{B}],\,\hat{B}] + \ldots . \qquad (22)$$

This is a version of the Baker–Campbell–Hausdorff relation treated in Sect. 1.3. Fortunately in our case the double commutator vanishes and the series in (22) terminates since $\hat{B} \equiv \hat{O}$ as well as $[\hat{A},\,\hat{B}] \equiv [J(x),\,\hat{O}]$ contain only derivative operators $\delta/\delta J$ that commute among each other according to (15). Using (21) and (22) we find the right-hand side of (1) becomes

$$\begin{aligned}J(x)W[J] &= J(x)\,\mathrm{e}^{\hat{O}} W_0[J] \\ &= \mathrm{e}^{\hat{O}}\left(J(x) - \frac{\hbar}{\mathrm{i}}\mathcal{L}'_{\text{int}}\left(\frac{\hbar}{\mathrm{i}}\frac{\delta}{\delta J(x)}\right)\right) W_0[J] . \end{aligned} \qquad (23)$$

This agrees with the left-hand side given by (14), thus proving that $e^{\hat{O}} W_0[J]$ satisfies the differential equation for the generating functional $W[J]$ of the interacting theory.

Exercise 12.3

12.6 Green's Functions in Momentum Space

When describing scattering processes it is often advantageous to work in momentum space. Accordingly we will now introduce the n-point Green's function in momentum space $G^{(n)}(p_1, \ldots, p_n)$, which is defined through a Fourier transformation. A little care is required since this function is overdetermined when expressed in terms of the set of momentum variables p_1 to p_n. As a consequence of translation invariance the total momentum is conserved in any process, thus putting a constraint on the n momentum variables. Therefore it is reasonable to define the momentum-space Green's function as follows:

$$G^{(n)}(p_1, \ldots, p_n) \left(\frac{\hbar}{2\pi}\right)^4 \delta^4(p_1 + \ldots + p_n)$$
$$= \int d^4 x_1 \ldots \int d^4 x_n \, e^{i(p_1 \cdot x_1 + \ldots + p_n \cdot x_n)/\hbar} G^{(n)}(x_1, \ldots, x_n) , \quad (12.102)$$

i.e., to split off the momentum-conserving delta function. Since space-time is homogeneous, $G^{(n)}(x_1, \ldots, x_n)$ will not depend on the absolute values of the coordinates but only on distances $x_i - x_j$.

Let us evaluated the *free two-point and four-point functions* in momentum space from (12.80) and (12.83). Going to center-of-mass and relative coordinates

$$\begin{aligned} x &= x_1 - x_2 &, \quad p &= \tfrac{1}{2}(p_1 - p_2) , \\ X &= \tfrac{1}{2}(x_1 + x_2) &, \quad P &= p_1 + p_2 , \end{aligned} \quad (12.103)$$

we find that the Fourier transform of $G_0^{(2)}(x_1, x_2) = i\hbar \Delta_F(x_1 - x_2)$ becomes

$$\int d^4 x_1 d^4 x_2 \, e^{i(p_1 \cdot x_1 + p_2 \cdot x_2)/\hbar} G_0^{(2)}(x_1, x_2)$$
$$= \left(\int d^4 X \, e^{iP \cdot X/\hbar}\right) \left(\int d^4 x \, e^{ip \cdot x/\hbar} i\hbar \Delta_F(x)\right)$$
$$= (2\pi\hbar)^4 \delta^4(P) \, i\hbar \Delta_F(p) . \quad (12.104)$$

As expected the free two-point Green's function in momentum space is just the Fourier-transformed Feynman propagator:

$$G_0^{(2)}(p, -p) = i\hbar \Delta_F(p) . \quad (12.105)$$

The minus sign in the second momentum argument arises from the fact that the two arguments are treated equally. All momenta are considered to *enter* the graph and they must sum to zero.

The calculation for $G_0^{(4)}(x_1, x_2, x_3, x_4)$ is very similar. Since (12.83) depends on coordinate differences $x_i - x_j$, transformations of the type (12.103) can be applied to each pair of coordinates and the integrals factorize. For example the first term in (12.83) leads to the Fourier transform

$$\int d^4x_1 d^4x_2 d^4x_3 d^4x_4 \, \Delta_F(x_1 - x_2) \Delta_F(x_3 - x_4) \tag{12.106}$$

$$= (2\pi\hbar)^4 \delta^4(p_1 + p_2)(2\pi\hbar)^4 \delta^4(p_3 + p_4) \, \Delta_F\left(\tfrac{1}{2}(p_1 - p_2)\right) \Delta_F\left(\tfrac{1}{2}(p_3 - p_4)\right).$$

Momentum conservation has lead to $p_2 = -p_1$ and $p_4 = -p_3$ separately since the particles do not interact. The other terms contributing to $G_0^{(4)}$ differ by permutations of the end points. The same delta function expressing conservation of total momentum can be extracted from all three graphs by using the identity $\delta(x)\delta(y) = \delta(x+y)\,\delta(x-y)$. This leads to the free four-point Green's function in momentum space defined according to (12.102):

$$G_0^{(4)}(p_1, p_2, p_3, p_4) \tag{12.107}$$

$$= (i\hbar)^2 \Big[(2\pi\hbar)^4 \delta^4(p_1 + p_2 - p_3 - p_4) \Delta_F\left(\tfrac{1}{2}(p_1 - p_2)\right) \Delta_F\left(\tfrac{1}{2}(p_3 - p_4)\right)$$

$$+ (2\pi\hbar)^4 \delta^4(p_1 + p_3 - p_2 - p_4) \Delta_F\left(\tfrac{1}{2}(p_1 - p_3)\right) \Delta_F\left(\tfrac{1}{2}(p_2 - p_4)\right)$$

$$+ (2\pi\hbar)^4 \delta^4(p_1 + p_4 - p_2 - p_3) \Delta_F\left(\tfrac{1}{2}(p_1 - p_4)\right) \Delta_F\left(\tfrac{1}{2}(p_2 - p_3)\right)\Big].$$

This result is valid under the constraint $p_1 + p_2 + p_3 + p_4 = 0$. Of course, also the connected Green's functions $G_c(x_1, \ldots, x_n)$ can be transformed into momentum space and (12.102) holds also for them.

EXAMPLE

12.4 The Perturbation Series for the ϕ^4 Theory

To gain acquaintance with functional techniques one has to carry out practical calculations. Complications arising from the presence of several coupled fields can be avoided if one studies a self-interacting field theory. As an example we will study the ϕ^4 model, which is defined by the Lagrangian

$$\mathcal{L} = \frac{\hbar^2}{2} \partial_\mu \phi \, \partial^\mu \phi - \frac{1}{2} m^2 \phi^2 - gV(\phi), \tag{1}$$

with the interaction term

$$gV(\phi) = \frac{g}{4!} \phi^4. \tag{2}$$

We have studied this theory already in the canonical-quantization framework (Sects. 8.7 and 9.4).

Let us explicitely evaluate the generating functional $W[J]$ and the two-point and four-point Green's functions in first-order perturbation theory. For this, according to (12.99), we need the functional

$$u_1[J] = W_0^{-1}[J] \left[\frac{-i}{\hbar} \int d^4x \, V\left(\frac{\hbar}{i} \frac{\delta}{\delta J(x)}\right)\right] W_0[J]$$

$$= W_0^{-1}[J] \frac{-i}{\hbar} \frac{1}{4!} \int d^4x \left(\frac{\hbar}{i}\right)^4 \left(\frac{\delta}{\delta J(x)}\right)^4 W_0[J], \tag{3}$$

where the free generating functional $W_0[J]$ is known analytically:

$$W_0[J] = \exp\left[-\frac{i}{2\hbar}\int d^4y d^4z\, J(y)\Delta_F(y-z)J(z)\right] \equiv \exp\left[-\frac{i}{2\hbar}(J,\Delta_F J)\right]. \quad (4)$$

A similar calculation was needed for the free Green's functions (12.4)), with the exception that the derivatives with respect to $J(x_i)$ had differing arguments x_i. The calculation is elementary, just making repeated use of the product and chain rules of differentiation. For the first derivative we obtain

$$\frac{\hbar}{i}\frac{\delta}{\delta J(x)}\exp\left[-\frac{i}{2\hbar}(J,\Delta_F J)\exp\right] \quad (5)$$

$$= -\int d^4y\, \Delta_F(x-y)J(y)\,\exp\left[-\frac{i}{2\hbar}(J,\Delta_F J)\exp\right].$$

The second derivative is given by

$$\left(\frac{\hbar}{i}\frac{\delta}{\delta J(x)}\right)^2 \exp\left[-\frac{i}{2\hbar}(J,\Delta_F J)\exp\right] \quad (6)$$

$$= \left[-\frac{\hbar}{i}\Delta_F(0) + \left(\int d^4y\, \Delta_F(x-y)J(y)\right)^2\right]\exp\left[-\frac{i}{2\hbar}(J,\Delta_F J)\exp\right],$$

and the third derivative is

$$\left(\frac{\hbar}{i}\frac{\delta}{\delta J(x)}\right)^3 \exp\left[-\frac{i}{2\hbar}(J,\Delta_F J)\exp\right]$$

$$= \left[3\frac{\hbar}{i}\Delta_F(0)\int d^4y\, \Delta_F(x-y)J(y) - \left(\int d^4y\, \Delta_F(x-y)J(y)\right)^3\right]$$

$$\times \exp\left[-\frac{i}{2\hbar}(J,\Delta_F J)\exp\right]. \quad (7)$$

Finally the fourth derivative becomes

$$\left(\frac{\hbar}{i}\frac{\delta}{\delta J(x)}\right)^4 \exp\left[-\frac{i}{2\hbar}(J,\Delta_F J)\exp\right]$$

$$= \left[3\left(\frac{\hbar}{i}\right)^2(\Delta_F(0))^2 - 6\frac{\hbar}{i}\Delta_F(0)\left(\int d^4y\, \Delta_F(x-y)J(y)\right)^2\right.$$

$$\left. + \left(\int d^4y\, \Delta_F(x-y)J(y)\right)^4\right]\exp\left[-\frac{i}{2\hbar}(J,\Delta_F J)\exp\right]. \quad (8)$$

These formulas can be made more intuitive by introducing a *graphic notation*. As usual each Feynman propagator $\Delta_F(x-y)$ is represented by a line $\overline{}_{xy}$. Furthermore the source terms $J(x)$ are represented by line endings according to $\bullet\!\!-\!\!-$. Table 12.1 shows the abbreviations used. Suitable powers of i and $\hbar$ have been included to simplify the equations written in graphical form.

Note that each vertex $\times$ with 4 lines and each source term $\bullet\!\!-\!\!-$ has its own coordinate value x attached and that an integration $\int d^4x$ over all coordinates is implied. In the compact graphical notation the functional (3) reads

Example 12.4

Example 12.4

Table 12.1. Graphical translation rules for the ϕ^4 theory.

Propagator	$i\hbar\Delta_F(x-y)$	$\overline{}_{xy}$
Loop	$i\hbar\Delta_F(0)$	◯
External source	$\frac{i}{\hbar}\int d^4x\, J(x)$	•——
ϕ^4 interaction	$-\frac{i}{\hbar}\int d^4x$	✕

$$u_1[J] = \frac{1}{4!}\left(3\,\infty + 6\,\text{─○─} + \text{✕}\right). \tag{9}$$

There are terms quadratic and quartic in J and the constant term ∞. As mentioned already in Sect. 12.5 the normalization of the generating functional to $W[0] = 1$ leads to the cancellation of contributions from unconnected vacuum bubbles.[6] Within first-order perturbation theory we have

$$W[J] = \frac{1 + gu_1[J] + O(g^2)}{1 + gu_1[0] + O(g^2)} W_0[J]$$

$$= \left(1 + g(u_1[J] - u_1[0]) + \ldots\right) W_0[J]. \tag{10}$$

The normalization term $u_1[0]$ contains only the vacuum graphs since the imposition of $J = 0$ eliminates the line endings. Therefore the the generating functional in first order has the following graphical representation

$$W[J] = \left[1 + \frac{g}{4!}\left(6\,\text{─○─} + \text{✕}\right) + \ldots\right] e^{\frac{1}{2}\text{─•─}}, \tag{11}$$

or, once again using the explicit notation

$$W[J] = \left[1 - \frac{i}{\hbar}\frac{g}{4!}\int d^4x \left(6i\hbar\Delta_F(0)\int d^4y_1 d^4y_2\, \Delta_F(x-y_1)\Delta_F(x-y_2)J(y_1)J(y_2)\right.\right.$$

$$\left.+ \int d^4y_1 d^4y_2 d^4y_3 d^4y_4\, \Delta_F(x-y_1)\Delta_F(x-y_2)\Delta_F(x-y_3)\Delta_F(x-y_4)\right.$$

$$\left.\left.\times J(y_1)J(y_2)J(y_3)J(y_4) + \ldots\right)\right]\exp\left[-\frac{i}{2\hbar}(J,\Delta_F J)\right]. \tag{12}$$

Now the *Green's functions of the interacting theory* in first-order perturbation theory can be evaluated. The two-point function $G(x_1, x_2)$ is given as the second-order functional derivative of (12):

$$G(x_1, x_2) = \left(\frac{\hbar}{i}\right)^2 \frac{\delta^2 W[J]}{\delta J(x_1)\delta J(x_2)}\bigg|_{J=0}. \tag{13}$$

---

[6] Within the canonical formalism the cancellation of unconnected graphs was discussed in Exercise 8.3. Strictly speaking one should admit that this is more of a formal manipulation since the normalization factor $\mathcal{N}$ is a divergent expression.

12.6 Green's Functions in Momentum Space

The functional derivatives can be conveniently performed by using the graphical notation. Differentiating with respec to $J(x_i)$ simply amounts to the replacement of a line ending by an open (external) line which carries the coordinate index x_i:

$$\frac{\hbar}{i}\frac{\delta}{\delta J(x_i)}\;\bullet\!\!-\!\!-\;=\;\overline{}_{x_i}\,. \tag{14}$$

For the remainder of the calculation, only the product and chain rules are needed. The first derivative of the functional (11) becomes

$$\frac{\hbar}{i}\frac{\delta}{\delta J(x_1)}W[J] = \frac{\hbar}{i}\frac{\delta}{\delta J(x_1)}\left[1+\frac{g}{4!}\left(6\,\text{◯}\!\!-\!\!\bullet + \text{✕}\right)\right]e^{\frac{1}{2}\bullet\!\!-\!\!\bullet}$$

$$= \left[\frac{g}{4!}\left(6\cdot 2\,{}_{x_1}\text{◯}\!\!-\!\!\bullet + 4\,{}_{x_1}\text{✕}\right) + \left(1+\frac{g}{4!}\left(6\,\text{◯}\!\!-\!\!\bullet + \text{✕}\right)\right)\,\overline{}_{x_1}\right]e^{\frac{1}{2}\bullet\!\!-\!\!\bullet}$$

$$= \left[\,\overline{}_{x_1} + \frac{g}{4!}\left(12\,{}_{x_1}\text{◯}\!\!-\!\!\bullet + 4\,{}_{x_1}\text{✕} + 6\,\text{◯}\!\!-\!\!\bullet\,{}_{x_1}\!\!-\!\!\bullet + \text{✕}\,{}_{x_1}\!\!-\!\!\bullet\right)\right]e^{\frac{1}{2}\bullet\!\!-\!\!\bullet} \tag{15}$$

Each of the terms contains at least one end point $\bullet\!\!-\!\!-$ (which corresponds to a factor $J(x)$) so that the one-point Green's function does not exist, because (15) vanishes in the limiting case $J\to 0$. The second derivative of $W[J]$ already looks rather complex. In graphical notation it is

$$\left(\frac{\hbar}{i}\right)^2 \frac{\delta^2}{\delta J(x_1)\delta J(x_2)} W[J] =$$

$$\Bigg\{\overline{}_{x_1\,x_2} + \overline{}_{x_1}\!\!\bullet\;\overline{}_{x_2}\!\!\bullet$$

$$+\frac{g}{4!}\bigg[12\,{}_{x_1}\text{◯}\!\!-\!\!{}_{x_2} + 12\,{}_{x_2}^{x_1}\text{✕} + 12\,\text{◯}\!\!-\!\!{}_{x_2}\,\overline{}_{x_1}\!\!\bullet + 6\,\text{◯}\!\!-\!\!\bullet\,\overline{}_{x_1\,x_2}$$

$$+4\,{}_{x_2}\text{✕}\,\overline{}_{x_1}\!\!\bullet + \text{✕}\,\overline{}_{x_1\,x_2} + 12\,{}_{x_1}\text{◯}\!\!-\!\!\bullet\,\overline{}_{x_2}\!\!\bullet + 4\,{}_{x_1}\text{✕}\,\overline{}_{x_2}\!\!\bullet$$

$$+6\,\text{◯}\!\!-\!\!\bullet\,\overline{}_{x_1}\!\!\bullet\,\overline{}_{x_2}\!\!\bullet + \text{✕}\,\overline{}_{x_1}\!\!\bullet\,\overline{}_{x_2}\!\!\bullet\bigg]\Bigg\}e^{\frac{1}{2}\bullet\!\!-\!\!\bullet}\,. \tag{16}$$

In the limit $J\to 0$ only two terms remain. The two-point Green's function of the ϕ^4 theory according to (13) in first-order perturbation theory is

$$G^{(2)}(x_1,x_2) = \overline{}_{x_1\,x_2} + \frac{1}{2}g\,\text{◯}\!\!-\!\!\bullet\,\overline{}_{x_1\,x_2} + O(g^2)\,, \tag{17}$$

or written explicitly

$$\begin{aligned}G^{(2)}(x_1,x_2) &= i\hbar\Delta_F(x_1-x_2) \\ &\quad -\frac{1}{2}\hbar^2 g\,\Delta_F(0)\int d^4x\,\Delta_F(x_1-x)\Delta_F(x-x_2) + O(g^2)\,.\end{aligned} \tag{18}$$

Example 12.4

Example 12.4 The influence of the interaction on the Green's function in lowest order consists in the appearance of the loop graph ⟲ (the *tadpole graph*). Since $\Delta_F(0)$ is a singular expression this correction is divergent. This is a problem regularly encountered in field theory and it is remedied by the renormalization procedure. To be specific, corrections to $G^{(2)}(x_1, x_2)$ lead to a *mass renormalization* of the boson field.

Remark on the tadpole graph: In the canonical quantization formalism there was a simple way to exclude graphs with loops containing a single vertex, i.e., the normal-ordering prescription applied to the interaction operator. These contributions, however, can also be kept and dealt with later at the stage of mass renormalization.

The evaluation of the *four-point Green's function*

$$G^{(4)}(x_1, x_2, x_3, x_4) = \left(\frac{\hbar}{i}\right)^4 \frac{\delta^4}{\delta J(x_1)\delta J(x_2)\delta J(x_3)\delta J(x_4)} W[J]\bigg|_{J=0} \quad (19)$$

is rather laborious. The next differentiation step of (16) already produced 30 different terms and this number gets inflated even more when the fourth derivative is formed. Fortunately we are only interested in those contributions that survive in the limit $J \to 0$, i.e., that do not contain any line endings •——• . Thus the interesting terms in the third derivative are those which contain exactly one line ending, a condition which considerably reduces the number of terms to be evaluated. Furthermore we do not have to repeat the calculation for the noninteracting contribution (the first term in (11); cf. section 12.4).

Let us first investigate the contribution of •⟲• in the generating functional (11). The terms of interest having one line ending at the stage of the third derivative are obtained by differentiating (16):

$$\left(\frac{\hbar}{i}\right)^3 \frac{\delta^3}{\delta J(x_1)\delta J(x_2)\delta J(x_3)} W'[J] = \frac{g}{4!} 12 \Big[\underset{x_1}{\bigcirc}\!\!-\!\!\underset{x_2\ x_3}{\bullet} + \underset{x_1}{\bigcirc}\!\!-\!\!\underset{x_3\ x_2}{\bullet}$$

$$+ \underset{x_2}{\bigcirc}\!\!-\!\!\underset{x_3\ x_1}{\bullet} + \underset{x_1}{\bigcirc}\!\!-\!\!\underset{x_2\ x_3}{\bullet} + \underset{x_2}{\bigcirc}\!\!-\!\!\underset{x_1\ x_3}{\bullet}$$

$$+ \underset{x_3}{\bigcirc}\!\!-\!\!\underset{x_1\ x_2}{\bullet} \Big] e^{\frac{1}{2}\bullet\!-\!\bullet} + \ldots . \quad (20)$$

The omitted terms contain at least two line endings, i.e., two factors $J(x_i)$. The step of forming the fourth derivative now is trivial. Since subsequently J is put to zero only the terms

$$\left(\frac{\hbar}{i}\right)^4 \frac{\delta^4}{\delta J(x_1)\delta J(x_2)\delta J(x_3)\delta J(x_4)} W'[J]\bigg|_{J=0}$$

$$= \frac{g}{4!} 12 \Big[\underset{x_2}{\bigcirc}\!\!-\!\!\underset{x_3\ x_4\ x_1}{\bullet} + \underset{x_2}{\bigcirc}\!\!-\!\!\underset{x_4\ x_3\ x_1}{\bullet} + \underset{x_3}{\bigcirc}\!\!-\!\!\underset{x_4\ x_2\ x_1}{\bullet}$$

12.6 Green's Functions in Momentum Space   397

Example 12.4

$$+ \underset{x_2}{\bigcirc} \underset{x_1\ x_3\ x_4}{---} + \underset{x_3}{\bigcirc} \underset{x_1\ x_2\ x_4}{---} + \underset{x_4}{\bigcirc} \underset{x_1\ x_2\ x_3}{---} \Bigg] \tag{21}$$

survive. The same type of graph consisting of two disjoint parts repeatedly contributes with the various possible permutations of the coordinates x_i.

The calculation of the third contribution in (11) containing the factor $\times$ proceeds similarly. From (16) only a single term is obtained,

$$\left(\frac{\hbar}{i}\right)^3 \frac{\delta^3}{\delta J(x_1)\delta J(x_2)\delta J(x_3)} W''[J] = \frac{g}{4!} 24\, {}^{x_1}\!\!\times_{\!x_3}^{\!x_2}\, e^{\frac{1}{2}\bullet\!\!-\!\!\bullet} + \ldots , \tag{22}$$

which simply produces

$$\left(\frac{\hbar}{i}\right)^4 \frac{\delta^4}{\delta J(x_1)\delta J(x_2)\delta J(x_3)\delta J(x_4)} W''[J]\Bigg|_{J=0} = \frac{g}{4!} 24\, {}^{x_2}_{x_1}\!\!\times_{\!x_3}^{\!x_4}. \tag{23}$$

Collecting the results (12.84b), (21), and (23), we can write down the *four-point function of ϕ^4 theory in first-order perturbation theory*:

$$G^{(4)}(x_1, x_2, x_3, x_4) = \Big(\equiv + |\ | + \times \Big) \tag{24}$$

$$+ \frac{1}{2} g \Big(\underset{}{\overset{\bigcirc}{-\!\!-}} + \overline{\underset{\bigcirc}{}} + \mathsf{q}\ | + |\ \mathsf{p} + \overset{\bigcirc}{\times} + \underset{\bigcirc}{\times} \Big) + g \times .$$

The structure of this result is quite plausible. In addition to the free four-point function all possible propagators with a self-energy insertion $-\!\!\bigcirc\!\!-$ are encountered. In addition there is a single "true" interaction graph $\times$.

The numerical factors associated with the various graphs (they are called *symmetry factors*) can be deduced from combinatorial reasoning. For example, the factor $\frac{1}{2}$ of the self-energy graphs is obtained as follows. Each of the six self-energy graphs in the second line of (24) contains a four-vertex connected to two external lines characterized by certain coordinates x_i, x_j. There are $4 \cdot 3$ ways to connect them to the four legs of the vertex. The remaining two lines are connected among each other via the Feynman propagator, which is unambiguous. If we take into account the normalization factor associated with the coupling constant this consideration leads to the symmetry factor $S = 4 \cdot 3/4! = 1/2$. The symmetry factor of the scattering graph $\times$ is $S = 4!/4! = 1$, because there are 4! alternative ways to connect the four external lines to the four legs of the vertex.

The rules for calculating Feynman graphs derived using the functional method are in full agreement with our earlier results based on the canonical formalism obtained in Chap. 8. The perturbation expansion that resulted from the Dyson series and Wick's theorem is now the outcome of repeatedly differentiating the vacuum functional $W[J]$ with respect to the fictitious source current J. It is very gratifying to see how two completely different routes lead to the same goal.

Example 12.4

The four-point function in (24) is quite messy because of the presence of products of disjoint two-point functions. If one considers, on the other hand, the *connected Green's function* $G_c^4(x_1, x_2, x_3, x_4)$ only the last term of (24) will contribute according to the general arguments given in Exercise 12.5. Here we will check this out explicitly for our example of the ϕ^4 model in first-order perturbation theory. Let us take the generating functional $\frac{i}{\hbar}Z[J] = \ln W[J]$ and insert the result of (11):

$$\frac{i}{\hbar}Z[J] = \ln \exp\left(\frac{1}{2}\text{\textemdash}\right) + \ln\left[1 + \frac{g}{4!}\left(6\,\bigcirc\!\!\!\text{\textemdash} + \times\right)\right] + O(g^2)$$

$$= \frac{1}{2}\text{\textemdash} + \frac{g}{4!}\left(6\,\bigcirc\!\!\!\text{\textemdash} + \times\right) + O(g^2) \,. \tag{25}$$

To obtain the connected four-point function according to (12.88) we have to form the fourth derivative with respect to J. Obviously the rule (14) produces only a single nonvanishing contribution:

$$G_c^{(4)}(x_1, x_2, x_3, x_4) = \left(\frac{\hbar}{i}\right)^3 \frac{\delta^4 Z[J]}{\delta J(x_1)\delta J(x_2)\delta J(x_3)\delta J(x_4)}$$

$$= 6g \times + O(g^2) \,. \tag{26}$$

The result has become as simple as this because after we take the logarithm, the term $W_0[J] = e^{\frac{1}{2}\text{\textemdash}}$ enters additively instead of being a multiplicative factor. Therefore upon differentiation the product rule does not get invoked and no "mixed terms" arise.

EXERCISE

12.5 Connected Green's Functions

Problem. Prove the following general relation between ordinary and connected n-point Green's functions:

(a) $G^{(2)}(x_1, x_2) = G_c^{(2)}(x_1, x_2) \,,$ $\hspace{2cm} (1)$

(b) $G^{(4)}(x_1, x_2, x_3, x_4)$
$$= G_c^{(4)}(x_1, x_2, x_3, x_4) + G^{(2)}(x_1, x_2)G^{(2)}(x_3, x_4)$$
$$+ G^{(2)}(x_1, x_3)G^{(2)}(x_2, x_4) + G^{(2)}(x_1, x_4)G^{(2)}(x_2, x_3) \,. \tag{2}$$

How can this relation be extended to the case $n > 4$?

Solution. (a) We insert the relation (12.86) into the definition (12.88) and take the derivative of the logarithm:

$$G_c^{(2)}(x_1, x_2) = \left(\frac{\hbar}{i}\right)\frac{\delta^2 Z[J]}{\delta J(x_1)\delta J(x_2)}\bigg|_{J=0} = \left(\frac{\hbar}{i}\right)^2 \frac{\delta^2 \ln W[J]}{\delta J(x_1)\delta J(x_2)}\bigg|_{J=0}$$

$$= \left(\frac{\hbar}{i}\right)^2 \frac{\delta}{\delta J(x_1)}\left[\frac{1}{W[J]}\frac{\delta W[J]}{\delta J(x_2)}\right]_{J=0}$$

$$= \left(\frac{\hbar}{i}\right)^2 \left[-\frac{1}{W^2}\frac{\delta W}{\delta J(x_1)}\frac{\delta W}{\delta J(x_2)} + \frac{1}{W}\frac{\delta^2 W}{\delta J(x_1)\delta J(x_2)}\right]_{J=0}. \quad (3)$$

Using the normalization condition for the generating functional and the fact that the one-point function vanishes,

$$W[0] = 1 \quad \text{and} \quad G^{(1)}(x) = \left(\frac{\hbar}{i}\right)\frac{\delta W}{\delta J(x)}\bigg|_{J=0} = 0, \quad (4)$$

we find that both Green's functions agree:

$$G_c^{(2)}(x_1, x_2) = \left(\frac{\hbar}{i}\right)^2 \frac{\delta^2 W}{\delta J(x_1)\delta J(x_2)}\bigg|_{J=0} = G^{(2)}(x_1, x_2). \quad (5)$$

(b) Since the derivatives of the extra terms in (3) do not vanish at $J = 0$ the functions $G^{(n)}$ and $G_c^{(n)}$ will not agree at higher orders of n. For the third derivative of the functional $Z[J]$ we find

$$\frac{i}{\hbar}\frac{\delta^3 Z}{\delta J(x_1)\delta J(x_2)\delta J(x_3)} = \frac{2}{W^3}\frac{\delta W}{\delta J(x_1)}\frac{\delta W}{\delta J(x_2)}\frac{\delta W}{\delta J(x_3)}$$
$$-\frac{1}{W^2}\left(\frac{\delta W}{\delta J(x_1)}\frac{\delta^2 W}{\delta J(x_2)\delta J(x_3)} + \frac{\delta W}{\delta J(x_2)}\frac{\delta^2 W}{\delta J(x_1)\delta J(x_3)}\right.$$
$$\left.+\frac{\delta W}{\delta J(x_3)}\frac{\delta^2 W}{\delta J(x_1)\delta J(x_2)}\right)$$
$$+\frac{1}{W}\frac{\delta^3 W}{\delta J(x_1)\delta J(x_2)\delta J(x_3)}. \quad (6)$$

When taking the fourth derivative of $Z[J]$ a large number of terms is obtained. We write down only those that survive in the limit $J = 0$:

$$\frac{i}{\hbar}\frac{\delta^4 Z}{\delta J(x_1)\delta J(x_2)\delta J(x_3)\delta J(x_4)}\bigg|_{J=0} = \left[-\left(\frac{\delta^2 W}{\delta J(x_1)\delta J(x_2)}\frac{\delta^2 W}{\delta J(x_3)\delta J(x_4)}\right.\right.$$
$$\left.+\frac{\delta^2 W}{\delta J(x_1)\delta J(x_3)}\frac{\delta^2 W}{\delta J(x_2)\delta J(x_4)} + \frac{\delta^2 W}{\delta J(x_1)\delta J(x_4)}\frac{\delta^2 W}{\delta J(x_2)\delta J(x_3)}\right)$$
$$\left.+\frac{\delta^4 W}{\delta J(x_1)\delta J(x_2)\delta J(x_3)\delta J(x_4)}\right]_{J=0}. \quad (7)$$

This is just equation (2).

Denoting the connected Green's function by a bubble marked with the letter C, we can write (1) and (2) graphically as

$$-\!\bigcirc\!- \;=\; -\!\text{\textcircled{C}}\!- \;, \quad (8)$$

$$\text{\scriptsize4-point} \;=\; \text{\textcircled{C}} + \text{(perm.)} + \cdots + \cdots$$
$$=\; \text{\textcircled{C}} + \sum_{\text{Permut.}} \cdots \quad (9)$$

Exercise 12.5

where the summation extends over all possible associations of the coordinates x_i to the lines of the graph.

The general connection between ordinary and connected Green's functions of order n looks rather complicated:

$$\begin{aligned}
G^{(n)}(x_1,\ldots,x_n) &= G_c^{(n)}(x_1,\ldots,x_n) \\
&+ \sum_\mu \sum_{\text{Perm.}} G_c^{(\mu)}(x_{\sigma_1},\ldots,x_{\sigma_\mu}) G_c^{(n-\mu)}(x_{\sigma_{\mu+1}},\ldots,x_{\sigma_n}) \\
&+ \sum_{\mu,\nu} \sum_{\text{Perm.}} G_c^{(\mu)}(x_{\sigma_1},\ldots,x_{\sigma_\mu}) G_c^{(\nu)}(x_{\sigma_{\mu+1}},\ldots,x_{\sigma_{\mu+\nu}}) \\
&\quad \times G_c^{(n-\mu-\nu)}(x_{\sigma_{\mu+\nu+1}},\ldots,x_{\sigma_n}) \\
&+ \ldots \,.
\end{aligned} \qquad (10)$$

Thus in addition to the connected Green's function $G_c^{(n)}$ all possible products of disjoint Green's functions of lower order are present, subject to the condition that the number of variables x_i, i.e., the "legs" of the graph, sums up to n. For example in the case $n=6$, (10) becomes

$$\Xi\bigcirc\Xi \;=\; \Xi\bigcirc\!\!c\!\!\bigcirc\Xi \;+\; \sum_{\text{Permut.}} \left(\Xi\bigcirc\!\!c\!\!\bigcirc\!\!\!-\!\!\!\bigcirc\!\!\!-\; +\; \Xi\bigcirc\!\!c\!\!\bigcirc\!\!-\!\!\bigcirc\!\!c\!\!\bigcirc\!\!-\; +\; \Xi\bigcirc\!\!-\!\!\bigcirc\!\!-\!\!\bigcirc\!\!- \right) \,. \qquad (11)$$

Remark. If the interaction Lagrangian $\mathcal{L}_{\text{int}}$ is an even function of ϕ, as is the case in the ϕ^4 model, there are no odd n-point functions and the intermediate term in (11) drops out.

12.7 One-Particle Irreducible Graphs and the Effective Action

Up to now we have introduced two kinds of n-point function, $G^{(n)}(x_1,\ldots,x_n)$ and $G_c^{(n)}(x_1,\ldots,x_n)$. The latter function has the advantage that it does not contain contributions that arise trivially from the multiplication of n-point functions of lower order. One can take a second step and further subdivide the set of connected graphs, ending up with a "hard core" of graphs having a nontrivial topology. The following examples from ϕ^4 theory illustrates that even the connected graphs can still have a certain redundancy:

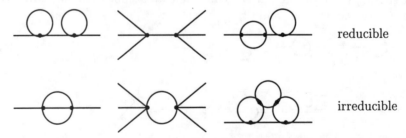

The diagrams in the first row can be split into two disjoint parts by cutting a single internal line, thus they are called *reducible*. The examples on the

12.7 One-Particle Irreducible Graphs and the Effective Action

second row are *irreducible*, or to be more precise *one-particle irreducible*, which is abbreviated as 1PI, since they cannot be dissected in this way. The 1PI property in a sense is a more rigorous variant of connectedness: the graph has to be connected in a "nontrivial" way. The irreducible graphs play an important role for the systematic construction of perturbation theory in higher orders. Every higher-order graph can be obtained in a unique way by taking irreducible graphs and free propagators $G_0^{(2)}$ as building blocks.

We define an *irreducible vertex function* which is an n-point function $\Gamma^{(n)}(x_1, \ldots, x_n)$ consisting of the sum of all irreducible graphs. An alternative notation is $G_p^{(n)}(x_1, \ldots, x_n)$ where p stands for "proper". Again it is possible to construct a generating functional from which the irreducible vertex functions can be obtained. The connection with the functionals studied so far is less straightforward, though.

Although the reason for the next step is not immediately obvious we now introduce the derivative of the generating functional $Z[J]$ with respect to the source $J(x)$ as a new variable $\phi_c(x)$:

$$\phi_c(x) = \frac{\delta Z[J]}{\delta J(x)} . \qquad (12.108)$$

The meaning of this variable becomes evident if one starts from the path-integral representation of the vacuum functional $W[J]$,

$$W[J] = e^{\frac{i}{\hbar}Z[J]} = \frac{\int \mathcal{D}\phi \, \exp\left[\frac{i}{\hbar}\int d^4x\, (\mathcal{L} + J\phi)\right]}{\int \mathcal{D}\phi \, \exp\left[\frac{i}{\hbar}\int d^4x\, \mathcal{L}\exp\right]} , \qquad (12.109)$$

which leads to

$$\begin{aligned}\phi_c(x) &= \frac{\delta Z[J]}{\delta J(x)} = \frac{\hbar}{i}\frac{1}{W[J]}\frac{\delta W[J]}{\delta J(x)} = \frac{\int \mathcal{D}\phi\, \exp\left[\frac{i}{\hbar}\int d^4x'\,(\mathcal{L}+J\phi)\right]\phi(x)}{\int \mathcal{D}\phi\, \exp\left[\frac{i}{\hbar}\int d^4x'\,(\mathcal{L}+J\phi)\right]} \\ &= \left(\frac{\langle 0|\hat{\phi}(x)|0\rangle}{\langle 0|0\rangle}\right)_J .\end{aligned} \qquad (12.110)$$

Here the interpretation of the path integral as a vacuum expectation value was introduced (cf. Sect. 12.1). Thus $\phi_c(x)$ is the normalized vacuum expectation value of the field operator $\hat{\phi}(x)$ in the presence of the source J; one also uses the names *mean field* or *classical field*. Note that $\phi_c(x)$ is both an ordinary function of x and a functional of J; one might use the notation $\phi_c(x, J]$. In the limit of vanishing perturbation $J \to 0$, under normal circumstances the classical field will also go to zero, $\phi_c \to 0$. Important exceptions to this rule are theories exhibiting spontaneous symmetry breaking, which are characterized by a nonzero value of the vacuum expectation value of the field.[7]

Now we assume that the connection between ϕ_c and J can be inverted so that the source J can be expressed as a functional of the classical field ϕ_c (at the same time being a function of x), $J = J(x, \phi_c]$. Then J can be eliminated and be replaced by ϕ_c as an independent variable. This is achieved by a (functional) *Legendre transformation* familiar from Hamiltonian mechanics or thermodynamics. We define the new generating functional

---

[7] See, e.g., D. Bailin, A. Love, *Introduction to Gauge Field Theory* (Adam Hilger, Bristol 1986), Chap. 13

$$\Gamma[\phi_c] = Z[J] - \int d^4x\, J(x)\phi_c(x)\,. \tag{12.111}$$

This functional is constructed in such a way that it has no explicit dependence on J, $\delta\Gamma/\delta J = 0$. For reasons that will become clear later, $\Gamma[\phi_c]$ is called the *effective action*. The functional derivative of the effective action $\Gamma[\phi_c]$ with respect to the classical field ϕ_c takes a simple form:

$$\frac{\delta\Gamma[\phi_c]}{\delta\phi_c(x)} = \frac{\delta Z}{\delta\phi_c(x)} - \int d^4y\, \frac{\delta J(y)}{\delta\phi_c(x)}\phi_c(y) - J(x)\,. \tag{12.112}$$

With the use of (12.108) the functional chain rule leads to

$$\frac{\delta Z}{\delta\phi_c(x)} = \int d^4y\, \frac{\delta Z}{\delta J(y)}\frac{\delta J(y)}{\delta\phi_c(x)} = \int d^4y\, \phi_c(y)\frac{\delta J(y)}{\delta\phi_c(x)}\,. \tag{12.113}$$

Therefore the first two terms in (12.112) cancel and we are left with

$$\frac{\delta\Gamma[\phi_c]}{\delta\phi_c(x)} = -J(x)\,. \tag{12.114}$$

Note that the relation between $\Gamma[\phi_c]$ and J is very similar to that between $Z[J]$ and ϕ_c given in (12.108).

The newly defined functional $\Gamma[\phi_c]$ is now taken as the generating functional for a new class of n-point functions, which will turn out to be the *irreducible vertex functions* $\Gamma^{(n)}(x_1, \ldots, x_n)$ introduced above. In keeping with (12.74) and (12.87) the Volterra expansion of the effective action reads

$$\Gamma[\phi_c] = \sum_n \int d^4x_1 \ldots \int d^4x_n\, \frac{1}{n!}\, \Gamma^{(n)}(x_1, \ldots, x_n)\, \phi_c(x_1)\cdots\phi_c(x_n)\,. \tag{12.115}$$

Now we investigate the relation of the newly defined functions to the ordinary and to the connected n-point Green's functions $G^{(n)}$ and $G_c^{(n)}$. The integral kernels in (12.115) are determined by

$$\Gamma^{(n)}(x_1, \ldots, x_n) = \left.\frac{\delta^n \Gamma[\phi_c]}{\delta\phi_c(x_1)\ldots\delta\phi_c(x_n)}\right|_{\phi_c=0}\,. \tag{12.116}$$

Let us start with the *two-point vertex function*

$$\Gamma^{(2)}(x_1, x_2) = \left.\frac{\delta^2 \Gamma[\phi_c]}{\delta\phi_c(x_1)\,\delta\phi_c(x_2)}\right|_{\phi_c=0}\,. \tag{12.117}$$

This object turns out to be the *inverse of the two-point Green's function* $G^{(2)}(x_1, x_2)$. From the functional chain rule and using (12.108) and (12.114) we find

$$\begin{aligned}\delta^4(x_1 - x_2) &= \frac{\delta\phi_c(x_1)}{\delta\phi_c(x_2)} = \int d^4x\, \frac{\delta\phi_c(x_1)}{\delta J(x)}\frac{\delta J(x)}{\delta\phi_c(x_2)} \\ &= -\int d^4x\, \frac{\delta^2 Z}{\delta J(x_1)\delta J(x)}\frac{\delta^2\Gamma}{\delta\phi_c(x)\delta\phi_c(x_2)}\,.\end{aligned} \tag{12.118}$$

Taken at $J = \phi_c = 0$ this implies

$$-\int d^4x\, \frac{i}{\hbar}G_c^{(2)}(x_1, x)\, \Gamma^{(2)}(x, x_2) = \delta^4(x_1 - x_2)\,. \tag{12.119}$$

12.7 One-Particle Irreducible Graphs and the Effective Action

Thus $\Gamma^{(2)}$ and $G_c^{(2)}$ (and also $G^{(2)}$; cf. (12.5a)) are inverse to each other (in the sense of a "matrix multiplication with continuous indices"). Equation (12.119) as an integral equation for $\Gamma^{(2)}$ is solved by the inverse propagator $G^{(2)-1}$ so that we can identify

$$\Gamma^{(2)}(x_1, x_2) = i\hbar G^{(2)-1}(x_1, x_2) \,. \tag{12.120}$$

This result becomes particulary simple for the noninteracting theory. Here $G^{(2)}(x_1, x_2)$ is just the Feynman propagator, multiplied by $i\hbar$, and therefore

$$\Gamma_0^{(2)}(x_1, x_2) = \Delta_F^{-1}(x_1 - x_2) \,. \tag{12.121}$$

While the Feynman propagator in coordinate space is a complicated function, in momentum space Δ_F is the simple multiplication operator $1/(p^2 - m^2 + i\epsilon)$, which can be trivially inverted. Introducing the Fourier-transformed vertex function $\Gamma^{(2)}(x_1, x_2)$ as in Sect. 12.6,

$$\Gamma^{(n)}(p_1, \ldots, p_n)(2\pi\hbar)^4 \delta^4(p_1 + \ldots + p_n)$$
$$= \int d^4x_1 \ldots d^4x_n \, e^{i(p_1 \cdot x_1 + \ldots + p_n \cdot x_n)/\hbar} \, \Gamma^{(n)}(x_1, \ldots, x_n) \,, \tag{12.122}$$

then according to (12.121) we simply have

$$\Gamma^{(2)}(p, -p) = \Delta_F^{-1}(p) = p^2 - m^2 \,. \tag{12.123}$$

Now we evaluate the *irreducible three-point vertex function*

$$\Gamma^{(3)}(x_1, x_2, x_3) = \left. \frac{\delta^3 \Gamma[\phi_c]}{\delta\phi_c(x_1)\delta\phi_c(x_2)\delta\phi_c(x_3)} \right|_{\phi_c=0} \,. \tag{12.124}$$

$\Gamma^{(3)}$ can be obtained by forming the functional derivative of the identity (12.118), taking into account the product and chain rules:

$$0 = \frac{\delta}{\delta J(x_3)} \int d^4x \, \frac{\delta^2 Z}{\delta J(x_1)\delta J(x)} \, \frac{\delta^2 \Gamma}{\delta\phi_c(x)\delta\phi_c(x_2')} \tag{12.125}$$

$$= \int d^4x \, \frac{\delta^3 Z}{\delta J(x_3)\delta J(x_1)\delta J(x)} \, \frac{\delta^2 \Gamma}{\delta\phi_c(x)\delta\phi_c(x_2')}$$

$$+ \int d^4x_1' \int d^4x_3' \, \frac{\delta^2 Z}{\delta J(x_1)\delta J(x_1')} \, \frac{\delta^3 \Gamma}{\delta\phi_c(x_1')\delta\phi_c(x_3')\delta\phi_c(x_2')} \, \frac{\delta\phi_c(x_3')}{\delta J(x_3)} \,.$$

The equation relates the three-point vertex function with the three-point Green's function. Both are integrated over and both are multiplied by second derivatives either of Γ with respect to ϕ_c or of Z with respect to J (the last term in (12.125) is also of this type because of (12.108)) With the help of the inverse operator provided by (12.118), (12.125) can be solved for one of the three-point functions. Let us multiply the equation by $\delta^2 Z/\delta J(x_2')\delta J(x_2)$ and integrate over x_2'. This leads to

$$\frac{\delta^3 Z}{\delta J(x_1)\delta J(x_2)\delta J(x_3)} = \int d^4x_1' d^4x_2' d^4x_3' \, \frac{\delta^2 Z}{\delta J(x_1)\delta J(x_1')} \, \frac{\delta^2 Z}{\delta J(x_2)\delta J(x_2')}$$
$$\times \frac{\delta^2 Z}{\delta J(x_3)\delta J(x_3')} \, \frac{\delta^3 \Gamma}{\delta\phi_c(x_1')\delta\phi_c(x_2')\delta\phi_c(x_3')} \,. \tag{12.126}$$

In the limit $J = \phi_c = 0$ and with the help of (12.88) and (12.116) this reduces to

$$G_c^{(3)}(x_1, x_2, x_3) = \frac{i}{\hbar} \int d^4x_1' d^4x_2' d^4x_3' \; G^{(2)}(x_1, x_1') G^{(2)}(x_2, x_2')$$
$$\times G^{(2)}(x_3, x_3') \, \Gamma^{(3)}(x_1', x_2', x_3') \; . \qquad (12.127)$$

Thus the connected three-point Green's function $G_c^{(3)}$ is obtained by appending propagators to the irreducible three-point vertex function $\Gamma^{(3)}$, one propagator for each of the external "legs". Therefore $\Gamma^{(n)}$ also is called an *amputated Green's function*.

The graphical representation of (12.127) looks like

$$\text{(diagram)} \qquad (12.128)$$

where $\;\text{---}\!\circ\!\text{---}\;$ describes the two-point Green's function, and on the right-hand side an integration over the internal coordinates (x_1', x_2', x_3') is implied. This shows that the amputated function Γ does not contain information on how the field evolves from the last point of interaction x_i' to the "observation point" x_i. This has to be accounted for separately by appending the propagators $G^{(2)}(x_i, x_i')$. These propagators contain effects of the self-interaction but there is no interaction with other particles. The "essential" physics of a scattering process therefore is contained in the amputated n-point function.

With the help of (12.118), (12.125) of course can also be solved for the other three-point function. Suitably renaming the coordinates, we obtain

$$\frac{\delta^3 \Gamma}{\delta\phi_c(x_1)\delta\phi_c(x_2)\delta\phi_c(x_3)} = -\int d^4x_1' d^4x_2' d^4x_3' \frac{\delta^2 \Gamma}{\delta\phi_c(x_1)\delta\phi_c(x_1')}$$
$$\times \frac{\delta^2 \Gamma}{\delta\phi_c(x_2)\delta\phi_c(x_2')} \frac{\delta^2 \Gamma}{\delta\phi_c(x_3)\delta\phi_c(x_3')} \frac{\delta^3 Z}{\delta J(x_1')\delta J(x_2')\delta J(x_3')} \; . \qquad (12.129)$$

With the use of (12.120) the irreducible three-point vertex function is obtained from the three-point Green's function through

$$\Gamma^{(3)}(x_1, x_2, x_3) = \frac{\hbar}{i} \int d^4x_1' d^4x_2' d^4x_3' \; G^{(2)-1}(x_1, x_1') G^{(2)-1}(x_2, x_2')$$
$$\times G^{(2)-1}(x_3, x_3') G_c^{(3)}(x_1', x_2', x_3') \; . \qquad (12.130)$$

The inverse propagator on the right-hand side serves to amputate the legs of $G_c^{(3)}$. This operation is graphically represented by marking the respective external lines with a cross. Then (12.130) can be written as

$$\frac{i}{\hbar} \; \text{(diagram)} \qquad (12.131)$$

It is obvious that this is equivalent to (12.128).

12.7 One-Particle Irreducible Graphs and the Effective Action

As we have seen, the connection between $\Gamma^{(n)}$ and $G_c^{(n)}$ in the case of $n = 3$ is rather simple, the only difference refers to the treatment of external lines. More profound differences are to be found at higher orders n. Let us investigate the connection between the *four-point functions*.

To obtain an expression for $\Gamma^{(4)}$ we differentiate (12.126) once more with respect to J. This gives rise to four different terms under the integral:

$$\frac{\delta^4 Z}{\delta J(x_1)\delta J(x_2)\delta J(x_3)\delta J(x_4)} = \int d^4x_2' d^4x_3' d^4x_4'$$

$$\left[\frac{\delta^2 Z}{\delta J(x_2)\delta J(x_2')}\frac{\delta^2 Z}{\delta J(x_3)\delta J(x_3')}\frac{\delta^2 Z}{\delta J(x_4)\delta J(x_4')}\right.$$

$$\times \frac{\delta^4 \Gamma}{\delta J(x_1)\delta\phi_c(x_2')\delta\phi_c(x_3')\delta\phi_c(x_4')}$$

$$+ \frac{\delta^3 Z}{\delta J(x_1)\delta J(x_2)\delta J(x_2')}\frac{\delta^2 Z}{\delta J(x_3)\delta J(x_3')}\frac{\delta^2 Z}{\delta J(x_4)\delta J(x_4')}$$

$$\times \frac{\delta^3 \Gamma}{\delta\phi_c(x_2')\delta\phi_c(x_3')\delta\phi_c(x_4')}$$

$$\left. + 2\text{ terms}\right]. \tag{12.132}$$

The derivative of Γ with respect to J in the first term can be replaced by a ϕ_c derivative through

$$\frac{\delta}{\delta J(x_1)} = \int d^4x_1' \frac{\delta\phi(x_1')}{\delta J(x_1)}\frac{\delta}{\delta\phi_c(x_1')} = \int d^4x_1' \frac{\delta^2 Z}{\delta J(x_1)\delta J(x_1')}\frac{\delta}{\delta\phi_c(x_1')}. \tag{12.133}$$

The right-hand side of (12.132) contains the irreducible vertex functions $\Gamma^{(3)}$ and $\Gamma^{(4)}$ and the Green's functions $G_c^{(3)}$. Now we employ (12.127) and replace $G_c^{(3)}$ by the vertex function $\Gamma^{(3)}$. In this way the last three terms become six-dimensional integrals. In the first contribution we rename x_2' as y and use x_1', x_2', y' as the new integration variables. Similar bookkeeping is done for the other contributions. Finally we arrive at

$$\frac{\delta^4 Z}{\delta J(x_1)\delta J(x_2)\delta J(x_3)\delta J(x_4)}$$

$$= \int d^4x_1' d^4x_2' d^4x_3' d^4x_4' \frac{\delta^2 Z}{\delta J(x_1)\delta J(x_1')}\frac{\delta^2 Z}{\delta J(x_2)\delta J(x_2')}$$

$$\times \frac{\delta^2 Z}{\delta J(x_3)\delta J(x_3')}\frac{\delta^2 Z}{\delta J(x_4)\delta J(x_4')}\frac{\delta^4 \Gamma}{\delta J(x_1')\delta J(x_2')\delta J(x_3')\delta J(x_4')}$$

$$+ \int d^4x_1' d^4x_2' d^4x_3' d^4x_4' d^4y d^4y' \left[\frac{\delta^2 Z}{\delta J(x_1)\delta J(x_1')}\frac{\delta^2 Z}{\delta J(x_2)\delta J(x_2')}\right.$$

$$\times \frac{\delta^2 Z}{\delta J(x_3)\delta J(x_3')}\frac{\delta^2 Z}{\delta J(x_4)\delta J(x_4')}\frac{\delta^2 Z}{\delta J(y')\delta J(y)}\frac{\delta^3 \Gamma}{\delta\phi_c(x_1')\delta\phi_c(x_2')\delta\phi_c(y')}$$

$$\left. \times \frac{\delta^3 \Gamma}{\delta\phi_c(y)\delta\phi_c(x_3')\delta\phi_c(x_4')} + 2\text{ terms}\right]. \tag{12.134}$$

Thus we have succeeded in expressing the connected four-point Green's function $G_c^{(4)}$ in terms of a combination of propagators and irreducible vertex functions $\Gamma^{(3)}$ and $\Gamma^{(4)}$:

$$\begin{aligned}G_c^{(4)}(x_1,x_2,x_3,x_4) &= \frac{i}{\hbar}\int d^4x_1' d^4x_2' d^4x_3' d^4x_4'\, G^{(2)}(x_1,x_1')G^{(2)}(x_2,x_2')\\ &\quad\times G^{(2)}(x_3,x_3')G^{(2)}(x_4,x_4')\,\Gamma^{(4)}(x_1',x_2',x_3',x_4')\\ &\quad+\left(\frac{i}{\hbar}\right)^2\int d^4x_1' d^4x_2' d^4x_3' d^4x_4'\, d^4y d^4y'\, G^{(2)}(x_1,x_1')G^{(2)}(x_2,x_2')\\ &\quad\times G^{(2)}(x_3,x_3')G^{(2)}(x_4,x_4')G^{(2)}(y,y')\\ &\quad\times\Big(\Gamma^{(3)}(x_1',x_2',y')\Gamma^{(3)}(y,x_3',x_4') + \Gamma^{(3)}(x_1',x_3',y')\Gamma^{(3)}(y,x_2',x_4')\\ &\quad+ \Gamma^{(3)}(x_3',x_2',y')\Gamma^{(3)}(y,x_1',x_4')\Big)\,.\end{aligned} \qquad (12.135)$$

The first term has the same structure as in (12.127), i.e., the amputated vertex function $\Gamma^{(4)}$ has four propagators attached as legs. There are additional terms, however, which consist of three-point vertex functions and an additional internal propagator.

As usual the rather intricate equation, (12.135), looks much more benign when written in graphical form:

$$\begin{array}{c}\includegraphics{}\end{array} \qquad (12.136)$$

It is evident that $\Gamma^{(4)}$ is justly called a one-particle irreducible vertex function. $G_c^{(4)}$ contains the whole set of connected graphs with four external lines. To obtain $\Gamma^{(4)}$, the last three terms in (12.136) have to be subtracted. These clearly are one-particle reducible since they can be split into two parts by cutting the internal line.

The relation (12.135) has been obtained by repeatedly taking functional derivatives of (12.129). In exactly the same way one can derive the connection between higher-order n-point functions. As in (12.136) the connected Green's function $G_c^{(n)}$ can always be constructed from irreducible vertex functions $\Gamma^{(m)}$, where in general all orders $3 \leq m \leq n$ will contribute. Of course, if n is large there will be many contributing graphs.

In certain cases simplifications will occur if some of the graphs do not exist. One such example is the familiar ϕ^4 theory. Since all interaction vertices have four legs, clearly no n-point function with odd n will exist. Then the last three terms in (12.132) drop out. In this special case $G_c^{(4)}$ agrees with $\Gamma^{(4)}$, except for the appending of external propagators.

As mentioned earlier the functional $\Gamma[\phi_c]$ is called the *effective action*. To elucidate the reason for this choice of name we take a look at the noninteracting theory and evaluate the functional $\Gamma_0[\phi_c]$ explicitly. In Sect. 12.4 we have been able to calculate the generating functional $W_0[J]$ of the free scalar field:

$$Z_0[J] = \frac{\hbar}{i} \ln W_0[J] = -\frac{1}{2} \int d^4x\, d^4x'\, J(x) \Delta_F(x-x') J(x') \,. \qquad (12.137)$$

Accordingly the "classical field" defined in (12.108) reads

$$\phi_c(x) = \frac{\delta Z_0[J]}{\delta J(x)} = -\int d^4x'\, \Delta_F(x-x') J(x') \,. \qquad (12.138)$$

The Feynman propagator Δ_F is a Green's function of the Klein–Gordon operator $(\Box + m^2)\Delta_F(x-x') = -\delta^4(x-x')$. Therefore (12.138) confirms that ϕ_c, which was christened the "classical field", indeed is a *solution of the classical field equation*, which is the Klein–Gordon equation, with $J(x)$ as an inhomogeneous source:

$$(\Box + m^2)\phi_c(x) = J(x) \,. \qquad (12.139)$$

The free functional $\Gamma_0[\phi_c]$ follows from the Legendre transformation

$$\begin{aligned}\Gamma_0[\phi_c] &= Z_0[J] - \int d^4x\, J(x)\phi_c(x) \\ &= \frac{1}{2} \int d^4x\, d^4x'\, J(x) \Delta_F(x-x') J(x') \,. \end{aligned} \qquad (12.140)$$

With the help of the field equation (12.139), the source $J(x)$ can be expressed in terms of the classical field. After repeated integration by parts we find

$$\begin{aligned}\Gamma_0[\phi_c] &= \frac{1}{2} \int d^4x\, d^4x'\, \phi_c(x)(\overleftarrow{\Box}+m^2)\Delta_F(x-x')(\overrightarrow{\Box}'+m^2)\phi_c(x') \\ &= \frac{1}{2} \int d^4x\, d^4x'\, \phi_c(x)(\overrightarrow{\Box}+m^2)\Delta_F(x-x')(\overrightarrow{\Box}'+m^2)\phi_c(x') \\ &= -\frac{1}{2} \int d^4x\, \phi_c(x)(\Box + m^2)\phi_c(x) \\ &= \frac{1}{2} \int d^4x\, \left(\partial_\mu \phi_c \partial^\mu \phi_c - m^2 \phi_c^2\right) \,. \end{aligned} \qquad (12.141)$$

This just is the *action integral* $S[\phi_c]$ of the free scalar field theory. Equation (12.141) is a very special kind of functional, being given by a simple integral over a density function (the Lagrange density) which depends on the *local* value of $\phi_c(x)$.

In the case of interacting field theories, $\Gamma[\phi_c]$ in general will no longer agree with the classical action. Rather the influence of quantum fluctuations will transform $\Gamma[\phi_c]$ into a highly *nonlocal* functional characterized by an integrand that depends on the simultaneous values of $\phi_c(x)$ at many different points x (cf. the general expression (12.115)). Formally it is always possible to write the functional as a local expression "plus corrections" by expanding all fields $\phi_c(x')$ into a power series around a common point x. The effective action of the interacting theory can thus be written as

$$\Gamma[\phi_c] = \int d^4x \left(-U(\phi_c) + \frac{1}{2}(\partial_\mu \phi_c)(\partial^\mu \phi_c) F(\phi_c) + \text{higher deriv.}\right). \qquad (12.142)$$

Here $U(\phi_c)$ and $F(\phi_c)$ are local functions (not functionals) of the field ϕ_c. In addition there are terms that contain derivatives of arbitrarily high order. The term containing no gradient, $U(\phi_c)$, is called the *effective potential*. This effective potential consists of the mass term $\frac{1}{2}m^2\phi_c^2$ and the interaction potential

$V(\phi_c)$ (for example $-\frac{g}{4!}\phi_c^4$) plus an extra term which originates from quantum fluctuations (the loop graphs). The influence of quantum fluctuations can be studied by expanding the effective potential with respect to powers of $\hbar$.

The effective action can be taken as a starting point for the discussion of nonperturbative effects such as spontaneous symmetry breaking. It is an important tool for the investigation of the properties of quantum field theories.

12.8 Path Integrals for Fermion Fields

In the previous sections the theory of scalar quantum fields $\hat{\phi}(x)$ has been reformulated in terms of path integrals. It is only natural to ask whether such a formulation can also be developed for fermion fields $\hat{\psi}(x)$. Any attempt to work this out is immediately confronted with a characteristic problem: the fermion field operators are anticommuting objects. In the canonical formalism they obey the equal-time anticommutation relations ($x_0 = y_0$)

$$\{\hat{\psi}_\alpha(x), \hat{\psi}_\beta^\dagger(y)\} = \delta_{\alpha\beta}\delta^3(\boldsymbol{x}-\boldsymbol{y}), \tag{12.143a}$$

$$\{\hat{\psi}_\alpha(x), \hat{\psi}_\beta(y)\} = \{\hat{\psi}_\alpha^\dagger(x), \hat{\psi}_\beta^\dagger(y)\} = 0. \tag{12.143b}$$

It is not possible to go back from the field operators $\hat{\psi}(x)$ to ordinary classical spinor fields $\psi(x)$ since in this way the Pauli principle gets lost. If one wants to defy this obstacle and to set up a path-integral representation for fermion fields nevertheless, the space of ordinary (real or complex) numbers has to be left and *anticommuting numbers* have to be introduced instead. Indeed this notion is not new to mathematics since anticommuting numbers were introduced around the middle of the 19th century by the mathematician Hermann Grassmann. These objects play an important role in the theory of exterior differential forms.[8]

The set of all anticommuting numbers (or *Grassmann variables*) is called the *Grassmann algebra* or exterior algebra. Each element of this algebra can be expressed in terms of n generators Θ_i. The index i runs from 1 to n where n is the dimension of the algebra (later we will consider the generalization to infinite dimension). The decisive relation defining the structure of the algebra is the anticommutation relation

$$\{\Theta_i, \Theta_j\} = 0 \tag{12.144}$$

for all i and j. As a particular consequence of this condition the square and all higher powers of a generator are seen to vanish,

$$\Theta_i^2 = 0. \tag{12.145}$$

Therefore any element of a finite-dimensional Grassmann algebra can be expanded into a *finite sum* containing products of generators:

$$g(\Theta) = g^{(0)} + \sum_i g_i^{(1)}\Theta_i + \sum_{i_1<i_2} g_{i_1 i_2}^{(2)}\Theta_{i_1}\Theta_{i_2} + \ldots + g^{(n)}\Theta_1\Theta_2\cdots\Theta_n, \tag{12.146}$$

---

[8] See e.g. Y. Choquet-Bruhat, C. deWitt-Morette: M. Dillard-Bleick: *Analysis, Manifolds and Physics* (North Holland, Amsterdam 1982).

where the coefficients $g^{(k)}$ are ordinary c numbers.

The indices in the sum have been arranged in ascending order, $i_1 < i_2 < \ldots < i_n$. Because of the property $\Theta_i\Theta_j = -\Theta_j\Theta_i$ the number of independent products of p generators is reduced to $\binom{n}{p}$. For example the last term $(n = p)$ in (12.146) in fact does not contain a summation at all since there is only one possible ordered combination of indices. The dimension of the Grassmann algebra, i.e., the maximum number of linearly independent terms in the basis expansion (12.146) is 2^n. (This is evident if one represents each term in the sum by a binary number with n digits, where 1(0) at the ith place signals the presence (absence) of the generator Θ_i. There are 2^n binary numbers of length n.)

In order to formulate field theory using the Grassmann algebra, certain tools from analysis have to be available. Indeed it is possible to define the operations of integration and differentiation involving Grassmann variables, although these turn out to be rather abstract concepts that do not share all the features familiar from analysis in the real or complex domain.

Let us first consider a simple Grassmann algebra of order $n = 2$ with the generators Θ_1 and Θ_2 having the basis $\{1, \Theta_1, \Theta_2, \Theta_1\Theta_2\}$.

The rules for *differentiation* with respect to the generators are

$$\frac{\mathrm{d}}{\mathrm{d}\Theta_i} 1 = 0 \quad , \quad \frac{\mathrm{d}}{\mathrm{d}\Theta_i}\Theta_j = \delta_{ij} \,, \qquad (12.147a)$$

$$\frac{\mathrm{d}}{\mathrm{d}\Theta_i}\Theta_1\Theta_2 = \delta_{i1}\Theta_2 - \delta_{i2}\Theta_1 \,. \qquad (12.147b)$$

Note the minus sign in the product rule (12.147b). The general rule for differentiation of a product is given by

$$\frac{\mathrm{d}}{\mathrm{d}\Theta_j}\Theta_{i_1}\Theta_{i_2}\ldots\Theta_{i_m} = \delta_{ji_1}\Theta_{i_2}\cdots\Theta_{i_m} - \delta_{ji_2}\Theta_{i_1}\Theta_{i_3}\ldots\Theta_{i_m} + \ldots$$
$$+ (-1)^{m-1}\delta_{ji_m}\Theta_{i_1}\Theta_{i_2}\cdots\Theta_{i_{m-1}} \,. \qquad (12.148)$$

The respective factor Θ_{i_k} is anticommuted to the left until the derivative operator can be directly applied.[9] Since differentiation is supposed to be a *linear* operator equation, (12.146) is sufficient to determine the derivative of any Grassmann-valued function. Using (12.148) one may prove the following properties concerning the anticommutators of differential operators:

$$\left\{\frac{\mathrm{d}}{\mathrm{d}\Theta_i}, \Theta_j\right\} = \delta_{ij} \,, \qquad (12.149a)$$

$$\left\{\frac{\mathrm{d}}{\mathrm{d}\Theta_i}, \frac{\mathrm{d}}{\mathrm{d}\Theta_j}\right\} = 0 \,. \qquad (12.149b)$$

In particular, all the higher derivatives with respect to the same generator Θ_i vanish. This is no surprise since according to (12.145) all the functions in existence are linear at most.

When trying to define an *integration over Grassmann variables*[10] we have to relinquish many of the familiar properties. The usual construction of Rie-

---

[9] Note that we have defined a *left-sided* derivative but one can similarly introduce a right-sided derivative. Then some of the signs in (12.148) are changed.

[10] F.A. Berezin: *The Method of Second Quantization* (Academic Press, New York 1966).

mann's sum cannot be extended to the Grassmann case. Also an attempt to introduce an indefinite integral as the inverse of differentiation is bound to fail. This is illustrated by the fact that according to (12.149b) the second derivative of any function with respect to a Grassmann variable vanishes, so that the inverse operation does not exist.

We will be content with a formal definition that extends some of the most basic properties of integration to Grassmann variables.

Let us start with a single Grassmann variable Θ. The integration over a function $f(\Theta)$ is constructed such that it produces an ordinary (real or complex) number. This mapping is constructed as a linear functional:

$$\int d\Theta \left(af(\Theta) + bg(\Theta) \right) = a \int d\Theta \, f(\Theta) + b \int d\Theta \, g(\Theta) \, . \tag{12.150}$$

This operation is uniquely fixed if the values of the integral of the constant function 1 and the linear function Θ are known. We postulate

$$\int d\Theta \, 1 \; = \; 0 \, , \tag{12.151a}$$

$$\int d\Theta \, \Theta \; = \; 1 \, . \tag{12.151b}$$

These are postulates that cannot be derived. Some intuitive understanding can be gained by comparing (12.151) with the rules for definite integrals $\int_{-\infty}^{\infty} dx \, f(x)$ of ordinary functions $f(x)$ that drop to zero at infinity. In this case the integral over the derivative of a function (a total differential) vanishes:

$$\int_{-\infty}^{\infty} dx \, \frac{df}{dx} = f(\infty) - f(-\infty) = 0 \, , \tag{12.152}$$

which agrees with the rule (12.151a) since $1 = d\Theta/d\Theta$ is such a derivative. The integrand in (12.151b), on the other hand, cannot be represented as a total differential since there is no "quadratic" Grassmann-valued function whose derivative would be Θ. The sole nonvanishing integral $\int d\Theta \, \Theta$ arbitrarily is assigned the value 1. This is a convenient normalization condition which amounts to defining a "scale" for the Grassmann variable.

One also can arrive at the integration rules (12.151) by postulating the *translation invariance* of the integration. For an arbitrary function $f(\Theta) = f_1 + f_2 \Theta$ we have

$$\begin{aligned} \int d\Theta \, f(\Theta + \eta) &= \int d\Theta \left[f_1 + f_2(\Theta + \eta) \right] = \int d\Theta \left[f_1 + f_2 \Theta \right] + \int d\Theta \, f_2 \eta \\ &= \int d\Theta \, f(\Theta) + \left[\int d\Theta \, 1 \right] f_2 \eta = \int d\Theta \, f(\Theta) \, . \end{aligned} \tag{12.153}$$

The most striking property of the definition (12.151) is its complete agreement with (12.147). *For Grassmann variables the operations of integration and differentiation are the same!* This has its roots in the fact that there are only constant or linear functions and that, roughly speaking, rising and lowering the "power" amount to the same.

Generalizing the integration rules to the case of a multidimensinal Grassmann algebra with the generators Θ_i poses no problem. For all $i = 1, \ldots, n$ one postulates

$$\int d\Theta_i\, 1 = 0, \tag{12.154a}$$

$$\int d\Theta_i\, \Theta_i = 1. \tag{12.154b}$$

Multiple integrals are reduced to repeated ordinary integrals in the usual way. The anticommutation rules for differentials $d\Theta_i$ are useful in this context:

$$\{d\Theta_i, d\Theta_j\} = 0, \tag{12.155a}$$

$$\{d\Theta_i, \Theta_j\} = \delta_{ij}. \tag{12.155b}$$

Because Grassmann integration and differentiation are equivalent, these rules agree with equations (12.149).

Given a function $g(\Theta)$ of Grassmann variables the coefficients in the sum representation (12.146) can be obtained by a projection. In particular, the highest term $g^{(n)}$ is obtained by n-fold integration over all the variables:

$$\int d\Theta_n \ldots d\Theta_1\, g(\Theta) = g^{(n)}. \tag{12.156}$$

The other coefficients $g^{(k)}_{i_1 \ldots i_k}$ follow similarly if extra factors Θ_{i_m} are inserted into the integral for all the complementary indices i_m not contained in the set $(i_1, \ldots, i_k)$.

To illustrate (12.156), let us write down the case $n = k = 2$ in detail:

$$\int d\Theta_2 \int d\Theta_1\, g(\Theta_1, \Theta_2) \tag{12.157}$$

$$= \int d\Theta_2 \int d\Theta_1 \left(g^{(0)} + g_1^{(1)} \Theta_1 + g_2^{(1)} \Theta_2 + g_{12}^{(2)} \Theta_1 \Theta_2 \right)$$

$$= g^{(0)} \left(\int d\Theta_2\, 1 \right) \left(\int d\Theta_1\, 1 \right) + g_1^{(1)} \left(\int d\Theta_2\, 1 \right) \left(\int d\Theta_1\, \Theta_1 \right)$$

$$- g_2^{(1)} \left(\int d\Theta_2\, \Theta_2 \right) \left(\int d\Theta_1\, 1 \right) + g_{12}^{(2)} \left(\int d\Theta_2\, \Theta_2 \right) \left(\int d\Theta_1\, \Theta_1 \right)$$

$$= g_{12}^{(2)}.$$

An important aspect when working with integrals is their behavior under *variable transformations*. Let us start with the example of a linear transformation in one dimension,

$$\Theta' = \eta + a\Theta, \tag{12.158}$$

where η is an anticommuting and a an ordinary number. For the integral of a function $g(\Theta')$ we find

$$\int d\Theta'\, g(\Theta') = g^{(1)}. \tag{12.159}$$

On the other hand,

$$\int d\Theta\, g(\Theta') = \int d\Theta \left(g^{(0)} + g^{(1)} \Theta' \right) = \int d\Theta \left(g^{(0)} + g^{(1)} \eta + g^{(1)} a\Theta \right)$$

$$= g^{(1)} a. \tag{12.160}$$

Comparing these results we deduce

$$\int d\Theta' \, g(\Theta') = \frac{1}{a} \int d\Theta \, g(\Theta'(\Theta)) = \int d\Theta \left(\frac{d\Theta'}{d\Theta}\right)^{-1} g(\Theta'(\Theta)) \,. \qquad (12.161)$$

The constant factor $1/a$ was moved under the integral in the form $(d\Theta'/d\Theta)^{-1}$ in order to emphasize the analogy to the transformation rule, which holds for the integration of an ordinary function $g(x)$:

$$\int dx' \, g(x') = \int dx \, \frac{dx'}{dx} \, g(x'(x)) \,. \qquad (12.162)$$

Thus Grassman integrals exhibit just the *opposite behavior under transformations* when compared to ordinary integrals! This peculiar property of course will also be found at higher dimensions. For an n-dimensional Grassmann algebra the linear transformation

$$\Theta'_i = \sum_{j=1}^{n} a_{ij}\Theta_j + \eta_i \qquad (12.163)$$

leads to the following transformation law for the integral:

$$\int d\Theta'_n \ldots d\Theta'_1 \, g(\Theta') = \int d\Theta_n \ldots d\Theta_1 \left[\det\left(\frac{d\Theta'}{d\Theta}\right)\right]^{-1} g(\Theta'(\Theta)) \,. \qquad (12.164)$$

Here the *inverse of the Jacobian determinant* of the transformation $\Theta \to \Theta'$ is encountered.

While the general proof of (12.164) is not trivial we can easily check its validity for the special case of the product function $g(\Theta) = \Theta_1 \cdots \Theta_n$. The normalization condition (12.154b) implies

$$\int d\Theta_n \ldots d\Theta_1 \, \Theta_1 \cdots \Theta_n = \int d\Theta'_n \ldots d\Theta'_1 \, \Theta'_1 \cdots \Theta'_n = 1 \,. \qquad (12.165)$$

The product of the Grassmann generators Θ'_i transforms with the determinant of the transformation matrix A

$$\Theta'_1 \Theta'_2 \cdots \Theta'_n = \sum_{i_1 \ldots i_n} a_{1i_1} a_{2i_2} \cdots a_{ni_n} \Theta_{i_1} \Theta_{i_2} \cdots \Theta_{i_n} = \det A \, \Theta_1 \Theta_2 \cdots \Theta_n \qquad (12.166)$$

since all permutations of the indices i_k contribute to the sum (with alternating sign). Therefore the volume element has to transform as the inverse of this determinant,

$$d\Theta'_n \cdots d\Theta'_2 \, d\Theta'_1 = \left(\det A\right)^{-1} d\Theta_n \cdots d\Theta_2 \, d\Theta_1 \,, \qquad (12.167)$$

in order to satisfy (12.165). This is the opposite of the transformation law for the volume element of ordinary numbers:

$$dx'_1 \cdots dx'_n = \det A \, dx_1 \cdots dx_n \,. \qquad (12.168)$$

In order to develop the path-integral formalism for fermion fields we will have to solve *Gaussian integrals*. Fortunately this can be done (and the calculation in principle is very simple since Grassmann-valued functions at worst can be linear, causing the series expansion of the exponential function to terminate).

Let us solve the following Gaussian integral in n dimensions:

$$I_n = \int d\Theta_1 \cdots d\Theta_n \, e^{-\frac{1}{2}\Theta^T A \Theta} \,. \qquad (12.169)$$

The exponent contains the bilinear form

$$\Theta^{\mathrm{T}} A \Theta = \sum_{i,j} \Theta_i A_{ij} \Theta_j , \tag{12.170}$$

where, because of the anticommuting property of the Grassmann generators Θ_i, only the *antisymmetric* part of the matrix A has to be taken into account. Remember that in the n-dimensional Gaussian integral over bosonic variables (c numbers) A was a symmetric matrix.

The integral (12.169) can be calculated by Taylor-expanding the exponential function and using the fundamental integral relations (12.154). It is obvious that only those terms that contain a product of all n generators Θ_i make a contribution to the integral, each generator being present exactly once. This is possible only if n is an even number; for n odd we have $I_n = 0$. Furthermore it is obvious that it is the term of order $n/2$ that will contribute to the integral:

$$I_n = \frac{1}{(n/2)!} \int \mathrm{d}\Theta_1 \cdots \mathrm{d}\Theta_n \left(-\frac{1}{2}\Theta^{\mathrm{T}} A \Theta\right)^{n/2} . \tag{12.171}$$

Because of the integration rule (12.154) those terms in the integrand are sorted out for which the indices of all n factors Θ_i are different. This leads to the product of $n/2$ matrix elements A_{ij} involving all index combinations and alternating signs. The result turns out to be the square root of the determinant of the matrix A (cf. Exercise 12.6):[11]

$$\int \mathrm{d}\Theta_n \cdots \mathrm{d}\Theta_1 \, \mathrm{e}^{-\frac{1}{2}\Theta^{\mathrm{T}} A \Theta} = \left(\det A\right)^{1/2} = \exp\left(\frac{1}{2}\operatorname{Tr}\ln A\right) . \tag{12.172}$$

Compared with the corresponding Gaussian integral formula for c numbers, (2) in Exercise 11.3, the result for Grassmann variables looks very similar. However, the determinant now stands in the numerator instead of the denominator! This reminds us of the inverse Jacobian encountered for the transformation of the volume element of Grassmann variables.

We remark that (12.172) immediately teaches us that the Gaussian integral vanishes for odd dimensions n. To wit, the determinant of an antisymmetric matrix of odd dimension is zero (check this, e.g., for the special case $n = 3$).

It is not surprising that the Gaussian integral formula can be generalized to the case of general bilinear forms in the exponent:

$$\int \mathrm{d}\Theta_1 \cdots \mathrm{d}\Theta_n \, \mathrm{e}^{-\frac{1}{2}\Theta^{\mathrm{T}} A \Theta + \rho^{\mathrm{T}}\Theta} = \left(\det A\right)^{\frac{1}{2}} \mathrm{e}^{-\frac{1}{2}\rho^{\mathrm{T}} A^{-1} \rho} . \tag{12.173}$$

Here ρ is an n-component vector of Grassmann variables. Equation (12.173) can be confirmed by translating the integration variable, $\Theta' = \Theta + A^{-1}\rho$, as in (2) in exercise 11.3.

Having in mind the description of Dirac fields we now introduce *complex Grassmann variables*. Let us start with two disjoint sets of Grassmann variables $\Theta_1, \ldots, \Theta_n$ and $\Theta_1^*, \ldots, \Theta_n^*$, which are all mutually anticommuting

$$\{\Theta_i, \Theta_j\} = \{\Theta_i^*, \Theta_j^*\} = \{\Theta_i, \Theta_j^*\} = 0 . \tag{12.174}$$

---

[11] Such an antisymmetric sum over products of $n/2$ elements of an antisymmetric matrix of even order is known in mathematics as a "Pfaffian".

Taken together, these generators form a $2n$-dimensional Grassmann algebra. The two sets are put into a relation by introducing the conjugation operation (which also is called "involution") according to

$$\begin{align}
(\Theta_i)^* &= \Theta_i^*, \\
(\Theta_i^*)^* &= \Theta_i, \\
(\Theta_{i_1}\Theta_{i_2}\cdots\Theta_{i_m})^* &= \Theta_{i_m}^*\cdots\Theta_{i_2}^*\Theta_{i_1}^*, \\
(\lambda\Theta_i)^* &= \lambda^*\Theta_i^*,
\end{align} \tag{12.175}$$

where λ is a complex number. Upon differentiation and integration, the Θ_i and Θ_i^* are treated as independent variables.

The Gaussian integral for complex Grassmann variables instead of (12.170) now contains the bilinear form

$$\Theta^\dagger A \Theta = \sum_{i,j} \Theta_i^* A_{ij} \Theta_j \tag{12.176}$$

in the exponent. The Gaussian integral formula reads

$$\int \mathrm{d}\Theta_1^* \cdots \mathrm{d}\Theta_n^* \, \mathrm{d}\Theta_1 \cdots \mathrm{d}\Theta_n \, e^{-\Theta^\dagger A \Theta} = \det A . \tag{12.177}$$

The absense of the square root, when this is compared with (12.172), is not surprising since the integration now extends over $2n$ instead of n variables. Equation (12.177) has the obvious generalization

$$\int \mathrm{d}\Theta_1^* \cdots \mathrm{d}\Theta_n^* \, \mathrm{d}\Theta_1 \cdots \mathrm{d}\Theta_n \, \exp\left(-\Theta^\dagger A \Theta + \Theta^\dagger \rho + \rho^\dagger \Theta\right)$$
$$= \det A \, \exp\left(-\rho^\dagger A^{-1} \rho\right) . \tag{12.178}$$

These equations are most easily proved for anihermitean matrices $A^\dagger = -A$ but their validity is more general.

To apply Grassmann variables in field theory we have to introduce *anticommuting fields*, i.e., to study the continuum limit $\Theta_i \to \Theta(x)$. The field variables $\Theta(x)$ are the generators of an *infinite-dimensional Grassmann algebra*, the elements of which all are anticommuting (i.e., taken at all points in space time). At least on a formal level this generalization presents no problems, the rules for the discrete finite-dimensional case can be simply copied. The fundamental relation is

$$\{\Theta(x), \Theta(y)\} = 0 \quad \text{for all } x, y . \tag{12.179}$$

Ordinary *differentiation* is translated to the functional derivative

$$\frac{\delta \Theta(x)}{\delta \Theta(y)} = \delta^4(x-y) \tag{12.180}$$

and

$$\left\{\frac{\delta}{\delta\Theta(x)}, \Theta(y)\right\} = \delta^4(x-y) , \tag{12.181a}$$

$$\left\{\frac{\delta}{\delta\Theta(x)}, \frac{\delta}{\delta\Theta(y)}\right\} = 0 . \tag{12.181b}$$

An arbitrary element of the algebra can be expanded like

$$g(\Theta) = g^{(0)} + \int dx_1 \, g^{(1)}(x_1)\Theta(x_1) + \ldots +$$
$$+ \int dx_1 \cdots dx_n \, g^{(n)}(x_1, \ldots, x_n) \, \Theta(x_1) \cdots \Theta(x_n) + \ldots , \quad (12.182)$$

where the coefficient functions $g^{(n)}(x_1, \ldots, x_n)$ are antisymmetric with respect to the interchange of the arguments x_i. If the coefficients $g^{(n)}$ depend on x, (12.182) forms the representation of a continuous Grassmann-valued function $g(\Theta, x)$.

In keeping with (12.151) we postulate the following *integration rules*:

$$\int d\Theta(x) \, 1 = 0, \quad (12.183a)$$

$$\int d\Theta(x) \, \Theta(x) = 1. \quad (12.183b)$$

Once the integration over $\Theta(x)$ is defined, it is possible to extend the concept of the *path integral* to Grassmann fields $\psi(x)$. The construction in Sect. 12.1 did not make use of any special properties of the integration over field variables which might restrict the validity to ordinary c numbers.[12]

For field-theoretical applications of the path integral over fermion fields only one integral formula is needed (and available). The *Gaussian integral* is given by

$$\int \mathcal{D}\bar{\psi} \int \mathcal{D}\psi \exp\left[-\int d^4x' d^4x \, \bar{\psi}(x')A(x',x)\psi(x)\right.$$
$$\left. + \int d^4x \, (\bar{\psi}(x)\rho(x) + \bar{\rho}(x)\psi(x))\right]$$
$$= \det A \, \exp\left[\int d^4x' d^4x \, \bar{\rho}(x')A^{-1}(x',x)\rho(x)\right]. \quad (12.184)$$

This is sufficient for all practical purposes since the Lagrangian of realistic field theories always consists of bilinear combinations of spinor fields.

EXERCISE

12.6 Grassmann Integration

Problem. (a) Prove the validity of the transformation law for Grassmann integration (12.164):

$$\int d\Theta'_n \ldots d\Theta'_1 \, g(\Theta') = \int d\Theta_n \ldots d\Theta_1 \left[\det\left(\frac{\partial \Theta'_i}{\partial \Theta_j}\right)\right]^{-1} g(\Theta'(\Theta)). \quad (1)$$

Hint: Use complete induction and employ the following relation between the determinants of an $n \times n$ and an $(n-1) \times (n-1)$ matrix:

---

[12] Nevertheless one has to admit that the Grassmann path integral is a rather formal concept for which an intuitive interpretation is quite remote. Of course, the ultimate justification of the construction is its success, i.e., its ability to derive the correct results for generating functionals, Green's functions, etc., incorporating the right fermionic properties.

Exercise 12.6

$$\det(a_{ij})_{n\times n} = a_{nn} \det(a_{ij} - a_{in}a_{nj}a_{nn}^{-1})_{n-1 \times n-1} \,. \tag{2}$$

(b) Derive the Gaussian integral formula (12.161) for Grassmann variables,

$$\int d\Theta_1 \cdots d\Theta_n\, e^{-\frac{1}{2}\Theta^{\mathrm{T}} A \Theta} = (\det A)^{1/2} \,, \tag{3}$$

where A ia a real antisymmetric matrix of even dimension.

Hint. A unitary transformation can be used to bring the matrix into the following form

$$A' = \begin{pmatrix} 0 & \lambda_1 & & & \\ -\lambda_1 & 0 & & & \\ & & 0 & \lambda_3 & \\ & & -\lambda_3 & 0 & \\ & & & & \ddots \\ & & & & & \ddots \end{pmatrix}, \tag{4}$$

where only the two lines adjacent to the main diagonal are nonzero.

Solution. (a) The identity is fulfilled for the special case $n = 1$: take the general function $g(\Theta) = g^{(0)} + g^{(1)}\Theta$ and the transformation $\Theta' = a\Theta + b$. Then the integrals over Θ' and over Θ read

$$\int d\Theta'\, g(\Theta') = \int d\Theta'\, (g^{(0)} + g^{(1)}\Theta') = g^{(1)} \,, \tag{5a}$$

$$\int d\Theta\, g(\Theta'(\Theta)) = \int d\Theta\, (g^{(0)} + g^{(1)}(a\Theta + b)) = a\, g^{(1)} \,. \tag{5b}$$

Therefore we find as claimed

$$\int d\Theta'\, g(\Theta') = \int d\Theta\, \frac{1}{a} g(\Theta'(\Theta)) = \int d\Theta \left[\frac{d\Theta'}{d\Theta}\right]^{-1} g(\Theta'(\Theta)) \,. \tag{6}$$

As the second step of the proof by induction we assume that the identity holds for a certain value of n:

$$\int d\Theta'_{n-1} \ldots d\Theta'_1\, g(\Theta') = \int d\Theta_{n-1} \ldots d\Theta_1 \left[\det\left(\frac{\partial \Theta'_i}{\partial \Theta_j}\right)_{n-1 \times n-1}\right]^{-1} g(\Theta'(\Theta)) \,. \tag{7}$$

Now an additional variable Θ'_n is added and integrated over:

$$\int d\Theta'_n \int d\Theta'_{n-1} \ldots d\Theta'_1\, g(\Theta') \tag{8}$$

$$= \int d\Theta'_n \int d\Theta_{n-1} \ldots d\Theta_1 \left[\det\left(\frac{\partial \Theta'_i}{\partial \Theta_j}\right)_{n-1 \times n-1}\right]^{-1} g(\Theta') \,.$$

$$= \int d\Theta_n \int d\Theta_{n-1} \ldots d\Theta_1 \left(\frac{\partial \Theta'_n}{\partial \Theta_n}\right)^{-1} \left[\det\left(\frac{\partial \Theta'_i}{\partial \Theta_j}\right)_{n-1 \times n-1}\right]^{-1} g(\Theta') \,.$$

For the integration over Θ'_n the one-dimensional formula (6) has been used. This is admissible only if the functional determinant (8) is evaluated under the constraint of keeping the value of Θ' fixed. In this way Θ_n becomes a *dependent variable*. If the $\Theta_1, \ldots, \Theta_n$ are to be used as independent variables then differentiation poduces an extra term:

$$\frac{\partial \Theta'_i}{\partial \Theta_j}\bigg|_{\Theta'_n} = \frac{\partial \Theta'_i}{\partial \Theta_j}\bigg|_{\Theta_n} + \frac{\partial \Theta_n}{\partial \Theta_j}\bigg|_{\Theta'_n} \frac{\partial \Theta'_i}{\partial \Theta_n}, \tag{9}$$

where $i, j = 1, \ldots, n-1$. The variable Θ_n depends implicitly on Θ_j through the transformation relation $\Theta'_n = \Theta'_n(\Theta_1, \ldots, \Theta_n)$. We can use the implicit-function differentiation law

$$\frac{\partial \Theta'_n}{\partial \Theta_j} + \frac{\partial \Theta_n}{\partial \Theta_j}\bigg|_{\Theta'_n} \frac{\partial \Theta'_n}{\partial \Theta_n} = 0. \tag{10}$$

Then (9) becomes

$$\frac{\partial \Theta'_i}{\partial \Theta_j}\bigg|_{\Theta'_n} = \frac{\partial \Theta'_i}{\partial \Theta_j}\bigg|_{\Theta_n} - \frac{\partial \Theta'_i}{\partial \Theta_n} \frac{\partial \Theta'_n}{\partial \Theta_j} \left(\frac{\partial \Theta'_n}{\partial \Theta_n}\right)^{-1}. \tag{11}$$

This expression just agrees with the argument of the determinant in equation (2), identifying $a_{ij} := \partial \Theta'_i / \partial \Theta_j$. Therefore we can use

$$\frac{\partial \Theta'_n}{\partial \Theta_n} \det\left(\frac{\partial \Theta'_i}{\partial \Theta_j}\bigg|_{\Theta'_n}\right)_{n-1 \times n-1} = \det\left(\frac{\partial \Theta'_i}{\partial \Theta_j}\right)_{n \times n} \tag{12}$$

and the proof of (1) is completed.

(b) The transition from A to the standard form A' proceeds in two steps. First we note that the matrix iA is hermitean since $A^T = -A$ and $A^* = A$ imply $(iA)^\dagger = -iA^T = iA$. The matrix iA can be diagonalized by using a unitary transformation and it has real eigenvalues λ_i,

$$A_d = U\, iA\, U^\dagger. \tag{13}$$

The antisymmetry of A implies that the eigenvalues come in pairs, with opposite sign. This can be shown by transposing the matrix in the secular equation, which determines the eigenvalues:

$$\det(iA - \lambda \mathbb{1}) = 0 \quad \to \quad \det(iA - \lambda \mathbb{1})^T = \det(-iA - \lambda \mathbb{1}) = -\det(iA + \lambda \mathbb{1})$$
$$= 0, \tag{14}$$

which shows that λ and $-\lambda$ simultaneously are eigenvalues. If we sort the eigensolutions in ascending order the matrix A_d takes the form

$$A_d = \begin{pmatrix} \lambda_1 & 0 & & & \\ 0 & -\lambda_1 & & & \\ & & \lambda_3 & 0 & \\ & & 0 & -\lambda_3 & \\ & & & & \ddots \\ & & & & & \ddots \end{pmatrix}. \tag{15}$$

A second unitary transformation R can be constructed which leads from (15) to (4). As an ansatz for R we assume it to be a block matrix

Exercise 12.6

$$R = \begin{pmatrix} R_2 & & \\ & R_2 & \\ & & \ddots \end{pmatrix}. \tag{16}$$

R_2 is a 2×2 matrix constructed in such a way that it "flips" the diagonal matrix It is easily verified that R_2 leads to the intended "flipping" of the diagonal matrix:

$$R_2 \begin{pmatrix} 1 & 0 \\ 0 & -1 \end{pmatrix} R_2^\dagger = i \begin{pmatrix} 0 & 1 \\ -1 & 0 \end{pmatrix}. \tag{17}$$

The explicit form of R_2 is easily found:

$$R_2 = \frac{1}{\sqrt{2}} \begin{pmatrix} i & 1 \\ 1 & i \end{pmatrix}, \quad R_2^\dagger = \frac{1}{\sqrt{2}} \begin{pmatrix} -i & 1 \\ 1 & -i \end{pmatrix} = R_2^{-1}. \tag{18}$$

Thus we have succeeded in constructing the sought-after transformation:

$$A' = -iR\, A_d\, R^\dagger = -iR U\, i A\, U^\dagger R^\dagger = R U\, A\, U^\dagger R^\dagger. \tag{19}$$

Now the transformation of variables

$$\Theta' = R U\, \Theta \tag{20}$$

suggests itself. It leads to a substantial simplification of the bilinear form in the exponent of the Gaussian integral:

$$\begin{aligned}\frac{1}{2}\Theta^T A \Theta &= \frac{1}{2}\Theta'^T A' \Theta' \\ &= \frac{1}{2}(\Theta'_1 \lambda_1 \Theta'_2 - \Theta'_2 \lambda_1 \Theta'_1 + \Theta'_3 \lambda_3 \Theta'_4 - \Theta'_4 \lambda_3 \Theta'_3 + \ldots) \\ &= \lambda_1 \Theta'_1 \Theta'_2 + \lambda_3 \Theta'_3 \Theta'_4 + \ldots + \lambda_{n-1}\Theta'_{n-1}\Theta'_n\,. \end{aligned} \tag{21}$$

The Jacobian determinant of the transformation (20) is 1 since RU is unitary. Thus we obtain for the Gaussian integral

$$\int d\Theta_n \cdots d\Theta_1\, e^{-\frac{1}{2}\Theta^T A \Theta} \tag{22}$$

$$= \int d\Theta_n \cdots d\Theta_1 \left[\det\left(\frac{\partial \Theta_i}{\partial \Theta_j}\right)\right]^{-1} e^{-\frac{1}{2}\Theta'^T A' \Theta'}$$

$$= \frac{1}{(n/2)!}\int d\Theta'_n \cdots d\Theta'_1\, (-)^{n/2}\left(\lambda_1 \Theta'_1 \Theta'_2 + \lambda_3 \Theta'_3 \Theta'_4 + \ldots + \lambda_{n-1}\Theta'_{n-1}\Theta'_n\right)^{n/2}$$

since this is the only term resulting from the power-series expansion of the exponential function that contributes, as explained in (12.171). When we multiplying out the power of the polynomial there are $(n/2)!$ terms that share the property that all the indices are different:

$$\int d\Theta_n \cdots d\Theta_1\, e^{-\frac{1}{2}\Theta^T A \Theta}$$

$$= (-)^{n/2}\left(\lambda_1 \lambda_3 \cdots \lambda_{n-1}\right)\int d\Theta'_n \cdots d\Theta'_1\, \Theta'_1 \Theta'_2 \cdots \Theta'_n$$

$$= (-)^{n/2}\left(\lambda_1 \lambda_3 \cdots \lambda_{n-1}\right) = (-)^{n/2}\sqrt{\det A}\,. \tag{23}$$

The last step made use of the fact that the determinant of A according to (13) is

$$\det A = \det U^{-1} \det(iA_d) \det U = i^n \det A_d = (-)^{n/2} \lambda_1 \lambda_2 \lambda_3 \cdots \lambda_n$$
$$= (-)^{n/2} \lambda_1(-\lambda_1)\lambda_3(-\lambda_3)\cdots\lambda_{n-1}(-\lambda_{n-1}) = \lambda_1^2 \lambda_3^2 \cdots \lambda_{n-1}^2 . \quad (24)$$

This proves (3). The sign factor $(-)^{n/2}$ originates from the inverted ordering of the integration variables.

Exercise 12.6

12.9 Generating Functional and Green's Function for Fermion Fields

If Grassmann variables are employed to enforce Fermi statistics it is rather straightforward to extend the results of the previous sections to the fermionic case.

Instead of the scalar field $\phi(x)$ we now have to deal with a four-component spinor field $\psi(x)$. The *action of the free theory* is given by

$$W = \int dt \int d^3x \, \mathcal{L}(\psi, \bar\psi) = \int d^4x \, \bar\psi(x)(i\gamma^\mu \partial_\mu - m)\psi(x) . \qquad (12.185)$$

The field $\psi(x)$ satisfies the free Dirac equation

$$(i\gamma^\mu \partial_\mu - m)\psi(x) = 0 . \qquad (12.186)$$

Within the framework of canonical field quantization, field operators $\hat\psi(x)$ and $\hat\pi(x) = i\hat\psi^\dagger(x)$ are introduced for which the equal-time anticommutation relations

$$\{\hat\psi(t,\boldsymbol{x}), \hat\psi(t,\boldsymbol{y})\} = \{\hat\psi^\dagger(t,\boldsymbol{x}), \hat\psi^\dagger(t,\boldsymbol{y})\} = 0 , \qquad (12.187a)$$

$$\{\hat\psi(t,\boldsymbol{x}), \hat\psi^\dagger(t,\boldsymbol{y})\} = \delta^3(\boldsymbol{x}-\boldsymbol{y})\,\mathbb{1} \qquad (12.187b)$$

are postulated, as discussed in detail in Chap. 4.

The quantities of interest are the *Green's functions* or *n*-point functions

$$G^{(2n)}(y_1,\ldots,y_n; x_1,\ldots,x_n) = \langle 0|T(\hat\psi(y_n)\cdots\hat\psi(y_1)\hat{\bar\psi}(x_1)\cdots\hat{\bar\psi}(x_n))|0\rangle , \qquad (12.188)$$

which we encountered in Chap. 9 in connection with the LSZ formalism. In (12.188) the factors $\hat\psi(y_i)$ have been written in descending order to ensure that the "direct term" of $G_0^{(2n)}$ has a positive sign.

As shown in Fig. 12.5 there are n incoming fermions (which are created at $x_1,\ldots,x_n$) and n outgoing fermions (which are annihilated at $y_1,\ldots,y_n$). Only Green's functions of even order will exist since the Lagrangian is always constructed from bilinear combinations of the type $\bar\psi\hat O\psi$. Thus each part of a Feynman graph contains an equal number of incoming and outgoing fermion lines (counting as usual outgoing antiparticles as incoming particles and vice versa).

The essential difference when comparing fermionic and bosonic Green's functions is the emergence of a *minus sign* upon the exchange of fermions:

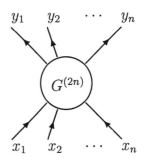

Fig. 12.5. The Green's function for fermions

$$G^{(2n)}(y_1, y_2, \ldots, y_n; x_1, x_2, \ldots, x_n) = -G^{(2n)}(y_2, y_1, \ldots, y_n; x_1, x_2, \ldots, x_n)$$
$$= -G^{(2n)}(y_1, y_2, \ldots, y_n; x_2, x_1, \ldots, x_n) \qquad (12.189)$$

etc. Within the path integral formalism this can be achieved only by using Grassmann variables. As in the bosonic case the fermionic Green's functions can also be obtained from a *generating functional*. For this one forms the functional derivative with respect to an auxiliary variable which couples to the field $\psi(x)$. In the scalar case an additional interaction term $J(x)\phi(x)$ was introduced. For Dirac fields two such terms are required since both ψ and $\bar\psi$ are present in the Lagrangian. We introduce the auxiliary fields $\eta(x)$ and $\bar\eta(x)$ with the hermitean coupling $\bar\eta\psi + \bar\psi\eta$. The generating functional of the free Dirac field is then taken as

$$W_0[\eta, \bar\eta] = \mathcal{N} \int \mathcal{D}\bar\psi \int \mathcal{D}\psi \exp\left[\frac{i}{\hbar} \int d^4x \left(\bar\psi(x)(i\hbar\gamma^\mu\partial_\mu - m)\psi(x) + \bar\eta(x)\psi(x) + \bar\psi(x)\eta(x)\right)\right]. \qquad (12.190)$$

The auxiliary variables $\eta(x)$ and $\bar\eta(x)$ are anticommuting Grassmann fields and at the same time they are four-component Dirac spinors, so that the Lagrangian is a Lorentz scalar. All of the fields $\psi(x), \bar\psi(x), \eta(x)$, and $\bar\eta(x)$ are assumed to anticommute at all space-time points and for all Dirac components. The rules for functional differentiation are

$$\left\{\frac{\delta}{\delta\bar\eta(x)}, \bar\eta(y)\right\} = \left\{\frac{\delta}{\delta\eta(x)}, \eta(y)\right\} = \delta^4(x-y), \qquad (12.191\text{a})$$

$$\left\{\frac{\delta}{\delta\bar\eta}, \eta\right\} = \left\{\frac{\delta}{\delta\eta}, \bar\eta\right\} = \left\{\frac{\delta}{\delta\bar\eta}, \frac{\delta}{\delta\bar\eta}\right\} = \left\{\frac{\delta}{\delta\eta}, \frac{\delta}{\delta\eta}\right\} = 0. \qquad (12.191\text{b})$$

The Green's function (12.188) can now be obtained by forming derivatives of the generating functional[13]

$$G^{(2n)}(y_1, \ldots, y_n; x_1, \ldots, x_n)$$
$$= \left(\frac{\hbar}{i}\right)^{2n} \frac{\delta^{2n} W[\eta, \bar\eta]}{\delta\eta(x_n)\cdots\delta\eta(x_1)\delta\bar\eta(y_1)\cdots\delta\bar\eta(y_n)}\bigg|_{\eta=\bar\eta=0}. \qquad (12.192)$$

The derivative is evaluated at $\eta = \bar\eta = 0$, since η and $\bar\eta$ are auxiliary fields, which have to drop out in the final expression. The connection between the expression (12.192) for the Green's function and the corresponding definition based on the vacuum expectation value of the time-ordered product of field operators (12.188) rests on the ordering of the time arguments in the path integral. Exactly the same argument as in the scalar case can be employed (see Sect. 12.4). The anticommuting property of the Grassmann variables guarantees that the correct minus sign emerges in the time-ordered product

---

[13] The sign of the n-point function of course is a matter of convention. In (12.192) the order of differentiations was chosen such that we get agreement with (12.188). This is not utterly trivial since the Grassmann derivatives $\delta/\delta\eta(x)$ and $\delta/\delta\bar\eta(x)$ anticommute with the field variables $\psi(x)$ and $\bar\psi(x)$. One can show, however, that there is an even number of commutations when we carry out the differentiations of (12.192) and write the result in the form (12.188).

$T(\bar\psi(y_1)\cdots\bar\psi(x_n))$. Clearly the Green's function of (12.192) satisfies the anticommutation property (12.189), which is an immediate consequence of the differentiation rule (12.191b).

12.10 Generating Functional and Feynman Propagator for the Free Dirac Field

The generating functional $W_0[\eta,\bar\eta]$ of the *free spinor field* can be easily evaluated in closed form since according to (12.190) it is defined by a Gaussian path integral. The calculation proceeds as in Sect. 12.3 where the scalar field was studied. In the following, however, we will use Minkowski coordinates for simplicity. We will ignore questions of convergence, appealing to the confidence that such problems can be avoided if the euclidian formulation of the theory (imaginary time coordinates) is used.

We want to calculate the generating functional expressed by the path integral (12.190). This is just the general Gaussian Grassmann integral solved in (12.184). To use the result we identify

$$\rho(x) = \frac{\mathrm{i}}{\hbar}\eta(x) \quad,\quad \bar\rho(x) = \frac{\mathrm{i}}{\hbar}\bar\eta(x)\,, \tag{12.193a}$$

and

$$A(x',x) = -\frac{\mathrm{i}}{\hbar}\big(-\mathrm{i}\hbar\gamma\cdot\partial - m\big)\delta^4(x'-x)\,. \tag{12.193b}$$

Here the extra minus sign in the Dirac differential operator results from an integration by parts:

$$\int\mathrm{d}^4x'\int\mathrm{d}^4x\,\bar\psi(x')\left[(-\mathrm{i}\hbar\gamma\cdot\partial - m)\delta^4(x'-x)\right]\psi(x)$$
$$= \int\mathrm{d}^4x'\int\mathrm{d}^4x\,\bar\psi(x')\left[(+\mathrm{i}\hbar\gamma\cdot\partial - m)\psi(x)\right]\delta^4(x'-x)$$
$$= \int\mathrm{d}^4x\,\bar\psi(x)\big(\mathrm{i}\hbar\gamma\cdot\partial - m\big)\psi(x)\,. \tag{12.194}$$

The *noninteracting generating functional* then reads

$$W_0[\eta,\bar\eta] = \mathcal{N}\det A\,\exp\left[-\frac{1}{\hbar^2}\int\mathrm{d}^4x'\mathrm{d}^4x\,\bar\eta(x')A^{-1}(x',x)\eta(x)\right]\,. \tag{12.195}$$

In order to invert the integration kernel $A(x',x)$ we can switch to momentum space as in Sect. 12.3. Equation (12.193b) can be written as

$$\begin{aligned}A(x',x) &= -\frac{\mathrm{i}}{\hbar}(-\mathrm{i}\hbar\gamma\cdot\partial - m)\int\frac{\mathrm{d}^4p}{(2\pi\hbar)^4}\,\mathrm{e}^{-\mathrm{i}p\cdot(x'-x)/\hbar}\\ &= -\frac{\mathrm{i}}{\hbar}\int\frac{\mathrm{d}^4p}{(2\pi\hbar)^4}\,\mathrm{e}^{-\mathrm{i}p\cdot(x'-x)/\hbar}(\gamma\cdot p - m)\,.\end{aligned} \tag{12.196}$$

Accordingly the inverse integral kernel has the momentum-space representation

$$A^{-1}(x',x) = \mathrm{i}\hbar\int\frac{\mathrm{d}^4p}{(2\pi\hbar)^4}\,\mathrm{e}^{-\mathrm{i}p\cdot(x'-x)/\hbar}\frac{1}{\gamma\cdot p - m}\,, \tag{12.197}$$

which immediately can be verified by evaluating

$$\int d^4x'\, A^{-1}(x'',x')A(x',x) = \delta^4(x''-x)\,. \tag{12.198}$$

To give a unique meaning to the integral (12.197) the Feynman contour will be used to circumvent the poles at $p^2 = m^2$ or, equivalently, the mass m is given a small negative imaginary part. Then (12.197) just coincides with the *Feynman propagator for Dirac fields* that was introduced in Sect. 5.4

$$A^{-1}(x',x) = i\hbar S_F(x'-x)\,. \tag{12.199}$$

It is related to the spin-0 Feynman propagator through

$$S_F(x'-x) = (i\hbar\gamma\cdot\partial_{x'} + m)\,\Delta_F(x'-x)\,. \tag{12.200}$$

The *free generating functional* then reads

$$W_0[\eta,\bar\eta] = \exp\left[-\frac{i}{\hbar}\int d^4x'd^4x\,\bar\eta(x')S_F(x'-x)\eta(x)\right]\,. \tag{12.201}$$

It is very similar to the Klein–Gordon functional (12.69). To arrive at (12.201) we have identified the (formally divergent) expression det A with $\mathcal{N}^{-1}$ in order to satisfy the normalization condition $W_0[0,0] = 1$.

The explicit expression (12.201) for the free generating functional again allows the construction of the *free n-point functions*

$$G_0^{(2n)}(y_1,\ldots,y_n;x_1,\ldots,x_n) = \left(\frac{\hbar}{i}\right)^{2n}\frac{\delta^{2n}W_0[\eta,\bar\eta]}{\delta\eta(x_n)\cdots\delta\eta(x_1)\delta\bar\eta(y_1)\cdots\delta\bar\eta(y_n)}\bigg|_{\eta=\bar\eta=0}. \tag{12.202}$$

The free *two-point function* of course is just the Feynman propagator:

$$\begin{aligned}
G_0^{(2)}(y;x) &= \left(\frac{\hbar}{i}\right)^2 \frac{\delta^2 W_0[\eta,\bar\eta]}{\delta\eta(x)\delta\bar\eta(y)}\bigg|_0 \\
&= \left(\frac{\hbar}{i}\right)^2\left(\frac{-i}{\hbar}\right)\frac{\delta}{\delta\eta(x)}\int d^4x'\,S_F(y-x')\eta(x')\,e^{-\frac{i}{\hbar}(\bar\eta,S_F\eta)}\bigg|_0 \\
&= i\hbar\,S_F(y-x)\,.
\end{aligned} \tag{12.203}$$

The exponent in (12.201) has been abbreviated by $(\bar\eta, S_F\eta)$. Equation (12.203) corresponds to the spin-0 result (12.80).

With some computational effort the n-point functions of higher order can be constructed. The free *four-point function* for fermions reads

$$\begin{aligned}
G_0^{(4)}(y_1,y_2;x_1,x_2) &= \left(\frac{\hbar}{i}\right)^4 \frac{\delta^4 W_0[\eta,\bar\eta]}{\delta\eta(x_2)\delta\eta(x_1)\delta\bar\eta(y_1)\delta\bar\eta(y_2)}\bigg|_{\eta=\bar\eta=0} \\
&= \left(\frac{\hbar}{i}\right)^4\left(\frac{-i}{\hbar}\right)\frac{\delta^3}{\delta\eta(x_2)\delta\eta(x_1)\delta\bar\eta(y_1)}\int d^4x\,S_F(y_2-x)\eta(x)\,e^{-\frac{i}{\hbar}(\bar\eta,S_F\eta)}\bigg|_0 \\
&= -\left(\frac{\hbar}{i}\right)^4\left(\frac{-i}{\hbar}\right)^2 \frac{\delta^2}{\delta\eta(x_2)\delta\eta(x_1)}\int d^4x\,S_F(y_2-x)\eta(x) \\
&\quad \times \int d^4x'\,S_F(y_1-x')\eta(x')\,e^{-\frac{i}{\hbar}(\bar\eta,S_F\eta)}\bigg|_0
\end{aligned}$$

12.10 Generating Functional and Feynman Propagator for the Free Dirac Field

$$
\begin{aligned}
&= -\left(\frac{\hbar}{i}\right)^4 \left(\frac{-i}{\hbar}\right)^2 \frac{\delta}{\delta\eta(x_2)} \Bigg[S_F(y_2 - x_1) \int d^4x' \, S_F(y_1 - x')\eta(x') \\
&\quad - \int d^4x \, S_F(y_2 - x)\eta(x) \, S_F(y_1 - x_1) + \ldots \Bigg] e^{-\frac{i}{\hbar}(\bar\eta, S_F \eta)}\bigg|_0 \\
&= -\left(\frac{\hbar}{i}\right)^4 \left(\frac{-i}{\hbar}\right)^2 \Big[S_F(y_2 - x_1)S_F(y_1 - x_2) - S_F(y_2 - x_2)S_F(y_1 - x_1) \Big] \\
&= (i\hbar)^2 \Big[S_F(y_1 - x_1)S_F(y_2 - x_2) - S_F(y_2 - x_1)S_F(y_1 - x_2) \Big] . \quad (12.204)
\end{aligned}
$$

If the Feynman propagator is represented by a directed line

$$
i\hbar S_F(y - x) = \underset{y \quad\quad x}{\longleftarrow} , \quad (12.205)
$$

this equation can be written in graphical form:

$$
G_0^{(4)}(y_1, y_2; x_1, x_2) = \begin{array}{c} y_1 \longleftarrow x_1 \\ y_2 \longleftarrow x_2 \end{array} - \begin{array}{c} y_1 x_1 \\ \times \\ y_2 x_2 \end{array} . \quad (12.206)
$$

This result differs in two aspects from the four-point function of the (neutral) Klein–Gordon field, equation (12.84b). First, the propagators here have a sense of direction, such that the points x_i and y_i cannot be interchanged. Therefore (12.204) contains only two possible permutations. Second, the fermionic character of the Dirac field gives a *minus sign* to the exchange graph in (12.206). This stems from the anticommuting property of the Grassmann-valued source fields η and $\bar\eta$.

To describe a nontrivial field theory an interaction Lagrangian $\mathcal{L}_{\text{int}}$ has to be specified. In general this interaction term can take on many different forms. If we adopt only the condition that $\mathcal{L}_{\text{int}}$ has to be a Lorentz scalar we know that it must be constructed from bilinear covariants $\bar\psi \hat O \psi$ (omitting possible "flavor" indices). For example we could have a quadrilinear self-interaction

$$
\mathcal{L}_{\text{int}} = g(\bar\psi\psi)^2 \quad \text{oder} \quad \mathcal{L}_{\text{int}} = g(\bar\psi\gamma_\mu\psi)(\bar\psi\gamma^\mu\psi) . \quad (12.207)
$$

This interaction is very similar to the ϕ^4 theory and the resulting perturbation series would produce graphs of the same topology. From a deeper point of view, however, theories involving fermionic self-interactions turn out to be *nonrenormalizable* and thus are ruled out as fundamental theories. Nevertheless couplings of the type (12.207) can still be used as *effective interactions* approximately valid in a low-energy approximation to a more fundamental theory. This is the case for Fermi's theory of weak interaction, which contains the product of four different spin-$\frac{1}{2}$ fields (two quarks, neutrino, and electron). Because of the large mass and short range of the intermediate vector bosons Z and W, which are responsible for the weak interaction, the four-fermion point-like interaction is a good approximation.

The criterion of renormalizability turns out to be very restrictive: only such terms in $\mathcal{L}_{\text{int}}$ are allowed that are a product of two spinor fields and one boson field. This criterion allows the following theories:

$$\begin{aligned}
\mathcal{L}_{\text{int}} &= g\,\bar\psi\psi\phi\,, & \phi &\text{ scalar}\,, \\
\mathcal{L}_{\text{int}} &= g\,\mathrm{i}\bar\psi\gamma_5\psi\phi\,, & \phi &\text{ pseudoscalar}\,, \\
\mathcal{L}_{\text{int}} &= g\,\bar\psi\gamma_\mu\psi A^\mu\,, & A^\mu &\text{ vector}\,, \\
\mathcal{L}_{\text{int}} &= g\,\bar\psi\gamma_\mu\gamma_5\psi A^\mu\,, & A^\mu &\text{ axialvector}\,.
\end{aligned}$$

The couplings to spin-1 fields are of fundamental importance for quantum electrodynamics (coupling to the photon field) and the theory of weak interaction (coupling to intermediate vector bosons). In Example 12.7 we will treat the coupling to a scalar boson field (the *Yukawa coupling*) which is somewhat simpler. The Yukawa coupling had originally been introduced to describe the strong interaction between nucleons via the exchange of mesons.

EXAMPLE

12.7 Yukawa Coupling

In the following we study the coupling between a spin-$\tfrac{1}{2}$ field and a spin-0 field in the form

$$\mathcal{L}_{\text{int}}(\bar\psi,\psi,\phi) = g\,\bar\psi(x)\psi(x)\phi(x)\,. \tag{1}$$

We evaluate the generating functional $W[\eta,\bar\eta,J]$ in second-order perturbation theory. The result will be used to obtain several n-point functions of the Yukawa model.

As in Sect. 12.5 the following representation of the generating functional can be used as a starting point for developing the perturbation expansion:

$$W[\eta,\bar\eta,J] = \mathcal{N}\,\exp\left[\frac{\mathrm{i}}{\hbar}\mathcal{L}_{\text{int}}\!\left(\frac{\hbar}{\mathrm{i}}\frac{\delta}{\delta\eta},\frac{\hbar}{\mathrm{i}}\frac{\delta}{\delta\bar\eta},\frac{\hbar}{\mathrm{i}}\frac{\delta}{\delta J}\right)\right]W_0[\eta,\bar\eta,J]\,. \tag{2}$$

This expression results from the path-integral representation

$$W[\eta,\bar\eta,J] = \mathcal{N}\int\mathcal{D}\phi\,\mathcal{D}\bar\psi\,\mathcal{D}\psi\,\exp\Big[\frac{\mathrm{i}}{\hbar}\Big(\mathcal{L}_0(\phi) + \mathcal{L}_0(\bar\psi,\psi) \\ + J\phi + \bar\psi\eta + \bar\eta\psi + \mathcal{L}_{\text{int}}(\bar\psi,\psi,\phi)\Big)\Big]\,. \tag{3}$$

Equation (2) follows if in the interaction Lagrangian the fields $\psi,\bar\psi$, and ϕ are replaced by functional derivatives with respect to the source fields $\bar\eta,\eta$, and J. Then $\mathcal{L}_{\text{int}}$ can be moved before the integral.

The generating functional of the free theory is simply given by the product

$$\begin{aligned}
W_0[\eta,\bar\eta,J] &= \exp\left[-\frac{\mathrm{i}}{\hbar}\int\mathrm{d}^4x\,\mathrm{d}^4y\,\bar\eta(x)S_{\mathrm{F}}(x-y)\eta(y)\right] \\
&\quad \times \exp\left[-\frac{\mathrm{i}}{2\hbar}\int\mathrm{d}^4x\,\mathrm{d}^4y\,J(x)\Delta_{\mathrm{F}}(x-y)J(y)\right]\,.
\end{aligned} \tag{4}$$

The expansion of the exponential function in (2) up to second order leads to

$$W[\eta,\bar\eta,J] = \mathcal{N}\left[1 + \frac{\mathrm{i}}{\hbar}g\int\mathrm{d}^4x\left(\frac{\hbar}{\mathrm{i}}\right)^3\frac{\delta^3}{\delta\eta(x)\delta\bar\eta(x)\delta J(x)}\right.$$

12.10 Generating Functional and Feynman Propagator for the Free Dirac Field

Example 12.7

$$+ \left(\frac{i}{\hbar}\right)^2 \frac{g^2}{2!} \int d^4y \left(\frac{\hbar}{i}\right)^3 \frac{\delta^3}{\delta\eta(y)\delta\bar\eta(y)\delta J(y)} \int d^4x \left(\frac{\hbar}{i}\right)^3 \frac{\delta^3}{\delta\eta(x)\delta\bar\eta(x)\delta J(x)}$$
$$+ \cdots \Bigg] W_0[\eta,\bar\eta,J] \,. \tag{5}$$

Since the differentiation refers to *different fields* the calculation here will be somewhat simplified (compared, e.g., to scalar ϕ^3 theory which topologically would produce similar graphs) since we do not have to apply the product rule as often. Nevertheless the derivation of the terms in second and higher order is quite laborious. To keep track of the calculation we make use of the graphical notation as in Example 12.4. The "dictionary" for translation between algebraic and graphical notation is found in Table 12.2.

Table 12.2. Graphical translation rules for the Yukawa theory.

Propagators	$i\hbar\Delta_F(x-y)$	x --- y
	$i\hbar S_F(x-y)$	$x \longleftarrow y$
External sources	$\frac{i}{\hbar}\int d^4x\, J(x)$	$\circ$ ---
	$\frac{i}{\hbar}\int d^4x\, \bar\eta(x)$	$\bullet\!\!\longleftarrow$
	$\frac{i}{\hbar}\int d^4x\, \eta(x)$	$\longrightarrow\!\!\bullet$
Interaction	$\frac{i}{\hbar}\int d^4x$	(vertex)

The free generating functional (4) can be written as

$$W_0 = e^{\bullet\!\!\leftarrow\!\!\bullet} \, e^{\frac{1}{2}\circ\text{---}\circ} \,. \tag{6}$$

Functional differentiation has the effect that one of the end circles is removed and is replaced by an external coordinate:

$$\frac{\hbar}{i}\frac{\delta}{\delta J(x)} \circ\text{---}\circ = 2\,\circ\text{---}\,x \,,$$

$$\frac{\hbar}{i}\frac{\delta}{\delta\bar\eta(x)} \bullet\!\!\leftarrow\!\!\bullet = {}_x\!\!\leftarrow\!\!\bullet \,,$$

$$\frac{\hbar}{i}\frac{\delta}{\delta\eta(x)} \bullet\!\!\leftarrow\!\!\bullet = -\bullet\!\!\leftarrow\,_x \,. \tag{7}$$

Note the minus sign in the last equation, which results from the anticommutation of $\delta/\delta\eta(x)$ and $\bar\eta(x')$. In general when working with Grassmann variables the following sign rule has to be taken into account: each time a fermionic differential operator is moved past a factor $\bullet\!\!\leftarrow$ or $\longrightarrow\!\!\bullet$ a minus sign is generated.

Now we successively evaluate the required derivatives of $W_0[\eta,\bar\eta,J]$:

Example 12.7

$$\frac{\hbar}{i}\frac{\delta W_0}{\delta J(x)} = \text{\small(diagram)}_x W_0, \qquad (8)$$

$$\left(\frac{\hbar}{i}\right)^2 \frac{\delta^2 W_0}{\delta\bar\eta(x)\delta J(x)} = \text{\small(diagram)}_{x\ x} W_0 = \text{\small(diagram)}_x W_0, \qquad (9)$$

where the line endings carrying the same coordinate index x have been glued together. The third differentiation leads to

$$\frac{i}{\hbar}\left(\frac{\hbar}{i}\right)^3 \frac{\delta^3 W_0}{\delta\eta(x)\delta\bar\eta(x)\delta J(x)} = \left[\text{\small(diagram)}_{x\ x} + (-1)^2 \text{\small(diagram)}_{x\ \ x}\right] W_0$$

$$= \left[\text{\small(diagram)} - \text{\small(diagram)}\right] W_0. \qquad (10)$$

Here the sign rule has been used. In the last step an extra minus sign has arisen from the reordering of the factors η and $\bar\eta$. Equation (10) has been multiplied with the factor $\frac{i}{\hbar}$ standing in front of $\mathcal{L}_{\text{int}}$. In this way (associating the factor $\frac{i}{\hbar}$ with each vertex —⃖|⃗— according to the table) all factors are absorbed in the graphical notation. Written out algebraically, (10) reads

$$\frac{i}{\hbar}\left(\frac{\hbar}{i}\right)^3 \frac{\delta^3 W_0}{\delta\eta(x)\delta\bar\eta(x)\delta J(x)} = \Bigg[S_F(x-x)\int d^4x'\, \Delta_F(x-x')J(x')$$

$$-\frac{-i}{\hbar}\int d^4x'\, \bar\eta(x')S_F(x'-x)\int d^4x'\, S_F(x-x')\eta(x')$$

$$\times \int d^4x'\, \Delta_F(x-x')J(x') \Bigg] W_0. \qquad (11)$$

This constitutes the contribution of first order to the generating functional (5). The result is not particularly rewarding, however, since it does not describe true interaction (scattering) processes. Therefore we have to push the calculation to the order $O(g^2)$. Thus (10) has to be differentiated three more times. We find

$$\left(\frac{\hbar}{i}\right)\frac{\delta}{\delta J(y)}\cdots = \left[\text{\small(diagram)}_y - \text{\small(diagram)}_y + \left(\text{\small(diagram)} - \text{\small(diagram)}\right)_y\right] W_0 \qquad (12)$$

and thus

$$\left(\frac{\hbar}{i}\right)^2 \frac{\delta^2}{\delta\bar\eta(y)\delta J(y)}\cdots = \Bigg[-\text{\small(diagram)}_y - \text{\small(diagram)}_y + \text{\small(diagram)}_y \qquad (13)$$

$$-\text{\small(diagram)}_y + \text{\small(diagram)} \ \text{\small(diagram)}_y - \text{\small(diagram)}_y \ \text{\small(diagram)}_y\Bigg] W_0.$$

After the reordering and collecting of terms the third derivative leads to

$$\frac{i}{\hbar}\left(\frac{\hbar}{i}\right)^3 \frac{\delta^3}{\delta\eta(y)\delta\bar\eta(y)\delta J(y)}\cdots = \Bigg[-\text{\small(diagram)} + \text{\small(diagram)} - \text{\small(diagram)}$$

12.10 Generating Functional and Feynman Propagator for the Free Dirac Field

Example 12.7

$$+ 2 \cdots - 2 \cdots + 2 \cdots + \cdots$$

$$+ \left(\cdots - \cdots \right)^2 \Bigg] W_0 \,. \tag{14}$$

Using Table 12.2 we can easily translate this result back to the algebraic form. The general structure of (14) might have been guessed without an explicit calculation. All topologically possible graphs with two vertices are present. There are pure "vacuum bubble" graphs and graphs with two or four external sources. In addition there is a contribution that consists of the product of two unconnected first-order graphs according to (10). Some of the contributions carry the symmetry factor 2.

The properties described so far would also apply to the case of two coupled Klein–Gordon fields ϕ_1 and ϕ_2 interacting via $\mathcal{L}_{\text{int}} = g\phi_1^2\phi_2$ instead of (1). The minus signs, however, are characteristic for the fermions. We note the rule:

- Graphs involving a *fermion loop* are multiplied by the factor -1.

This is a consequence of Fermi statistics as we already noted in connection with the canonical formalism (cf. Sect. 8.6). In the framework of the functional method the minus sign arises from an anticommutation of Grassmann variables.

The generating functional $W[\eta, \bar{\eta}, J]$ in second order is obtained by inserting (14) and (10) into (5). The normalization constant $\mathcal{N}$ is fixed by the condition $W[0,0,0] = 1$. In Excercise 12.2 we have found the general connection

$$\begin{aligned} W[J] &= \frac{1 + gu_1[J] + g^2 u_2[J] + \cdots}{1 + gu_1[0] + g^2 u_2[0] + \cdots} W_0[J] \\ &= \left(1 + gw_1[J] + g^2 w_2[J] + \cdots\right) W_0[J] \,, \end{aligned} \tag{15}$$

where

$$\begin{aligned} w_1[J] &= u_1[J] - u_1[0] \,, \\ w_2[J] &= \bigl(u_2[J] - u_2[0]\bigr) - \bigl(u_1[J] - u_1[0]\bigr) u_1[0] \,, \\ &\cdots \end{aligned} \tag{16}$$

Now we have fermion sources η and $\bar{\eta}$ in addition to the scalar source J, but this does not affect the validity of these equations. In the present case, (16) is particularly simple since the Yukawa theory does not allow for vacuum bubbles in first order (in contrast to, e.g., the ϕ^4 theory), and thus $u_1[0] = 0$. The normalization factor $\mathcal{N}$ has the effect only of cancelling the unconnected "vacuum bubbles" in $u_2[J]$. Thus the first two graphs in (14) are then removed. The correctly normalized generating functional now reads

Example 12.7

$$W[\eta,\bar\eta,J] = \Big\{ 1 + g\Big(\,\text{[diagram]} - \text{[diagram]}\,\Big)$$

$$+ \frac{g^2}{2}\Big[-\text{[diagram]} + 2\,\text{[diagram]} - 2\,\text{[diagram]} + 2\,\text{[diagram]}$$

$$+ \text{[diagram]} + \Big(-\text{[diagram]} + \text{[diagram]}\Big)^2 \Big]$$

$$+ O(g^3) \Big\} W_0[\eta,\bar\eta,J]\,. \tag{17}$$

By differentiating this functional with respect to the sources the interacting n-point functions can be evaluated. The *two-point Green's function for fermions* in order $O(g^2)$ reads

$$G_F^{(2)} = \left(\frac{\hbar}{i}\right)^2 \frac{\delta^2 W}{\delta\eta(x)\delta\bar\eta(y)}\bigg|_0$$

$$= \;\text{[diagram]}\; + g^2\Big(\,\text{[diagram]} - \text{[diagram]}\,\Big) + O(g^3)\,. \tag{18}$$

Thus the free fermion propagator is modified by a *self-energy* contribution. Explicitly we have

$$G_F^{(2)}(y;x) = i\hbar S_F(y-x)$$

$$+ g^2(i\hbar)^4 \left(\frac{i}{\hbar}\right)^2 \int d^4x''\,d^4x'\, S_F(y-x'')S_F(x''-x')S_F(x'-x)\Delta_F(x''-x')$$

$$+ O(g^3)\,. \tag{19}$$

This integral is divergent and has to be regularized. Subsequently the theory has to renormalized. Equation (18) contains a second radiative correction, namely the "tadpole" involving a closed fermion line. According to the argument presented in connection with the ϕ^4 theory this contribution has no physical consequences and can be simply left out. This is what we did in (19).

Similarly we obtain the *two-point Green's function for bosons*

$$G_B^{(2)} = \left(\frac{\hbar}{i}\right)^2 \frac{\delta^2 W}{\delta J(x_1)\delta J(x_2)}\bigg|_0$$

$$= \;\text{[diagram]}\; + g^2\Big(-\text{[diagram]} + \text{[diagram]}\Big) + O(g^3) \tag{20}$$

or

$$G_F^{(2)}(x_1,x_2) = i\hbar\Delta_F(x_1-x_2)$$

Example 12.7

$$-g^2(i\hbar)^4\left(\frac{i}{\hbar}\right)^2 \int d^4x'' \, d^4x' \, \Delta_F(x_1-x'')S_F(x''-x')S_F(x'-x'')\Delta_F(x'-x_2)$$

$$+O(g^3), \tag{21}$$

where the irrelevant tadpole contribution again was left out. Note the negative sign of the *vacuum polarization* graph.

In addition to the radiative corrections of the propagators, as expressed in (19) and (21), the Yukawa coupling of course also describes true scattering interactions. Therefore we will briefly discuss the possible four-point functions.

The interaction of two fermions is described by

$$G_F^{(4)}(y_1, y_2; x_1, x_2) = \left(\frac{\hbar}{i}\right)^4 \frac{\delta^4 W}{\delta\eta(x_2)\delta\eta(x_1)\delta\bar{\eta}(y_1)\delta\bar{\eta}(y_2)}\bigg|_0. \tag{22}$$

Using (17) we easily find the *four-point function of the fermion field*

$$G_F^{(4)}(y_1, y_2; x_1, x_2) = G_{0F}^{(4)}$$

$$+g^2\left[\begin{array}{c}\text{diagram}\end{array} + \begin{array}{c}\text{diagram}\end{array} + \begin{array}{c}\text{diagram}\end{array} - \text{exchange } (x_1 \leftrightarrow x_2)\right]$$

$$+O(g^3). \tag{23}$$

In addition to the disconnected graphs there is a true scattering graph of order $O(g^2)$ which describes the exchange of a virtual boson.

However, *four-point function of the boson field*,

$$G_B^{(4)}(x_1, x_2, x_3, x_4) = \left(\frac{\hbar}{i}\right)^4 \frac{\delta^4 W}{\delta J(x_1)\delta J(x_2)\delta J(x_3)\delta J(x_4)}\bigg|_0, \tag{24}$$

in the same order contains only trivial disconnected contributions:

$$G_B^{(4)}(x_1, x_2, x_3, x_4) = G_{0B}^{(4)} + g^2\left(\begin{array}{c}\text{diagram}\end{array} + 5 \text{ permut.}\right) + O(g^3). \tag{25}$$

Here one has to go to the fourth order of scattering theory to find an interaction. We do not want to work out the fourth-order generating functional $W[\eta, \bar{\eta}, J]$ in detail. It is rather easy, however, to obtain that part responsible for the boson–boson interaction.

The g^4 contribution to $W[\eta, \bar{\eta}, J]$ is obtained if the differential operator $(i/\hbar)\mathcal{L}_{\text{int}}\big((\hbar/i)\delta/\delta\bar{\eta}, (\hbar/i)\delta/\delta\eta, (\hbar/i)\delta/\delta J\big)$ is applied twice to the g^2 term in (17). We are interested only in that part that contains four bosonic and no fermionic source. Terms of this type can only result from ⊶⊷⊷⊷ and from (⊶⊷⊷)^2. After carrying out the differentiation we find that the resulting expressions can be combined into the "box diagram"

$$W_{\text{Box}}[\eta, \bar{\eta}, J] = -\frac{g^4}{4!}\,6\;\;\text{[box diagram]}. \tag{26}$$

The minus sign has its origin in the closed fermion loop. Equation (26) produces a g^4 contribution to the bosonic four-point function

Example 12.7

$$G_B^{(4)}(x_1, x_2, x_3, x_4) = \cdots - g^4 \left(\begin{array}{c} \text{[box diagram with } x_1, x_2, x_3, x_4\text{]} \end{array} + \text{5 permut.} \right). \tag{27}$$

In the formation of the derivatives of (26), in all there are $4! = 24$ permutations of the coordinates. $3! = 6$ different configurations of the coordinates lead to nonequivalent diagrams which have to be added up coherently in (27). The remaining factor 4 (which may be interpreted as the possible association of a chosen coordinate, say x_1 with each of the four corners of the box) has cancelled with the factor present in (26), $\frac{6}{4!}4 = 1$. In contrast with the ϕ^n theories, see Example 8.7, here we do not encounter extra symmetry factors.

The box diagram leads to a new type of *self-interaction of the boson field*, which was not present in the original Lagrangian and which has its origin in quantum fluctuations.[14]

The induced interaction graphs correspond exactly to those of the ϕ^4 theory. There is a fundamental difference, however, in that here we do not have a point interaction. At sufficiently high values of the momentum transfer (the scale is set by the fermion mass) the scattering process will resolve the "internal structure" of the vertex and there will be deviations from pure ϕ^4 theory.

Finally we take notice of the *two-fermion-two-boson Green's function*

$$G_{FB}^{(4)}(x_1, x_2; y; x) = \left(\frac{\hbar}{i}\right)^4 \frac{\delta W}{\delta J(x_1)\delta J(x_2)\delta\eta(x)\delta\bar{\eta}(y)}\bigg|_0, \tag{28}$$

which according to (17) in lowest order is given by

$$G_{FB}^{(4)}(x_1, x_2; y; x) = \text{[diagram]} + g^2 \left(\text{[diagram with } x_1, x_2\text{]} + \text{[diagram with } x_2, x_1\text{]} \right) + O(g^3). \tag{29}$$

This is the scalar equivalent of the Compton diagram of QED.

As discussed in Sect. 12.4 for the spin-0 case we can introduce the *irreducible Green's functions* and their generating functional $Z[\eta, \bar{\eta}, J] = \frac{\hbar}{i} \ln W[\eta, \bar{\eta}, J]$. According to (9) in Exercise 12.2, (17) gets transformed into

$$\frac{i}{\hbar}Z[\eta, \bar{\eta}, J] = \frac{1}{2}\text{[diagram]} + g\left(\text{[diagram]} - \text{[diagram]}\right)$$

$$+ \frac{g^2}{2}\left(-\text{[diagram]} + 2\text{[diagram]} - 2\text{[diagram]} + 2\text{[diagram]}\right.$$

$$\left. + \text{[diagram]}\right) + O(g^3). \tag{30}$$

---

[14] The analog effect in quantum electrodynamics (which contains a vector instead of our scalar field) is the *Delbrück graph* which describe a very weak (because of the factor $\alpha^4 \simeq 10^{-9}$) photon–photon interaction; cf. Example 8.5.

In this way the nonconnected diagrams are eliminated. For example the *irreducible fermionic four-point function* is reduced to the true scattering graphs and its exchange partner

$$G^{(4)}_{\text{cF}}(y_1, y_2; x_1, x_2) = \left(\frac{\hbar}{i}\right)^3 \frac{\delta^4 Z}{\delta\eta(x_2)\delta\eta(x_1)\delta\bar\eta(y_1)\delta\bar\eta(y_2)}\bigg|_0 \qquad (31)$$

$$= G^{(4)}_{\text{0F}} + g^2 \left(\begin{array}{c} \text{diagram} \end{array} - \begin{array}{c} \text{diagram} \end{array} \right) + O(g^3).$$

Example 12.7

Index

Action, 4, 32, 344
- effective, 402, 406
- euclidian, 374
Adiabatic switching, 221, 360
Advanced Green's function, 113
Advanced propagator, 273
Amputated Green's function, 404
Amputation of external lines, 286, 298
Angular momentum
- Klein–Gordon field, 83
- orbital, 45, 85, 120
- spin, 45, 120
Angular-momentum operator
- Dirac field, 131
- Klein–Gordon field, 85–91, 93
Angular-momentum tensor, 45, 85, 99
- Dirac field, 119
- electromagnetic, 148
- modified, 48
Annihilation operator, 9, 61, 67, 77, 83, 127, 162, 214, 276
Anticommutation relations, 65, 125, 408
Anticommuting numbers, 408
Antilinear operator, 307
Antiparticles, 94, 127
Antiunitary operator, 325
Antiunitary transformation, 307
Asymptotic completeness, 271
Asymptotic condition
- for field operator, 275, 276
Axial gauge, 144
Axiomatic quantum field theory, 269

Baker–Campbell–Hausdorff relation, 25, 27, 215, 390
Bare mass, 270, 272, 295
Basis, 59, 60, 76
Basis function, 12
Belinfante tensor, 47, 148
Bethe–Salpeter equation, 320
Bilinear covariants, 328, 330
Bilinear forms, 103, 139
Bose particles
- quantization, 58
Bose–Einstein statistics, 64, 68, 76, 123, 130, 323, 329

Boson loop, 265
Boson–boson scattering, 298
Box diagram, 242, 429
Box normalization, 79, 127, 179
Branching ratio, 333

Canonical commutation relations, 9, 215, 269
- Dirac field, 123
- electromagnetic field, 176
- Klein–Gordon field, 76, 92, 102
- Proca field, 159
- Schrödinger field, 58
Canonical energy-momentum tensor, 43
Canonical quantization, 58, 177
- Coulomb gauge, 196
- electromagnetic field, 176
- massive vector field, 158
Canonical transformation, 6, 128
Canonically conjugate field, 34, 172, 251
- Dirac, 118
- Klein–Gordon, 75
- Proca, 158
- Schrödinger, 57
Canonically conjugate momentum, 5
Casimier effect, 82
Chain rule for functional derivatives, 37
Charge
- Bose field, 93
- Dirac field, 120
- Noether, 46
Charge conjugation, 301
- Dirac field, 313
- scalar field, 305, 309
- vector field, 319
Charge operator
- Dirac, 130
Charge parity, 305, 320
- positronium, 322
Circular polarization, 161, 166
Classical field, 401, 407
Classical field theory, 31–51
Classical limit, 344
Classical path, 346
Closed loops, 297
Coherent state, 10, 26

Commutation relations, *see* Equal-time commutation relations
- general, Dirac field, 138–139
- general, electromagnetic field, 205
- invariant, 100

Completeness, 63, 339
Completeness relation, 7, 13
- polarization vectors, 155, 157

Complex rotation, 358, 359
Compton scattering, 239, 247–249, 430
- of positron, 248
- Spin 0, 259

Connected Green's function, 383, 387, 398–400
Conservation laws, 43
- in classical field theory, 39

Constraint condition, 141, 153
Continuity equation, 42, 142, 143, 146
Continuous matrix, 376, 403
Continuum limit, 11
Continuum threshold, 281
Contracted normal product, 230
Contraction, 228
Convergence
- strong, 274
- weak, 274, 275, 278, 284

Coordinate operator, 338
Coordinate-space representation, 8
Correlations, 74
Coulomb gauge, 144–145, 171, 196, 197, 208, 319
Coulomb interaction, 69, 187, 200, 202
Coulomb photon, 187
Coulomb potential, 145, 203
Coulomb propagator, 186
Counter term, 296
Coupled fields
- Maxwell and Dirac, 150
- scalar electrodynamics, 251
- Yukawa model, 424

Covariant commutation relation, 160
Covariant notation, 33
Covariant photon propagator, 187
Covariant quantization, 171
- Proca field, 160

CPT theorem, 326–331
Creation operator, 9, 61, 67, 77, 83, 127, 162
Cross section, 267
Current
- conserved, 42, 46, 47
- Dirac, 120

Current conservation, 143, 146, 153
Current operator, 139
- symmetrized, 134

Degeneracy factor, 268
Degrees of freedom

- Lorentz gauge, 144
- photon field, 171
- Proca field, 158
- vector field, 141

Delbrück graph, 250, 430
Delbrück scattering, *see* Photon–photon scattering
$\Delta(x)$, 100, 109–115, 160, 167
$\Delta_1(x)$, 104, 105, 109–115
$\Delta_F(x)$, 106, 109–115
Dimension
- Dirac field, 117
- Klein–Gordon field, 75

Dirac equation, 117, 124, 129, 288
Dirac field, 117
Dirac picture, *see* Interaction picture
Dirac sea, 126, 130
Dirac unit spinor, 124, 128, 137
Dirac's hole picture, 126
Disconnected graphs, 243
Discrete symmetry, 301
Discretization, 32
Dispersion relation
- linear chain, 14
- photon, 156
- relativistic, 77

Dual field-strength tensor, 142
Dyson operator, 215, 292
Dyson product, 217
Dyson propagator, 113

Effective action, 402, 406
Effective potential, 407
Electron–electron scattering, 240, 245–247
Electron–positron
- pair annihilation, 239
- pair creation, 239
- scattering, 240

Energy shift, 219, 220, 222, 224
Energy-momentum tensor, 43
- Dirac field, 119
- electromagnetic field, 146, 174
- Proca field, 153
- symmetrical, 43
- symmetrized, 47, 49, 147, 152

Equal-time anticommutation relation, 123
Equal-time commutation relations, 269
- electromagnetic field, 176
- Klein–Gordon field, 76, 92, 102
- Proca field, 159
- Schrödinger field, 58

Equation of continuity, 42, 142, 143, 146
ETCR, *see* Equal-time commutation relations
Euclidian action, 374
Euclidian coordinates, 359, 361, 372

Euclidian field theory, 371–375
Euler–Lagrange equation, 5, 32, 33, 57, 118, 149, 152, 172
Exchange graph, 239, 246, 260
Exchange term, 73
External current, 146, 148
External fermion line, 236
External forces, 20
External photon line, 236
External potential, 69
External source, 360, 367, 384, 420

Fermi level, 71
Fermi particles
– quantization, 65
Fermi's Lagrangian, 172
Fermi–Dirac statistics, 68, 104, 123, 130, 246
Fermion loop, 241, 427
Fermionic quantization, 65, 104, 159
Feynmal rules
– scalar electrodynamics, 260
Feynman contour, 107, 422
Feynman diagrams, 236, 239, 382
Feynman gauge, 173, 185, 189
Feynman kernel, 339, 345, 350, 360, 366, 371
– euclidian, 359
Feynman path integral, 344, 360, 362, 367, 370
Feynman propagator, 188, 230, 296
– Coulomb gauge, 199
– Dirac, 132–134
– euclidian, 378
– gauge dependence, 189
– Klein–Gordon, 106–109, 188
– momentum space, 108, 134, 170, 185
– path-integral representation, 375–380
– photons, 185–187, 195
– Proca, 167, 169, 188
– scalar, 112–115
– spectral representation, 280
Feynman rules
– ϕ^4 theory, 262
– in coordinate space, 241
– in momentum space, 242
– quantum electrodynamics, 233–243
Field
– classical, 31
Field strenght
– electric, 143
– magnetic, 143
Field-strength tensor
– dual, 142
– electromagnetic, 142, 152
Fock space, 61, 67, 79, 83, 130, 182, 270
– photons, 179, 182
Fock state
– normalization, 68
Four-boson vertex, 263
Four-current
– electromagnetic, 142
Four-momentum, 43, 119
Four-point function, 382, 391, 405, 422
– ϕ^4 theory, 297, 397
– bosons, 429
– fermions, 429, 431
Fourier decomposition, 59, 92
– discrete, 12
– photon field, 177
– Proca field operator, 160
Fourier expansion, 78
Fréchet derivative, see Functional derivative
Functional, 31, 37
Functional derivative, 31, 35–39
– chain rule, 37
– product rule, 37
Furry's theorem, 320

Gauge, see Coulomb, Fermi, Feynman, Landau, Lorentz gauge
– noncovariant, 171
Gauge-covariant derivative, 151
Gauge-fixing parameter, 173
Gauge fixing, 143, 172, 185, 189
Gauge invariance, 143, 153, 182, 210
Gauge transformation, 143, 146, 147, 171, 183
Gaussian integral, 343, 345, 351, 353–356, 370, 375
– Grassmann, 412, 415
Gaussian units, 142
Gell-Mann–Low theorem, 219–225
Generating functional
– fermion fields, 419–421
– for Green's function, 380
– for irreducible Green's function, 383
– free Dirac field, 421–424
– interacting, 384
– perturbation series, 385–387
Generator, 49, 98
– Grassmann algebra, 408
– Poincaré group, 50
Gradient coupling, 251, 253
Grassmann algebra, 408, 409, 414
Grassmann integration, 415–419
Grassmann variable, 408
– complex, 413
Green's function, 145
– amputated, 404
– connected, 383, 387, 398–400
– higher-order, 382
– irreducible, 383, 430
– Klein–Gordon, 108
– momentum space, 391–392

- multi-particle, 283, 286, 287, 290
- pole structure, 287
- Schrödinger equation, 346

Ground state
- nondegenerate, 359

Gupta–Bleuler method, 180–183

Hamilton density, 34
Hamilton equations, 5, 34, 36
Hamilton formalism
- classical fields, 34

Hamilton function, see Hamiltonian
Hamilton's principle, 4, 32, 344
Hamiltonian, 5
- classical field theory, 34
- Dirac, 119, 126, 129
- electromagnetic field, 173
- harmonic oscillator, 9
- interaction, 291
- Klein–Gordon, 75
- linear chain, 12
- normal-ordered, 81, 92, 164
- Proca, 159, 164
- Schrödinger field, 60

Harmonic oscillator, 8–10, 14, 80
Hartree-Fock approximation, 69–74
Heisenberg picture, 8, 60, 212, 213, 215, 216, 252, 269, 275, 338
Heisenberg's equation, 8, 58, 123, 159, 291, 338
- four-dimensional, 97

Heisenberg's quantization rule, 6
Helicity, 131, 136, 154, 165, 166, 199, 319
Hermitean operator, 6, 96
Hilbert space, 6
- photons, 183

Hole picture, 126, 130

Imaginary time, 343, 359, 371, 372
In field, 270–276
Indefinite metric, 176, 180, 181
Infinitesimal generator, 44
- electromagnetic field, 148

Infinitesimal rotation, 44
Infinitesimal transformation, 40
- Dirac field, 119

Initial-value problem, 17
Instantaneous Coulomb potential, 145
Integration
- Grassmann variables, 409, 415–419

Interacting fields, 211, 272, 291
Interacting ground state, 22
Interaction picture, 211–216, 221, 252
Internal fermion line, 236
Internal photon line, 236
Internal symmetries, 46
Interpolating field, 272, 291

Intrinsic parity, 302, 312, 321
Invariance
- S matrix, 324

Invariant amplitude, 267
Invariant commutation relations, 100
- Proca field, 167

Irreducible, see One-particle irreducible
Irreducible Green's function, 383, 430
Irreducible vertex function, 402

Jordan–Wigner quantization, 65

Kinetic energy, 16
Klein–Gordon equation, 75, 76, 153, 168, 275
Klein–Gordon field
- charged, 91
- neutral, 75

Ladder operators, 9
Lagrange density, 33
- Dirac field, 117
- electromagnetic field, 145
- Klein–Gordon field, 75, 92
- Schrödinger field, 57

Lagrange function, see Lagrangian
Lagrange functional, 31
Lagrange multiplier, 172
Lagrangian, 4
- ϕ^4 theory, 261
- classical field theory, 31
- Dirac, 117
- electromagnetic, 149–150
- -- with gauge fixing, 172, 173
- linear chain, 11
- Lorentz scalar, 329
- Proca, 152
- scalar electrodynamics, 251
- symmetrized Dirac, 120–122

Landau gauge, 185
Landau–Yang rule, 323
Lattice field theory, 374
Legendre transformation, 5, 34, 401, 407
Lehmann–Källen representation, 278–282
Lehmann–Symanzik–Zimmermann formalism, see LSZ formalism
Lie algebra, 45, 149
Lifetime, 332
- of antiparticle, 333

Light cone, 101, 115, 206, 207
Linear chain
- classical treatment, 10–15
- quantum treatment, 18–20

Linear polarization, 161, 166
Local field theory, 33
Longitudinal photon, 181, 186

Longitudinal polarization state, 141, 145, 166, 178
Longitudinal polarization vector, 154, 157
Lorentz boost, 44
Lorentz condition, 153, 168, 172, 177, 180, 183
Lorentz gauge, 144, 172, 175, 210
Lorentz invariance, 44
Lorentz scalar, 33
Lorentz transformation, 97, 148, 272
- infinitesimal generators, 44, 97
Lorentz–Heaviside units, 142
LSZ formalism, 269, 282–288
LSZ reduction formula
- for spin-$\frac{1}{2}$ particles, 290
- for spin-0 particles, 286
- in momentum space, 286

Many-body Schrödinger equation, 64, 66
Mass
- bare, 270
- counter term, 296
- renormalization, 238, 295–297, 396
Massive vector bosons, 152
Massless vector field, 156
Matrix representation, 44
Maxwell equations, 141, 143
Maxwell field, 154
Mean field, 401
Microcausality, 103, 104, 115, 139, 159, 207
Minimal coupling, 151
Møller scattering, 245–247
Momentum, 43
Momentum eigenstates, 7
Momentum operator
- Dirac field, 131, 135
- Klein–Gordon field, 82, 93
- Proca field, 164
Monte Carlo integration, 374
Moving basis, 339, 361, 366
Multi-phonon state, 20, 25
Multi-photon state, 320

n-point function, 283, 359, 363, 367, 380
- connected, 383
- free, 381, 422
- perturbation series, 293
- perturbation theory, 290
Negative-energy continuum, 124, 126
Neumann series, 217, 292
Newton's equation of motion, 4
Noether charge, 46
Noether's theorem, 40–43, 82, 85
Noncovariant gauge, 171
Norm
- euclidian, 371
- one-photon state, 179
Norm convergence, 274
Normal coordinates, 12, 15–18
Normal-dependent terms, 169, 170, 252, 253, 259
- cancellation, 258
Normal ordering, 26, 81, 132, 164, 237, 253, 254
Normal product, 226
Normalization, 61, 68, 77, 363
- box, 79, 127
- continuum, 79
- polarization vectors, 155
Number operator
- for holes, 127

Occupation-number representation, 61
One-particle irreducible, 400
Orbital angular momentum, 45
Orthogonality, 14
- polarization vectors, 154, 162, 177
Orthopositronium, 322
Oscillator, see Harmonic oscillator
Out field, 270–276

Pair annihilation, 237
Pair creation, 237
Parapositronium, 322
Parity
- charge, 305, 320
- intrinsic, 302, 312, 321
Particle-antiparticle symmetry, 331
Particle-number operator, 60
- localized, 62
Path integral, 339
- coordinate-space, 344, 352
- Feynman, 344, 360, 362, 367, 370
- for fermion fields, 408–415
- for free particle, 345
- for vacuum functional, 362
- Grassmann fields, 415
- in field theory, 366, 368–371
- in phase space, 342, 351
- multi-dimensional, 350
- notation, 342
- time-ordered product, 357
- Wiener, 342
Pauli–Jordan function, 100, 109–115, 138, 160, 167, 256, 277
- spectral representation, 280
Pauli matrix, 120
Pauli principle, 65, 67, 129, 130
PCT theorem, see CPT theorem
Periodic boundary conditions, 12
Perturbation
- stationary, 221
Perturbation operator, 220

Perturbation series
- ϕ^4 theory, 392–398
- generating functional, 385–387
- S operator, 220, 234
- time-evolution operator, 218, 292
- two-point function, 295
- Yukawa model, 424

Perturbation theory, 211, 225
Pfaffian, 413
Phase factor, 304
- divergent, 221, 224, 245
Phase transformation, 46, 93
ϕ^4 theory
- triviality, 262
ϕ^4 theory, 261–267, 295–299, 392–398
Phonons, 20
Photon
- absorption, 237
- emission, 237
- longitudinal, 181
- scalar, 181
- virtual, 183, 187
Photon field
- energy density, 175
Photon propagator
- covariant, 187
Photon–photon scattering, 249–251
Plane waves, 79
- Dirac field, 288
Plane-wave basis, 76
Plane-wave expansion, 131
- Dirac field, 124, 125
- electromagnetic field, 177
- Klein–Gordon field, 77
- vector field, 154
Poincaré
- algebra, 51, 98
- group, 49, 50
- transformation, 96, 272, 324
Poisson brackets, 5, 15, 36
- in classical field theory, 35
Poisson distribution, 26
Poisson equation, 200
Polarization
- circular, 161, 166
- linear, 161, 166
Polarization state
- longitudinal, 141, 145, 166, 178
- massive vector field, 156
- photon, 177
Polarization vector, 154, 199
- completeness relation, 155, 156
- longitudinal, 154, 157
- time-like, 155, 157
- transverse, 154, 156
Position operator, 7
Positron–positron scattering, 240
Positronium, 312, 320–323

- selection rule, 323
Potential
- external, 69
Potential energy, 4, 16
Poynting vector, 148
Principle of least action, 4
Proca equation, 152
Proca field, 152, 154
- Feynman propagator, 167, 169
- invariant commutation relation, 167
Product rule for functional derivatives, 37
Projection operator, 133
- transverse, 145, 197
Propagator, see Feynman propagator
- advanced, 273
- retarded, 273, 277
Pseudo photon, 182, 183

QED, see Quantum electrodynamics
Quantization
- of a mechanical system, 6
Quantum electrodynamics, 150, 320
- Feynman rules, 233–243
Quantum mechanics, 6

Radiation gauge, 197
Random walk, 342
Reciprocity theorem, 325
Reducible, 400
Renormalization, 231
- coupling constant, 266, 299
- mass, 238, 295–297, 396
Renormalization constant, 273, 281, 299
Resonance, 281
Retarded Green's function, 113
Retarded propagator, 273, 277

S matrix, see Scattering matrix
S matrix
- invariance, 324
S matrix element, 220, 243, 267, 272, 283, 286, 288, 298
S operator
- perturbation series, 220, 234
- second order, 237, 238
- unitarity, 271
Scalar electrodynamics, 251–261
Scalar field theory, 261, 384
Scalar photon, 178, 180, 181, 186
Scalar product, 6, 162
- Klein–Gordon, 78
Scattering cross section, 267, 326
Scattering matrix, 219–220
Schrödinger equation, 20, 57, 59, 339, 345
Schrödinger picture, 212, 213, 338, 357
Schrödinger wave function, 7

Schwinger function, *see* Pauli–Jordan function
Seagull graph, 254
Second quantization, 57
Selection rule, 325
- decay of positronium, 323
Self-energy, 240, 254, 264
- Yukawa model, 428
Self-interaction, 73, 262, 273, 430
$S_F(x)$, 132
Shift operator, 22, 24
Single-particle
- basis, 71
- energy, 73
- state, 272, 273, 281, 287
- wavefunction, 70
Singlet state, 322
Slater determinant, 68, 71
Smeared field operators, 275
Source term, 360, 367, 384, 420
Space inversion, 301
- Dirac field, 312
- scalar field, 301, 309
- vector field, 319
Space-like separations, 103
Spectral density, 279, 281
Spectral representation, *see* Lehmann–Källen representation
Spectrum
- Dirac equation, 126
- of harmonic oscillator, 9
Spherical basis, 165, 319
Spherical waves, 79
Spin, 45, 322
- electromagnetic field, 149
Spin operator, 131
- Proca field, 165
Spin-1 mesons, 141
Spin-statistics theorem, 130, 140, 159
Spinors, 117
State vector
- localized, 62
Stationary phase approximation, 344
Statistical mechanics, 374
Strong operator convergence, 274
Stückelberg, 109, 113, 249
Sum rule for spectral density, 281
Superposition principle, 342
Symmetrized product wave function, 65
Symmetry, 272
- discrete, 301
Symmetry factor, 261, 397
- ϕ^4 theory, 266
- quantum electrodynamics, 267
Symmetry transformation, 40, 43, 95
- internal, 46

Tadpole graph, 238, 396, 428

TCP theorem, *see* CPT theorem
Temporal gauge, 144
Tensors, 329
Three-point function, 382, 404
Time-evolution operator, 8, 215–219, 292, 340
- adiabatic, 221
- perturbation series, 218
Time-ordered exponential, 219, 292
Time-ordered product, 227, 293, 357, 367
- Grassmann fields, 420
- modified, 258
Time ordering, 217
Time reversal, 301, 325
- Dirac field, 315
- scalar field, 306
- vector field, 319
Tomonaga picture, *see* Interaction picture
Topologically equivalent graphs, 237
Trajectory, 340
- classical, 337
Transfer matrix, 340
Transformation
- field operators, 96
- infinitesimal, 40, 96
Transition amplitude, 339, 340
- infinitesimal, 341
Transition current, 187, 247
Transition matrix element, 357
Translation, 43, 272
Transversality condition, 197
- four-dimensional, 156
Transverse delta function, 197, 202–205
Transverse gauge, 144
Transverse photons, 198
Transverse polarization vector, 154
Transverse projection operator, 197, 208
Tree approximation, 299
Triplet state, 322
Two-body interaction, 69
Two-particle scattering, 225, 262
Two-point function, 357–359, 381, 391, 394
- ϕ^4 theory, 395
- bosons, 428
- fermions, 428
Two-point Green's function, 295

Uncertainty relation, 7, 8
Unitary representation, 96
Unitary transformation, 22, 96

Vacuum, 61, 129, 270, 279, 293
- charge, 131
- Dirac, 127

- energy, 81, 127
- fluctuations, 240, 293, 394
- polarization, 240
- stability, 244

Vacuum functional, 362, 368
- euclidian, 375

Vacuum persistence amplitude, 360–363
Vacuum–vacuum amplitude, 362
Variation, 40
- modified, 40
- total, 42

Variation of a functional, 31
Vector field
- massive, 154
- plane-wave expansion, 154

Vector potential, 143
- plane-wave decomposition, 161

Vertex
- electron-photon, 236
- four-boson, 263
- two-photon two-boson, 254

Vertex function
- amputated, 406
- irreducible, 402
- irreducible three-point, 403

Virtual field quanta, 272
Virtual phonon, 25
Virtual photon, 183, 187
Volterra series, 380, 402

Wave equation, 143, 173
Wave function, 62
Wave packet, 179, 275
Weak operator convergence, 274, 275, 278, 284
Weyl ordering, 341, 347–350
Wick rotation, 343, 359, 372
Wick's theorem, 225–231, 255, 293
- proof, 231–233
Wiener path integral, 342
Wightman function, 278

Yang–Feldman equation, 272, 276–278
Yukawa coupling, 424–431

Z, *see* Renormalization constant
Zero-point energy, 9, 20, 80, 127

Springer-Verlag and the Environment

We at Springer-Verlag firmly believe that an international science publisher has a special obligation to the environment, and our corporate policies consistently reflect this conviction.

We also expect our business partners – paper mills, printers, packaging manufacturers, etc. – to commit themselves to using environmentally friendly materials and production processes.

The paper in this book is made from low- or no-chlorine pulp and is acid free, in conformance with international standards for paper permanency.